IET POWER AND ENERGY SERIES 74

Power Electronic Converters and Systems

Other volumes in this series:

Volume 1	**Power Circuit Breaker Theory and Design** C.H. Flurscheim (Editor)
Volume 4	**Industrial Microwave Heating** A.C. Metaxas and R.J. Meredith
Volume 7	**Insulators for High Voltages** J.S.T. Looms
Volume 8	**Variable Frequency AC Motor Drive Systems** D. Finney
Volume 10	**SF_6 Switchgear** H.M. Ryan and G.R. Jones
Volume 11	**Conduction and Induction Heating** E.J. Davies
Volume 13	**Statistical Techniques for High Voltage Engineering** W. Hauschild and W. Mosch
Volume 14	**Uninterruptible Power Supplies** J. Platts and J.D. St Aubyn (Editors)
Volume 15	**Digital Protection for Power Systems** A.T. Johns and S.K. Salman
Volume 16	**Electricity Economics and Planning** T.W. Berrie
Volume 18	**Vacuum Switchgear** A. Greenwood
Volume 19	**Electrical Safety: a guide to causes and prevention of hazards** J. Maxwell Adams
Volume 21	**Electricity Distribution Network Design, 2nd Edition** E. Lakervi and E.J. Holmes
Volume 22	**Artificial Intelligence Techniques in Power Systems** K. Warwick, A.O. Ekwue and R. Aggarwal (Editors)
Volume 24	**Power System Commissioning and Maintenance Practice** K. Harker
Volume 25	**Engineers' Handbook of Industrial Microwave Heating** R.J. Meredith
Volume 26	**Small Electric Motors** H. Moczala *et al.*
Volume 27	**AC-DC Power System Analysis** J. Arrillaga and B.C. Smith
Volume 29	**High Voltage Direct Current Transmission, 2nd Edition** J. Arrillaga
Volume 30	**Flexible AC Transmission Systems (FACTS)** Y-H. Song (Editor)
Volume 31	**Embedded generation** N. Jenkins *et al.*
Volume 32	**High Voltage Engineering and Testing, 2nd Edition** H.M. Ryan (Editor)
Volume 33	**Overvoltage Protection of Low-Voltage Systems, Revised Edition** P. Hasse
Volume 36	**Voltage Quality in Electrical Power Systems** J. Schlabbach *et al.*
Volume 37	**Electrical Steels for Rotating Machines** P. Beckley
Volume 38	**The Electric Car: Development and future of battery, hybrid and fuel-cell cars** M. Westbrook
Volume 39	**Power Systems Electromagnetic Transients Simulation** J. Arrillaga and N. Watson
Volume 40	**Advances in High Voltage Engineering** M. Haddad and D. Warne
Volume 41	**Electrical Operation of Electrostatic Precipitators** K. Parker
Volume 43	**Thermal Power Plant Simulation and Control** D. Flynn
Volume 44	**Economic Evaluation of Projects in the Electricity Supply Industry** H. Khatib
Volume 45	**Propulsion Systems for Hybrid Vehicles** J. Miller
Volume 46	**Distribution Switchgear** S. Stewart
Volume 47	**Protection of Electricity Distribution Networks, 2nd Edition** J. Gers and E. Holmes
Volume 48	**Wood Pole Overhead Lines** B. Wareing
Volume 49	**Electric Fuses, 3rd Edition** A. Wright and G. Newbery
Volume 50	**Wind Power Integration: Connection and system operational aspects** B. Fox *et al.*
Volume 51	**Short Circuit Currents** J. Schlabbach
Volume 52	**Nuclear Power** J. Wood
Volume 53	**Condition Assessment of High Voltage Insulation in Power System Equipment** R.E. James and Q. Su
Volume 55	**Local Energy: Distributed generation of heat and power** J. Wood
Volume 56	**Condition Monitoring of Rotating Electrical Machines** P. Tavner, L. Ran, J. Penman and H. Sedding
Volume 57	**The Control Techniques Drives and Controls Handbook, 2nd Edition** B. Drury
Volume 58	**Lightning Protection** V. Cooray (Editor)
Volume 59	**Ultracapacitor Applications** J.M. Miller
Volume 62	**Lightning Electromagnetics** V. Cooray
Volume 63	**Energy Storage for Power Systems, 2nd Edition** A. Ter-Gazarian
Volume 65	**Protection of Electricity Distribution Networks, 3rd Edition** J. Gers
Volume 66	**High Voltage Engineering Testing, 3rd Edition** H. Ryan (Editor)
Volume 67	**Multicore Simulation of Power System Transients** F.M. Uriate
Volume 68	**Distribution System Analysis and Automation** J. Gers
Volume 69	**The Lightening Flash, 2nd Edition** V. Cooray (Editor)
Volume 70	**Economic Evaluation of Projects in the Electricity Supply Industry, 3rd Edition** H. Khatib
Volume 74	**Power Electronic Converters and Systems: Frontiers and applications** Andrzej M. Trzynadlowski (Editor)
Volume 76	**Power System Stability: Modelling, analysis and control** B. Om P. Malik
Volume 78	**Numerical Analysis of Power System Transients and Dynamics** A. Ametani (Editor)
Volume 79	**Vehicle-to-Grid: Linking electric vehicles to the smart grid** J. Lu and J. Hossain (Editors)
Volume 905	**Power system protection, 4 volumes**

Power Electronic Converters and Systems

Frontiers and applications

Edited by
Andrzej M. Trzynadlowski

The Institution of Engineering and Technology

Published by The Institution of Engineering and Technology, London, United Kingdom

The Institution of Engineering and Technology is registered as a Charity in England & Wales (no. 211014) and Scotland (no. SC038698).

First published 2015

The Institution of Engineering and Technology
Michael Faraday House
Six Hills Way, Stevenage
Herts, SG1 2AY, United Kingdom

www.theiet.org

British Library Cataloguing in Publication Data
A catalogue record for this product is available from the British Library

ISBN 978-1-84919-826-4 (hardback)
ISBN 978-1-84919-827-1 (PDF)

Typeset in India by MPS Limited
Printed in the UK by CPI Group (UK) Ltd, Croydon

This book is dedicated to those countless researchers and engineers around the world who diligently strive to maintain the high rate of progress in modern power electronics

Contents

Preface xvii

Part I Converters 1

1 Semiconductor power devices 3
- 1.1 Introduction 3
- 1.2 High-voltage SiC power devices 4
  - 1.2.1 Characterization of 15 kV SiC N-IGBTs 4
  - 1.2.2 Characterization of 10 kV SiC MOSFETs 9
- 1.3 Low-voltage SiC devices and its characteristics 11
  - 1.3.1 Low-voltage gate drive design 11
  - 1.3.2 Common-mode current minimization 12
- 1.4 Characterization of 1,200 V, 100 A SiC MOSFET 12
  - 1.4.1 1,200 V, 100 A SiC MOSFET device characterization without complementary device of the half-bridge module 12
  - 1.4.2 1,200 V, 100 A SiC MOSFET device characterization with complementary device of the half-bridge module 12
  - 1.4.3 Hard-switching characterization of 1,700 V SiC MOSFET [11] 14
  - 1.4.4 Performance comparison of MOSFET and IGBT 16
  - 1.4.5 Gate drive design and characterization of 1,200 V/45 A infineon SiC JFET module [12] 23
  - 1.4.6 SiC super-junction transistor characteristics 24
- 1.5 Zero voltage switching characterization of 12 kV SiC [14] 25
  - 1.5.1 ZVS turn-on characteristics 26
  - 1.5.2 ZVS turn-off characteristics 28
- 1.6 All SiC-based SST 34
- 1.7 Summary 39
- Acknowledgements 41
- References 41

2 Multilevel converters 43
- 2.1 Introduction 43
- 2.2 Basic concepts of multilevel converters 43
  - 2.2.1 One-branch converter 44
  - 2.2.2 Two branches, "H-bridge" converter 46
- 2.3 Electronic switches to implement the converters 47

2.3.1 NPC converter 47
2.3.2 FC converter 49
2.3.3 CHB converter 49
2.3.4 Combined topologies 53
2.4 Three-phase multilevel converters 54
2.4.1 Phase-to-phase and phase-to-neutral voltages 55
2.4.2 Space vector representation 56
2.5 Modulation strategies for multilevel converters 60
2.5.1 Voltage levels–based algorithms 61
2.5.2 Space vector–based algorithms 69
References 72
Further reading 73

3 Multi-input converters 75
3.1 Introduction 75
3.2 Realizing multi-input converter topologies 76
3.3 Multi-port converters 87
3.3.1 Synthesis of multi-port converters by extending multi-input topologies 87
3.3.2 Multi-port converters with dc link 89
3.3.3 Ac link multi-port power converters 92
3.4 Applications of multi-port power converters 97
3.4.1 Multi-port power converters for renewable energy systems 99
3.4.2 Application of multi-input converters in micro-grids 101
3.4.3 Multi-port converters for vehicular power systems 103
3.5 Summary 106
References 106

4 Modular converters 111
4.1 Introduction 111
4.2 Modular converter topologies and description 115
4.2.1 MMCC arrangements and SM topologies 116
4.2.2 Basic modular multilevel cascade converters 121
4.3 Control strategies 125
4.3.1 Voltage-balancing control 125
4.3.2 Circulating current control 128
4.4 Modulation techniques 130
4.4.1 High switching frequency techniques 130
4.4.2 Low switching frequency techniques 132
4.5 Operational issues in MMCCs 134
4.5.1 Fault-tolerant operation 134
4.5.2 Floating DC capacitor pre-charging procedures 136
4.6 Main applications 136
Acknowledgements 138
References 138

5 Matrix converters **147**
5.1 Introduction 147
5.2 Direct matrix converter 147
5.2.1 Circuit topology 147
5.2.2 Modulation techniques 147
5.3 Indirect matrix converter 159
5.3.1 Circuit topology 161
5.3.2 Modulation techniques 162
5.4 Technological issues of MCs 165
5.5 MC versus voltage back-to-back converter 167
5.6 Summary 167
References 168

6 Soft-switching converters **169**
6.1 Resonant converters 169
6.1.1 Second-order resonant converters 169
6.1.2 Resonant converters with three or more resonating elements 178
6.2 Quasi-resonant converters 181
6.2.1 Example 1 – half-wave ZCS-QRC 183
6.2.2 Example 2 – full-wave ZCS-QRC 185
6.2.3 Example 3 – half-wave ZVS-QRC 186
6.2.4 Example 4 – full-wave ZVS-QRC 188
6.2.5 The effect of parasitic oscillations in QRCs 189
6.3 Multi-resonant converters 190
6.4 Quasi-square-wave converters 191
6.5 Other types of ZVS and ZCS converters 196
References 200

7 *Z*-source converters **205**
7.1 Introduction 205
7.1.1 General overview 205
7.1.2 Basic principles 207
7.1.3 Modeling and control 209
7.2 Categories of impedance source power converters based on conversion functionality 215
7.2.1 DC–DC converter topologies 215
7.2.2 DC–AC inverter topologies 222
7.2.3 AC–AC converter topologies 226
7.2.4 AC–DC converter topologies 227
7.3 Impedance source network topologies 227
7.3.1 Non-transformer based 229
7.3.2 With transformer or magnetic coupling 233
7.4 Conclusions 236
References 237

8 Switching power supplies **245**
8.1 Introduction 245
8.2 Non-isolated converters topologies 246
8.2.1 Buck converter 246
8.2.2 Boost converter 249
8.2.3 Buck–boost converter 251
8.2.4 Integrated buck and boost converter 252
8.2.5 Power factor correction 253
8.3 Isolated converters topologies 254
8.3.1 Flyback converter 254
8.3.2 Forward converter 255
8.3.3 Half-bridge converter 256
8.3.4 Full-bridge converter 258
8.3.5 Rectifiers 259
8.4 Parasitics in DC–DC converters 262
8.5 Continuous and discontinuous conduction modes 264
8.6 Synchronous rectification 264
8.7 Bidirectional converters 265
8.8 Interleaving 266
8.9 Control principles 268
Further reading 272

9 Smart power electronic modules **273**
9.1 History 273
9.2 Technology background 276
9.2.1 IGBT device technologies and their performance 277
9.2.2 Gate driver technology 280
9.2.3 Packaging technologies 282
9.3 Basic usage 284
9.3.1 Protection 284
9.3.2 Bootstrap power supply 285
9.3.3 Digital interface 287
9.4 Reliability 287
9.5 Variety of products 288
9.6 Future usage and emerging solutions 290
9.6.1 Grid interface with multiple power modules 290
9.6.2 Matrix converter with SPM devices 295
9.6.3 Multilevel converter with SPM devices 296
9.6.4 A different direct converter with SPM devices 298
9.7 Conclusion 306
References 306

Part II Applications **311**

10 Permanent magnet synchronous motor drives **313**
10.1 Introduction 313
10.2 Trends in sensorless control of PMSM 314

10.2.1 Key factors for sensorless controls evaluation 315
10.2.2 A glance to HF injection methods 316
10.2.3 A HF sensorless technique for IPM and PMASR motors 318
10.2.4 A HF injection method for surface-mounted permanent magnet motors 319
10.2.5 A glance to extended EMF-based methods 320
10.2.6 Sliding mode observers for the extended EMF 321
10.3 Trends in MPC of PMSM 322
10.3.1 Key factors for MPC 323
10.3.2 Direct torque and flux control 323
10.3.3 An MPC method for IPM motors 324
10.3.4 A perspective on centralised MPC-based structures 325
10.4 Some hints about energy efficiency in PMSM drives 326
10.5 Final considerations 327
Acknowledgements 328
References 328

11 Induction motor drives 333
11.1 Induction motors 333
11.2 IM model 334
11.3 Variable frequency drives 335
11.3.1 Scalar control 335
11.3.2 Field-oriented control 335
11.3.3 Direct torque control 336
11.4 DTC schemes 336
11.4.1 Space vector modulation DTC 336
11.4.2 Feedback linearization and sliding mode DTC 337
11.4.3 Intelligent DTC schemes 340
11.5 IM speed estimation with Kalman filtering 342
11.5.1 High-speed operation 343
11.5.2 Low-speed operation 344
11.5.3 Implementation considerations 346
11.6 Switched reluctance sensorless drives 348
References 349

12 Wind energy systems 351
12.1 Introduction 351
12.1.1 Overview of the wind energy technology 351
12.1.2 Types of WT rotors 352
12.1.3 Generators for WTs and standard configurations of WECSs 354
12.2 Power electronic interfaces for variable speed WTs 361
12.2.1 Conventional power electronic building blocks 363
12.2.2 Common PEC topologies for WTs 366
12.2.3 Emerging PEC topologies for WTs 369
12.2.4 PEC topologies for high power WTs 371

12.3 Control algorithms for PECs 374
12.3.1 Maximum power point tracking 375
12.3.2 Control for DC/DC boost converters 377
12.3.3 Control for IS converters 378
12.3.4 Field oriented control 380
12.3.5 Direct torque control-space vector modulated 382
12.3.6 Voltage oriented control 382
12.3.7 Direct power control-space vector modulated 384
12.3.8 Single-phase grid converter control 385
12.3.9 Control for stand-alone mode of operation 386
Acknowledgements 388
References 388

13 Photovoltaic energy systems **395**
13.1 Introduction 395
13.1.1 Brief overview of photovoltaic generation 395
13.1.2 PV inverter circuit 395
13.1.3 Centralized PV plant 397
13.2 The technologies 399
13.2.1 State-of-the-art technologies 399
13.2.2 Reliability 404
13.3 The grid interface 409
13.3.1 Basic control of real and reactive power in a two-bus power system 409
13.4 The standards 414
13.4.1 Protection 417
13.4.2 Islanding 417
13.4.3 Power quality 419
13.4.4 Ancillary services 420
13.5 The field measurements 422
13.5.1 Intermittence in solar field results 422
13.5.2 LVRT test results of the 500 kW RX series 422
13.6 Summary and conclusions 423
Acknowledgements 424
References 424

14 Automotive energy systems **425**
14.1 Electric vehicle batteries 425
14.1.1 Introduction 425
14.2 EV charging 440
14.2.1 Plugged charging 440
14.3 Wireless charging 456
14.3.1 Introduction 456
14.3.2 Necessity of compensation for wireless charging 458
14.3.3 Analysis of series–series topology 461
14.3.4 Analysis of series–parallel topology 461

14.3.5 Peak efficiency of series–series and series–parallel topology 462
14.3.6 Control strategies for SS and SP topology 463
14.3.7 Advantages of EV wireless charging 464
References 464

15 Shipboard power systems 469
15.1 Shipboard power system topologies 469
15.2 Shipboard propulsion drives 472
15.2.1 Voltage link systems 474
15.2.2 Current link systems 477
15.2.3 Direct AC–AC conversion – cycloconverters 478
15.3 Power quality requirements in shipboard systems 479
15.4 Harmonic mitigation in shipboard systems 483
15.4.1 Harmonic cancellation 483
15.4.2 Reactive harmonic suppressors 485
15.4.3 Active filters 489
15.4.4 Case study 492
15.5 Frequency variation and converter control 495
15.6 Concepts for future shipboard power systems 495
15.6.1 Power electronics building block 495
15.6.2 Medium voltage DC integrated power system 496
Acknowledgements 498
References 498

16 Converters in power grid 501
16.1 Introduction 501
16.2 Power converter topologies 501
16.2.1 AC–DC converters 502
16.2.2 AC phase controllers 505
16.2.3 AC pulse width modulated controllers 505
16.2.4 Modular multilevel converter 508
16.3 Application examples of power converters in power grid 512
16.3.1 Shunt compensation 512
16.3.2 Series compensation 515
16.3.3 Shunt-series compensation 521
16.3.4 Series-series compensation 523
16.3.5 High-voltage direct current transmission 524
16.3.6 Low-frequency high-voltage ac transmission 526
16.3.7 Solid-state power transformer 528
16.4 Summary 529
References 530

17 Distributed generation and microgrids 535
17.1 Introduction 535
17.2 Distribution generators 536

17.2.1 Examples of DGs 536
17.2.2 Technical impacts due to DG 541
17.2.3 IEEE1547 544
17.3 Microgrid 545
17.3.1 DC and AC microgrids 546
17.3.2 Stand-alone microgrids 546
17.3.3 Grid-tied microgrids 546
17.3.4 Centralized control 547
17.3.5 Conventional droop control method 547
17.3.6 Local control 550
17.3.7 Multifunctionalities 551
References 553

18 Uninterruptible power supplies 557
18.1 Introduction 557
18.2 Topologies 557
18.2.1 On-line UPS systems 557
18.2.2 Off-Line UPS 558
18.2.3 Line-interactive UPS 560
18.2.4 Delta conversion UPS 561
18.2.5 Tri-mode UPS 561
18.2.6 Rotary UPS 562
18.2.7 Hybrid static and rotary UPS 563
18.2.8 Flywheels 564
18.2.9 DC UPS for pulse load with power leveling 565
18.2.10 Redundant bus 566
18.3 Controls for UPS systems 566
18.4 Applications 571
18.4.1 Desktop personal computers 572
18.4.2 Industrial systems 572
18.4.3 Data centers 572
18.4.4 Medical equipment 572
References 573

19 Wireless power transfer 577
19.1 Introduction 577
19.2 Basic principles and two fundamental concepts of WPT 579
19.2.1 Basic principles 579
19.2.2 Two fundamental concepts 580
19.3 Different forms of WPT systems 582
19.3.1 Two-coil systems 582
19.3.2 Three-coil systems 583
19.3.3 Four-coil systems 584
19.3.4 WPT systems with relay and domino resonators 585

19.4 Power electronics and control 585
19.4.1 Series and parallel tuned primary and secondary circuits 586
19.4.2 Practical circuit topologies 588
19.4.3 Control 592
19.5 Safety regulations 594
19.5.1 Electromagnetic compatibility 594
19.5.2 Human exposure 595
19.6 Conclusion 597
Acknowledgements 597
References 597

20 Advanced control of power electronic systems 601
20.1 Introduction 601
20.2 Brief overview of historic advanced nonlinear controllers for PES applications 601
20.3 Switching SBC [9–11] 604
20.3.1 SBC for standalone PES 604
20.3.2 SBC for networked PESs 610
Acknowledgements 617
References 617

Index 619

Preface

The era of modern power electronics began in the late 1950s when the silicon-controlled rectifier (SCR) was developed by General Electric Corporation. Most of the early applications of SCRs involved electric drives. In recent decades, power electronic converters spread to the electric grid, distributed generation systems, renewable energy sources, transportation, and a variety of industrial processes. Today's power electronics is sustaining a robust growth.

This book is intended as a reference for professionals who are already familiar with the fundamentals of power electronics. Consequently, in contrast to typical textbooks, no coverage of basic principles of electric power conditioning is provided. It is assumed that Readers do not need explanation of such terms as the rectifier, inverter, chopper, or pulse width modulation.

The content of the book is mostly focused on recent advances in power electronic converters and systems, but the technological progress in the area of the associated semiconductor devices cannot be overlooked. The traditional silicon-based semiconductor power switches, such as thyristors or IGBTs, are reaching limits of their highly impressive operating parameters and characteristics. However, a new era of the so-called wide bandgap (WBG) semiconductor devices has already begun, promising revolutionary enhancement of the existing power electronic circuits. The most advanced WBG technology is that of silicon carbide devices, which are described in details in Chapter 1.

Most of the first part of the book deals with those power electronic converters, which thanks to their unique properties, enjoy currently high interest of researchers. Thus, the subjects of Chapters 2–7 include multilevel, multi-input, modular, matrix, soft switching, and *Z*-source converters. Switching power supplies, explained in Chapter 8, provide high-quality power to electronic devices, including the ubiquitous laptops, tablets, and smart phones. Chapter 9 describes "smart" power electronic modules, which combine power and control circuits in the same package.

The second part of the book describes the most common applications of modern power electronics systems. Electric drives with synchronous and induction motors have always been in the mainstream of power electronics. Photovoltaic and most of the wind energy sources are interfaced with the power grid through power electronic converters. Recently, battery-fed electric cars, with a power electronic inverter driving an ac motor, have been gaining popularity. Hybrid cars, in which sophisticated gearing links an electric drive system with an internal combustion engine, have already entered the mainstream of automobile markets. Shipboard power systems progressively employ power electronics. Those topics are covered in Chapters 10–15.

Modern power grids increasingly use power electronic systems for energy conversion, control of the power flow, and stability enhancement. Integration of the renewable energy sources through distributed generation and microgrids would not be possible without those systems. The ubiquitous data and communication centers require uninterruptable power supplies to prevent catastrophic information loss due to power outages and disturbances. Wireless power transfer allows remote energizing of battery-fed devices and vehicles. All the sophisticated power electronic systems need correspondingly advanced control methods. Chapters 16–20 deal with the aforementioned issues.

The Editor wants to express his deep gratitude to the forty-four contributing Authors, all accomplished specialists in various areas of power electronics and its applications. Their collective expertise and efforts, supported by the most helpful Publisher's personnel, made this book a highly valuable source of engineering knowledge.

A.M. TRZYNADLOWSKI

Part I

Converters

Chapter 1
Semiconductor power devices

Arun Kadavelugu, Samir Hazra*, Sachin Madhusoodhanan*, Awneesh Tripathi*, Kasunaidu Vechalapu*, Ankan De*, Krishna Mainali*, Dhaval Patel*, Kamalesh Hatua* and Subhashish Bhattacharya**

1.1 Introduction

Power electronics systems are very important part of today's energy conversion systems. Power semiconductor devices are the basis for power electronics systems. In power electronics system, a power semiconductor device is used as a switch either in the ON or OFF state. Thus, power devices are optimized to be used in commutation mode (either ON or OFF). Common power devices are power diode, thyristor, bipolar junction transistor (BJT), metal–oxide–semiconductor field-effect transistor (MOSFET), insulated-gate bipolar transistor (IGBT), and different derivatives of them. Since the very first power device, power electronics technologies have been developed on silicon material. With more than 60 years of silicon technology development, the silicon power device fabrication process is very mature.

The demand for faster switching speed devices and higher voltage blocking capabilities is growing. Silicon-based power devices are limited in these performances due to inherent material properties. Silicon has low band gap energy, low critical electric field, low thermal conductivity, and limited switching frequency. Since early 1980s, research has been focused on wide band gap semiconductor materials for power devices to overcome these limitations of the silicon-based power devices. Now, silicon carbide (SiC) is considered to be most promising material for power devices. SiC is wide band gap semiconductor, with much lower leakage currents and higher operating temperatures compared to silicon power devices. It has high critical electric fields; device can have high doping concentrations with thin blocking layers giving lower specific on-resistance. SiC has higher electron saturation velocity which leads to higher operating frequencies. SiC also has higher thermal conductivity which improves the heat dissipation and helps in achieving higher power densities.

In this chapter, the characteristics of high-voltage SiC IGBT, SiC MOSFET, SiC junction gate field-effect transistor (JFET), and low-voltage SiC MOSFET are discussed.

*North Carolina State University, Raleigh, NC, USA

1.2 High-voltage SiC power devices

The several fold increase in critical electric field in comparison to silicon makes it feasible to build high-voltage unipolar (majority carrier) devices with extremely low resistance as predicted by Baliga's figure of merit [1]. The better thermal conductivity of SiC is another factor of considerable importance which is critical in heat removal process [2,3]. The high-temperature capability of SiC, due to its wider band gap, is another attractive feature.

These unique advantages of SiC led to development of MOSFETs beyond 10 kV [4,5]. The 10 kV SiC MOSFETs have been demonstrated on a single-phase 1 MVA soft-switched Solid State Power Substation by GE in 2010 [5]. The 13 kV SiC MOSFETs have been demonstrated on a 20 kVA single-phase solid-state transformer (SST) by FREEDM Systems Center at North Carolina State University (NCSU) [6,7].

However, similar to the case of silicon, as the voltage is scaled-up, IGBT is more efficient due to its bipolar physics with metal–oxide–semiconductor (MOS) gate structure. Therefore, significant research effort has been spent in the recent years to develop high-voltage 4H-SiC IGBTs. Due to difficulty in producing low-resistivity P+ substrate, the initial SiC IGBTs have been developed with P-channel (using N+ substrate) [8]. However, with advances in SiC technology, low-resistivity P+ substrates have been developed leading to the development of SiC N-IGBTs [9].

1.2.1 Characterization of 15 kV SiC N-IGBTs

The simplified cross-sectional view of the 15 kV SiC N-IGBT is shown in Figure 1.1. The drift layer is 140 μm thick with doping concentration of 2×10^{14} cm^{-3}. The IGBT used for the characterization has a buffer layer thickness of 5 μm, and its chip area is 0.84 cm × 0.84 cm, with an active area of 0.32 cm^2. The current-sense resistors and the thermistor integrated in the module for over current and temperature protection are seen in Figure 1.2.

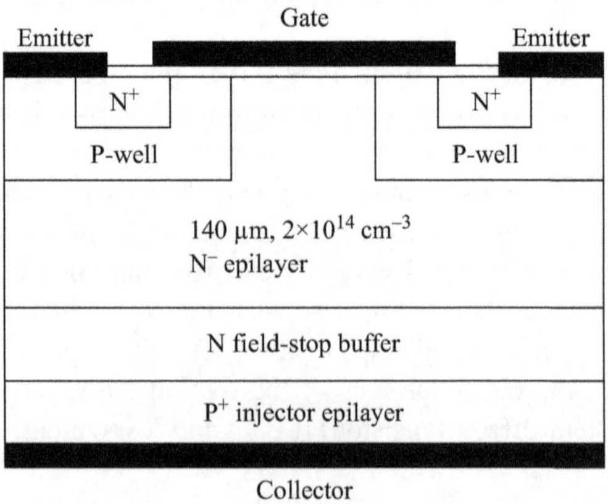

Figure 1.1 Simplified cross-sectional view of 15 kV SiC N-IGBT

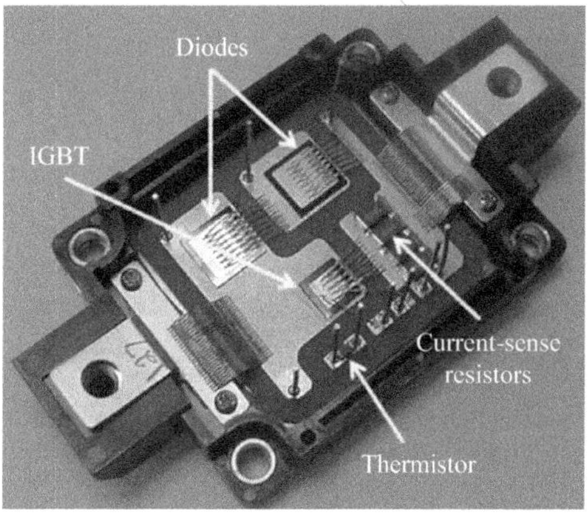

Figure 1.2 15 kV SiC N-IGBT co-pack module

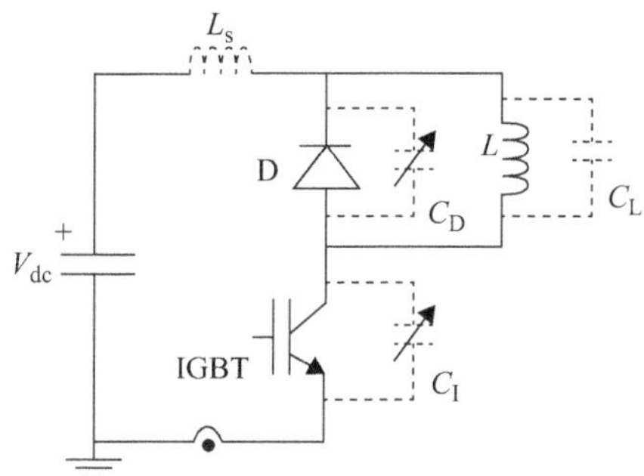

Figure 1.3 The double-pulse test circuit schematic with different parasitic components

The clamped inductive (or fully hard-switched) characteristics are most widely provided data for power semiconductor devices to cover majority of applications. The schematic of the double-pulse test circuit is shown in Figure 1.3. Figure 1.4 shows turn-on switching loss (E_{on}) and turn-off switching loss (E_{off}) at 25 °C, at 10 kV, 2–10 A. Unlike the turn-on loss, the turn-off loss is weakly dependent on the current. Figure 1.5 shows the turn-off transitions at 10 kV, 5 A, and 10 A at 25 °C. The duration of the 10 A transition is about 60% of that of the 5 A transition, resulting in very slight variation of energy loss from 5 to 10 A. The temperature dependency of the turn-off loss is shown in Figure 1.6. The energy loss is increased to a factor of three from 25 to 175 °C, due to increased injection at higher temperature resulting in significantly larger amounts of charge in the drift region. As shown in Figure 1.7, the larger amount of charge to be removed at higher temperature is slowing down the voltage rise (before punch through) which in consequence is resulting in higher energy loss.

The turn-on loss behavior at high temperatures is evaluated with higher gate resistance of 33 Ω (turn-off loss is evaluated with gate resistance of 10 Ω).

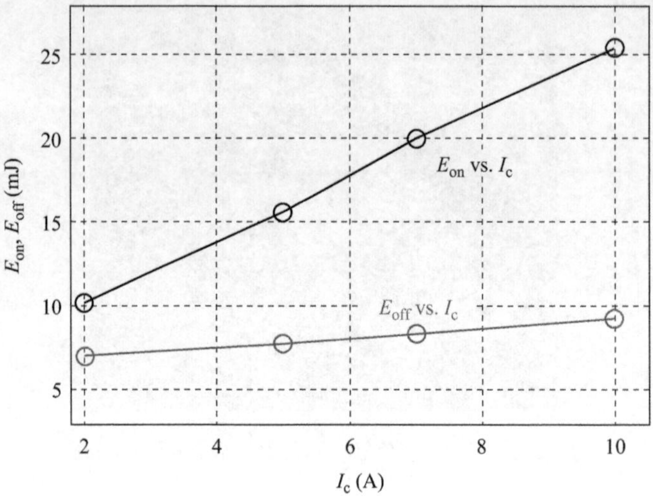

Figure 1.4 Turn-on and turn-off energy loss values with current variation at 10 kV and 25 °C, under clamped inductive load

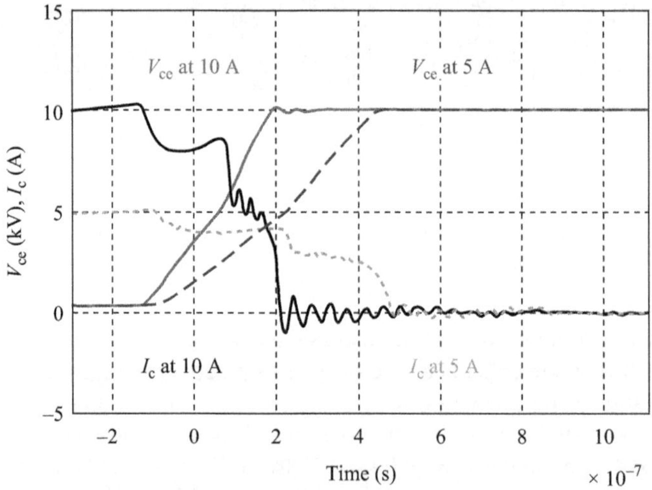

Figure 1.5 Turn-off switching transitions at 10 kV, 25 °C, 5 and 10 A, under clamped inductive load conditions

Based on the results shown in Figure 1.8, the turn-on energy loss is not drastically varying with temperature, unlike the turn-off energy loss.

The 15 kV SiC N-IGBT characteristics reported earlier has 5 μm thick field-stop buffer layer. The reduction in buffer layer thickness from 5 to 2 μm reduces the conduction drop from 7.2 to 6.0 V at 20 A. Both the IGBTs have been designed with same drift layer parameters and same buffer layer doping, except for its

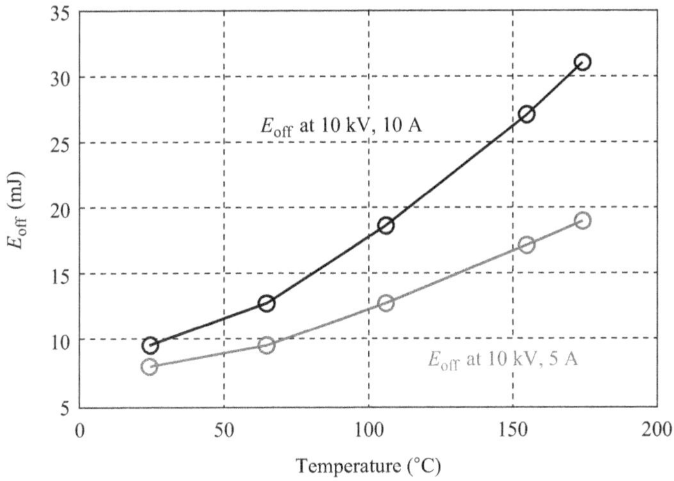

Figure 1.6 Turn-off switching loss variation with temperature at 5 and 10 A at 10 kV, under clamped inductive load conditions

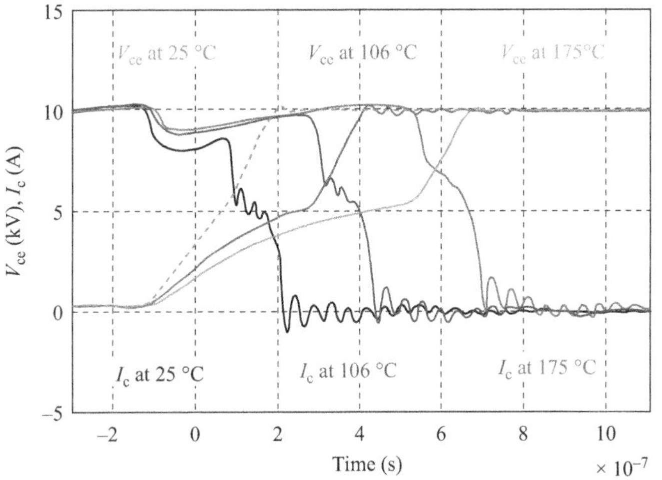

Figure 1.7 Turn-off switching transitions with temperature at 10 kV and 10 A, under clamped inductive load conditions

thickness. Figure 1.9 shows turn-off transition of 2 and 5 μm buffer layer IGBTs at 10 kV and 10 A. The turn-off transition of the 5 μm buffer layer IGBT is much faster than that of the 2 μm buffer layer IGBT, due to reduced injection.

The turn-on transitions of both the IGBTs at 10 kV, 8 A are shown in Figure 1.10. The 2 μm IGBT has significantly larger dv/dt above the punch-through voltage resulting in a large current spike due to discharge of the capacitance of the free-wheeling diode (FWD) of the clamped inductive test circuit. Table 1.1 summarizes

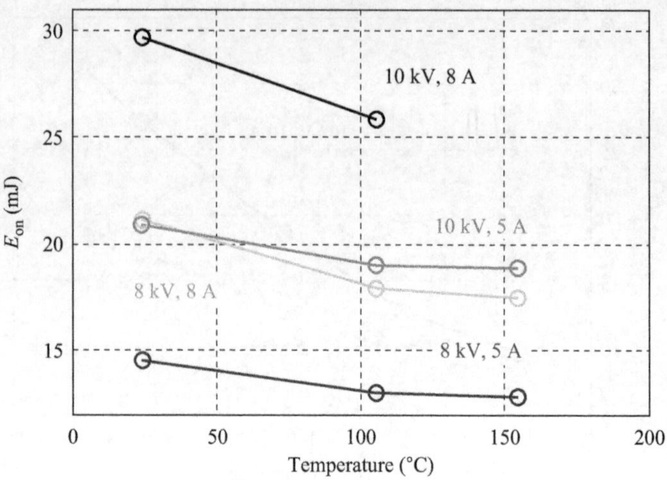

Figure 1.8 The turn-on energy loss variation of 15 kV N-IGBT with temperature with $R_{G(ON)} = 33\ \Omega$

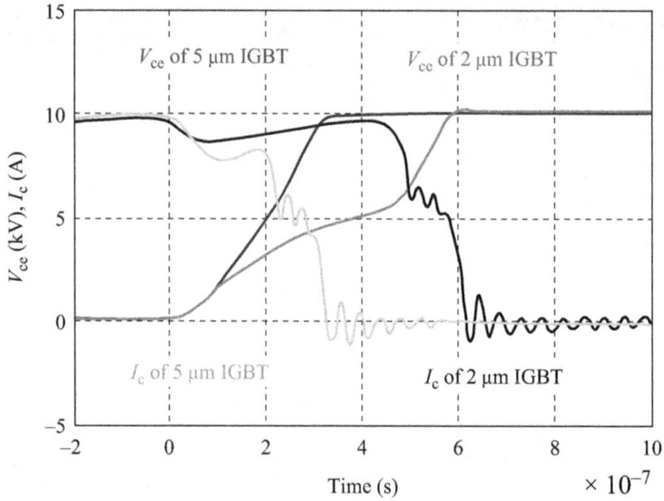

Figure 1.9 Turn-off transitions of 15 kV, 2 and 5 μm thick buffer layer, SiC IGBT at 10 kV, 10 A, 25 °C with $R_{G(OFF)}$ *of 10 Ω*

d*v*/d*t* values of both the devices at 25 °C with $R_{G(ON)} = 50\ \Omega$ and $R_{G(OFF)} = 10\ \Omega$. Table 1.2 summarizes energy loss values of both the IGBTs. Based on the data shown in these two tables, the thicker buffer layer IGBT has significantly lower turn-off loss and turn-on d*v*/d*t* (over punch-through voltage). However, the conduction drop is moderately increased due to the thicker buffer layer. Thus, the 5 μm devices are more suitable for high-frequency, high-voltage power conversion.

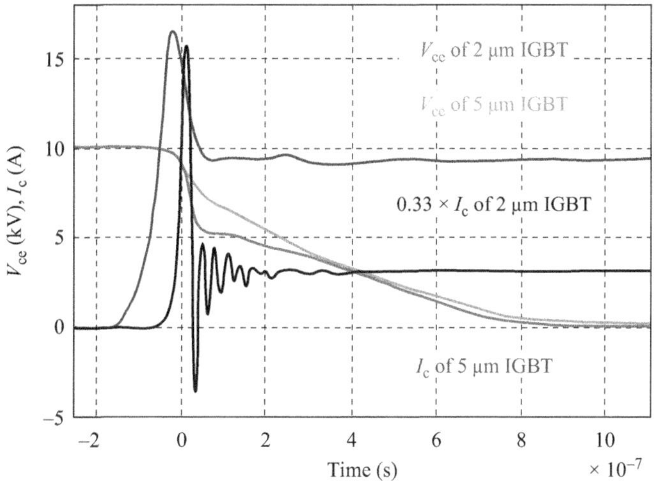

Figure 1.10 Turn-on transitions of 15 kV, 2 and 5 μm thick buffer layer, SiC IGBT at 10 kV, 8 A, 25 °C with $R_{G(ON)}$ *of 50 Ω (the 2 μm IGBT current is scaled down to 0.33× to accommodate the current spike)*

Table 1.1 dv/dt of 2 and 5 μm IGBTs at 10 kV, 8 A (turn-on) and 10 A (turn-off)

Parameter	**2 μm IGBT (kV/μs)**	**5 μm IGBT (kV/μs)**
Turn-on dv/dt (above PT[a])	135	39
Turn-on dv/dt (under PT)	7.3	9.0
Turn-off dv/dt (before PT)	12	30
Turn-off dv/dt (after PT)	39	47

[a]PT, punch through.

Table 1.2 Energy loss values of 15 kV, 2 and 5 μm SiC IGBTs

Parameter	**2 μm IGBT**	**5 μm IGBT**
Forward drop at 20 A, 25 °C ($V_{GE} = 20$ V)	5.52 V	7.12 V
Turn-off energy loss at 10 kV, 10 A and 25 °C ($R_{G(ON)} = 10$ Ω)	20.4 mJ	7.2 mJ

1.2.2 Characterization of 10 kV SiC MOSFETs

Cree Inc. has developed 10 kV, 10 A SiC MOSFET module with 10 kV, 10 A Junction Barrier Schottky (JBS) diode. At the room temperature, MOSFET has a 3 V threshold on gate voltage. The on state drop across the device Vdson is 4.1 V while conducting 10 A and gate voltage of 20 V, as shown in Figure 1.11. At the

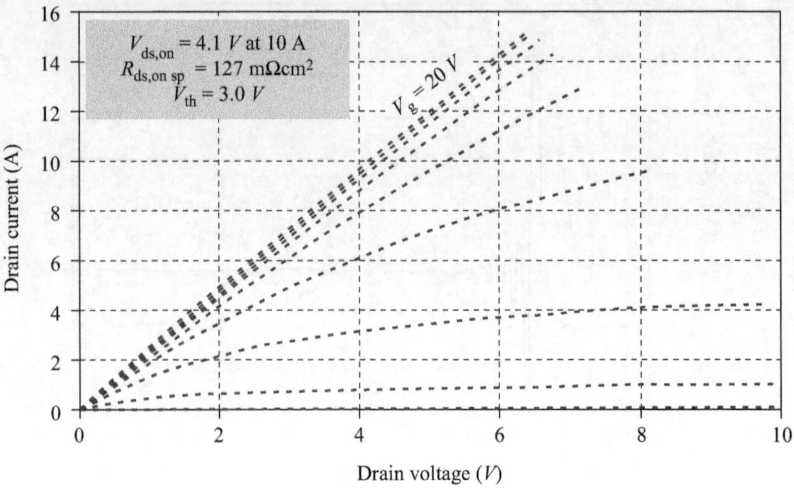

Figure 1.11 SiC 10 kV, 10 A MOSFET drain conductance: $R_{ds,on} = 400\ m\Omega$ *with* $V_g = 20\ V$ *[5]*

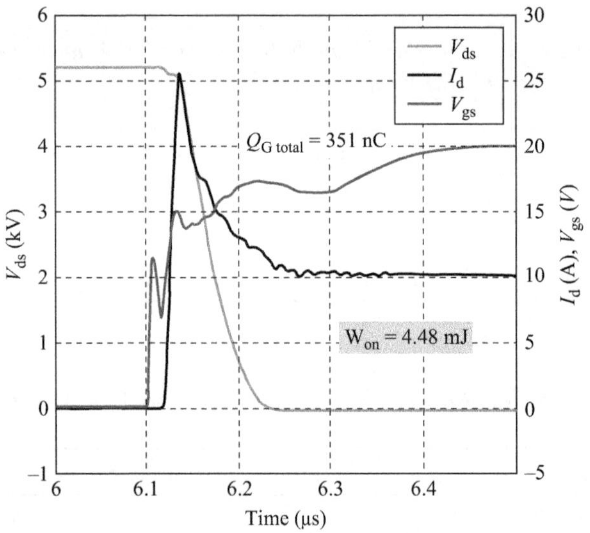

Figure 1.12 Turn-on waveform for SiC 10 kV, 10 A MOSFET [5]

room temperature, the JBS diode has knee voltage of 1.2 V and forward drop V_F of 4 V with 10 A current. Both the MOSFET and JBS diode have positive temperature coefficient for their on-resistance, thus it facilitates the parallel operation for higher current ratings. In the OFF state, both the devices have sub microampere leakage currents at 10 kV; and avalanche at 12 kV. The clamped inductive switching turn-on measurements are shown in Figure 1.12; it reveals the fast transition times for the MOSFET and no stored minority carrier charge in the JBS diode reverse

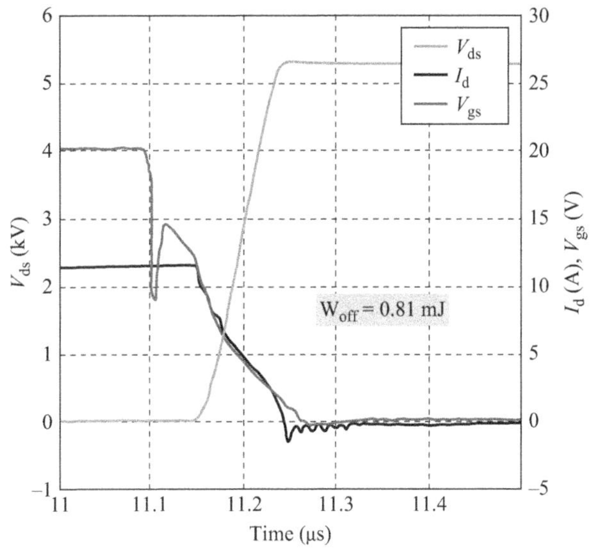

Figure 1.13 Turn-off waveform for SiC 10 kV, 10 A MOSFET [5]

recovery. The MOSFET completes the turn-on transient in 150 ns and requires 351 nC of gate charge. The turn-off transient is shown in Figure 1.13; the transition is fast and takes 144 ns without any tail current effects. The total switching power losses is much lesser than those of Si devices [5].

1.3 Low-voltage SiC devices and its characteristics

1.3.1 Low-voltage gate drive design

The gate driver is an essential subsystem of the power converter and influences the switching characteristics of the devices; and therefore the development of a suitable gate driver for SiC devices is discussed. The gate drive circuit has eight isolated channels to drive eight devices. Each channel is galvanically isolated at power supply side and optically isolated at signal side. The salient features of the gate drive are:

- Gate signal from control circuit is passed to the device through opto-coupler (HCPL-316J) based gate driver which provides optical isolation.
- Isolated power supply is used to provide power supply to the gate. Gate supply voltage for SiC MOSFET is +20/−5 V. The power supply levels can be changed.
- The gate driver can provide protection from current shoot through, by sensing drain to source voltage, V_{ds} of the device.
- Common-mode current injection in the control circuit through the coupling capacitor of isolated power supply is minimized by inserting common-mode choke.

The designed gate drive board is shown in Figure 1.14 with different sections of it highlighted.

Figure 1.14 Eight channel gate driver for SiC devices

1.3.2 Common-mode current minimization

Common-mode current is generated in the switched mode power converter due to presence of parasitic capacitance to ground. Due to faster switching of the SiC device, the dv/dt of converter common-mode voltage is much higher compared to Si device. Therefore, even a small coupling capacitor across either isolated power supply or across optically isolated driver, can cause significant amount of common-mode current to flow into control ground. To minimize the common-mode current flow into the control circuit, a common-mode choke is used at the input side of the isolated power supply. High impedance of common-mode choke can reduce the common-mode current peak substantially.

1.4 Characterization of 1,200 V, 100 A SiC MOSFET

1.4.1 1,200 V, 100 A SiC MOSFET device characterization without complementary device of the half-bridge module

Characterization of 1,200 V, 100 A SiC MOSFET is performed in two different ways. In the first case, the top device is taken out of the circuit by not connecting in to dc bus. Instead, a FWD is used. Turn-ON characteristics at 800 V and 100 A with gate resistance (R_g) of 15 Ω is shown in Figure 1.15, and the turn-ON energy loss (E_{on}) is measured to be 8.2 mJ. Turn-OFF characteristics are shown in Figure 1.16, and turn-OFF energy loss (E_{off}) is measured to be 5 mJ. During turn ON, device current is seen to be ringing due to L-C oscillation of FWD capacitance and stray inductance of the bus bar and device layout. Device voltage overshoot is observed during turn OFF because of the stray inductance.

1.4.2 1,200 V, 100 A SiC MOSFET device characterization with complementary device of the half-bridge module

In this section, device characterization, using complementary device in the module as FWD, has been discussed. From the turn-ON characteristics in Figure 1.17,

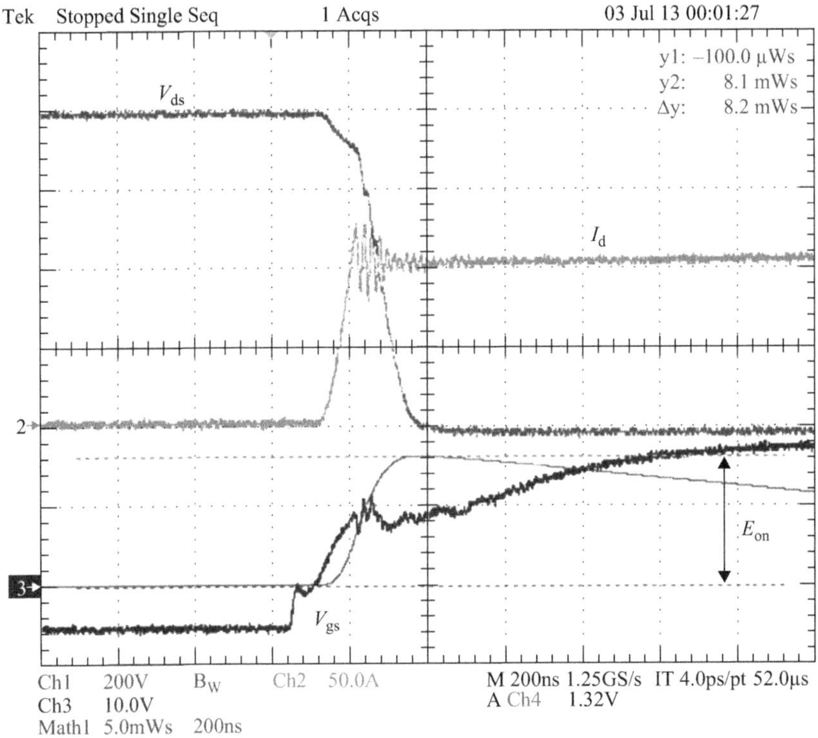

Figure 1.15 Turn-ON characteristics with $R_g = 15\ \Omega$ *and* $T_j = 125\ °C$*, measured* $E_{on} = 8.2$ *mJ, scale:* $V_{ds} \Rightarrow 200$ *V/div,* $I_d \Rightarrow 50$ *A/div,* $V_{gs} \Rightarrow 10$ *V/div, energy ⇒ 5 mJ/div, time ⇒ 200 ns/div*

it can be seen that the device current has a spike which is not present without the complementary device in the circuit as shown in Figure 1.15. This is due to the high drain to source capacitance (C_{ds}) of the top device. When diode of the top device free wheels, the voltage across the top device is the forward drop of the diode. At low voltage, C_{ds} is higher and the current spike is generated from high dv/dt. The turn-ON current spike magnitude can be related with dv/dt as

$$\Delta I_d = -C_{oss}\frac{\Delta V_{ds}}{\Delta t} \tag{1.1}$$

The current spike causes additional turn-ON loss which is measured as 10.2 mJ against 8.2 mJ in the case when the complementary device remains unused. However, turn-OFF characteristics with and without complementary device shown in Figures 1.16 and 1.18, respectively, show almost same amount of overshoot of switch voltage. This is due to the unchanged commutation inductance in the circuit. The voltage overshoot can be expressed as

$$\Delta V_{ds} = -L_{com}\frac{\Delta I_d}{\Delta t} \tag{1.2}$$

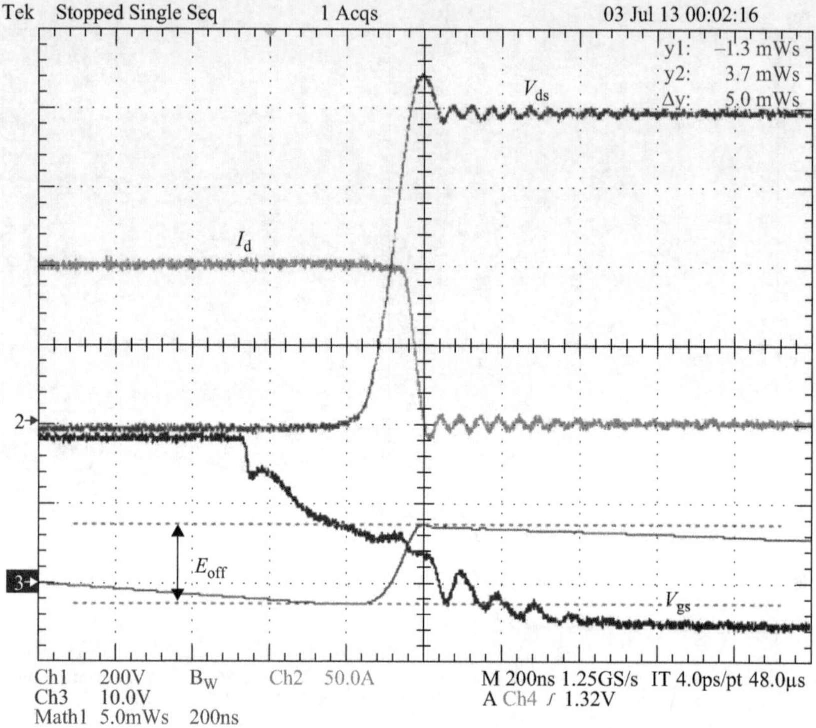

Figure 1.16 Turn-OFF characteristics with $R_g = 15\ \Omega$ *and* $T_j = 125\ °C$*, measured* $E_{off} = 5\ mJ$*, scale:* $V_{ds} \Rightarrow 800$ *V/div,* $I_d \Rightarrow 50$ *A/div,* $V_{gs} \Rightarrow 10$ *V/div, energy ⇒ 5 mJ/div, time ⇒ 200 ns/div*

The turn-ON and turn-OFF transients at different gate resistances are shown in Figures 1.19 and 1.20. Device voltages, device currents, and gate voltages are captured. It can be seen from Figure 1.19 that the delay and fall time of voltage decrease as the gate resistance is reduced. However, faster turn ON increases dv/dt and hence the device current spike is increased. From Figure 1.19, it can be seen that the dv/dt of the voltage fall at R_g of 5 Ω is 6 kV/μs.

During turn OFF, delay and fall time of current decrease with the decrease of gate resistance. Faster turn OFF increases di/dt and hence the device voltage oscillation and overshoot increase as seen from Figure 1.20. Switching losses versus device current is plotted in Figure 1.21.

1.4.3 Hard-switching characterization of 1,700 V SiC MOSFET [11]

1.4.3.1 SiC MOSFET switching

The 1,700 V, 50 A SiC MOSFET is tested with the dc bus voltage of 1,200 V and maximum switch current of 50 A. The gate voltage for SiC MOSFET is

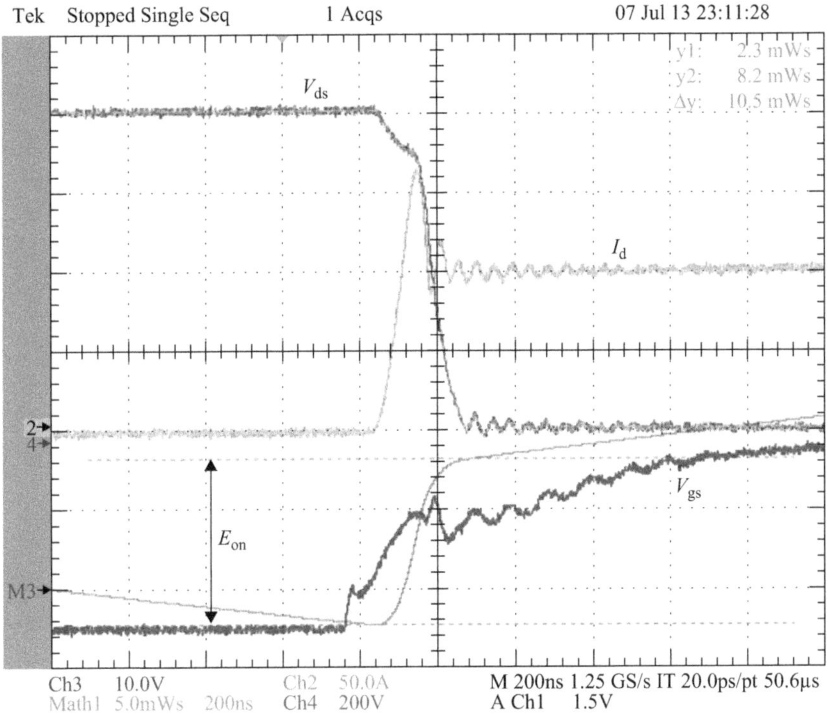

Figure 1.17 Module turn-ON characteristics at $R_g = 15\ \Omega$ *and* $T_j = 125\ °C$*, measured* $E_{on} = 10.5$ *mJ, scale:* $V_{ds} \Rightarrow 200$ *V/div,* $I_d \Rightarrow 50$ *A/div,* $V_{gs} \Rightarrow 10$ *V/div, energy ⇒ 5 mJ/div, time ⇒ 200 ns/div*

+20 V during turn-ON condition and −5 V during turn-OFF condition. In Figure 1.22(a), the turn-ON characteristics are shown using 20-Ω gate resistance at junction temperature of 125 °C. Switching energy is measured by integrating the product of device voltage (V_{ds}) and device current (I_d). Measured energy loss (E_{on}) during turn-ON transient is 3.38 mJ. Turn-OFF behavior of the device is shown in Figure 1.22(b) and measured turn-OFF loss (E_{off}) is 2.16 mJ. The maximum turn-ON and turn-OFF d*v*/d*t*s are found to be 23.7 and 24.5 kV/μs, respectively. The turn-ON and turn-OFF d*i*/d*t*s are 0.75 and 2.25 kA/μs, respectively.

1.4.3.2 Si IGBT switching

Characterization of the 1,700 V, 32 A Si IGBT is performed in the same test setup with the gate voltage of −15 to +15 V. The turn-ON and turn-OFF behaviors are shown in Figure 1.23 using gate resistance of 5 Ω at junction temperature of 125 °C. The energy loss during ON and OFF transients is 6.6 and 19.8 mJ, respectively. Turn-OFF loss of IGBT is much higher due to tailing effect of the current.

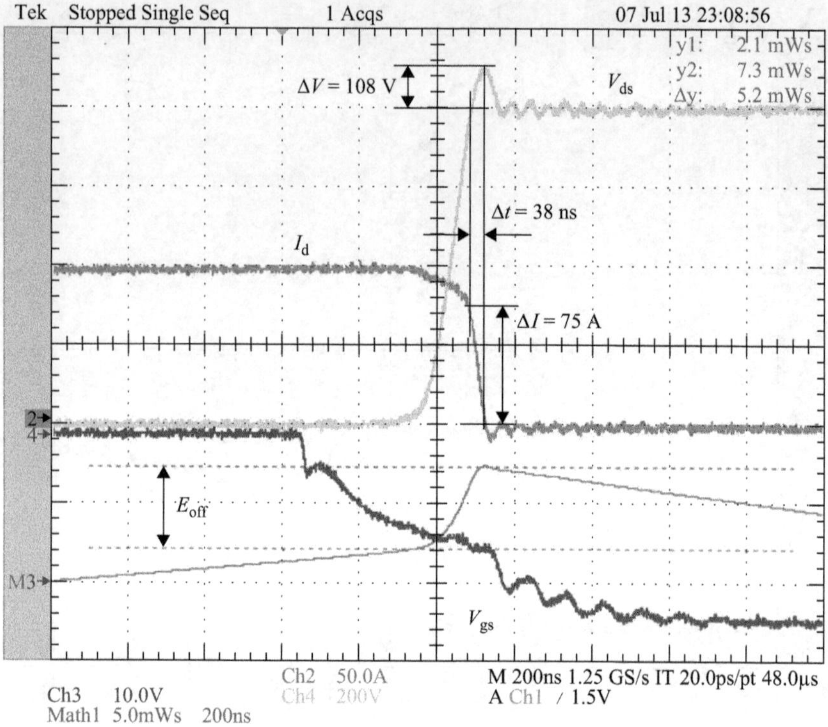

Figure 1.18 Module turn-OFF characteristics at $R_g = 15\ \Omega$ *and* $T_j = 125\ °C$*, measured* $E_{off} = 5.2$ *mJ, scale:* $V_{ds} \Rightarrow 200$ *V/div,* $I_d \Rightarrow 50$ *A/div,* $V_{gs} \Rightarrow 10$ *V/div, energy ⇒ 5 mJ/div, time ⇒ 200 ns/div*

1.4.3.3 BiMOSFET switching

BiMOSFET, a silicon device from IXYS Corporation, is constructed by keeping MOSFET and IGBT in its structure. Switching losses of the 1,700 V, 42 A BiMOSFET are evaluated using the same circuit setup with gate voltage of −15 and +15 V for OFF and ON conditions. Figure 1.24 shows the BiMOSFET turn-ON and turn-OFF characteristics using a gate resistance of 5 Ω at junction temperature of 125 °C. The turn-OFF tail current is observed, and the turn-OFF loss is much higher compared to turn-ON loss. Also, the switching loss found to be similar to the IGBT.

1.4.4 Performance comparison of MOSFET and IGBT

Variation of switching energy loss with different gate resistances is shown in Figure 1.25. Gate resistance has prominent effect on both turn-ON and turn-OFF energy losses of SiC MOSFET. The turn-OFF loss of the bipolar device is dominated by the tail current and it is not influenced by the change of the gate resistance as can

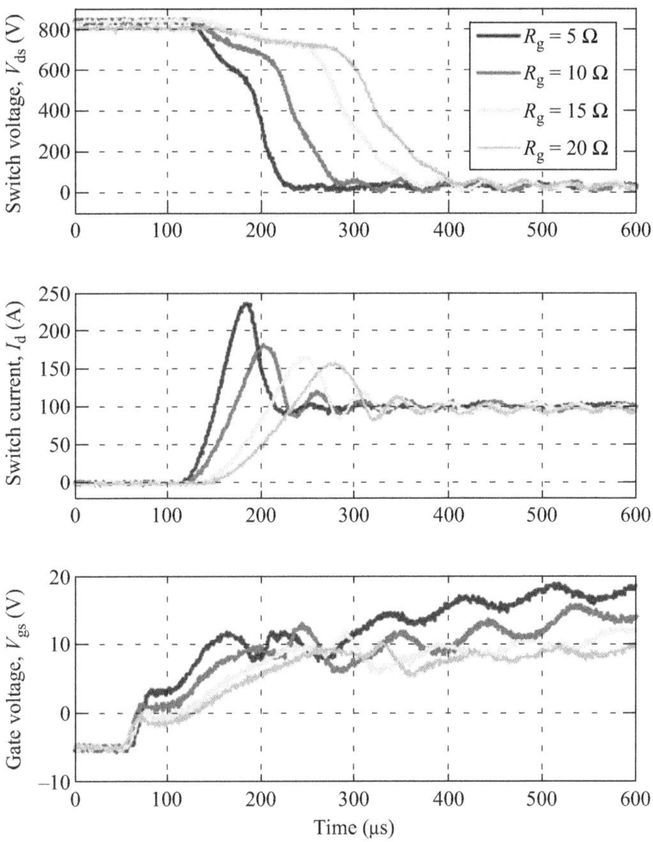

Figure 1.19 Turn-ON transient behavior with different gate resistances, R_g *at junction temperature,* $T_j = 125\ °C$

be seen in Figure 1.25. Turn-ON loss tends to decrease with faster switching by lowering the gate resistance. However, at lower gate resistances, the change becomes insignificant. Switching energy loss of three devices as a function of the device current at dc bus voltage of 1,200 V is characterized and given in Figure 1.26. It can be seen that the total loss of both BiMOSFET and IGBT is almost 5–6 times higher than that of SiC MOSFET. The devices are characterized at different junction temperatures to find out their switching loss dependency on junction temperature which is important for scaling up the power. The switching loss variation with junction temperature is plotted in Figure 1.27. Turn-ON loss of SiC MOSFET goes down slightly whereas its turn-OFF loss remains constant with the increase of junction temperature. As a result, the total switching loss (E_{sw}) decreases slightly with temperature elevation. Whereas, the increase of junction temperature causes steady increase of both turn-ON and turn-OFF losses of IGBT as well as BiMOSFET. The junction temperature of the devices is raised up to 150 °C.

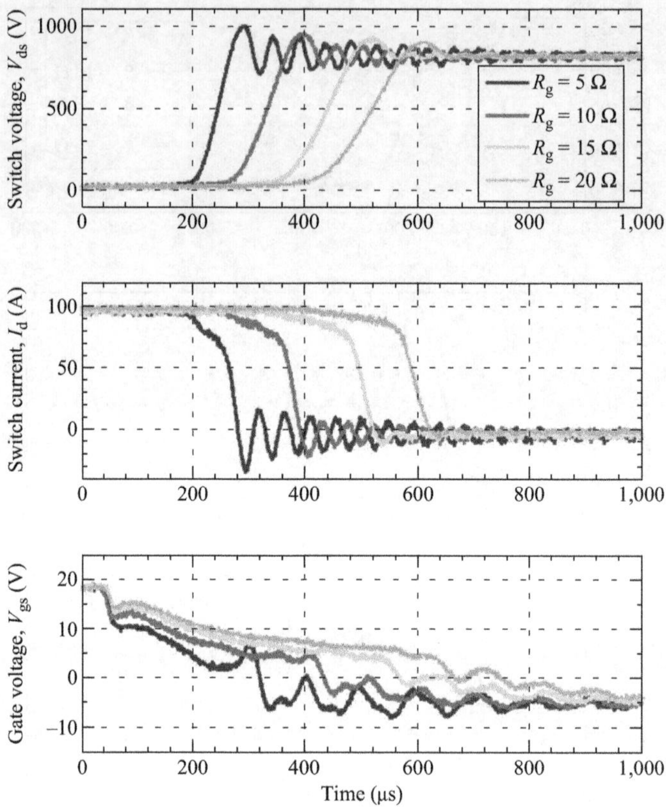

Figure 1.20 Turn-OFF transient behavior with different gate resistances, R_g *at junction temperature,* $T_j = 125\ ^\circ C$

1.4.4.1 Forward characterization

Generally, due to conductivity modulation of the drift region, the on-state voltage drop of Si IGBT is quite low. However, with the use of a wide band gap material like SiC, for the same break down voltage the drift region width of MOSFET is shorter. Hence, it offers considerably lower on-state voltage drop than the Si devices. The forward characteristics of the devices are given in Figure 1.28. The maximum current of characterization is limited by the curve tracer. Over the entire temperature range, the forward voltage drop of SiC MOSFET is found to be significantly less compared to IGBT and BiMOSFET. Therefore, the use of SiC MOSFET reduces the conduction loss substantially in the high-power converters.

1.4.4.2 Operating boundary in output power-switching frequency plane

The results from the device characterization are put together to evaluate the operating capabilities of these devices in a particular converter. A non-isolated dc–dc boost converter is selected to compute the limits of the output power and frequency

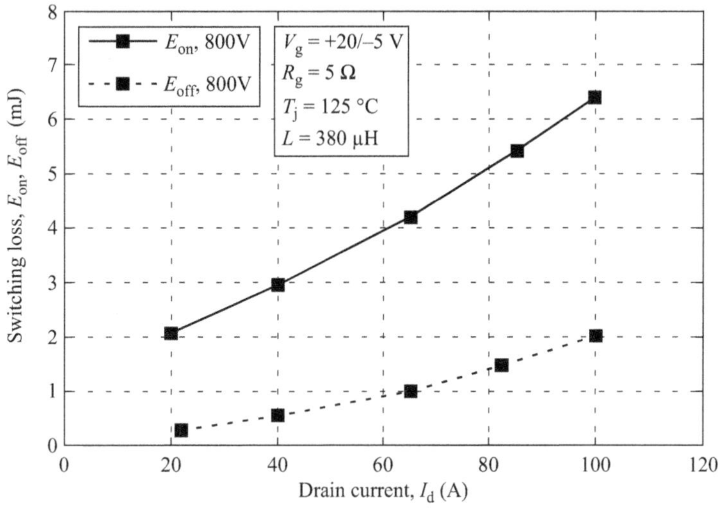

Figure 1.21 *E_{on} and E_{off} measured with* $R_g = 5\ \Omega$ *and* $T_j = 125\ °C$ *at different device currents,* I_d

of operation using each device. The thermal arrangements and gate resistances are assumed to be fixed for computing the operating points. For each device, the ON and OFF condition gate resistances are selected to be 5 and 10 Ω, respectively. It is assumed that the thermal arrangements can dissipate 150 W of heat at 125 °C of device junction temperature at a certain ambient temperature. Also, the output voltage of the converter is assumed to be 1,200 V while operating at 40% duty ratio (D). Now, the switching power loss, P_{sw} and conduction loss, P_{cond} are expressed as

$$P_{sw} = E_{sw} f_{sw} \tag{1.3}$$

$$P_{cond} = V_{fwd} I_{sw} D \tag{1.4}$$

where f_{sw}, E_{sw}, V_{fwd}, and I_{sw} are switching frequency, total switching loss, forward voltage drop, and switch current, respectively. In the analysis, ripple current is assumed to be small and insignificant. Switching loss and forward voltage drop can be expressed as the functions of device current with fixed junction temperature, ON and OFF gate resistances, and blocking voltage as

$$E_{sw} = f(I_{sw}, 125\ °C, 5\ \Omega, 10\ \Omega, 1,200\ V) = f_1(I_{sw}) \tag{1.5}$$

$$V_{fwd} = f(I_{sw}, 125\ °C) = f_2(I_{sw}) \tag{1.6}$$

E_{sw} and V_{fwd} can be found by curve fitting of the experimental results given in Figures 1.26 and 1.28. For SiC MOSFET, the switching loss is scaled appropriately for the lower gate resistance. By selecting a particular switch current, the power

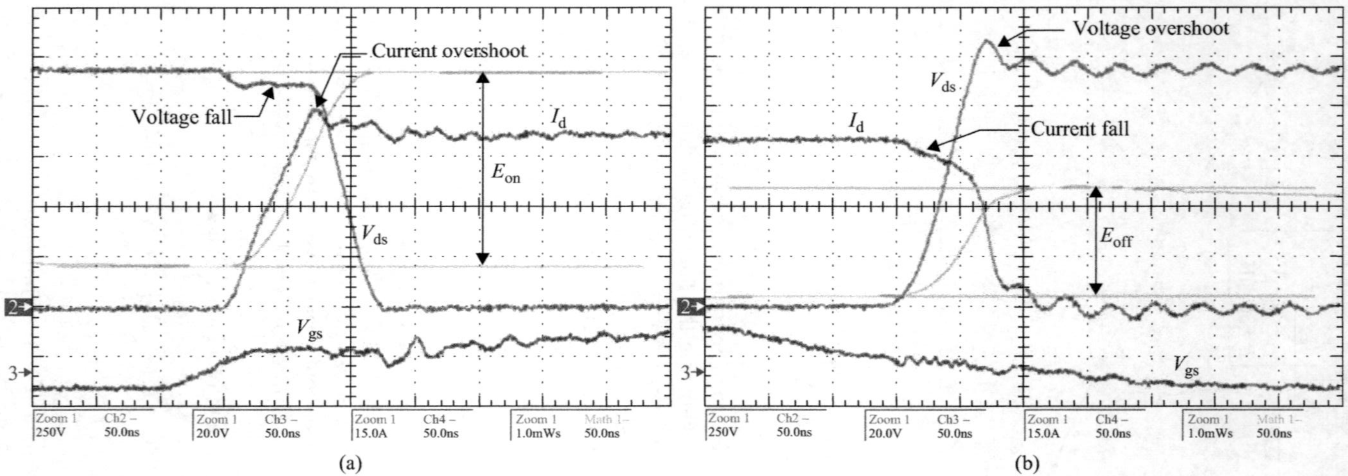

(a) (b)

Figure 1.22 Switching characteristics of SiC MOSFET: $R_g = 20\ \Omega$ *and* $T_j = 125\ °C$*; measured* $E_{on} = 3.88$ *mJ and* $E_{off} = 2.16$ *mJ; scale:* $V_{ds} \Rightarrow 250$ *V/div,* $I_d \Rightarrow 15$ *A/div,* $V_{gs} \Rightarrow 20$ *V/div, time* $\Rightarrow 50$ *ns/div. (a) Turn-ON. (b) Turn-OFF*

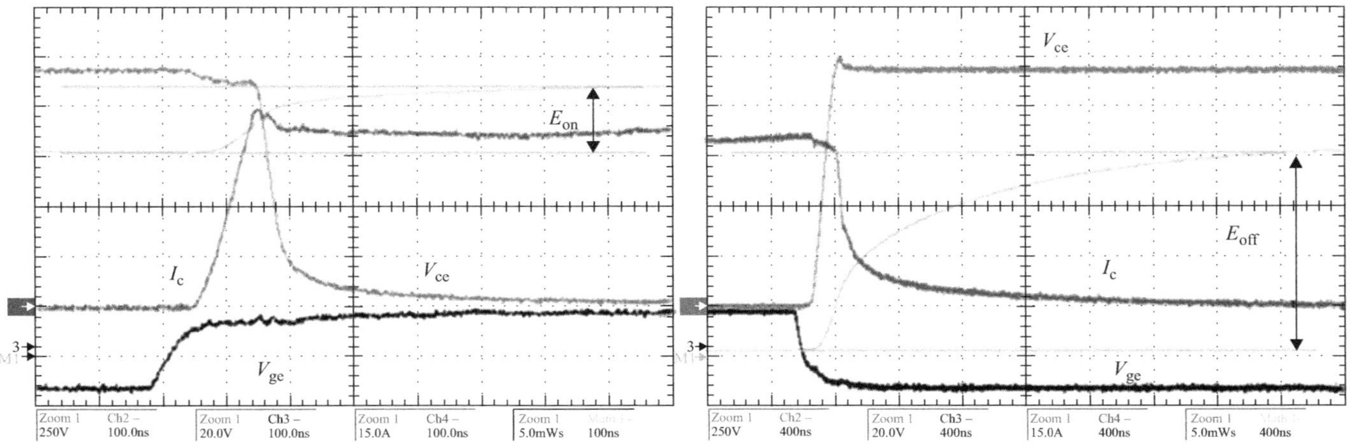

(a) (b)

Figure 1.23 Switching characteristics of Si IGBT: $R_g = 5\ \Omega$ *and* $T_j = 125\ °C$*; measured* $E_{on} = 6.6$ *mJ,* $E_{off} = 19.8$ *mJ; scale:* $V_{ce} \Rightarrow 250$ *V/div,* $I_c \Rightarrow 15$ *A/div,* $V_{ge} \Rightarrow 20$ *V/div, time* $\Rightarrow$ *100 ns/div (turn-ON), 400 ns/div (turn-OFF). (a) Turn-ON. (b) Turn-OFF*

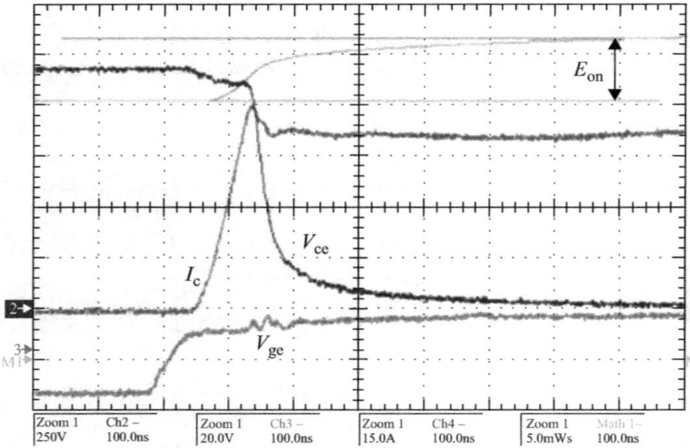

(a)

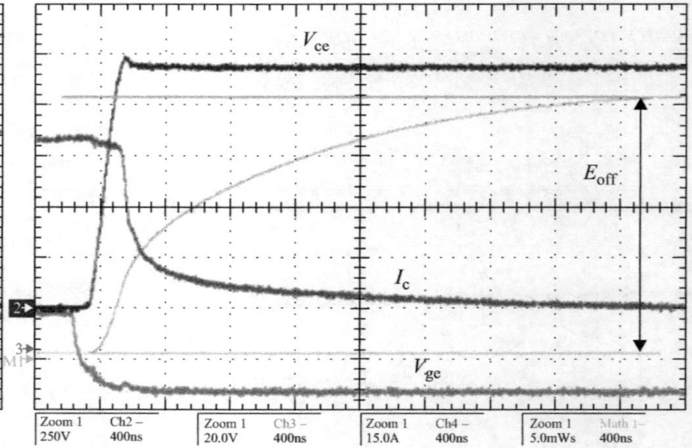

(b)

Figure 1.24 *Switching characteristics of Si BiMOSFET:* $R_g = 5\ \Omega$ *and* $T_j = 125\ °C$*; measured* $E_{on} = 6.1$ *mJ,* $E_{off} = 25.2$ *mJ; scale:* $V_{ce} \Rightarrow 250$ *V/div,* $I_c \Rightarrow 15$ *A/div,* $V_{ge} \Rightarrow 20$ *V/div, time* $\Rightarrow$ *100 ns/div (turn-ON), 400 ns/div (turn-OFF). (a) Turn-ON. (b) Turn-OFF*

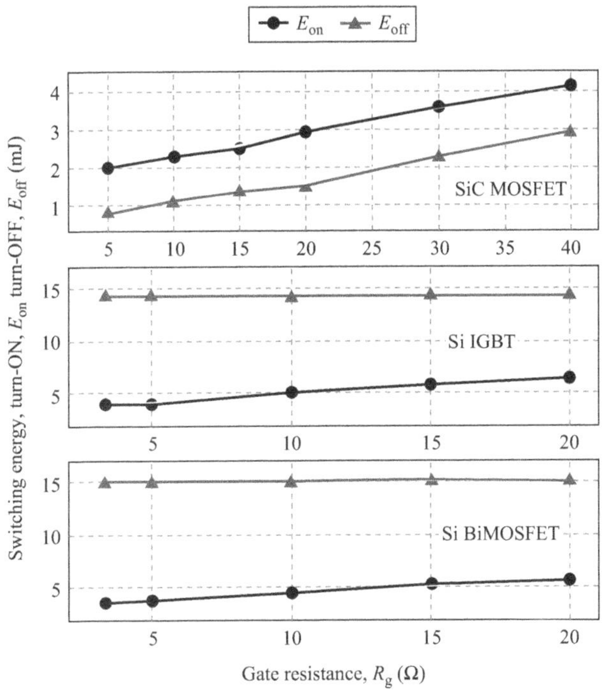

Figure 1.25 Switching loss variation with gate resistance, R_g*; junction temperature of SiC MOSFET,* $T_j = 125\,°C$*; junction temperature of IGBT and BiMOSFET,* $T_j = 25\,°C$*; dc bus voltage,* $V_{dc} = 1{,}200\ V$*; switch current = 40 A*

output and the corresponding switching frequency can be found from (1.7) and (1.8), respectively.

$$P_o = V_o I_{sw}(1 - D) - 150 \tag{1.7}$$

$$f_1(I_{sw})f_{sw} + f_2(I_{sw})I_{sw}D = 150 \tag{1.8}$$

For simplicity, losses except in the device are neglected for computation of power output. The operating boundary for each device is plotted in Figure 1.29. It can be seen that, over the entire region of operation, the SiC MOSFET can operate at much higher frequency while delivering the same power output. At 10 kW, the switching frequency multiplicity is almost ten; and at 30 kW, it is found to be around seven. Here, three 10-kW converters are demonstrated using these devices and compared for maximum possible switching frequency of operation.

1.4.5 Gate drive design and characterization of 1,200 V/45 A infineon SiC JFET module [12]

The 1,200 V/45 A Infineon SiC JFET module features two pairs of cascodes in series. However, the design approach is novel – the low-voltage MOSFET is p-type and the

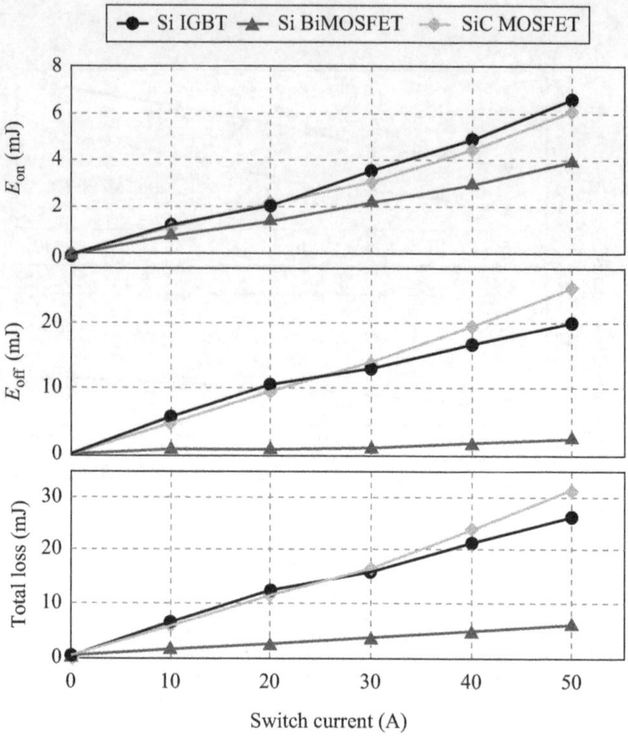

Figure 1.26 Switching loss comparison at different switch currents; gate resistance for IGBT and BiMOSFET, $R_g = 5\ \Omega$*; gate resistance for SiC MOSFET,* $R_g = 20\ \Omega$*;* $T_j = 125\ °C$ *and dc bus voltage,* $V_{dc} = 1{,}200\ V$

gate of the JFET is not tied to ground as in the classic case. The JFET and MOSFET in each pair have an anti-parallel diode across them, as shown in Figure 1.30. These anti-parallel diodes serve as freewheeling agents for the current when the JFET modules are tested under load.

The double-pulse test results with zero gate resistance are shown in Figure 1.31. Zoomed waveforms of the turn-on for $R_g = 0\ \Omega$ are shown in Figure 1.32. The turn-on current spike is appeared due to the capacitance charging of the complementary device.

With zero external gate resistance, the turn-ON and turn-OFF losses of JFET switching are 1.3 and 1.9 mJ, respectively.

1.4.6 SiC super-junction transistor characteristics

Super-junction transistor (SJT) is a bipolar junction transistor with higher gain, which reduces the base (gate) drive significantly. This device has very low switching energy loss and can operate with high switching frequency. The switching waveforms are given in [13], shown in Figure 1.33. SJT device can

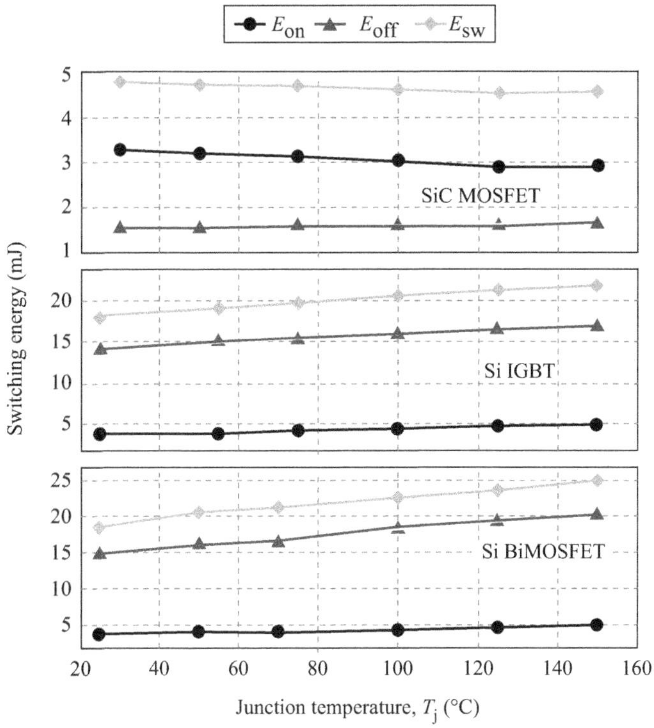

Figure 1.27 Switching loss variation with junction temperature, T_j*; gate resistance for IGBT and BiMOSFET,* $R_g = 5\ \Omega$*; gate resistance for SiC MOSFET,* $R_g = 20\ \Omega$*; dc bus voltage,* $V_{dc} = 1{,}200$ *V; switch current = 40 A*

operate at higher junction temperature without any significant rise of its switching loss.

1.5 Zero voltage switching characterization of 12 kV SiC [14]

Zero voltage switching (ZVS) technique is a proven method of reducing d*v*/d*t* stress and switching losses. In this section, the ZVS characteristics of the SiC IGBTs are illustrated.

The 2 μm buffer layer N-IGBT has about 20% lower forward drop than the 5 μm buffer layer IGBT. However, the 5 μm buffer layer IGBT has significantly lower turn-off loss and turn-on d*v*/d*t*. Therefore, the 5 μm buffer layer IGBT is a favorable candidate for high switching frequency applications, whereas, the 2 μm IGBT is favorable for high-current, low-frequency applications. The high d*v*/d*t* should be limited to minimize common-mode currents, stress on insulation, and parasitic ringing for reliable operation of power converter systems. To achieve that objective, ZVS performance of the SiC IGBTs is being investigated.

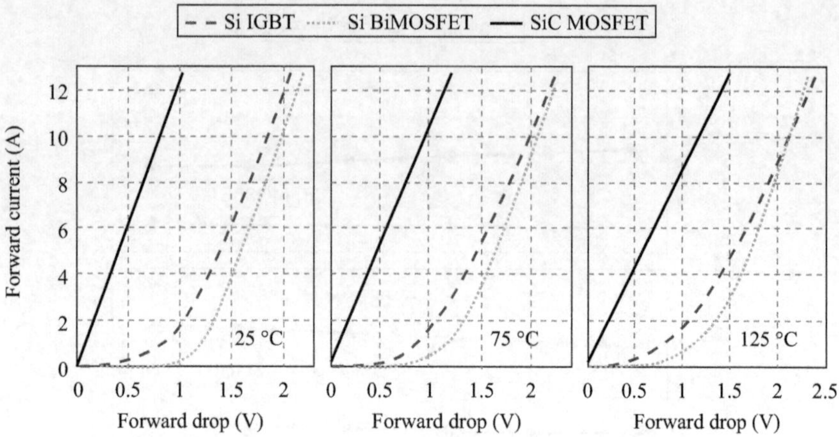

Figure 1.28 Forward characteristics of IGBT ($V_{ge} = 5$ V), BiMOSFET ($V_{ge} = 15$ V), and SiC MOSFET ($V_{gs} = 20$ V) at different junction temperatures

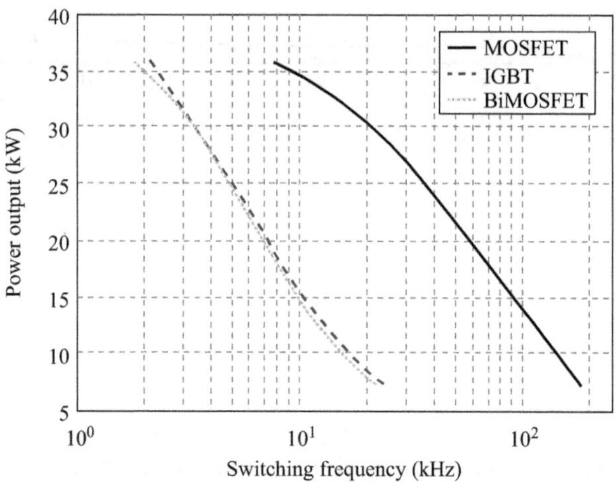

Figure 1.29 Boundary of converter operating points with 150 W of heat dissipation from the device for maximum junction temperature of 125 °C

1.5.1 ZVS turn-on characteristics

Under ZVS conditions, the turn-on gate bias is provided for the IGBT just before the circuit gets configured to conduct current through it. Despite formation of the MOS channel, the conductivity modulation lag (forward recovery) in the drift region results in higher forward drop (V_{ce}) across the IGBT, when the current starts to flow with reasonable di/dt. Once the drift layer is fully modulated, the V_{ce} of the IGBT settles down to steady-state saturation value. Consequently, higher di/dt

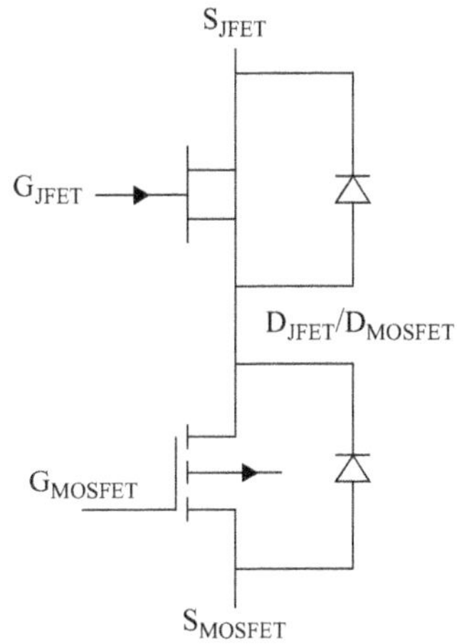

Figure 1.30 Novel cascode topology with P-MOSFET and anti-parallel diodes

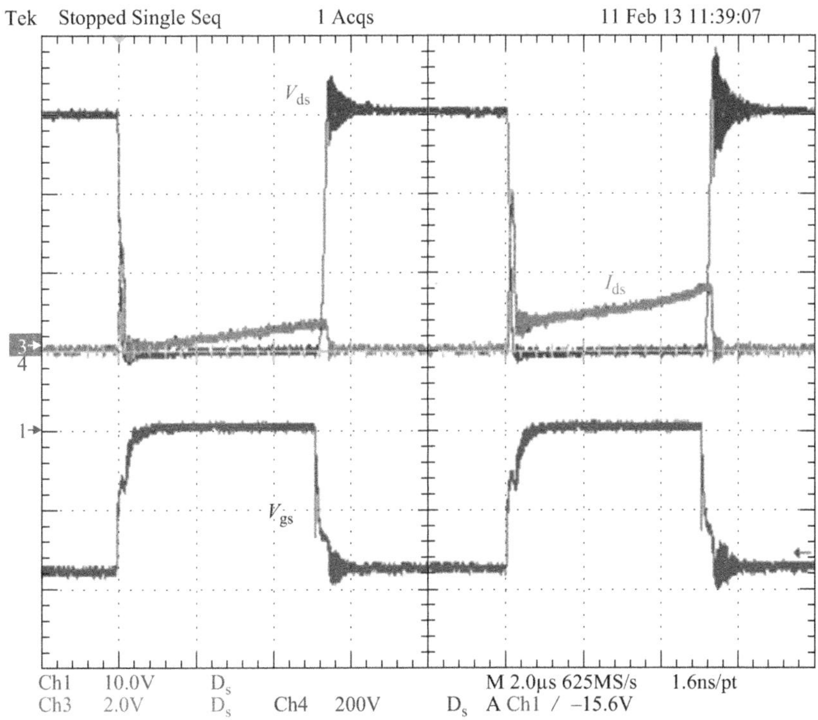

Figure 1.31 Switching waveforms – $V_{dc} = 600$ *V,* $R_g = 0\ \Omega$ *(*$V_{ds} = 200$ *V/div,* $I_{ds} = 40$ *A/div,* $V_{gs} = 10$ *V/div)*

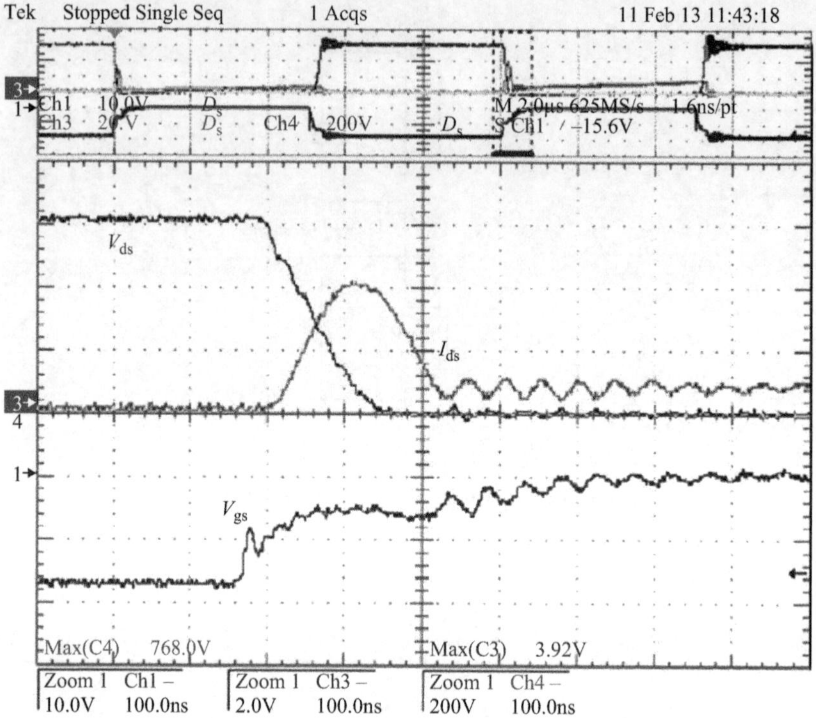

Figure 1.32 Zoomed switching waveforms – $V_{dc} = 600$ *V,* $R_g = 0\ \Omega$ *(*$V_{ds} = 200$ *V/div,* $I_{ds} = 40$ *A/div,* $V_{gs} = 10$ *V/div)*

results in higher voltage spike because of instant high current availability in the unmodulated drift region.

Figures 1.34 and 1.35 show the turn-on voltage spike under different di/dt conditions for the 2 and 5 μm buffer layer IGBTs, respectively. The voltage spike has increased from 6 V to over 15 V when the di/dt is increased from 1 to 10 A/μs for both IGBTs. Also, it can be noticed that the 5 μm IGBT has about 1 V higher steady-state voltage drop, as explained earlier. The reduced mobility at higher temperatures also increases the resistance of unmodulated drift that result in higher voltage spike. Figure 1.36 shows the effect of temperature on the voltage spike for the 5 μm buffer layer IGBT. Similar behavior is observed for the 2 μm buffer layer IGBT. The voltage spike is found to have negligible influence due to package inductance.

The ZVS turn-on and turn-off characteristics are obtained through experiments on the test circuit with 10 kV SiC MOSFET/JBS diode co-packs used in place of the switch 1 and also FWDs (with gate shorted).

1.5.2 ZVS turn-off characteristics

The turn-off characteristics at different temperatures, with and without a 1.1 nF external snubber capacitor are presented in this section along with the explanation of underlying physics.

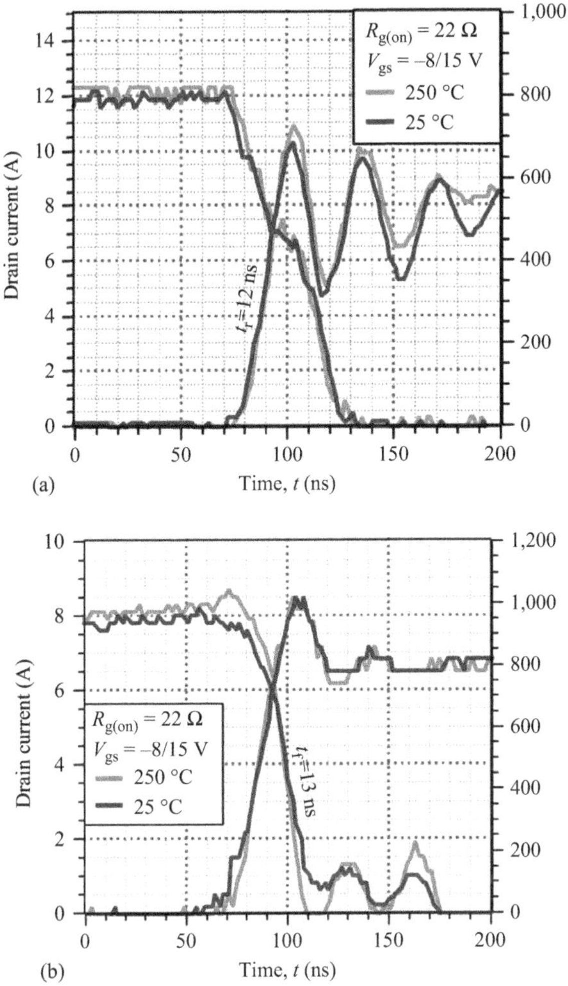

Figure 1.33 (a) Turn-ON and (b) turn-OFF drain current and voltage transients recorded for switching at 800 V and 8 A through a 3 mm² SiC SJT. There is no difference in switching speed between 25 and 250 °C, due to the unipolar nature of the SJT device design [13]

1.5.2.1 12 kV, 2 μm buffer layer N-IGBT

Figure 1.37 shows turn-off transition at 7 kV, 10 A at 25 and 150 °C without external snubber capacitor. The turn-off current has five different stages during the transition: a dip at the beginning of the transition resulting from transferring load inductor current into the IGBT capacitance (and capacitance of the FWD); the characteristic bump existing until the punch-through voltage (4.3 kV); a slow drop in current from punch-through voltage until the voltage reached dc bus value of 7 kV; a sudden drop in current once the voltage reached dc bus value of 7 kV; and,

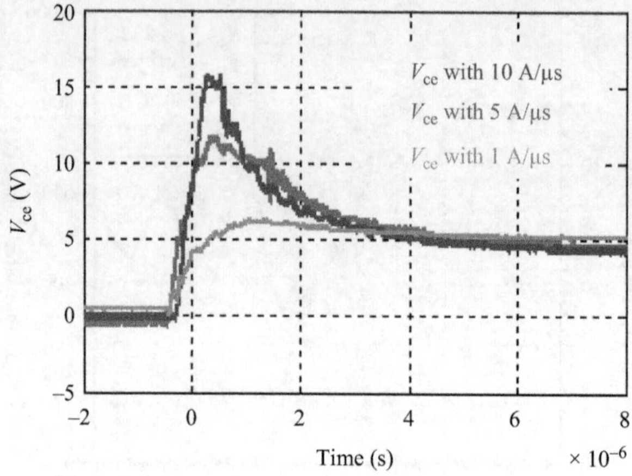

Figure 1.34 Turn-on voltage spike of 12 kV, 2 μm buffer layer SiC IGBT under different di/dt conditions at 150 °C

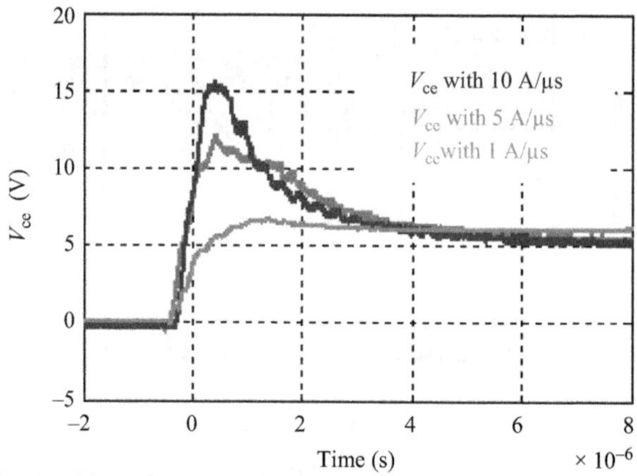

Figure 1.35 Turn-on voltage spike of 12 kV, 5 μm buffer layer SiC IGBT under different di/dt conditions at 150 °C

the current tail resulting from recombination of the stored charge. The voltage transition has two slopes, relatively slower one before the punch-through voltage, followed by faster one above the punch-through voltage. It is to be noted that the magnitude of collector–emitter capacitance, C_{CE}, of the IGBT is significantly different in the sweep-out phase (under 4.3 kV, removal of excess carriers), and punch-through phase (over 4.3 kV). Once the punch-through phase has been reached, the amount of minority carriers removed by the increase in V_{CE} becomes very small, resulting in much faster voltage rise, comparable to the ones in majority carrier devices.

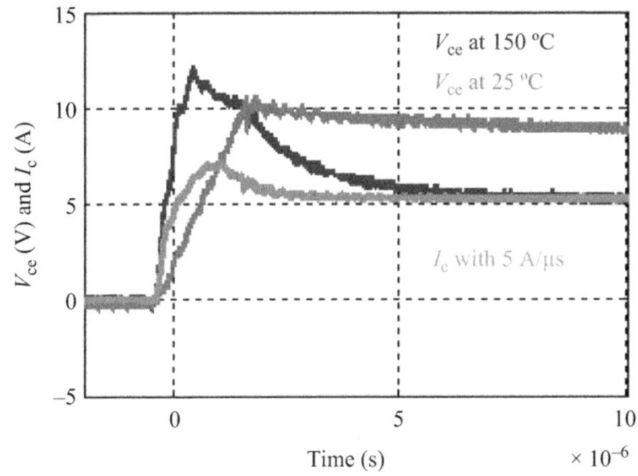

Figure 1.36 Turn-on voltage spike of 12 kV, 5 μm buffer layer SiC IGBT under di/dt of 5 A/μs at 25 and 150 °C

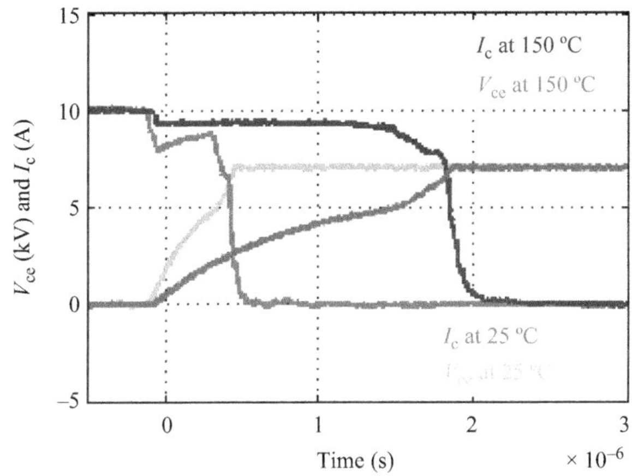

Figure 1.37 Turn-off transitions of 12 kV, 2 μm buffer layer SiC IGBT at 25 and 150 °C without an external snubber capacitor

At 150 °C, the magnitude of the characteristic current bump is higher and its duration is longer. The current bump is resulting from base drive current to the internal PNP transistor provided by the dv/dt, but with much larger magnitude due to deep punch-through design adopted for the SiC IGBT. The magnitude of the current bump, with voltage rise, is decided by the amount of base drive current dictated by dv/dt [$i = C \times (\mathrm{d}v/\mathrm{d}t)$], and the gain of the IGBT. At 150 °C, the dv/dt is lower than that at 25 °C. However, the increased injection at 150 °C has resulted in higher current gain of the internal BJT, which in turn, is causing higher current bump. There is no current tail at 25 °C, but the increased stored charge at 150 °C is responsible for the small tail current.

Figures 1.38 and 1.39 show turn-off transitions with a 1.1 nF external snubber capacitor at 25 and 150 °C. The external capacitor is providing reduced dv/dt, but for longer duration. Consequently, the magnitude of the current bump is lower and is existing for longer duration. The increase in magnitude of the current bump as the voltage is rising is due to trade-off between IGBT capacitance reduction with voltage, and the increase in gain of the IGBT resulting from reduced undepleted drift layer. Over 4.3 kV, the low-depletion capacitance has dominant effect in comparison to the increase in the IGBT gain. Therefore, the current is dropping once the IGBT reaches its punch-through voltage. The same explanation is valid for the case of 150 °C shown in Figure 1.39, except for the presence of short tail

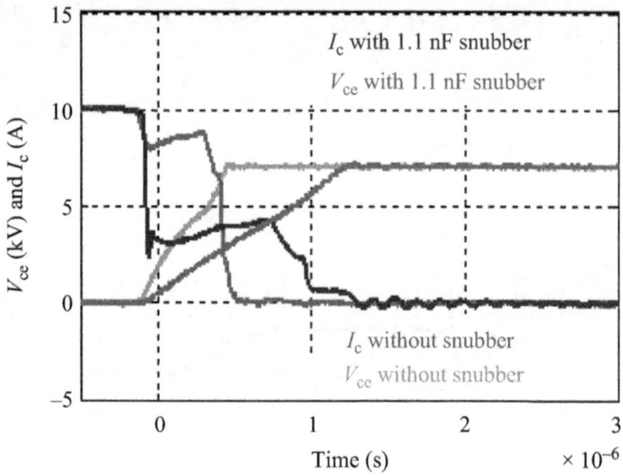

Figure 1.38 Turn-off transitions of 12 kV, 2 μm buffer layer SiC IGBT at 25 °C with and without the 1.1 nF external snubber capacitor

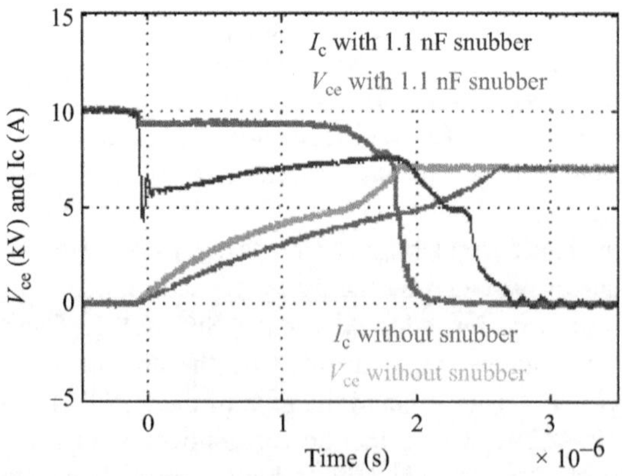

Figure 1.39 Turn-off transitions of 12 kV, 2 μm buffer layer SiC IGBT at 150 °C with and without the 1.1 nF external snubber capacitor

current for the case without the external snubber. In the presence of external snubber, the voltage rise is slower, providing enough time for charge recombination and no current tail is observed.

1.5.2.2 12 kV, 5 μm buffer layer N-IGBT

The turn-off characteristics of the 5 μm buffer layer IGBT without external snubber capacitor are shown in Figure 1.40. By comparing with the similar results of the 2 μm buffer layer IGBT (shown in Figure 1.37), it is apparent that the transitions of the 5 μm IGBT at 25 and 150 °C are significantly faster resulting in considerably lower switching loss. The 2 and 5 μm buffer layer IGBTs are identical, except for thickness of the buffer layer. The thicker buffer layer results in lower hole concentration at the junction of buffer and drift layers (due to reduced injection) and in the entire drift layer which causes faster rise of voltage. The lower gain of the internal PNP transistor (due to lower injection) is consequently resulting in current bump of smaller magnitude for the 5 μm buffer layer IGBT, as shown in Figure 1.40. The transitions with 1.1 nF external snubber capacitor at 25 and 150 °C are shown in Figures 1.41 and 1.42, respectively. The explanation of the transitions for the 2 μm buffer layer IGBT is completely valid for the 5 μm IGBT as well.

The 5 μm buffer layer IGBT, in comparison to the 2 μm IGBT, has lower magnitude and shorter duration of the turn-off current bump because of lower gain of the internal PNP transistor (due to reduced injection). Also, the turn-off energy loss of the 5 μm IGBT about 40%–50% of that of 2 μm IGBT, with and without the external snubber capacitor. In addition, the percentage turn-off loss reduction resulting from using external snubber capacitor is found to be about 10% more for the 5 μm IGBT. Considering these points, it is promising to explore further optimization of buffer layer parameters (e.g. thickness, doping concentration, and life time) as a promising future research direction for further efficient high-power devices.

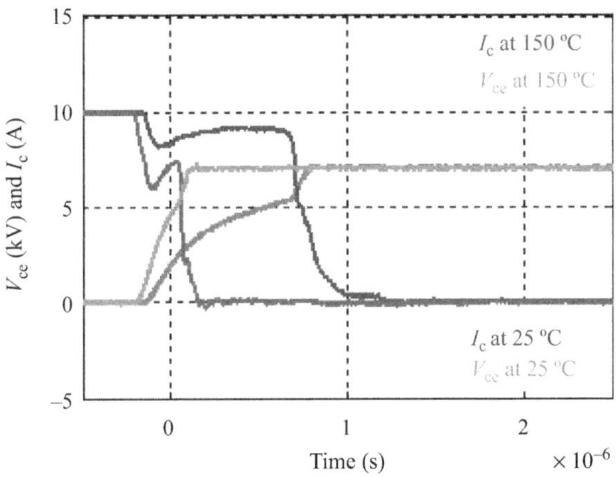

Figure 1.40 Turn-off transitions of 12 kV, 5 μm buffer layer SiC IGBT at 25 and 150 °C without an external snubber capacitor

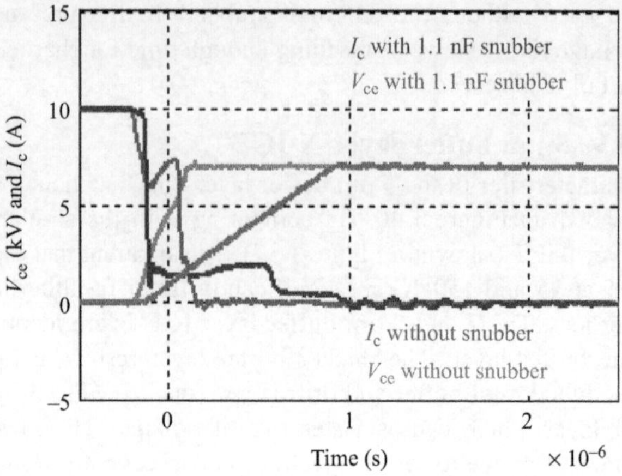

Figure 1.41 Turn-off transitions of 12 kV, 5 μm buffer layer SiC IGBT at 25 °C with and without the 1.1 nF external snubber capacitor

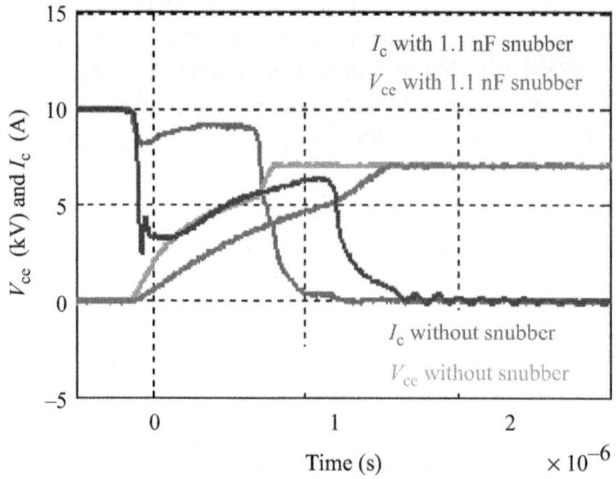

Figure 1.42 Turn-off transitions of 12 kV, 5 μm buffer layer SiC IGBT at 150 °C with and without the 1.1 nF external snubber capacitor

1.6 All SiC-based SST

An all SiC power device based three-phase SST topology is shown in Figure 1.43. This topology has been referred as "Transformerless Intelligent Power Substation" (TIPS) which interfaces the three phase, 13.8 kV and 480 V distribution grids [15]. TIPS has three power conversion stages. The active front-end converter (AFEC) is tied to medium voltage (MV) 13.8 kV distribution grid with a three-level neutral point clamped (3L-NPC) converter built using the 15 kV

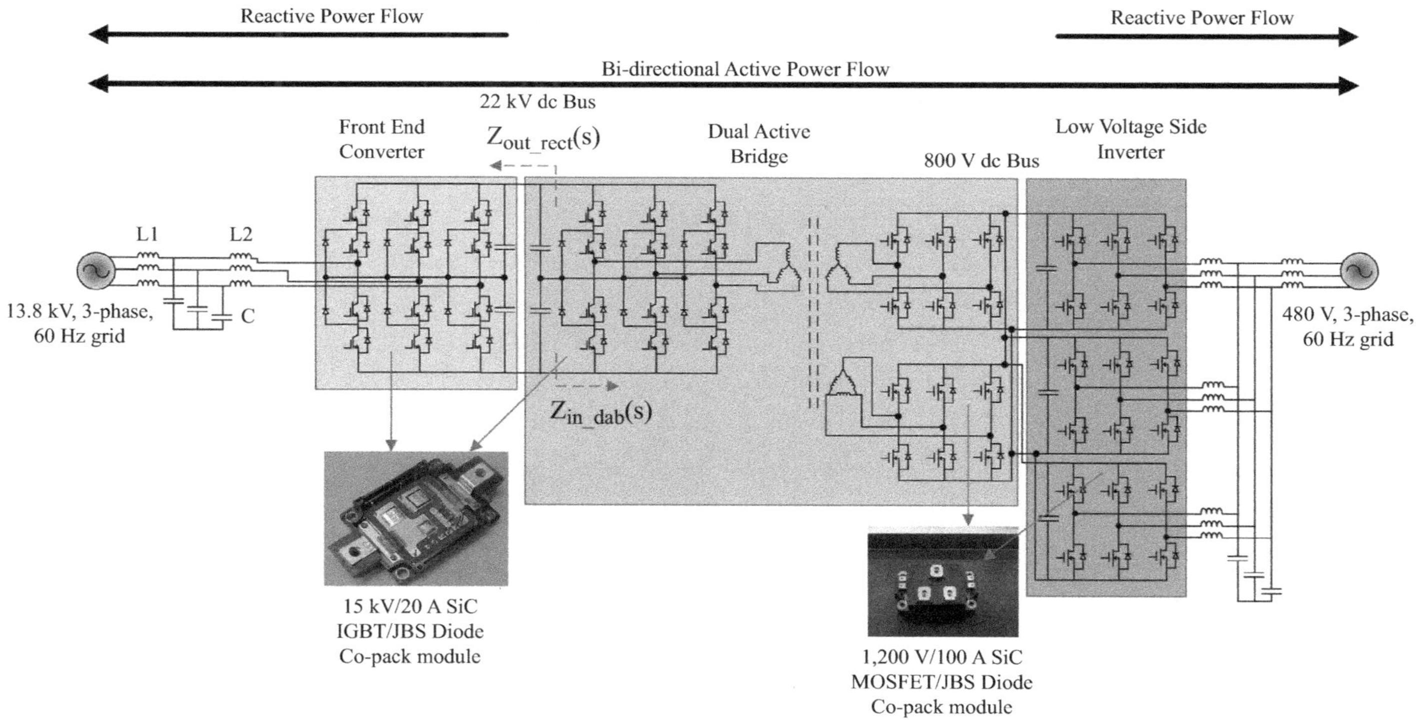

Figure 1.43 Schematic of the TIPS – a 13.8 kV–480 V grid-interfaced three-phase SST

SiC 0IGBTs. It is a PWM boost rectifier with 22 kV dc bus on the output, and is operated with hard-switched PWM at 3–5 kHz. The AFEC dc bus (22 kV) is converted to low voltage (LV) side dc bus (800 V) using a high-frequency dc link, enabled by a three-phase dual active bridge converter (DABC). The DABC is responsible for isolation and voltage conversion in the TIPS system. The DABC is ZVS soft-switched converter operated at 10 kHz. The MV side of DABC is also 3L-NPC converter built using 15 kV SiC IGBTs, and the LV side has 1,200 V SiC MOSFET half-bridge modules based two-level converter for each star- and delta-connected secondary winding. The output LV inverter stage is developed using 1,200 V SiC MOSFET modules with three interleaved 17–20 kHz, 35 kVA converters. The foremost advantage of the 15 kV SiC IGBT is simplified converter topologies for MV applications.

The complete hardware setup of the TIPS system developed at FREEDM Systems Center at NCSU is shown in Figure 1.44.

The 3L-NPC converters on the MV side of the TIPS have a modular structure with three poles for AFEC stage and three poles for primary side of the DABC. Each 3L-NPC pole has its own dc-link capacitor with a bus-bar connection to the 15 kV SiC IGBTs and 20 kV (2 × 10 kV) SiC JBS clamping diodes for low-stray inductance. The poles have been individually tested up to 10 kV dc input in inverter mode at 5 kHz and 7.5 kW before integrating them into the three-phase TIPS [15]. Figure 1.45 shows a 3L-NPC pole mounted on high-voltage bushings for reducing

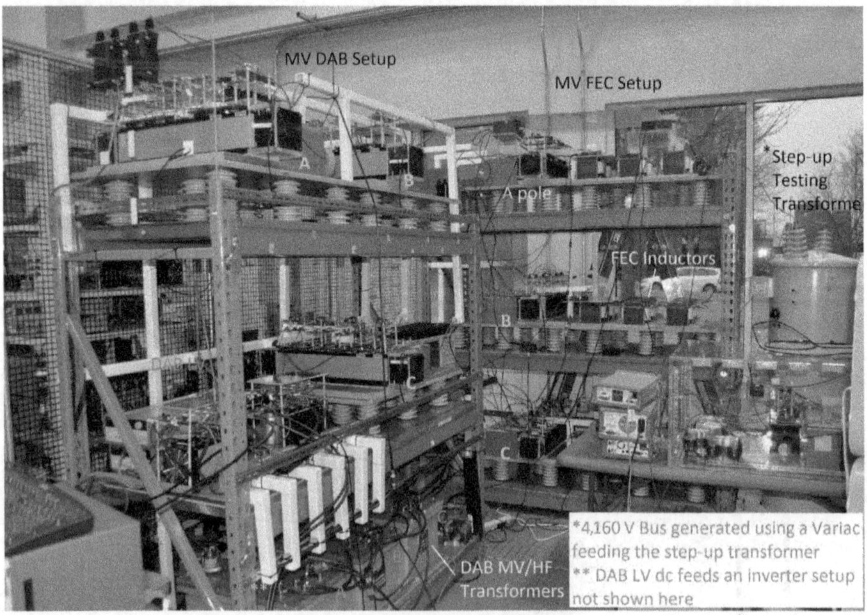

Figure 1.44 Experimental setup of TIPS system at FREEDM Systems Center at NCSU

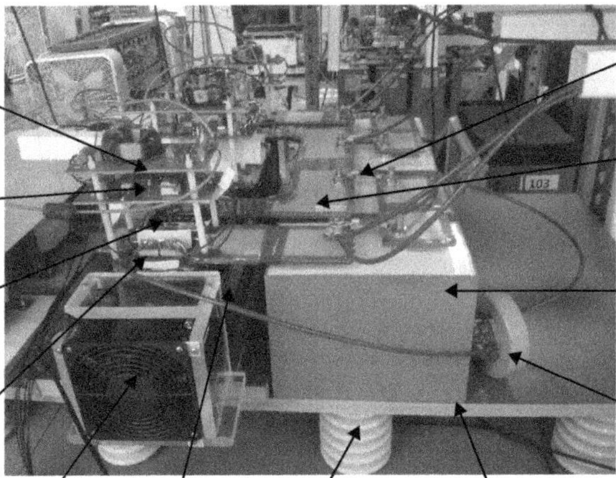

Figure 1.45 A pole of 3L-NPC converter

common-mode currents. The PWM signals are transmitted optically to minimize EMI issues and provide higher voltage isolation and also controller operation distant from the TIPS 3L-NPC converter poles.

Figures 1.46 and 1.47 show the test waveforms of the FEC and DABC under integrated mode, the complete TIPS in operation. The three-phase input currents are sinusoidal. The switching ripple in the current can be improved by designing filter inductors with very low capacitance. Also, the ringing currents in DABC currents, shown in Figure 1.47, are contributed by the parasitic coupling capacitances of the high-frequency transformer. Along with the SiC devices, improvements in the magnetic design have to be done to harness the benefits of SiC devices.

Two-level three-phase converters for the low-voltage side of the DABC and the inverter stage for 480 V grid interface are built with 1,200 V, 100 A SiC MOSFET half-bridge modules packaged by Powerex with dies from Cree. The V–I characteristics of the 1,200 V SiC MOSFET and its anti-parallel JBS diodes are given in [16]. Hard switching characterization of these 1,200 V SiC MOSFETs are performed to evaluate their turn-on and turn-off switching behaviors. Switching loss is measured as the function of switch current at different junction temperatures and gate resistances. Also, the switching dv/dts are characterized with varying gate resistances to select a suitable gate resistance to limit dv/dt to 15 kV/μs. With the experimentally evaluated switching energy loss data, the cooling system of the converters and its packaging are designed.

The gate driver is designed to provide high-peak current (up to 15 A if required) and shoot through capability tested up to 1 kV dc bus voltage. Since these gate drivers handle comparatively lower dv/dt (in the range of 15 kV/μs) compared to the MV side gate drivers, commercially available isolated dc–dc converters are used for

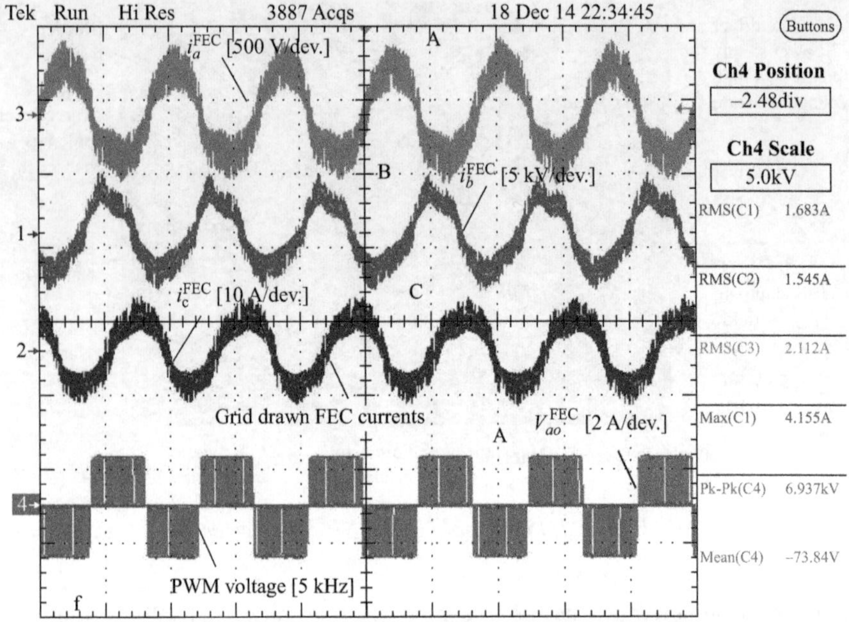

Figure 1.46 AFEC converter three-phase currents and phase A pole voltage under ac 3.42 kV grid tied TIPS system integration mode

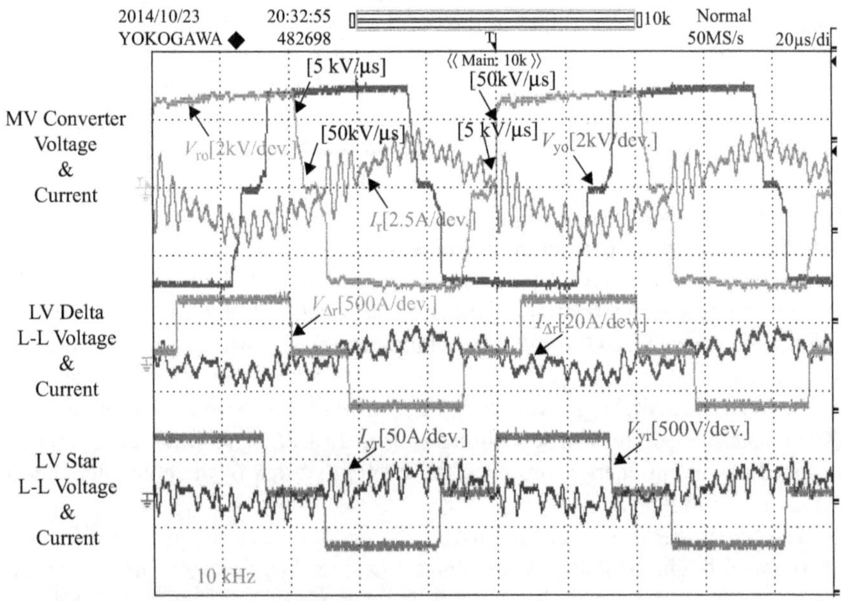

Figure 1.47 Experiment waveforms of DABC operation at 6 kV–400 V, 7.4 kW and 10 kHz switching frequency

Figure 1.48 The 1,200 V, 100 A, SiC half-bridge MOSFET module-based three-phase, two-level voltage source converter

isolated gate driver power supply. The packaged three-phase two-level converter is shown in Figure 1.48. All components of the converter including the 1,200 V, 100 A modules, sandwiched dc bus bar, gate drivers, heat sink with forced air cooling system, and dc bus capacitor are shown in Figure 1.48. Each converter has been demonstrated up to 50 kVA at 20 kHz switching frequency and 800 V dc bus voltage [10]. The converter test results at 35 kVA operation are shown in Figure 1.49.

At this point of time, the 15 kV SiC IGBT used in TIPS system are experimental devices and are not commercially available. With future commercialization of HV SiC devices, more industries, focusing on grid connected power electronics, will develop commercial SSTs applied to distributed renewable energy integration, MV motor drives, MV traction applications, MV shipboard power systems and disaster recovery transformers and many more applications.

1.7 Summary

Power electronics is a fundamental industry finding its usage everywhere where energy is used. Advances in power electronics devices would also impact all these applications and open avenues for more applications which were previously not possible. Silicon-based power devices are widely used in power electronics systems. With more demand for high switching frequency, high-power density, and higher blocking voltage, the application of the silicon-based power devices are limited. SiC, a wide band gap semiconductor has received tremendous interest from

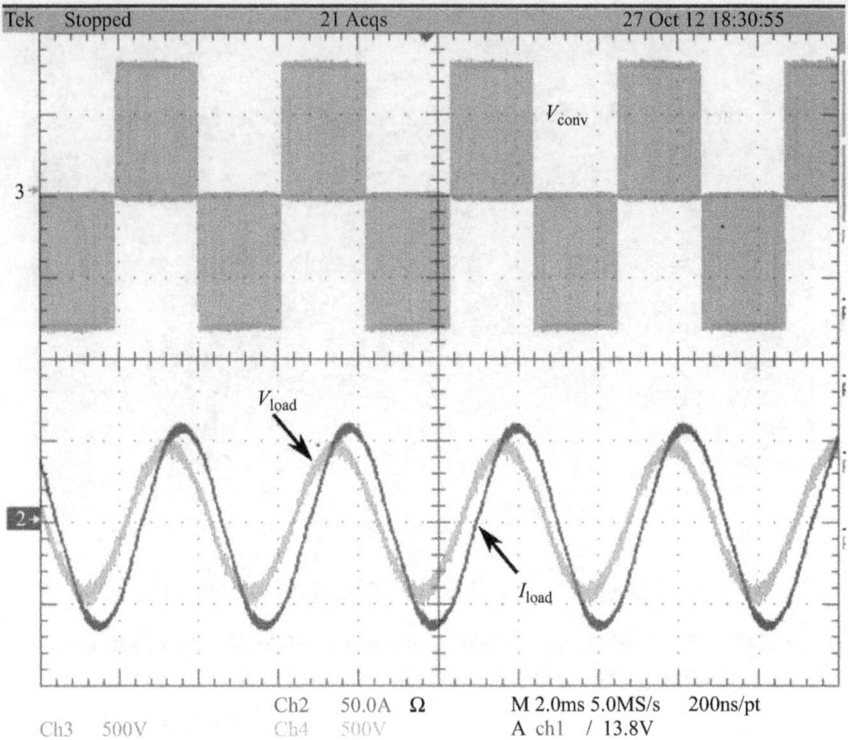

Figure 1.49 Experimental waveforms of 1,200 V SiC MOSFET-based converter operating at 35 kVA (line current, I_{load}*: 50 A/div; converter line voltage,* V_{conv}*: 500 V/div; load line voltage,* V_{load}*: 500 V/div; time: 8 ms/div)*

academia and industry in last two decades. Various manufacturers like Cree, GeneSiC, ROHM, Infineon, Panasonic, STMicroelectronics and others are making the commercial devices. The advantages of SiC over Si for power devices include lower losses for higher efficiency, higher switching frequencies to reduce passive components size, and higher breakdown voltages and high-temperature operating conditions. SiC power devices will have large applications on renewable energy generation, geothermal (down-hole drilling), automotive (hybrid/electric vehicles), transportation (aircraft, ships, and rail traction), military systems, space programs, industrial motor drives, and grid power processing applications.

The present high-power converters are used for mostly the MV drive applications and use most often the Si IGBTs. These IGBTs are typically limited to 6.5 kV blocking voltage. With the advancement in SiC high-voltage technology, higher blocking voltage SiC devices are expected to come to market and impact the MV applications. Mitsubishi Electric has recently announced 40% power saving with SiC power modules used in electric railway in Japan.

Acknowledgements

This work was supported by US Government through the DOE ARPA-E program under Contract No. DE-AR0000110. This work made use of FREEDM ERC shared facilities supported by NSF under Award No. EEC-0812121. The authors also thank CREE, POWEREX, and DOE POWER AMERICA for their support. Part of this work related to current source-based converter systems and testing was supported through Department of Energy (DOE) grant and by Varentec Inc.

References

[1] B. J. Baliga, *Fundamentals of power semiconductor devices*. New York: Springer-Verlag, 2008.
[2] A. Agarwal, M. Das, B. Hull, S. Krishnaswami, J. W. Palmour, J. Richmond, S. Ryu and J. Zhang, "Progress in silicon carbide power devices," Device Research Conference, 2006, pp. 155–158.
[3] L. Stevanovic, K. Matocha, Z. Stum, P. Losee, A. Gowda, J. Glaser and R. Beaupre, "Realizing the full potential of silicon carbide power devices," IEEE 12th Workshop on Control and Modeling for Power Electronics, June 2010, pp. 1–6.
[4] J. W. Palmour, L. Cheng, V. Pala, E. V. Brunt, D. J. Lichtenwalner, G. Y. Wang, J. Richmond, M. O'Loughlin, S. Ryu, S. T. Allen, A. A. Burk and C. Scozzie, "Silicon carbide power MOSFETs: breakthrough performance from 900 V up to 15 kV," IEEE International Symposium on Power Semiconductor Devices and ICs (ISPSD), 2014, pp. 79–82.
[5] M. K. Das, C. Capell, D. E. Grider, R. Raju, M. Schutten, J. Nasadoski, S. Leslie, J. Ostop and A. Hefner, "10 kV, 120 A SiC half H-bridge power MOSFET modules suitable for high frequency, medium voltage applications," Energy Conversion Congress and Exposition (ECCE), 2011, pp. 2689–2692.
[6] G. Wang, A. Huang and C. Li, "ZVS range extension of 10 A 15 kV SiC MOSFET based 20 kW dual active half bridge (DHB) DC–DC converter," IEEE Energy Conversion Congress and Exposition (ECCE), 2012, pp. 1533–1539.
[7] F. Wang, G. Wang, A. Huang, W. Yu and X. Ni, "Design and operation of a 3.6 kV high performance solid state transformer based on 13 kV SiC MOSFET and JBS diode," IEEE Energy Conversion Congress and Exposition (ECCE), 2014, pp. 4553–4560.
[8] Q. Zhang, C. Jonas, S. Ryu, A. Agarwal and J. Palmour, "Design and fabrications of high voltage IGBTs on 4H-SiC," IEEE International Symposium on Power Semiconductor Devices and IC's, ISPSD, 2006, pp. 1–4.
[9] B. J. Baliga, *Advanced high voltage power device concepts*. New York: Springer-Verlag, 2012, pp. 293–294.

[10] S. Hazra, S. Madhusoodhanan, G. K. Moghaddam, K. Hatua and S. Bhattacharya, "Design considerations and performance evaluation of 1200 V, 100 A SiC MOSFET based converter for high power density application," in Proceedings of the IEEE ECCE, Denver, CO, Sept. 2013.

[11] S. Hazra, A. De, L. Cheng, J. Palmour, M. Schupbach, B. A. Hull, S. Allen and S. Bhattacharya, "High switching performance of 1700 V, 50 A SiC power MOSFET over Si IGBT/BiMOSFET for advanced power conversion applications," Transactions on Power Electron. Early access.

[12] V. Baliga, S. Hazra, S. Singh, S. Roy, S. Bhattacharya, J. Paulakonis and S. Notani, "Device characterization and performance of 1200 V/45 A SiC JFET module," in Proceedings of the IEEE ECCE, Denver, CO, Sept. 2013, pp. 273–278.

[13] S. Sundaresan, R. Singh and R. W. Johnson, "Silicon carbide 'super' junction transistors operating at 500 °C," GeneSiC Semiconductor Inc. Dulles, VA.

[14] A. Kadavelugu, S. Bhattacharya, B. J. Baliga, S. H. Ryu, D. Grider and J. Palmour, "Zero voltage switching characterization of 12 kV SiC N-IGBTs," IEEE Power Semiconductor Devices and ICs Symposium (ISPSD), 2014, pp. 350–353.

[15] A. Kadavelugu, K. Mainali, D. Patel, S. Madhusoodhanan, A. Tripathi, K. Hatua, S. Bhattacharya, S.-H. Ryu, D. Grider and S. Leslie, "Medium voltage power converter design and demonstration using 15 kV SiC N-IGBTs," in Proceedings of the IEEE Applied Power Electronics Conference and Exposition (APEC), 2015, pp. 1396–1403.

[16] http://www.mskennedy.com/products/Bridges/MSK4804.prod

Chapter 2
Multilevel converters

*Juan W. Dixon, Ph.D.**

2.1 Introduction

This chapter introduces the concept and application of power converters using many voltage sources to increase the quality of output voltage and the power rating of those devices. They are called "multilevel converters" (or multilevel inverters) and their mission is to improve efficiency in industrial process and expand applications in new areas such as renewable energy conversion, power transmission and distribution, and transportation. Most of converters used today work with two or three levels of voltage and contribute with important part of total harmonic distortion (THD). The reduction of THD strongly depends on switching frequency of the power semiconductors and are limited to low and medium power conversion. These reasons have generated much interest on the topic of *Multilevel Converters* [1–5]. Multilevel converters make use of a series connection of switches, which allow the use of switches with reduced voltage ratings. These lower voltage switches have lower conduction losses and can operate at a higher frequency. Higher switching frequencies with many voltage levels result in higher quality voltage waveforms.

2.2 Basic concepts of multilevel converters

Multilevel converters comprise a new type of topologies that generate very much cleaner voltage and current waveforms, reducing significantly the amount of harmonics. Given the special topologies that can be implemented, they permit application in high-power conversion, not achievable with conventional two-level converters. The term "multilevel" defines topologies with more than one power source, and one of the first commercially available converters of this type is the three-level neutral point clamped (NPC) converter. Subsequently, other multilevel topologies, such as cascaded H-bridge (CHB) and flying capacitors (FCs), have been developed. CHBs can reach high output voltage and power levels and high reliability due to its modular topology. H-bridge converters can be combined with NPC topologies to improve energy conversion quality and reach higher levels of power.

*Professor Emeritus, Department of Electrical Engineering, Pontificia Universidad Católica de Chile, Vicuña Mackenna 4860, Santiago 7820436, Chile

2.2.1 One-branch converter

Multilevel converters include an arrangement of semiconductors and DC voltage sources required to generate a staircase output voltage waveform. Figure 2.1 shows the schematic diagram of voltage source inverters with different number of levels, which utilize only one branch of switches. As it is well known, a two-level inverter, like the one shown in Figure 2.1(a), generates an output voltage with two different levels: 0 and V_{DC}, while a three-level module, like the one shown in Figure 2.1(b), generates three different output voltages: 0, V_{DC}, and $2V_{DC}$. The different positions of the ideal switches are implemented with a number of semiconductors that are in direct relation with the number of levels.

It can be observed that the larger the number of DC sources, the greater the number of steps that can be generated and the number of sources is directly related with the number of levels through (2.1):

$$s = n - 1 \tag{2.1}$$

where s is the number of DC supplies connected in series and n is the number of voltage levels generated.

As it was already mentioned, multilevel converters are implemented with a number of DC sources to form a staircase AC waveform, which follows a given reference template. For example, with ten DC sources, an 11-level waveform can be obtained (five positive, five negatives, and zero with respect to the middle point between the ten sources). If the template is a sinusoidal waveform like the one shown in Figure 2.2, it is possible to obtain a waveform with very low THD.

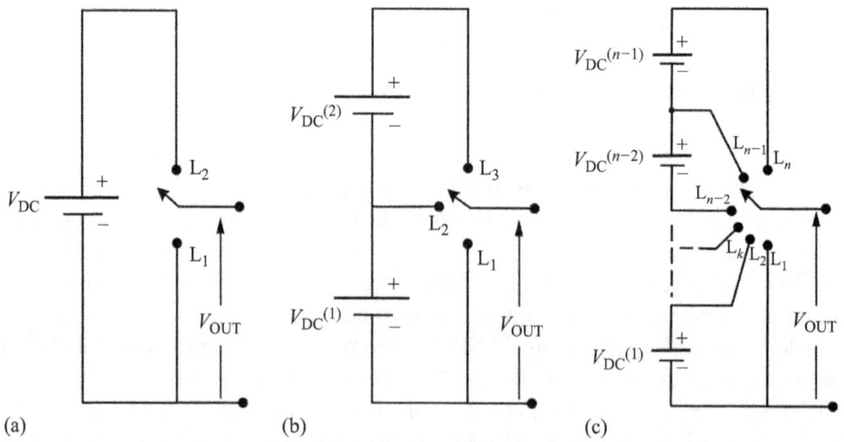

Figure 2.1 Basic multilevel converter topologies: (a) two levels, (b) three levels, and (c) n levels

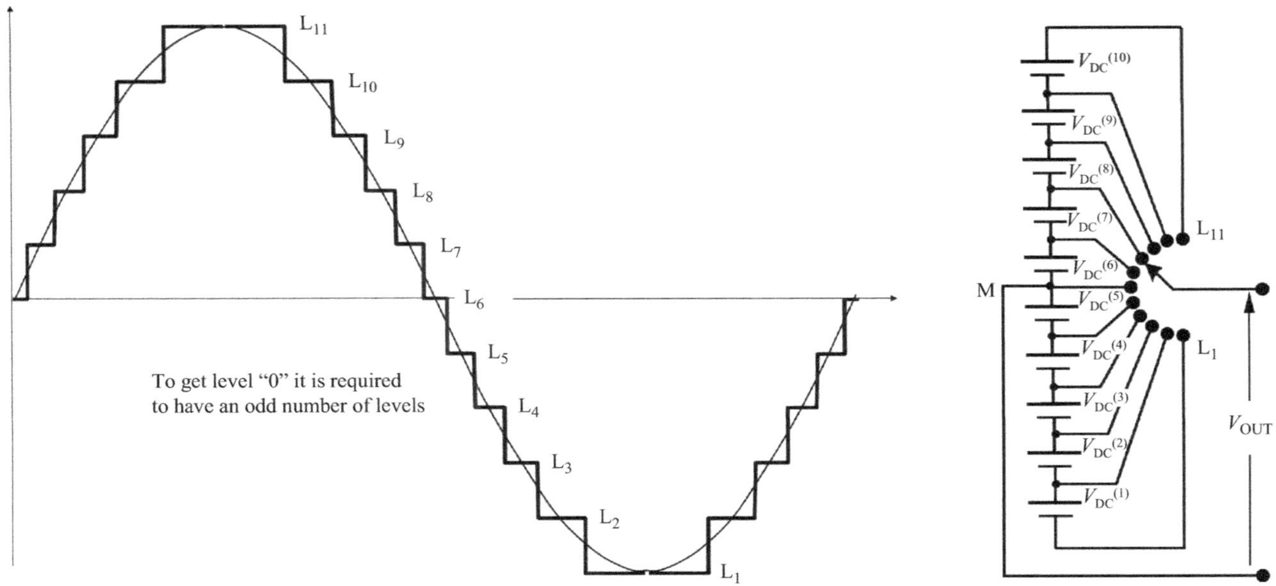

Figure 2.2 Voltage output waveform from an 11-level inverter

2.2.2 Two branches, "H-bridge" converter

To increase the number of levels, two branches of the previous topologies are combined to get the well-known "H-bridge" (HB) converter. It consists of connecting two of the previously discussed topologies using a common set of DC sources, which almost doubles the number of levels generated. This topology is displayed in Figure 2.3.

In this type of topology, the concept of "redundant levels" appears. The "H-bridge" topology has many redundant combinations to produce the same voltage level. As an example, the level "zero" can be generated with the load connected between L_1 and L_1', or in general, between L_k and L_k'. In a similar way, the level that produces $V_{OUT} = 2V_{DC}$ can be generated connecting L_3–L_1' or L_{k+2}–L_k'. Negative levels are generated in reverse way (e.g., $V_{OUT} = -2V_{DC}$ is obtained between positions L_k–L_{k+2}'). In general, all levels generated with inner switches of any of the two branches are redundant, that means, from L_2' to L_{n-1}' (or L_2 to L_{n-1}). Figure 2.4 shows, with black circles, all the positions (or switches) that can be removed, because they only generate redundant levels, and this removal can be done in whichever of the two branches.

When these switches are eliminated, only one redundant level remains: the L_1–L_1' because it generates the same level as L_n–L_n' ($V_{OUT} = 0$). The total number of redundant levels can be easily obtained from Figure 2.4, because each one of the $(n-2)$ removed switches produces n_R redundant levels and hence the total number of them is

$$n_R = n \cdot (n-2) + 1 \tag{2.2}$$

where n_R is the number of redundant levels and n is the number of voltage levels in one branch. Equation (2.2) includes the redundant L_1–L_1' (or else L_n–L_n').

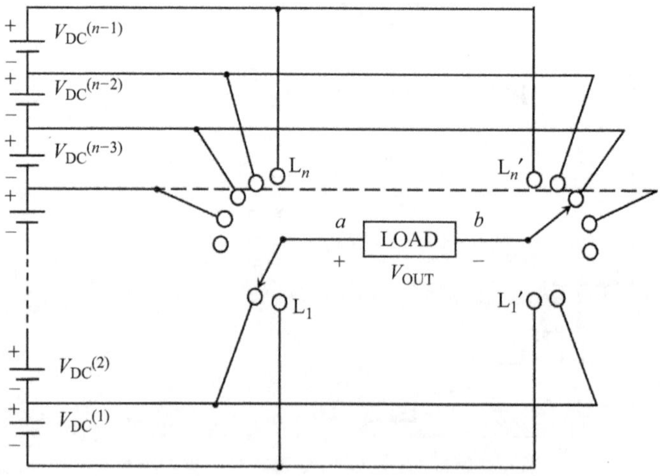

Figure 2.3 "H-bridge" inverter topology

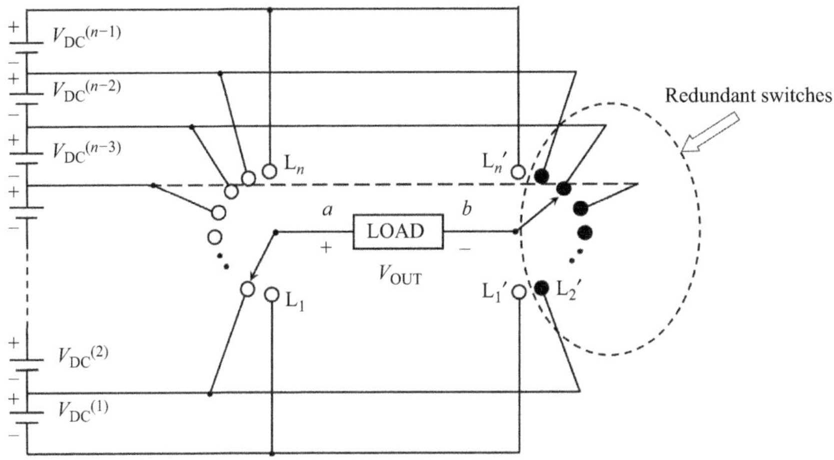

Figure 2.4 Redundant switches in an "H-bridge" converter

On the other hand, the total number of non-redundant levels reached with this topology is

$$n_{ab} = 2n - 1 \tag{2.3}$$

From (2.3), it can be observed that H-bridges only produce odd number of levels, which ensures the existence of $V_{OUT} = 0$ at the load.

2.3 Electronic switches to implement the converters

The above-described multilevel converters need electronic switches to operate, like metal oxide semiconductor field effect transistors, insulated gate bipolar transistors, or gate turn-off thyristors. These devices cannot replace the ideal switches shown in previous sections, because they have some limitations. They are unidirectional, are limited in switching speed and produce power losses. Besides, none of them can commutate from one power source to another; it is required the use of special topologies for multilevel implementation, which will be described in the next sections.

2.3.1 NPC converter

The NPC topology was invented in Japan by Akira Nabae *et al.* [6] at Nagaoka University of Technology in 1980. The NPC converter found widespread applications in the 1990s and gradually, the number of levels of the converter was increased. Currently, the NPC topology is the one most commonly used. Figure 2.5(a) shows an n-level topology and Figure 2.5(b) shows the well-known three-level topology.

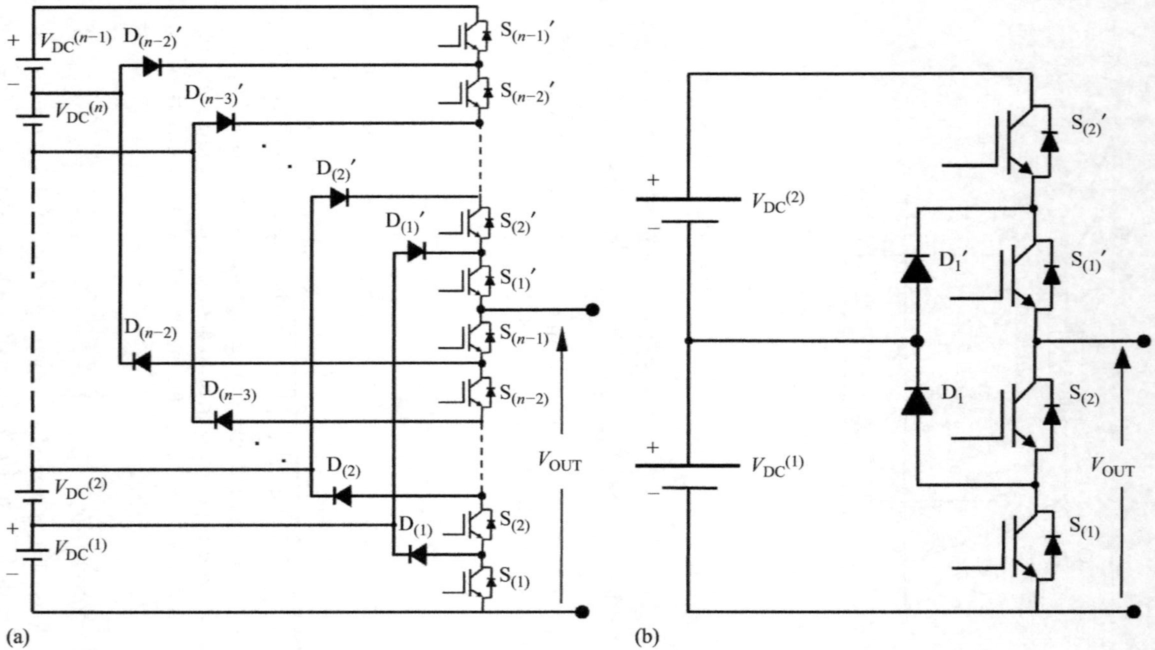

Figure 2.5 Neutral point clamped (NPC) topology. (a) n-level and (b) three-level

To explain how this type of topology works, Figure 2.6 shows a specific example using a seven-level NPC converter. The figure shows only four of the seven possible outputs: $V_{\mathrm{OUT}} = 0$ V, $V_{\mathrm{OUT}} = 2V_{\mathrm{DC}}$, $V_{\mathrm{OUT}} = 3V_{\mathrm{DC}}$, and $V_{\mathrm{OUT}} = 5V_{\mathrm{DC}}$. As can be seen, a chain of six transistors is moved together to get the desired level.

In general, if the required voltage at the output is kV_{DC}, the entire chain of transistors located between the diodes $D_{(k)}$ and $D_{(k)}'$ should be ON. This will allow the current to flow bidirectionally. In this way, all the DC sources located at the bottom of diodes $D_{(k)}$ and $D_{(k)}'$ will create the output voltage V_{OUT}. For example, if a $3V_{\mathrm{DC}}$ voltage is required at the output, then all transistors located between diodes $D_{(3)}$ and $D_{(3)}'$ must be switched ON:

$$V_{\mathrm{OUT}} = V_{\mathrm{DC}}^{(1)} + V_{\mathrm{DC}}^{(2)} + V_{\mathrm{DC}}^{(3)} = 3V_{\mathrm{DC}} \qquad (2.4)$$

2.3.2 FC converter

The FC voltage source converter was proposed about 15 years ago [7,8]. As with the NPC, the main characteristic of the FC technology is the connection of several semiconductors in series to share the voltage. However, instead of introducing clamping diodes to provide new nodes to connect the output, they are replaced by capacitors, which are connected as shown in Figure 2.7. These capacitors are pre-charged to certain voltage levels, always at a fraction of the main *DC* source [9]. When properly charged, they can generate intermediate voltage levels. Although three levels and up (with no theoretical limit) are possible, today the four-level FC is produced by one manufacturer of industrial medium voltage drives. The circuit configuration of the topology is depicted in Figure 2.7. Figure 2.7(a) shows an n-level topology and Figure 2.7(b) shows a three-level topology. The number of transistors required for a FC converter is the same as the NPC converter.

The FC topology can be presented in a more modular way, also known as multicell, as shown in Figure 2.8.

Figure 2.9 shows how the levels of a four-level FC converter are generated. Level 1 ($V_{\mathrm{OUT}} = 0$ volt) is generated by switching-on transistors $S_{(1)}$, $S_{(2)}$, and $S_{(3)}$; level 2 ($V_{\mathrm{OUT}} = V_{\mathrm{DC}}$) is obtained by switching-on transistors $S_{(2)}$, $S_{(3)}$, and $S_{(3)}'$; level 3 switching-on $S_{(3)}$, $S_{(2)}'$, and $S_{(3)}'$, and level 4 with $S_{(1)}'$, $S_{(2)}'$, and $S_{(3)}'$. Note that $V_{\mathrm{OUT}} = V_{\mathrm{DC}}$ is not generated using the voltage from C_1. Instead, the source $3V_{\mathrm{DC}} - Vc_2$ is utilized. In a similar way, the voltage $V_{\mathrm{OUT}} = 2V_{\mathrm{DC}}$ is obtained.

2.3.3 CHB converter

The CHB inverter uses a combination of many H-bridges in cascade, usually of the two-level type. There are two kinds of CHB converters: symmetric CHB (SCHB) and asymmetric CHB (ACHB). In the SCHB, each H-bridge uses power sources of the same value ($V_{\mathrm{DC}j} = V_{\mathrm{DC}k}$), while in the ACHB the power sources have different values ($V_{\mathrm{DC}j} \neq V_{\mathrm{DC}k}$). One of the advantages of the ACHB is that with the same number of H-bridges and power sources more voltage levels are obtained. Its drawback is that the H-bridges are not interchangeable.

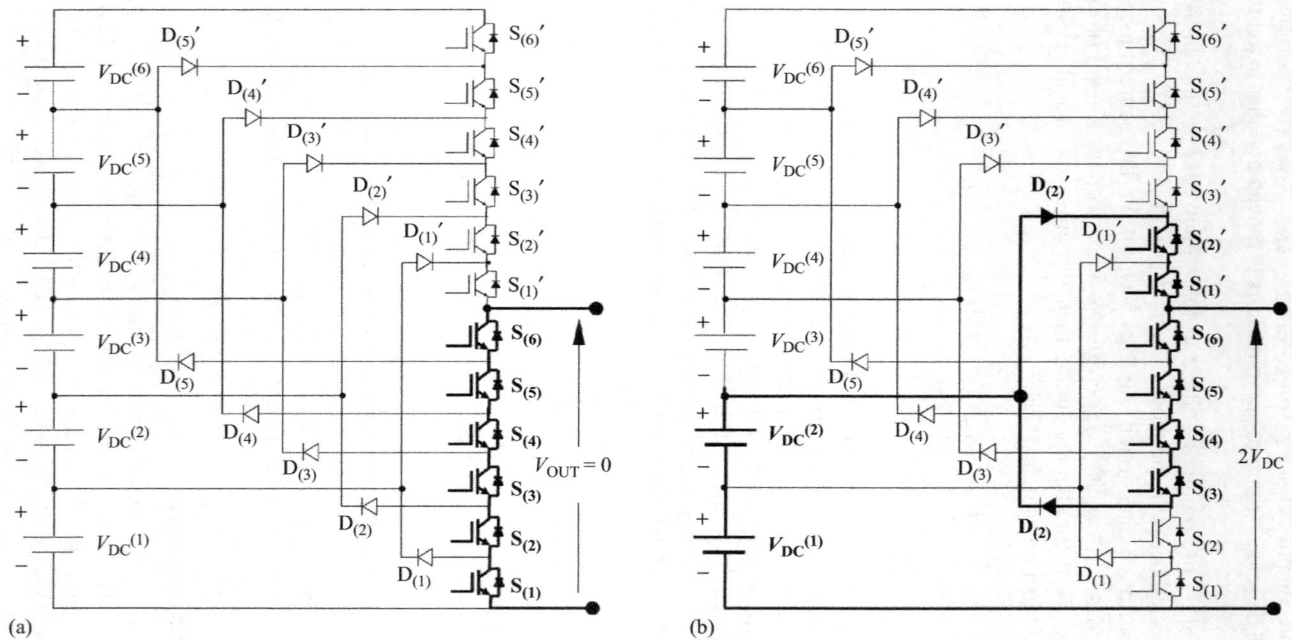

Figure 2.6 (*Continued*)

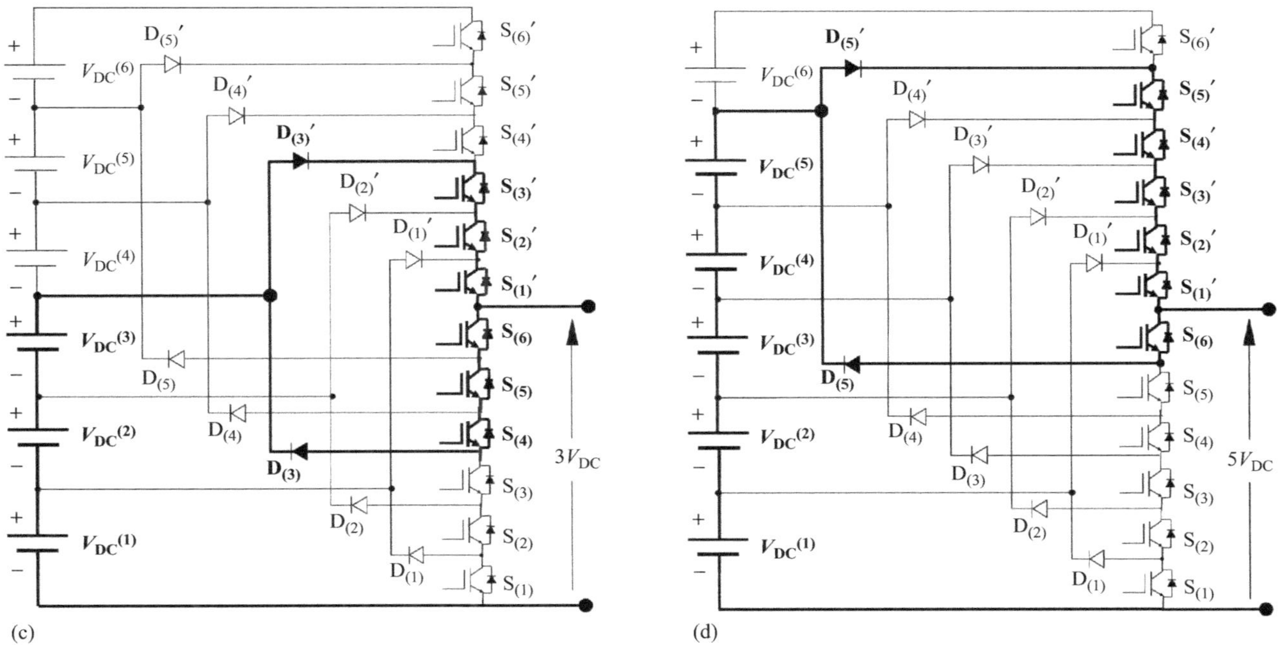

Figure 2.6 *Seven-level NPC topology. (a) $V_{OUT} = 0$, (b) $V_{OUT} = 2V_{DC}$, (c) $V_{OUT} = 3V_{DC}$, and (d) $V_{OUT} = 5V_{DC}$*

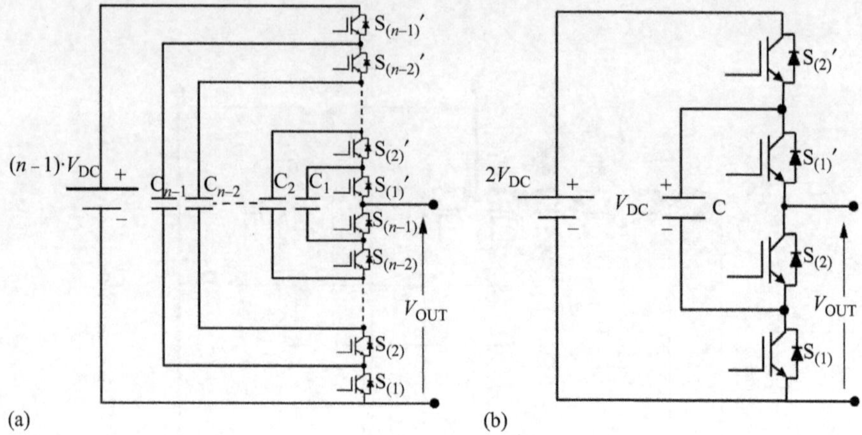

Figure 2.7 Flying capacitors (FC) topology. (a) n-level and (b) three-level scheme

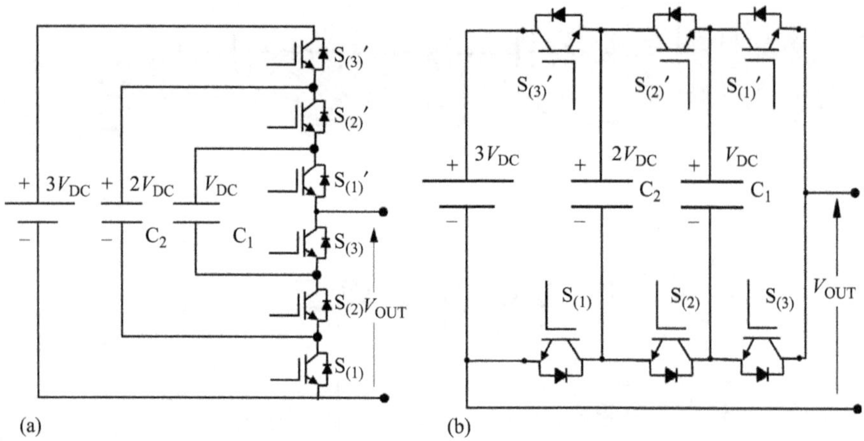

Figure 2.8(a) Four-level FC configuration and (b) multicell representation

Figure 2.10(a) shows a SCHB inverter with three H-bridges and Figure 2.10(b) shows an ACHB of the same type, but with the power sources scaled in power of 3. In the first case, only seven levels are obtained, while the second topology is able to generate 27 levels of voltage.

In general, ACHB converters produce more levels because they use redundant combination of switches from the SCHB converters. ACHB converters scaled in power of 3 optimize the number of levels because a two-level H-bridge can generate three different levels: "1", "0", and "−1". The total number of levels that an ACHB converter with *m* bridges scaled in power of 3 is 3^m. More details about other ACHB characteristics will be discussed later on.

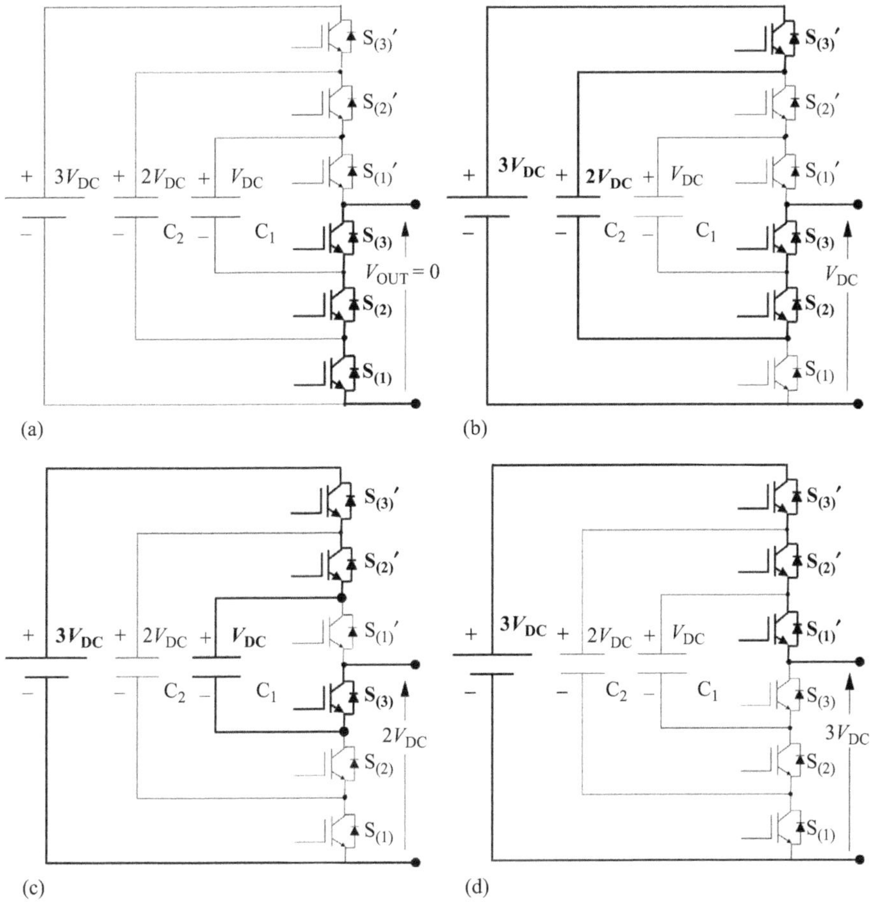

Figure 2.9 Voltage levels generation in a four-level FC converter. (a) $V_{OUT} = 0$, (b) $V_{OUT} = V_{DC}$, (c) $V_{OUT} = 2V_{DC}$, (c) $V_{OUT} = 3V_{DC}$

2.3.4 Combined topologies

Other topologies are obtained using combination of different converters. Figure 2.11 shows a simple two-level H-bridge which uses sub-cells to generate as many levels as the number of sub-cells. As each sub-cell can generate two different levels (level "0" and level "1"), the number of levels generated by the sub-cells is optimized when their power sources are scaled in power of 2. In this case, redundant levels do not exist, and with m sub-cells, 2^m positive levels are generated. The H-bridge is in charge of producing positive and negative half cycles, and then the total number of levels is given by (2.5).

$$n_{ab} = 2m^2 - 1 \tag{2.5}$$

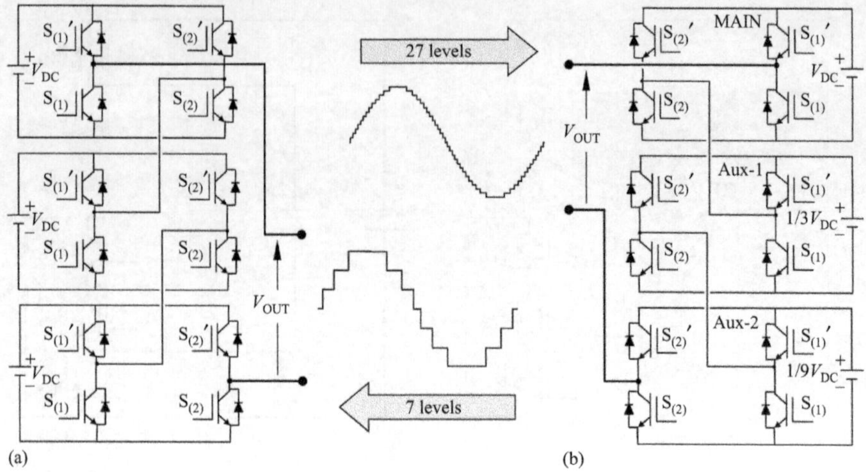

Figure 2.10 CHB converters using three two-level H-bridges. (a) SCHB type, (b) ACHB type with power supplies scaled in power of 3

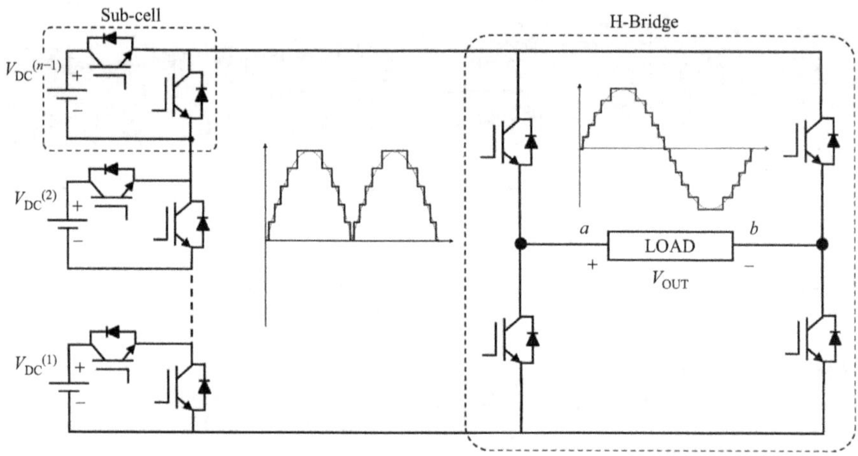

Figure 2.11 Multilevel inverter using sub-cells

Figure 2.12 shows a combined topology for an H-bridge, using one branch with a three-level converter and one branch with a two-level converter.

2.4 Three-phase multilevel converters

Most of the applications of static converters are for industrial purposes, which require three-phase topologies. Some of these applications are for electric machine drives, active power filters, reactive power compensation, renewable energy conversion, flexible AC transmission systems and voltage source high-voltage DC

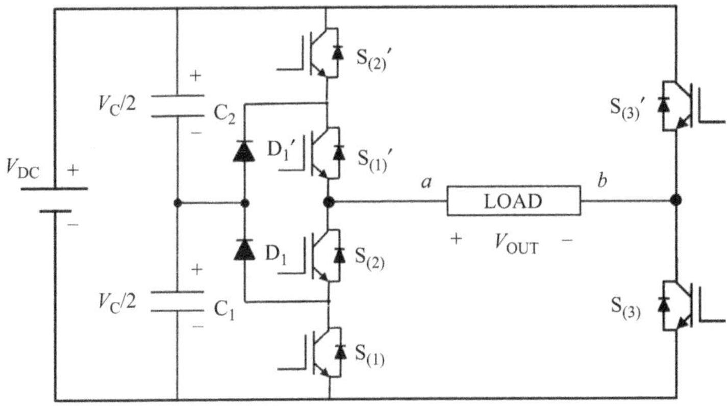

Figure 2.12 Combined H-bridge with NPC three-level and two-level converters

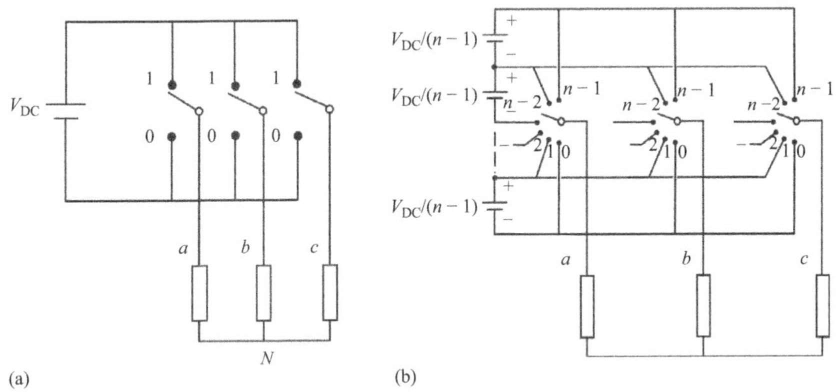

Figure 2.13 Three-phase converters. (a) Two-level and (b) n-level topologies

transmission. Multilevel inverters for most of those applications mean better performance, less harmonic distortion, higher levels of power conversion, and less power losses in electrical equipment. Figure 2.13 shows the schematic of a two-level and an n-level, three-phase converters.

2.4.1 Phase-to-phase and phase-to-neutral voltages

Generally, three-phase converters use the load in wye connection with neutral floating. This connection, shown in Figure 2.13, generates more voltage levels at the load because it corresponds to the phase-to-neutral voltage. As the phase-to-phase voltage have the same number of levels than an H-bridge, phase-to-neutral voltage can be obtained from the following relation:

$$n_{f-f} = n_{ab} = 2n - 1 \tag{2.6}$$

where n_{f-f} is the phase-to-phase voltage and n the number of levels in one branch of the inverter. Similarly, the number of levels in the phase-to-neutral connection with respect to phase-to-phase voltage is

$$n_{f-N} = 2n_{ab} - 1 \tag{2.7}$$

Replacing (2.6) in (2.7) it yields:

$$n_{f-N} = 2 \cdot (2n - 1) - 1 = 4n - 3 \tag{2.8}$$

where n_{f-N} is the number of levels in the phase-to-neutral voltage in whichever of the three phases, V_{aN}, V_{bN}, or V_{cN}.

2.4.2 Space vector representation

Three-phase systems can be analyzed using vector diagrams, which are especially useful for machine drives applications. The vector diagrams show, in a two-dimensional way, the instantaneous magnitude of each phase-to-neutral voltage at the load. Figure 2.14 shows the vector diagram of an ideal sinusoidal source and Figure 2.15 shows the vector diagram of a two-level converter. The composition of the three phases generates a rotating vector of voltage, which travels at an angular speed of ω rad/s and describes a perfect circle, whose magnitude is $V_T = 3/2V_{MAX}$, where V_{MAX} is the peak value of phase-to-neutral voltage.

Assuming that

$$\begin{aligned} v_a &= V_{MAX} \cdot \cos \omega \cdot t \\ t\, v_b &= V_{MAX} \cdot \cos(\omega \cdot t - 120°) \\ v_c &= V_{MAX} \cdot \cos(\omega \cdot t - 240°) \end{aligned} \tag{2.9}$$

where v_a, v_b, and v_c are the instantaneous voltages of phases a, b, and c, respectively, and ω is the angular frequency of the mains supply.

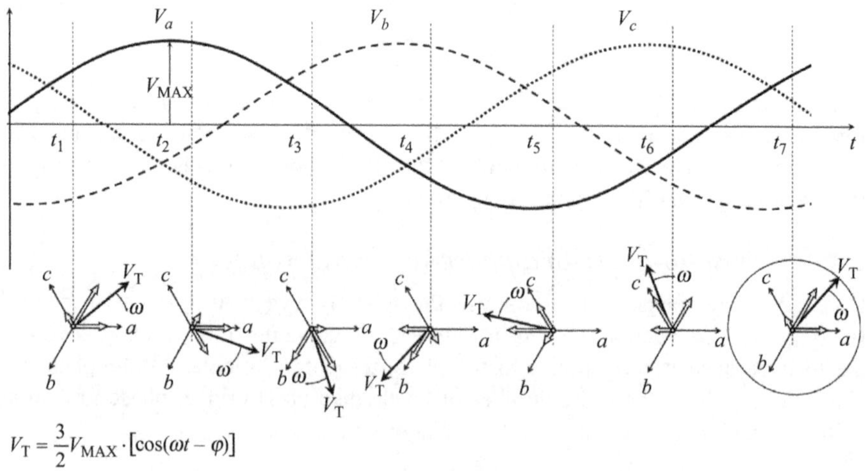

Figure 2.14 Vector diagram of a pure sinusoidal power source

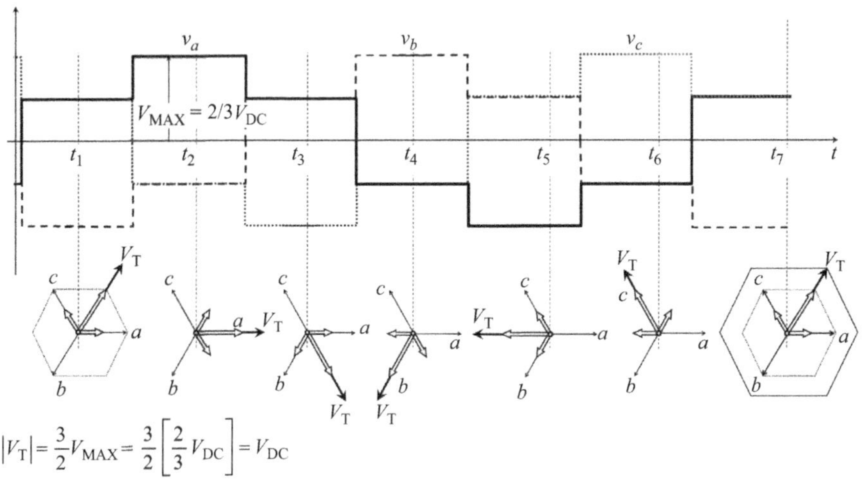

Figure 2.15 Phase-to-neutral voltages and vector diagram of a two-level inverter

Assuming that the vector V_T is being measured at a specific angle φ with respect to phase a:

$$V_T = v_a \cdot \cos\phi + v_b \cdot \cos(\phi - 120°) + v_c \cdot \cos(\phi - 240°) \tag{2.10}$$

Replacing (2.6) in (2.7):

$$\begin{aligned} V_T = V_{MAX}[&\cos(\omega t) \cdot \cos(\phi) + \cos(\omega t - 120°) \cdot \cos(\phi - 120°) \\ &+ \cos(\omega t - 240°) \cdot \cos(\phi - 240°)] \end{aligned} \tag{2.11}$$

It yields:

$$V_T = \frac{1}{2} V_{MAX} \begin{bmatrix} \cos(\omega t + \phi) + \cos(\omega t - \phi) + \cos(\omega t + \phi - 240°) + \cos(\omega t - \phi) \\ + \cos(\omega t + \phi - 120°) + \cos(\omega t - \phi) \end{bmatrix} \tag{2.12}$$

But

$$\cos(\omega t + \phi) + \cos(\omega t + \phi - 240°) + \cos(\omega t + \phi - 120°) = 0 \tag{2.13}$$

And finally,

$$V_T = \frac{3}{2} V_{MAX} \cdot [\cos(\omega t - \phi)] \tag{2.14}$$

Equation (2.14) represents a rotating vector that travels at ω rad/s and have a constant magnitude of $3/2 V_{MAX}$.

In the case of a two-level converter, the vector $\boldsymbol{V}_{\mathbf{T}}$ only has six positions, because each phase can only be at two different locations (two levels): "0" and "V_{DC}", where V_{DC} is the voltage of the power source of the converter. Besides, there are two

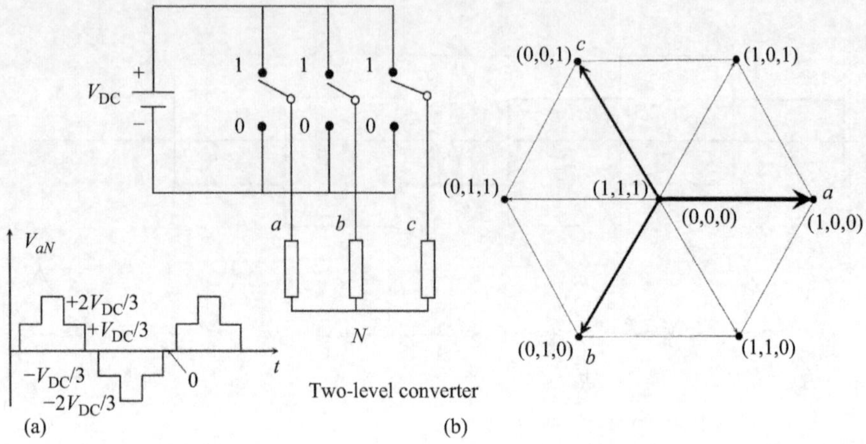

Figure 2.16 Two-level inverter. (a) Phase-to-neutral voltage and (b) space vector diagram

redundant vectors of magnitude "zero", which are generated with switches in position (0,0,0) or (1,1,1). Figure 2.16 shows the phase-to-neutral voltage waveform and vector diagram of a two-level converter, which can be associated with a six-sided polygon (hexagon). In this hexagon, each vertex represents one of the six vectors and at the center are located vectors (0,0,0) and (1,1,1). In this case, the magnitude of vector $\boldsymbol{V}_{\mathbf{T}} = V_{DC}$, where V_{DC} is the voltage source that feeds the converter.

$$|\boldsymbol{V}_{\mathrm{T}}| = \frac{3}{2} V_{\mathrm{MAX}} = \frac{3}{2} \cdot \left[\frac{2}{3} V_{\mathrm{DC}} \right] = V_{\mathrm{DC}} \tag{2.15}$$

When the number of levels is increased, more vectors with different magnitudes and angles appear. Figures 2.17 and 2.18 show similar diagrams for two-level and four-level converters, respectively. No matter the number of levels, the peak values of phase-to-neutral voltages are always $\pm\frac{2}{3}V_{DC}$ (assuming that $V_{DC} = \sum V_{DCi}$).

Each dot represents a new voltage vector that starts from the center of the hexagon. The number of vectors with magnitude "zero" is equal to the number of levels and hence, an n-level converter has n vectors at the origin: $(0,0,\ldots,0)$, $(1,1,\ldots,1)$, …, $(n-1, n-1, \ldots, n-1)$. There are no redundant levels at the periphery of the hexagon, but they increase by one unit when going to the center of the polygon. At the center there are $(n-1)$ redundant levels.

Figure 2.19 shows the space vector diagram of an n-level inverter. The total number of non-redundant vectors in an n-level inverter can easily be found from the space vector diagram. The hexagon can be decomposed in six sectors (six equilateral triangles in Figure 2.19), where each triangle has a number of vectors v_Δ given by

$$v_\Delta = \frac{n \cdot (n-1)}{2} \tag{2.16}$$

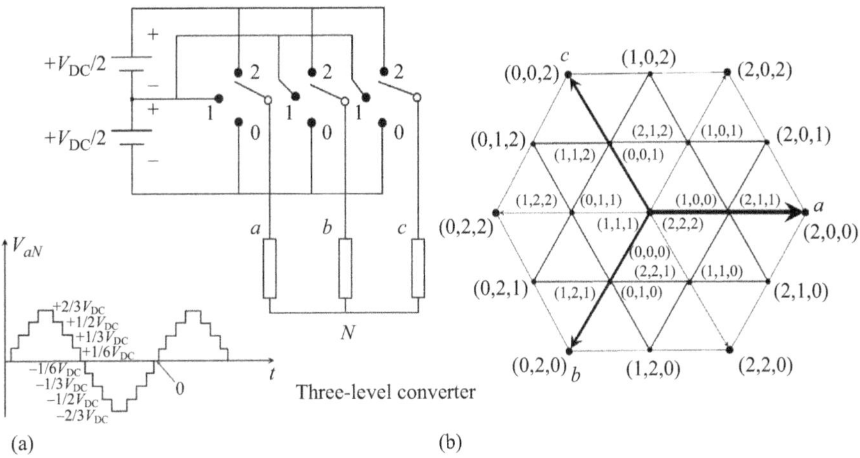

Figure 2.17 Three-level inverter. (a) Phase-to-neutral voltage and (b) space vector diagram

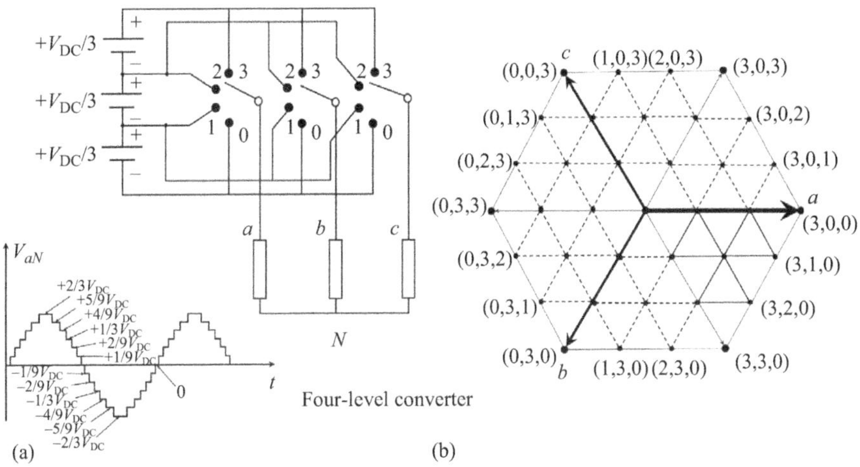

Figure 2.18 Four-level inverter. (a) Phase-to-neutral voltage and (b) space vector diagram

where n is the number of levels of the three-phase converter. Hence, the total number of non-redundant vectors is

$$v_{NR} = 6 \cdot \frac{n \cdot (n-1)}{2} + 1 = 3n \cdot (n-1) + 1 \tag{2.17}$$

The total number of vectors (redundant and non-redundant) may also be evaluated from Figure 2.19, and they correspond to all combination of switches of the three-phase inverter. The switches of phase a may be in n different positions for

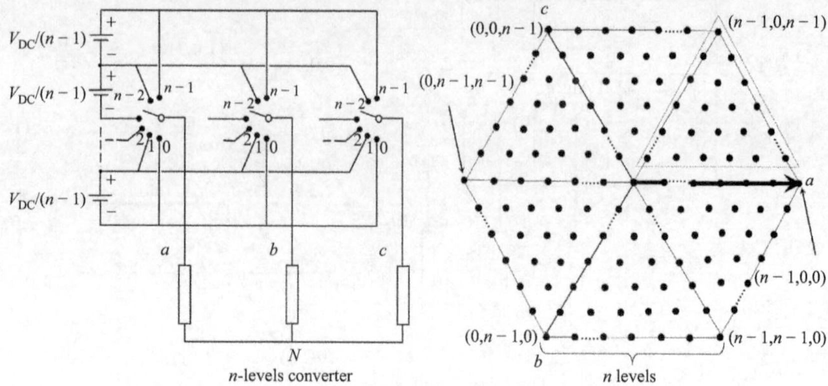

Figure 2.19 n-level inverter and space vector diagram

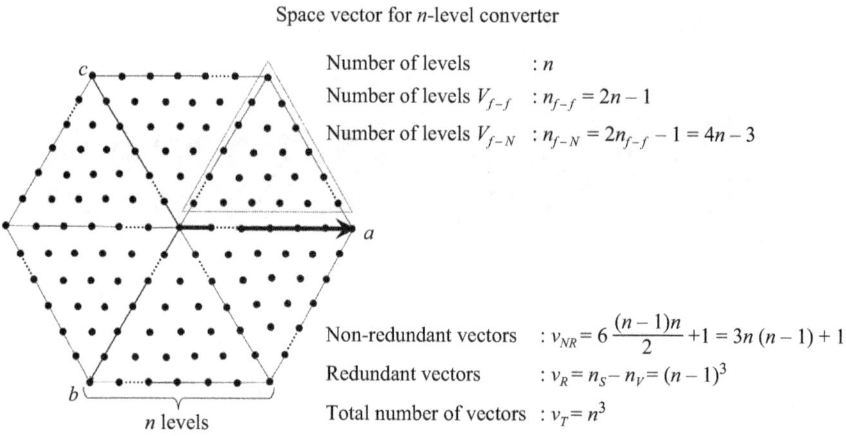

Figure 2.20 n-level space vector diagram and related equations

each one of the n positions of the phase b and also for each one of the n positions of phase c. As a result, the total number of vectors v_T is

$$v_T = n \cdot n \cdot n = n^3 \tag{2.18}$$

The total number of redundant vectors, v_R is obtained from (2.16) and (2.17)

$$v_R = v_T - v_{NR} = (n-1)^3 \tag{2.19}$$

Figure 2.20 resumes the results obtained in previous sections.

2.5 Modulation strategies for multilevel converters

With the development of multilevel inverter, topologies came the challenge to extend traditional modulation methods for two-level converters to the multilevel case. On the

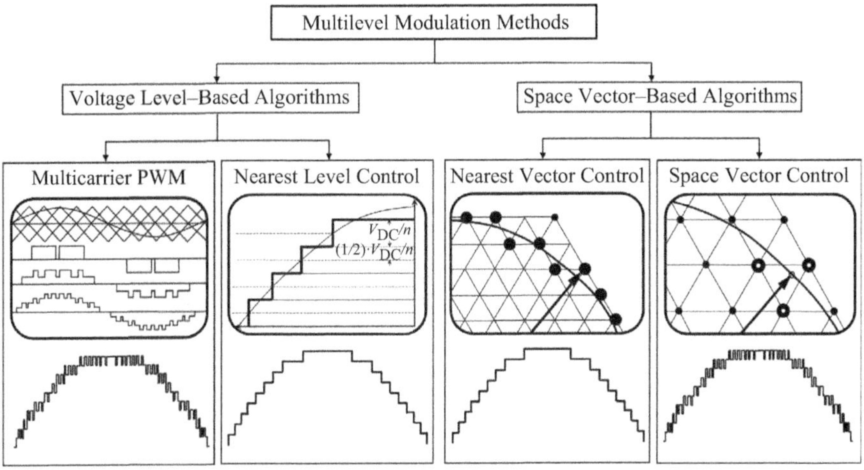

Figure 2.21 Modulation methods for multilevel converters

one hand, there is the inherent additional complexity of having more power electronics devices to control, and on the other hand, the possibility to take advantage of the extra degrees of freedom provided by the redundant switching states generated by these topologies. As a consequence, a large number of different modulation algorithms have been developed or adapted, each one with unique features and drawbacks, depending on the application. A classification of some of the modulation methods for multilevel inverters is presented in Figure 2.21. The modulation algorithms can be classified in to two different ways [10]: (a) the time domain, in which the method is based on the voltage level generation over a time frame, and (b) the space vector domain, in which the operating principle is based on the voltage vector generation.

2.5.1 Voltage levels–based algorithms

There are many methods that use this strategy, some of them based on traditional pulse width modulation (PWM) used in two-level topologies, in which a carrier signal is compared with a reference frame. In this case, with the increased number of levels, a bigger number of carriers comes out.

Other strategies that appear with the development of multilevel inverters are related with the direct utilization of a particular level, depending on the desired output. One of them is the so-called "nearest level control" or NLC, which is very simple to implement and has very low THD when the number of levels is larger than 9.

2.5.1.1 Multicarrier PWM

In this modulation method, each cell is modulated independently using PWM with the same reference signal. One of the methods, called "phase shift PWM" (PS-PWM), uses a phase shift across all the carrier signals of each cell in order to produce the stepped multilevel waveform. The lowest output voltage distortion is achieved with 180/*k* phase shifts between the carriers.

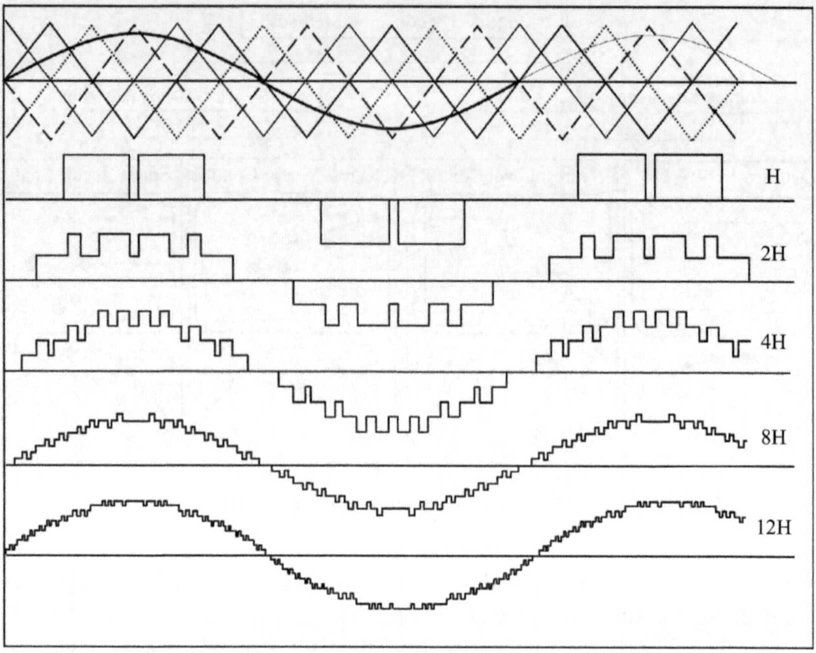

Figure 2.22 Twelve-cell (25 levels) PS-PWM waveform generation

The operating principle is illustrated in Figure 2.22 for a 12-level CHB example. Since the phase shift introduces a multiplicative effect, the amplitude of inverter output voltage has k times the amplitude of fundamental component of each cell. In a similar way, the frequency of the inverter output voltage switching pattern is k times the cell switching frequency. This is very useful, since the device switching frequency (hence the switching losses) is k times lower than the apparent switching frequency perceived by the load.

Since all the cells are controlled with the same reference and same carrier frequency (which has only been shifted), the power is evenly distributed among the cells, achieving current harmonic reduction. An example of application is illustrated in a back-to back, 50–60 Hz, system in Figure 2.23.

Another modulation method based on multicarrier is the "level shifted PWM" (LS-PWM). This method consists of an arrangement in vertical shifts instead of the phase shift used in previous method. Each carrier is set between two voltage levels, therefore the name level shifted. Since each carrier is associated with two levels, the carriers cover the whole amplitude range that can be generated by the converter. Figure 2.24 shows one phase of the typical NPC three-level converter, controlled with the LS-PWM method. As can be seen, each power source is controlled by a different triangular carrier.

Figure 2.25 shows the waveform generation for an 11-level converter. It can be implemented with any type of converter, NPC, FC, or CHB.

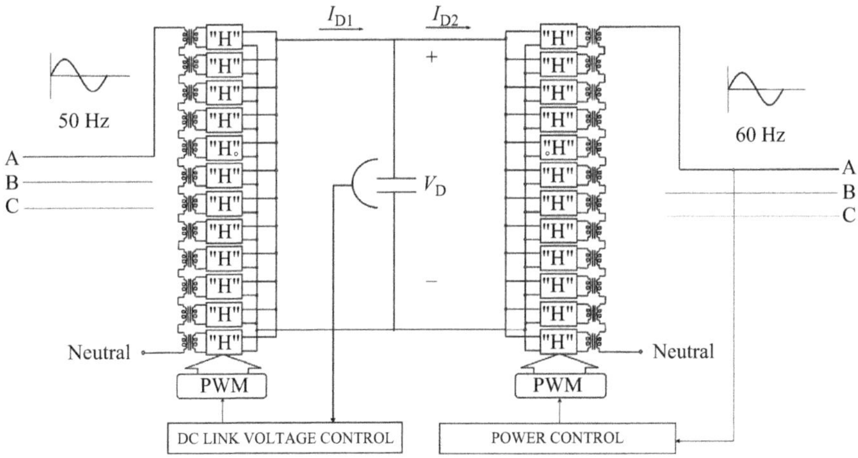

Figure 2.23 Back-to-back PS-PWM system using 12 cells

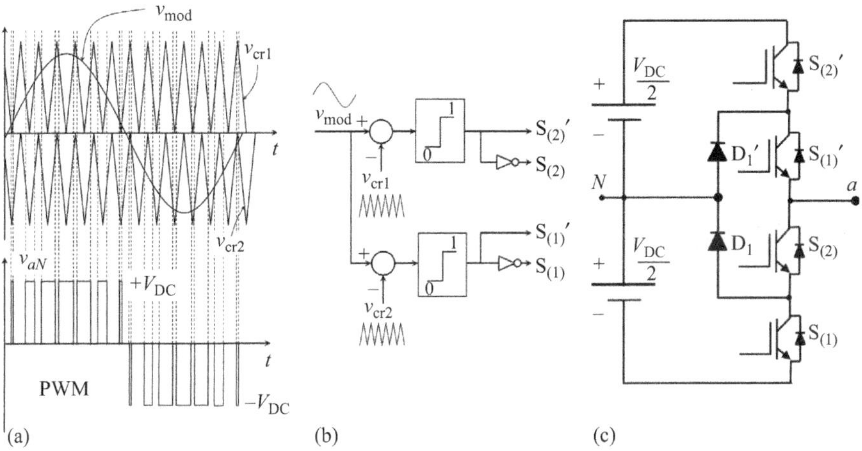

Figure 2.24 LS-PWM for NPC: (a) waveform generation, (b) control diagram, and (c) NPC

2.5.1.2 Nearest level control

The NLC also known as the rounding method consists of selecting the nearest voltage level that can be generated by the inverter to the desired output voltage reference. It can be expressed as

$$v_0 = V_{DC} \cdot \text{Round}\left(\frac{v_{REF}}{V_{DC}}\right) \tag{2.20}$$

where V_{DC} is the voltage of each power source and v_{REF} is the reference voltage, which normally is a sinusoidal waveform template. Figure 2.26 shows the waveform generation for an n-level inverter using the NLC strategy.

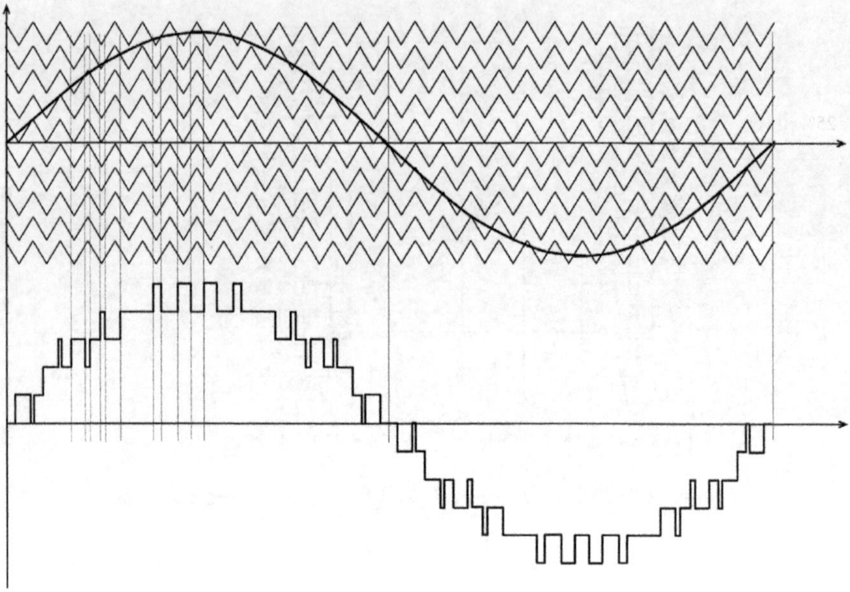

Figure 2.25 LS-PWM for and 11-level converter

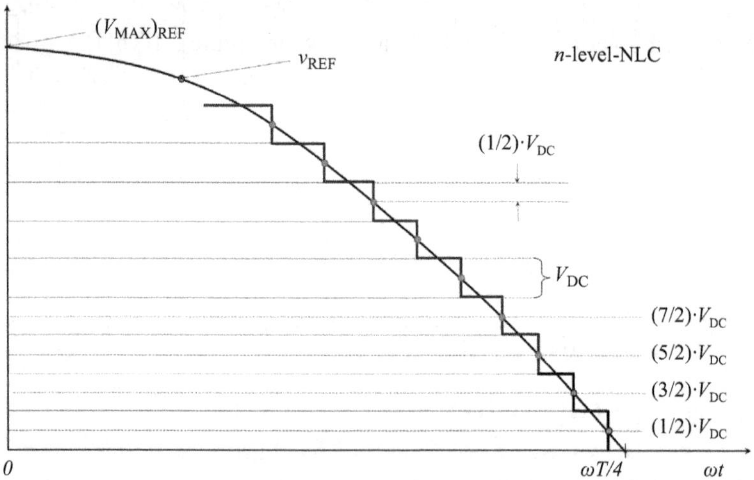

Figure 2.26 Waveform generation for an n-level inverter using the NLC strategy

As can be seen, the NLC method only produces one commutation between two voltage levels, which minimizes the switching frequency. However, the THD is large unless the number of levels is larger than 20 levels, because the THD of voltage becomes lesser than 4%. Figure 2.27 shows the THD of voltage as a function of number of levels for the NLC modulation.

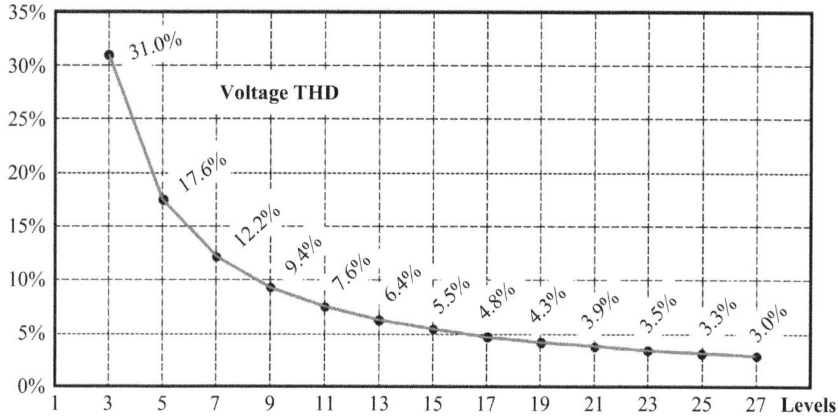

Figure 2.27 THD of voltage as a function of number of levels, using the NLC strategy

To generate more than 20 levels, at least ten HBs are needed. One option to reduce the number of bridges is using ACHBs, which were mentioned in Section 2.3.3. ACHBs use power sources with different voltages, and when power sources have voltages scaled in power of 3, the number of H-bridges required is minimized.

A 27-level inverter using ACHB was shown in Figure 2.10(b) (Section 2.3.3), in which the NLC modulation is utilized. This ACHB uses only three H-bridges and it is of interest doing a complete analysis of this particular topology. Figure 2.27 shows half cycle of the voltage waveform, using the NLC modulation in a 27-level converter. It is composed of three H-bridges whose power supplies are scaled in power of 3. The bigger bridge is the MAIN Bridge, and the ones located at the middle and at the bottom are called Aux-1 and Aux-2, respectively. This figure shows the voltage waveforms of each H-bridge (V_{MAIN}, $V_{\mathrm{Aux\text{-}1}}$, and $V_{\mathrm{Aux\text{-}2}}$) and the resultant voltage waveform (V_{LOAD}) that feeds one phase of the load. As can be seen in Figure 2.28(a), the MAIN Bridge switches at fundamental frequency, reducing power losses. The sinusoidal waveform $(V_{\mathrm{MAX}})_{\mathrm{REF}}$ shown in Figure 2.28(c) is the amplitude of the reference voltage that generates the NLC modulation and $(V^1_{\mathrm{MAX}})_{\mathrm{LOAD}}$ is the amplitude of the fundamental of V_{LOAD} (load voltage).

According to the analytical development given in [11,12], the value of $(V^1_{\mathrm{MAX}})_{\mathrm{LOAD}}$ is

$$(V^1_{\mathrm{MAX}})_{\mathrm{LOAD}} = \frac{4 \cdot V_{\mathrm{DC}}}{3^N \pi} \sum_{j=0}^{j=\frac{3^{N+1}}{2}-1} \int_{\omega t=0}^{\omega t=\cos^{-1}\left(\frac{2j+1}{3^{N+1}}\right)} \cos(\omega\, t)\, \mathrm{d}\omega\, t \qquad (2.21)$$

where N is the number of Aux Bridges and V_{DC} is DC voltage that feeds the MAIN Bridge.

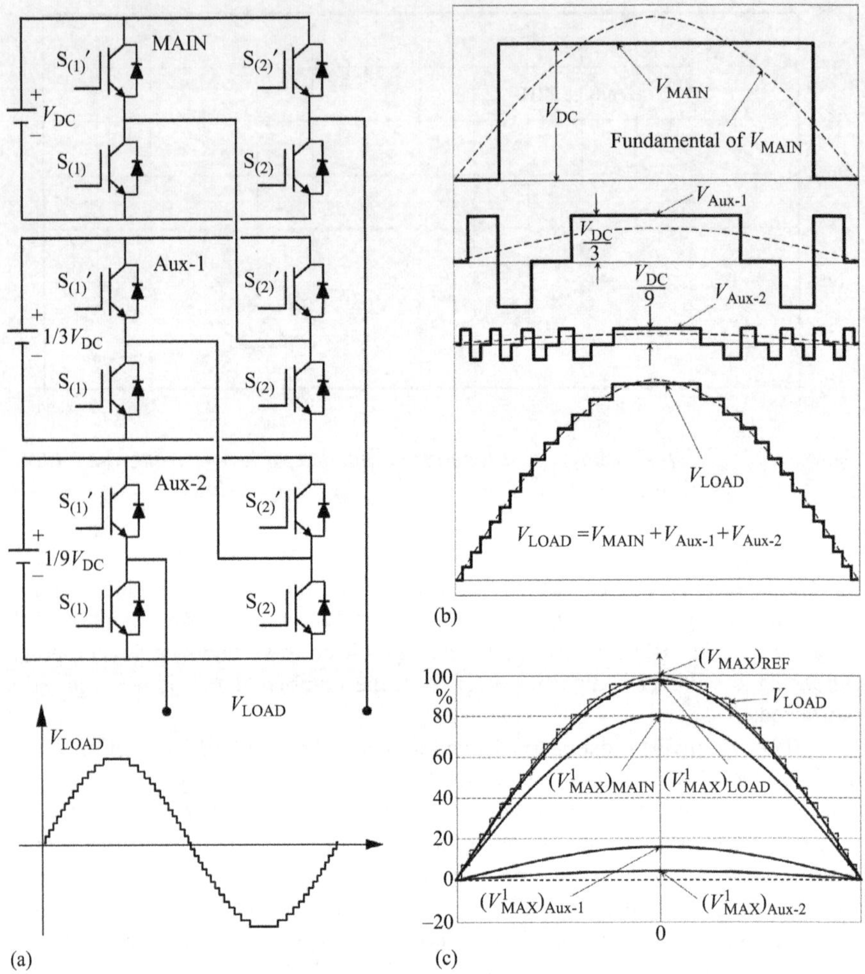

Figure 2.28 (a) 27-level ACHB and voltage V_{LOAD}, (b) voltage waveforms at each H-bridge, using the NLC modulation, and (c) fundamentals of voltages in each bridge and reference voltage

Equation (2.21) allows getting the values of $\left(V^1_{\mathrm{MAX}}\right)_{\mathrm{LOAD}}$ for whatever number of Aux Bridges. If the number of Aux Bridges is zero ($N = 0$), then the topology becomes a three-level inverter:

$$\left(V^1_{\mathrm{MAX}}\right)_{\mathrm{LOAD}}\Big|^{N=0} = \frac{4\cdot V_{\mathrm{DC}}}{\pi}\sum_{j=0}^{j=1}\int_{\omega t=0}^{\omega t=\cos^{-1}\left(\frac{2j+1}{3}\right)}\cos(\omega\, t)\,\mathrm{d}\omega\, t = 1.2\cdot V_{\mathrm{DC}} \tag{2.22}$$

Now, when the converter has only one Aux Bridge (Aux-1 or $N=1$), it becomes a nine-level device, and from (2.21):

$$\left(V_{\mathrm{MAX}}^{1}\right)_{\mathrm{LOAD}}\Big|^{N=1} = \frac{4 \cdot V_{\mathrm{DC}}}{3\pi} \sum_{j=0}^{j=4} \int_{\omega t=0}^{\omega t=\cos^{-1}\left(\frac{2j+1}{9}\right)} \cos(\omega\, t)\, \mathrm{d}\omega\, t = 1.44 \cdot V_{\mathrm{DC}} \tag{2.23}$$

And when the number of Aux Bridges is two ($N=2$), then it becomes a 27-level inverter, and the value of $\left(V_{\mathrm{MAX}}^{1}\right)_{\mathrm{LOAD}}$ in terms of V_{DC} is

$$\left(V_{\mathrm{MAX}}^{1}\right)_{\mathrm{LOAD}}\Big|^{N=2} = \frac{4 \cdot V_{\mathrm{DC}}}{9\pi} \sum_{j=0}^{j=13} \int_{\omega t=0}^{\omega t=\cos^{-1}\left(\frac{2j+1}{27}\right)} \cos(\omega\, t)\, \mathrm{d}\omega\, t = 1.49 \cdot V_{\mathrm{DC}} \tag{2.24}$$

Theoretically, if the number of Aux Bridges $\rightarrow \infty$:

$$\left(V_{\mathrm{MAX}}^{1}\right)_{\mathrm{LOAD}}\Big|^{N\rightarrow\infty} = \frac{4 \cdot V_{\mathrm{DC}}}{3^{N}\pi} \sum_{j=0}^{j=\infty} \int_{\omega t=0}^{\omega t=\cos^{-1}(0)} \cos(\omega\, t)\mathrm{d}\omega\, t = 1.5 \cdot V_{\mathrm{DC}} = \left(V_{\mathrm{MAX}}^{1}\right)_{\mathrm{REF}} \tag{2.25}$$

It can be noted that when $N \rightarrow \infty$:

$$\left(V_{\mathrm{MAX}}^{1}\right)_{\mathrm{LOAD}} = \left(V_{\mathrm{MAX}}^{1}\right)_{\mathrm{REF}} \tag{2.26}$$

The particular values of $\left(V_{\mathrm{MAX}}^{1}\right)_{\text{Aux-1}}$ and $\left(V_{\mathrm{MAX}}^{1}\right)_{\text{Aux-2}}$ can be obtained with

$$\left(V_{\mathrm{MAX}}^{1}\right)_{\text{Aux-1}} = \left(V_{\mathrm{MAX}}^{1}\right)_{\mathrm{LOAD}}\Big|^{N=1} - \left(V_{\mathrm{MAX}}^{1}\right)_{\mathrm{LOAD}}\Big|^{N=0} = 0.24 \cdot V_{\mathrm{DC}} \tag{2.27}$$

$$\left(V_{\mathrm{MAX}}^{1}\right)_{\text{Aux-2}} = \left(V_{\mathrm{MAX}}^{1}\right)_{\mathrm{LOAD}}\Big|^{N=2} - \left(V_{\mathrm{MAX}}^{1}\right)_{\mathrm{LOAD}}\Big|^{N=1} = 0.05 \cdot V_{\mathrm{DC}} \tag{2.28}$$

This mean, in terms of $\left(V_{\mathrm{MAX}}^{1}\right)_{\mathrm{LOAD}}$:

$$\frac{\left(V_{\mathrm{MAX}}^{1}\right)_{\mathrm{MAIN}}}{\left(V_{\mathrm{MAX}}^{1}\right)_{\mathrm{LOAD}}} = \frac{\left(V_{\mathrm{MAX}}^{1}\right)_{\mathrm{LOAD}}\Big|^{N-0}}{\left(V_{\mathrm{MAX}}^{1}\right)_{\mathrm{LOAD}}\Big|^{N=2}} = \frac{1.2}{1.49} = 0.81 = 81\% \tag{2.29}$$

$$\frac{\left(V_{\mathrm{MAX}}^{1}\right)_{\text{Aux-1}}}{\left(V_{\mathrm{MAX}}^{1}\right)_{\mathrm{LOAD}}} = \frac{0.24}{1.49} = 0.16 = 16\% \tag{2.30}$$

$$\frac{\left(V_{\mathrm{MAX}}^{1}\right)_{\text{Aux-2}}}{\left(V_{\mathrm{MAX}}^{1}\right)_{\mathrm{LOAD}}} = \frac{0.05}{1.49} = 0.03 = 3\% \tag{2.31}$$

Equations (2.29)–(2.31) show the percentage of voltage distribution between the H-bridges: MAIN, Aux-1, and Aux-2. As the H-bridges have a common output (they are series connected with the load), the currents and displacement factor are the same for all the H-bridges [11]. As a consequence, the percentage of power

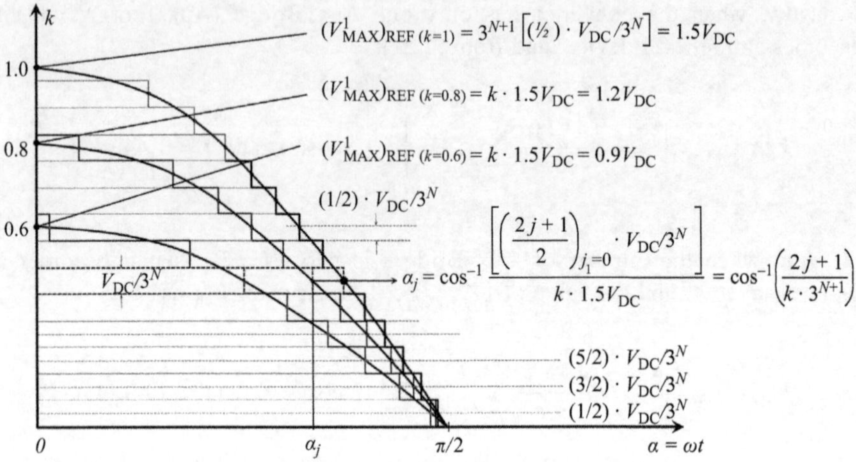

Figure 2.29 Reference voltage modulated in amplitude for N Aux Bridges

distribution is also the same as the voltage distribution. This means that over 80% of the power is controlled by the MAIN Bridge.

The previous analysis corresponds to the maximum reference voltage $(V^1_{MAX})_{REF}$ that can be applied to avoid distortion. This maximum value is equal to $1.5V_{DC}$ and can be modulated changing the amplitude of this reference as shown in Figure 2.29, by multiplying it by a factor k ($0 \leq k \leq 1$). With this form of modulation, the number of levels decreases when $(V_{MAX})_{REF(k)}$ decreases, as shown in Figure 2.4. Using (2.21), which correspond to factor $k = 1$, a generalized equation for different values of k is obtained.

$$(V^1_{MAX})_{LOAD}[k] = \frac{4 \cdot V_{DC}}{3^N \pi} \sum_{j=0}^{j=\frac{3^{N+1}-1}{2}} \int_{\alpha=0}^{\alpha=\cos^{-1}\left(\frac{2j+1}{k \cdot 3^{N+1}}\right)} \cos\alpha \, d\alpha \tag{2.32}$$

with $\alpha = \omega t$.

Using this equation for different values of N (number of Aux Bridges), the voltage variation for the MAIN and Aux Bridges as a function of k can be obtained. For example, for Aux-1:

$$\begin{aligned}(V^1_{MAX})_{Aux\text{-}1}[k] &= (V^1_{MAX})_{LOAD}[k]\Big|^{N=1} - (V^1_{MAX})_{LOAD}[k]\Big|^{N=0} \\ &= \frac{4 \cdot V_{DC}}{3\pi}\left[\sum_{j=0}^{j=4} \int_{\alpha=0}^{\alpha=\cos^{-1}\left(\frac{2j+1}{9k}\right)} \cos\alpha \, d\alpha - \sum_{j=0}^{j=1} \int_{\alpha=0}^{\alpha=\cos^{-1}\left(\frac{2j+1}{3k}\right)} \cos\alpha \, d\alpha\right]\end{aligned} \tag{2.33}$$

$$(V^1_{MAX})_{Aux\text{-}1}[k] = \frac{4 \cdot V_{DC}}{3\pi}\left\{\sum_{j=0}^{j=4} \int_{\alpha=0^\circ}^{\alpha=\cos^{-1}\left(\frac{2j+1}{9k}\right)} \cos\alpha \, d\alpha - 3 \int_{\alpha=0^\circ}^{\alpha=\cos^{-1}\left(\frac{1}{3k}\right)} \cos\alpha \, d\alpha\right\} \tag{2.34}$$

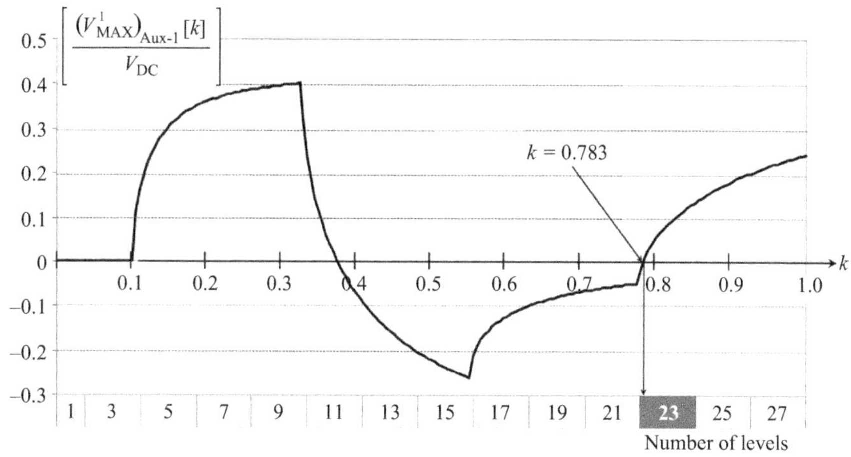

Figure 2.30 Amplitude of $(V^1_{MAX})_{Aux\text{-}1}$ as a function of k

Figure 2.30 is obtained by plotting (2.34) as a function of k.

As can be seen from Figure 2.30, Aux-1 does not give power to the load when $k = 0.783$, $k = 0.38$, and $k < 0.115$. Of particular interest is the first value ($k = 0.783$), because at this value the converter works with 23 levels and without the need to feed the Aux-1 Bridges. As a result, these H-bridges behave like *Series Active Power Filters* [13,14] because they do not transfer power to the load. If the same analysis is applied to Aux-2, it does not give power when $k = 0.796$. In this case, there are many more values of k where Aux-2 transfers zero power to the load.

2.5.2 *Space vector–based algorithms*

Two kinds of methods using space vector control are the most common: nearest vector control (NVC) and space vector modulation (SVM). The operation principles of each one are presented in the next sections.

2.5.2.1 Nearest vector control

NVC was first introduced in 2002 [15], as an alternative to selected harmonic elimination or staircase modulation, to provide a low-switching frequency modulation method, without the offline requirements and poor dynamic performance.

The basic idea is to take advantage of the high number of voltage vectors generated by a multilevel inverter, by simply approximating the reference to the closest voltage vector that can be generated. This principle is illustrated in Figure 2.31, where a nine-level space vector complex plane is illustrated. Each circle is one of the possible voltage vectors generated by the inverter. They are surrounded by hexagons that represent the boundary of the area closest to the corresponding vector. The circle is the voltage space vector reference trajectory through the plane. When the reference falls into a certain hexagon, the closest vector $\boldsymbol{V}_T$ will be generated by the inverter.

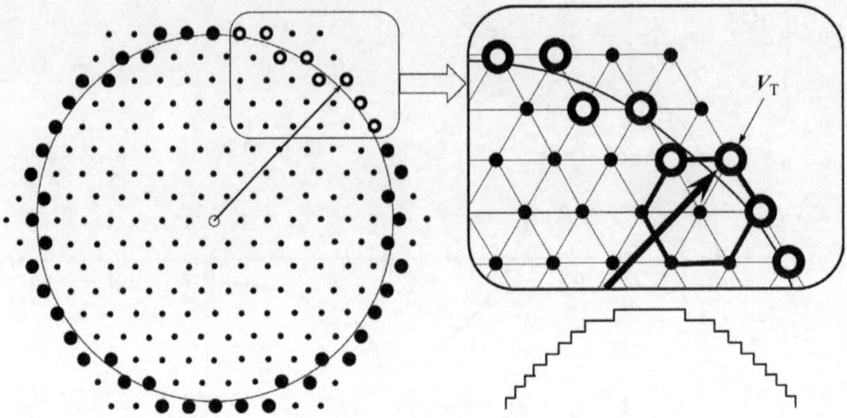

Figure 2.31 Nearest vector control (NVC)

2.5.2.2 Space vector modulation

The main idea behind SVM is to synthesize the rotating reference vector by a time average of a combination of the fixed vectors generated by the inverter. For a given reference vector magnitude and position (angle), the three closest vectors are used to generate the mean value over the modulation period. This strategy reduces distortion in voltage by reducing the dv/dts. Consider for example, the reference vector v_r is illustrated in Figure 2.32. The mean value of the inverter voltage can be expressed as

$$v_r = \frac{1}{T_M}\left(v_i \cdot t_i + v_j \cdot t_j + v_k \cdot t_k\right) \tag{2.35}$$

where t_i, t_j, and t_k are the times in which the fixed vectors v_i, v_j, and v_k are applied to the load, with the restriction

$$T_M = t_i + t_j + t_k \tag{2.36}$$

where T_M is the modulation period.

2.5.2.3 Special topologies for electric machine drives

Despite space vector–based algorithms are a technology useful for any application, they are especially applicable to electrical machines. Vector control permits a dynamic behavior not reachable with other technologies, because they allow changing vector position in a matter of milliseconds. Besides, different topologies of multilevel inverters can be combined to increase the size and number of vectors. Figure 2.33 shows one of these topologies, in which a two-level converter is combined with a three-level converter using H-bridge topologies. As these converters are fed with independent power supplies, their voltages can be selected using different magnitudes. It is interesting to mention that the H-bridges can operate with capacitors instead of power sources, using an adequate control

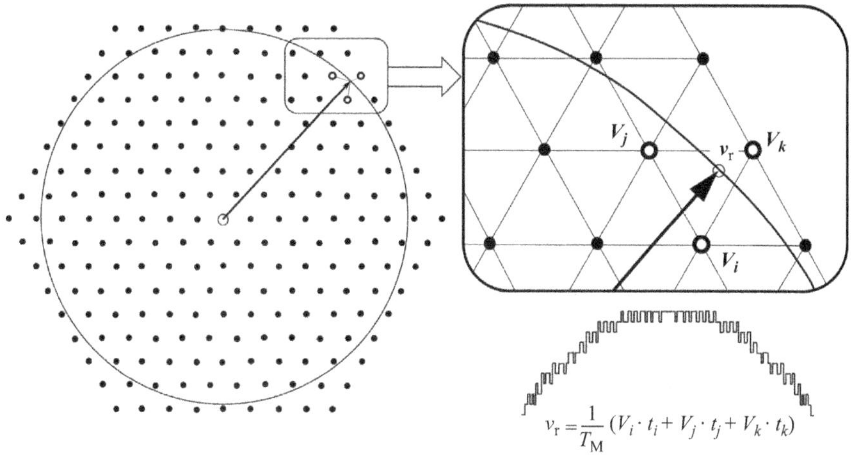

Figure 2.32 Space vector modulation (SVM)

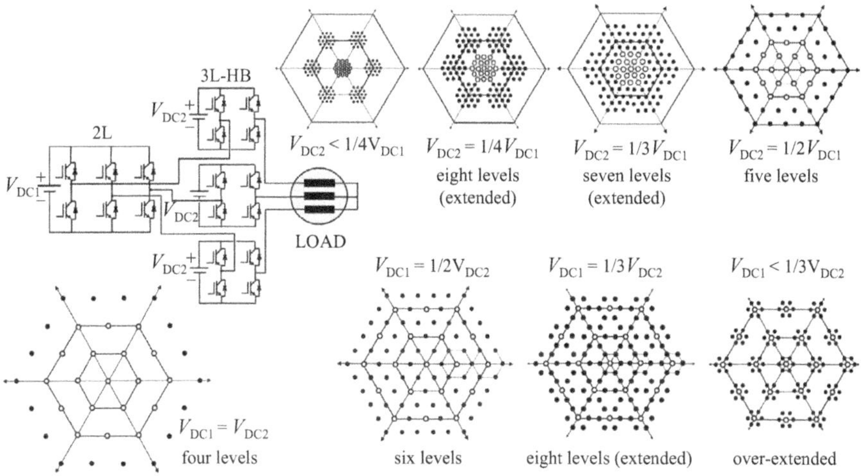

Figure 2.33 Combined topology using conventional two-level with three-level HB. White circles are the basic hexagon of the HB three-level converter

strategy. The figure shows how the space vector changes when relative amplitudes of V_{DC1} and V_{DC2} change. In the hexagons at the top of the figure, V_{DC1} is kept constant and in the hexagons at the bottom, V_{DC2} is kept constant.

Figure 2.34 shows a combined topology using three different converter schemes: (a) two three-level NPC, (b) one three-level NPC with one three-level HB, and (c) two three-level HBs. As in Figure 2.32, the space vector diagram can be modified changing the voltage relation between the first and the second converter. The number of levels increases until the smaller hexagons do not interact

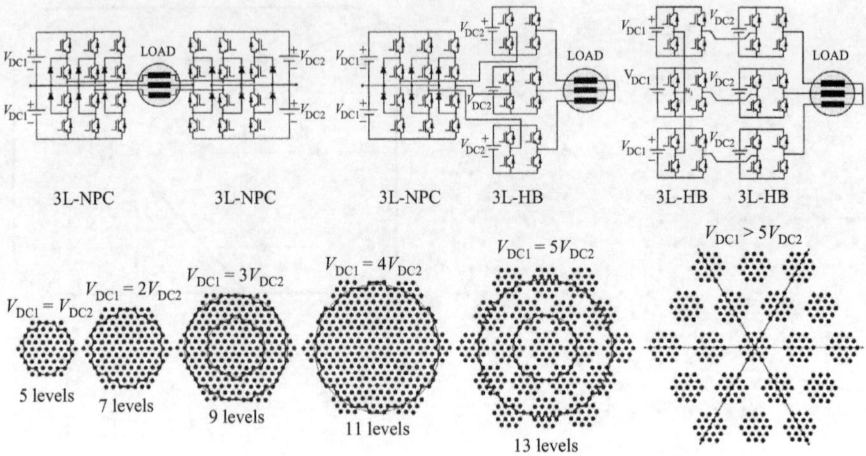

Figure 2.34 Combined topologies using two three-level converters

with the larger one. In this case, until 13 levels are obtainable, but some vectors, due to over extension of each hexagon, are not obtainable. Up to a limit of nine levels can be reached in this case, to have all vectors within the hexagon. With more than nine levels, some vectors cannot be generated (extended and over-extended vector diagrams) [16].

It is possible to get other combined topologies using more than two converters, and each one with different values of voltage supplies. Besides, as was mentioned before, with an adequate strategy some converters can work with capacitors instead of power sources. In this case, these converters do not supply any power but they increase the number of vectors improving the quality of the output voltage. When some converters work using capacitors, they work as series active power filters.

References

[1] Tolbert L., Peng F., Habetler T., 'Multilevel Converters for Large Electric Drives'. *IEEE Transactions on Industry Applications*. 1999, vol. 35(1), pp. 36–44.

[2] Franquelo L., Rodriguez J., Leon J., Kouro S., Portillo R., Prats M., 'The Age of Multilevel Converters Arrives'. *IEEE Industrial Electronics Magazine*. 2008, vol. 2(2), pp. 28–39.

[3] Rodríguez J., Lai J.-S., Zheng Peng F., 'Multilevel Inverters: A Survey of Topologies, Controls, and Applications'. *IEEE Transactions on Power Electronics*. 2002, vol. 49(4), pp. 724–737.

[4] Lai J.-S., Peng F. Z., 'Multilevel Converters – A New Breed of Power Converters'. *IEEE Transactions on Industry Applications*. 1996, vol. 32(3), pp. 509–517.

[5] Dixon J., Bretón A., Ríos F., Rodríguez J., Pontt J., Pérez M., 'High-Power Machine Drive, Using Non-Redundant 27-Level Inverters and Active Front End Rectifiers'. *IEEE Transactions on Power Electronics*. 2007, vol. 22(6), pp. 2527–2533.
[6] Nabae A., Takahashi I., Akagi H., 'A New Neutral-Point-Clamped PWM Inverter'. *IEEE Transactions on Industry Applications*. 1981, vol. IA-17(5), pp. 518–523.
[7] Meynard T. A., Foch H., 'Electronic device for electrical energy conversion between a voltage source and a current source by means of controllable switching cells' US patent 5,737,201, Apr. 1998 (priority Jul. 1991).
[8] Fazel S. S., Bernet S., Krug D., Jalili K., 'Design and Comparison of 4-kV Neutral-Point-Clamped, Flying-Capacitor, and Series-Connected H-Bridge Multilevel Converters'. *IEEE Transactions on Industrial Applications*. 2007, vol. 43(4), pp. 1032–1040.
[9] Wilkinson R. H., Meynard T. A., du Toit Mouton H., 'Natural Balance of Multicell Converters: The Two-Cell Case'. *IEEE Transactions on Power Electronics*. 2006, vol. 21(6), pp. 1649–1657.
[10] Kouro S., *Multilevel Converter Performance Improvements*. Ph.D. Thesis, Universidad Técnica Federico Santa María, Valparaíso, Chile. 2008.
[11] Dixon J., Pereda J., Castillo C., Bosch S., 'Asymmetrical Multilevel Inverter for Traction Drives, Using Only One DC Supply'. *IEEE Transactions on Vehicular Technology*. 2011, vol. 59(8), pp. 3736–3743.
[12] Dixon J., Pereda J., 'High Frequency Link: A Solution for Using Only One DC Source in Asymmetric Cascaded Multilevel Inverters'. *IEEE Transactions on Power Electronics*. 2011, vol. 58(9), pp. 3884–3892.
[13] Dixon J., Pereda J., '23-Level Inverter with Only One Power Supply, Using Series Active Filters'. *IEEE Transactions on Vehicular Technology*. 2012, vol. 61(3), pp. 1043–1051.
[14] Varschavsky A., Dixon J., Rotella M., Morán L., 'Cascaded Nine-Level Inverter for Hybrid Series Active Power Filter, Using Industrial Controller'. *IEEE Transactions on Industrial Electronics*. 2010, vol. 57(8), pp. 271–277.
[15] Rodríguez J., Morán L., Correa P., Silva C., 'A Vector Control Technique for Medium Voltage Multilevel Inverters'. *IEEE Transactions on Industrial Electronics*. 2002, vol. 49(4), pp. 882–888.
[16] Pereda J., Dixon J., 'Cascaded Multilevel Converters: Optimal Asymmetries and Floating Capacitor Control'. *IEEE Transactions on Industrial Electronics*. 2013, vol. 60(11), pp. 4784–4793.

Further reading

[1] Dixon J., Morán L., 'High-Level Multistep Inverter Optimization Using a Minimum Number of Power Transistors'. *IEEE Transactions on Power Electronics*. 2006, vol. 21(2), pp. 330–337.

[2] Franquelo L. G., Prats M. M., Portillo R., León J. I., Carrasco J. M., Galván E. *et al.*, 'Three Dimensional Space Vector Modulation Algorithm for Four Leg Multilevel Converters Using abc Coordinates'. *IEEE Transactions on Industrial Electronics*. 2006, vol. 53(2), pp. 458–466.

[3] Prats M. M., León J. I., Portillo R., Carrasco J. M., Franquelo L. G., 'A Novel Space Vector Algorithm for Multilevel Converters Based on Geometrical Considerations Using a New Sequence Control Technique'. *Journal of Circuits, Systems and Computers*. 2004, vol. 13(4), pp. 845–861.

[4] Rech C., Pinheiro J. R., 'Hybrid Multilevel Converters: Unified Analysis and Design Considerations'. *IEEE Transactions on Industrial Electronics*. 2007, vol. 54(2), pp. 1092–1104.

[5] McGrath B. P., Holmes D. G., 'Multicarrier PWM Strategies for Multilevel Inverters', *IEEE Transactions on Industrial Electronics*. 2002, vol. 49(4), pp. 858–867.

[6] Meynard T., Foch H., Thomas P., Cournault J., Jakob R., Nahrstaedt M., 'Multicell Converters: Basic Concepts and Industry Applications'. *IEEE Transactions on Industrial Electronics.* 2002, vol. 49(5), pp. 955–964.

[7] Escalante M., Vannier J. C., Arzondé A., 'Flying Capacitor Multilevel Inverters and DTC Motor Drive Applications'. *IEEE Transactions on Industrial Electronics.* 2002, vol. 49(4), pp. 805–815.

[8] Pou J., Boroyevich D., Pindado R., 'New Feedforward Space-Vector PWM Method to Obtain Balanced AC Output Voltages in a Three-Level Neutral-Point-Clamped Converter'. *IEEE Transactions on Industrial Electronics.* 2002, vol. 49(50), pp. 1026–1034.

[9] Corzine K., Baker J., 'Multilevel Voltage Source Duty Cycle Modulation: Analysis and Implementation'. *IEEE Transactions on Industrial Electronics*. 2002, vol. 49(5), pp. 1009–1016.

Chapter 3
Multi-input converters

Babak Farhangi and Hamid Toliyat**

3.1 Introduction

Conventional unidirectional power electronic converters have a single input and a single output. These converters are time-variant, nonlinear, single-input, single-output electric systems [1–3]. In contrast, in hybrid systems or integrated power systems, more than one source may supply the load. Such systems may be realized through a set of single-input, single-output power converters from each source to the load. In this case, the converters may be coupled to the load in a parallel scheme through a capacitive filter or in a series scheme through an inductive filter [4,5]. If the separated power conversion paths combine into one power converter, they form a multi-input power converter. This approach can potentially reduce the component count. Moreover, the multi-input converters may be designed for improving functionality and reliability of the system.

In multiple-input, single-output converters, each source is conditioned through unidirectional power converter stages [5–21]. In this case, the sources are not able to exchange energy between them. When bidirectional sources or energy storage elements exist in the system, bidirectional power conditioning stages are needed to condition each input. The sources may freely exchange power together. Such systems are called multi-port systems as each port can be either a load or a source to the system [22–30]. In this situation, a group of sources supply a group of loads in a multi-input, multi-output system. Multiple-input, multiple-output systems can also interconnect a group of unidirectional sources to a group of active and passive loads [31,32]. The multi-input converters may be dc–dc, dc–ac, ac–dc or hybrid dc and ac systems, according to the nature of the loads and the sources. The galvanic isolation between the sources and loads depends on the application requirements.

Several emerging applications utilize multi-input converters. Renewable energy systems have been the subject of interest in academia and industries. The renewable sources, such as photovoltaic sources and wind turbines, are not constantly available. A hybrid renewable system may benefit from multiple energy sources in an integrated power system or a micro-grid. This will improve the

*Texas A&M University, College Station, TX, USA

versatility of the system. Moreover, energy storage elements, such as batteries or flywheels, introduce bidirectional ports to the system. Integrated power systems and micro-grids are implemented in a small physical space; they are appealing applications for multi-input converters [7,12,19,20,24,27,33]. More specifically, hybrid electric and fully electric vehicles are mobile systems that host multi-input converters. Battery-powered vehicles, grid-connected automobiles, all electric ships and airplanes, and satellites need multi-input vehicular systems for land, sea, air, and space applications [4,20,22,23,34–40].

This chapter is dedicated to recent advances in multi-input converters. The novel applications for multi-input converters need customized power converter topologies. These topologies need to be synthesized considering the power conditioning requirements of the sources and the loads. The techniques recently proposed for synthesizing multi-input power converter topologies will be introduced in Section 3.2. The trivial multi-input topologies including dc link and ac link configurations will be discussed in Section 3.3. The major applications of the multi-input converters are reviewed in Sections 3.4 and 3.5. Section 3.5 belongs to the renewable energy systems, and Section 3.6 belongs to the vehicular power systems. The specific requirements of the mentioned applications are discussed from a designer point of view in these two sections. Finally, this chapter is summarized in Section 3.7.

3.2 Realizing multi-input converter topologies

Multi-input converters have emerging applications, and it is desired to develop novel topologies customized for each application. The fundamental concepts for developing single-input, single-output power converters can be extended to the synthesis of multi-input converters. Fundamentally, power converters consist of switching networks and passive elements between the sources and loads [1,2]. In other words, between each source and load there is graph that contains time-variant components, such as switches and diodes, and passive elements, such as inductors and capacitors. By cascading switching stages and passive networks, a multi-stage converter is formed. While there are different definitions for power conversion stages, the definition used in this chapter is the number of switching networks separated by passive elements between sources and loads determines the stages count. According to this definition, basic buck and boost converters are single-stage converters, and so is the differential synchronous buck converter, the h-bridge converter. In contrast, the z-source converter is not a single-stage converter because a diode is separated from the other switches by the z-source network [34,35,41]. Extending the multi-stage switching power conversion networks, it is possible to connect several sources through switching networks and passive elements to form a multi-input power converter. If each converter is directly connected to the load, it is a separated solution. In contrast, a multi-input power converter is realized if there is a common power conversion path in between the multiple sources and loads.

Assuming each source is either a voltage source or a current source, it is preferred to connect each source through a switching stage to the power converter network. If the ideal voltage sources are in parallel, they have to be identical, or they would violate Kirchhoff's Voltage Law (KVL). Alternatively, if the current sources are in series, they have to be identical, or they would violate Kirchhoff's Current Law (KCL). If a number of voltage sources are in series or a number of current sources are in parallel, they will add up according to KVL and KCL and represent as one unified source to the power converter network. It is not possible to condition these sources individually, which does not satisfy the goal of developing multi-input converters. The conceptual theory presented in this section describes the fundamentals of realizing multi-input power converters.

In [1], the fundamental theory of power converters as repetitive switching networks is applied to the analytical synthesis of single-input, single-output switching power converters; moreover, the properties of different switching power converters are discussed. The buck and the boost converters are two basic converters, and several converters can be derived by cascade or differential connection of these two basic converters as shown in Figure 3.1. As an example, buck–boost converter is derived by cascading buck and boost converters and simplifying the passive elements in between. The h-bridge converter can be derived by differential connection of two buck converters. The boost converter is actually dual of the buck converter [42]; also, it can be realized by the inversion of the source and the load in the buck converter. Not all power converters can be derived similarly from buck and boost converters. The single-ended primary inductor converter (SEPIC) and Watkins–Johnson converters are two examples shown in [1]. These two topologies are shown in Figure 3.2. In both converters, the passive network between the switches is more complicated than a simple inductor or a single capacitor; a transformer, which is a network of inductors and capacitors, is used within the sources and switching networks. A generalized approach is proposed to realize single-input, single-output power converters using the state space equations in [1]. This method uses connection matrices for describing the power converters' graphs.

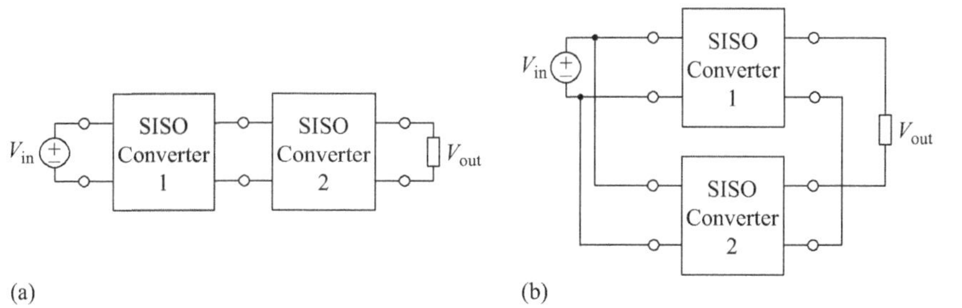

Figure 3.1 Converters can be derived from the existing converters. (a) The cascade connection of the converters and (b) the differential connection of the converters

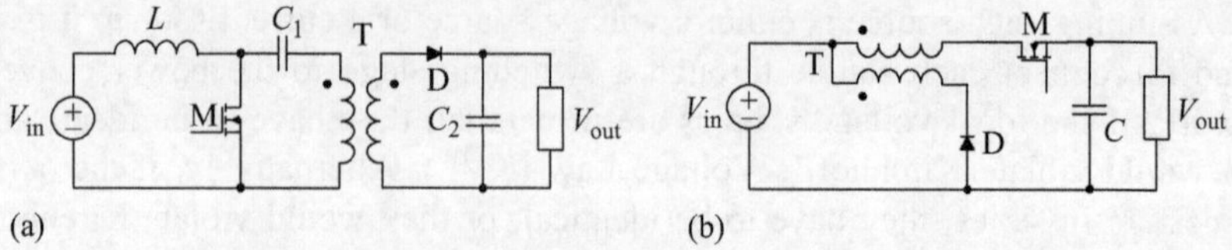

Figure 3.2 SEPIC and Watkins–Johnson topologies cannot be derived from buck and boost converters. (a) SEPIC and (b) Watkins–Johnson converter

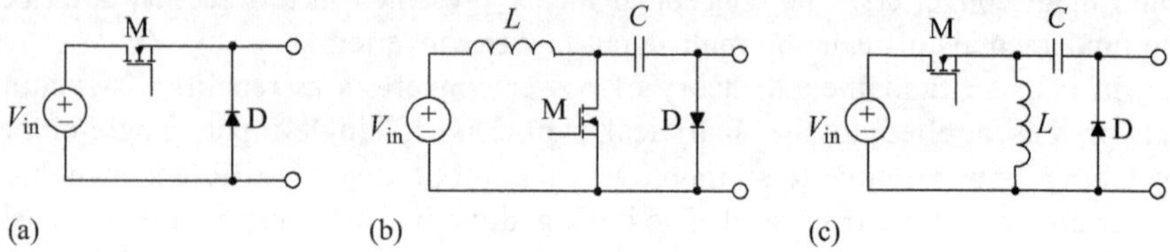

Figure 3.3 Examples of voltage input cells to introduce new voltage sources into existing dc–dc converters. The configurations are inspired from: (a) buck converter, (b) Cuk converter, and (c) Zeta converter. These input cells are connected in series with an inductor of the existing converters

In [3], the single-input, single-output dc–dc power converters are realized by defining various converter cells. Each converter cell is a three terminal graph of switches, inductors, and capacitors. Rotating the cell between the input and output generates different converters with the same converter cell. A conceptual method extending the existing single-input, single-output power converters to multi-input single-output converters is proposed in [6]. This method is limited to the multiple-input, single-output topologies and only allows unidirectional power flow out of the sources. Only non-isolated dc–dc converters can be derived using this method. Three networks consisting of a switch, a diode, and reactive elements are considered for connection of the voltage sources to the existing power converters as shown in Figure 3.3. These connection schemes are inspired from existing power converter topologies, specifically, buck, Cuk and Zeta converters. The source connection schemes shown in Figure 3.3 are called pulsating voltage source cells (PVSCs) in [14].

The conceptual idea of [14] is synthesizing new multi-input converters by connecting the voltage source cells in series with a current buffer or a current sink (an inductor) in one of six existing power converters, specifically, the buck, boost, buck–boost, Cuk, Zeta, and SEPIC converters. Similarly, pulsating current source cells (PCSCs) are inspired from boost, buck–boost, and SEPIC converters as shown in Figure 3.4. These networks have to be connected in parallel with a voltage buffer or a voltage sink (a capacitor) in one of the six existing converters. This method is addressed as the nested approach in this chapter. The nested approach for realizing multi-input converters is adding a new source through a power conversion stage to

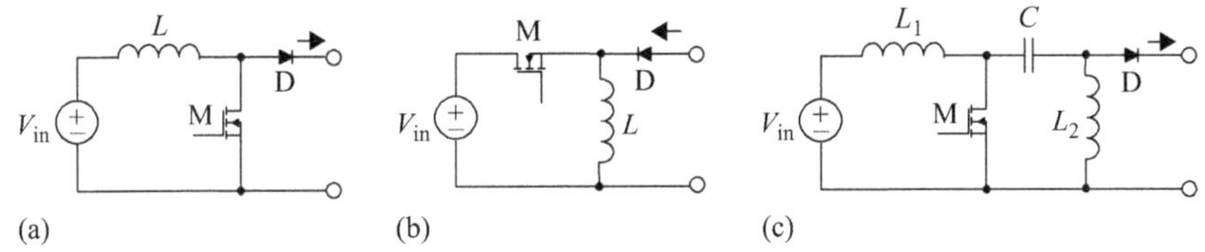

Figure 3.4 Examples of current input cells to introduce new voltage sources into existing dc–dc converters, these configurations are inspired from: (a) boost converter, (b) buck–boost converter, and (c) SEPIC. These networks are connected in series with an inductor of the existing converters

an existing power converter topology. The method can be repetitively applied to realize a more complicated multi-input topology.

Acronyms, such as PVSC and PCSC, are repetitively used in [6,14]. In this chapter, PVSCs are called voltage input cells, and PCSCs are called current input cells. Alternatively, the network that connects a source to the power converter circuit is called the source's power conditioning stage. This is based on the general theory of power converters as receptive networks of switching and passive elements, which was introduced in beginning of this chapter.

The source and its power conditioning network are combined into one input cell according to [6,14]; hence, the cell has only one free terminal. This will limit the realized topologies, as the power conditioning stage cannot be altered. In contrast, in [3], a three terminal network of switches and passive elements was defined as a power converter cell. The power converter cells defined in [3] do not include any source or load (input or output). The three terminal power converter cells can rotate between the input and output and form alternative power converter topologies; whereas, the input cells that are defined in [6,14] are fixed and cannot rotate.

The voltage input cells need to be connected within a mesh containing an output sink. Alternatively, the current input cells need to form a mesh with an output sink. This will assure all the inputs can transfer power to the load separately; otherwise, the sources have to work together to supply the load simultaneously. Such configurations are called quasi-multi-input converters in [6].

An example of the nested extension of the existing dc–dc converters is illustrated for a dual-input, single-output dc–dc converter for renewable energy applications in [19]. A voltage input cell is introduced to the buck–boost converter as shown in Figure 3.5. This topology saves an inductor and an output capacitor when compared with two individual buck or buck–boost converters. The multi-input converter of Figure 3.5 saves one inductor compared with two separate dc–dc converters supplying a common load.

The steady-state transfer function of this converter is formulated in (3.1), where d_1 and d_2 are the duty cycles of M_1 and M_2, respectively. If d_2 is zero, V_{s1} supplies the load through a buck converter with duty cycle of d_1. If d_2 is not zero and d_1 is zero, V_{s2} supplies the load through a buck–boost converter. Finally, both voltages can supply the load according to (3.1), when both d_1 and d_2 are not zero.

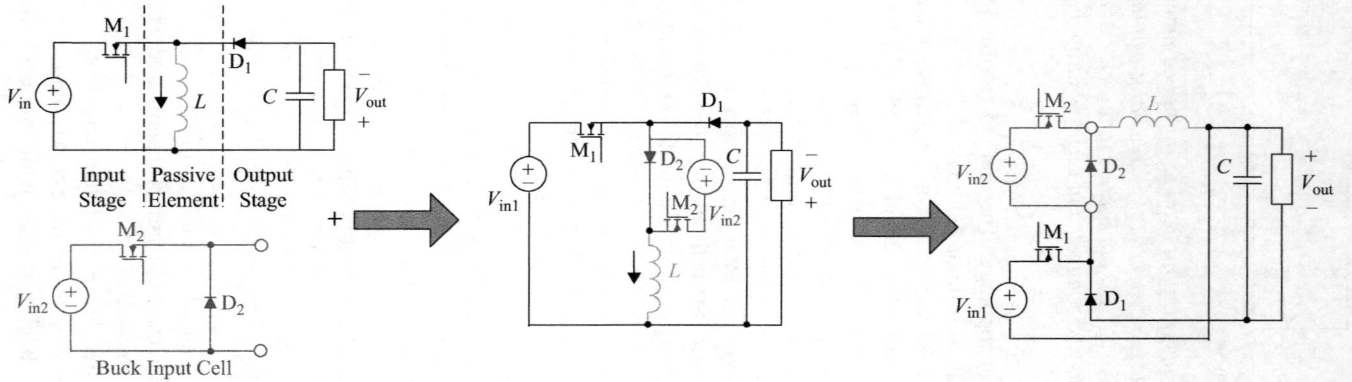

Figure 3.5 Extending a buck–boost converter to a dual-input single-output converter through a buck type voltage input cell

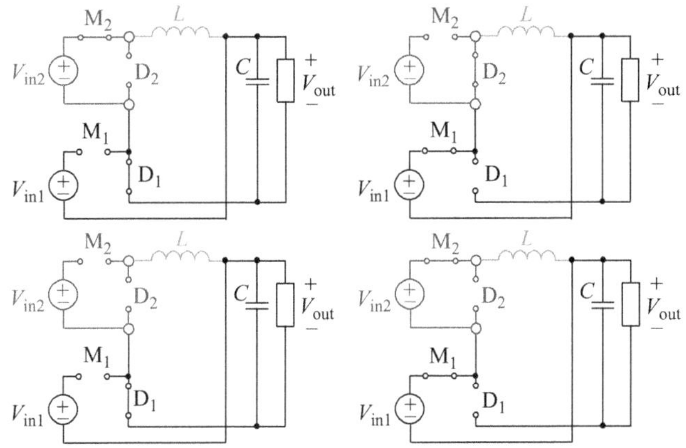

Figure 3.6 Switching states of dual-input converter as shown in Figure 3.5

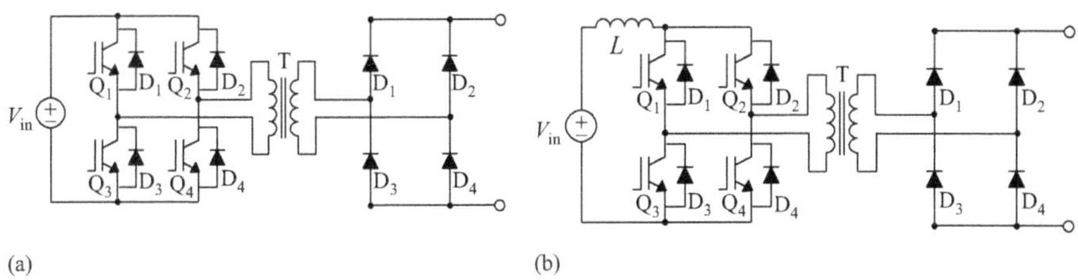

Figure 3.7 Examples of the isolated input cells: (a) full bridge type network for a voltage input cell and (b) current source full bridge network as a current input cell

There are two switches and four switching states for this converter, as shown in Figure 3.6.

$$V_o = \frac{d_1}{1-d_1} V_{\text{in1}} + \frac{d_2}{1-d_1} V_{\text{in2}} \tag{3.1}$$

The idea of building multi-input converters by introducing input cells is extended in [14] by introducing isolation transformers to the input cells. Examples of the isolated input cells are shown in Figure 3.7. The approach proposed in [14] is a modular approach. In addition to source cells, filter cells are defined. The multi-input converters are formed by series and parallel connections of the input cells and filter cells to the load. The voltage input cells are connected in series in a loop containing an inductive capacitive (LC) filter cell and the source, as shown in Figure 3.8. The current input cells are connected in parallel with a capacitive filter to the source. The parallel connection of the voltage input cells and the series connection of the current input cells are discussed, but only the parallel connection of the voltage input cells is recommended in [14]. Compared to the nested approach discussed earlier, this approach can realize fewer power conversion graphs. Similarly to the nested approach, realized multi-input converters will be unidirectional and single output.

Looking at Figure 3.7, the whole isolated full bridge converter is selected as the input voltage cell. When a number of similar cells are connected to an LC filter, only an output inductor is saved compared to the case in which separate power converters were used for different sources. This criticism is partially addressed in the same reference by employing common rectifiers or transformers in the multi-input converter as shown in Figure 3.8. Initially, an example of cascaded full bridge converters connected to the single load through the LC filter is shown. In this realization, each input has a dedicated high-frequency inverter, an isolation transformer, and a rectifier. The outputs are cascaded. Next, the rectifier stage is shared between both sources through the secondary of the isolation transformer. Inputs excite the primaries of the isolation transformer, the secondary supplies a common rectifier. A winding, a transformer core, and four diodes are reduced comparing to the cascaded full bridges. The third alternative of this topology cascades the outputs of h-bridges and eliminates one more winding. In this case, the input voltages need to be isolated. Both sources share the same turn ratio to the output side winding when they share the same primary winding. This simplification is not seen in the fundamental synthesizing rules suggested by Yan *et al.* [14]. The described shortcoming, presented in Figure 3.8, suggests the power conditioning stages include too many components and the suggested connection schemes are oversimplified.

The hybrid input cells by combining known power input cells according to the rules suggested for the nested approach are discussed in [14]. It will make it possible to realize more complicated power conversion graphs; whereas, the input cells will be even larger. An additional observation from the example shown in Figure 3.8 is that, the more features that are introduced into the power converter, the more effective a multi-input topology becomes. Adding an isolation transformer and rectifier stages provided more flexibility for combining power converter stages between the multiple sources and the loads.

In [14], the filter cells were defined to condition the load. A second look at the example shown in Figure 3.8 suggests more common elements can be gathered into the load-side cells in multi-input converters. This is the approach suggested in [5] for synthesizing multi-input power converters by extending single-input, single-output power converters. In addition to the input cells, the common output cell is defined. The multiple-input, single-output converter is divided into input cells and the common output cell. Multiple-input cells will be employed for conditioning additional inputs in multi-input converters. The common output cell is the common portion of the power converter network between all the input cells and the load. In order to achieve the minimum component count in the multi-input converter, the maximum possible number of components should be moved into the common output cell. At the same time, each input needs to be controlled individually. The following rules are developed to achieve the described multi-input power converter:

"***Rule 1:*** *The input cell must contain at least one forward conducting and bidirectional blocking switch.*

Rule 2: *Only one switch terminal can be connected to the input cell terminals. This assures the input cell will not have a redundant switch between several input cells.*

Rule 3: *If the input cell has a center capacitor, this capacitor average voltage should be independent from the source voltage.*

Rule 4: *The input source cannot directly be connected to both of the input cell terminals."*

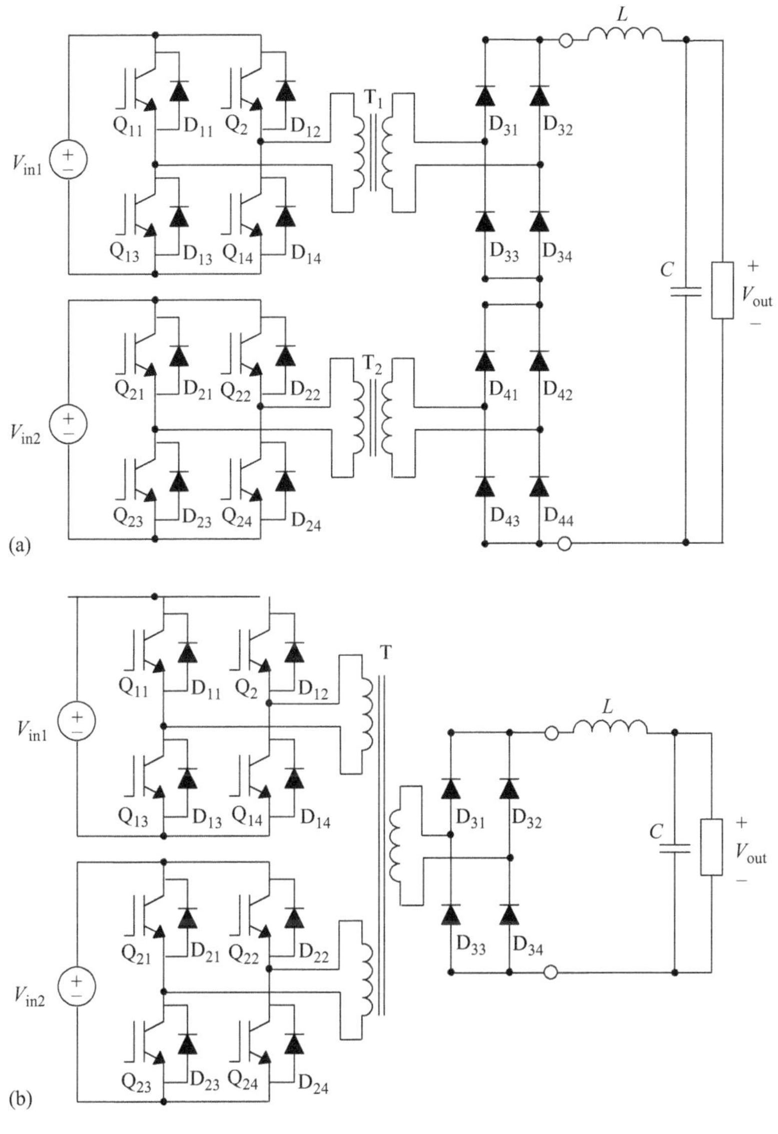

Figure 3.8 (Continued)

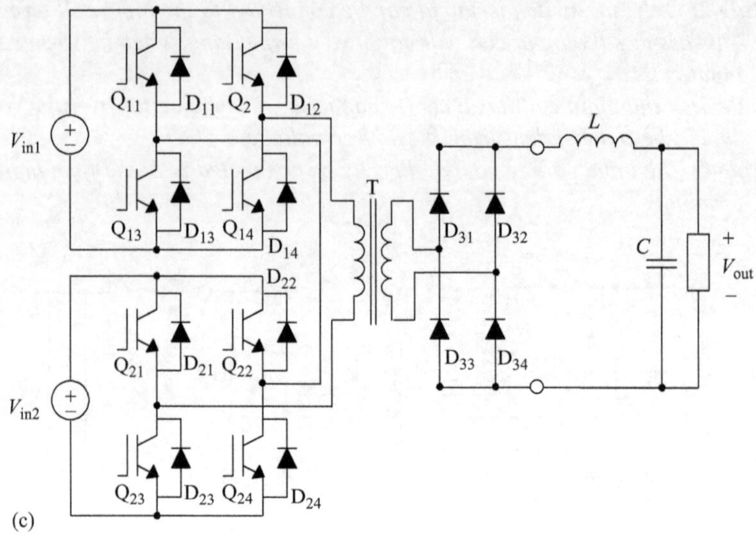

(c)

Figure 3.8 Simplifying isolated multi-input converters by employing common transformer or rectifier stages: (a) two cascaded full bridge converters supplying a load, (b) separated primary windings supply the common rectifier, and (c) the cascaded ac voltage is applied to the common primary winding and rectified by a common rectifying stage

The various converter cells listed in [3] are evaluated as potential candidates according to the mentioned rules. This results into four input cells; buck, buck–boost, G1, and G6 satisfy these rules as feasible input cells. G1 and G6 topologies are introduced in [3]. These topologies are shown in Figure 3.9. If these input cells draw pulsating current from the source, a source conditioning filter can be employed as shown in Figure 3.10. More feasible input cells can qualify if the third rule is relaxed by moving the middle capacitor to the input cells from the common stage; accordingly, additional cells, which are shown in Figure 3.11, will be feasible. Non-minimal multi-input power converters are realized from the input cells shown in Figure 3.11.

Two examples of multi-input converters realized by following the rules mentioned in [5] are shown in Figures 3.12 and 3.13. Figure 3.12 shows a multi-input converter extended from the G6 topology shown in Figure 3.9. It has the minimum possible components in the input cell. The multi-input topology shown in Figure 3.13 is based on the SEPIC topology. The middle capacitor filter is distributed among all the input cells. Although the approach proposed in [5] systematically leads to multi-input converters with minimal components, the derived topologies are limited to those with the identical input cells. It does not discuss if the common output stages can be shared among several different input cells.

This section provided an overview of the proposed methods for realizing multi-input converters. The fundamental procedure suggested by each reference briefly presented and the limitations of each method were discussed. All the methods were

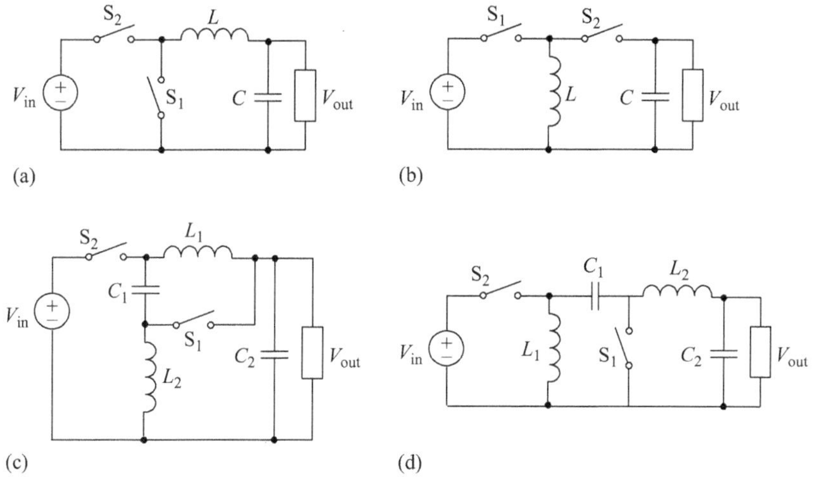

Figure 3.9 Feasible single-input, single-output topologies to be extended into multi-input, single-output topologies according to the rules indicated by Kwasinski [5], (a) buck, (b) buck–boost, (c) G1, and (d) G6

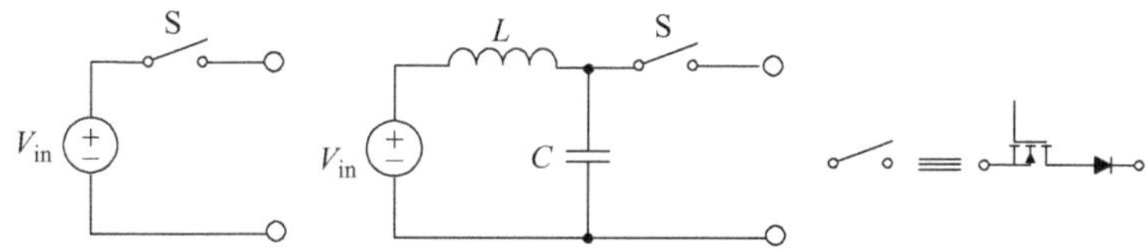

Figure 3.10 Employing current source conditioning filter (L and C) into the buck type input cell

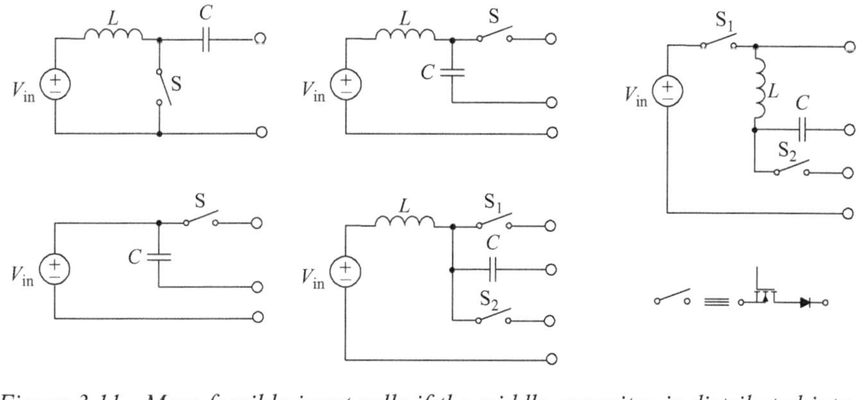

Figure 3.11 More feasible input cells if the middle capacitor is distributed into the non-minimal input cell

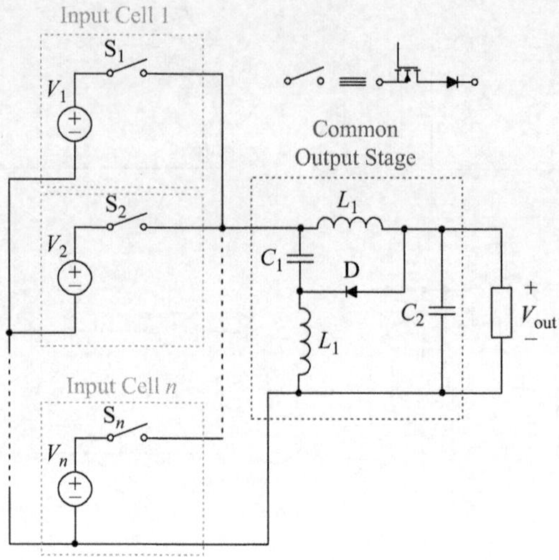

Figure 3.12 A multiple-input topology extending the G6 converter as shown in Figure 3.9

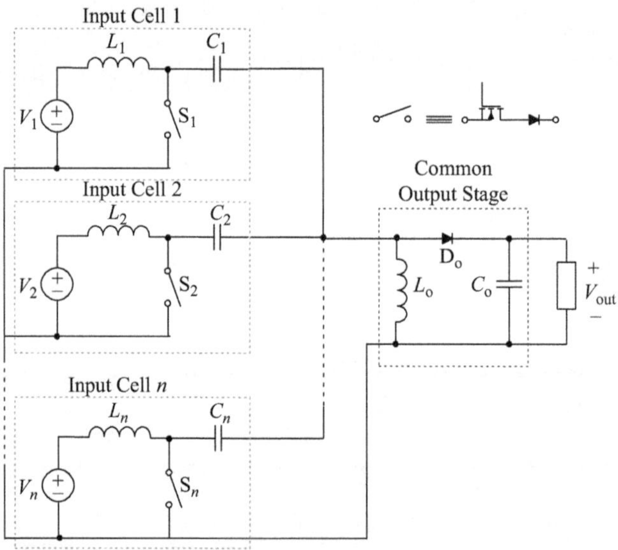

Figure 3.13 A multiple-input topology extending SEPIC

able to creatively realize new multi-input topologies; however, none of the discussed methods were able to address all the requirements of the practical applications. In contrary, the trivial rules of extending the power converters through dc and ac links remain as two flexible methods that can be applied to several multi-input, multi-output power converter systems. The multi-input converters formed around a link will be discussed in the next section that is dedicated to realization of the multi-port converters.

3.3 Multi-port converters

Multi-input converters only enable unidirectional power conversion paths from sources to the load. The ac grid and batteries are examples of sources and energy storage elements that accept bidirectional power flow. As discussed in Section 3.1, multi-port converters can provide bidirectional power flow for the bidirectional components. Multi-port converters can be synthesized similarly to multi-input converters. Multi-port converters may also be realized through dc link and ac link configurations. All these methods will be presented in this section.

3.3.1 Synthesis of multi-port converters by extending multi-input topologies

In Section 3.2, extending multi-input, single-output converters from the single-input, single-output converters was discussed. In [25], an approach for extending double-input, single-output converters to three-port converters is proposed. These three-port converters have one bidirectional port, one unidirectional source, and one unidirectional load. An example of this system is a renewable system with an energy storage element. When the source and the energy storage element supply the load, the converter is a double-input single-output converter. Alternatively, the converter is a single-input double-output converter when the source supplies the load and the energy storage element.

In [25], adding a unidirectional power conversion path from the bidirectional port (energy storage element) to the input (source) in order to realize a three-port converter is suggested. The procedure is applied to a boost dual-input converter, which was realized in [6], as shown in Figure 3.14. Initially, a boost path is added; then, the converter is simplified in a few steps to the final topology. Only one inductor, three active switches, and three diodes exist in the final power converter topology. The simplification process is called topology optimization in [25]. Although conceptual directions are given for topology optimization, this procedure needs to be manually customized for each power converter topology.

The relationship between the outputs and the input in the dual-output mode is written in (3.2) for steady-state conditions. The outputs are not decoupled and need to be controlled together through a central controller. The output is related to the inputs according to (3.3) during the dual-input mode. The steady-state relationship between the inputs and outputs in (3.2) and (3.3) shows the control variables are coupled. In [27,29,30], a decoupling method is suggested for a converter with two inputs, one bidirectional port, and one dc load.

$$V_{\mathrm{in}} = V_{\mathrm{o}}(1 - D_1) - V_{\mathrm{b}}D_2 \tag{3.2}$$

$$V_{\mathrm{o}} = \frac{V_{\mathrm{in}}(1 - D_3) + D_3 V_{\mathrm{b}}}{1 - D_1} \tag{3.3}$$

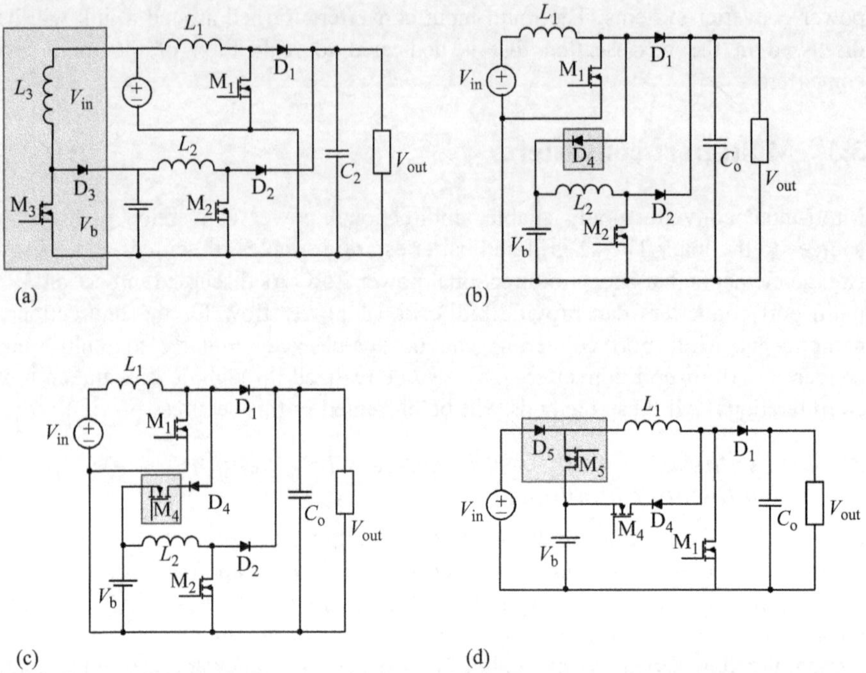

Figure 3.14 Extension of a three-port converter from the dual-input boost converter as described in [25]. (a) A power conversion path is added from the source to the battery, (b) L_1–M_1 and L_3–M_3 paths are merged, (c) adding M_4 enables independent power flow from V_{in} to V_b and V_{out}, (d) two boost paths are paralleled in the previous sub-figure (L_1–M_1–D_1 and L_2–M_2–D_2), these two paths are merged and M_5–D_5 are placed in order to regulate V_{in} output and block reverse current into V_{in}

If a three-port converter has two bidirectional ports, following a similar concept will result in four unidirectional power conversion paths or two unidirectional paths and one bidirectional path. If all the components of a three-port converter need bidirectional power flow, then six unidirectional paths or three bidirectional paths are needed in the converter. In a four-port converter, up to six bidirectional paths or twelve unidirectional paths are required. This is illustrated in Figure 3.15. Generally, in an n-port converter the required number of power conversion paths can be calculated as k_n in (3.4).

$$k_{n(\max)} = \begin{cases} \dfrac{n(n-1)}{2} & \text{Bidirectional paths} \\ n(n-1) & \text{Unidirectional paths} \end{cases} \tag{3.4}$$

If each power conversion path is implemented with a diode and an active switch, then the n-port converter needs $2n(n-1)$ switches; whereas, the number of

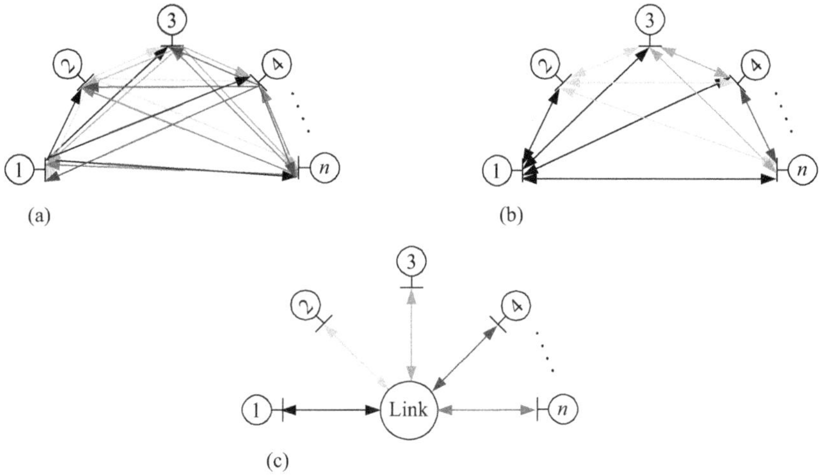

Figure 3.15 *Power conversion paths in multi-port converters; the number of paths for an n-port converter is: (a) n(n − 1) unidirectional paths, (b) n(n − 1)/2 bidirectional paths, and (c) 2n unidirectional paths or n bidirectional paths to the link*

switches is $2n$ if each power conversion path is connected to a dc link. This analysis suggests the benefit of developing a multi-port converter around a dc or ac link. This benefit becomes more significant as the number of ports grows. Multi-port converters formed around a dc link will be discussed in the next subsection.

3.3.2 Multi-port converters with dc link

Developing multi-port power converters around a dc link is a trivial approach. The capacitive dc link integrates multiple voltage source power converters into one multi-port power converter. All the loads, sources, and energy storage elements are connected to the dc link through a power conditioning stage. The capacitive dc link is a parallel dc link as all the power conditioning stages will be in parallel with the link. The dc link decouples the controller function of each stage from rest of the power converter according to KVL.

In [28], the implementation and control of a five-port power converter around a dc link is considered. The sources include a fuel cell, a solar panel, and a low-power wind turbine. Three unidirectional boost converters have been coupled to the dc link. The wind turbine generator has an ac output and is conditioned through a rectifier prior connecting to the dc–dc stage. A battery is connected through a bidirectional boost (buck) converter. A bidirectional h-bridge inverter (rectifier) is the grid's power conditioning stage. This topology includes six legs in parallel with the dc link. The general topology of a similar multi-port system formed around a dc link is shown in Figure 3.16.

In the topology of Figure 3.16, each source is conditioned through an inductor. The inductors currents and the dc link voltage are the state variables of the system.

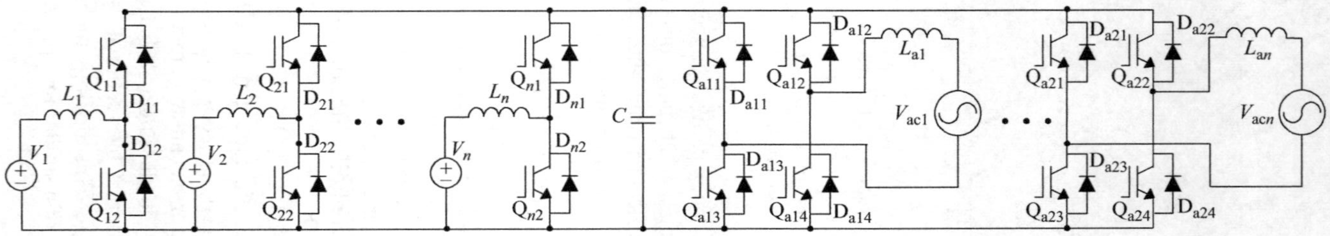

Figure 3.16 Multi-port power converter formed around the capacitive dc link

Thus the system described in [28] is a fifth-order system. The average and small signal models of the system are developed; moreover, the control system is developed in [28]. The system is partitioned into the sources and loads. The main control goal is regulating the dc link voltage. One converter may regulate the dc link voltage, and the other converters will exchange power accordingly. Similarly to the single-input, single-output converters with a dc link, the sources' and the loads' power flows are synchronized if the dc link voltage is regulated.

In contrast to the capacitive dc link, the inductive dc link can be employed in a power converter system. In this case, all the power converters are in series with an inductor. An example is shown in [31,32]. The general topology of this multi-input, multi-output power converter is shown in Figure 3.17. Each source is connected through a forward conducting bidirectional blocking switch similar to the switches used in [5]. In this chapter, the described switches are realized with a MOSFET in series with a diode. The diode is needed to avoid paralleling different voltage sources when the switches are on simultaneously. The small signal model of the system is derived by the state space averaging method for a two-input, two-output system. The controller for the same system is designed. This system is evaluated by implementing a three-input, two-output prototype. Although the inductor L in Figure 3.17 serves as an inductive dc link, the converter is essentially an extension of single-input, single-output converters to multi-input, multi-output converters.

This topology is a limited multi-input, multi-output topology formed around an inductive dc link. When several sources are connected to the dc link inductor, only the one with the highest voltage transfers energy. The sources with lower voltages will be blocked by the series diodes. Only sequential power transfer is possible with this topology. The accommodation of bidirectional components such as batteries is not discussed in [31,32]. One solution can be connecting the battery to the switching cells on both sides of the series dc inductor. In the proposed switching scheme, all the input switches are initially turned on. The switches will be turned off from the highest input source to the lowest input source sequentially.

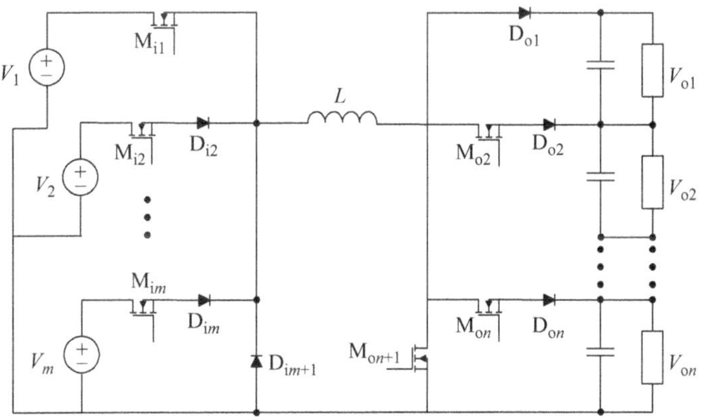

Figure 3.17 Multiple-input multiple-output converter formed around the inductive dc link

Similarly, the output switches are turned off sequentially. The source and load power conditioning stages are oversimplified in [32]. Looking at the source cells, only one switch is used. This will limit the controllability of the sources. The flexibility of the system is lost in exchange for simplicity when compared to Figure 3.16.

3.3.3 Ac link multi-port power converters

When galvanic isolation is required between the system components, an isolation transformer needs to be placed between the ports. Moreover, the transformer improves power converter utilization when the voltages of either side are very different. In order to reduce the transformer's volume and weight, the switching frequency needs to be increased. This is especially important for portable applications, such as vehicular applications. Ac link converters are ideal topologies for accommodating isolation transformers. Besides galvanic isolation, ac link converters provide benefits such as soft switching, resonant power conversion, and multiple-quadrant power conversion. The ac link current and voltage are alternating and have zero crossings. Switching at zero crossings enables zero voltage switching and (or) zero current switching.

Similarly to the dc link implementation, the ac link is implemented by one or more passive components and an isolation transformer, when galvanic isolation is required. In contrast to dc link approach, the dc bias is avoided for the ac link components. This is mandatory for proper transformer utilization. The ac link components may be integrated into the isolation transformer. The electric model of the transformer consists of the leakage inductance and magnetizing inductance. If the leakage inductance is utilized as the ac link, this is a series inductive ac link. If the magnetizing inductance is utilized as the ac link, this is a parallel ac link.

The dual active bridge (DAB) converter is an example of a single-input, single-output, series ac link topology. The DAB topology was introduced in the early 1990s [43]. The two-quadrant DAB is shown in Figure 3.18. A four-quadrant DAB is investigated in [44] as shown in Figure 3.19. The equivalent circuit of a DAB is shown in Figure 3.20. The equivalent circuit includes two equivalent ac sources which are connected though an inductor, L, in series. The ac inductor, L, shown in Figure 3.20, is the key component for transferring power between the ac sources.

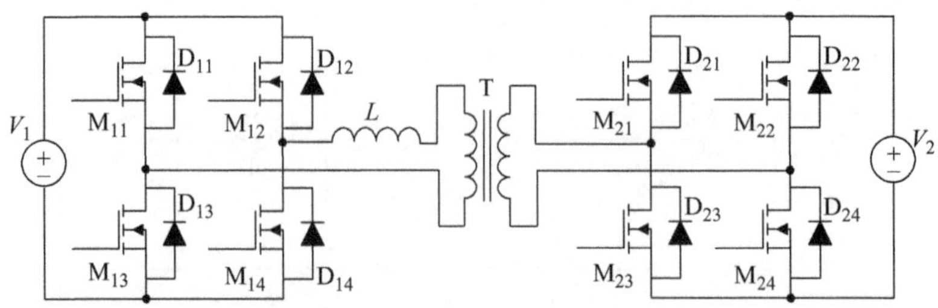

Figure 3.18 A two-quadrant DAB is a series ac link dc–dc converter

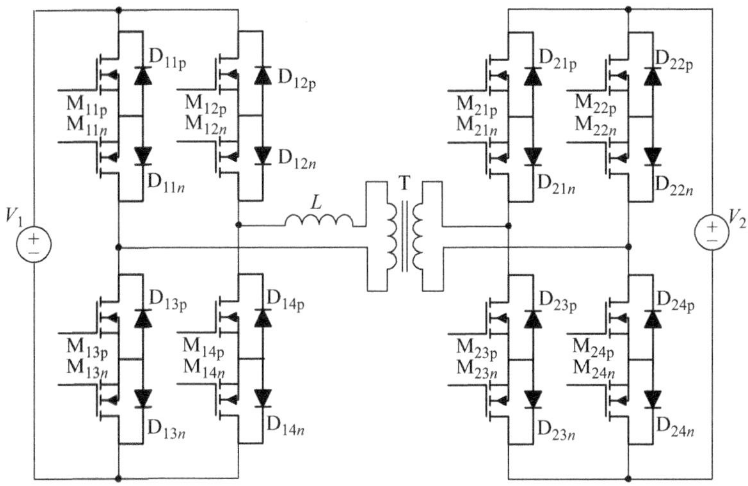

Figure 3.19 A four-quadrant DAB is implemented with four-quadrant switches

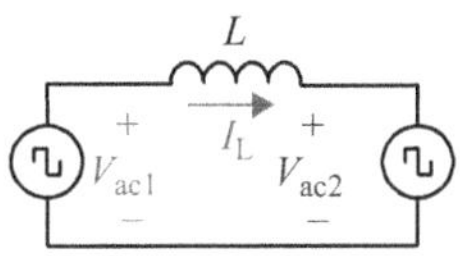

Figure 3.20 DAB equivalent circuit

The power flow is controlled by controlling the phase shift between the primary-side bridge's ac voltage and the secondary-side bridge's ac voltage. This modulation method is called phase shift modulation [43]. Phase shift modulation enables maximum power transfer between the bridges.

The flyback converter and the converter proposed in [26] are examples of parallel ac link converters. A single-input, single-output, parallel ac link converter is shown in Figure 3.21. The modulation method proposed in [45] can be applied to the power converter as shown in Figure 3.21. This modulation method results in a triangular ac link current waveform.

In order to compare the ac link topologies, the ac link utilization factor (UF) is defined in (3.5) [39]. The ac link UF is the ratio of the average power to the maximum power transferred through the ac link. The ac link is cascaded with the power path; thus, this definition determines the power utilization of the converter; furthermore, it evaluates the power density. In (3.5), i_{link} is the link current and v_{sec} is the transformer's secondary-side voltage.

$$\text{UF} = \frac{\frac{1}{T_s}\int_{T_s}(i_{\text{link}}(t)\cdot v_{\text{sec}}(t))\mathrm{d}t}{i_{\text{link(max)}}\cdot v_{\text{sec(max)}}} \tag{3.5}$$

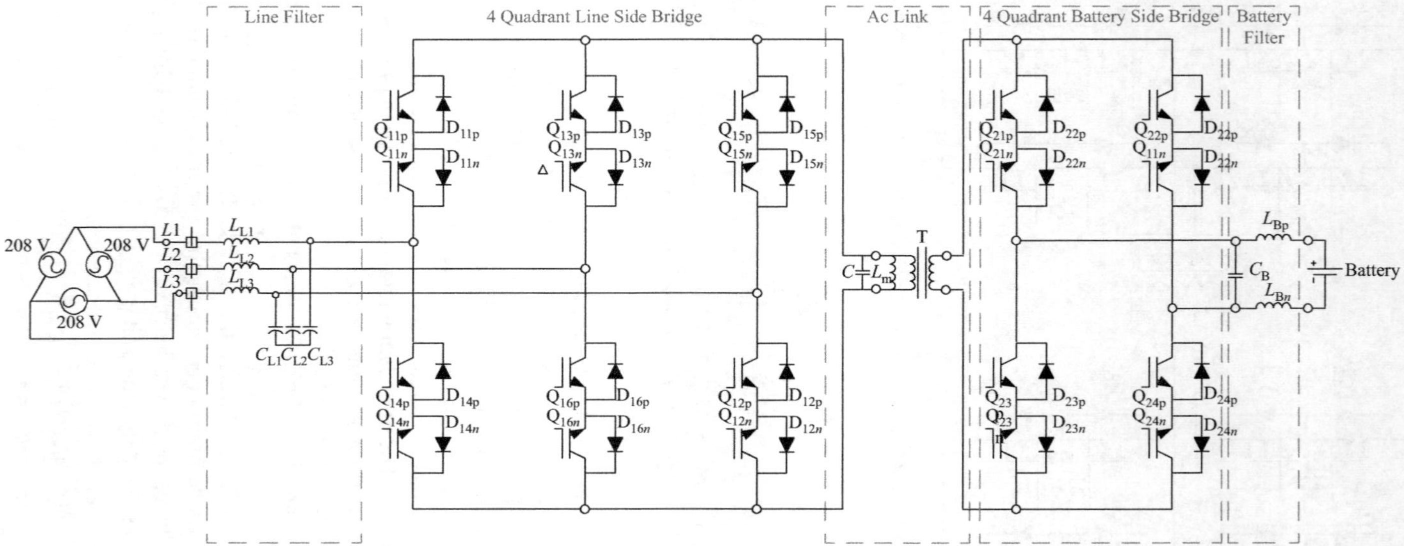

Figure 3.21 Single-input single-output ac–dc converter based on a parallel ac link converter

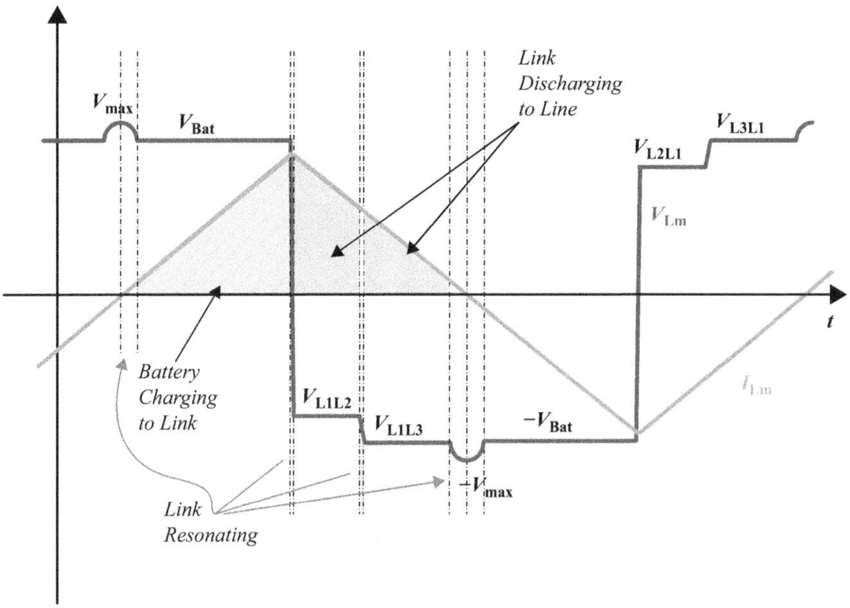

Figure 3.22 Ac link waveforms of the parallel ac link converter as shown in Figure 3.21

The ac link UF will be calculated for the discussed topologies by considering the ac link waveforms. The diagrams of Figures 3.22 and 3.23 are the ac link waveforms of parallel and series ac link topologies, respectively. In the parallel ac link topology, the link is parallel with the power conversion path. All the transferred power is charged into the link inductor and then discharges to the output port alternatively during each switching cycle. In contrast, the series ac link is in series with the power conversion path. With phase shift modulation, square waveform voltages are applied to each side of the link, with a phase shift. Only the difference of primary and secondary sides' voltages will store energy into the link. Thus, the ac link inductance causes less circulating power for the series configuration with the phase shift modulation.

The proposed ac link UF criterion supports the previous statement about the series ac link configuration's excellence. Following (3.5), it suggests the maximum UF for a square waveform, which is 1. The phase shift modulation with low-phase shift is the closest to a unity UF. On the other hand, UF is less than 0.5 for a triangular link current. Since only half of the switching cycle is dedicated to each side in parallel series ac link converter, the UF is less than 0.25 for this configuration. The UF for the discussed topologies is approximately calculated as follows. The parallel ac link converter's UF is estimated as (3.6). If the resonating cycles are neglected, the UF is 0.25 for this topology. Similarly, the series ac link converter UF is calculated in (3.7). In (3.7), PS is the phase shift between the bridges in radians. The phase shift is always below 180°; hence, the UF is always more than 0.5. In practice, the phase shift is controlled

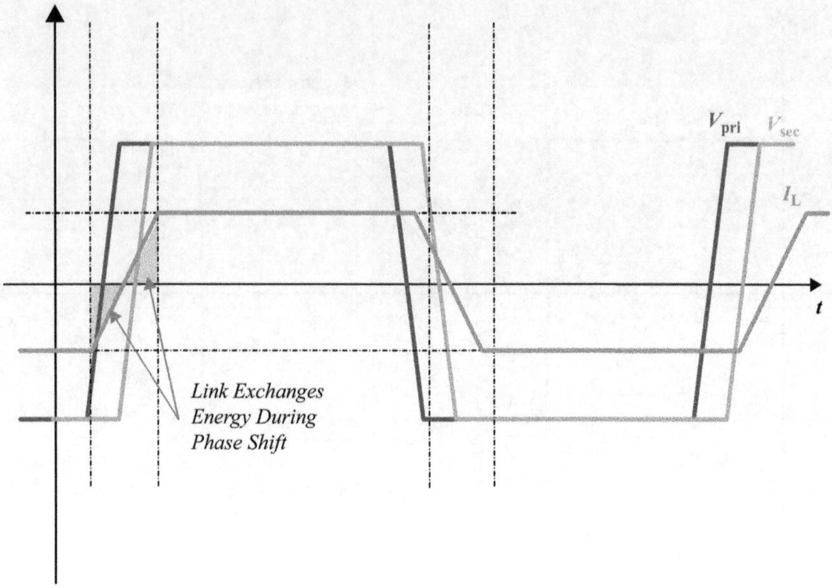

Figure 3.23 Ac link waveforms of the phase shift modulated DAB as the series ac link converter

below 90° in order to avoid excessive reactive power. Thus, the UF for series ac link topology would be more than 0.75.

$$\mathrm{UF}_{\mathrm{Par}} \approx \frac{\frac{1}{Ts} \cdot V_{\mathrm{Bat}} \cdot I_{L\max} \cdot \frac{T_s}{4}}{V_{\mathrm{Bat}} \cdot I_{L\max}} = \frac{1}{4} \tag{3.6}$$

$$\mathrm{UF}_{\mathrm{Ser}} \approx 1 - \frac{\mathrm{PS}}{2\pi} \tag{3.7}$$

When galvanic isolation is required, the magnetizing inductance will be the ac link. In other words, the energy needs to be stored in the magnetizing inductance, and the magnetizing inductance acts as an ac inductor. The magnetizing inductance easily saturates, hence an air gap needs to be integrated into the core. This reduces the magnetic coupling and increases the leakage flux. The leakage inductance is not utilized in this topology and leads to a low efficiency for the magnetic components.

Both the series and parallel ac link topologies can be extended to multi-port configurations. This can be achieved by introducing more windings around the transformer's core. In [26], an example of a multi-port parallel ac link converter is presented. In [46–50], examples of multi-active bridge converters are suggested. Figure 3.24 shows a triple-active bridge converter as an example of a multi-active bridge topology. Alternatively, Figure 3.25 shows a system based on the three-port, parallel, high-frequency ac link topology.

The multi-port variations of these two topologies inherit the characteristics that were discussed earlier in this subsection. Additionally, the multi-active bridge is more flexible for power flow control between the ports. All the ports may

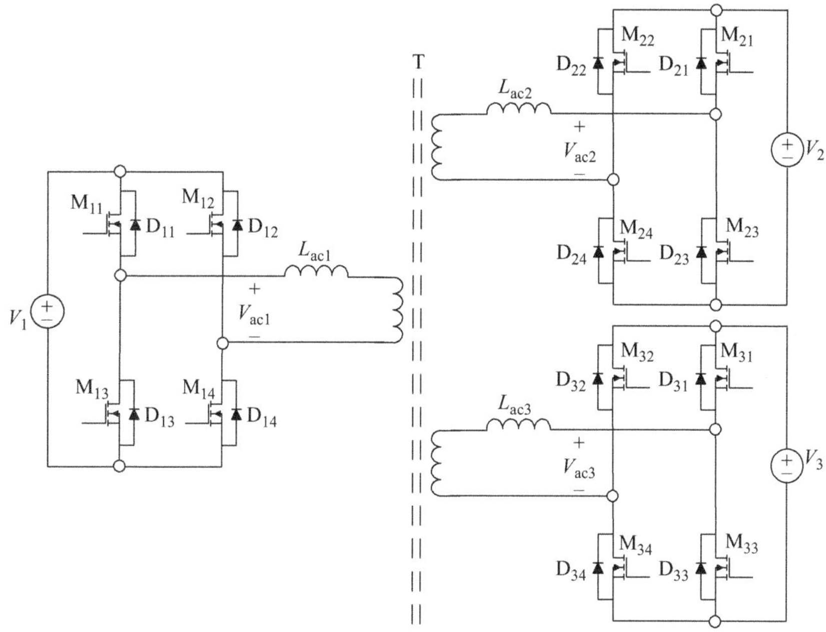

Figure 3.24 MOSFET realization of the triple-active bridge

simultaneously increase or decrease the magnetic flux in the magnetic circuit. On the other hand, in the parallel ac link multi-port converter, the link is charged and discharged sequentially by the ports, and it is not possible for several ports to charge or discharge the link at the same time. As a result, the UF can potentially approach the unity for the multi-active bridge topology, but the UF will be less than 0.25 for any of the sources that are attached to the parallel ac link converter. Indeed, the more sources that are added, the lower UF each has from the ac link. This translates to higher power density and more flexibility for the series ac link configuration that utilizes the phase shift modulation. In conclusion, the fundamental comparison between the isolated multi-port topologies suggested that the multi-active bridge topology is the superior topology to interconnect all isolated sources.

3.4 Applications of multi-port power converters

Different methods for realizing multi-port power converter topologies were introduced in the previous sections. In this section, the major applications of multi-port power converters will be reviewed. Renewable energy systems, micro-grids, and vehicular power systems are among the major applications for multi-port power converters. These applications have been the subject of interest in recent publications. The following subsections are dedicated to the multi-port converters proposed for the applications mentioned above. In each subsection, a brief introduction to the application will be provided and a few examples of recent works will be reviewed.

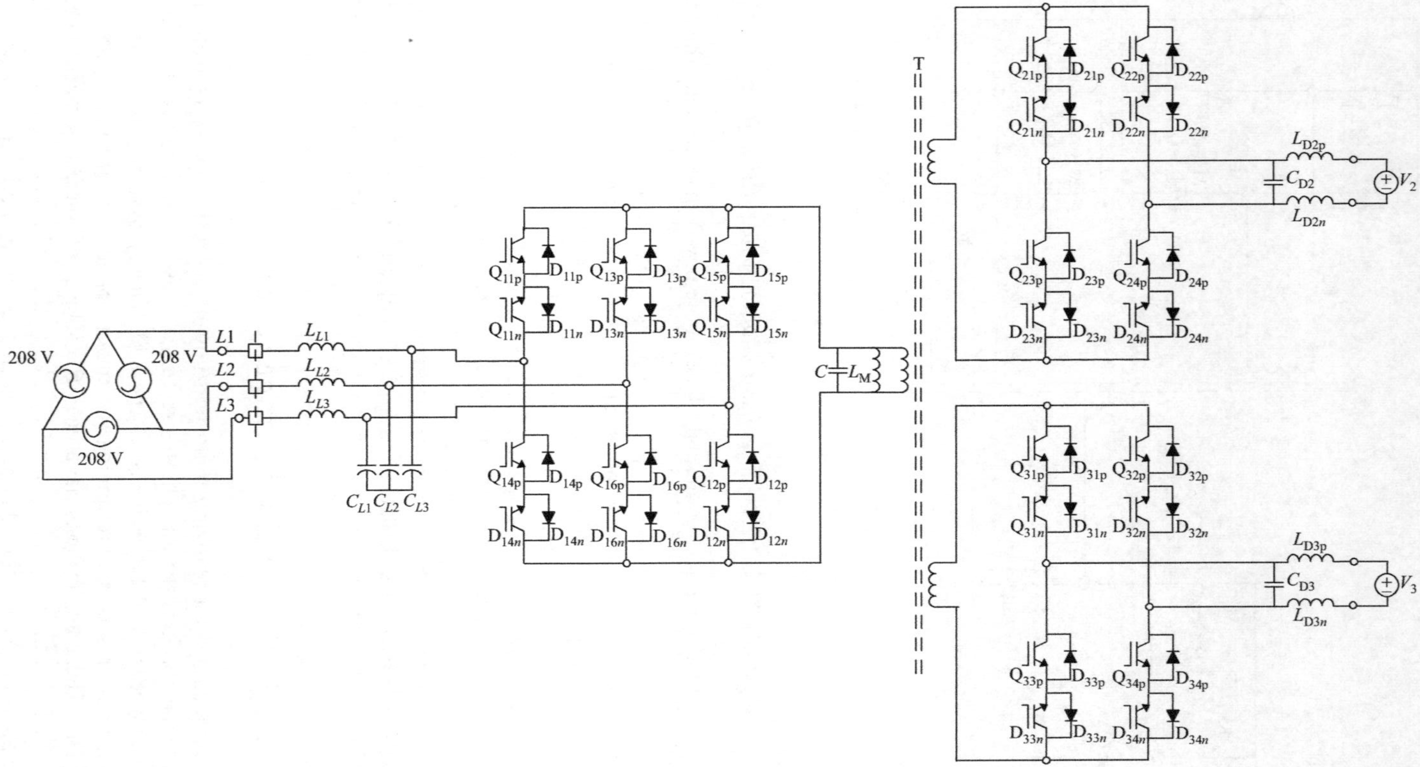

Figure 3.25 Three-port parallel high-frequency ac link converter

3.4.1 Multi-port power converters for renewable energy systems

Distributed generation is possible due to renewable energy resources. Renewable energy resources, including photovoltaic panels and wind turbines, are programmable by the load. A combination of multiple renewable sources and energy storage elements increases the availability of renewable power in grid-connected and stand-alone systems [7,12,19,20,24,27,33].

A hybrid renewable energy system or renewable backup system includes a renewable source, an energy storage system, and loads. Renewable backup systems are being commercialized. A photovoltaic backup system includes a photovoltaic plant, a battery pack, and interconnections to the ac grid and local ac and dc loads. A solution practiced in industry includes separate commercial converters that are all connected to the low-frequency ac grid [33]. In contrast, a multi-port converter can integrate all these sources and loads into one power conditioner [7,19,24,25].

Figure 3.26 shows a multi-port power converter proposed in [7] for renewable energy systems. The core of this topology is a non-isolated multi-port converter, which is realized with two switching legs (half-bridges). The battery can be conditioned by both switching legs. The ac load is supplied through the differential connection of the switching legs. Each leg is supplied by a number of loads connected through a dc link. This solution is a multi-port power converter extended through two dc links. The authors discuss different operating modes and suggest the state-feedback control method for the converter. The battery does not have a dedicated power conditioning cell in this system. It balances the supply and demand in this system. If the source provides more power than the load, the battery is charged. Alternatively, the battery is discharged to supply to the load if the source power is less than the demanded load power. In Figure 3.26, the battery does not have a dedicated power conditioning stage. The switches are reduced at the cost of less flexibility in the power conditioning of the battery.

The solution used in [33] is a bidirectional, high frequency, isolated inverter that connects the battery to the low frequency (50 Hz/60 Hz) ac grid. This solution aims to form energy storage systems from blocks that are all connected to the terrestrial ac grid. In contrast, a more integrated photovoltaic backup system is developed in [24]. In this system, both the photovoltaic array and the batteries are connected to the dc link. The photovoltaic array is conditioned through a non-isolated boost converter. The battery is conditioned through a bidirectional push–pull forward isolated dc–dc converter. The dc link is also connected to the grid through a bidirectional ac–dc converter. These photovoltaic backup systems are compared in Figure 3.27.

It is possible for more than one source to need to be isolated from rest of the system. In this case, the sources may use a multi-input converter. In [12], multiple renewable sources are conditioned by the source cells as shown in Figures 3.3 and 3.4, which were suggested by Yuan-Chuan and Yaow-Ming [6]. The output of the resulting converter is cascaded with an isolated dc–dc converter. This approach includes a non-isolated dc–dc stage and an isolated dc–dc stage, which is illustrated in Figure 3.28. It has one more stage when compared with the approach proposed in

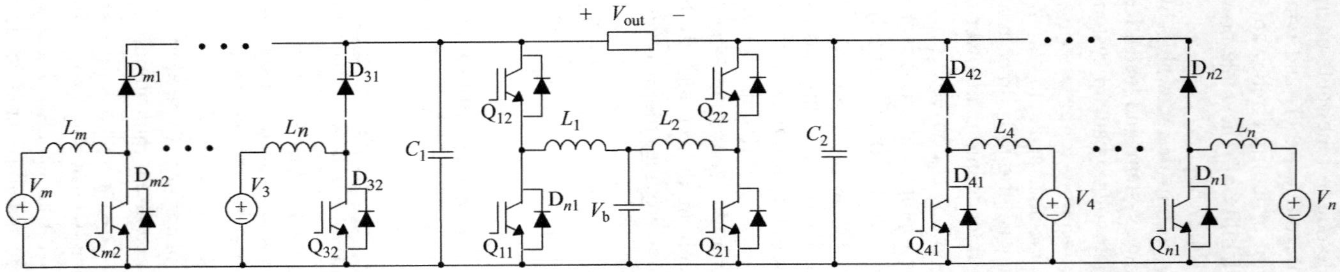

Figure 3.26 A multi-port power converter proposed for renewable energy systems

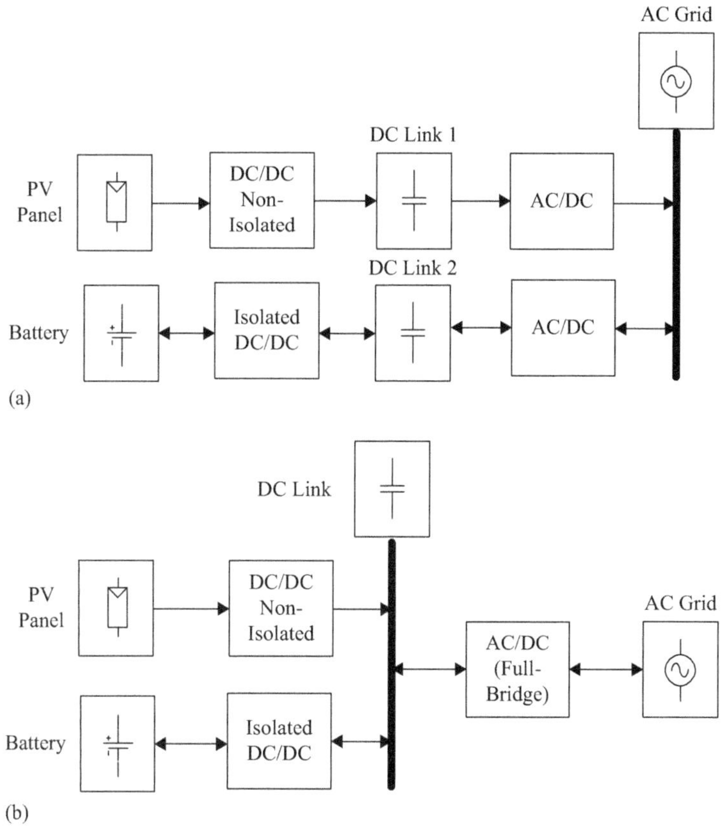

Figure 3.27 Photovoltaic backup systems, (a) the system is formed around the ac grid and (b) the sources are integrated through the dc link

[14], which is shown in Figure 3.8. However, fewer switches are used to condition each dc source.

The idea of a multi-port converter cascaded with an isolated dc–dc converter is also illustrated in [29]. A four-port converter is proposed as shown in Figure 3.29. Two unidirectional inputs, which are solar and wind sources, are conditioned. Additionally, a bidirectional battery is integrated into the system. The converter has an isolated dc source which is connected through an LC filter. The isolated dc–dc converter is a half-bridge converter with a synchronous rectifier, which is integrated into the multi-port converter. This four-port system is controlled through a decoupling network in [29]. The three-port variation of this converter is proposed for a satellite power system in [30]; this is an example of multi-port power converters in space power systems.

3.4.2 Application of multi-input converters in micro-grids

Micro-grids are areas in a grid that may either work independently or in conjunction with the main grid. Each micro-grid needs to be supplied by distributed

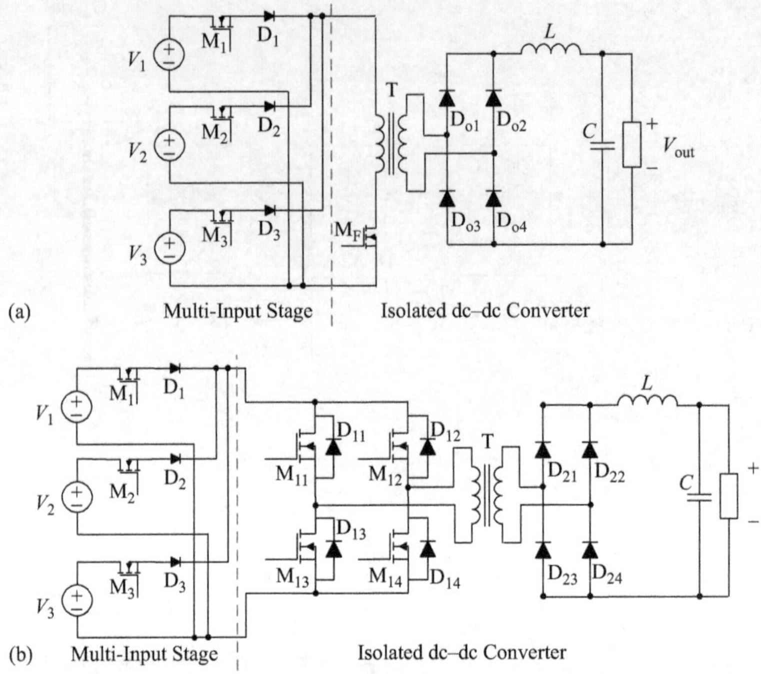

Figure 3.28 Multi-input converters cascaded with an isolated dc–dc converter, (a) three buck input cells in parallel are cascaded with the flyback converter and (b) three buck input cells in parallel are cascaded with the full-bridge converter

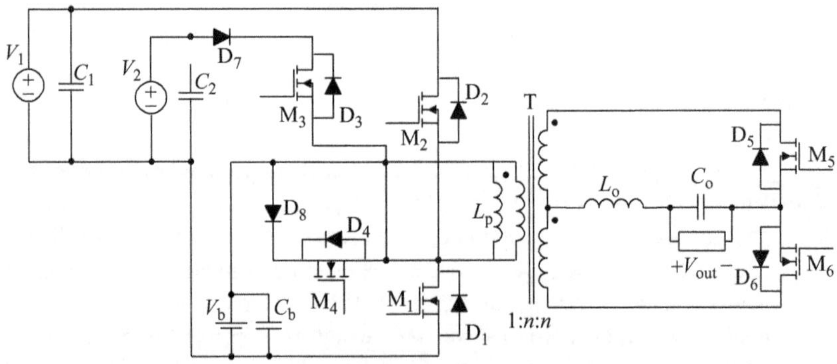

Figure 3.29 A four-port converter built with a multi-port converter and an isolated half-bridge converter for renewable energy applications

generation and backed-up with energy storage. The distributed sources can be renewable sources, which were discussed in the previous subsection. Small diesel generators are examples of the distributed generators that are not renewable.

Multiple-input converters enable the integration of multiple distributed resources and energy storage elements into the micro-grids through cost-effective and efficient solutions. In bipolar dc micro-grids, a positive and a negative voltages are distributed [16,51]. A double-input single-output, a single-input double-output, or a triple-port converter is needed to interface unipolar and bipolar dc micro-grids. This is an example of how multi-input converters can be used as the building blocks of futuristic dc micro-grids.

A telecommunications power system is an example of a micro-grid that is implemented with a multi-input, non-isolated positive buck–boost converter in [15]. This converter is formed around an inductive dc link similarly to the converter as shown in Figure 3.17; in contrast, the multiple-input positive buck–boost converter does not condition multiple outputs. Alternatively, bidirectional power flow through the dc link inductor is allowed [17]. The switches are unidirectional conducting and bidirectional blocking in [15]. Bidirectional switches are utilized in [17] to condition several current sources.

3.4.3 Multi-port converters for vehicular power systems

Hybrid electric and fully electric vehicles include several electric components gathered in a mobile electric power system [52]. If all the sources and loads are interconnected within a vehicle, a vehicular integrated power system is formed [23]. The integrated power system concept was specifically investigated for the shipboard power systems [4]. An integrated shipboard power system requires less reserve power and offers more reliability than a segregated shipboard power system.

Multi-port power converters are attractive topologies for vehicular power systems, because multi-port power converters can reduce the component count and increase the power density of the converter. Vehicular power systems are mobile, and power density matters for such applications. Multi-port power converters can be designed for interconnecting several components in a vehicular system. This can improve the reliability and reduce the required power capacity of the system as used in integrated shipboard power systems. This subsection provides a few examples of multi-port power converters that have been proposed for battery-powered vehicles.

A conceptual block-diagram of a battery-powered vehicle is shown in Figure 3.30 [23]. This system includes two dc buses, a high-voltage dc bus and a low-voltage dc bus. The voltage of the high-voltage bus is determined by the high-voltage battery of the vehicle. The traction drive is connected to this battery. This battery is charged from the ac grid. Additionally, the battery's power can be transferred to the ac grid for vehicle-to-grid operation [38,39,53,54]. The low-voltage bus is determined by the low-voltage battery in the vehicle. The control system and the cabin loads are connected to this low-voltage bus. The low-voltage battery is charged from the high-voltage bus. Renewable sources may be connected to the low-voltage or the high-voltage dc bus. The low-voltage components are grounded to the chassis. The high-voltage components are referenced to the chassis through a high impedance grounding network. The

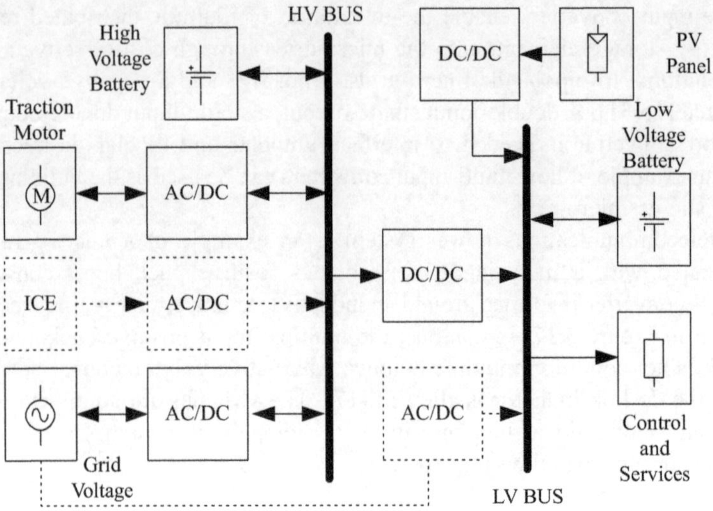

Figure 3.30 Conceptual single-line block diagram of a conventional vehicular power system

high-voltage components need to be isolated from the low-voltage components due to safety standards [38].

In [36], a hybrid battery and ultra-capacitor energy storage system are investigated for battery-powered vehicles. The authors indicate that the multi-input converter is expensive for this particular application. They have placed the ultra-capacitor as the dc link and interconnected the battery and the ultra-capacitor through a dc–dc converter. The power flow between the energy storage elements is controlled by adjusting the ultra-capacitor's voltage. In addition to the battery and ultra-capacitor, a fuel cell is integrated into the vehicle's propulsion system in [20]. A multiple-input topology based on the dc link connection is utilized. Each source is conditioned through a synchronous buck–boost converter. All the dc–dc converters are connected to the dc link that supplies the drive's inverter in [20].

A charger that is able to charge both the low-voltage and the high-voltage batteries in the vehicle is an attractive product. In [55], a unidirectional auxiliary power conditioning pass is added to a two-stage non-isolated vehicular utility interface. The shortcomings are that the auxiliary mode requires the power converter to be reconfigured and there is no means to transfer energy from the low-voltage battery to the high-voltage battery. In [56], an isolated three-stage charger connects to the high-voltage dc bus, and an auxiliary unidirectional dc–dc converter interconnects the low-voltage battery bus to the high-voltage dc bus. In order to avoid the mentioned shortcomings, a multi-port series ac link converter is proposed for interconnecting the low-voltage battery and the grid side converter in [23].

The system shown in Figure 3.31 interconnects all the sources and loads of a vehicle into a vehicular integrated power system [23]. The core of this vehicular integrated power system is a multi-port isolated converter.

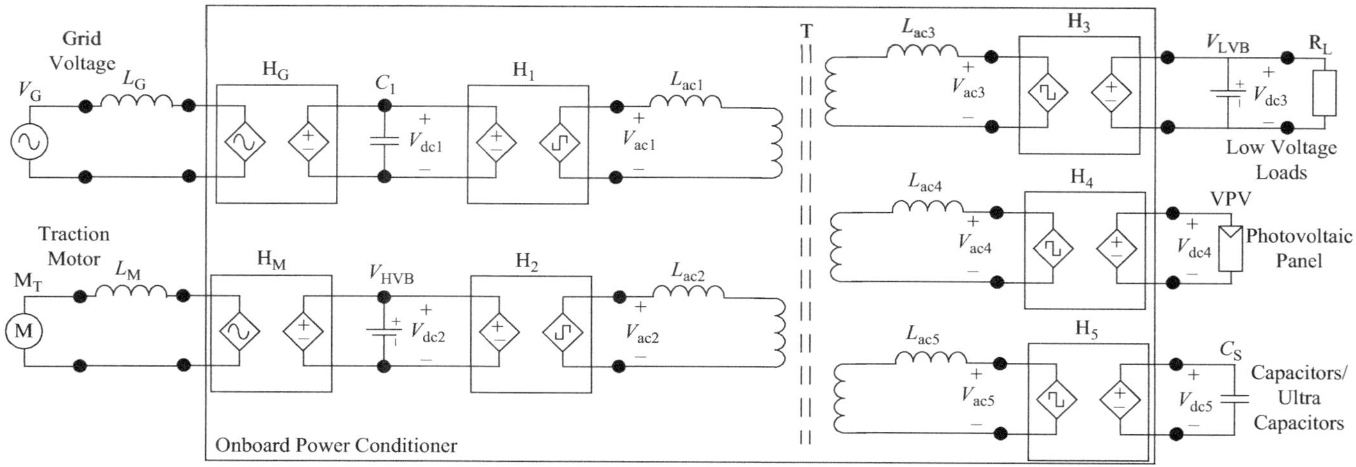

Figure 3.31 A vehicular integrated power system realized with ac–dc power conditioning stages connected through multi-port series ac link converter

The isolated components are conditioned through a series ac link, which offers high-power density power conversion. Several dc links extend the system where needed. The design procedure for the magnetic components of this system is presented in [22,37].

3.5 Summary

Recent advances in multi-input and multi-port converters were reviewed in this chapter. Multi-input converters bring new features to the emerging applications such as renewable energy systems, micro-grids, and vehicular systems. Synthesis of multi-input converters by extending existing single-input, single-output converters was presented. Additionally, realizing multi-port systems through dc link and ac link converters was discussed. The examples of recent multi-input and multi-port power converters in several applications were presented. Multi-port converters can be developed in order to reduce the component count. Alternatively, multi-port converters can offer more flexibility to the power conversion system. Both aspects of realizing multi-port converters were discussed. This chapter provided an overview to realizing the proper multiple-input topologies for several applications. Furthermore, the proper control techniques and design procedures were referenced from the latest works. Finally, potential applications of the multi-input converters were evaluated for the researchers, developers, and entrepreneurs.

References

[1] R. W. Erickson, "Synthesis of Switched-Mode Converters," in *IEEE Power Electronics Specialists Conference*, 1983, pp. 9–22.

[2] P. Wood, *Switching Power Converters*. New York: Van Nostrand Reinhold Co., 1981.

[3] R. Tymerski and V. Vorperian, "Generation and Classification of PWM DC-to-DC Converters," *Aerospace and Electronic Systems, IEEE Transactions on*, vol. 24, pp. 743–754, 1988.

[4] B. Farhangi and K. Butler-Purry, "Transient Study of DC Zonal Electrical Distribution System in Next Generation Shipboard Integrated Power Systems Using PSCAD^{TM}," in *North American Power Symposium (NAPS), 2009*, 2009, pp. 1–8.

[5] A. Kwasinski, "Identification of Feasible Topologies for Multiple-Input DC–DC Converters," *Power Electronics, IEEE Transactions on*, vol. 24, pp. 856–861, 2009.

[6] L. Yuan-Chuan and C. Yaow-Ming, "A Systematic Approach to Synthesizing Multi-Input DC–DC Converters," *Power Electronics, IEEE Transactions on*, vol. 24, pp. 116–127, 2009.

[7] S. Danyali, S. H. Hosseini, and G. B. Gharehpetian, "New Extendable Single-Stage Multi-Input DC–DC/AC Boost Converter," *Power Electronics, IEEE Transactions on*, vol. 29, pp. 775–788, 2014.

[8] L. W. Zhou, B. X. Zhu, and Q. M. Luo, "High Step-Up Converter with Capacity of Multiple Input," *Power Electronics, IET*, vol. 5, pp. 524–531, 2012.
[9] Z. Ruichen and A. Kwasinski, "Analysis of Decentralized Controller for Multiple-Input converters," in *Applied Power Electronics Conference and Exposition (APEC), 2012 Twenty-Seventh Annual IEEE*, 2012, pp. 1853–1860.
[10] C. N. Onwuchekwa and A. Kwasinski, "A Modified-Time-Sharing Switching Technique for Multiple-Input DC–DC Converters," *Power Electronics, IEEE Transactions on*, vol. 27, pp. 4492–4502, 2012.
[11] W. Rong-Jong, L. Chung-You, L. Jun-Jie, and C. Yung-Ruei, "Newly Designed ZVS Multi-Input Converter," *Industrial Electronics, IEEE Transactions on*, vol. 58, pp. 555–566, 2011.
[12] W. Qin, Z. Jie, R. Xinbo, and J. Ke, "Isolated Single Primary Winding Multiple-Input Converters," *Power Electronics, IEEE Transactions on*, vol. 26, pp. 3435–3442, 2011.
[13] A. I. Bratcu, I. Munteanu, S. Bacha, D. Picault, and B. Raison, "Cascaded DC–DC Converter Photovoltaic Systems: Power Optimization Issues," *Industrial Electronics, IEEE Transactions on*, vol. 58, pp. 403–411, 2011.
[14] L. Yan, R. Xinbo, Y. Dongsheng, L. Fuxin, and C. K. Tse, "Synthesis of Multiple-Input DC/DC Converters," *Power Electronics, IEEE Transactions on*, vol. 25, pp. 2372–2385, 2010.
[15] O. C. Onar, O. H. A. Shirazi, and A. Khaligh, "Grid Interaction Operation of a Telecommunications Power System with a Novel Topology for Multiple-Input Buck–Boost Converter," *Power Delivery, IEEE Transactions on*, vol. 25, pp. 2633–2645, 2010.
[16] H. Kakigano, Y. Miura, and T. Ise, "Low-Voltage Bipolar-Type DC Microgrid for Super High Quality Distribution," *Power Electronics, IEEE Transactions on*, vol. 25, pp. 3066–3075, 2010.
[17] A. Khaligh, J. Cao, and L. Young-Joo, "A Multiple-Input DC–DC Converter Topology," *Power Electronics, IEEE Transactions on*, vol. 24, pp. 862–868, 2009.
[18] A. Khaligh, "A Multiple-Input DC–DC Positive Buck–Boost Converter Topology," in *Applied Power Electronics Conference and Exposition, 2008. APEC 2008. Twenty-Third Annual IEEE*, 2008, pp. 1522–1526.
[19] C. Yaow-Ming, L. Yuan-Chuan, and L. Sheng-Hsien, "Double-Input PWM DC/DC Converter for High-/Low-Voltage Sources," *Industrial Electronics, IEEE Transactions on*, vol. 53, pp. 1538–1545, 2006.
[20] L. Solero, A. Lidozzi, and J. A. Pomilio, "Design of Multiple-Input Power Converter for Hybrid Vehicles," *Power Electronics, IEEE Transactions on*, vol. 20, pp. 1007–1016, 2005.
[21] G. R. Walker and P. C. Sernia, "Cascaded DC–DC Converter Connection of Photovoltaic Modules," *Power Electronics, IEEE Transactions on*, vol. 19, pp. 1130–1139, 2004.

[22] B. Farhangi and H. A. Toliyat, "Modeling and Analyzing Multi-Port Isolation Transformer Capacitive Components for Onboard Vehicular Power Conditioners," *Industrial Electronics, IEEE Transactions on*, vol. 62, pp. 3134–3142, 2015.

[23] B. Farhangi and H. Toliyat, "A Novel Vehicular Integrated Power System Realized with Multi-Port Series AC Link Converter," in *Applied Power Electronics Conference and Exposition, 2015. APEC 2015. 30th Annual IEEE*, 2015, pp. 1353–1359.

[24] M. Moosavi, S. Farhangi, H. Iman-Eini, and A. Haddadi, "An LCL-Based Interface Connecting Photovoltaic Back-Up Inverter to Load and Grid," in *Power Electronics, Drive Systems and Technologies Conference (PEDSTC), 2013 4th*, 2013, pp. 465–470.

[25] W. Hongfei, S. Kai, D. Shun, and X. Yan, "Topology Derivation of Nonisolated Three-Port DC–DC Converters from DIC and DOC," *Power Electronics, IEEE Transactions on*, vol. 28, pp. 3297–3307, 2013.

[26] M. Amirabadi, H. A. Toliyat, and W. C. Alexander, "A Multiport AC Link PV Inverter with Reduced Size and Weight for Stand-Alone Application," *Industry Applications, IEEE Transactions on*, vol. 49, pp. 2217–2228, 2013.

[27] F. Nejabatkhah, S. Danyali, S. H. Hosseini, M. Sabahi, and S. M. Niapour, "Modeling and Control of a New Three-Input DC–DC Boost Converter for Hybrid PV/FC/Battery Power System," *Power Electronics, IEEE Transactions on*, vol. 27, pp. 2309–2324, 2012.

[28] J. Wei and B. Fahimi, "Multiport Power Electronic Interface – Concept, Modeling, and Design," *Power Electronics, IEEE Transactions on*, vol. 26, pp. 1890–1900, 2011.

[29] Q. Zhijun, O. Abdel-Rahman, and I. Batarseh, "An Integrated Four-Port DC/DC Converter for Renewable Energy Applications," *Power Electronics, IEEE Transactions on*, vol. 25, pp. 1877–1887, 2010.

[30] Q. Zhijun, O. Abdel-Rahman, H. Al-Atrash, and I. Batarseh, "Modeling and Control of Three-Port DC/DC Converter Interface for Satellite Applications," *Power Electronics, IEEE Transactions on*, vol. 25, pp. 637–649, 2010.

[31] H. Behjati and A. Davoudi, "A Multiple-Input Multiple-Output DC–DC Converter," *Industry Applications, IEEE Transactions on*, vol. 49, pp. 1464–1479, 2013.

[32] H. Behjati and A. Davoudi, "Power Budgeting Between Diversified Energy Sources and Loads Using a Multiple-Input Multiple-Output DC–DC Converter," *Industry Applications, IEEE Transactions on*, vol. 49, pp. 2761–2772, 2013.

[33] A. Falk, "Bidirectional Battery Power Inverter," ed: Google Patents, 2014.

[34] B. Farhangi and S. Farhangi, "Application of Z-Source Converter in Photovoltaic Grid-Connected Transformer-Less Inverter," *Electrical Power Quality and Utilisation, Journal*, vol. 12, pp. 41–45, 2006.

[35] B. Farhangi and S. Farhangi, "Comparison of Z-Source and Boost–Buck Inverter Topologies as a Single Phase Transformer-Less Photovoltaic

Grid-Connected Power Conditioner," in *Power Electronics Specialists Conference, 2006. PESC'06. 37th IEEE*, 2006, pp. 74–79.

[36] J. Cao and A. Emadi, "A New Battery/Ultra-Capacitor Hybrid Energy Storage System for Electric, Hybrid, and Plug-In Hybrid Electric Vehicles," *Power Electronics, IEEE Transactions on*, vol. 27, pp. 122–132, 2012.

[37] B. Farhangi and H. A. Toliyat, "Modeling Isolation Transformer Capacitive Components in a Dual Active Bridge Power Conditioner," in *Energy Conversion Congress and Exposition (ECCE), 2013 IEEE*, 2013, pp. 5476–5480.

[38] B. Farhangi, H. A. Toliyat, and A. Balaster, "High Impedance Grounding for Onboard Plug-In Hybrid Electric Vehicle Chargers," in *Power Engineering, Energy and Electrical Drives (POWERENG), 2013 Fourth International Conference on*, 2013, pp. 609–613.

[39] B. Farhangi, "Power Conditioning for Plug-In Hybrid Electric Vehicles," Ph. D. Dissertation, Texas A&M University, College Station, TX, 2014.

[40] B. Farhangi and H. A. Toliyat, "Piecewise Linear Modeling of Snubberless Dual Active Bridge Commutation," in *Energy Conversion Congress and Exposition (ECCE), 2014 IEEE*, 2014, pp. 2065–2071.

[41] P. Fang Zheng, "Z-Source Inverter," *Industry Applications, IEEE Transactions on*, vol. 39, pp. 504–510, 2003.

[42] R. P. Severns and G. Bloom, *Modern DC-to-DC Switchmode Power Converter Circuits*. New York: Van Nostrand Reinhold Co., 1985.

[43] R. W. A. A. De Doncker, D. M. Divan, and M. H. Kheraluwala, "A Three-Phase Soft-Switched High-Power-Density DC/DC Converter for High-Power Applications," *Industry Applications, IEEE Transactions on*, vol. 27, pp. 63–73, 1991.

[44] Q. Hengsi and J. W. Kimball, "Solid-State Transformer Architecture Using AC–AC Dual-Active-Bridge Converter," *Industrial Electronics, IEEE Transactions on*, vol. 60, pp. 3720–3730, 2013.

[45] M. Amirabadi, A. Balakrishnan, H. A. Toliyat, and W. C. Alexander, "High-Frequency AC-Link PV Inverter," *Industrial Electronics, IEEE Transactions on*, vol. 61, pp. 281–291, 2014.

[46] T. Haimin, A. Kotsopoulos, J. L. Duarte, and M. A. M. Hendrix, "Transformer-Coupled Multiport ZVS Bidirectional DC–DC Converter with Wide Input Range," *Power Electronics, IEEE Transactions on*, vol. 23, pp. 771–781, 2008.

[47] M. Michon, J. L. Duarte, M. Hendrix, and M. G. Simoes, "A Three-Port Bi-Directional Converter for Hybrid Fuel Cell Systems," in *Power Electronics Specialists Conference, 2004. PESC 04. 2004 IEEE 35th Annual*, vol. 6, pp. 4736–4742, 2004.

[48] C. Zhao, S. D. Round, and J. W. Kolar, "An Isolated Three-Port Bidirectional DC–DC Converter with Decoupled Power Flow Management," *Power Electronics, IEEE Transactions on*, vol. 23, pp. 2443–2453, 2008.

[49] K. Sung-young, S. Hong-Seok, and N. Kwanghee, "Idling Port Isolation Control of Three-Port Bidirectional Converter for EVs," *Power Electronics, IEEE Transactions on*, vol. 27, pp. 2495–2506, 2012.

[50] F. Jauch and J. Biela, "An Innovative Bidirectional Isolated Multi-Port Converter with Multi-Phase AC Ports and DC Ports," in *Power Electronics and Applications (EPE), 2013 15th European Conference on*, 2013, pp. 1–7.

[51] A. Kwasinski, "Quantitative Evaluation of DC Microgrids Availability: Effects of System Architecture and Converter Topology Design Choices," *Power Electronics, IEEE Transactions on*, vol. 26, pp. 835–851, 2011.

[52] A. Emadi, S. S. Williamson, and A. Khaligh, "Power Electronics Intensive Solutions for Advanced Electric, Hybrid Electric, and Fuel Cell Vehicular Power Systems," *Power Electronics, IEEE Transactions on*, vol. 21, pp. 567–577, 2006.

[53] W. Kempton and J. Tomić, "Vehicle-to-Grid Power Implementation: From Stabilizing the Grid to Supporting Large-Scale Renewable Energy," *Journal of Power Sources*, vol. 144, pp. 280–294, 2005.

[54] B. Farhangi, "A Novel Modified Deadbeat Controller for Vehicle to Grid Application," in *Power Electronic & Drive Systems & Technologies Conference (PEDSTC), 2015 6th*, 2015, pp. 47–52.

[55] J. G. Pinto, V. Monteiro, H. Goncalves, and J. L. Afonso, "On-Board Reconfigurable Battery Charger for Electric Vehicles with Traction-to-Auxiliary Mode," *Vehicular Technology, IEEE Transactions on,* vol. 63, pp. 1104–1116, 2014.

[56] M. Pahlevaninezhad, D. Hamza, and P. K. Jain, "An Improved Layout Strategy for Common-Mode EMI Suppression Applicable to High-Frequency Planar Transformers in High-Power DC/DC Converters Used for Electric Vehicles," *Power Electronics, IEEE Transactions on*, vol. 29, pp. 1211–1228, 2014.

Chapter 4
Modular converters

Italo Roger F. M. P. da Silva, Cursino B. Jacobina* and Alexandre C. Oliveira**

4.1 Introduction

In the last two decades, voltage source (VS) topologies based on modular structure have been used in various applications of power electronics, such as power conditioning system (PCS) for renewable energy sources; battery energy storage system (BESS) for power leveling; active power line conditioner (APLC) system for harmonic minimization and reactive power compensation; adjustable speed drive (ASD) system, and high-voltage direct current (HVDC) system. The similarity of modular converters is the unlimited capability of combining identical low- or medium-power subsystems to achieve a system with higher power ratings. Mainly due to the high degree of modularity, modular topologies provide high output energy quality, high reliability, high efficiency, easy maintenance, and cost-weight-volume reduction. All these features are due to the series connection of submodules (SMs), whereas they are identical with the same rated power and can be seen as power cells with similar circuit topology and controlled by the same control and modulation schemes. Thus, for high-power large-scale modular structures, in which the number of power cells per arm (m) are usually more than ten, designing the SMs with standard low- or medium-voltage technology devices has a significant impact on converter efficiency, since it is possible to obtain a high number of voltage levels, allowing an expressive reduction in average switching frequency without compromising the power quality.

Since the mid-1990s, the cascade H-bridge (CHB) topology has been the subject of many studies due to its modularity feature [1,2]. Intrinsically, the modular concept has already been used, since the CHB topology is designed from multiple single-phase converters, which are simple and mature building blocks of power electronics systems. Likewise, a few years later, a novel topology also based on single-phase converters was proposed by Marquardt *et al.* [3]. At first, it was named modular multilevel converter (M^2 LC) [4] and, unlike the CHB, this topology is composed of half-bridge (HB) inverters connected in series, providing a common DC-link, which made it very attractive for back-to-back (BTB)

*Federal University of Campina Grande, Campina Grande, Paraíba, Brazil applications. Therefore, the terminology "modular converters" has become increasingly popular, but it has been leading to a vast group of topologies, whose structure is based on the connection of power cells or SMs. So, the concept of what power cell and SM means is important to define a reasonable classification of modular converters. In this case, both the power cell and SM are considered an arrangement of a switching devices, DC capacitors or sources, based on the same electrical circuit. However, an SM should be designed with the same power semiconductor technology, so that thermal components have an equal project for each SM as well. This ensures the modularity and the redundant operation, once a defective SM can be bypassed without interrupting the energy transfer [5].

From Figure 4.1, it is possible to understand that all SMs are power cells, but the reverse is not true. The modular structure allows the use of different kinds of converter topologies as SMs, besides the conventional HB and full-bridge (FB) configurations, such as neutral-point clamped (NPC), flying capacitor (FC), and neutral-point piloted (NPP), which lead to the advanced modular topologies. From these advanced solutions, it is possible to reach high output waveform quality with less SMs, and better fault tolerance, or even to reduce the SM switch-count, extending to unidirectional topologies. Although the use of advanced SM topologies can provide some benefits, it also results in loss of modularity though, since less SMs are needed to obtain the desired output waveform. It is important to mention that taking as basis the duality principle, current source (CS) circuits can be derived from VS ones. Nami *et al.* [6] investigated different CS modular topologies for HVDC systems. Notwithstanding, this study focuses on the main modular topologies, whose SMs are based on VSs, hence the CS-based SMs will not be discussed.

In the literature, "modular converters" are usually included in "multicell", "chain link", and "cascaded" topologies [6–9]. However, recently, the search for modular converters focuses on a particular family of multilevel topologies named modular multilevel cascade converter (MMCC) [9]. The basic topologies of the MMCC family are based on HB- and FB-SM topologies. For instance, the star- and delta-connected CHB converters are detached in single-star bridge-cells (SSBCs) and single-delta bridge-cells (SDBCs), respectively, as well as the M^2 LC, MMC, and M2C which are other nomenclatures of the double-star chopper-cells (DSCCs). The terminology adopted for the MMCC family follows the arrangement of the arm-clusters, which are a set of SMs per arm, and the circuit configuration used as SM. For this reason, it is possible to classify the multilevel converters according to the diagram illustrated by Figure 4.2 based on the classification proposed in [9].

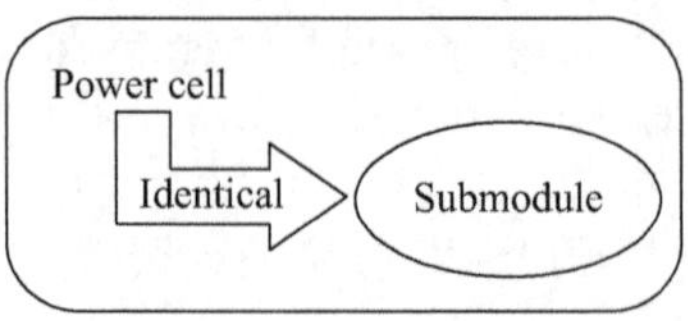

Figure 4.1 SM definition

As it can be seen, the FC and the MMCC topologies are included in multicell topologies, since they are composed by a set of power cells. Nevertheless, to be considered a modular structure, besides the series connection of SMs, the scaling to different voltage and power levels should be achieved only by varying the number of SMs (with no additional central components). Figure 4.3(a) and (b) illustrates one phase-leg of a FC converter and one phase-leg of a DSCC converter, respectively. Since the DSCC topology is composed of two sets of SMs per phase-leg, it has double the number of power cells per phase (M) in comparison to the FC. In order to better understand the proposed classification, both topologies are taken as a basis of comparison between multicell and modular topologies. Therefore, Figure 4.4 illustrates the parameter M as a function of the voltage and current levels. As shown in Figure 4.4(a), the voltage level in each power cell of the FC converter is unequal and increases proportionately with the expansion of M, while in Figure 4.4(b), the modular structure ensures an equal voltage distribution, even if M is enlarged. As a result, the FC and its derivations, such as stacked FC, cannot meet these requirements. Also, the asymmetric CHB (ACHB) cannot be considered as MMCC, because in medium- and high-power applications, different switching device technologies are required, leading to different thermal designs for each power cell and consequently eliminating the modularity benefit. Further details

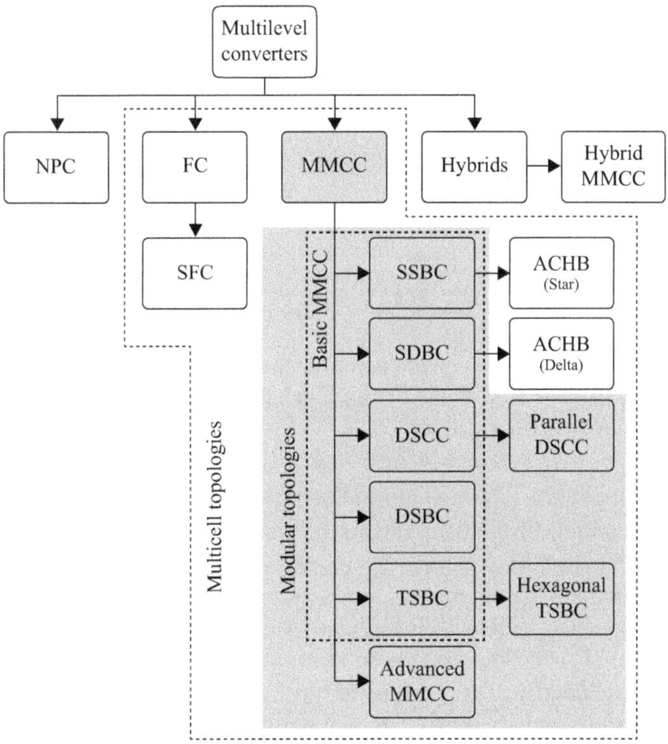

Figure 4.2 Multilevel converter classification

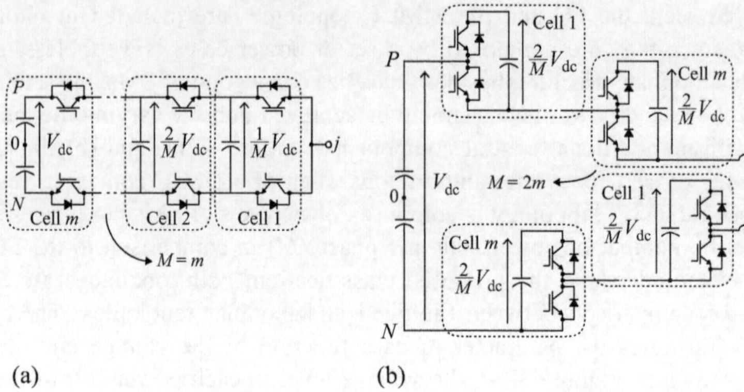

Figure 4.3 Multilevel converter circuit configuration. (a) One phase-leg of FC converter and (b) one phase-leg of DSCC converter

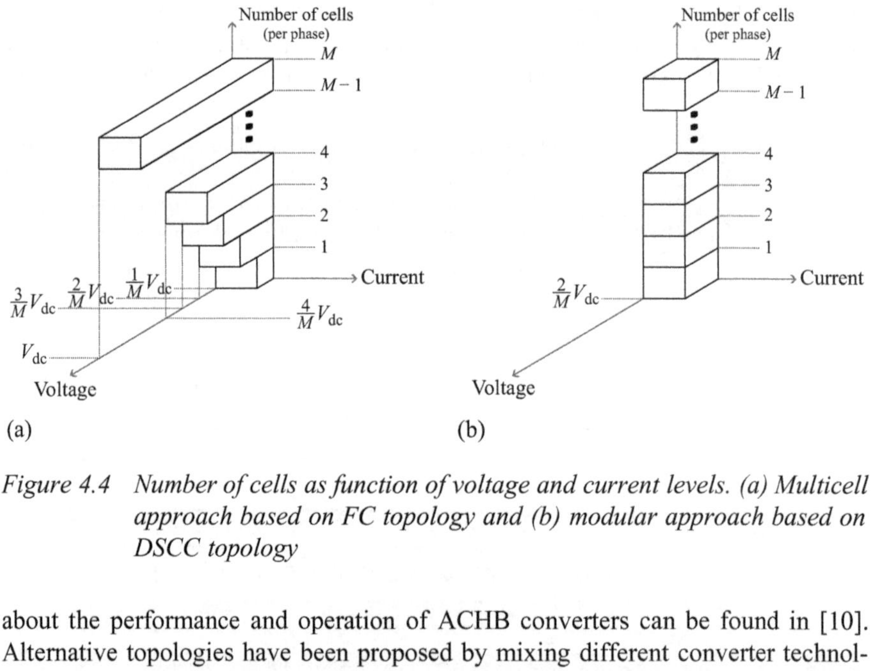

Figure 4.4 Number of cells as function of voltage and current levels. (a) Multicell approach based on FC topology and (b) modular approach based on DSCC topology

about the performance and operation of ACHB converters can be found in [10]. Alternative topologies have been proposed by mixing different converter technologies. Denoted hybrid MMCC, the topologies of this set present a certain degree of modularity; although they improve some aspect of the MMCC topologies, they lose modularity, reducing the reliability of the whole converter, since their design is not totally modular [6,11]. Thus, as can be seen in Figure 4.2, hybrid MMCC topologies are not included in the MMCC family. All in all, the modular topologies correspond to the darkened boxes in the classification diagram shown in Figure 4.2.

This chapter aims to provide a comprehensive review on modular topologies based on the proposed classification of multilevel converters. The review is conducted by discussing the most relevant characteristics of the new MMCC family, concerning the circuit configurations, control strategies, modulation techniques, operational issues, and main applications. The new MMCC family not only includes the basic modular topologies, but also covers the latest topology developments, such as the advanced modular topologies and the derivation of basic ones. Notwithstanding, the object of study is focused on the basic MMCC topologies, whose advantages have made them a competitive solution for industry and the utility systems in recent years, especially with the high demand of sustainable energy supply. The rest of this chapter is organized as follows. Section 4.2 describes the MMCC circuit configuration, presenting a complete review of the different MMCC arrangements, and several SM topologies that can be used in the design of converters. Section 4.3 presents the main contributions on the voltage-balancing and circulating current control strategies. Next, the modulation techniques are approached in Section 4.4. Section 4.5 presents the latest developments on operational issues, regarding the fault-tolerant operation and pre-charging procedures of the floating DC capacitors. Finally, a brief overview of the main applications is addressed in Section 4.6.

4.2 Modular converter topologies and description

The contribution that multilevel converters have been providing to the power electronic in high-power applications is well known and with proven performance [12,13]. Performed by the proper switching states, multilevel converters can provide a quasi-sinusoidal output waveforms with low harmonic contents, which reduce or even eliminate the need of passive filters. Also, they can reduce the voltage stress across semiconductor devices, demanding smaller cooling systems. As a result, the power quality has been improved, but to achieve high number of voltage levels and expand the power capacity of the multilevel converters, the complexity of design, the capacitor voltage unbalance problems, and the computational efforts have to be considered. With this in mind, the NPC and FC topologies, for example, have some problems with excessive number of clamping diodes and bulk capacitors, respectively, when the number of levels is high. In this way, the modular topologies arise with a different approach of stepped output waveform synthesis.

Figure 4.5(a) and (b), respectively, illustrates the conventional multilevel solution (based on NPC topology) and the modular solution (based on DSCC topology), wherein the stepped voltages are obtained by selecting different voltage levels from a single DC-link voltage (V_{dc}), resulting in the output phase voltage (v_j), where j is the converter phase, as shown in Figure 4.5(c).

The main difference between both solutions is in the arrangement proposal of the capacitive DC-link. In the conventional solution as presented in Figure 4.5(a), the central DC-link capacitor is required for the DC energy storage, while the modular solution provides a distributed DC energy storage from the floating DC capacitors, without the need of a common DC-link capacitor between the DC

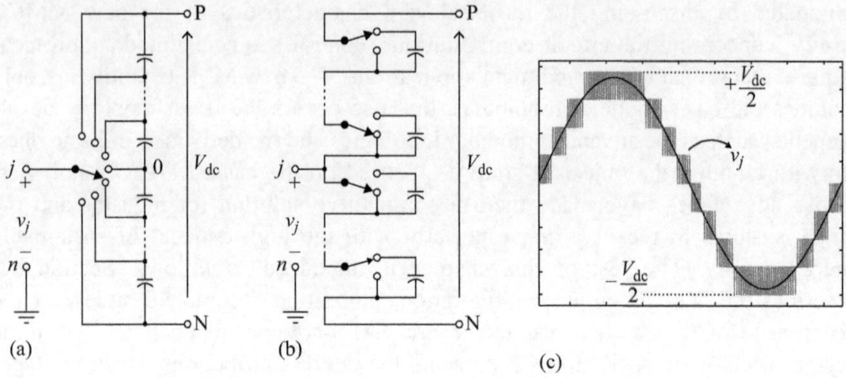

Figure 4.5 Multilevel converter solutions. (a) Conventional solution based on NPC topology, (b) modular solution based on DSCC topology, and (c) nine-level output waveform

terminals (P, N) according to Figure 4.5(b). This increases the overall converter reliability against extremely high surge currents and subsequent damage, if short circuits occur at the DC terminals [14]. Also, the modular solution transmits the idea of easy failure management, maintenance, and consistent redundancy concept. Although the mentioned modular solution is based on the DSCC topology (see Figure 4.3(b)), the main idea of separated DC-link capacitors and their advantages can be widespread for the other modular topologies.

4.2.1 MMCC arrangements and SM topologies

The MMCC topologies are basically composed of series-connected SMs, forming a cluster of SMs in the arm, or just arm-cluster. The combination of arm-clusters can originate different basic groups of possible modular phase-leg arrangements: single-star (SS), single-delta (SD), double-star (DS), and triple-star (TS). The initial letter defines the number of the set of the three arm-clusters (single, double, or triple) and the subsequent letter represents the disposition (star or delta) of these arm-clusters. To simplify, both the SS and SD arrangements can generally be named Sx (x = S or D). Figure 4.6(a) and (b) illustrates MMCC topologies based on SS and SD arrangements, respectively. The SS arrangement in particular can be divided into positive SS (P-SS) and negative SS (N-SS) subconverters, which depend on the connections of the arm-clusters in relation to the common terminals (P, N). As a result, the combination of P-SS and N-SS subconverters originates the DS arrangement, as shown in Figure 4.7. Besides, the parallel connection of three N-SS or P-SS subconverters originates the TS arrangement, as shown in Figure 4.8.

The MMCC family members composed by FB- or HB-SMs are considered as basic MMCC topologies, while the advanced MMCC ones are those in which the SMs are based on the circuits unlike the FB and HB. Hence, depending on the SM topology, several MMCC topologies can be created. Figure 4.9 shows various SM circuit configurations proposed over the past two decades. The output voltage of

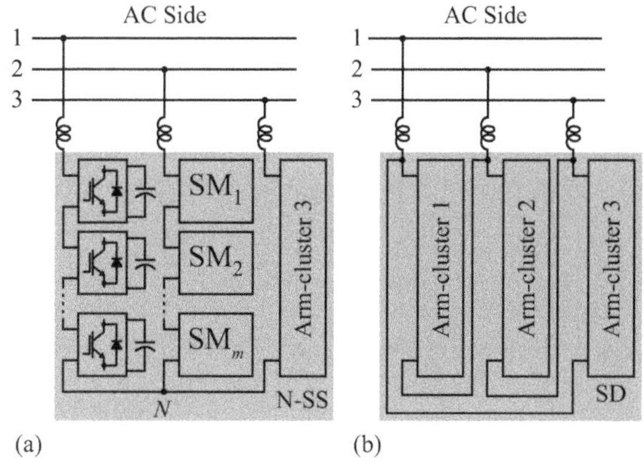

Figure 4.6 Sx arrangements. (a) SS arrangement and (b) SD arrangement

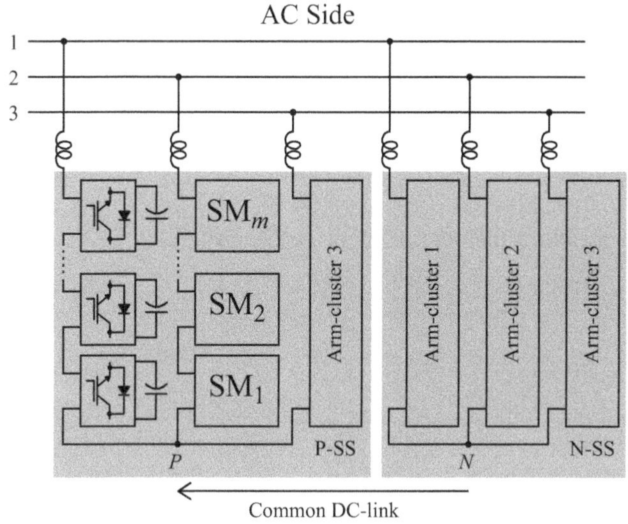

Figure 4.7 DS arrangement

each SM (v_m) in relation to the floating DC capacitor voltage (v_C) is described in Table 4.1. The FB-SM, as in Figure 4.9(a), can provide a symmetrical three-level bipolar voltage [8], unlike the HB-SM, as in Figure 4.9(b), which can achieve only two unipolar voltage levels [3,4]. The HB-SM is only used in the DS arrangement, because it can only provide unipolar voltages. Otherwise, bipolar SMs are interesting to cut-off arm currents in any direction by impressing the appropriate polarity of terminal voltages in the arm [15]. For instance, in case of MMCC topologies with DC-link availability and a subsequent DC short circuit, FB-SM offers better DC fault-handling capability than the HB-SM, since reverse voltage can be generated from the capacitor voltage to block the AC side currents when all

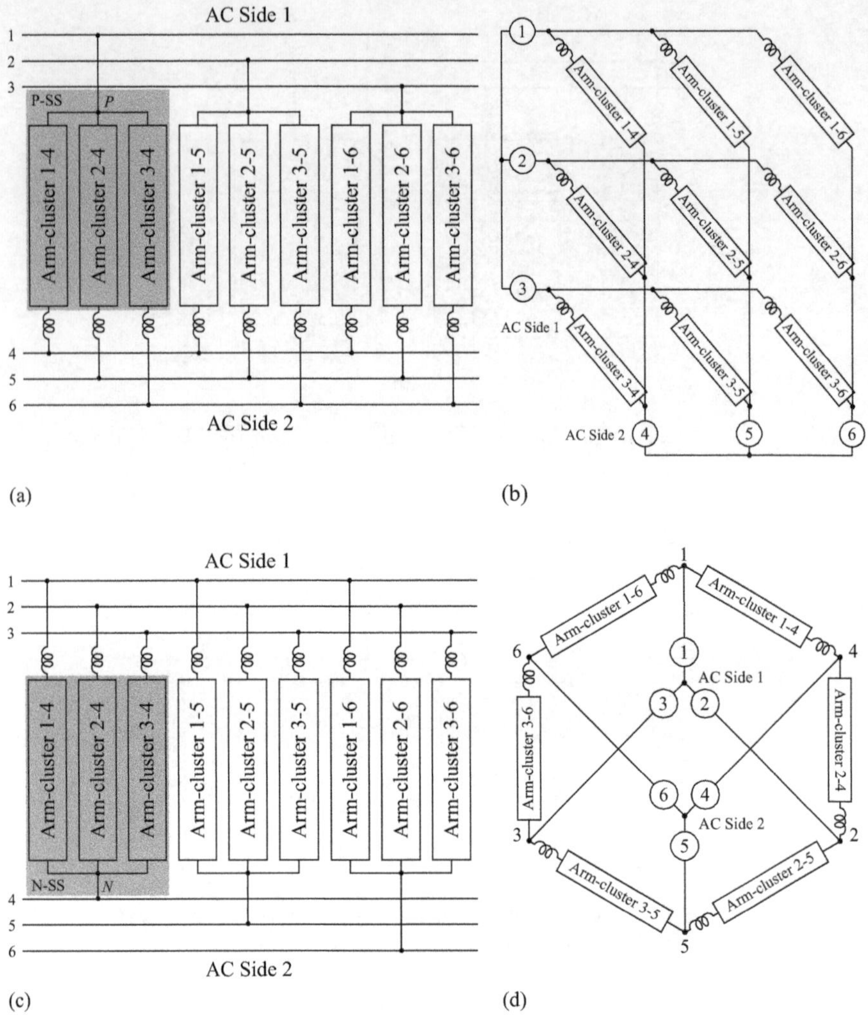

Figure 4.8 TS arrangements. (a) TS arrangement with P-SS subconverters, (b) typical representation of the TS arrangement with P-SS subconverters, (c) TS arrangement with N-SS subconverter, and (d) hexagonal configuration based on TS arrangement

switches of the FB-SM are off, potentially driving the fault current to zero [5]. Since the HB-SM cannot provide DC fault-handling capability, protective thyristors should be equipped parallel to the endangered diode to bypass the fault current.

Meanwhile, using SMs based on multilevel topologies allows for reducing the dimensions of the whole converter at the expense of complicating the SM technology. Well-known multilevel topologies such as the three-level NPC and FC

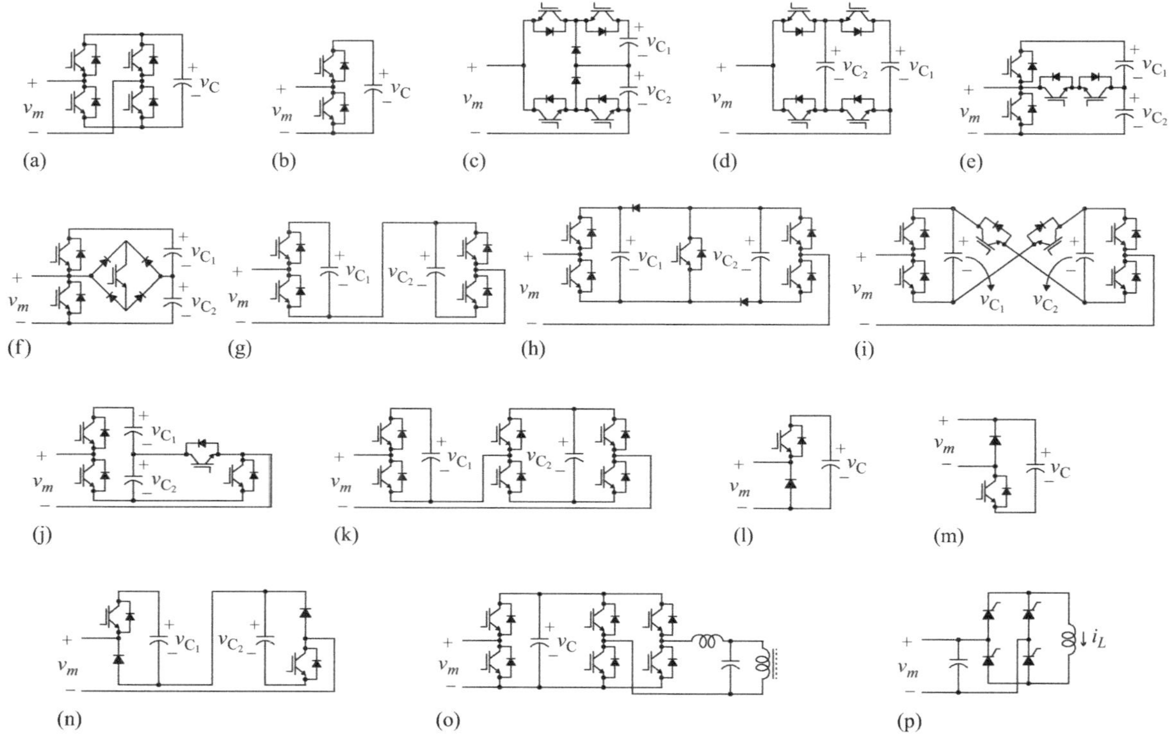

Figure 4.9 *SM topologies. (a) FB-SM, (b) HB-SM, (c) NPC-SM, (d) FC-SM, (e) NPP1-SM, (f) NPP2-SM, (g) TC-SM, (h) CD-SM, (i) CCC-SM, (j) ACD-SM, (k) MC-SM, (l) RC1-SM, (m) RC2-SM, (n) RC3-SM, (o) HB-SM with IPT capability, and (p) CS-based FB-SM*

Table 4.1 Output voltage for the VS-based SM topologies

Acronym	Number of voltage levels	Output voltage of each SM (v_m)
FB	3	$-v_C, 0, v_C$
HB	2	$0, v_C$
NPC	3	$0, v_{C_1}, v_{C_2}, (v_{C_1} + v_{C_2})$
FC1	3	$0, v_{C_2}, (v_{C_1} - v_{C_2}), v_{C_1}$
NPP1	3	$0, v_{C_1}, (v_{C_1} + v_{C_2})$
NPP2	3	$0, v_{C_1}, (v_{C_1} + v_{C_2})$
TC	3	$0, v_{C_1}, v_{C_2}, (v_{C_1} + v_{C_2})$
CD2	4	$-\overline{v}_C, 0, \overline{v}_C, v_{C_1}, v_{C_2}, (v_{C_1} + v_{C_2})$
CCC	5	$-(v_{C_1} + v_{C_2}), -v_{C_1}, -v_{C_2}, 0, v_{C_1}, v_{C_2}, (v_{C_1} + v_{C_2})$
ACD	4	$-v_{C_2}, 0, v_{C_1}, v_{C_2}, (v_{C_1} + v_{C_2})$
MC	4	$-v_{C_2}, (v_{C_1} - v_{C_2}), 0, v_{C_1}, v_{C_2}, (v_{C_1} + v_{C_2})$
RC1	2	$0, v_C$
RC2	2	$0, v_C$
RC3	3	$0, v_{C_1}, (v_{C_1} + v_{C_2})$

[1] $v_{C_1} = 2v_{C_2}$.
[2] $\overline{v}_C = \{v_{C_1}, v_{C_2}\}$, during a fault (FB mode).

have been applied as SMs [16], as shown in Figure 4.9(c) and (d), respectively. Both the NPC- and FC-SM can synthesize three-level unipolar voltage, although using two NPC-SM or FC-SM in parallel, forming a bridge, it is possible to obtain symmetrical bipolar voltages too [17]. Notwithstanding, these solutions imply in modularity reduction, not to mention the inherent issues of FC and NPC topologies. Following this reasoning, Sahoo *et al.* [18] proposed two new multilevel SM topologies based on the NPP topology, as shown in Figure 4.9(e) and (f). The first, named NPP-type 1 (NPP1), is based on the conventional circuit of the NPP converter, and the second one, named NPP-type 2 (NPP2), is an alternative solution for the NPP1-SM, although working in a similar way. Likewise, the NPC- and FC-SM, both NPP1- and NPP2-SM can synthesize three-level unipolar voltage.

Other SM topologies have been proposed to improve the power density and the fault-handling capacity, as well as to achieve more output voltage levels. The combination of two HB-SMs connected by the terminals of the floating DC capacitors can result in four different SM concepts: twin-cells (TCs) [19], clamp-double (CD) [14], cross-connected cells (CCCs) [20], and asymmetric CD (ACD) [6], as shown in Figure 4.9(g)–(j), respectively. Besides, the combination of HB- and FB-SMs originates a mixed cell (MC) [6], as shown in Figure 4.9(k). In normal operation, the CD-SM represents an equivalent of two HBs, just like the TC-SM, both operating in the same manner and synthesizing three-level unipolar voltage. However, in a DC fault condition, the CD-SM operates as the FB-SM, due to the opening of the intermediate switch between the two HBs, which results in voltage clamping and energy absorption. During the voltage clamping, the CD-SM can provide reverse voltage to clear a DC fault current as well as to ensure minimized over-voltage [14]. The ACD-SM is another way of doubling the commutation cells and achieving reverse voltage. Therefore, just like the CD-SM, both the ACD- and

MC-SM can synthesize asymmetrical four-level bipolar voltage, and the CCC-SM can synthesize symmetrical five-level bipolar voltage [6].

When the bidirectional power flow is not necessary, unidirectional topologies can be a suitable alternative for energy conversion systems, reducing the switch-count and the overall cost, as well as leading to a simpler design, compared to full-controlled structures. The MMCC concept can be extended to unidirectional topologies by using SMs based on reduced cells (RCs). Two unidirectional SM topologies resulted from the HB-SM, named RC-type 1 (RC1) and RC-type 2 (RC2), as shown, respectively, in Figure 4.9(l) and (m), and another from the TC-SM, named RC-type 3 (RC3) as shown in Figure 4.9(n). Nevertheless, these can switch the output voltage only in case of positive current flows through the SM, whereas if the negative current flows, the output voltage remains constant. Both RC1 and RC2 are two-level unipolar SMs, while RC3 can reach one more voltage level. Similar to bridge-based NPC- and FC-SM, the bridge-based RC-SM can synthesize symmetrical bipolar voltage.

Furthermore, additional benefits can be achieved by connecting different subsystems parallel to each SM. For example, the inductive power transfer (IPT) technology was used to keep the floating DC capacitors in the HB-SMs balanced within tight bounds [21]. As shown in Figure 4.9(o), an IPT-based converter was parallel connected to a HB-SM to exchange energy among the floating DC capacitors and keep them regulated around the nominal value. Also, to improve the fault-tolerant operation and to achieve better DC voltage regulation, boost rectifiers were used parallel-connected to FB-SMs [22]. As can be noticed, however, these solutions increase the cost and the converter switch-count when compared to the original circuit. Furthermore, from the duality concept, SM topologies based on CSs can be designed for enhancing the fault-handling capability [6]. Figure 4.9(p) illustrates a dual version of the FB-SM presented in [23]. The CS-based SMs can be applied to transmit a higher power compared to the standard VS-based SMs; however, the losses are increased and large footprints are required for the DC inductors [6].

4.2.2 Basic modular multilevel cascade converters

Despite most of the SM topologies presented in Section 4.2.1 having been implemented to improve the performance of systems based on single-star or -delta bridge-cells (S*x*BC) and DSCC topologies, motivated by their increasing popularity amongst academic and industrial community, other SM topologies can also be extended for the remaining MMCC family members. From now on, this study will be directed to the basic MMCC topologies: S*x*BC, DSCC, double-star bridge-cells (DSBCs), and triple-star bridge-cells (TSBCs), since the FB- and HB-SM are the default power cells of the MMCC family members. The modular design of the MMCC topologies, theoretically, allows for expanding the cascading SMs to an unlimited number, without increasing the complexity of the circuit, which makes the MMCC topologies easily scalable to different power and voltage levels. As a result, low voltage steps can be generated, which lead to higher output wave quality, lower harmonic content, as well as lower electromagnetic interference emission. Thus, small or even no passive filters are required, which results in a space-saving and cost-effective system. Hence, using only industrial proven-technology low-voltage components, MMCC topologies are suitable for a very wide range of high-power applications.

Each arm-cluster in MMCC topologies is followed by inductors, which make the arm currents be continuous shaped. Besides, they have an important role in limiting the rate of fault current through the phase-leg. The use of a single-coupled inductor is very common in each phase-leg of DS-based topologies, which results in size and weight reduction compared to two non-coupled inductors. In all basic MMCC topologies, except in the SSBC, the arm-clusters are connected in a way that inner current loops are formed, wherein the arm currents flow carrying a component defined as a circulating current that is needed to keep the floating DC capacitors energized. The voltage imbalance among the arm-clusters originate this current, whose main function is to exchange active power between the DC-link and the arm-clusters, or even among the arm-clusters themselves. Additionally, AC components provided from the AC side flow through each SM, and due to this, MMCC topologies are subjected to voltage fluctuations in the floating DC capacitors. Hence, the circulating current may also contain undesired harmonics. Therefore, circulating currents can be controlled to improve the performance, to reduce the losses and energy fluctuations and even to reach the operation stability in conditions previously not possible.

However, the primary purpose of controlling the circulating currents is to achieve the average voltage balance among the arm-clusters. The voltage of each floating DC capacitor must be regulated, so that the converter synthesizes the correct voltage level. Both the voltage-balancing and circulating current control strategies, as well as the modulation techniques, will be discussed in Sections 4.3 and 4.4, respectively. Considering the five basic MMCC topologies, the choice of the converter depends mainly on its application, semiconductor ratings and energy storage requirements, because each topology has its own characteristics in the terms of performance and viability [9]. In Sections 4.2.2.1–4.2.2.4, the main characteristics of each basic modular topology will be approached, comparing each one by taking into account physical and technical constraints and suitability for some application segments. Further considerations about the applicability of the basic MMCC family members in the current literature scenario will be discussed in Section 4.6.

4.2.2.1 SxBC topologies

The first works about S*x*BC topologies were presented in [1,2]. Both S*x*BC topologies are arranged with one arm-cluster per phase-leg (i.e., $M = m$), then no common DC-link is available. Therefore, when S*x*BC topologies are used in direct DC-to-AC power inversion, such as in ASD applications, isolated DC sources for each SM are required, which usually are provided by rectifiers fed by phase-shifting transformers, i.e., multipulse transformers [13]. For regenerative applications, the use of S*x*BC converters is not usual, owing to the high number of devices required by active front-end rectifiers used to improve the input power quality. On the other hand, they are very suitable for grid-connected applications intended to enhance power quality, like reactive power and harmonic compensation [24,25], in which only floating DC capacitors can be used instead of isolated DC sources [8]. The SSBC has the capability of controlling positive sequence by leading and lagging reactive power [9]; however, at unbalanced conditions, it is not suitable for compensation of negative-sequence power without increasing the converter voltage ratings, because it should exchange both negative-sequence active and reactive power with the grid [26]. In principle, the first exchange results in charge/discharge of the floating DC capacitors, which is against the basic operation of a STATic synchronous COMpensator (STATCOM) [27], and the second is not feasible, since the SSBC has no circulating current unlike the SDBC [9]. However, several studies have implemented control strategies based on the zero-sequence voltage injection to achieve the desired operating range for practical unbalanced conditions up to some extent, since the increase of the maximum attainable output voltage will raise the cost [28,29]. So, the power converter rating and the degree of unbalance (ratio between the negative-sequence current and the positive-sequence current components) must be considered in the SSBC design when used to compensate negative-sequence reactive power [28]. Nevertheless, an active power control can also be implemented to achieve the state-of-charge (SOC) in SSBC-based BESS [30,31], as well as to control low-frequency active power in SDBC-based STATCOM intended for flicker compensation [32]. Hagiwara *et al.* [32] proved that the SDBC can control at the same time positive- and negative-sequence reactive power beyond the low-frequency active power, considering a symmetrical and sinusoidal three-phase supply. However, in case of serious unbalanced conditions (e.g., unbalanced load under asymmetrical grid fault), large circulating current is required to balance the floating DC capacitors, so the capability of compensating negative-sequence reactive power is reduced, and the SDBC is not able to compensate it sufficiently.

4.2.2.2 DSCC topology

The DSCC topology was first intended to be applied in HVDC transmission systems, whereas the modular design provides high scalability, the DS arrangement can make a common DC-link available and the internal arm inductors limit both parasitic and fault currents. But the DSCC also has found applications in medium-voltage high-power ASD by AC-to-AC indirect power conversion as well as in improvement of the power quality in grid-connected systems. In the DSCC, the arm-clusters are composed by a stack of HB-SMs, which produces unipolar voltages with a phase difference of 180° between the arm-clusters of the same phase-leg. Furthermore, the common DC-link capacitor can be removed in many applications, such as BTB and grid-connected systems, because the energy storage in the DSCC is distributed among the floating DC capacitors. Regarding the floating DC capacitor voltage, its ripple is one of the most important issues in the DSCC, because it is inversely proportional to the frequency of the load current and directly proportional to the magnitude of the load current as well [33]. Hence, the

DSCC has restrictions to drive motors at low frequency, especially for constant-torque loads, wherein the ripple magnitude will have direct influence from the frequency. Conversely, for driving quadratic-torque loads, this issue is less severe. Nevertheless, using the appropriate control strategy based on common-mode voltage and AC circulating current injection, the DSCC can operate in a wide range of frequency, even at start-up and low-speed operation for constant-torque motor drive [34–36]. Just as the SDBC, the DSCC (and also the DSBC) has the capability of exchanging negative-sequence reactive power with the grid due to the presence of circulating currents [9]. Unlike the SDBC, the DSCC is able to achieve full negative-sequence reactive power compensation with a small amount of DC circulating current, as well as being able to be applied to compensate unbalanced non-linear loads even under asymmetrical grid fault conditions [37]. Besides, to deal with large-current requirements, a parallel configuration of the DSCC was proposed by Pirouz and Bina [38] named extended MMC (EMMC). Initially, the EMMC or parallel DSCC has separated DC-links, but a version of the parallel DSCC sharing the same DC-link was presented in [39].

4.2.2.3 DSBC topology

The DSBC topology is an interesting derivation of the DSCC. Its control and modulation techniques are very similar to the DSCC ones, but the DSBC presents more flexibility than DSCC topology, practically speaking. By replacing the HB-SMs with FB ones, it is possible to synthesize bipolar voltages, providing the DSBC with an additional capability for bucking and boosting the DC-link voltage in power rectification or inversion [9]. As a result, the DSBC can produce an AC voltage independently from DC-link voltage, which also means that it carries out AC-to-AC direct power conversion when a single-phase AC VS is connected to the DC-link terminals [40–42]. Furthermore, depending on the DC-link voltage utilization, the number of voltages levels can also vary [43]. Summarizing, the DSBC can operate not only as DC-to-AC converter [43], such as the DSCC, but also as a single-phase-to-three-phase and single-phase-to-single-phase direct AC-to-AC converter. When the DSBC operates as single-phase-to-three-phase AC-to-AC converter, the input and output frequencies must not be the same for proper operation [42]. Hence, this topology is very suitable to distributed PCSs for renewable energy resources, in which the DC-link voltage varies over a wide range [9].

4.2.2.4 TSBC topology

The TSBC topology consists of a set of three SSBCs, each of which is denoted by a subconverter (P-SS or N-SS). Also known as modular multilevel matrix converter (M3C), the first approach about this topology was presented by Erickson and Al-Naseem [44]. The TSBC is suitable for three-phase-to-three-phase AC-to-AC direct power conversion, especially for medium-voltage high-power motor drives at low frequency and regenerative braking [45–48]. Just like the DSBC when used for single-phase-to-three-phase AC-to-AC conversion, the TSBC are not recommended for synchronous frequency operation, due to the serious increasing ripple in the floating DC capacitors. Also, without appropriately controlling the circulating currents, further voltage fluctuations in the floating DC capacitors increase, whereas the output frequency is reduced below one-third of the input frequency [45,46]. To overcome the synchronous operation problem and achieve a broad range frequency operation, a control method was proposed in [47]. The operation has become possible when the input and output frequencies are close or equal, albeit with restrictions, because this method enlarged the ripple in the floating DC capacitors at low-frequency range, compared to the previous methods proposed in [45,46]. Besides, since some energy variations may still occur in lower frequencies, the semiconductor ratings have to be designed to support them [48]. Therefore, the control method to achieve synchronous operation commits the operation at lower frequencies. Furthermore, an alternative topology of the TSBC based on hexagonal configuration was presented in [49], called "hexverter". It consists of a reduced version of the TSBC, since one arm-cluster is counted out in each subconverter, totaling six arm-cluster for the whole topology, as shown in Figure 4.8(d). The hexverter or hexagonal TSBC presents similar performance in comparison to the conventional one, but with the advantage of having fewer SMs. Comparisons between DSCC and TSBC topologies for three-phase AC-to-AC conversion was conducted by Ilves *et al.* [48]. Both the DSCC and TSBC topologies can be applied for AC-to-AC conversion in a broad range frequency, presenting better performance and stable operation with proper control technique. Considering that, the DSCC-based BTB system is found to be the most suitable solution for synchronous frequency operation as well as at higher frequencies, while the TSBC topologies can provide superior performance and efficiency than DSCC at low nominal frequency applications.

4.3 Control strategies

4.3.1 Voltage-balancing control

Under practical circumstances, there are voltage differences due to losses and other non-ideal conditions among the phase-legs, resulting in non-zero active power flowing into the phase-legs, since the capacitors can be floating in MMCC topologies. Therefore, it is convenient to implement voltage-balancing control strategies to regulate a constant voltage level. In order to assist in the floating DC capacitors regulation, as well as to achieve high-performance and to improve the operational range, circulating current control strategies have been implemented. Both the voltage-balancing and circulating current control strategies are usually based on the overall control schemes illustrated by Figures 4.10 and 4.11, one for each individual regulation scheme of the floating DC capacitors.

Generally, voltage-balancing control is implemented by controlling the average voltage per phase and also by simply regulating the voltage of each SM. The averaging voltage control, also known as cluster-balancing control, is a closed-loop strategy that regulates the average value of the sum of all M floating DC capacitors, ensuring the energy balance among the converter legs [50–53]. The individual regulation can be implemented in two basic ways: feeding back each SM capacitor

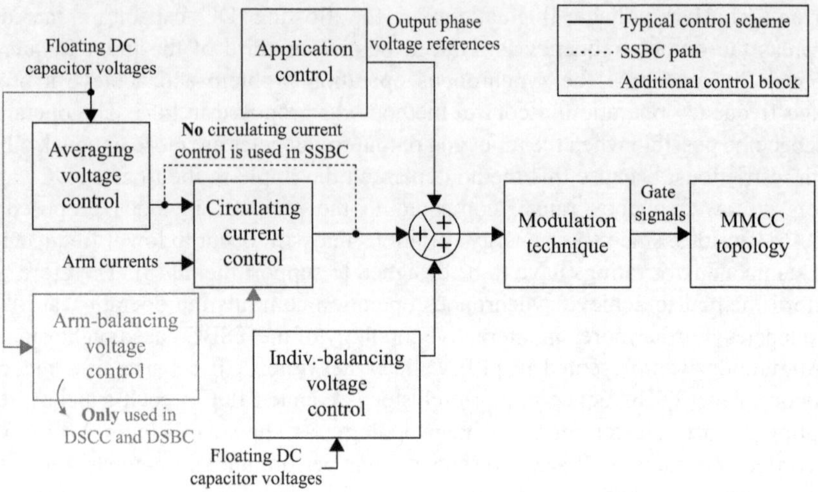

Figure 4.10 Block diagram of the overall control scheme based on individual-balancing voltage control principle

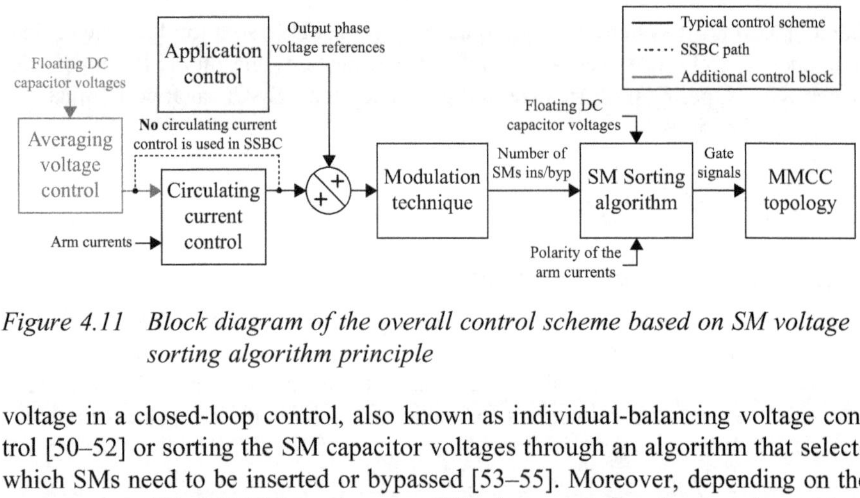

Figure 4.11 Block diagram of the overall control scheme based on SM voltage sorting algorithm principle

voltage in a closed-loop control, also known as individual-balancing voltage control [50–52] or sorting the SM capacitor voltages through an algorithm that selects which SMs need to be inserted or bypassed [53–55]. Moreover, depending on the arrangement of the MMCC topology, modifications can be implemented to achieve the desired voltage balance, e.g., since the SSBC converter has no circulating current [9], adding a zero-sequence voltage [51,52] or slightly shifting the switching pattern [24,56], can either be used to compensate the converter losses and maintain the SMs balanced. Thus, as can be seen in Figures 4.10 and 4.11, the block of the circulating current control is despised (by a dashed line path) in the control scheme for SSBC converter. On the other hand, SDBC, DSCC, DSBC, and TSBC converters have circulating currents, and the averaging voltage control is usually combined to an inner circulating current control. The circulating current control strategies will be broached in the Section 4.3.2.

4.3.1.1 Individual-balancing voltage control

Peng *et al.* [1] introduced a voltage-balancing voltage control for S*x*BC topologies linked to the selective harmonic elimination (SHE) technique by using the averaging voltage control. A slight shifting in the SHE switching pattern was used to guarantee the balance of the individual voltages by a closed-loop controller. Furthermore, this voltage-balancing strategy was optimized in [24] by swapping the pulse patterns among the SMs per arm instead of fixed ones. From this, to keep all SMs balanced, only a single floating DC capacitor needs to be measured and fed back. Fujii *et al.* [57] implemented the averaging voltage control for SDBC topology with m SMs using unipolar sinusoidal pulse-width modulation (PWM) with phase-shifted carriers (PSCs) while the individual-balancing voltage control was implemented by feeding back $m - 1$ SMs. Akagi *et al.* [51] presented a voltage-balancing strategy for SSBC topology similar to that proposed in [57]; however, the individual-balancing control fed back all of m floating DC capacitors, as well as an inner decoupled current loop for independent active and reactive power control was introduced in the averaging voltage control [30]. Considering an eventual SM number expansion, the delay time of this PSC-PWM voltage-balancing control should also be considered [52]. Thereafter, other works have been published based on PSC-PWM voltage-balancing control (averaging and individual-balancing voltage controls) for all basic MMCC topologies, as follow: SSBC [51,52], SDBC [32], DSCC [58], DSBC [42,43], and TSBC [47]. Hagiwara *et al.* [50,59] introduced the arm-balancing voltage control to balance the voltage between the converter arms for DS-based topologies, since they have two arms per phase-leg, and make possible the voltage-balancing control proposed in [58] stable in four-quadrant operation.

4.3.1.2 SM voltage sorting algorithm

The first SM voltage sorting algorithms were developed for DSCC topology [53–55]. In these algorithms, the SMs are inserted or bypassed to produce the required voltage level, according to the voltage magnitude of each floating DC capacitor sorted in ascending (or descending) order and the direction of the arm currents. The basic principle of the SM voltage sorting algorithms was described in [4,40]. Following this reasoning, Ängquist *et al.* [60] presented the open-loop control based on the estimation of the stored energy using the SM sorting algorithm principle to minimize the impact of delays in the measurement and communication systems of the converter, which can be a critical factor in large-scale MMCC topologies [61]. The open-loop scheme does not use any voltage feedback controllers to balance the floating DC capacitors; however, the voltage of each SM capacitor is still measured and then compared directly in the sorting process [60]. Nevertheless, Ilves *et al.* [62] proposed an SM sorting algorithm based on round-robin schedule, wherein the floating DC capacitors do not need to be measured to achieve the voltage balance. To do so, the converter must be controlled insomuch that the average energy transfer is null. Furthermore, a predictive SM voltage sorting algorithm, aiming to reduce the capacitor voltage ripple under low switching frequency modulation was proposed by Ilves *et al.* [63], since in conventional voltage-balancing strategies, there is a trade-off between switching frequency and the capacitor voltage ripple. As can be noticed, SM voltage sorting algorithms do not control the voltages of the floating DC capacitors, but instead all voltages remain within a small voltage band [54] by selecting the inserted or bypassed SMs, which, in turn, depends on the sampling frequency [61,64]. Even though most the sorting algorithms are implemented for DSCC topology, they also have been implemented for the remaining MMCC topologies, such as S*x*BC [65], DSBC [41], TSBC [49], and advanced MMCC topologies [18,66]. Furthermore, over last the years, several studies have proposed SM voltage sorting algorithms associated to different modulation techniques, improving the switching frequency efficiency [54,61–64,67–70] and the quality of the output waveforms [71,72], as well as reducing the number of sensors [62,63,73] and the algorithm complexity [64,73].

4.3.2 Circulating current control

After the DSCC topology was proposed, special attention has been paid to circulating current control strategies. Before that, just the SDBC had been proposed, among the MMCC topologies which have circulating currents, and in this case the circulating current control had not been fully explored and detailed as reported by the authors in [32]. Thus, the most of the papers dedicated to circulating current control have been used in DSCC-based systems. The mathematical model of the DSCC has been thoroughly studied by several papers [74–79], both under the context of the dynamic behavior and the steady state to develop active and passive solutions to regulate the circulating currents.

Under ideal conditions, the circulating currents in the DSCC topology consist of a portion of the DC-link current (or even itself in case of single phase-leg) for energy transfer between the DC-bus and arm-clusters. While in the DSBC topology, an additional fundamental-frequency component is present to exchange active power between the upper and lower arm-clusters in the same phase-leg, and in the SDBC and TSBC topologies, the circulating currents are composed of only the fundamental-frequency component, since no DC-bus is available in both of them. As reported in [9], circulating currents have an important role in exchanging active power among the arm-clusters of the MMCC topologies and have no influence on the AC side signals as well. However, in real conditions, oscillating components rise in the circulating currents, degrading the waveform of the arm currents. This study considers these oscillating components as circulating harmonics. In balanced systems, these components are mostly composed by low-order harmonics at negative-sequence double-line frequency, but under unbalanced voltage conditions, zero- and positive-sequence components should be considered too [80]. The modulation and control schemes, as well as the operating converter point, have significant influence on the appearance of harmonics in circulating currents [54], not to mention that resonances between the arm inductances and SM capacitances can contribute to the increase of these harmonics [81].

Circulating harmonics have direct impact on the ripple of the floating DC capacitors, since they distort the arm currents, and thereby if not controlled, they can not only compromise the converter stability, but also affect the rating of capacitors and semiconductor devices, and increase the total power losses. Thus, controlling the circulating currents is one of the most important issues in MMCC topologies. Besides, suppressing undesired circulating harmonics is essential to achieve high performance and efficiency of the converter, in which it can be ensured that only active power is exchanged inside the converter, although the injection of selective harmonics in the circulating current can be used to overcome some operational issues, like reducing voltage fluctuations in the floating DC capacitors.

The primary purpose of the circulating current control strategies is to achieve the energy balance of the floating DC capacitors during the dynamic state and maintain these capacitors regulated in the steady state operation. Thus, considering an ideal scenario for a DSCC-based system, wherein the circulating current is purely DC, just a proportional-integral (PI) controller would be sufficient to perform a satisfactory dynamic response [58]. However, this controller cannot eliminate circulating harmonics due to its limited gains in other harmonics frequencies. This encouraged several studies to find a suitable solution to eliminate or suppress these harmonics to an acceptable magnitude threshold. The literature is full of control schemes to control/suppress the circulating harmonics covering the various kinds of closed-loop and open-loop strategies, which will be defined henceforward as active methods. Passive methods, conversely, involve physical modifications, which may be used to minimize the circulating harmonics.

Ilves *et al.* [75] proved that the value of the arm inductances and SM capacitances can be calculated to avoid a specific resonant harmonic frequency. Since each SM capacitance is calculated according to energy storage requirements and maximum permissible fluctuation magnitude, modifications in the arm inductors have been evaluated for DSCC converters [74,75]. Tu *et al.* [74] developed a theoretical analysis which resulted in the arm inductances as function of the second-order circulating harmonic and the fault current rise rate as well. This harmonic order can be minimized to a certain limit by increasing the arm inductances; however, this solution enlarges the overall size and weight of the converter, not to mention that high voltage across the arm inductors can compromise performance. In contrast, the use of coupled inductors is an interesting and easy alternative not only to suppress circulating harmonic currents, but also to reduce the size, weight, and cost of the magnetic core [58]. Other passive solution was proposed in [75], which is based on the insertion of a tuned *LC* filter for elimination of the second-order circulating harmonic. The passive methods are limited solutions for circulating harmonic elimination, because they can be designed just to suppress undesired low-order circulating harmonics to some extent. Hence, an effective way to use these methods is allying them to active methods.

Regarding the active suppress method based on closed-loop controllers for the DSCC, the control strategy proposed by Hagiwara and Akagi [58] had some restrictions concerning the point operation, which were overcome in [50,59] by inserting the arm-balancing voltage control to the voltage-balancing control strategy. In doing that, the circulating current controller was changed to a proportional (P) controller and there was a soft reduction of the circulating harmonics. Clearly, these works were not fully concerned about the circulating harmonic elimination, but actually, they aimed to achieve stable four-quadrant operation. Hence, Tu *et al.*

[67] used PI controllers to suppress second-order circulating harmonics in a negative-sequence rotational *dq* coordinate. This control strategy can perform an effective circulating harmonic suppression; however, it is restricted to three-phase systems, since it is not applicable to multiphase or single-phase systems, and only under balanced conditions, whereas in real conditions other even-order harmonics are also presented in circulating currents [81]. Therefore, proportional-resonant (PR) controllers were used to suppress other even-order circulating harmonics, which are applicable to both single-phase and multi-phase systems [76]. As described in [81], the circulating harmonics can be represented as a series of even-order harmonics, then resonant transfer functions are tuned in the harmonic orders that need to be suppressed, and due to this, the implementation of PR controllers becomes difficult when multiple harmonics have to be eliminated.

As presented in [77], the minimization of circulating harmonics above the tenth order by using multi-resonant controllers has been ineffective. Thus, a plug-in repetitive plus PI control strategy was proposed in [77,78]. The repetitive control scheme based on the internal mode principle has shown a good steady-state performance in circulating harmonic suppression, inasmuch as it can provide higher gains at fundamental frequency and its multiples. To overcome the slow dynamic response of the repetitive controller, a PI controller has been combined with it. Some drawbacks of repetitive control are the requirement of accurate control parameters, e.g., sampling frequency and number of samples in a cycle, and the noise sensitivity at intermediate frequencies. As aforementioned in Section 4.3.1.2, open-loop control strategies have been studied aiming to reduce the number of closed-loop controllers. Although the results have showed an effective dynamic performance and suppression of the circulating harmonics, this open-loop method requires precise analytical model and parameters. Harnefors *et al.* [79] included a circulating current control in the open-loop control proposed in [60] (with slight modifications) derived on the basis of a continuous-variable dynamic model. The circulating current control in question uses a P controller (referred to as "active resistance") to regulate the internal arm voltage and a feedforward path of the circulating current reference. Under another perspective, the circulating currents can be controlled to reduce voltage fluctuations in the floating DC capacitors by injecting zero-sequence components [34–36,45–48]. As investigated in [36], it is clear that there is a trade-off between mitigation of the voltage fluctuations and the reduction of the peak current. Nevertheless, the validity in using these control strategies is the capability of operating over a wide frequency range for both the DSCC and the TSBC.

4.4 Modulation techniques

4.4.1 High switching frequency techniques

High switching frequency techniques can be highlighted in carrier-based and space-vector (SV) techniques, as illustrated in Figure 4.12. For the MMCC family, level-shifted carrier (LSC) and PSC are the most common carrier-based PWM

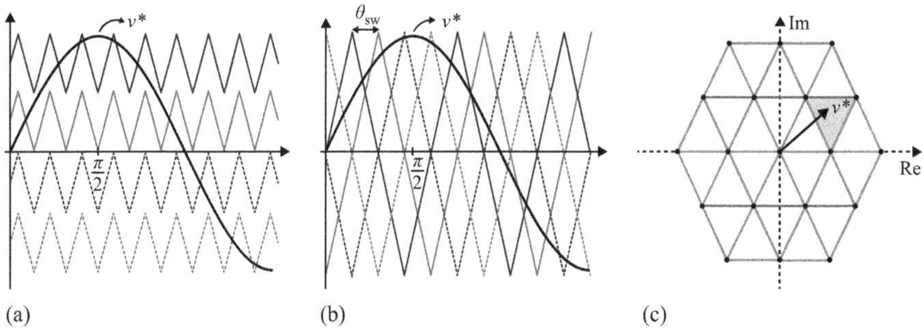

Figure 4.12 Basic principle of high-frequency modulation techniques. (a) LSC-PWM technique, (b) PSC-PWM technique, and (c) SV-PWM technique

techniques [18,82], particularly due to their easy implementation. Both LSC- and PSC-PWM techniques are based on multiple carrier arrangements. In LSC-PWM technique, m carriers are stacked on top of each other, so that each voltage level corresponds to one carrier, as shown in Figure 4.12(a). Since each carrier is associated to a certain voltage level, this technique produces unequal power distribution across the SM terminals.

Thus, some modifications can be implemented to overcome this problem and equalize the voltage distribution, such as: (i) simple alternance in the level between the contiguous carriers each switching frequency [83]; (ii) signal rotation [82,84]; (iii) carrier rotation each fundamental frequency [82,85]; and (iv) continuous carrier rotation at the end of each carrier cycle [86]. For DSCC topology, LSC-PWM produces $m + 1$ pole voltage levels [55] with the possibility of expanding up to $2m + 1$ voltage levels by controlling the number of inserted SMs (N_{on}) (or number of on-state SMs per arm) [54,69,87] and/or doubling the number of carriers compared to the conventional [88].

As can be seen in Figure 4.12(b), the conventional PSC-PWM technique is based on the phase shifting of carriers, in which low-order harmonics can be suppressed for multilevel converters [67]. Naturally, this technique provides equal power distribution, since all SMs are commanded by the same voltage reference (v^*) and all the carriers are on the same level with phase-shifting such that they are equally spaced in one cycle [18]. The PSC-PWM technique requires m carriers for SxBC and TSBC topologies and $2m$ carriers for DSCC topology to synthesize $2m + 1$ pole voltage levels, so that the phase-shifted angle (θ_{sw}) is defined by 180°/m [83,84]. However, it is possible to reduce the number of carriers to a half for DSCC topology by modifying the conventional technique [89] and controlling the number N_{on} as well [67,73,87]. For DSBC topology, $2m$ carriers are also required to use PSC-PWM, but the number of voltage levels can vary depending on the DC-link voltage utilization [43].

Therefore, considering that both LSC- and PSC-PWM techniques have the same number of switching transitions, the first technique provides a better line voltage harmonic spectrum under both normal and over-modulation conditions [5,82,90]. Also, LSC-PWM technique produces less semiconductor losses in comparison with the conventional PSC [87]. The combination of both techniques was investigated in [18,91]. Besides, when the conventional PSC-PWM is used in MMCC topologies, feedback voltage-balancing control strategies are usually implemented to ensure the energy balance among the converter phases and individually in each SM [50,51,58]. Conversely, the implementation of the LSC-PWM technique is a bit more complex than the PSC-PWM, due to the appropriate modifications to ensure equal power distribution for all SMs.

Continuing with high switching frequency techniques, SV-PWM is also widely applied in MMCC topologies [92–94], especially if the converter does not synthesize a high number of voltage levels. In comparison with carrier-based techniques, SV-PWM provides higher DC-link voltage utilization [93] and has more flexibility in optimizing the switching pattern design [94], since additional switching states can be selected to reduce the output voltage distortion, besides the fact that the proper selection of the switching states avoids unnecessary switching transitions [92]. For multilevel converters, the SV-PWM technique can produce a large number of switching states represented by vectors, which are derived from the voltage reference mapping, as shown in Figure 4.12(c). Then, as the level number increases, a significant number of redundancy state sequences are generated, which make the computational implementation extremely complex. However, there are studies focused to simplify the SV-PWM implementation [93,94].

High switching frequency techniques can fulfil the total harmonic distortion (THD) level standards with low-order harmonic distortion in the output waveforms, covering the entire modulation range. Harmonic sidebands of the carrier frequency are displaced to higher frequencies, which makes the filtering easier by passive filters [84]. Nevertheless, the increased number of commutation per cycle still leads to high switching losses. In three-phase systems, zero-sequence voltage injection technique has been implemented in high-frequency PWM techniques to extend the linearity range of the voltage utilization, so that the switching losses can be reduced and the output waveform quality can be improved. For this reason, high-frequency PWM techniques are very suitable for MMCC topologies with a low number of SMs per arm, usually less than ten, since the THD has not been the major issue for large-scale MMCC topologies, but instead the computational effort and algorithm complexity of these PWM techniques are, not to mention that the voltage-balancing strategies also have to be considered. Therefore, the implementation of low switching frequency techniques has been taking advantage of the high modularity degree of MMCC topologies to synthesize good quality output waveforms with minimal switching commutation, thus improving the overall efficiency and reliability of the converter.

4.4.2 Low switching frequency techniques

The implementation of low switching frequency techniques in large-scale MMCC topologies has been an attractive manner to combine quality of converter output waveform, closely related to the uniform sampling frequency, with semiconductor losses closer to the thyristor technology [61]. For high-power applications, low switching frequencies are considered to be those below 1 kHz [12]. So, it is possible to detach low switching frequency techniques in those whose frequency is based on the fundamental, such as the nearest vector control (NVC) and nearest level control (NLC), and those whose frequency is above of the fundamental frequency and below 1 kHz, such as the SHE technique and its variations.

The SHE technique has been applied in conventional VS converters (VSCs) providing the elimination of undesired low order harmonics by switching angles (α_m) per quarter of the fundamental period [12], as can be seen in Figure 4.13(a). These angles are calculated off-line and stored in lookup tables for various modulation indices, and that is why this technique is also known as "programmed technique". In MMCC topologies with FB-SMs, the output waveform synthesized by the SHE technique usually has a stair-like shape, wherein each SM is associated to a switching angle at fundamental frequency, providing $2m + 1$ pole voltage levels [1,12]. Thus, to improve the THD levels even in higher harmonic orders, the active harmonic elimination (AHE) technique was proposed in [95], in which new switching angles were added corresponding to the high-order harmonic frequencies. In case of the DSCC, SHE-based techniques have also been developed at fundamental frequency [62]. Konstantinou *et al.* [72] proposed a SHE-based technique, considering a significant number (up to 17) of switching angles per quarter-wave, providing high-quality output waveforms. Also, depending on the number of inserted SMs per arm, the DSCC can reach $m + 1$ or $2m + 1$ pole voltage levels, which can lead to the absence or presence of high-frequency components in the DC-link voltage, respectively.

Even though the SHE and SHE-based techniques can provide low THD levels in the output waveforms for MMCC topologies, the computational difficulty increases significantly, due to the need to pre calculate a high number of non-linear equations [70]. Moreover, the poor dynamic response makes these techniques not feasible for real projects. Rodríguez *et al.* [92] proposed the NVC technique, which

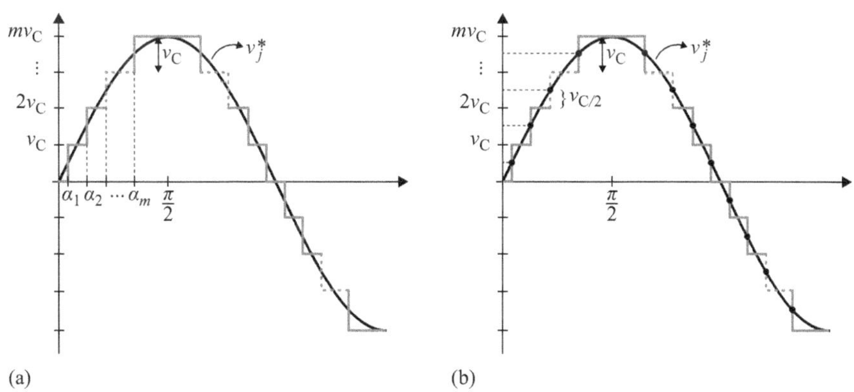

Figure 4.13 Basic principle of low-frequency modulation techniques. (a) SHE technique and (b) NLC technique

is a version of the SV-PWM for low switching frequency. However, for large-scale MMCC topologies, finding the nearest vector is more difficult than calculating the nearest voltage level [12]. Therefore, the NLC has been considered the most efficient technique for large-scale MMCC topologies commanded at fundamental frequency [61,64], as shown in Figure 4.13(b). The NLC technique is a result of a simple mathematical operation in time-domain, producing the pulse patterns by finding the nearest voltage level. Due to its simplicity, several papers have implemented the NLC combined to SM voltage sorting algorithms to drive the DSCC [61,64,70,96]. By doing so, the NLC technique returns the number N_{on} that is needed to generate the desired output voltage level. Once the number of SMs is large, the sampling frequency of a digital implementation must be considered, because if the sampling frequency is not high enough to ensure that all SMs can produce the required voltage level, the voltage step will increase and quality of the output voltage waveform can be degraded [61]. Similar to the SHE technique, the NLC technique can synthesize $2m + 1$ and $m + 1$ pole voltage levels with a staircase-shaped waveform for MMCC topologies, respectively, based on FB- and HB-SM.

4.5 Operational issues in MMCCs

4.5.1 Fault-tolerant operation

Fault-tolerant operation is essential to increase system reliability. For MMCC topologies, the faults can be highlighted in external and internal faults. The external faults can occur at the AC side or DC side (if the converter has a common DC-link), and the internal faults can be comprehended by defects occurred in the SMs or short-circuit at the floating DC capacitors.

4.5.1.1 External faults

In case of AC-side short circuits, several studies have been investigating the MMCC topologies to maintain the operation even in asymmetrical grid fault conditions, due to their good fault ride-through (FRT) capability without additional hardware cost [29,37,55,80]. In case of DC-link short circuit, fast interruption of the DC fault current is a very important factor to avoid irreparable damage to the system. This is a significant issue in HVDC systems, and so, three solutions are commonly implemented to extinguish the DC fault current: open the AC side circuit breakers or DC side circuit breakers and use advanced SM topologies. Opening the AC side circuit breakers is known to be worst solution of the three aforementioned, because it may require enough time to damage the components involved [5,97]. When HB-SMs are used, the first procedure to initiate the DC-link fault protection is bypassing all SMs soon after detecting the fault to avoid the converter feeding the DC fault current. The basic solution to this procedure is to connect press-pack thyristor to protect the free-wheeling diodes from over-current; however, this provides a one-way path to the DC fault current, making its fast interruption difficult to accomplish, inasmuch as the DC arc cannot be extinguished until the AC side circuit breaker is opened. Hence, Li *et al.* [97] proposed the use of double-thyristor switches to provide bidirectional path for DC non-permanent fault current, eliminating the diode-bridge rectifier mode. In doing so, the DC arc can naturally be extinguished and the converter can be restarted to normal operation after a few fundamental cycles. Although the double-thyristor strategy represents a simple, reliable, and cost-effective protection scheme, the DC fault current cannot be extinguished faster than employing DC side circuit breakers or using CD-SM [5,97]. Besides, using a DC side circuit breaker can quickly extinguish DC fault current, however, it is costly and its technology is not yet mature [15,97].

At last, advanced SM topologies can be used to achieve the high-performance fault-tolerant operation [14,20,66]. SM topologies that provide bipolar voltages, like FB- and CCC-SM, can impose the reverse voltage polarity to block the AC side currents during the DC short-circuit. However, in applications like HVDC transmission, the reverse voltage has no utility in normal operation and the high number of semiconductor devices increases the design cost and the losses (around 70% more than the HB-SM) [15]. On the other hand, the CD-SM operates as HB-SM with only an extra current path, but during the DC short-circuit, it operates as FB-SM producing reverse voltage polarity by the two capacitors to clear the DC fault current. In comparison with HB-SM, the CD-SM topology has around 35% more losses, due to the extra current path [15]. Nevertheless, in the literature, the CD-SM is considered one of the best SM topologies for MMCC-based HVDC system, which can simultaneously offer high reliability and low losses. Whereas CD-SM has more semiconductor devices than the HB-SM and slightly increases the conduction losses, it is important to evaluate the overall cost of the system, since using HB-SMs in the current technology scenario may require bulky and expensive DC side circuit breaker to achieve good performance in DC fault clearance like the CD-SM.

4.5.1.2 Internal faults

In order to maintain the operation, faulty SMs must be bypassed and the system can be carried out compensating their operation by redundant SMs [68,98] or by software [85,99]. The redundancy process is a reliable way to ensure the functionality of the MMCC topologies without decreasing the output power quality. The basic procedure to use the redundant SMs is to design the converter with additional SMs, which are pre charged and in stand-by mode. In the event of failure, the faulty SM is shorted out by highly reliable high-speed bypass switch and replaced by the redundant SM. Son *et al.* [68] investigated the use of spare SMs to guarantee a fail-safe operation and reinforce the redundancy process. As presented in [68], the spare SMs are not only bypassed in principle, like the redundant ones, but also disconnected from the system, i.e., these SMs are totally discharged. The spare SMs are connected to the system when the number of fault SM exceeds the redundancy, which may lead to same voltage transients, due to the capacitor charging procedure. Konstantinou *et al.* [98] proposed the use of redundant SMs, not always bypassed, but instead included in the normal operation as active redundant SMs. In this way, the SMs in the same arm are swapped and selected by a voltage algorithm. As a result, a fault-tolerant operation could be achieved, reducing the floating DC capacitor ripple as well. Another way to achieve fail-safe operation in case of SM failure is based on software modifications in the control and modulation strategies. The procedure of bypassing the faulty SM occurs similarly to the redundancy process. Hence, Maharjan *et al.* [99] proposed a technique based on the fundamental-frequency zero-sequence voltage injection, named "neutral shift". The zero-sequence component is added in each healthy SM to compensate the operation of the faulty SM and maintain the line voltages balanced. From another perspective, Shen *et al.* [85] proposed a modification (based on the carrier rotation) in the LSC-PWM technique to maintain the system operating under a SM failure. While redundant solutions basically improve system reliability by hardware backup, though increasing the cost, the software solutions may lead to higher voltage requirement than the rated value, not to mention that the phase voltages can have the waveform quality degraded.

4.5.2 Floating DC capacitor pre-charging procedures

Before the MMCC topologies start the operation, the floating DC capacitors are generally under de-energized condition. Thus, to avoid large inrush currents at the start-up, capacitors must be pre-charged up to their nominal value or near it. Besides, after a DC fault (in case of topologies with DC-link availability) the re-charging procedure has to be considered for the same reason. Hence, pre-charging or black-start solutions have been implemented to ensure a fast, soft, and failsafe start-up [4,66,100,101]. Most solutions are based on current-limiting resistor [66,100] or auxiliary DC VS [4,101]. A simple pre-charging procedure can be accomplished by using an AC side current-limiting resistor in parallel to a magnetic contactor, wherein both are in series with AC circuit breaker. From this procedure, all floating DC capacitors are charged up to a value near the rated voltage from the supply currents that flow through the current-limiting resistor. Once the capacitors are pre-charged, the AC circuit breaker is turned on and the magnetic contactor bypasses the current-limiting resistor. Although this procedure is simple, there is no control of the time of charging. It is not an efficient solution and can lead to bulky resistances in high-power systems [7]. Another common solution uses an auxiliary DC VS and DC circuit breaker to charge the capacitors in a staggered sequence one by one through a constant time up to their rated value [4]. However, this solution is not cost-effective, due to the need of a costly DC circuit breaker [102]. The start-up procedure based on current-limiting resistors was expanded to different controlled schemes [66,100,102]. For instance, Li *et al.* [102] proposed a closed-loop pre-charging control producing a constant charging current from either the DC side or the AC side main voltage. By doing so, no auxiliary power supply is required, the charging time is reduced, and inrush current is eliminated.

4.6 Main applications

The benefits and prominent features of MMCC topologies have attracted a lot of interest in all applications segments, especially those requiring high output power quality. S*x*BC topologies have been widely used as APLCs in high-voltage transmission grid, known as flexible AC transmission system (FACTS) devices, or in distribution systems, known as custom power (CP) devices. Among the APLC systems, STATCOM is the most implemented device based on S*x*BC topologies, due to their simplest layout, since isolated DC sources are not required [27–29,32,51,56,57,65]. Nevertheless, depending on the connection among the arm-clusters, both the star- and delta-type topologies present some issues for reactive power and harmonic compensation. The SSBC is suitable for positive-sequence reactive power compensation [51,56,65]; however, the compensation of negative-sequence reactive power is restricted by voltage requirements [28]. Conversely, the SDBC has the additional capability of compensating negative-sequence reactive power [27,32,57], although it is suitable up to some extent of unbalanced conditions. Hence, Hagiwara *et al.* [59] have investigated the negative-sequence reactive power compensation using a DSCC-based STATCOM. Besides, a DSCC-based distribution STATCOM (D-STATCOM) was applied for full compensation under a more critical scenario with unbalanced non-linear load and asymmetrical grid voltages [37]. The harmonic filtering was performed not only by SSBC-based active power filter (APF) [24,25], but also by DSCC-based APF [103]. Furthermore, Ota *et al.* [29] investigated FRT capability of a SSBC-based STATCOM, and Sotoodeh and Miller [104] applied a DSCC-based D-STATCOM for regulating active and reactive power transferred to the grid by a wind energy conversion system (WECS). As reported in [9], the DSBC can fulfil the requirements for PCSs for renewable energy sources, since DC-link voltage can vary over a wide range without interfering with the AC voltage. Therefore, Thitichaiworakorn *et al.* [43] have investigated the DSBC to be applied in WECSs. On the other hand, the interface between the renewable energy sources and the grid was also achieved by S*x*BC-based photovoltaic systems (PVSs) [105], S*x*BC-based WECSs [22], as well as by DSCC-based PVSs [69]. Under the same perspective, BESSs intended for active-power levelling have been investigated based on the SSBC [30,31] and DSCC [106]. Regarding the ASDs, the S*x*BC can achieve a high-performance operation, however, the lack of a DC-link requires the use of isolated DC sources, which results in complex front ends [13]. Unlike the S*x*BC, the DSCC has been arousing interest in BTB systems, especially in those requiring regenerative applications. In this case, the common DC-link capacitor is dispensable [89], and the AC-to-AC conversion is carried out indirectly. The direct AC-to-AC conversion can be achieved by TSBC, which is suitable for ASDs requiring regenerative braking [46,47], and also by the DSBC as investigated by Glinka and Marquardt [40]. The TSBC is practical under asynchronous operation [44,45]. Whereas the DSCC does not have restrictions to drive with the input and output frequencies synchronized, it has issues when operating at lower frequencies [33]. However, by using the proper control strategy, not only the problems of the TSBC with synchronous operation, but also the difficulty of the DSCC in driving at low speeds can be overcome [48]. Hence, the DSCC has also been become a competitive alternative for ASD [34–36,45,48]. The advent of modular converter technology has resulted in the fourth generation of VSC transmission [6]. Since the DSCC was introduced in [3], several papers have been published, evaluating it for HVDC transmission [4–7,55,64,68,96,97]. A summary of some commercialized DSCC-based HVDC systems and projects underway can be found in [7]. From another perspective, the DSBC has also been investigated for HVDC systems [5,6,14], due to its better capability of DC fault-handling than the DSCC.

Acknowledgements

This work was supported in part by the Coordination for the Improvement of Higher Education Personnel (CAPES), in part by the Postgraduate Program in Electrical Engineering (PPgEE)/COPELE, Department of Electrical Engineering (DEE), Federal University of Campina Grande (UFCG). The authors also acknowledge the support of Dra Camila Seibel Gehrke, who generously helped us in the text editing and for her valuable suggestions.

References

[1] Fang Z. Peng, Jih-Sheng Lai, John W. McKeever, and James VanCoevering. A Multilevel Voltage-Source Inverter with Separate DC Sources for Static VAr Generation. *IEEE Trans. Ind. Appl.*, 32(5):1130–1138, 1996.

[2] Jih-Sheng Lai and Fang Z. Peng. Multilevel Converters – A New Breed of Power Converters. *IEEE Trans. Ind. Appl.*, 32(3):509–517, 1996.

[3] Rainer Marquardt, Anton Lesnicar, and Jürgen Hildinger. Modulares Stromrichterkonzept für Netzkup-plungsanwendung bei hohen Spannungen. In *ETG Conf.*, Bad Nauheim, 2002.

[4] Anton Lesnicar and Rainer Marquardt. An Innovative Modular Multilevel Converter Topology Suitable for a Wide Power Range. In *2003 IEEE Power Tech Conf. Proc.*, volume 3, pages 272–277, Bologna, Italy, IEEE, 2003.

[5] Suman Debnath, Jiangchao Qin, Behrooz Bahrani, Maryam Saeedifard, and Peter Barbosa. Operation, Control, and Applications of the Modular Multilevel Converter: A Review. *IEEE Trans. Power Electron.*, 30(1):37–53, January 2015.

[6] Alireza Nami, Jiaqi Liang, Frans Dijkhuizen, and Georgios D. Demetriades. Modular Multilevel Converters for HVDC Applications: Review on Converter Cells and Functionalities. *IEEE Trans. Power Electron.*, 30(1):18–36, January 2015.

[7] Marcelo A. Perez, Steffen Bernet, José Rodríguez, Samir Kouro, and Ricardo Lizana. Circuit Topologies, Modelling, Control Schemes and Applications of Modular Multilevel Converters. *IEEE Trans. Power Electron.*, 30(1):4–17, 2015.

[8] Mariusz Malinowski, K. Gopakumar, José Rodríguez, and Marcelo A. Perez. A Survey on Cascaded Multi-Level Inverters. *IEEE Trans. Ind. Electron.*, 57(7):2197–2206, July 2010.

[9] Hirofumi Akagi. Classification, Terminology, and Application of the Modular Multilevel Cascade Converter (MMCC). *IEEE Trans. Power Electron.*, 26(11):3119–3130, November 2011.

[10] José Rodríguez, Steffen Bernet, Bin Wu, Jorge O. Pontt, and Samir Kouro. Multilevel Voltage-Source-Converter Topologies for Industrial Medium-Voltage Drives. *IEEE Trans. Ind. Electron.*, 54(6):2930–2945, 2007.

[11] Jun Wang, Rolando Burgos, and Dushan Boroyevich. A Survey on the Modular Multilevel Converters Modeling, Modulation and Controls. In *2013*

IEEE Energy Convers. Congr. Expo., pages 3984–3991, IEEE, September 2013.

[12] José Rodríguez, Leopoldo G. Franquelo, Samir Kouro, Jose I. Leon, Ramon C. Portillo, Maria A. M. Prats, and Marcelo A. Perez. Multilevel Converters: An Enabling Technology for High-Power Applications. *Proc. IEEE*, 97(11):1786–1817, November 2009.

[13] Samir Kouro, José Rodríguez, Bin Wu, Steffen Bernet, and Marcelo A. Perez. Powering the Future of Industry: High-Power Adjustable Speed Drive Topologies. *IEEE Ind. Appl. Mag.*, 18(4):26–39, July 2012.

[14] Rainer Marquardt. Modular Multilevel Converter: An Universal Concept for HVDC-Networks and Extended DC-Bus-Applications. In *2010 IEEE Int. Power Electron. Conf.*, pages 502–507, Sapporo, June 2010.

[15] Noman Ahmed, Staffan Norrga, Hans-Peter Nee, Arif Haider, Dirk Van Hertem, Lidong Zhang, and Lennart Harnefors. HVDC SuperGrids with Modular Multilevel Converters the Power Transmission Backbone of the Future. In *Int. Multi-Conference Syst. Signals Devices*, pages 1–7, March 2012.

[16] Estibaliz Solas, Gonzalo Abad, Jon A. Barrena, Sergio Aurtenetxea, Ainhoa Carcar, and Ludwik Zajac. Modular Multilevel Converter with Different Submodule Concepts Part II: Experimental Validation and Comparison for HVDC Application. *IEEE Trans. Ind. Electron.*, 60(10):4536–4545, October 2013.

[17] José Rodríguez, Jih-Sheng Lai, and Fang Z. Peng. Multilevel Inverters: A Survey of Topologies, Controls, and Applications. *IEEE Trans. Ind. Electron.*, 49(4):724–738, August 2002.

[18] Ashish K. Sahoo, Ruben Otero-De-Leon, Visweshwar Chandrasekaran, and Ned Mohan. New 3-Level Submodules for a Modular Multilevel Converter Based HVDC System with Advanced Features. *IECON'13 – 39th Annu. Conf. IEEE Ind. Electron. Soc.*, pages 6269–6274, November 2013.

[19] Marc Hiller, Dietmar Krug, Rainer Sommer, and Steffen Rohner. A New Highly Modular Medium Voltage Converter Topology for Industrial Drive Applications. In *EPE'09 – 13th Eur. Conf. Power Electron. Appl.*, pages 1–10, 2009.

[20] Alireza Nami, Liwei Wang, Frans Dijkhuizen, and Anshuman Shukla. Five Level Cross Connected Cell for Cascaded Converters. *EPE'13 – 15th Eur. Conf. Power Electron. Appl.*, pages 1–9, September 2013.

[21] Baljit S. Riar, Udaya K. Madawala, and Duleepa J. Thrimawithana. Analysis and Control of a Three-Phase Modular Multi-Level Converter Based on Inductive Power Transfer technology (M2LC-IPT). In *2013 IEEE Int. Conf. Ind. Technol.*, pages 475–480, February 2013.

[22] Max A. Parker, Li Ran, and Stephen J. Finney. Distributed Control of a Fault-Tolerant Modular Multilevel Inverter for Direct-Drive Wind Turbine Grid Interfacing. *IEEE Trans. Ind. Electron.*, 60(2):509–522, February 2013.

[23] Marcelo A. Perez, Ricardo Lizana, Camilo Azocar, José Rodríguez, and Bin Wu. Modular Multilevel Cascaded Converter based on Current Source H-Bridges Cells. In *IECON'12 – 38th Annu. Conf. IEEE Ind. Electron. Soc.*, pages 3443–3448, Montreal, QC, 2012.

[24] Fang Z. Peng, John W. McKeever, and Donald J. Adams. A Power Line Conditioner Using Cascade Multilevel Inverters for Distribution Systems. *IEEE Trans. Ind. Appl.*, 34(6):1293–1298, 1998.

[25] Chen Junling, Li Yaohua, Wang Ping, Yin Zhizhu, and Dong Zuyi. A Closed-Loop Selective Harmonic Compensation with Capacitor Voltage Balancing Control of Cascaded Multilevel Inverter for High-Power Active Power Filters. In *2008 IEEE Power Electron.Spec. Conf.*, pages 569–573, IEEE, June 2008.

[26] Fang Z. Peng, Jr. Ott, George W., and Donald J. Adams. Harmonic and Reactive Power Compensation Based on the Generalized Instantaneous Reactive Power Theory for Three-Phase Four-Wire Systems. *IEEE Trans. Power Electron.*, 13(6):1174–1181, 1998.

[27] Fang Z. Peng and Jin Wang. A Universal STATCOM with Delta-Connected Cascade Multilevel Inverter. In *2004 IEEE Power Electron. Spec. Conf.*, pages 3529–3533, Aachen, Germany, 2004. IEEE.

[28] Qiang Song and Wenhua Liu. Control of a Cascade STATCOM with Star Configuration under Unbalanced Conditions. *IEEE Trans. Power Electron.*, 24(1):45–58, 2009.

[29] Joao I. Y. Ota, Yuji Shibano, and Hirofumi Akagi. Low-Voltage-Ride-Through (LVRT) Capability of a Phase-Shifted-PWM STATCOM Using the Modular Multilevel Cascade Converter Based on Single-Star Bridge-Cells (MMCC-SSBC). In *2013 IEEE Energy Convers. Congr. Expo.*, pages 3062–3069. IEEE, September 2013.

[30] Laxman Maharjan, Shigenori Inoue, and Hirofumi Akagi. A Transformerless Energy Storage System Based on a Cascade Multilevel PWM Converter with Star Configuration. *IEEE Trans. Ind. Appl.*, 44(5):1621–1630, September 2008.

[31] Laxman Maharjan, Shigenori Inoue, Hirofumi Akagi, and Jun Asakura. State-of-Charge (SOC)-Balancing Control of a Battery Energy Storage System Based on a Cascade PWM Converter. *IEEE Trans. Power Electron.*, 24(6):1628–1636, June 2009.

[32] Makoto Hagiwara, Ryo Maeda, and Hirofumi Akagi. Negative-Sequence Reactive-Power Control by a PWM STATCOM Based on a Modular Multilevel Cascade Converter (MMCC-SDBC). *IEEE Trans. Ind. Appl.*, 48(2): 720–729, 2012.

[33] Makoto Hagiwara, Kazutoshi Nishimura, and Hirofumi Akagi. A Medium-Voltage Motor Drive with a Modular Multilevel PWM Inverter. *IEEE Trans. Power Electron.*, 25(7):1786–1799, July 2010.

[34] Antonios Antonopoulos, Lennart Ängquist, Staffan Norrga, Kalle Ilves, Lennart Harnefors, and Hans-Peter Nee. Modular Multilevel Converter AC Motor Drives with Constant Torque from Zero to Nominal Speed. *IEEE Trans. Ind. Appl.*, 50(3):1982–1993, May 2014.

[35] Makoto Hagiwara, Isamu Hasegawa, and Hirofumi Akagi. Start-Up and Low-Speed Operation of an Electric Motor Driven by a Modular Multilevel Cascade Inverter. *IEEE Trans. Ind. Appl.*, 49(4):1556–1565, July 2013.

[36] Yuhei Okazaki, Hitoshi Matsui, Makoto Hagiwara, and Hirofumi Akagi. Research Trends of Modular Multilevel Cascade Inverter (MMCI-DSCC)-Based Medium-Voltage Motor Drives in a Low-Speed Range. In *2014 IEEE Int. Power Electron. Conf. (IPEC/ECCE Asia)*, pages 1586–1593. IEEE, May 2014.

[37] Sixing Du and Jinjun Liu. A Study on DC Voltage Control for Chopper-Cell-Based Modular Multilevel Converters in D-STATCOM Application. *IEEE Trans. Power Deliv.*, 28(4):2030–2038, October 2013.

[38] Hassan M. Pirouz and Mohammad T. Bina. A Transformerless Medium-Voltage STATCOM Topology Based on Extended Modular Multilevel Converters. *IEEE Trans. Power Electron.*, 26(5):1534–1545, May 2011.

[39] Feng Gao, Decun Niu, Hao Tian, Chunjuan Jia, Nan Li, and Yong Zhao. Control of Parallel-Connected Modular Multilevel Converters. *IEEE Trans. Power Electron.*, 30(1):372–386, January 2015.

[40] Martin Glinka and Rainer Marquardt. A New AC/AC Multilevel Converter Family. *IEEE Trans. Ind. Electron.*, 52(3):662–669, June 2005.

[41] Marcelo A. Perez, José Rodríguez, Esteban J. Fuentes, and Felix Kammerer. Predictive Control of ACAC Modular Multilevel Converters. *IEEE Trans. Ind. Electron.*, 59(7):2832–2839, July 2012.

[42] Nuntawat Thitichaiworakorn, Makoto Hagiwara, and Hirofumi Akagi. A Single-Phase to Three-Phase Direct AC/AC Modular Multilevel Cascade Converter Based on Double-Star Bridge-Cells (MMCC-DSBC). In *2013 IEEE 1st Int. Futur. Energy Electron. Conf.*, pages 476–481. IEEE, November 2013.

[43] Nuntawat Thitichaiworakorn, Makoto Hagiwara, and Hirofumi Akagi. Experimental Verification of a Modular Multilevel Cascade Inverter Based on Double-Star Bridge Cells. *IEEE Trans. Ind. Appl.*, 50(1):509–519, January 2014.

[44] Robert W. Erickson and Osama A. Al-Naseem. A New Family of Matrix Converters. In *IECON'01 – 27th Annu. Conf. IEEE Ind. Electron. Soc.*, volume 2, pages 1515–1520, IEEE, 2001.

[45] Arthur J. Korn, Manfred Winkelnkemper, Peter K. Steimer, and Johann W. Kolar. Direct Modular Multilevel Converter for Gearless Low-Speed Drives. In *EPE'11 – 14th Eur. Conf. Power Electron. Appl.*, pages 1–7, Birmingham, IEEE, 2011.

[46] Wataru Kawamura, Makoto Hagiwara, and Hirofumi Akagi. Control and Experiment of a Modular Multilevel Cascade Converter Based on Triple-Star Bridge Cells. *IEEE Trans. Ind. Appl.*, 50(5):3536–3548, September 2014.

[47] Wataru Kawamura, Makoto Hagiwara, and Hirofumi Akagi. A Broad Range of Frequency Control for the Modular Multilevel Cascade Converter Based on Triple-Star Bridge-Cells (MMCC-TSBC). In *2013 IEEE Energy Convers. Congr. Expo.*, pages 4014–4021, IEEE, September 2013.

[48] Kalle Ilves, Luca Bessegato, and Staffan Norrga. Comparison of Cascaded Multilevel Converter Topologies for AC/AC Conversion. In *2014 IEEE Int. Power Electron. Conf. (IPEC/ECCE Asia)*, pages 1087–1094, IEEE, May 2014.

[49] Lennart Baruschka and Axel Mertens. A New Three-Phase AC/AC Modular Multilevel Converter with Six Branches in Hexagonal Configuration. *IEEE Trans. Ind. Appl.*, 49(3):1400–1410, May 2013.

[50] Makoto Hagiwara, Ryo Maeda, and Hirofumi Akagi. Control and Analysis of the Modular Multilevel Cascade Converter Based on Double-Star Chopper-Cells (MMCC-DSCC). *IEEE Trans. Power Electron.*, 26(6):1649–1658, June 2011.

[51] Hirofumi Akagi, Shigenori Inoue, and Tsurugi Yoshii. Control and Performance of a Transformerless Cascade PWM STATCOM with Star Configuration. *IEEE Trans. Ind. Appl.*, 43(4):1041–1049, 2007.

[52] Joao I. Y. Ota, Yuji Shibano, Naoto Niimura, and Hirofumi Akagi. Current Control of a Phase-Shifted-PWM STATCOM Using the Modular Multilevel Cascade Converter Based on Single-Star Bridge-Cells (MMCC-SSBC). In *2013 IEEE Energy Convers. Congr. Expo.*, pages 420–427, IEEE, September 2013.

[53] Estibaliz Solas, Gonzalo Abad, Jon A. Barrena, A. Carear, and Sergio Aurtenetxea. Modulation of Modular Multilevel Converter for HVDC Application. In *2010 IEEE 14th Int. Power Electron. Motion Control Conf.*, pages T2-84–T2-89, IEEE, September 2010.

[54] Steffen Rohner, Steffen Bernet, Marc Hiller, and Rainer Sommer. Modulation, Losses, and Semiconductor Requirements of Modular Multilevel Converters. *IEEE Trans. Ind. Electron.*, 57(8):2633–2642, August 2010.

[55] Maryam Saeedifard and Reza Iravani. Dynamic Performance of a Modular Multilevel Back-to-Back HVDC System. *IEEE Trans. Power Deliv.*, 25(4):2903–2912, October 2010.

[56] Yu Liu, Alex Q. Huang, Wenchao Song, Subhashish Bhattacharya, and Guojun Tan. Small-Signal Model Based Control Strategy for Balancing Individual DC Capacitor Voltages in Cascade Multilevel Inverter-Based STATCOM. *IEEE Trans. Ind. Electron.*, 56(6):2259–2269, 2009.

[57] Kansuke Fujii, Rik W. De Doncker, and Shigeo Konishi. A Novel DC-Link Voltage Control of PWM-Switched Cascade Cell Multi-Level Inverter Applied to STATCOM. *2005 IEEE 40th Annu. Meet. Ind. Appl. Soc.*, 2: 961–967, 2005.

[58] Makoto Hagiwara and Hirofumi Akagi. Control and Experiment of Pulsewidth-Modulated Modular Multilevel Converters. IEEE Trans. Power Electron., 24(7):1737–1746, July 2009.

[59] Makoto Hagiwara, Ryo Maeda, and Hirofumi Akagi. Negative-Sequence Reactive-Power Control by the Modular Multilevel Cascade Converter Based on Double-Star Chopper-Cells (MMCC-DSCC). In *2010 IEEE Energy Convers. Congr. Expo.*, volume 1, pages 3949–3954, Atlanta, GA, 2010.

[60] Lennart Ängquist, Antonios Antonopoulos, Daniel Siemaszko, Kalle Ilves, Michail Vasiladiotis, and HansPeter Nee. Open-Loop Control of Modular Multilevel Converters Using Estimation of Stored Energy. *IEEE Trans. Ind. Appl.*, 47(6):2516–2524, November 2011.

[61] Qingrui Tu and Zheng Xu. Impact of Sampling Frequency on Harmonic Distortion for Modular Multilevel Converter. *IEEE Trans. Power Deliv.*, 26(1):298–306, July 2011.

[62] Kalle Ilves, Antonios Antonopoulos, Staffan Norrga, and Hans-Peter Nee. A New Modulation Method for the Modular Multilevel Converter Allowing Fundamental Switching Frequency. *IEEE Trans. Power Electron.*, 27(8): 3482–3494, August 2012.

[63] Kalle Ilves, Lennart Harnefors, Staffan Norrga, and Hans-Peter Nee. Predictive Sorting Algorithm for Modular Multilevel Converters Minimizing the Spread in the Submodule Capacitor Voltages. *IEEE Trans. Power Electron.*, 30(1):440–449, January 2015.

[64] Jiangchao Qin and Maryam Saeedifard. Reduced Switching-Frequency Voltage-Balancing Strategies for Modular Multilevel HVDC Converters. *IEEE Trans. Power Deliv.*, 28(4):2403–2410, October 2013.

[65] Christopher D. Townsend, Terrence J. Summers, John Vodden, Alan J. Watson, Robert E. Betz, and Jon C. Clare. Optimization of Switching Losses and Capacitor Voltage Ripple Using Model Predictive Control of a Cascaded H-Bridge Multilevel StatCom. *IEEE Trans. Power Electron.*, 28(7):3077–3087, July 2013.

[66] Yinglin Xue, Zheng Xu, and Geng Tang. Self-Start Control with Grouping Sequentially Precharge for the C-MMC-Based HVDC System. *IEEE Trans. Power Deliv.*, 29(1):187–198, February 2014.

[67] Qingrui Tu, Zheng Xu, and Lie Xu. Reduced Switching-Frequency Modulation and Circulating Current Suppression for Modular Multilevel Converters. *IEEE Trans. Power Deliv.*, 26(3):2009–2017, July 2011.

[68] Gum Tae Son, Hee-Jin Lee, Tae Sik Nam, Yong-Ho Chung, Uk-Hwa Lee, Seung-Taek Baek, Kyeon Hur, and Jung-Wook Park. Design and Control of a Modular Multilevel HVDC Converter with Redundant Power Modules for Noninterruptible Energy Transfer. *IEEE Trans. Power Deliv.*, 27(3):1611–1619, 2012.

[69] Jun Mei, Bailu Xiao, Ke Shen, Leon M. Tolbert, and Jianyong Zheng. Modular Multilevel Inverter with New Modulation Method and Its Application to Photovoltaic Grid-Connected Generator. *IEEE Trans. Power Electron.*, 28(11):5063–5073, November 2013.

[70] Sixing Du, Jinjun Liu, and Teng Liu. Modulation and Closed-Loop-Based DC Capacitor Voltage Control for MMC with Fundamental Switching Frequency. *IEEE Trans. Power Electron.*, 30(1):327–338, January 2015.

[71] Zixin Li, Ping Wang, Haibin Zhu, Zunfang Chu, and Yaohua Li. An Improved Pulse Width Modulation Method for Chopper-Cell-Based Modular Multilevel Converters. *IEEE Trans. Power Electron.*, 27(8):3472–3481, August 2012.

[72] Georgios S. Konstantinou, Mihai Ciobotaru, and Vassilios G. Agelidis. Selective Harmonic Elimination Pulsewidth Modulation of Modular Multilevel Converters. *IET Power Electron.*, 6(1):96–107, January 2013.

[73] Fujin Deng and Zhe Chen. A Control Method for Voltage Balancing in Modular Multilevel Converters. *IEEE Trans. Power Electron.*, 29(1):66–76, January 2014.

[74] Qingrui Tu, Zheng Xu, Hongyang Huang, and Jing Zhang. Parameter Design Principle of the Arm Inductor in Modular Multilevel Converter Based HVDC. In *2010 Int. Conf. Power Syst. Technol.*, pages 1–6. IEEE, October 2010.

[75] Kalle Ilves, Staffan Norrga, Lennart Harnefors, and Hans-Peter Nee. Analysis of Arm Current Harmonics in Modular Multilevel Converters with Main-Circuit Filters. In *Int. Multi-Conference Syst. Signals Devices*, pages 1–6. IEEE, March 2012.

[76] Zixin Li, Ping Wang, Zunfang Chu, Haibin Zhu, Yongjie Luo, and Yaohua Li. An Inner Current Suppressing Method for Modular Multilevel Converters. *IEEE Trans. Power Electron.*, 28(11):4873–4879, November 2013.

[77] Ming Zhang, Long Huang, Wenxi Yao, and Zhengyu Lu. Circulating Harmonic Current Elimination of a CPS-PWM-Based Modular Multilevel Converter with a Plug-In Repetitive Controller. *IEEE Trans. Power Electron.*, 29(4):2083–2097, April 2014.

[78] Liqun He, Kai Zhang, Jian Xiong, and Shengfang Fan. A Repetitive Control Scheme for Harmonic Suppression of Circulating Current in Modular Multilevel Converters. *IEEE Trans. Power Electron.*, 30(1):471–481, January 2015.

[79] Lennart Harnefors, Antonios Antonopoulos, Staffan Norrga, Lennart Ängquist, and Hans-Peter Nee. Dynamic Analysis of Modular Multilevel Converters. *IEEE Trans. Ind. Electron.*, 60(7):2526–2537, July 2013.

[80] Ji-Woo Moon, Chun-Sung Kim, Jung-Woo Park, Dea-Wook Kang, and Jang-Mok Kim. Circulating Current Control in MMC under the Unbalanced Voltage. *IEEE Trans. Power Deliv.*, 28(3):1952–1959, 2013.

[81] Kalle Ilves, Antonios Antonopoulos, Staffan Norrga, and Hans-Peter Nee. Steady-State Analysis of Interaction between Harmonic Components of Arm and Line Quantities of Modular Multilevel Converters. *IEEE Trans. Power Electron.*, 27(1):57–68, January 2012.

[82] S. Sedghi, A. Dastfan, and A. Ahmadyfard. A New Multilevel Carrier Based Pulse Width Modulation Method for Modular Multilevel Inverter. In ICPE*'11 – 8th Int. Conf. Power Electron. – ECCE Asia*, pages 1432–1439. IEEE, May 2011.

[83] Mauricio Angulo, Pablo Lezana, Samir Kouro, José Rodríguez, and Bin Wu. Level-Shifted PWM for Cascaded Multilevel Inverters with Even Power Distribution. In *2007 IEEE Power Electron. Spec. Conf.*, pages 2373–2378. IEEE, 2007.

[84] Georgios S. Konstantinou and Vassilios G. Agelidis. Performance Evaluation of Half-Bridge Cascaded Multilevel Converters Operated with Multicarrier Sinusoidal PWM Techniques. In *2009 IEEE 4th Conf. Ind. Electron. Appl.*, pages 3399–3404. IEEE, May 2009.

[85] Ke Shen, Bailu Xiao, Jun Mei, Leon M. Tolbert, Jianze Wang, Xingguo Cai, and Yanchao Ji. A Modulation Reconfiguration Based Fault-Tolerant

Control Scheme for Modular Multilevel Converters. In *APEC'13 – Twenty-Eighth Annu. IEEE Appl. Power Electron. Conf. Expo.*, pages 3251–3255. IEEE, March 2013.

[86] Dharmavarapu Sreenivasarao, Pramod Agarwal, and Biswarup Das. Performance Evaluation of Carrier Rotation Strategy in Level-Shifted Pulse-Width Modulation Technique. *IET Power Electron.*, 7(3):667–680, March 2014.

[87] Emmanuel K. Amankwah, Jon C. Clare, Patrick W. Wheeler, and Alan J. Watson. Multi Carrier PWM of the Modular Multilevel VSC for Medium Voltage Applications. In *APEC'12 – Twenty-Seventh Annu. IEEE Appl. Power Electron. Conf. Expo.*, pages 2398–2406. IEEE, February 2012.

[88] Georgios S. Konstantinou, Mihai Ciobotaru, and Vassilios G. Agelidis. Analysis of Multi-Carrier PWM Methods for Back-to-Back HVDC Systems Based on Modular Multilevel Converters. In *IECON'11 – 37th Annu. Conf. IEEE Ind. Electron. Soc.*, pages 4391–4396. IEEE, November 2011.

[89] Italo R. F. M. P. Da Silva, Alexandre C. Oliveira, and Cursino B. Jacobina. Single-Phase AC-AC Double-Star Chopper-Cells (DSCC) Converter without Common DC-Link Capacitor. In *2014 IEEE Energy Convers. Congr. Expo.*, pages 548–555, Pittsburgh, PA, IEEE, 2014.

[90] Xiaojie Shi, Zhiqiang Wang, Leon M. Tolbert, and Fred Wang. A Comparison of Phase Disposition and Phase Shift PWM Strategies for Modular Multilevel Converters. In *2013 IEEE Energy Convers. Congr. Expo.*, pages 4089–4096, September 2013.

[91] Roozbeh Naderi and Abdolreza Rahmati. Phase-Shifted Carrier PWM Technique for General Cascaded Inverters. *IEEE Trans. Power Electron.*, 23(3):1257–1269, 2008.

[92] José Rodríguez, Luis Moran, Pablo Correa, and Cesar Silva. A Vector Control Technique for Medium-Voltage Multilevel Inverters. *IEEE Trans. Ind. Electron.*, 49(4):882–888, August 2002.

[93] F. J. Wu, K. Zhao, and L. Sun. Simplified Multilevel Space Vector Pulse-Width Modulation Scheme Based on Two-Level Space Vector Pulse-Width Modulation. *IET Power Electron.*, 5(5):609, 2012.

[94] Irfan Ahmed and Vijay B. Borghate. Simplified Space Vector Modulation Technique for Seven-Level Cascaded H-Bridge Inverter. *IET Power Electron.*, 7(3):604–613, March 2014.

[95] Leon M. Tolbert and John N. Chiasson. Active Harmonic Elimination for Multilevel Converters. *IEEE Trans. Power Electron.*, 21(2):459–469, March 2006.

[96] Prafullachandra M. Meshram and Vijay B. Borghate. A Simplified Nearest Level Control (NLC) Voltage Balancing Method for Modular Multilevel Converter (MMC). *IEEE Trans. Power Electron.*, 30(1):450–462, January 2015.

[97] Xiaoqian Li, Qiang Song, Wenhua Liu, Hong Rao, Shukai Xu, and Licheng Li. Protection of Nonpermanent Faults on DC Overhead Lines in MMC-Based HVDC Systems. *IEEE Trans. Power Deliv.*, 28(1):483–490, January 2013.

[98] Georgios S. Konstantinou, Josep Pou, Salvador Ceballos, and Vassilios G. Agelidis. Active Redundant Sub-Module Configuration in Modular Multilevel Converters. *IEEE Trans. Power Deliv.*, 28(4):2333–2341, October 2013.

[99] Laxman Maharjan, Tsukasa Yamagishi, Hirofumi Akagi, and Jun Asakura. Fault-Tolerant Operation of a Battery-Energy-Storage System Based on a Multilevel Cascade PWM Converter with Star Configuration. *IEEE Trans. Power Electron.*, 25(9):2386–2396, September 2010.

[100] Keyan Shi, Feifei Shen, Dong Lv, Ping Lin, Min Chen, and Dehong Xu. A Novel Start-Up Scheme for Modular Multilevel Converter. In *2012 IEEE Energy Convers. Congr. Expo.*, pages 4180–4187. IEEE, September 2012.

[101] Jianzhong Xu, Chengyong Zhao, Baoshun Zhang, and Li Lu. New Precharge and Submodule Capacitor Voltage Balancing Topologies of Modular Multilevel Converter for VSC-HVDC Application. In *2011 IEEE Asia-Pacific Power Energy Eng. Conf.*, pages 1–4, IEEE, March 2011.

[102] Binbin Li, Dandan Xu, Yi Zhang, Rongfeng Yang, Gaolin Wang, Wei Wang, and Dianguo Xu. Closed-Loop Precharge Control of Modular Multilevel Converters during Start-Up Processes. *IEEE Trans. Power Electron.*, 30(2):524–531, February 2015.

[103] Italo R. F. M. P. Da Silva, Alexandre C. Oliveira, Cursino B. Jacobina, Camila S. Gehrke, and Victor F. M. B. Melo. Active Power Line Conditioner based on Modular Multilevel Cascade Converter – Double-Star Chopper-Cells. In *IECON'13 – 39th Annu. Conf. IEEE Ind. Electron. Soc.*, pages 8510–8515, Vienna, Austria, November 2013.

[104] Pedram Sotoodeh and Ruth D. Miller. Design and Implementation of an 11-Level Inverter with FACTS Capability for Distributed Energy Systems. *IEEE J. Emerg. Sel. Top. Power Electron.*, 2(1):87–96, March 2014.

[105] E. Villanueva, Pablo Correa, José Rodríguez, and M. Pacas. Control of a Single-Phase Cascaded H-Bridge Multilevel Inverter for Grid-Connected Photovoltaic Systems. *IEEE Trans. Ind. Electron.*, 56(11):4399–4406, November 2009.

[106] Makoto Hagiwara and Hirofumi Akagi. Experiment and Simulation of a Modular PushPull PWM Converter for a Battery Energy Storage System. *IEEE Trans. Ind. Appl.*, 50(2):1131–1140, March 2014.

Chapter 5
Matrix converters

*Mehdi Farasat**

5.1 Introduction

Voltage and/or current back-to-back converters are traditionally used to interface an ac source with an ac load. An energy storage element is used to couple the dc-link of the front-end ac–dc rectifier to the back-end dc–ac inverter. A matrix converter (MC), however enables ac–ac conversion without any intermediate energy storage element. Conventional MCs, known as direct matric converters (DMCs), are single-stage converters that connect an m-phase voltage source to an n-phase output load through an $m \times n$ array of bidirectional switches. On the other hand, an indirect matrix converter (IMC) requires separate stages for the voltage and current conversion. In this chapter, the most popular MC topologies along with their control and modulation strategies are presented. A brief discussion on the technological and practical issues facing MCs, and a comparative assessment of their performance with the voltage back-to-back converters is given.

5.2 Direct matrix converter

5.2.1 Circuit topology

Figure 5.1 depicts the circuit topology of a three-phase to three-phase DMC. The circuit consists of nine bidirectional switches which enable direct connection of a three-phase input source to a three-phase load. The LC filter is employed to compensate the input current harmonics, so that sinusoidal currents are drawn from the source. Furthermore, the LC filter mitigates the overvoltages caused by the input current commutation.

5.2.2 Modulation techniques

The modulation techniques can be categorized into scalar and pulse width modulation (PWM) techniques. The PWM techniques in turn are divided into

*Louisiana State University, LA, USA.

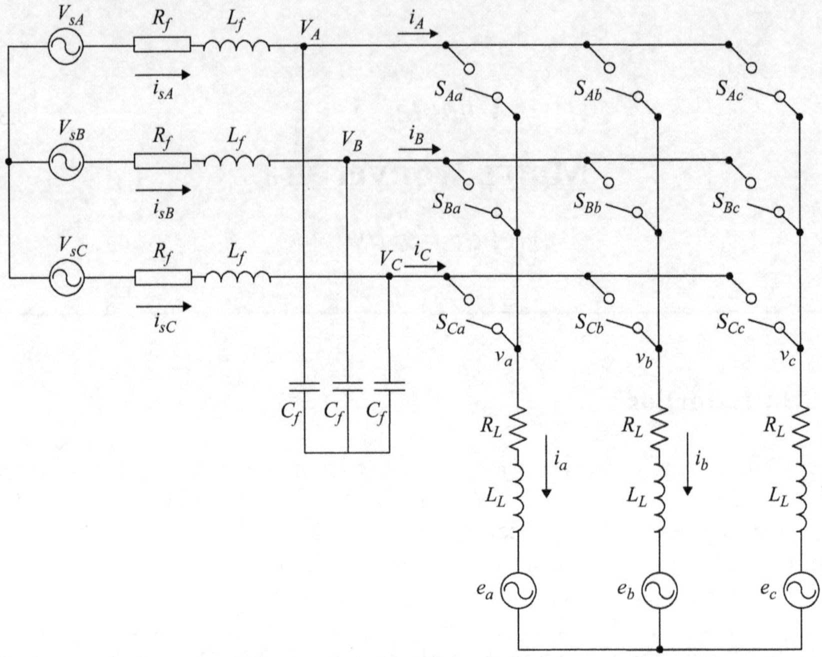

Figure 5.1 Three-phase direct matrix converter topology

carrier-based PWM and space vector modulation (SVM) methods. The detailed description of the scalar and SVM techniques is presented in the following subsections. Recently, predictive control technique has been applied to the MCs. This technique is discussed, as well.

5.2.2.1 Scalar technique

In this method, a desired set of three-phase output voltages are synthesized from the three-phase input voltages by sequential piecewise sampling [1]

$$\begin{bmatrix} \overline{v}_a \\ \overline{v}_b \\ \overline{v}_c \end{bmatrix} = \frac{1}{T_s} \begin{bmatrix} t_{Aa} & t_{Ba} & t_{Ca} \\ t_{Ab} & t_{Bb} & t_{Cb} \\ t_{Ca} & t_{Cb} & t_{Cc} \end{bmatrix} \begin{bmatrix} V_A \\ V_B \\ V_C \end{bmatrix} \tag{5.1}$$

where $\overline{v}_j; j = \{a, b, c\}$ is the average value of the output phase voltage, T_s is the sampling period, and t_{ij}; $i \in \{A, B, C\}; j \in \{a, b, c\}$ is the time during which switch S_{ij} is on (see Figure 5.1). The following duty ratios can be defined assuming $T_s = t_{Aj} + t_{Bj} + t_{Cj}$

$$m_{Aj}(t) = \frac{t_{Aj}}{T_s}; \quad m_{Bj}(t) = \frac{t_{Bj}}{T_s}; \quad m_{Cj}(t) = \frac{t_{Cj}}{T_s} \tag{5.2}$$

Therefore, (5.1) can be rewritten as

$$\underbrace{\begin{bmatrix} \bar{v}_a \\ \bar{v}_b \\ \bar{v}_c \end{bmatrix}}_{\bar{\boldsymbol{v}}_{\text{out}}(t)} = \underbrace{\begin{bmatrix} m_{Aa}(t) & m_{Ba}(t) & m_{Ca}(t) \\ m_{Ab}(t) & m_{Bb}(t) & m_{Cb}(t) \\ m_{Ca}(t) & m_{Cb}(t) & m_{Cc}(t) \end{bmatrix}}_{\boldsymbol{M}(t)} \underbrace{\begin{bmatrix} V_A \\ V_B \\ V_C \end{bmatrix}}_{\boldsymbol{v}_{\text{in}}(t)} \tag{5.3}$$

where $\bar{\boldsymbol{v}}_{\text{out}}(t)$ represents the average output voltage vector, $\boldsymbol{v}_{\text{in}}(t)$ is the input voltage vector, and $\boldsymbol{M}(t)$ is the transfer matrix.

Following the same procedure, the equation that describes the average input current vector, $\bar{\boldsymbol{i}}_{\text{in}}(t)$, in terms of the transfer matrix and output current vector, $i_{\text{out}}(t)$, can be derived as follows

$$\bar{\boldsymbol{i}}_{\text{in}}(t) = \boldsymbol{M}^{\text{T}}(t)\boldsymbol{i}_{\text{out}}(t) \tag{5.4}$$

Here, $\boldsymbol{M}^{\text{T}}(t)$ is the transpose of the transfer matrix.

Consider the three-phase input voltages and the output currents as follows

$$\boldsymbol{v}_{\text{in}}(t) = V_{\text{in}} \begin{bmatrix} \cos(\omega_{\text{in}}t) \\ \cos\left(\omega_{\text{in}}t - \dfrac{2\pi}{3}\right) \\ \cos\left(\omega_{\text{in}}t - \dfrac{4\pi}{3}\right) \end{bmatrix} \tag{5.5}$$

$$\boldsymbol{i}_{\text{out}}(t) = I_{\text{out}} \begin{bmatrix} \cos(\omega_{\text{out}}t + \phi_{\text{out}}) \\ \cos\left(\omega_{\text{out}}t + \phi_{\text{out}} - \dfrac{2\pi}{3}\right) \\ \cos\left(\omega_{\text{out}}t + \phi_{\text{out}} - \dfrac{4\pi}{3}\right) \end{bmatrix} \tag{5.6}$$

where ω_{in} and ω_{out} are the source and load frequencies, respectively, V_{in} and I_{out} are the input voltage and output current amplitudes, respectively, and ϕ_{out} is the load phase angle. The input currents can be written as

$$\boldsymbol{i}_{\text{in}}(t) = I_{\text{in}} \begin{bmatrix} \cos(\omega_{\text{in}}t + \phi_{\text{in}}) \\ \cos\left(\omega_{\text{in}}t + \phi_{\text{in}} - \dfrac{2\pi}{3}\right) \\ \cos\left(\omega_{\text{in}}t + \phi_{\text{in}} - \dfrac{4\pi}{3}\right) \end{bmatrix} \tag{5.7}$$

with I_{in} and ϕ_{in} denoting the input current amplitude and input displacement angle, respectively.

Assume that the output voltages are described by the following equation

$$\boldsymbol{v}_{\text{out}}(t) = qV_{\text{in}} \begin{bmatrix} \cos(\omega_{\text{out}}t) \\ \cos\left(\omega_{\text{out}}t - \dfrac{2\pi}{3}\right) \\ \cos\left(\omega_{\text{out}}t - \dfrac{4\pi}{3}\right) \end{bmatrix} \tag{5.8}$$

Here, q is the voltage gain of the MC.

Venturini proposed the transfer matrix that yields output voltages as (5.8) with the following elements [1]

$$m_{ij}(t) = \frac{1}{3}\left[1 + \frac{2v_i\bar{v}_j}{V_{\text{in}}^2}\right] \tag{5.9}$$

This solution is of little practical significance, since the voltage gain of the MC cannot exceed 50%. This is because the reference output voltage to be synthesized must remain within an envelope formed by the input voltages [2]. In order to enlarge the area within the input voltage envelope, common-mode voltages are added to the reference output voltages

$$\boldsymbol{v}_{\text{out}}(t) = qV_{\text{in}} \begin{bmatrix} \cos(\omega_{\text{out}}t) - \dfrac{1}{6}\cos(3\omega_{\text{out}}t) + \dfrac{1}{2\sqrt{3}}\cos(3\omega_{\text{in}}t) \\ \cos\left(\omega_{\text{out}}t - \dfrac{2\pi}{3}\right) - \dfrac{1}{6}\cos(3\omega_{\text{out}}t) + \dfrac{1}{2\sqrt{3}}\cos(3\omega_{\text{in}}t) \\ \cos\left(\omega_{\text{out}}t - \dfrac{4\pi}{3}\right) - \dfrac{1}{6}\cos(3\omega_{\text{out}}t) + \dfrac{1}{2\sqrt{3}}\cos(3\omega_{\text{in}}t) \end{bmatrix} \tag{5.10}$$

The transfer matrix that yields output voltages as (5.10) is given as

$$m_{ij}(t) = \frac{1}{3}\left[1 + \frac{2v_i\bar{v}_j}{V_{\text{in}}^2} + \frac{4q}{3\sqrt{3}}\sin(\omega_{\text{in}}t + \beta_i)\sin(3\omega_{\text{in}}t)\right] \tag{5.11}$$

where $\beta_i = 0, \frac{2\pi}{3}, \frac{4\pi}{3}$ for $i = \{A, B, C\}$ [1]. This method yields the maximum voltage gain of $\frac{\sqrt{3}}{2}$. Typical waveforms of the output voltage and current are shown in Figure 5.2.

Another scalar method proposed by Roy yields the same voltage gain of $\frac{\sqrt{3}}{2}$. In this technique, the modulation duty ratios are expressed as follows [2]

$$m_{ij}(t) = \frac{1}{3}\left[1 + \frac{2v_i\bar{v}_j}{V_{\text{in}}^2} + \frac{2}{3}\sin(\omega_{\text{in}}t + \beta_i)\sin(3\omega_{\text{in}}t)\right] \tag{5.12}$$

Comparing (5.11) and (5.12), it can be concluded that the only difference is that the term q is used in Venturini method and is fixed at its maximum in Roy's

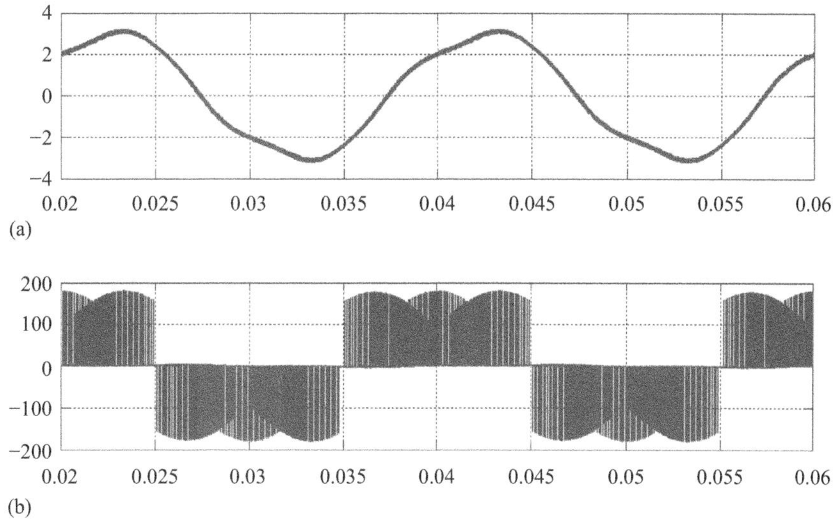

Figure 5.2 (a) Output phase voltage, (b) Output phase current of DMC with Venturini's method

method. The two methods result in the same output voltage quality except at low switching frequencies where the Venturini method is superior [2].

5.2.2.2 Space vector modulation

The MC is typically supplied from a voltage source. Therefore, the input terminals should not be short circuited. On the other hand, the output current should never be interrupted as the load is mostly inductive. Based on these considerations, there are 27 switching states available, given in Table 5.1. The available switching states produce three groups of output vectors, as follows:

Group I: Output voltage vectors with constant amplitude and rotating at the input frequency. These vectors are generated by connecting each output phase to a different input phase. Hence, there are six vectors in this group.

Group II: Output stationary vectors with variable amplitude. This group can be classified into three subsets with each subset consisting of six vectors which are generated by connecting two output phases to the same input phase.

Group III: Zero output vectors. This group consists of three vectors that are generated by simultaneous connection of the three output phases to the same input phase.

In the SVM, only the Groups II and III vectors are used. The SVM is based on the indirect transfer function approach which emulates the back-to-back voltage source rectification (VSR) and voltage source inversion (VSI), as shown in Figure 5.3 [3]. Here, V_{pn} represents a fictitious dc-link voltage. In the following subsections, VSR and VSI are reviewed, and the steps to combine those methods for control of the MC are explained.

Table 5.1 Switching states of the DMC

	No.	a	b	c	S_{Aa}	S_{Ab}	S_{Ac}	S_{Ba}	S_{Bb}	S_{Bc}	S_{Ca}	S_{Cb}	S_{Cc}	v_{ab}	v_{bc}	v_{ca}	i_A	i_B	i_C
Group III	1	A	A	A	1	0	0	1	0	0	1	0	0	0	0	0	0	0	0
	2	B	B	B	0	1	0	0	1	0	0	1	0	0	0	0	0	0	0
	3	C	C	C	0	0	1	0	0	1	0	0	1	0	0	0	0	0	0
Group I	4	A	C	C	1	0	0	0	0	1	0	0	1	$-V_{CA}$	0	V_{CA}	i_a	0	$-i_a$
	5	B	C	C	0	1	0	0	0	1	0	0	1	V_{BC}	0	$-V_{BC}$	0	i_a	$-i_a$
	6	B	A	A	0	1	0	1	0	0	1	0	0	$-V_{AB}$	0	V_{AB}	$-i_a$	i_a	0
	7	C	A	A	0	0	1	1	0	0	1	0	0	V_{CA}	0	$-V_{CA}$	$-i_a$	0	i_a
	8	C	B	B	0	0	1	0	1	0	0	1	0	$-V_{BC}$	0	V_{BC}	0	$-i_a$	i_a
	9	A	B	B	1	0	0	0	1	0	0	1	0	V_{AB}	0	$-V_{AB}$	i_a	$-i_a$	0
Group II	10	C	A	C	0	0	1	1	0	0	0	0	1	V_{CA}	$-V_{CA}$	0	i_b	0	$-i_b$
	11	C	B	C	0	0	1	0	1	0	0	0	1	$-V_{BC}$	V_{BC}	0	0	i_b	$-i_b$
	12	A	B	A	1	0	0	0	1	0	0	0	1	V_{AB}	$-V_{AB}$	0	$-i_b$	i_b	0
	13	A	C	A	1	0	0	0	0	1	1	0	0	$-V_{CA}$	V_{CA}	0	$-i_b$	0	i_b
	14	B	C	B	0	1	0	0	0	1	0	1	0	V_{BC}	$-V_{BC}$	0	0	$-i_b$	i_b
	15	B	A	B	0	1	0	1	0	0	0	1	0	$-V_{AB}$	V_{AB}	0	i_b	$-i_b$	0
	16	C	C	A	0	0	1	0	0	1	1	0	0	0	V_{CA}	$-V_{CA}$	i_c	0	$-i_c$
	17	C	C	B	0	0	1	0	0	1	0	1	0	0	$-V_{BC}$	V_{BC}	0	i_c	$-i_c$
	18	A	A	B	1	0	0	1	0	0	0	1	0	0	V_{AB}	$-V_{AB}$	$-i_c$	i_c	0
	19	A	A	C	1	0	0	1	0	0	0	0	1	0	$-V_{CA}$	V_{CA}	$-i_c$	0	i_c
	20	B	B	C	0	1	0	0	1	0	0	0	1	0	V_{BC}	$-V_{BC}$	0	$-i_c$	i_c
	21	B	B	A	0	1	0	0	1	0	1	0	0	0	$-V_{AB}$	V_{AB}	i_c	$-i_c$	0
	22	A	B	C	1	0	0	0	1	0	0	0	1	V_{AB}	V_{BC}	V_{CA}	i_a	i_b	i_c
	23	A	C	B	1	0	0	0	0	1	0	1	0	$-V_{CA}$	$-V_{BC}$	$-V_{AB}$	i_a	i_c	i_b
	24	B	A	C	0	1	0	1	0	0	0	0	1	$-V_{AB}$	$-V_{CA}$	$-V_{BC}$	i_b	i_a	i_c
	25	B	C	A	0	1	0	0	0	1	1	0	0	V_{BC}	V_{CA}	V_{AB}	i_c	i_a	i_b
	26	C	A	B	0	0	1	1	0	0	0	1	0	V_{CA}	V_{AB}	V_{BC}	i_b	i_c	i_a
	27	C	B	A	0	0	1	0	0	1	1	0	0	$-V_{BC}$	$-V_{AB}$	$-V_{CA}$	i_c	i_b	i_a

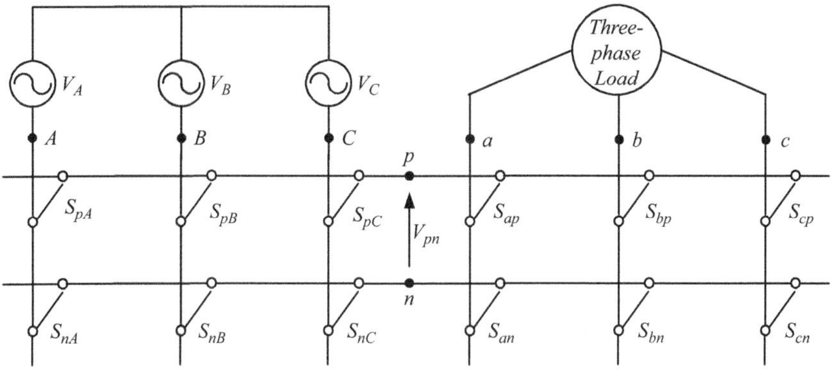

Figure 5.3 Emulation of VSR–VSI conversion

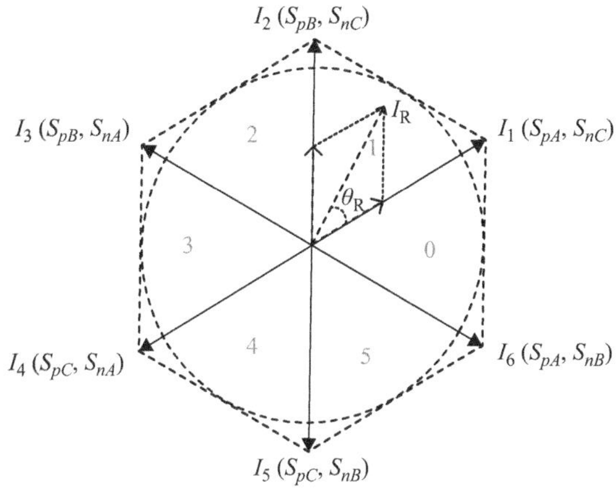

Figure 5.4 Diagram of input current space vectors

5.2.2.2.1 Voltage source rectification

Based on the available switching states, there are six active and two zero current vectors, as shown in Figure 5.4. Location of the reference vector within one of the 60° sectors defines the framing vectors I_λ and I_δ. For example, in Figure 5.4, the current reference angle is in the first sector.

The reference current vector is synthesized from the adjacent vectors and a zero vector. Denoting the local angle of the reference current by θ_R, the duty ratios of the active, d_λ, d_δ, and zero, d_{0R}, vectors are calculated as

$$d_\lambda = m_c \sin\left(\frac{\pi}{3} - \theta_R\right) \tag{5.13}$$

$$d_\delta = m_c \sin(\theta_R) \tag{5.14}$$

$$d_{0R} = 1 - (d_\lambda + d_\delta) \tag{5.15}$$

where $0 \leq m_c \leq 1$ is the current modulation index [4].

5.2.2.2.2 Voltage source inversion

Similar to VSR, the reference voltage vector is synthesized from six active and two zero voltage vectors, as shown in Figure 5.5. Location of the reference vector within one of the 60° sectors defines the framing vectors V_α and V_β. For example, in Figure 5.5, the reference vector of output voltage is in the zero sector.

Denoting the local angle of the reference voltage by θ_i, the duty ratios of the active, d_α, d_β, and zero, d_{0i}, vectors are calculated as

$$d_\alpha = m_{\rm v} \sin\left(\frac{\pi}{3} - \theta_i\right) \tag{5.16}$$

$$d_\beta = m_{\rm v} \sin(\theta_i) \tag{5.17}$$

$$d_{0i} = 1 - \left(d_\alpha + d_\beta\right) \tag{5.18}$$

where $0 \leq m_v \leq 1$ is the voltage modulation index [4].

5.2.2.2.3 Combined VSR and VSI

The time-averaged power outputs from the rectifying stage and the inversion stage are equal. This allows combining the two modulation strategies. Assuming the input displacement angle of zero, the combined duty ratios are given by

$$d_{\lambda\alpha} = d_\lambda * d_\alpha = m_{\rm c} m_{\rm v} \sin\left(\frac{\pi}{3} - \theta_{\rm R}\right) \sin\left(\frac{\pi}{3} - \theta_i\right) = T_{\lambda\alpha}/T_{\rm s} \tag{5.19}$$

$$d_{\lambda\beta} = d_\lambda * d_\beta = m_{\rm c} m_{\rm v} \sin\left(\frac{\pi}{3} - \theta_{\rm R}\right) \sin(\theta_i) = T_{\lambda\beta}/T_{\rm s} \tag{5.20}$$

$$d_{\delta\alpha} = d_\delta * d_\alpha = m_{\rm c} m_{\rm v} \sin(\theta_{\rm R}) \sin\left(\frac{\pi}{3} - \theta_i\right) = T_{\delta\alpha}/T_{\rm s} \tag{5.21}$$

$$d_{\delta\beta} = d_\delta * d_\beta = m_{\rm c} m_{\rm v} \sin(\theta_{\rm R}) \sin(\theta_i) = T_{\delta\beta}/T_{\rm s} \tag{5.22}$$

$$d_0 = 1 - \left(d_{\lambda\alpha} + d_{\lambda\beta} + d_{\delta\alpha} + d_{\delta\beta}\right) = T_0/T_{\rm s} \tag{5.23}$$

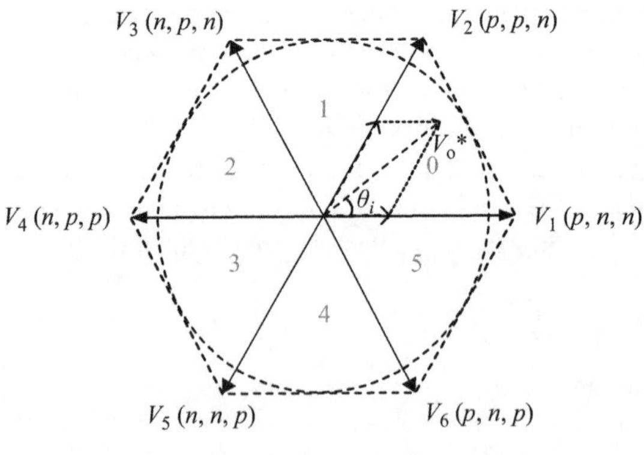

Figure 5.5 Diagram of output voltage space vectors

Final step is to determine the switching sequence within one switching period. The selection of zero vectors in the sequence is important. Those zero vectors which require switches to change state only once during a switching period are normally selected. A commonly used sequence is $d_{\lambda\alpha}d_{\lambda\beta}d_{\delta\beta}d_{\delta\alpha}d_0$.

The relationship between the voltage modulation index and fundamental component of the output voltage, $\hat{V}_{\text{out}}$, can be expressed as follows

$$m_{\text{v}} = \frac{\hat{V}_{\text{out}}}{V_{pn}} \tag{5.24}$$

Neglecting the losses in the rectification stage, the input power to the MC, P_{in}, equals the fictitious dc-link power, P_{dc}

$$P_{\text{in}} = P_{\text{dc}} \Rightarrow \frac{\sqrt{3}}{2}\hat{V}_{\text{in}}\hat{I}_{\text{in}} \cos\phi_{\text{in}} = V_{pn}I_p \tag{5.25}$$

where $\hat{V}_{\text{in}}$ and $\hat{I}_{\text{in}}$ represent the fundamental input line voltage and current components, respectively, and I_p is the dc-link current. The current modulation index relates the fundamental input current to the dc-link current as follows

$$m_{\text{c}} = \frac{\hat{I}_{\text{in}}}{I_p} \tag{5.26}$$

Using (5.25) and (5.26), it can be written

$$V_{pn} = \frac{\sqrt{3}}{2}\hat{V}_{\text{in}}m_{\text{c}} \cos\phi_{\text{in}} \tag{5.27}$$

Substituting (5.27) into (5.24), the following equation can be derived to express the relation between the input and the average output voltages of the MC

$$\hat{V}_{\text{out}} = \frac{\sqrt{3}}{2}m_{\text{v}}m_{\text{c}}\hat{V}_{\text{in}} \cos\phi_{\text{in}} \tag{5.28}$$

Assuming a unity input power factor, i.e. $\cos\phi_{\text{in}} = 1$, the maximum voltage ratio at maximum voltage and current modulation is $\sqrt{3}/2$ [4]. Figure 5.6 depicts the typical output voltage and current waveforms.

5.2.2.3 Predictive control technique

The main characteristics of predictive control are the use of a model of the system for predicting the future behavior of the controlled variables. This information is used by the controller to obtain the optimal actuation, here the switching states of the converter, to minimize a cost function which describes the desired behavior of the system [5]. The most desired objectives that must be achieved are control of the input power factor and the load current.

The model of the MC for predictive current control considers (5.3) and (5.4), which relate the input and output currents and voltages. However, the load and input filter models should be derived, as well.

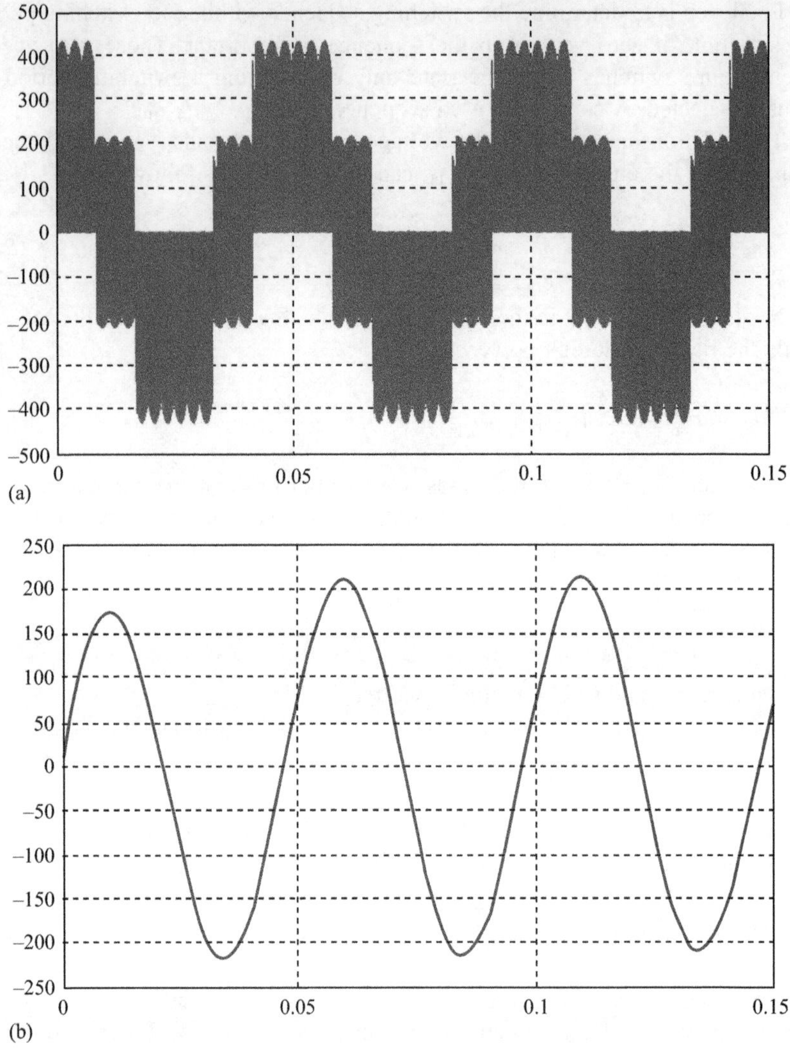

Figure 5.6 (a) Output phase voltage, (b) Output phase current of DMC with SVM

The load model gives the value of the load current in the next sampling time for each of the 27 switching states of the MC. The equation for the *R–L–E* load in Figure 5.1 is

$$L_{\mathrm{L}} \frac{\mathrm{d}\boldsymbol{i}_{\mathrm{out}}(t)}{\mathrm{d}t} = \boldsymbol{v}_{\mathrm{out}}(t) - R_{\mathrm{L}}\boldsymbol{i}_{\mathrm{out}}(t) - \boldsymbol{e}(t) \tag{5.29}$$

Discretizing the above equation yields the value of the load current in the next sampling time, $\boldsymbol{i}_{\text{out}}(k+1)$, as follows

$$\boldsymbol{i}^{p}_{\text{out}}(k+1) = \left(1 - \frac{R_{\text{L}}T_{\text{sam}}}{L_{\text{L}}}\right)\boldsymbol{i}_{\text{out}}(k) + \frac{T_{\text{sam}}}{L_{\text{L}}}(\boldsymbol{v}_{\text{out}}(k) - \hat{e}(k)) \tag{5.30}$$

where T_{sam} represents the sampling time.

The filter model can be described by the following equation in the matrix form (see Figure 5.1)

$$\underbrace{\begin{bmatrix} \dot{\boldsymbol{v}}_{\text{in}}(t) \\ \dot{\boldsymbol{i}}_{\text{s}}(t) \end{bmatrix}}_{\dot{x}(t)} = \underbrace{\begin{bmatrix} 0 & 1/C_{\text{f}} \\ -1/L_{\text{f}} & -R_{\text{f}}/L_{\text{f}} \end{bmatrix}}_{A} \underbrace{\begin{bmatrix} \boldsymbol{v}_{\text{in}}(t) \\ \boldsymbol{i}_{\text{s}}(t) \end{bmatrix}}_{x(t)} + \underbrace{\begin{bmatrix} 0 & -1/C_{\text{f}} \\ 1/L_{\text{f}} & 0 \end{bmatrix}}_{B} \underbrace{\begin{bmatrix} \boldsymbol{v}_{\text{s}}(t) \\ \boldsymbol{i}_{\text{in}}(t) \end{bmatrix}}_{u(t)} \tag{5.31}$$

where

$$\boldsymbol{v}_{\text{s}}(t) = \frac{2}{3}\left(V_{\text{s}A} + e^{j120^\circ}V_{\text{s}B} + e^{j240^\circ}V_{\text{s}C}\right) \tag{5.32}$$

$$\boldsymbol{i}_{\text{s}}(t) = \frac{2}{3}\left(i_{\text{s}A} + e^{j120^\circ}i_{\text{s}B} + e^{j240^\circ}i_{\text{s}C}\right) \tag{5.33}$$

For the MC predictive control, the value of the source current in the next sampling time, $\boldsymbol{i}_{\mathbf{s}}(k+1)$, is required, which can be calculated as follows

$$\boldsymbol{i}_{\text{s}}(k+1) = B_d(2,1)\boldsymbol{v}_{\text{s}}(k) + A_d(2,1)\boldsymbol{v}_{\text{in}}(k) + A_d(2,2)\boldsymbol{i}_{\text{s}}(k) + B_d(2,2)\boldsymbol{i}_{\text{in}}(k) \tag{5.34}$$

where

$$A_d = e^{AT_{\text{sam}}} = L^{-1}\left\{(s\boldsymbol{I} - A)^{-1}\right\}|_{t=T_{\text{sam}}} \tag{5.35}$$

$$B_d = \int_{\tau=0}^{T_{\text{sam}}} e^{A(T_{\text{sam}}-\tau)}B\,d\tau = A^{-1}(A - \boldsymbol{I})B \tag{5.36}$$

with $\boldsymbol{I}$ represents the identity matrix.

The most commonly used cost function, g, improves the input power factor by minimizing the instantaneous reactive power, and controls the load current to follow the reference with acceptable accuracy

$$g = \left|i^{*}_{\text{out}_\alpha}(k+1) - i^{p}_{\text{out}_\alpha}(k+1)\right| + \left|i^{*}_{\text{out}_\beta}(k+1) - i^{p}_{\text{out}_\beta}(k+1)\right| + C\left|Q^{*} - Q^{p}(k+1)\right| \tag{5.37}$$

Here, $i^{*}_{\text{out}_\alpha}$ and $i^{*}_{\text{out}_\beta}$ are the $\alpha-\beta$ components of the reference load current, $i^{p}_{\text{out}_\alpha}$ and $i^{p}_{\text{out}_\beta}$ are the $\alpha-\beta$ components of the predicted load current calculated by (5.30), and C is the weighting factor. The values of the instantaneous reactive power along with the predicted reactive power, $Q^{p}(k+1)$, can be calculated by the following equations

$$Q = \left(v_{\text{s}\beta}i_{\text{s}\alpha} - v_{\text{s}\alpha}i_{\text{s}\beta}\right) \tag{5.38}$$

$$Q^p(k+1) = v_{s\beta}(k+1)i_{s\alpha}(k+1) - v_{s\alpha}(k+1)i_{s\beta}(k+1) \tag{5.39}$$

Subscripts α and β represent the $\alpha - \beta$ components of the corresponding vectors [2,5].

The weighting factor, C, is the only parameter that should be adjusted in the predictive control. However, there is no analytical method to determine the weighting factor for a system with certain operational constraints and specific structures of the cost function. Figure 5.7 depicts the block diagram of the predictive current control strategy with minimization of the input reactive power.

Figure 5.8 illustrates the behavior of the model predictive control scheme for the DMC when the cost function has a value of $C = 0$ for the weighting factor. The output current is sinusoidal. However, the input current is highly distorted. In order to improve the quality of the input current as well as the input power factor, the weighting factor is increased in Figure 5.9.

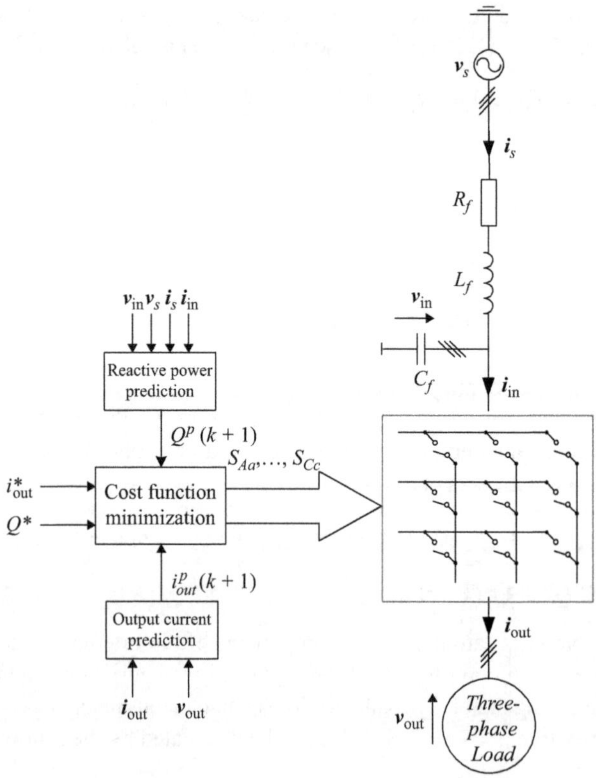

Figure 5.7 Block diagram of the predictive current and reactive power control

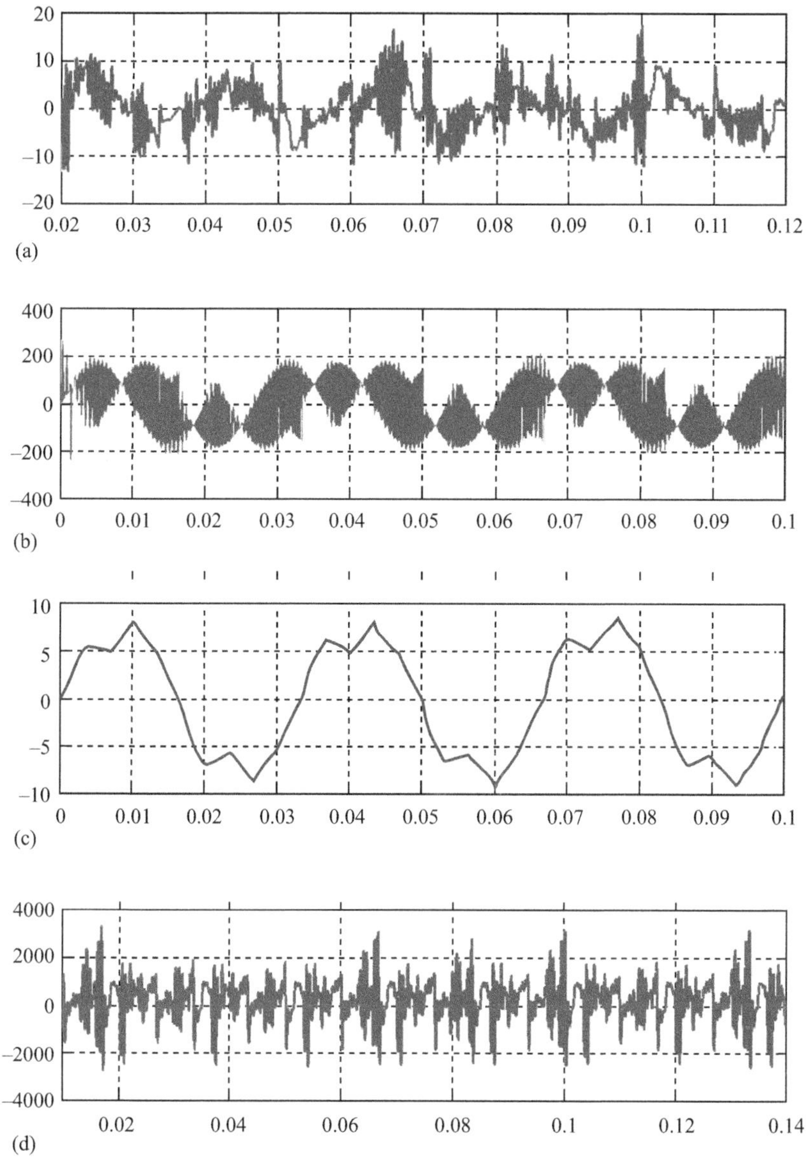

Figure 5.8 Model predictive control of DMC without reactive power control. (a) Input phase current, (b) Output phase voltage, (c) Output phase current, (d) Input reactive power

5.3 Indirect matrix converter

The IMC, shown in Figure 5.10, is the physical implementation of the VSR and VSI idea which is used to control the DMC. In this type of MC, a conventional voltage-source-type inverter is supplied from a current-source-type rectifier, which

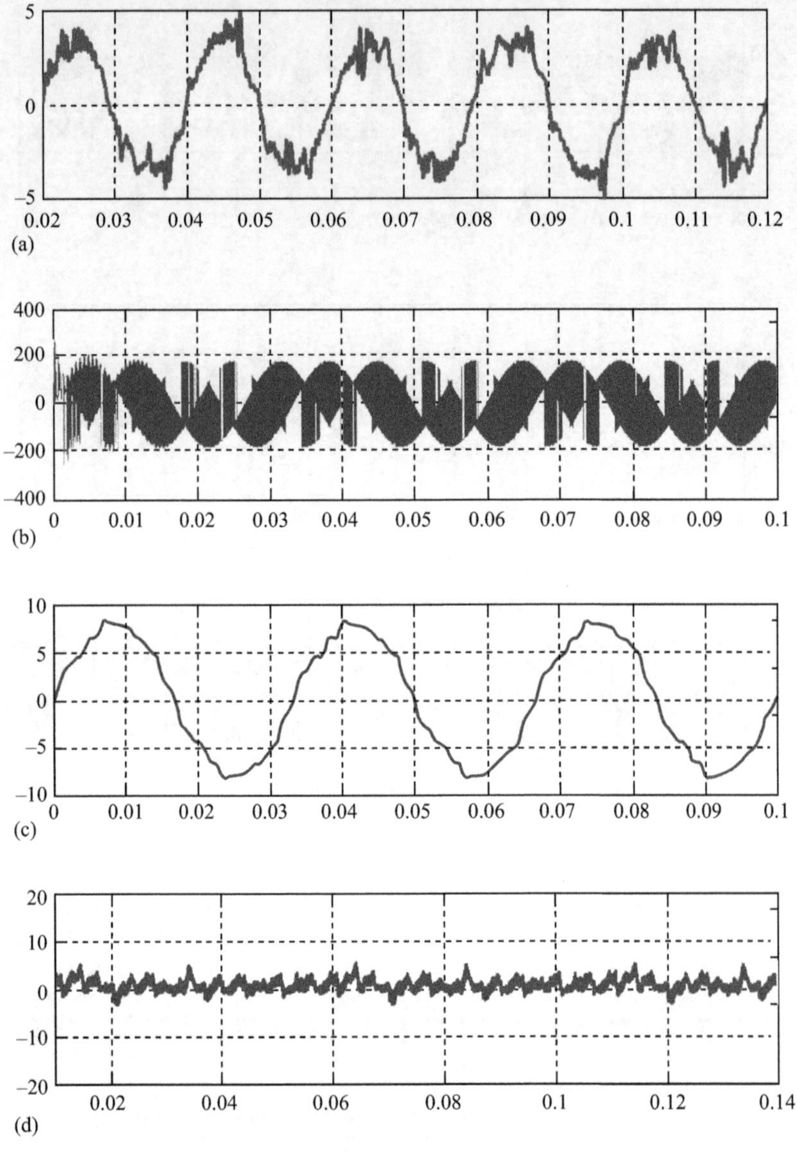

Figure 5.9 Model predictive control of DMC with reactive power control. (a) Input phase current, (b) Output phase voltage, (c) Output phase current, (d) Input reactive power

is able to operate with positive and negative dc currents, and provides bipolar dc-link voltage [6]. Similar to the DMC, the IMC employs 18 unipolar switches and 18 diodes. Sparse and ultra-sparse matrix converters (USMCs) are two types of IMCs with reduced number of switching devices. In the following subsections, the circuit topology and modulation techniques of these converters are explained.

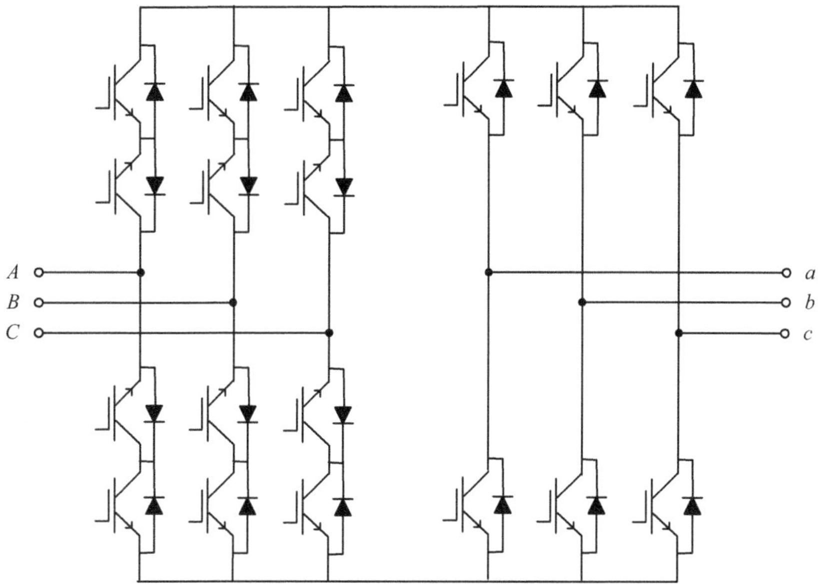

Figure 5.10 Circuit topology of the indirect matrix converter

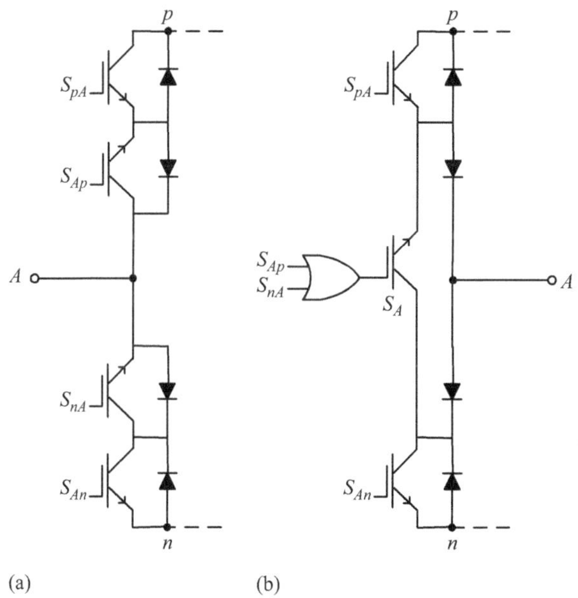

Figure 5.11 Structure of a bride leg of the input rectifier stage, (a) IMC and (b) SMC

5.3.1 Circuit topology

The sparse matrix converter (SMC) is derived from the fact that the voltage-source-type inverter does not require a bipolar dc-link voltage to operate. Figure 5.11(a) depicts one leg of the IMC. Assuming that S_{Ap} and S_{pA} are conducting, it is obvious

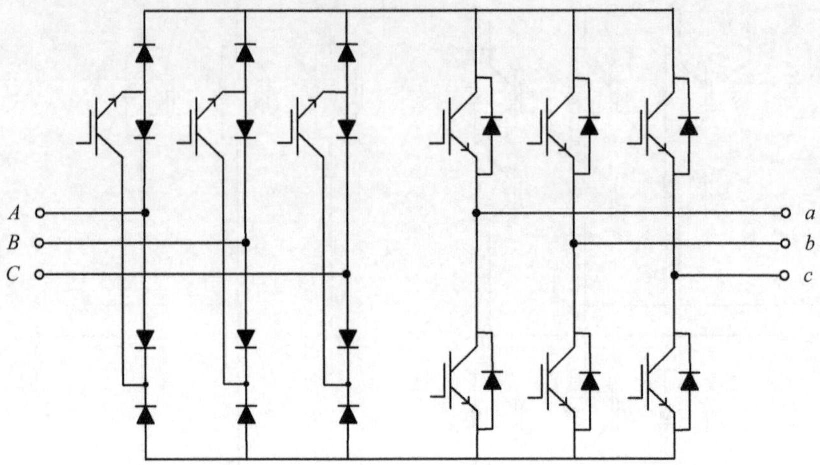

Figure 5.12 Circuit topology of the ultra-sparse matrix converter

that for a positive dc-link polarity, $V_{pn} > 0$, S_{An} is blocking voltage, while for a negative dc link, S_{nA} is blocking voltage. By restricting the operation to $V_{pn} > 0$, blocking of S_{nA} within the conduction interval of S_{Ap} is not necessary. Therefore, the two transistors can be combined to form a single transistor which connects *a* to *p* or *a* to *n*, as shown in Figure 5.11(b). The resulting leg topology provides bidirectional current flow for $V_{pn} > 0$. Therefore, the functionality of the SMC is realized by employing 15 switches and 18 diodes. It should be mentioned that despite the reduction in number of switches, the controllability and the operating range of the SMC are not restricted compared to the DMC [6,7].

Further reduction in the number of switches can be achieved by omitting S_{pi} and S_{in} switches in the rectifier stage. However, the operating range of the resulting topology is restricted to unidirectional power flow $\left(V_{pn} > 0, i > 0\right)$. Furthermore, the controllability of the phase displacement of input voltage and input current fundamental is limited to $\pm\pi/6$. The same restriction exists for the phase displacement of load current and load voltage fundamental. This topology, shown in Figure 5.12, consists of nine switches and 18 diodes and is known as USMC [6,7]. Due to significant reduction in the number of switches, this converter is of high practical interest.

5.3.2 Modulation techniques

5.3.2.1 Space vector modulation

The modulation concept explained in this section is applicable to both SMC and USMC. The rectifier stage should provide a maximum voltage for the inverter stage in order to achieve a maximum voltage gain. Therefore, in every $\pi/3$-wide interval, a phase voltage with the highest absolute value is clamped to the positive or negative dc-link bus. Consequently, the required operating condition $V_{pn} > 0$ of the SMC and USMC is inherently satisfied. The following discussions are limited to the $0 < \omega_{\text{in}}t < \pi/6$ interval where phase a remains clamped to the positive dc bus p assuming the following input voltages

$$\begin{aligned} V_A &= V_{\text{in}}\cos(\omega_{\text{in}}t) \\ V_B &= V_{\text{in}}\cos\left(\omega_{\text{in}}t - \frac{2\pi}{3}\right) \\ V_C &= V_{\text{in}}\cos\left(\omega_{\text{in}}t - \frac{4\pi}{3}\right) \end{aligned} \tag{5.40}$$

It is also assumed that the average value of the dc-link current, I, remains constant in each rectifier switching state. In this interval, the dc-link voltage, V_{pn}, is formed by segments of the input line-to-line voltages V_{AB} and V_{AC}. Therefore, the output voltage formed by the inverter shows two different levels within each pulse half period. In order to simplify the modulation scheme, the switching of the rectifier stage should occur during the freewheeling interval of the inverter and the output voltage zero vector should be generated by operating only the inverter stage with freewheeling intervals. At the rectifier input stage during $0 < \omega_{\text{in}}t < \pi/6$ interval, this can be achieved by

$$d_{ab} + d_{ac} = 1 \tag{5.41}$$

where d_{ab} and d_{ac} are the relative on-time of the switching states that will result in a dc-link voltage of V_{AB} and V_{AC}, respectively. The average input phase currents in this interval are given as follows

$$\bar{i}_a = (d_{ab} + d_{ac})I; \quad \bar{i}_b = -d_{ab}I; \quad \bar{i}_c = -d_{ac}I \tag{5.42}$$

In order to achieve unity input power factor, the average value of an input phase current during a pulse period must be proportional to the average value of the corresponding input phase voltage

$$d_{ab} = -\frac{\bar{i}_b}{\bar{i}_a} = -\frac{\bar{v}_b}{\bar{v}_a}; \quad d_{ac} = -\frac{\bar{i}_c}{\bar{i}_a} = -\frac{\bar{v}_c}{\bar{v}_a} \tag{5.43}$$

Equation (5.43) is derived assuming that $\bar{v}_a + \bar{v}_b + \bar{v}_c = 0$.

At the inverter output, a voltage space vector v^*_{out} of given amplitude $|v^*_{\text{out}}|$ and phase $\phi^*_{\text{out}} = \omega^*_{\text{out}}t$ has to be formed over half a pulse period, $T_{\text{p}}/2$. The following discussions are limited to the $0 < \phi_{\text{out}} < \pi/6$ interval. In this interval, the output voltage is synthesized by active voltage space vectors (1 0 0) and (1 1 0) and zero voltage space vectors (1 1 1) or (0 0 0). In the time intervals $T_{ac} = d_{ac}T_{\text{p}}/2$ and $T_{ab} = d_{ab}T_{\text{p}}/2$, the dc-link voltage is V_{AC} and V_{AB}, respectively. For calculating the on-times of the active switching states, the local average value of the dc-link voltage can be employed. The following equations express the on-times of the output voltage space vectors in the $0 < \phi_{\text{out}} < \pi/6$ interval

$$T_{100(ac)} = -\frac{2}{3\sqrt{3}}T_{\text{p}}\frac{V^*_{\text{out}}}{V^2_{\text{in}}}V_c\cos\left(\phi^*_{\text{out}} + \frac{\pi}{6}\right) \tag{5.44}$$

$$T_{100(ab)} = -\frac{2}{3\sqrt{3}} T_p \frac{V_{out}^*}{V_{in}^2} V_b \cos\left(\phi_{out}^* + \frac{\pi}{6}\right) \tag{5.45}$$

$$T_{110(ac)} = -\frac{2}{3\sqrt{3}} T_p \frac{V_{out}^*}{V_{in}^2} V_c \sin\phi_{out}^* \tag{5.46}$$

$$T_{110(ab)} = -\frac{2}{3\sqrt{3}} T_p \frac{V_{out}^*}{V_{in}^2} V_b \sin\phi_{out}^* \tag{5.47}$$

The voltage transfer ratio of the SMC and UMC is given by

$$M = \frac{V_{out}^*}{V_{in}} \leq \frac{\sqrt{3}}{2} \tag{5.48}$$

The maximum voltage transfer ratio is only available for unity input power factor, i.e. $\cos\phi_{in} = 1$ [6]. Figure 5.13 [6] depicts the formation of the dc-link voltage and dc-link current along with the switching states of the rectifier and inverter stages for $0 < \omega_{in}t < \pi/6$ and $0 < \omega_{out}t < \pi/6$. In this figure, S_a, S_b, and S_c are the switching states of the upper switches in each leg of the output inverter stage. Figure 5.14 shows the typical output voltage and current waveforms of an USMC.

5.3.2.2 Predictive control

The predictive control of SMC and USMC is conceptually very similar to that of the DMC. However, in addition to minimizing the line side current error and reactive power, the dc-link voltage of the SMC and USMC must be maintained positive for proper operation. Therefore, the cost function should be modified as follows

$$g = Q^{k+1} + C.\Delta i_1^{k+1} + h^{k+1} \tag{5.49}$$

where

$$\Delta i_1^{k+1} = \left|i_{out_\alpha}^*(k+1) - i_{out_\alpha}^p(k+1)\right| + \left|i_{out_\beta}^*(k+1) - i_{out_\beta}^p(k+1)\right|$$

$$Q^{k+1} = |Q^* - Q^p(k+1)|$$

$$h^{k+1} = \begin{cases} 0, & V_{pn}^p(k+1) > 0 \\ N, & V_{pn}^p(k+1) < 0 \end{cases}$$

Here, N is the maximum positive number that can be generated by the arithmetic unit of the controller [8].

The weighting factor C can be adjusted empirically such that the output current has no noticeable deviations from the reference and in the meantime, the input currents are not highly distorted.

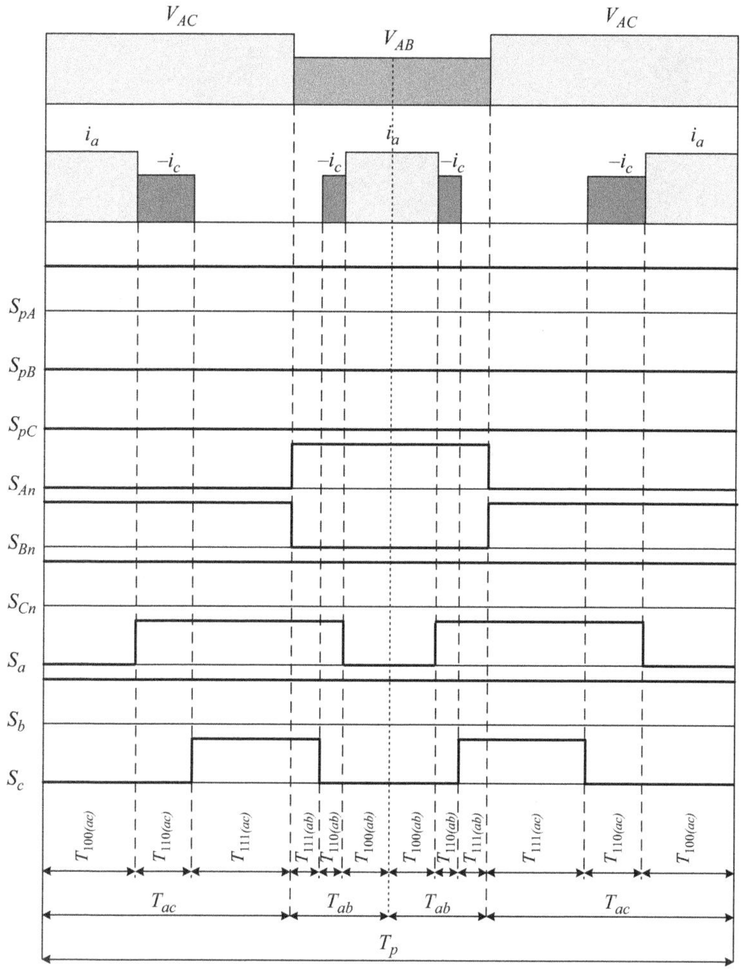

Figure 5.13 Top to bottom: dc-link voltage, dc-link current, switching states of the rectifier stage, and switching states of the inverter stage

5.4 Technological issues of MCs

The MC topology is based on the bidirectional semiconductor switches to both conduct current in each direction and block voltage in both directions. However, such a device is currently not commercially available. Therefore, insulated gate bipolar transistor (IGBT) devices and diodes are mostly used to create the power circuit. The reverse blocking IGBT has been reportedly used in the MC topology since anti-parallel diodes can be eliminated from the converter. Other arrangements include the diode bridge arrangement which uses only one active device, and the common emitter and common collector arrangements which use two active devices and two diodes, as shown in Figure 5.15 [9].

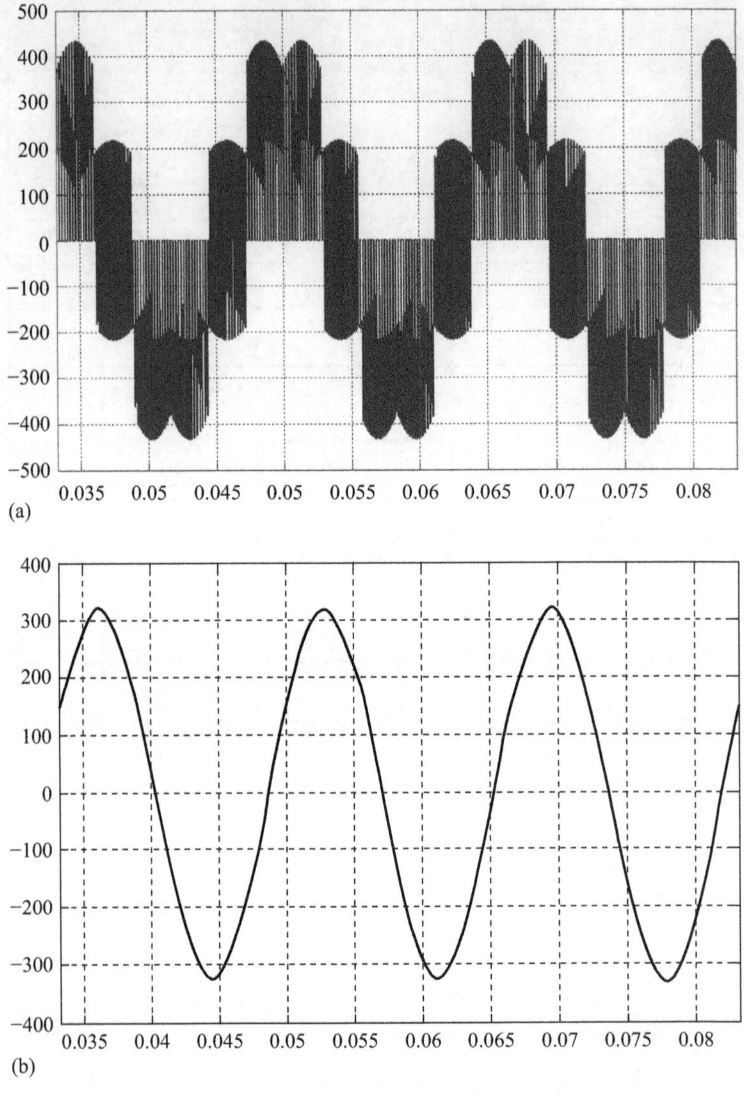

Figure 5.14 (a) Output phase voltage, (b) Output phase Current of USMC with SVM

Similar to any converter, a simple LC filter is often used at the input of the MC in order to reduce the switching harmonics of the input current. However, the LC circuit creates overvoltages during the transient operation, especially the power-up procedure of the MC. Some passive solutions such as connecting damping resistors in parallel with the input inductors have been proposed to reduce overvoltages. Overvoltages can also occur due to an open circuit of the load as there are no freewheeling paths within the MC circuit for the inductive load [9]. Several

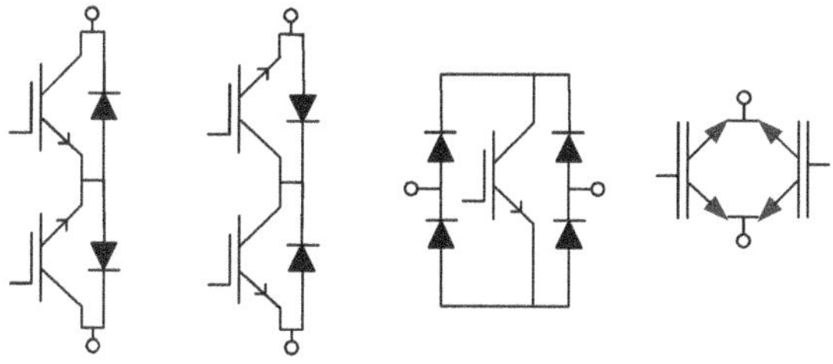

Figure 5.15 Possible bidirectional switch arrangements. (a) Common emitter, (b) common collector, (c) diode bridge, and (d) reverse blocking IGBT

schemes, such as passive and active clamp circuits, have been proposed and successfully applied to protect the MCs against these overvoltages.

5.5 MC versus voltage back-to-back converter

Within a certain switching frequency range, the DMC and IMC enable higher maximum power density and maximum power-to-mass ratio at a higher efficiency compared to the voltage back-to-back converter. The back-to-back converter requires a significantly smaller semiconductor chip area compared to the DMC and IMC for switching frequencies below a specific switching frequency, whereas above that frequency, the back-to-back converter requires the largest chip area, followed by the DMC and IMC.

Alas, those advantages are outweighed by the low-voltage transfer ratio, constrained input reactive power compensation, dependency of control of the input currents on the output currents, and restriction on single-phase operation of MCs. Consequently, the MC is not suitable for applications where a bidirectional, low-voltage, and low-power ac–ac converter system is required. The voltage back-to-back converter clearly is the preferred choice for such requirements [10].

5.6 Summary

Due to intensive research on MCs during the past two decades, these converters are finding wide variety of applications in the power converter industry, as to date, two drive manufacturers, i.e. Yaskawa and Fuji Electric Systems, offer MC products. Features such as lack of large and fragile dc-link components make these converters suitable for deep-sea remotely operated vehicle applications where extreme undersea pressure is present. Furthermore, due to the higher power density and the elevated temperature capability, MCs are employed in the aircraft industry. The direct and indirect matrix converters have also been applied to doubly fed induction generator based wind energy systems with the advantage of controlling a high-power generator with a relatively low power, four-quadrant power converter. MCs are finding applications in the power supply generation area, as well, where a fixed voltage and frequency power supplies are implemented from variable frequency diesel generators.

References

[1] P. W. Wheeler, J. Rodriguez, J. C. Clare, L. Empringham, and A. Weinstein, "Matrix converters: a technology review," *IEEE Trans. Ind. Electron.*, vol. 49, no. 2, pp. 276–287, Apr. 2002.

[2] J. Rodriguez, M. Rivera, J. W. Kolar, and P. W. Wheeler, "A review of control and modulation methods for matrix converters," *IEEE Trans. Ind. Electron.*, vol. 59, no. 1, pp. 58–70, Jan. 2012.

[3] L. Huber, and D. Borojevic, "Space vector modulated three-phase to three-phase matrix converter with input power factor correction," *IEEE Trans. Ind. Appl.*, vol. 31, no. 6, pp. 1234–1246, Nov./Dec. 1995.

[4] E. Karaman, M. Farasat, and A. M. Trzynadlowski, "A comparative study of series and cascaded *Z*-source matrix converters," *IEEE Trans. Ind. Electron.*, vol. 61, no. 10, pp. 5164–5173, Oct. 2014.

[5] J. Rodriguez, and P. Cortes, "Predictive Control of Power Converters and Electric Drives," John Wiley, Chichester, 2012.

[6] J. W. Kolar, F. Schafmeister, S. D. Round, and H. Ertl, "Novel three-phase AC–AC sparse matrix converters," *IEEE Trans. Power Electron.*, vol. 22, no. 5, pp. 1649–1661, Sep. 2007.

[7] J. W. Kolar, T. Friedli, J. Rodriguez, and P. W. Wheeler, "Review of three-phase PWM AC–AC converter topologies," *IEEE Trans. Ind. Electron.*, vol. 58, no. 11, pp. 4988–5006, Nov. 2011.

[8] P. Correa, J. Rodriguez, M. Rivera, J. R. Espinoza, and J. W. Kolar, "Predictive control of an indirect matrix converter," *IEEE Trans. Ind. Electron.*, vol. 56, no. 6, pp. 1847–1853, Jun. 2009.

[9] L. Empringham, J. W. Kolar, J. Rodriguez, P. W. Wheeler, and J. C. Clare, "Technological issues and industrial application of matrix converters: a review," *IEEE Trans. Ind. Electron.*, vol. 60, no. 10, pp. 4260–4271, Oct. 2013.

[10] T. Friedli, J. W. Kolar, J. Rodriguez, and P. W. Wheeler, "Comparative evaluation of three-phase AC–AC matrix converter and voltage DC-link back-to-back converter systems," *IEEE Trans. Ind. Electron.*, vol. 59, no. 12, pp. 4487–4510, Dec. 2012.

Chapter 6
Soft-switching converters

*Mahshid Amirabadi**

Switch-mode power converters are used in a wide variety of applications. In most applications, it is desirable to design high-frequency switching power converters to increase their power density. However, the switching losses are increased by increasing the frequency. Moreover, large dv/dt and di/dt produce electromagnetic interference.

To minimize the problems associated with the high-switching frequencies of power converters, several soft-switching techniques have been developed. In this chapter, these techniques will be reviewed.

6.1 Resonant converters

As mentioned earlier, power density of the converters is limited by their switching losses. Resonant converters were among the early solutions that were proposed for addressing this limitation. In resonant converters, as shown in Figure 6.1, a resonant tank network interfaces two switch networks. In case of a dc–dc converter, the switch networks are an inverter and a rectifier. In this converter, switches can benefit from zero-current turn-off or zero-voltage turn-on; therefore, switching losses are minimized, and the switching frequency may be increased to very high values. The resonant tank network may be formed by two or more resonating elements. In Section 6.1.1, the performance and specifications of the second-order resonant converters, in which the resonant tank network is formed by an inductor and a capacitor, will be studied. Section 6.1.2 studies resonant converters that use three or more resonating elements.

6.1.1 Second-order resonant converters

Resonant tank network in a second-order resonant converter is formed by a resonating inductor–capacitor pair. Figure 6.2 shows possible combinations for a resonant tank network that contain only one inductor and one capacitor [1]. Similar to other types of converters, resonant converters obey the following rules:

- Current sources, including inductors, cannot be placed in series.
- Voltage sources, including capacitors, cannot be placed in parallel.

*Northeastern University, Boston, MA, USA

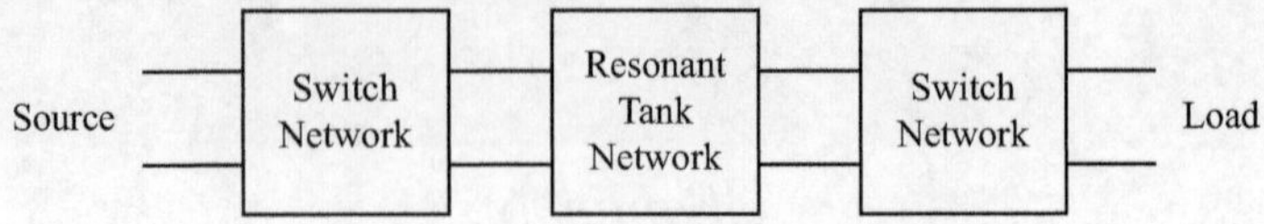

Figure 6.1 Resonant converters

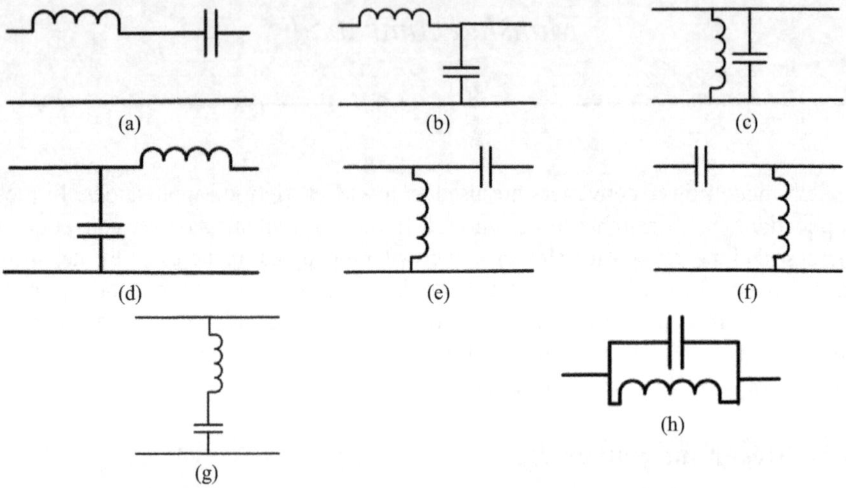

Figure 6.2 Resonant tank networks with two resonating elements (adapted from [1])

These two rules should be taken into consideration when developing resonant converters.

Among all the combinations that are illustrated in Figure 6.2, resonant tank networks shown in Figure 6.2(a) and (b) have attracted more attention. These resonant tanks, which are called series resonant tank and parallel load resonant tank, form the traditional series resonant converters (SRCs) and parallel resonant converters (PRCs).

The series resonant tank network can only interface a voltage source and a voltage sink; thus it can be implemented by a voltage source inverter and a current source rectifier. Similarly, the parallel resonant network interfaces a voltage source inverter and voltage source rectifier. Figure 6.3 illustrates the SRC and PRC using a full-bridge inverter and a diode rectifier.

6.1.1.1 Dc–dc SRCs

To study the performance of the SRC, its equivalent circuit, as depicted in Figure 6.4, will be used [2]. The output of the full-bridge inverter is a square-wave voltage. To be able to use sinusoidal analysis, fundamental component of this voltage is considered. The rectifier is assumed to be ideal; hence, the load voltage, V_o, and the load current, I_o, are pure dc. Since the rectifier is connected to a voltage

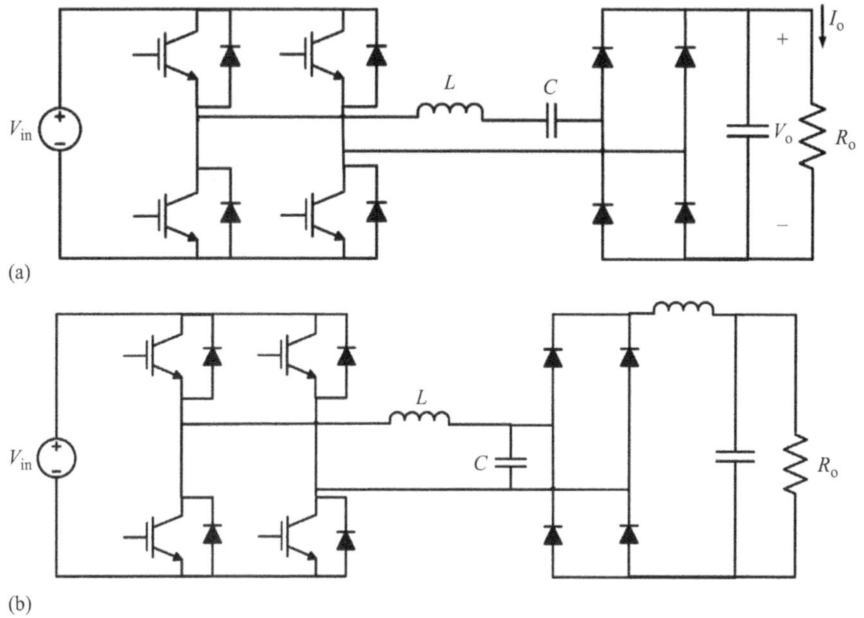

Figure 6.3 Dc–dc series and parallel resonant converters. (a) Series resonant converter (SRC) and (b) parallel resonant converter (PRC)

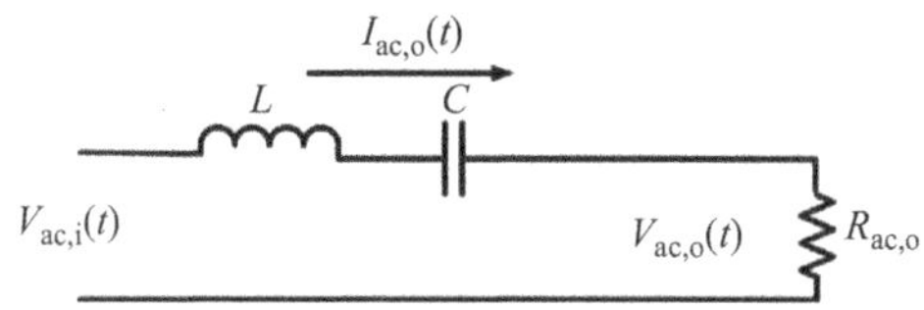

Figure 6.4 Equivalent circuit used for analyzing the SRC

sink with pure dc voltage, the voltage seen at the input of the rectifier is a square wave. The amplitude of the fundamental component of this voltage is:

$$V_{ac,o,peak} = \frac{4}{\pi} V_o \tag{6.1}$$

The current seen at the output of the rectifier is the rectified sinusoidal current passing through the resonant circuit formed by the series inductor L and Capacitor C. The average of the rectified current is equal to the load current (I_o). Therefore, the amplitude of the sinusoidal current is as follows:

$$I_{ac,o,peak} = \frac{\pi}{2} I_o \tag{6.2}$$

The equivalent load resistance, $R_{ac,o}$, can be determined as follows [2]:

$$R_{ac,o} = \frac{V_{ac,o,rms}}{I_{ac,o,rms}} = \frac{8}{\pi^2}\frac{V_o}{I_o} = \frac{8}{\pi^2} R_o \tag{6.3}$$

Using the sinusoidal analysis, voltage gain ratio of this converter is equal to:

$$\frac{V_{ac,o}}{V_{ac,i}} = \frac{R_{ac,o}}{R_{ac,o} + j(X_L - X_C)} = \frac{R_{ac,o}C\omega_s}{R_{ac,o}C\omega_s + j\left(LC(\omega_s)^2 - 1\right)} = \frac{1}{1 + j\frac{\left((\omega_s)^2 - (\omega_o)^2\right)L}{R_{ac,o}C\omega_s}} \tag{6.4}$$

where ω_s and ω_o are the switching and the resonant angular frequencies, and:

$$f_o = \frac{\omega_o}{2\pi} = \frac{1}{2\pi\sqrt{LC}} \tag{6.5}$$

$$f_s = \frac{\omega_s}{2\pi} \tag{6.6}$$

By defining Q as:

$$Q = \frac{\omega_o L}{R_o} \tag{6.7}$$

We can simplify (6.4) as follows:

$$\frac{V_o}{V_{in}} = \frac{V_{ac,o}}{V_{ac,i}} = \frac{1}{1 + j\frac{\pi^2}{8}Q\left[\frac{\omega_s}{\omega_o} - \frac{\omega_o}{\omega_s}\right]} \tag{6.8}$$

$$\left|\frac{V_o}{V_{in}}\right| = \left|\frac{V_{ac,o,rms}}{V_{ac,i,rms}}\right| = \frac{1}{\sqrt{1 + \left(\frac{\pi^2}{8}Q\left[\frac{\omega_s}{\omega_o} - \frac{\omega_o}{\omega_s}\right]\right)^2}} \tag{6.9}$$

Figure 6.5 depicts the voltage gain ratio versus the normalized frequency for different values of Q. This figure shows that the SRC can only step down the voltage. At any specific load, we may regulate the voltage by changing the

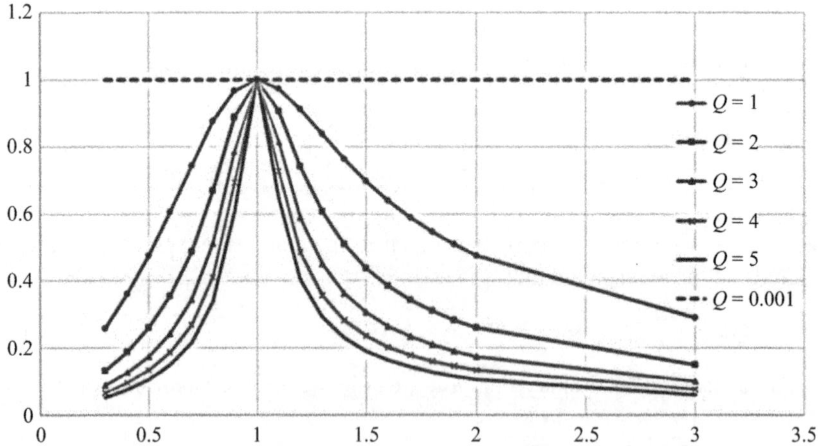

Figure 6.5 Voltage gain versus normalized frequency (f_s/f_o)

switching frequency. However, as seen in Figure 6.5, at light loads, voltage gain is almost fixed. Therefore, one of the major drawbacks of the SRC is its poor voltage regulation at light loads or no-load condition.

The current in the resonant tank of a SRC is equal to:

$$I_{ac,o} = \frac{V_{ac,o}}{R_{ac}} = \frac{1}{R_{ac} + j(X_L - X_C)} = \frac{\frac{1}{R_{ac,}} V_{ac,i}}{1 + j\frac{\left((\omega_s)^2 - (\omega_o)^2\right)L}{R_{ac,o} C \omega_s}} \tag{6.10}$$

According to (6.10), the phase shift between $I_{ac,o}$ and $V_{ac,i}$ is as follows:

$$\theta = -\tan^{-1}\left(\frac{\left((\omega_s)^2 - (\omega_o)^2\right)L}{R_{ac} C \omega_s}\right) \tag{6.11}$$

This implies that depending on ω_s and ω_o, the current passing a switch may be leading or lagging the voltage across that switch. Figure 6.6 depicts the current and voltage of a switch at above resonance and below resonance frequencies of operation. As seen in these figures, operating below the resonating frequency results in zero-current turn-off; whereas, operating above the resonating frequency allows the switches to benefit from zero-voltage turn-on.

The peak of the sinusoidal current passing the resonant tank, which is equal to the maximum switch current, may be calculated as follows:

$$I_{ac,o,peak} = \frac{\frac{\pi^2}{8R_o} V_{ac,i,peak}}{\sqrt{1 + \frac{\pi^4}{64} Q^2 \left[\frac{\omega_s}{\omega_o} - \frac{\omega_o}{\omega_s}\right]^2}} = \frac{\frac{\pi}{2R_o} V_{in}}{\sqrt{1 + \frac{\pi^4}{64} Q^2 \left[\frac{\omega_s}{\omega_o} - \frac{\omega_o}{\omega_s}\right]^2}} \tag{6.12}$$

It can be seen that the current passing the switches is proportional to the load. Therefore, the conduction losses will be lower at light loads, and the efficiency is high at these loads.

Important features of the SRCs may be summarized as follows:

- The SRC is a step-down converter.
- At light loads, voltage regulation is very poor.
- Operating below the resonating frequency results in zero-current turn-off of the switches; whereas, operating above the resonating frequencies allows the switches to benefit from zero-voltage turn-on.
- At light loads, the efficiency of the converter is high.

6.1.1.2 Dc–dc PRCs

Similar to a SRC, the equivalent circuit of a PRC is developed as shown in Figure 6.7 [2]. The input side of the equivalent circuit is similar to that of the SRC; however, the load side is a current sink; therefore, the equivalent resistance is not the same as that of the SRC. Assuming the rectifier is ideal, the load current, I_o, and load voltage, V_o, are pure dc. Therefore, the current seen at the input of the rectifier

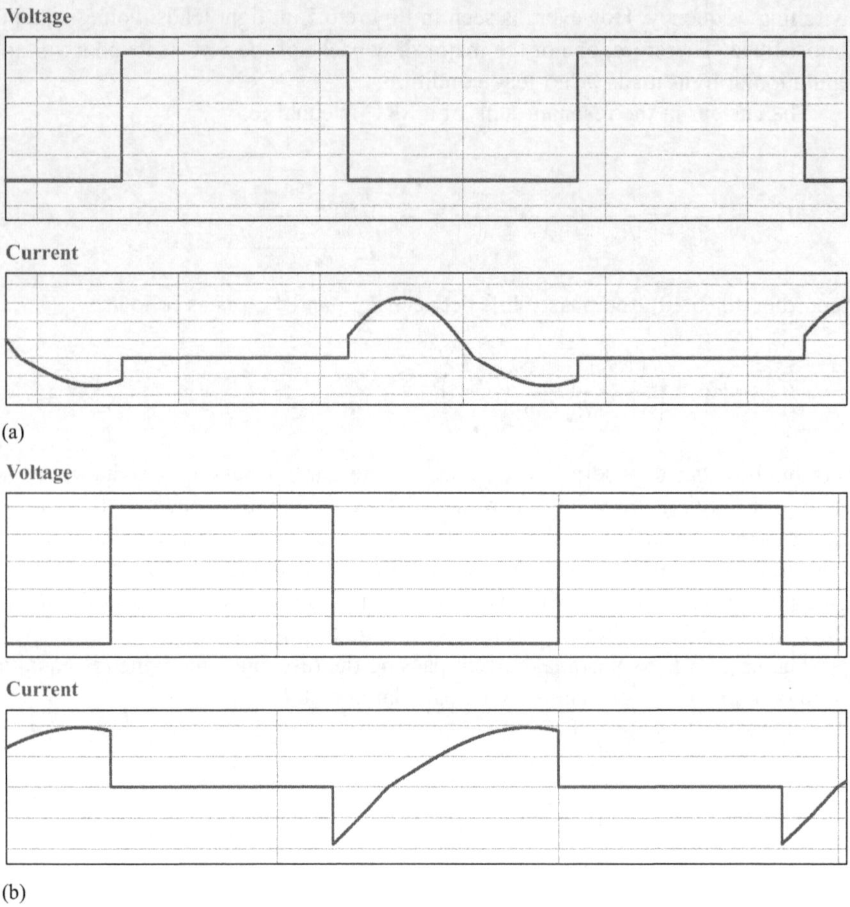

Figure 6.6 Current and voltage of a switch in SRC. (a) $f_s < f_o$ and (b) $f_s > f_o$

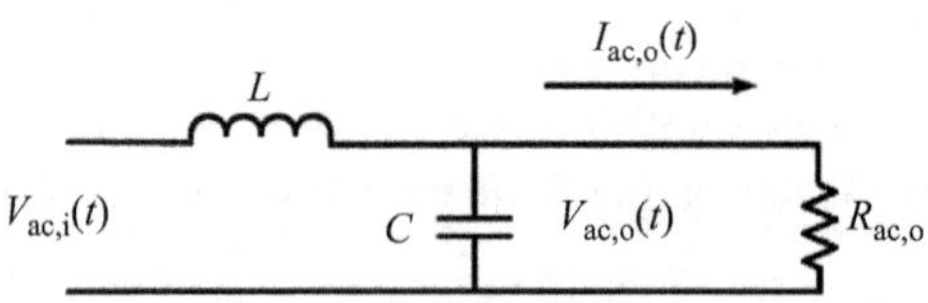

Figure 6.7 Equivalent circuit of a PRC

is a square wave that varies between I_o and $-I_o$, and the peak of the fundamental component of this current is equal to:

$$I_{ac,o,peak} = \frac{4}{\pi} I_o \tag{6.13}$$

The average of the rectified voltage is equal to V_o; thus, the peak of the sinusoidal voltage seen at the input of the rectifier is as follows:

$$V_{ac,o,peak} = \frac{\pi}{2} V_o \tag{6.14}$$

The equivalent load resistance in the PRC can be calculated as follows:

$$R_{ac,o} = \frac{V_{ac,o,rms}}{I_{ac,o,rms}} = \frac{8}{\pi^2}\frac{V_o}{I_o} = \frac{\pi^2}{8} R_o \tag{6.15}$$

Using the sinusoidal analysis, the voltage gain in this converter can be derived as follows:

$$\frac{V_{ac,o}}{V_{ac,i}} = \frac{1}{\left[1 - \left(\frac{\omega_s}{\omega_o}\right)^2\right] + j\frac{8}{\pi^2}\left(\frac{\omega_s}{\omega_o}\right)\frac{1}{Q}} \tag{6.16}$$

$$\left|\frac{V_o}{V_{in}}\right| = \frac{8}{\pi^2}\frac{V_{ac,o,peak}}{V_{ac,i,peak}} = \frac{1}{\sqrt{\frac{\pi^4}{64}\left[1 - \left(\frac{\omega_s}{\omega_o}\right)^2\right]^2 + \left(\left(\frac{\omega_s}{\omega_o}\right)\frac{1}{Q}\right)^2}} \tag{6.17}$$

where ω_s and ω_o are the switching and the resonant angular frequencies, and:

$$f_o = \frac{\omega_o}{2\pi} = \frac{1}{2\pi\sqrt{LC}} \tag{6.18}$$

$$f_s = \frac{\omega_s}{2\pi} \tag{6.19}$$

$$Q = \frac{R_o}{\omega_o L_o} \tag{6.20}$$

Figure 6.8 depicts the voltage gain of the PRC versus the normalized frequency for different values of Q. Unlike SRC, the PRC is capable of stepping up the voltage. If Q is more than 2, maximum voltage gain occurs at resonant frequency. When the converter operates at frequencies below this frequency, switches will be turned on at zero voltage; whereas, when they operate at frequencies above this point, they will be turned off at zero current.

The peak value of the inductor current, which is equal to the maximum current passing through the conducting switches of the inverter, can be calculated as follows:

$$I_{L,o,peak} = \frac{\frac{4}{\pi}\frac{V_{in}}{R_o}\sqrt{\left(1 + \frac{\pi^4 Q^2}{64}\left(\frac{\omega_s}{\omega_o}\right)^2\right)}}{\sqrt{\frac{\pi^4}{64}\left[1 - \left(\frac{\omega_s}{\omega_o}\right)^2\right]^2 + \left(\left(\frac{\omega_s}{\omega_o}\right)\frac{1}{Q}\right)^2}} \tag{6.21}$$

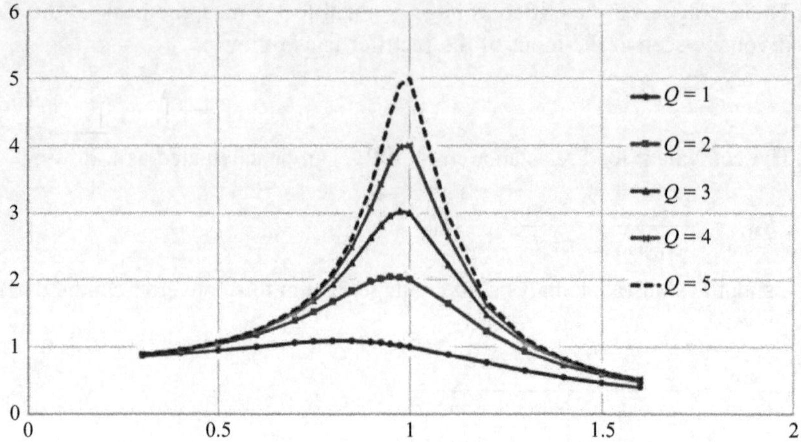

Figure 6.8 Voltage gain of a PRC versus normalized frequency

By increasing Q, the maximum current of the switches becomes equal to a fixed non-zero value:

$$I_{L,o,peak} = \frac{4}{\pi}\frac{V_{in}}{L}\frac{\omega_s}{\left((\omega_o)^2 - (\omega_s)^2\right)} \tag{6.22}$$

This implies that in this converter, the current passing the switches, and consequently the conduction losses are noticeable at light-load or no-load condition.

Important features of the PRCs may be summarized as follows:

- The PRC can step up the voltage.
- Voltage can be regulated at light loads.
- Operating above the resonating frequency results in zero-current turn-off of the switches; whereas, operating below the resonating frequency allows the switches to benefit from zero-voltage turn-on.
- At light loads, the efficiency of the converter is low.

6.1.1.3 Poly-phase series resonant ac-link converters

Resonant converters shown in Figure 6.3 convert dc voltage to dc. By extending the principles of the operation of these converters, dc–ac, ac–dc, or multi-phase resonant converters may be developed.

In [3], a SRC with bipolar output voltage was introduced and the principles of the operation of the dc-to-single-phase ac SRC was studied. Dc-to-poly-phase ac SRC was investigated in [4]. Poly-phase ac–ac SRCs were proposed and studied in [5–7]. Figure 6.9 depicts a three-phase ac–ac SRC, in which the switches benefit from zero-current turn-off. Therefore, thyristors can be used in this converter. In a three-phase ac–ac SRC, two of the three input phases, and two of the output phases will be connected to the resonating circuit at each moment. Therefore, the

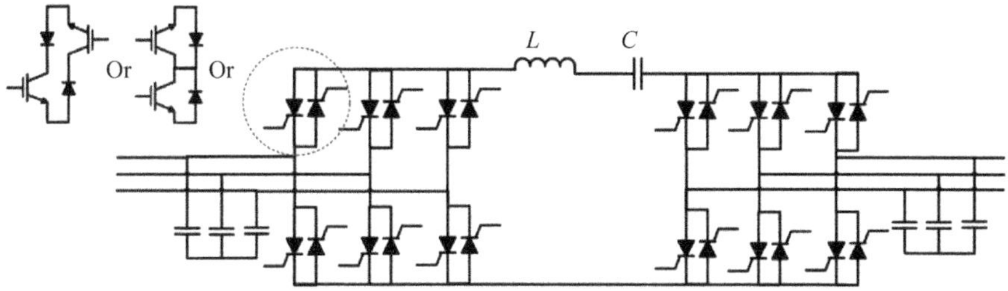

Figure 6.9 Three-phase ac–ac series ac-link resonant converter

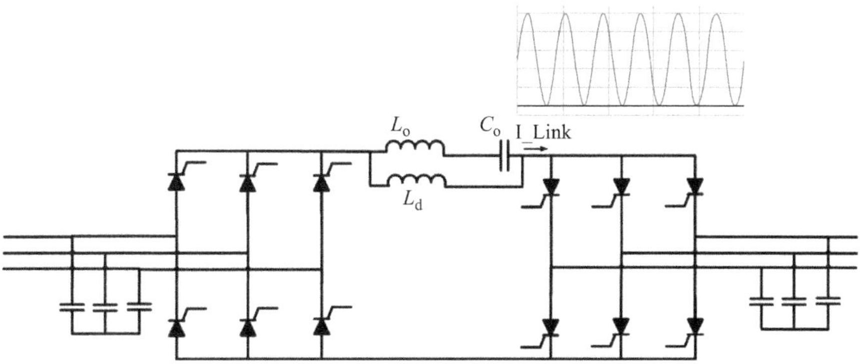

Figure 6.10 Three-phase ac–ac series dc-link resonant converter

performance of the circuit will be similar to that of the dc–dc converter. In this converter, bidirectional-conducting bidirectional-blocking switches are required.

6.1.1.4 Poly-phase series resonant dc-link converters

One of the limitations of the series ac-link resonant converters was the large number of switches. As seen in Figure 6.9, for a three-phase ac–ac series ac-link converter, 24 active switches are required. To address this problem, a modified configuration, called series resonant dc-link converter, was proposed in [8,9]. This configuration is depicted in Figure 6.10. As shown in this figure, another inductor is added to the link such that the current of the link is sinusoidal with a dc offset. This inductor (L_d) that controls the dc bias of the link inductor is much larger than L_o; therefore, it will not be involved in the resonance. Since the link current is always positive in this converter, only 12 active switches are required.

6.1.1.5 Poly-phase parallel resonant ac-link converters

Another type of three-phase ac–ac resonant link converter was proposed in [10,11]. This converter uses the resonant tank network shown in Figure 6.3(c), which is a parallel LC pair. Therefore, this converter is called parallel resonant ac-link converter. This converter is dual of the series resonant ac-link converter, and interfaces current sources and current sinks, as shown in Figure 6.11.

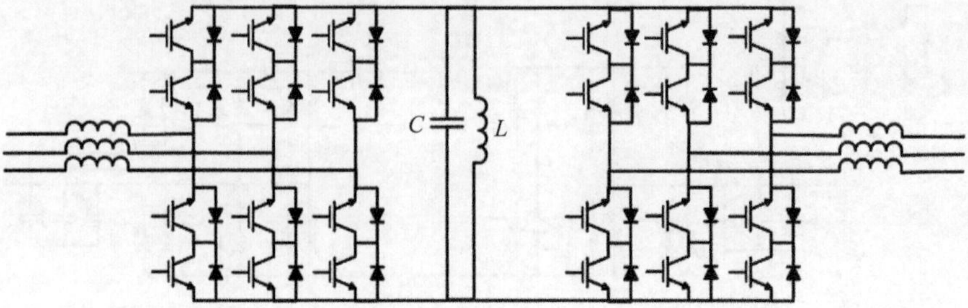

Figure 6.11 Three-phase ac–ac parallel ac-link resonant converter

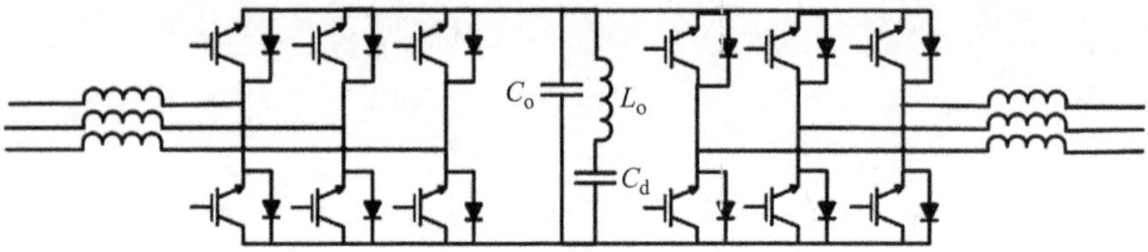

Figure 6.12 Three-phase ac–ac parallel dc-link resonant converter

6.1.1.6 Poly-phase parallel resonant dc-link converters

Similar to the series resonant ac-link converter, the parallel resonant ac-link converter requires bidirectional-blocking bidirectional-conducting switches. Therefore, 24 active switches are required for a three-phase ac–ac converter. To reduce the number of switches to 12, the parallel resonant dc-link converter, shown in Figure 6.12, was proposed in [12]. In this converter, another capacitor, C_d in Figure 6.12, is added to the link to add a dc bias to the link voltage. This capacitor is not involved in resonance. The link voltage will be positive in this converter.

6.1.2 Resonant converters with three or more resonating elements

Both SRC and PRC have limitations that can affect the overall performance of the system. To overcome these shortcomings, hybrid resonant converters, in which the resonant tank is formed by three or more resonating elements, were proposed.

There are numerous combinations for forming a resonant tank network with three elements. A number of these combinations are shown in Figure 6.13. Among these combinations, the LCC resonant tank that is formed by an inductor (L) and two capacitors (C) as shown in Figure 6.13(a), is the most common solution.

Figure 6.14 depicts the LCC resonant converter. To analyze this converter, its equivalent circuit, shown in Figure 6.15, is used. The output side of this converter is similar to that of the PRC; therefore, the equivalent resistance is equal to:

$$R_{ac,o} = \frac{V_{ac,o,rms}}{I_{ac,o,rms}} = \frac{8}{\pi^2}\frac{V_o}{I_o} = \frac{8}{\pi^2}R_o \tag{6.23}$$

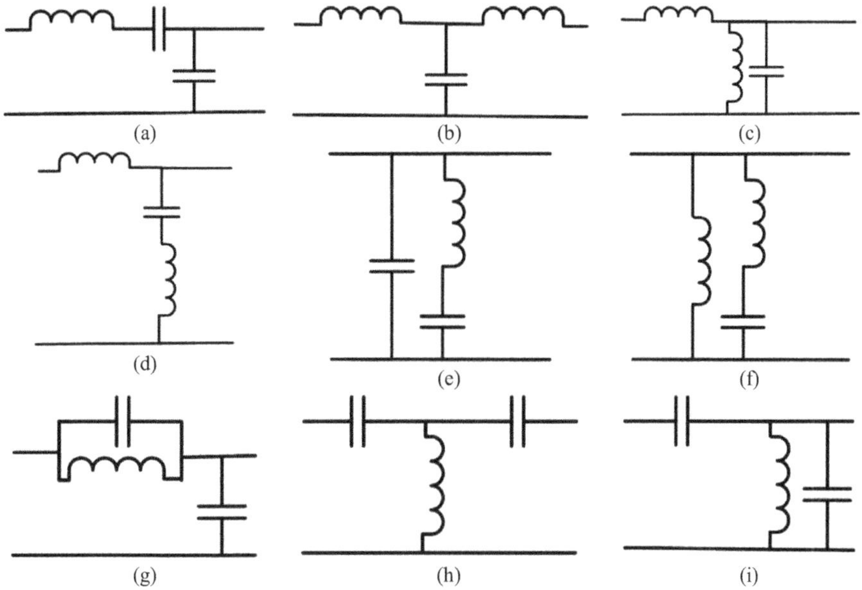

Figure 6.13 Examples of the third-order resonant tank network

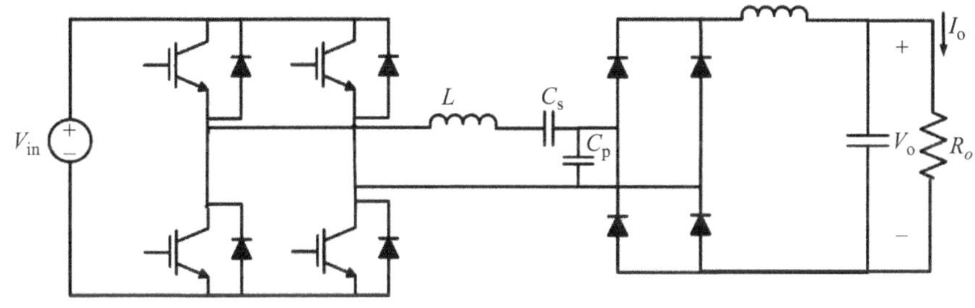

Figure 6.14 LCC resonant converter

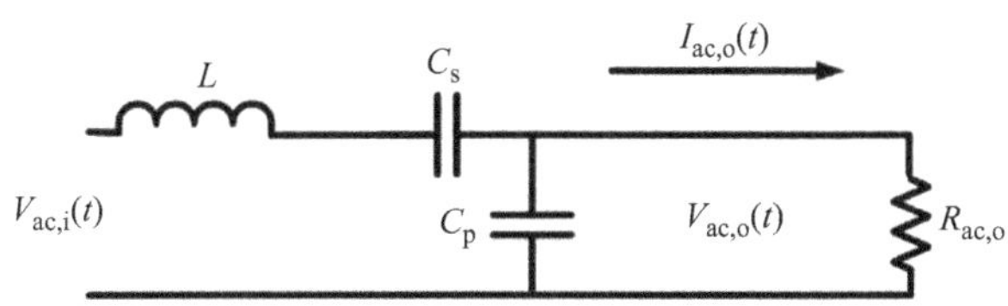

Figure 6.15 Equivalent circuit of the LCC resonant converter

Using the sinusoidal analysis, it can be shown that the voltage gain of this converter is as follows:

$$\frac{V_{ac,o}}{V_{ac,i}} = \frac{1}{\left[1+\frac{C_p}{C_s}-LC_p(\omega_s)^2\right]+j\frac{Q\pi^2}{8}\left[\frac{\omega_s}{\omega_{o1}}-\frac{\omega_{o1}}{\omega_s}\right]} \tag{6.24}$$

$$\left|\frac{V_o}{V_{in}}\right| = \frac{1}{\sqrt{\frac{\pi^4}{64}\left[1+\frac{C_p}{C_s}-LC_P(\omega_s)^2\right]^2+Q^2\left[\frac{\omega_s}{\omega_{o1}}-\frac{\omega_{o1}}{\omega_s}\right]^2}} \tag{6.25}$$

where

$$f_{o1} = \frac{\omega_{o1}}{2\pi} = \frac{1}{2\pi\sqrt{LC_s}} \tag{6.26}$$

$$f_{o2} = \frac{\omega_{o2}}{2\pi} = \frac{1}{2\pi\sqrt{\frac{LC_sC_p}{C_p+C_s}}} \tag{6.27}$$

$$f_s = \frac{\omega_s}{2\pi} \tag{6.28}$$

$$Q = \frac{\omega_o L}{R_o} \tag{6.29}$$

Voltage gain of the LCC resonant converter versus normalized frequency is depicted in Figure 6.16. In this figure, it is assumed that C_s and C_p are equal. The maximum gain of the LCC resonant converter can occur at f_{o1} or f_{o2}.

For frequencies above f_{o2}, switches benefit from zero voltage switching (ZVS); whereas, for frequencies below f_{o1}, the switches have zero current switching (ZCS).

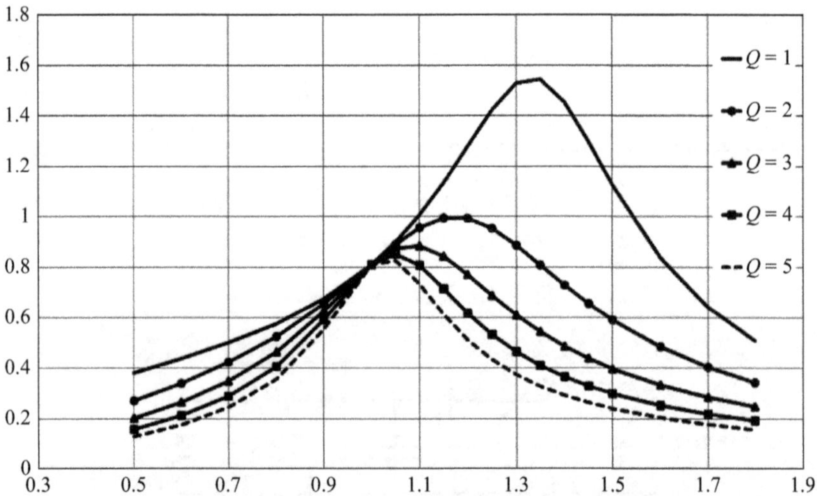

Figure 6.16 Voltage gain of an LCC resonant converter

For frequencies between f_{o1} and f_{o2}, the load value determines whether the switches will benefit from ZVS or ZCS [13].

The LCC resonant converter has superior performance compared to SRC or PRC. It can both step up and step down the voltage. Moreover, at light loads, its voltage can be regulated and its efficiency is high.

6.2 Quasi-resonant converters

Another category of the converters that are capable of operating at high frequencies and benefit from the soft switching are quasi-resonant converters (QRCs). In these converters, each switch is substituted by a "resonant switch" that includes a switch, an inductor, and a capacitor. QRCs are classified as follows:

- Zero Current Switching Quasi-Resonant Converters (ZCS-QRCs)
- Zero Voltage Switching Quasi-Resonant Converters (ZVS-QRCs)

Each of these converters can be half-wave or full-wave. To benefit from ZCS, the inductor should be placed in series with the switch. The possible resonant switch arrangements for half-wave and full-wave configurations are shown in Figures 6.17 and 6.18, respectively. As seen in these figures, bidirectional-conducting switch needs to be used for full-wave configuration. Figures 6.19 and 6.20 show possible resonant switch configurations in order for the switches to benefit from ZVS [14,15]. Full-wave configuration in ZVS-QRCs requires bidirectional-blocking switches.

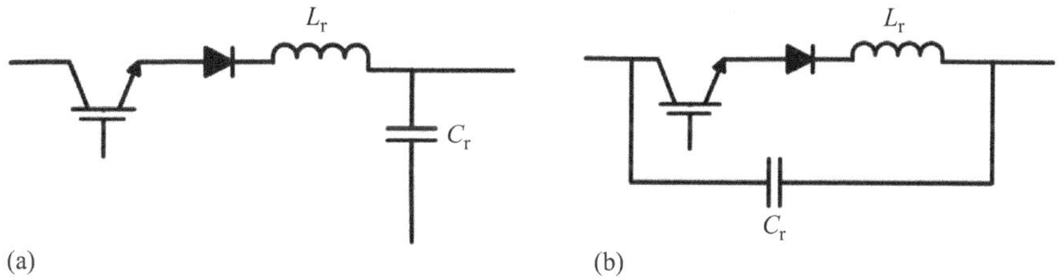

Figure 6.17 Resonant switch for half-wave ZCS-QRC

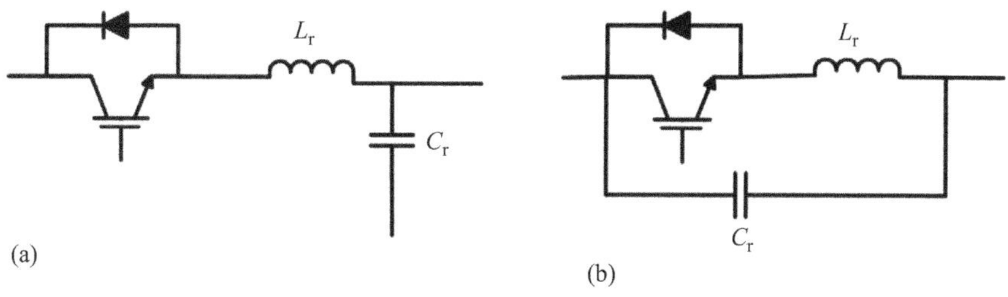

Figure 6.18 Resonant switch for full-wave ZCS-QRC

A wide variety of soft-switching power converters can be developed by using the resonant switches in the Pulse Width Modulated (PWM) converters [14].

In a ZCS-QRC, when the switch is on, an LC resonant circuit is formed; therefore, L_r, C_r, switch, and possibly a voltage source will be in series. In a ZVS-QRC, during the time that the freewheeling diode conducts the LC resonant circuit will be resonating; therefore, the diode, L_r, C_r, and possibly a voltage source will be in series. Figure 6.21 shows the equivalent circuit of a ZCS-QRC and a ZVS-QRC when voltage sources are shorted and current sources are open circuit.

In this section, performance of the quasi-resonant buck–boost converters will be studied.

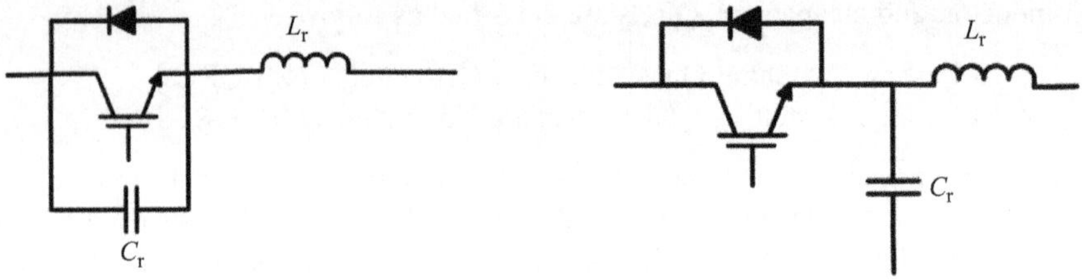

(a) (b)

Figure 6.19 Resonant switch for half-wave ZVS-QRC

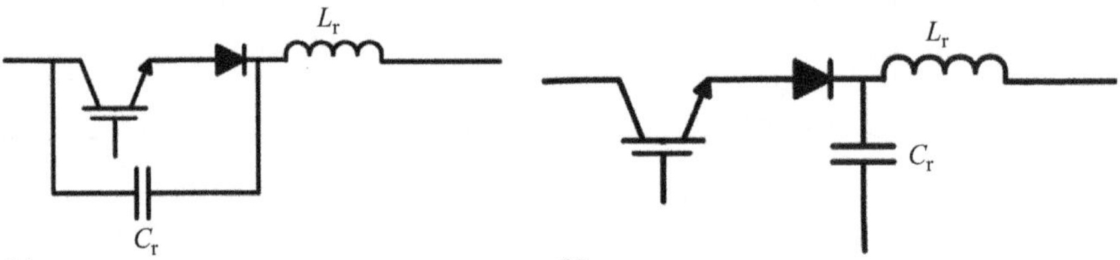

(a) (b)

Figure 6.20 Resonant switch for full-wave ZVS-QRC

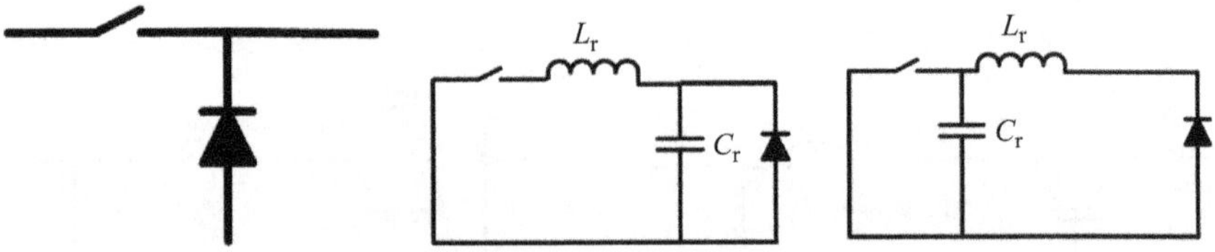

(a) (b) (c)

Figure 6.21 (a) PWM switch network, (b) ZCS-QRC circuit when voltage sources are shorted and current sources are open circuit, (c) ZVS-QRC circuit when voltage sources are shorted and current sources are open circuit

6.2.1 Example 1 – half-wave ZCS-QRC

Figure 6.22 depicts the schematic of a half-wave ZCS buck–boost QRC. L_r and C_r are the resonating components, and L_d is the energy storage component that transfers power from input to the output. The resonance inductance, L_r, should be much smaller than the link inductance, L_d.

There are four modes of operation in each switching cycle of this converter. Figure 6.23 shows the behavior of the converter in each mode, along with the current and voltage waveforms. Mode 1 is initiated when the switch is turned on. Before this mode, the diode was conducting and the voltage across C_r was equal to "$-V_o$". In a conventional buck–boost converter, once the switch is turned on, the diode stops conducting and the current of the switch becomes equal to the link inductor current (I_{Ld}). However, in this converter L_r does not allow the current of the switch have a sharp change. Thus, during mode 1, both the diode and switch will be conducting. The current of L_r increases linearly during this mode:

$$V_{in} + V_o = L_r \frac{d(i_{Lr}(t))}{dt} = L_r \frac{i_{Lr}(t)}{t} \tag{6.30}$$

This mode continues until the current of the switch becomes equal to I_{Ld}. At this moment, the diode stops conducting. Duration of this mode, t_1, may be calculated as follows:

$$t_1 = L_r \frac{I_{Ld}}{V_{in} + V_o} \tag{6.31}$$

During mode 2, C_r and L_r resonate until the current of L_r, which is equal to the switch current, becomes zero. The following equations describe the behavior of the circuit during this mode:

$$V_{cr}(t) = -(V_{in} + V_o)\cos\omega_o t + V_{in} \tag{6.32}$$

$$i_{Lr}(t) = (V_{in} + V_o)\omega_o C_r \sin\omega_o t + I_{Ld} \tag{6.33}$$

where

$$\omega_o = \frac{1}{\sqrt{L_r C_r}} \tag{6.34}$$

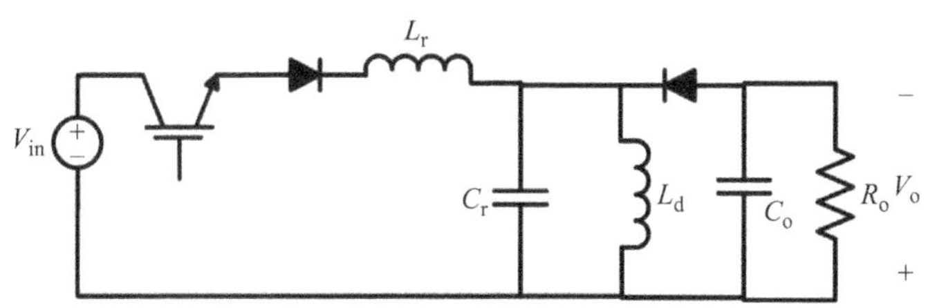

Figure 6.22 Half-wave ZCS buck–boost QRC

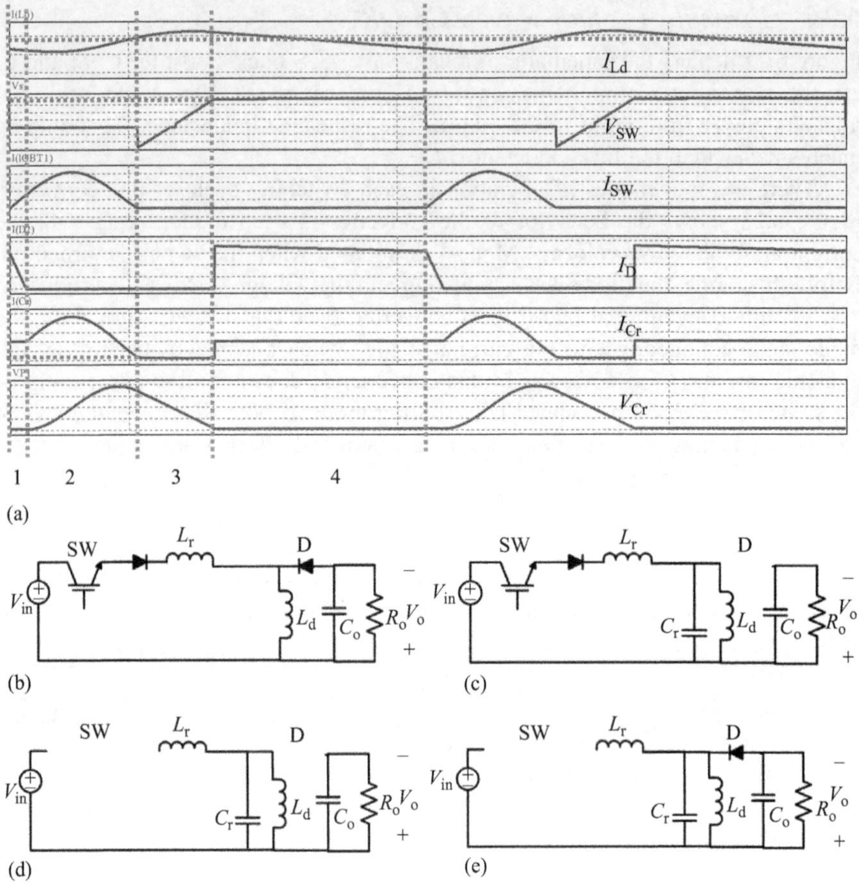

Figure 6.23 Modes of operation in half-wave ZCS buck–boost QRC. (a) Waveforms, (b) mode 1, (c) mode 2, (d) mode 3, and (e) mode 4

In the above equations, $V_{cr}(t)$ and $i_{Lr}(t)$ are the voltage across C_r and the current of L_r, respectively.

This mode is over when current of L_r becomes zero. Duration of mode 2, t_2, may be calculated as follows:

$$t_2 = \frac{\alpha}{\omega_o} \tag{6.35}$$

where

$$\alpha = \sin^{-1}\left(\frac{-I_{Ld}}{(V_{in} + V_o)\omega_o C_r}\right) \tag{6.36}$$

It should be noted that for the half-wave operation, α is in the third quadrant:

$$\pi \leq \alpha \leq \frac{3\pi}{2} \tag{6.37}$$

Therefore

$$\sin\alpha < 0 \tag{6.38}$$

$$\cos\alpha < 0 \tag{6.39}$$

The voltage across C_r at the end of this mode is equal to:

$$V_{cr}(t_2) = -(V_{in} + V_o)\cos\alpha + V_{in} \tag{6.40}$$

The maximum current of the switch is as follows:

$$I_{Lr,max} = (V_{in} + V_o)\omega_o C_r + I_{Ld} \tag{6.41}$$

During mode 3, the current of the switch remains zero, and the voltage across C_r decreases. This results in zero-current turn-off. During this mode, the voltage of C_r decreases linearly:

$$I_{Ld} = C_r \frac{V_{cr}(t_2)}{t_3} \tag{6.42}$$

$$t_3 = C_r \frac{-(V_{in} + V_o)\cos\alpha + V_{in}}{I_{Ld}} \tag{6.43}$$

When the voltage across the capacitor becomes equal to "$-V_o$", the diode becomes forward biased, and mode 4 during which the diode conducts is initiated. Duration of this mode may be found by:

$$t_4 = \frac{1}{f_s} - t_1 - t_2 - t_3 \tag{6.44}$$

where f_s is the switching frequency.

6.2.2 *Example 2 – full-wave ZCS-QRC*

Figure 6.24 depicts a full-wave ZCS buck–boost QRC. The performance of this converter is similar to that of half-wave ZCS buck–boost QRC; however, since the diode that was in series with the switch is removed and an antiparallel diode is added, the switch can handle bidirectional current and its current becomes negative during mode 2. Therefore, the duration of mode 2 (t_2) will be different than that of

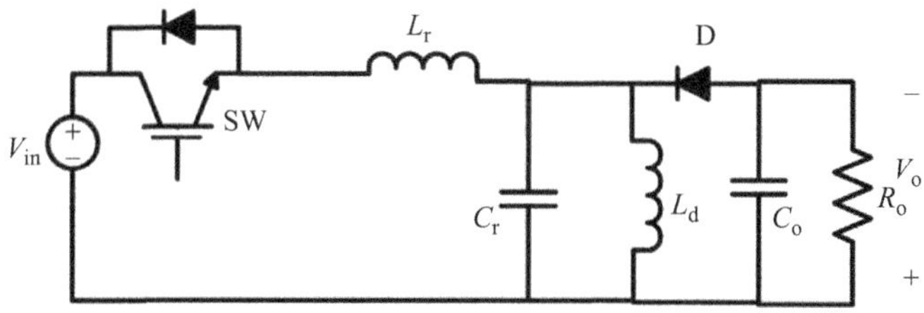

Figure 6.24 Full-wave ZCS buck–boost QRC

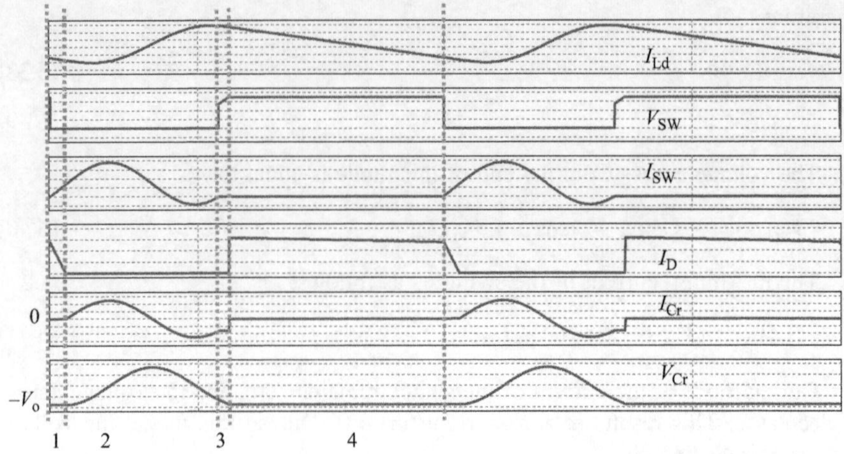

Figure 6.25 Waveforms in full-wave ZCS buck–boost QRC

the previous example. Equations (6.35) and (6.36) may still be used; however, α is in the fourth quadrant in this case:

$$\frac{3\pi}{2} \le \alpha \le 2\pi \tag{6.45}$$

Therefore

$$\sin\alpha < 0 \tag{6.46}$$

$$\cos\alpha > 0 \tag{6.47}$$

Figure 6.25 shows the current and voltage waveforms in this case. If the duty cycle of the gate signal is increased the resonant switch will be capable of conducting multiple cycles [16].

6.2.3 Example 3 – half-wave ZVS-QRC

In converters shown in Figures 6.22 and 6.24, the switch could benefit from zero-current turn-off. In order for the switch to be turned on at zero voltage, the configuration of the resonant switch should be changed. Figure 6.26 shows a half-wave ZVS buck–boost QRC. Again, there are four modes of operation in this converter. Figure 6.27 shows the behavior of the circuit during each mode along with the current and voltage waveforms.

Before turning off the switch, the current of L_r is equal to I_{Ld}, the diode is turned off, and the voltage across C_r is 0. Once the switch is turned off, mode 1, during which the voltage across C_r will be increasing, starts. Freewheeling diode (diode "D") remains reverse biased during this mode. When the voltage across C_r becomes equal to $V_{in} + V_o$, the diode becomes forward biased, and it starts to conduct (mode 2). During mode 2, L_r and C_r resonate. When the voltage across C_r

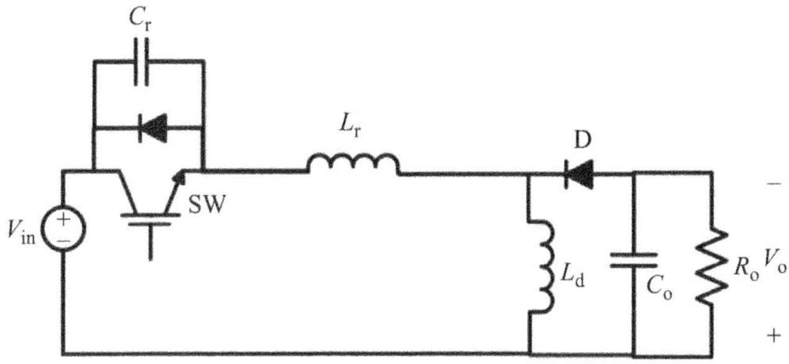

Figure 6.26 Half-wave ZVS buck–boost QRC

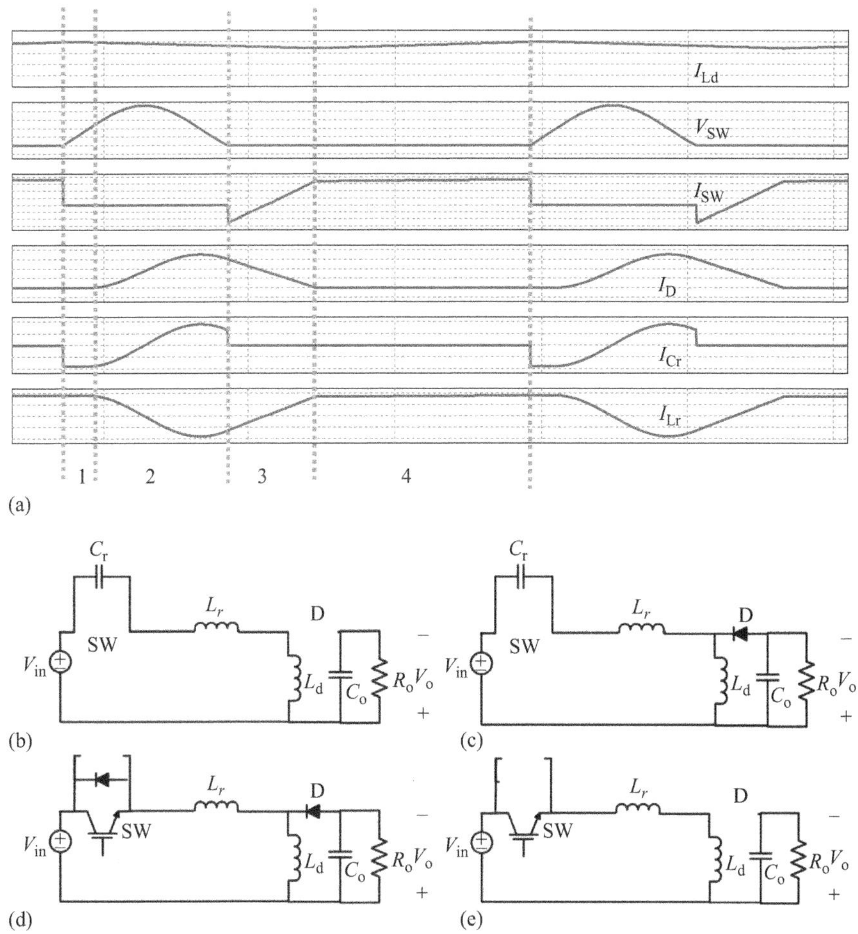

Figure 6.27 Modes of operation in half-wave ZVS buck–boost QRC. (a) Waveforms, (b) mode 1, (c) mode 2, (d) mode 3, and (e) mode 4

becomes zero the anti-parallel diode becomes forward biased and the voltage across C_r remains zero. This initiates mode 3, during which the freewheeling and the antiparallel diodes conduct. The switch should be turned on during this mode before the current of L_r becomes positive. Once the polarity of I_{Lr} changes, the anti-parallel diode stops to conduct and the switch starts to conduct. The voltage across the switch at the turn-on moment is zero. Mode 4 is initiated when the current of the freewheeling diode becomes zero.

6.2.4 Example 4 – full-wave ZVS-QRC

If switch SW in Figure 6.26 is substituted by a bidirectional-blocking switch, as shown in Figure 6.28, the voltage across the switch will be allowed to become negative during mode 2. The full-wave ZVS-QRC and the corresponding waveforms are depicted in Figures 6.28 and 6.29, respectively.

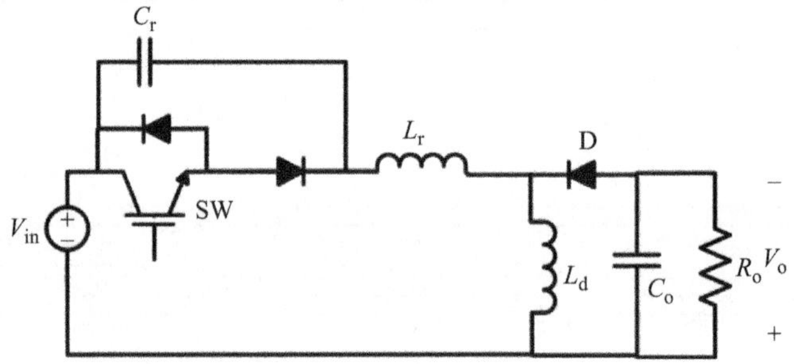

Figure 6.28 Full-wave ZVS buck–boost QRC

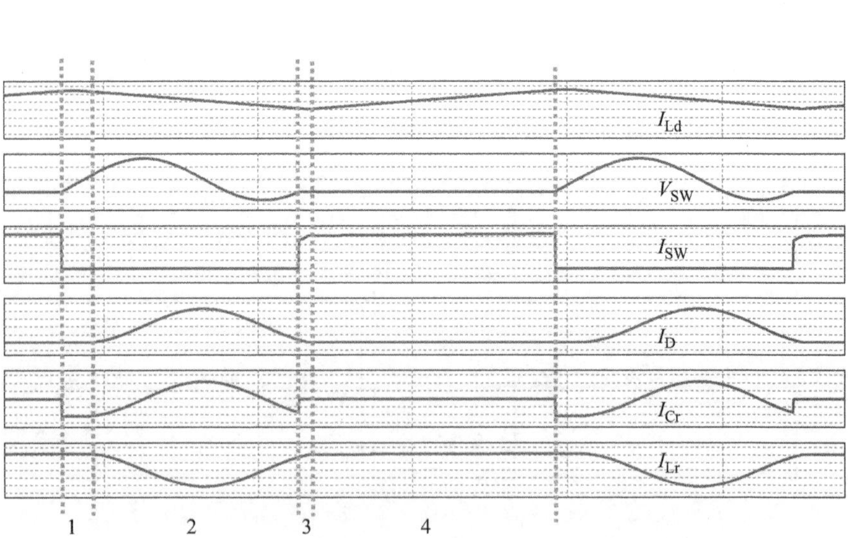

Figure 6.29 Waveforms in full-wave ZVS buck–boost QRC

6.2.5 The effect of parasitic oscillations in QRCs

Parasitic elements can affect the performance of the QRCs. It can be shown that the output capacitance of the active switch deteriorates the performance of the ZCS-QRCs; whereas, the junction capacitance of the diode affects the performance of the ZVS-QRCs. Figures 6.30 and 6.31 depict the current and voltage across the active switch and diode in a full-wave ZCS-QRC and a full-wave ZVS-QRC when the parasitic capacitors are modeled [15].

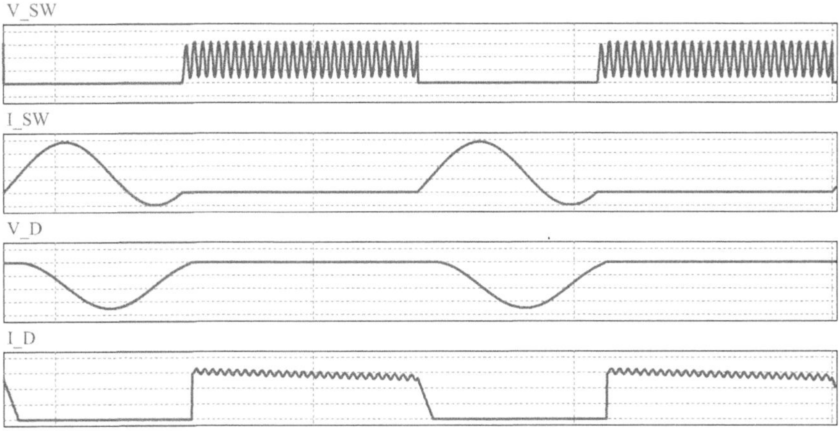

Figure 6.30 Voltage across the active switch, current of the switch, diode voltage, and diode current in a full-wave ZCS-QRC when parasitic elements are present

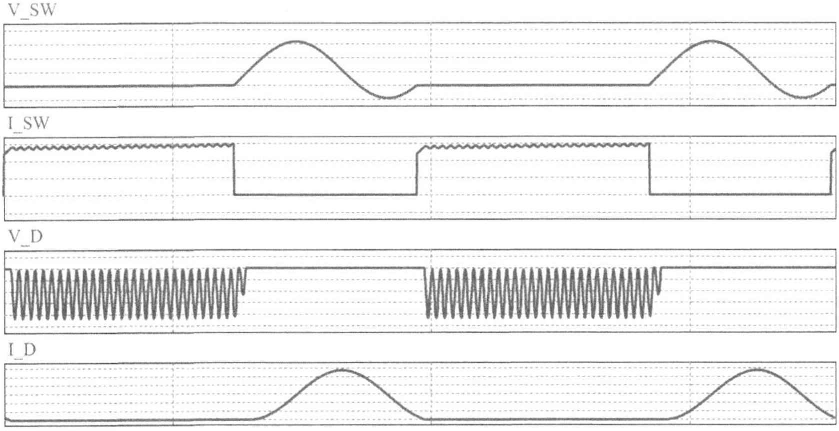

Figure 6.31 Voltage across the active switch, current of the switch, diode voltage, and diode current in a full-wave ZVS-QRC when parasitic elements are present

6.3 Multi-resonant converters

In a ZCS-QRC, the active switch benefits from zero-current turn-off and the freewheeling diode starts to conduct at zero voltage. Therefore, the active switch has a hard turn-on, and is exposed to large values of dv/dt when it is turned on. On the other hand, in a ZVS-QRC, the active switch is turned on at zero voltage and benefits from low dv/dt; whereas, the freewheeling diode is exposed to high-voltage stress.

Multi-resonant converters (MRCs) were proposed to allow both the active switch and the diode benefit from zero-voltage turn-on or to allow them both benefit from zero-current turn-off [15,17–21].

In a MRC, the PWM switch network is replaced by a multi-resonant switch network as shown in Figure 6.32.

A half-wave ZVS buck–boost MRC is shown in Figure 6.33. There are four modes of operation in this converter. Figure 6.34 shows the behavior of the converter in each mode along with the waveforms. As seen in this figure, both the diode and active switch can benefit from low dv/dt.

Another advantage of the ZVS-MRC is that in this converter the output capacitance of the active switch and the junction capacitance of the diode will not affect the performance of the converter.

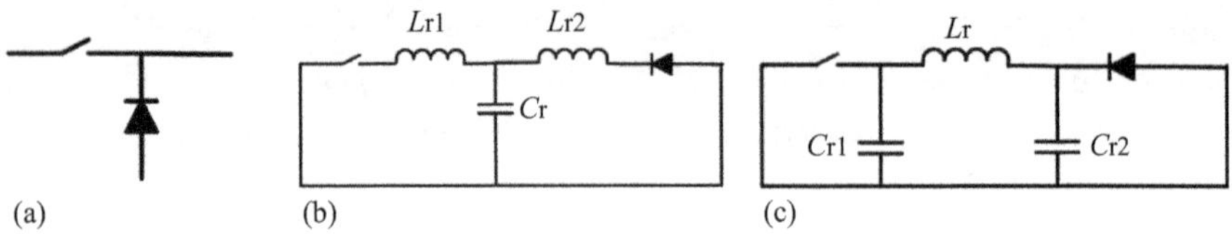

Figure 6.32 (a) PWM switch network, (b) ZCS-MRC circuit when voltage sources are shorted and current sources are open circuit, (c) ZVS-MRC circuit when voltage sources are shorted and current sources are open circuit

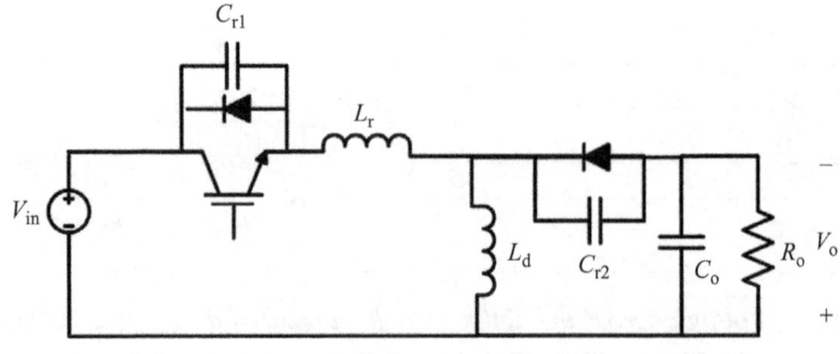

Figure 6.33 A half-wave ZVS buck–boost MRC

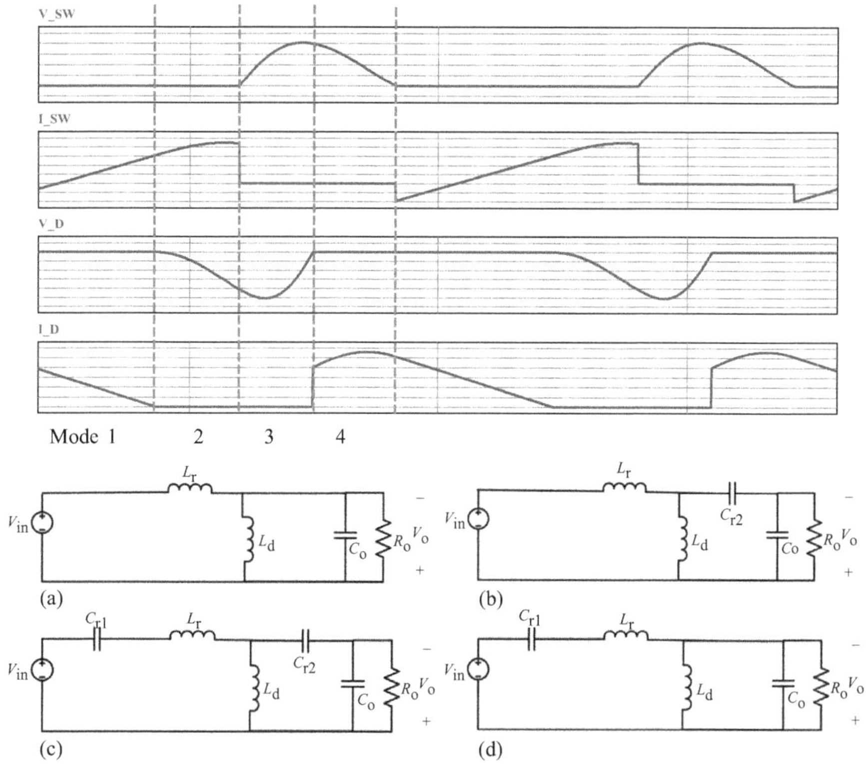

Figure 6.34 Principles of the operation of the ZVS buck–boost MRC. (a) Mode 1, (b) mode 2, (c) mode 3, and (d) mode 4

6.4 Quasi-square-wave converters

MRCs offer several advantages over QRCs; however, in both converters, switches and diodes have higher current or voltage ratings compared to their corresponding PWM configurations. To address this concern, quasi-square-wave converters (QSWCs) were proposed in [22,23].

In a QSWC, the PWM switch network is substituted by a quasi-square wave resonant switch network as shown in Figure 6.35. T sections need to be placed between a voltage source and a voltage sink to offer ZVS; whereas, the Pi sections can be placed between a current source and a current sink to offer ZCS. Several new converters can be formed using T and Pi sections. One of the advantages of using T or Pi sections is that the energy storage elements that exist in a switch network can be used as a resonant tank element, as well.

In this part, the behavior of a ZVS buck–boost QSWC will be studied. The schematic of this converter is depicted in Figure 6.36(a). Section T_b is used in this converter and the link inductor plays the role of the resonating inductor, as well. As shown in Figure 6.36(b)–(d), there are four modes of operation in this converter.

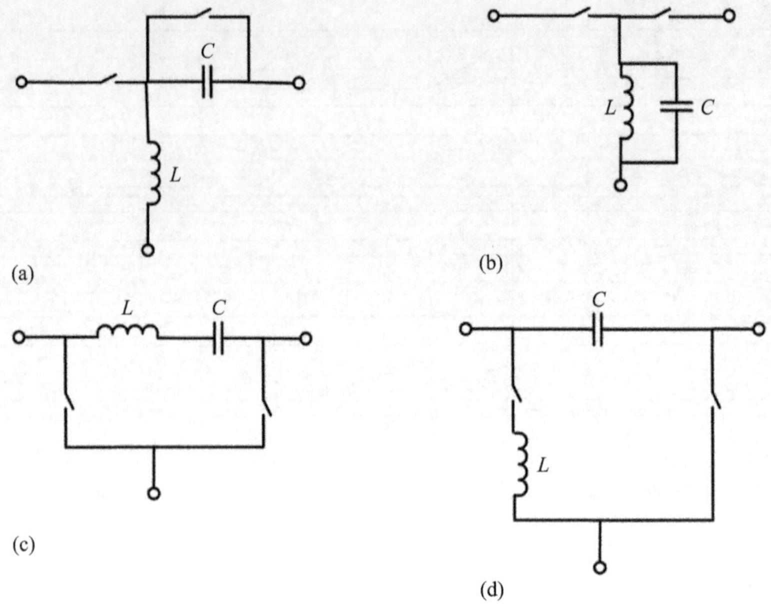

Figure 6.35 T and Pi sections used in QSWCs. (a) T_a, (b) T_b, (c) Pi_a, and (d) Pi_b (adapted from [23])

During the first mode, the inductor is charged through input voltage source. When the switch is turned off, mode 2, which is a resonating mode starts. During mode 2, the LC link resonates, and this results in soft turn off of the switch. When the voltage across the inductor becomes equal to the output voltage, which is negative, diode starts to conduct and the inductor will be discharged into the output (mode 3). Once the inductor is fully discharged, mode 4 starts. Mode 4 is another resonating mode that guaranties ZVS in this converter. The converter should be designed such that the resonating capacitor will not affect the performance of the converter during modes 1 and 3. If the impedance of the capacitor is much higher than that of the inductor, no current will flow through the capacitor during modes 1 and 3.

The principles of the operation of the QSWCs have been extended to three-phase ac–ac configuration, as well. In [24], a bidirectional three-phase ac–ac converter that could benefit from ZVS of the switches was proposed. In [24], gate turn-off thyristors were employed; however, any other forward conducting bidirectional-blocking switch can be used. Figure 6.37 shows the schematic of this converter, which is also called partial resonant link converter. Since there are three phases at the input of this converter, in each cycle one input phase pair should be selected for charging the link inductor. Similarly, one output phase pair is selected for being charged from the link inductor. There are different methods for choosing these phase pairs. In [24], the output phase pair whose voltage has the maximum error from its reference is chosen. The current and voltage of the link inductor are shown in Figure 6.37. These waveforms are similar to the waveforms of the dc–dc

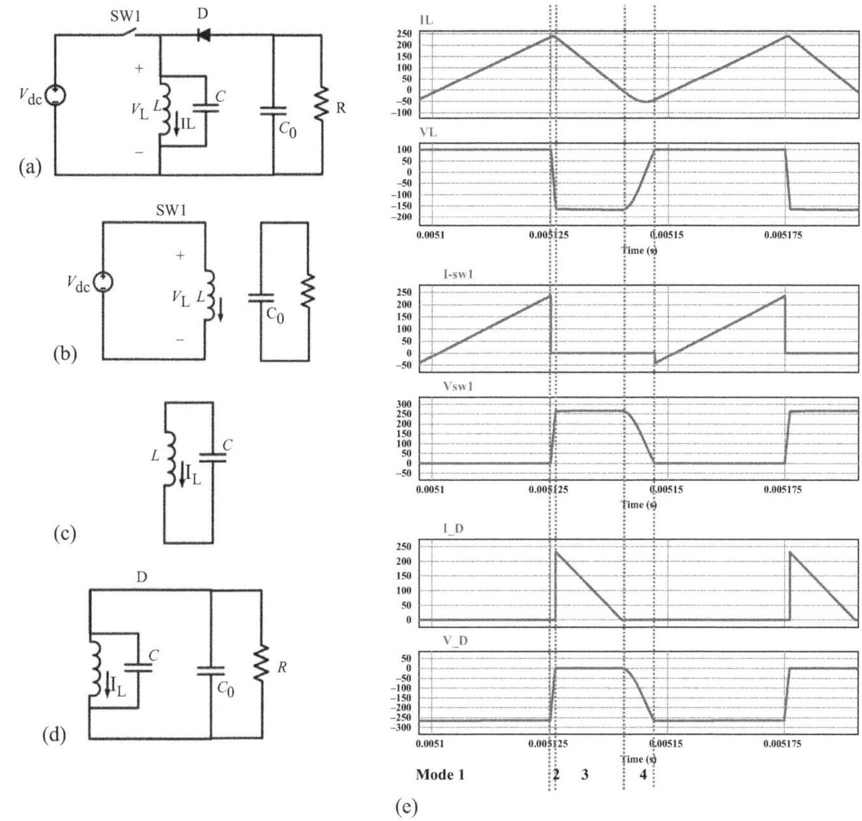

Figure 6.36 Buck–boost QSWC (a) configuration, (b) behavior during mode 1, (c) behavior during modes 2 and 4, (d) behavior during mode 3, and (e) waveforms (inductor current, inductor voltage, switch current, switch voltage, diode current, and diode voltage)

converter shown in Figure 6.36; however, since the output-side switches are turned off before the link current becomes zero, mode 4 is longer.

To increase the utilization of the inductor, in [25] the partial resonant ac-link converter, also called parallel ac-link universal converter or soft-switching ac-link buck–boost converter, was proposed. This converter operates similar to the converter shown in Figure 6.37; however, by doubling the number of switches it is capable of charging and discharging of the link inductor in both positive and negative directions. This not only results in better utilization of the inductor, but also minimizes the duration of mode 4. Although resonating modes are required for providing ZVS, their duration is preferred to be as short as possible. The ac-link feature of the converter proposed in [25] doubles the number of modes when compared to the converter proposed in [24]. Another modification that was proposed in [25], is dividing each charging and discharging mode in a three-phase ac-to-ac configuration into two modes, such that all the input and output phases are

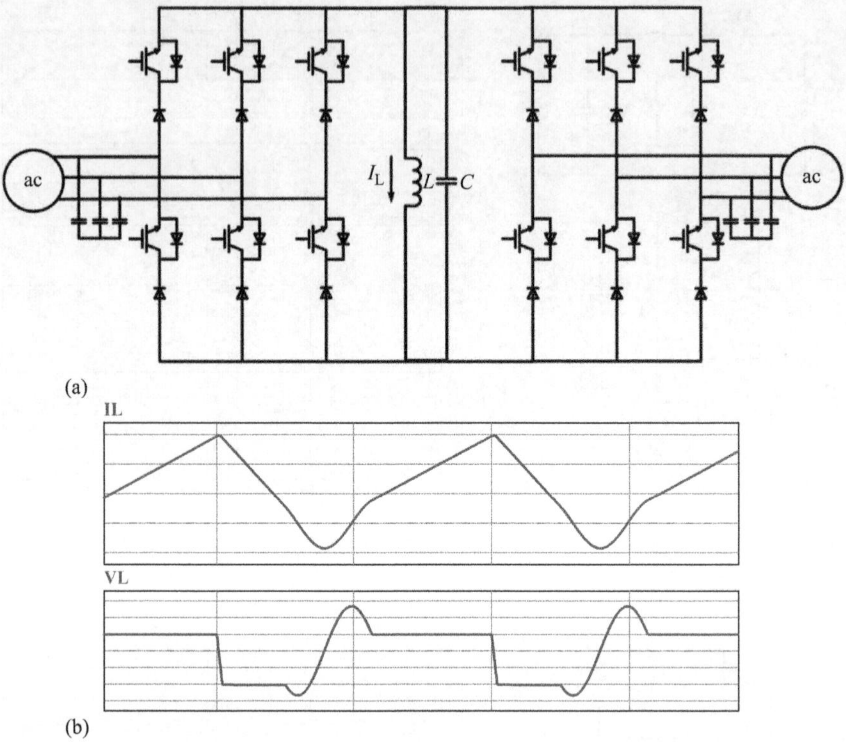

Figure 6.37 ZVS three-phase ac-to-ac buck–boost converter proposed in [24] (a) configuration and (b) link current and voltage

involved in charging and discharging of the link during each cycle. This results in a much smaller total harmonic distortion, and reduces the size of the required filter elements. Therefore, there will be 16 modes of operation in this converter. This converter is depicted in Figure 6.38(a), and the waveforms corresponding to this converter are shown in Figure 6.38(b). The order of input phase pairs for charging the link inductor, and the order of output phase pairs for discharging the link are determined such that the resonating time is minimized. Therefore, the input phase pair having higher voltage charges the link before the phase pair having a lower voltage. However, the output phase pair having lower voltage discharges the link before the output phase pair that has higher voltage.

The principles of the operation of this converter when used as dc-to-three-phase ac, three-phase ac-to-ac, and hybrid converters are described in [26–28]. One of the advantages of this converter is that regardless of the form, frequency, or number of phases of the input or output, it can provide galvanic isolation by the addition of a single-phase high-frequency transformer to the link. In dc-link converters, which are formed by a three inverter and a three-phase rectifier, galvanic isolation is provided by a bulky three-phase low-frequency transformer. Therefore, the parallel ac-link universal converter has a very high power density. The main problem associated with this converter is the large number of switches.

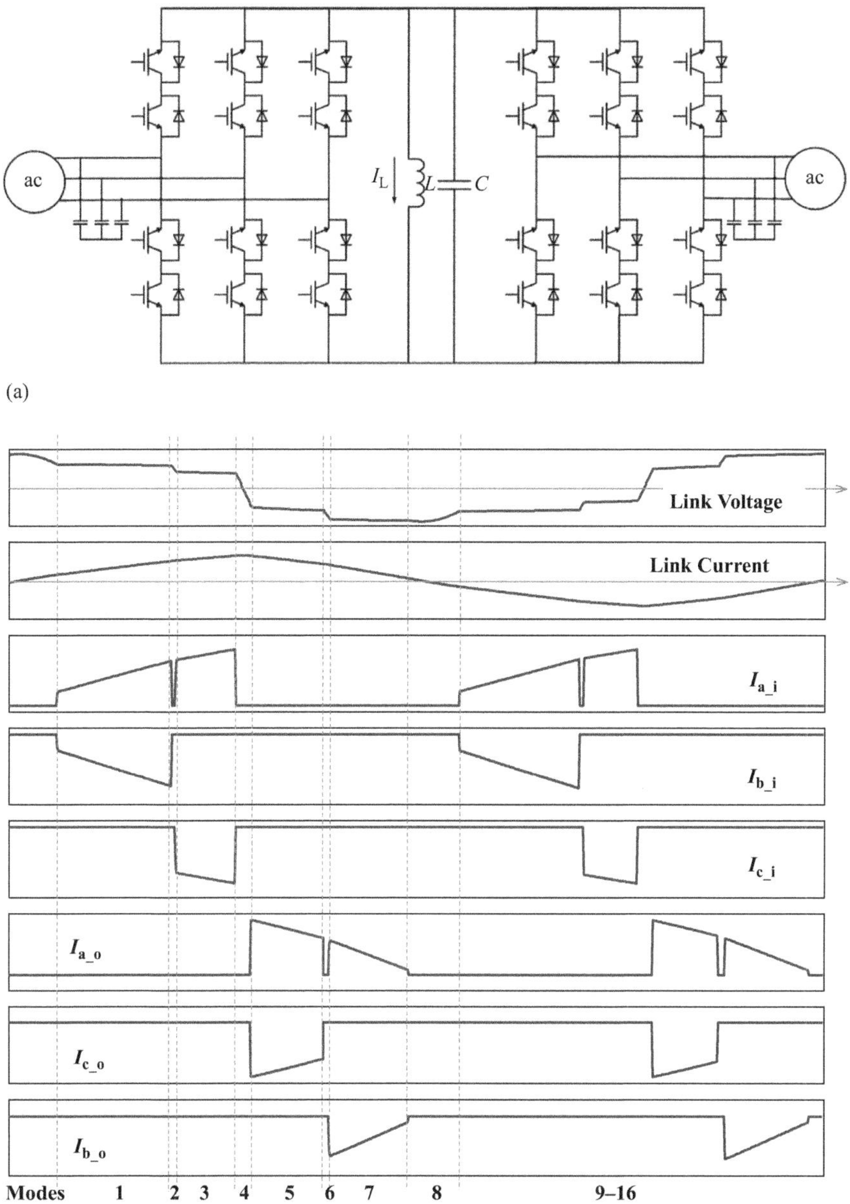

(b)

Figure 6.38 ZVS ac-link three-phase ac-to-ac buck–boost converter proposed in [25] (a) configuration and (b) link voltage, link current, and unfiltered phase currents

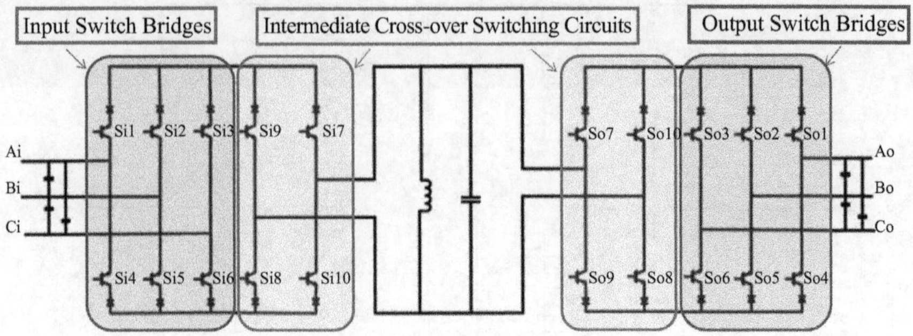

Figure 6.39 Sparse ac-link buck–boost converter

In [29], a reduced switch configuration, called sparse ac-link buck–boost converter, was proposed. The principles of the operation of this converter is similar to that of the converter introduced in [25]; however, unidirectional switches are used at the input and output switch bridges. To allow the link to be charged and discharged in both positive and negative directions, and have an alternating current, one switch-bridge at the input side and another switch-bridge at the output side are added. As depicted in Figure 6.39, these switch bridges each contain four switches.

In [30,31], another soft-switching converter was introduced, which was an extension of a soft-switching Cuk converter. In this converter, which is called series ac-link universal converter, all the switches can benefit from zero-current turn-off and soft turn-on. As shown in Figure 6.40, this converter is dual of the parallel ac-link universal converter.

6.5 Other types of ZVS and ZCS converters

The main advantage of the ZVS-QSWCs is providing zero-voltage turn-on without increasing the voltage stress over devices. In a ZVS-QSWC, the performance of the converter during the switching transitions is similar to that of the resonant converters; whereas, at other modes the converter operates similar to a PWM converter. This feature is very desirable. Several other soft-switching topologies have been proposed with a similar feature. Among these converters are pseudo-resonant full-bridge DC–DC converter [32], phase-shifted ZVS full-bridge DC–DC converter [33,34], and the auxiliary resonant commutated bridge converter [35,36].

A class of zero voltage transition (ZVT) converters can be developed by substituting the PWM switch network by a zero-voltage-resonant-transition (ZVRT) switch composite as shown in Figure 6.41 [37].

The principles of the operation of the converters developed by this method are similar to those of the QSWCs. During the voltage transition, the converter operates similar to a resonant converter to benefit from the soft switching; whereas, during the normal operation, the converter operates similar to a PWM converter. Figure 6.42 shows a ZVT buck–boost converter developed by this method along with the corresponding waveforms [13,37].

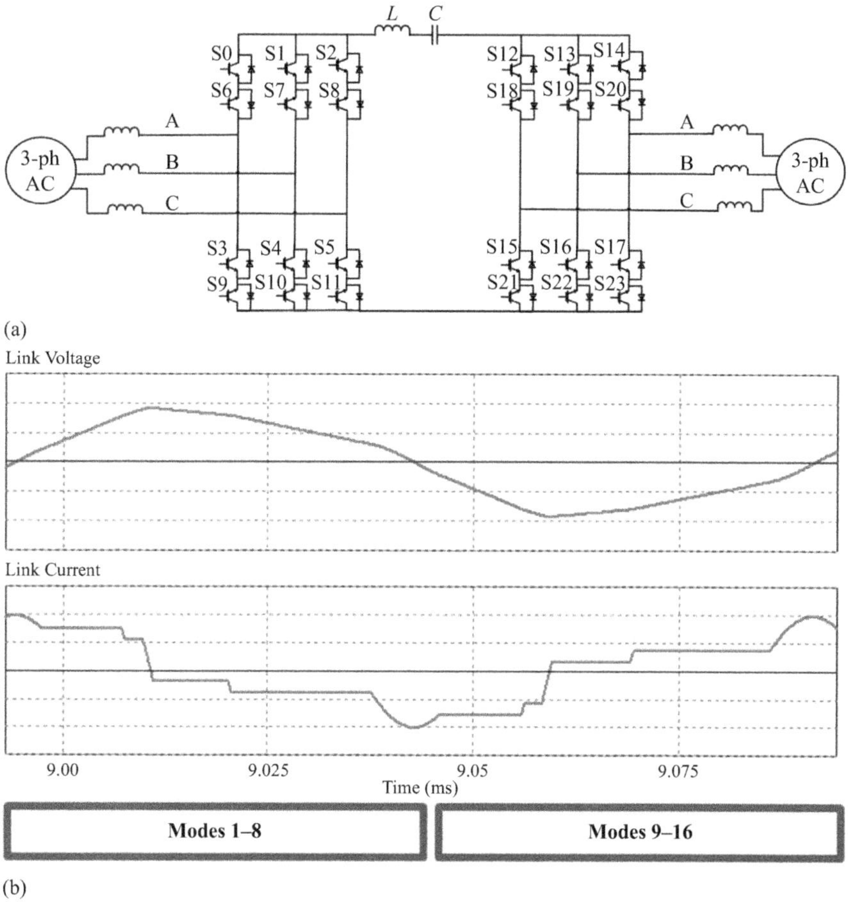

Figure 6.40 ZCS ac-link three-phase ac-to-ac Cuk converter proposed in [30,31] (a) configuration (b) and link voltage and link current

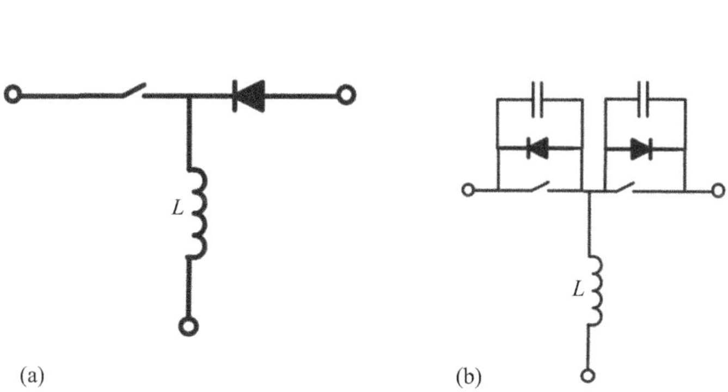

Figure 6.41 (a) PWM switch network and (b) zero-voltage-resonant-transition switch composite (adapted from [37])

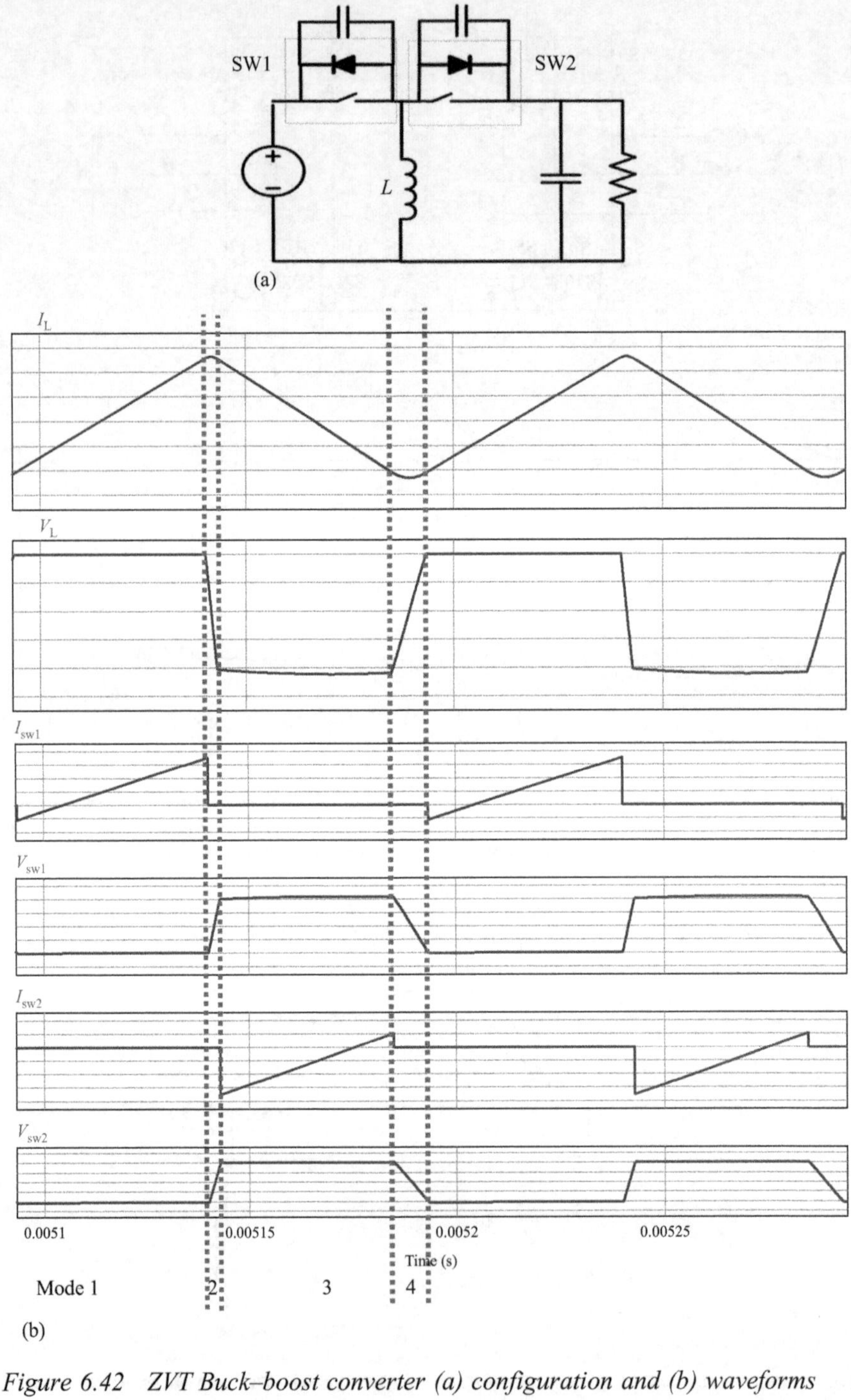

Figure 6.42 ZVT Buck–boost converter (a) configuration and (b) waveforms (inductor current, inductor voltage, switch sw1 current, switch sw1 voltage, switch sw2 current, and switch sw2 voltage)

Figure 6.43 shows another soft-switching converter that is developed using the switch composite shown in Figure 6.41 [38]. This topology, which is a ZVS full-bridge converter, is of specific importance, and is well suited for high power applications. In this converter, the two legs of the switch bridge operate with phase shift, and this allows the switches to benefit from the ZVS. The leakage inductance of the transformer (L) can play the role of the resonating inductor.

Another class of ZVT converters is formed by adding an auxiliary switch that allows controlling the path of current such that a resonating element is involved only during the switching transition. Figure 6.44 shows a ZVT converter that uses an auxiliary switch [39–42]. One problem with this converter is that the auxiliary switch has a hard switching.

The concept of resonant pole, which is shown in Figure 6.45(a), was also proposed to allow the switches benefit from ZVS without increasing their voltage stress. This converter operates similar to a PRC; however, the voltage of the link is clamped to the input voltage. This implies that before an active switch conducts, the corresponding anti-parallel diode conducts. The pseudo-resonant converter, which is formed by two resonant poles, is depicted in Figure 6.45 [32].

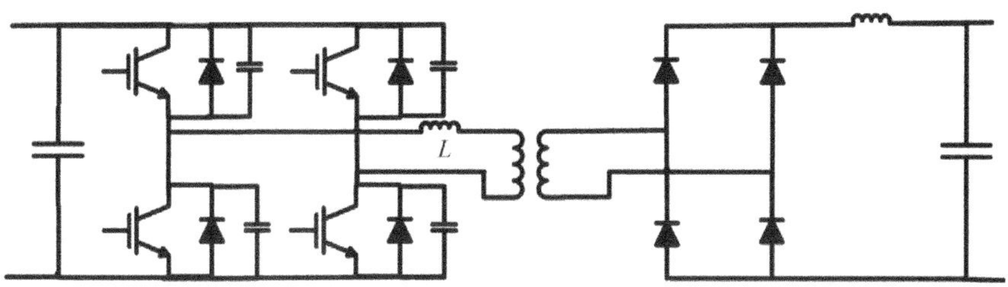

Figure 6.43 ZVS-full-bridge converter

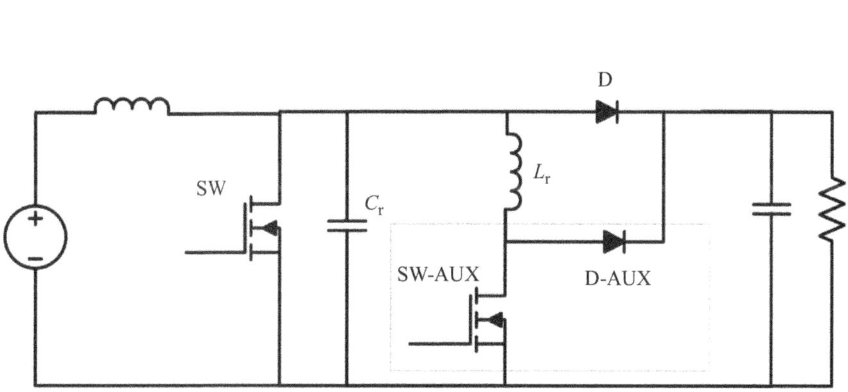

Figure 6.44 ZVT boost converters with an auxiliary switch

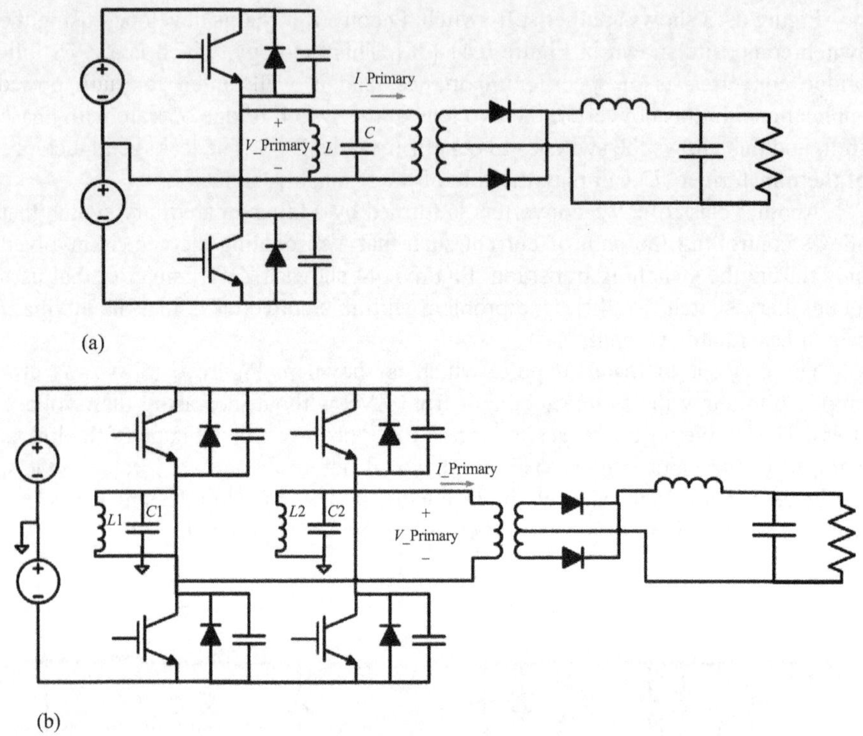

Figure 6.45 (a) Resonant pole circuit and (b) pseudo-resonant converter formed by two resonant pole circuits (adapted from [32])

References

[1] I. Batarseh, "Resonant converter topologies with three and four energy storage elements," *Power Electronics, IEEE Transactions on*, vol. 9, pp. 64–73, 1994.

[2] R. L. Steigerwald, "A comparison of half-bridge resonant converter topologies," *Power Electronics, IEEE Transactions on*, vol. 3, pp. 174–182, 1988.

[3] J. B. Klaassens, "DC-to-AC series-resonant converter system with high internal frequency generating synthesized waveforms for multikilowatt power levels," *Power Electronics, IEEE Transactions on*, vol. PE-1, pp. 9–20, 1986.

[4] J. B. Klaassens, "DC–AC series-resonant converter system with high internal frequency generating multiphase AC waveforms for multikilowatt power levels," *Power Electronics, IEEE Transactions on*, vol. PE-2, pp. 247–256, 1987.

[5] J. B. Klaassens, "Power conditioning with a switched series-resonant circuit operating at a high internal frequency," in *Telecommunications Energy*

Conference, 1987. INTELEC'87. The Ninth International, 1987, pp. 193–203.

[6] H. K. Lauw, J. B. Klaassens, N. G. Butler, and D. B. Seely, "Variable-speed generation with the series-resonant converter," *Energy Conversion, IEEE Transactions on*, vol. 3, pp. 755–764, 1988.

[7] J. B. Klaassens and F. de Beer, "Three-phase AC-to-AC series-resonant power converter with a reduced number of thyristors," in *Power Electronics Specialists Conference, 1989. PESC'89 Record, 20th Annual IEEE*, 1989, pp. 376–384, vol. 1.

[8] T. A. Lipo, "Recent progress in the development in solid-state AC motor drives," *Power Electronics, IEEE Transactions on*, vol. 3, pp. 105–117, 1988.

[9] Y. Murai and T. A. Lipo, "High frequency series resonant DC link power conversion," in *Industry Applications Society Annual Meeting, 1988, Conference Record of the 1988 IEEE*, 1988, pp. 772–779, vol. 1.

[10] P. K. Sood, T. A. Lipo, and I. G. Hansen, "A versatile power converter for high-frequency link systems," *Power Electronics, IEEE Transactions on*, vol. 3, pp. 383–390, 1988.

[11] P. K. Sood and T. A. Lipo, "Power conversion distribution system using a high-frequency AC link," *Industry Applications, IEEE Transactions on*, vol. 24, pp. 288–300, 1988.

[12] D. M. Divan, "The resonant DC link converter-a new concept in static power conversion," *Industry Applications, IEEE Transactions on*, vol. 25, pp. 317–325, 1989.

[13] R. W. Erickson and D. Maksimovic, *Fundamentals of Power Electronics*: Springer, New York, 2001.

[14] F. C. Lee, "High-frequency quasi-resonant converter technologies," *Proceedings of the IEEE*, vol. 76, pp. 377–390, 1988.

[15] F. C. Lee, "High-frequency quasi-resonant and multi-resonant converter technologies," in *Industrial Electronics Society, 1988. IECON'88. Proceedings, 14 Annual Conference of*, 1988, pp. 509–521.

[16] K.-H. Liu, R. Oruganti, and F. C. Y. Lee, "Quasi-resonant converters-topologies and characteristics," *Power Electronics, IEEE Transactions on*, vol. PE-2, pp. 62–71, 1987.

[17] W. A. Tabisz and F. C. Lee, "Zero-voltage-switching multi-resonant technique-a novel approach to improve performance of high frequency quasi-resonant converters," in *Power Electronics Specialists Conference, 1988. PESC'88 Record, 19th Annual IEEE*, 1988, pp. 9–17, vol. 1.

[18] F. H. Choo, B. R. Duggal, V. Rajagopalan, and A. Chandrasekaran, "Multi-resonant DC to AC converter," in *Industry Applications Society Annual Meeting, 1989, Conference Record of the 1989 IEEE*, 1989, pp. 1010–1014, vol. 1.

[19] M. M. Jovanovic and F. C. Lee, "Mode analysis of half-bridge zero-voltage-switched multi-resonant converter," in *Power Electronics Specialists Conference, 1989. PESC'89 Record, 20th Annual IEEE*, 1989, pp. 252–261, vol. 1.

[20] W. A. Tabisz and F. C. Lee, "DC analysis and design of zero-voltage-switched multi-resonant converters," in *Power Electronics Specialists Conference, 1989. PESC'89 Record, 20th Annual IEEE*, 1989, pp. 243–251, vol. 1.

[21] R. Farrington, M. M. Jovanovic, and F. C. Lee, "Constant-frequency zero-voltage-switched multi-resonant converters: analysis, design, and experimental results," in *Power Electronics Specialists Conference, 1990. PESC'90 Record, 21st Annual IEEE*, 1990, pp. 197–205.

[22] K. D. Ngo, "Generalization of resonant switches and quasi-resonant DC–DC converters," in *PESC'87-Annual IEEE Power Electronics Specialists Conference*, 1987, pp. 395–403.

[23] V. Vorperian, "Quasi-square-wave converters: topologies and analysis," *Power Electronics, IEEE Transactions on*, vol. 3, pp. 183–191, 1988.

[24] I. D. Kim and C. Gyu-Hyeong, "New bilateral zero voltage switching AC/AC converter using high frequency partial-resonant link," in *Industrial Electronics Society, 1990. IECON'90, 16th Annual Conference of IEEE*, 1990, pp. 857–862, vol. 2.

[25] W. C. Alexander, "Universal power converter," U.S. Patent 2008/0013351A1, Jan. 17, 2008.

[26] M. Amirabadi, A. Balakrishnan, H. A. Toliyat, and W. C. Alexander, "High-frequency AC-link PV inverter," *Industrial Electronics, IEEE Transactions on*, vol. 61, pp. 281–291, 2014.

[27] M. Amirabadi, B. Jeihoon, H. A. Toliyat, and W. C. Alexander, "Soft-switching AC-link three-phase AC–AC buck-boost converter," *Industrial Electronics, IEEE Transactions on*, vol. 62, pp. 3–14, 2015.

[28] M. Amirabadi, H. A. Toliyat, and W. C. Alexander, "A multiport AC link PV inverter with reduced size and weight for stand-alone application," *Industry Applications, IEEE Transactions on*, vol. 49, pp. 2217–2228, 2013.

[29] M. Amirabadi, B. Jeihoon, and H. A. Toliyat, "Sparse AC-link buck-boost inverter," *Power Electronics, IEEE Transactions on*, vol. 29, pp. 3942–3953, 2014.

[30] M. Amirabadi, J. Baek, and H. Toliyat, "Bidirectional soft switching series AC-link inverter," *Industry Applications, IEEE Transactions on*, vol. 51, pp. 2312–2320, 2015.

[31] M. Amirabadi and H. A. Toliyat, "A highly reliable converter for wind power generation application," in *Applied Power Electronics Conference and Exposition (APEC), 2013 Twenty-Eighth Annual IEEE*, 2013, pp. 1117–1123.

[32] O. D. Patterson and D. M. Divan, "Pseudo-resonant full bridge DC/DC converter," *Power Electronics, IEEE Transactions on*, vol. 6, pp. 671–678, 1991.

[33] R. A. Fisher, K. D. T. Ngo, and M. H. Kuo, "A 500 kHz, 250-W DC–DC converter with multiple outputs controlled by phase-shifted PWM and magnetic amplifiers," in *High Frequency Power Converter Conference*, 1988, pp. 100–110.

[34] R. Redl, L. Balogh, and D. W. Edwards, "Optimum ZVS full-bridge DC/DC converter with PWM phase-shift control: analysis, design considerations, and experimental results," in *Applied Power Electronics Conference and Exposition, 1994. APEC'94. Conference Proceedings 1994, Ninth Annual*, 1994, pp. 159–165, vol. 1.

[35] R. W. De Doncker and J. P. Lyons, "The auxiliary resonant commutated pole converter," in *Industry Applications Society Annual Meeting, 1990, Conference Record of the 1990 IEEE*, 1990, pp. 1228–1235, vol. 2.

[36] W. McMurray, "Resonant snubbers with auxiliary switches," in *Industry Applications Society Annual Meeting, 1989, Conference Record of the 1989 IEEE*, 1989, pp. 289–834, vol. 1.

[37] C. P. Henze, H. C. Martin, and D. W. Parsley, "Zero-voltage switching in high frequency power converters using pulse width modulation," in *Applied Power Electronics Conference and Exposition, 1988. APEC'88. Conference Proceedings 1988, Third Annual IEEE*, 1988, pp. 33–40.

[38] J. A. Sabate, V. Vlatkovic, R. B. Ridley, F. Lee, and B. H. Cho, "Design considerations for high-voltage high-power full-bridge zero-voltage-switched PWM converter," in *Applied Power Electronics Conference and Exposition, 1990. APEC'90, Conference Proceedings 1990, Fifth Annual*, 1990, pp. 275–284.

[39] G. Hua, C. S. Leu, and F. C. Lee, "Novel zero-voltage-transition PWM converters," in *Power Electronics Specialists Conference, 1992. PESC'92 Record, 23rd Annual IEEE*, 1992, pp. 55–61, vol. 1.

[40] G. Hua, C. S. Leu, J. Yimin, and F. C. Y. Lee, "Novel zero-voltage-transition PWM converters," *Power Electronics, IEEE Transactions on*, vol. 9, pp. 213–219, 1994.

[41] G. Hua, E. X. Yang, J. Yimin, and F. C. Lee, "Novel zero-current-transition PWM converters," in *Power Electronics Specialists Conference, 1993. PESC'93 Record, 24th Annual IEEE*, 1993, pp. 538–544.

[42] G. Hua, E. X. Yang, J. Yimin, and F. C. Lee, "Novel zero-current-transition PWM converters," *Power Electronics, IEEE Transactions on*, vol. 9, pp. 601–606, 1994.

Chapter 7
Z-source converters

Yam P. Siwakoti, Frede Blaabjerg* and Poh Chiang Loh**

7.1 Introduction

7.1.1 General overview

Impedance-sourced networks provide an efficient means of power conversion between source and load in a wide range of electric power conversion applications (dc–dc, dc–ac, ac–dc, and ac–ac). Various topologies and control methods using different impedance source networks have been presented in the literature, e.g. for adjustable-speed drives, uninterruptible power supply, distributed generation (fuel cell, photovoltaic (PV), wind, etc.), battery or super-capacitor energy storage, electric vehicles, distributed dc power systems, avionics, flywheel energy storage systems, electronic loads, dc circuit breaker, and many more [1–13]. A variety of converter topologies with buck, boost, buck–boost, unidirectional, bidirectional, isolated as well as non-isolated converters are possible by proper implementation of the impedance source network with various switching devices, topologies, and configurations. Figure 7.1 shows the general configuration of an impedance source network for electric power conversion.

The impedance source network was originally invented to overcome the limitations of the voltage source inverter (VSI) and current source inverter (CSI) topologies, which are commonly used in electric power conversion [1]. The ac output voltage of the VSI is limited to be below the input voltage, i.e. the VSI is a buck type inverter which cannot serve the need of distributed generation and ac drives alone. It requires an additional dc–dc boost converter to obtain a desired ac output, which increases system cost and lowers the efficiency. In addition, the switching devices are vulnerable to electromagnetic interference as misgating-on causes a short circuit across the inverter bridge and destroys the switching devices. The dead time introduced in such cases causes waveform distortion at the output. On the other hand, in the case of the CSI, the output voltage cannot be less than the input voltage. For applications where a wide voltage range is desirable, an additional dc–dc buck converter is needed. In addition, the upper and the lower switches

*Aalborg University, Aalborg, Denmark

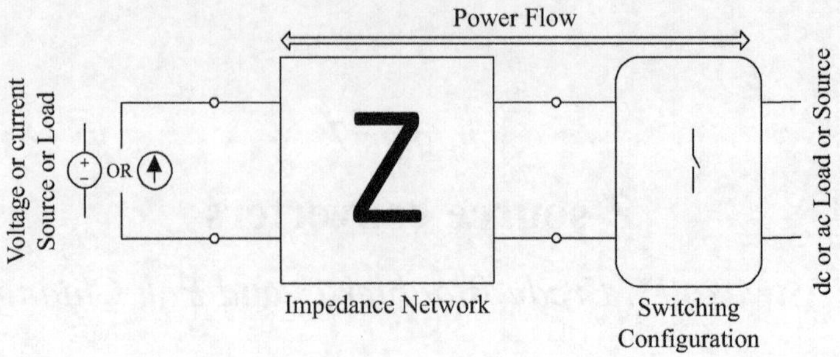

Figure 7.1 A general circuit configuration of an impedance-sourced network for power electronic conversion

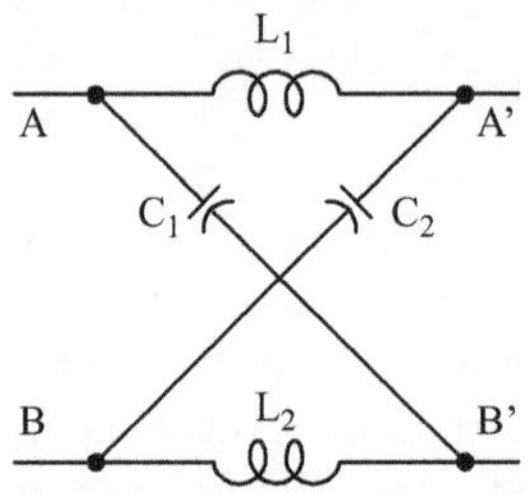

Figure 7.2 Basic Z-source impedance network with two inductors and capacitors

of the inverter have to be gated on and maintained on at any time. Otherwise, an open circuit of the dc inductor would occur and destroy the devices.

The impedance-sourced converter overcomes the above-mentioned conceptual and theoretical barriers and limitations of the classical VSI and CSI and provides a novel power-conversion concept. The major advantage of this topology is that it can operate as V-source or an I-source depending on the application and needs, and the output voltage can be varied from 0 to ∞. Since the publication of the first impedance source network, called a "Z-source network", in year 2002 [1], many modified topologies with improved modulation and control strategies have been proposed and published to improve the performance in various applications [14]. Figure 7.2 shows the basic Z-source impedance network, which consists of inductors L_1 and L_2 and capacitors C_1 and C_2 connected at both ends (Z-shape) and acts as a buffer between load and source (voltage source or current source).

According to the conversion functionality, impedance (Z-) source power converter can be divided into four main categories: dc–dc converters, dc–ac inverters, ac–ac converters, and ac–dc rectifiers. A further breakdown leads to two-level and multilevel, ac–ac and matrix converters, and non-isolated and isolated dc–dc converters. From the Z-source network topology standpoint, it can be voltage-fed or current-fed. Further, impedance networks can be divided based on the magnetics used in the impedance source network, i.e. non-transformer based and transformer or coupled inductor based. Each topology has its own unique features and applications to which it is best suited. There is no one-size-fits-all solution. It is expected that new *Z*-source topologies will continuously be put forth to meet and improve converter performance in different applications.

7.1.2 Basic principles

The concept of the impedance source network can be applied to any dc-to-dc, ac-to-ac, ac-to-dc, and dc-to-ac power conversion. The dc source and/or load can be a voltage or current source and/or a load. A *Z*-source impedance network is used as an example to briefly illustrate the operating principle and control of the impedance source network. Figure 7.3 shows the circuit diagram of the *Z*-source converter and its equivalent circuit during active and shoot-through states. During the shoot-through state, the output terminals of the impedance networks A′ and B′ are short circuited by a switch or a combination of switches which will, in turn, cause diode D in the network to reverse-bias as shown in Figure 7.3(a). The energy stored in the inductor and capacitor during this shoot-through state is transferred to the load during the next active state, in which the diode D is returned to conduction. The switching circuit viewed from the dc side during the active state is equivalent to a current source as shown in Figure 7.3(b). Averaging of these two switching states

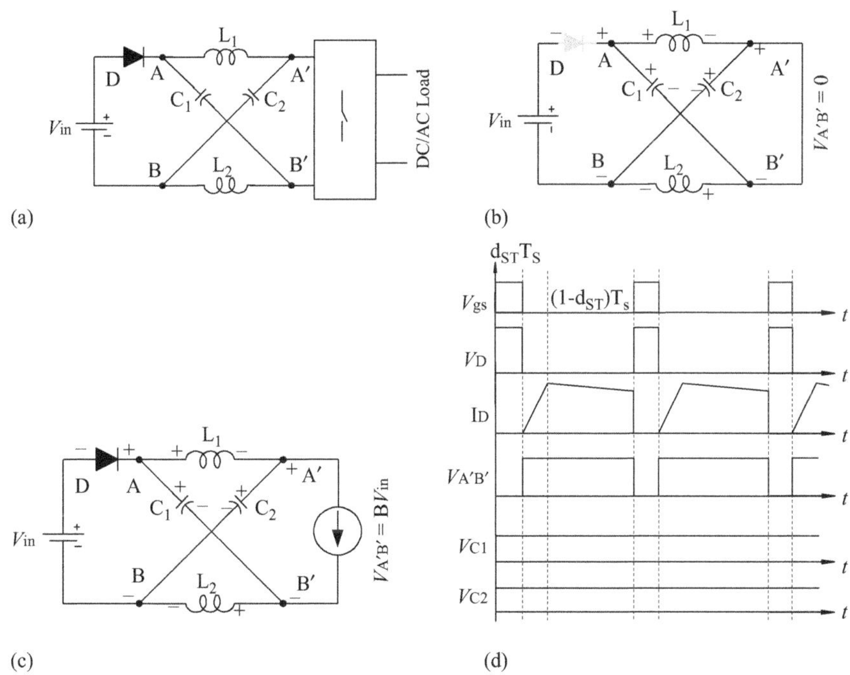

Figure 7.3 Voltage-fed Z-source converter: (a) circuit diagram, (b) equivalent circuit at a shoot-through, (c) equivalent circuit in the active state, (d) current and voltage waveforms

results in an expression to compute the peak dc-link voltage $\hat{v}_{A'B'}$ across terminals A′ and B′, in terms of its input voltage V_{in} as

$$\hat{v}_{A'B'} = \frac{1}{1 - \beta d_{ST}} V_{in} \tag{7.1}$$

$$= BV_{in} \tag{7.2}$$

where d_{ST} is the fractional shoot-through time assumed in a switching period, and $\beta \geq 2$ is a factor determined by the impedance network chosen, e.g. for Z-source inverter and quasi-Z-source inverter (ZSI and qZSI), $\beta = 2$. Equating the denominator of the boost factor (B) to zero then results in the permissible range of d_{ST} as $0 \leq d_{ST} < \frac{1}{\beta}$, whose upper limit corresponds to an infinite gain. Waveforms of current and voltage considering single switch at the dc-link (A′ and B′) are illustrated in Figure 7.3(c).

A three-phase voltage-fed Z-source inverter, as shown in Figure 7.4, is used as an example to briefly illustrate the operating principle as described in Figure 7.3. The three-phase Z-source inverter bridge has nine permissible switching states (six active states, two zero states, and one shoot-through state), unlike the traditional three-phase VSI, which has eight (six active states and two zero states). During zero states, the upper three or lower three switches of the inverter bridge are turned on simultaneously, thus shorting the output terminals of the inverter and producing zero voltage across the load. During one of the six active states, the dc voltage is impressed across the load, positively or negatively. However, during the shoot-through states, the load terminals are shorted through both the upper and lower devices of any phase leg, any two-phase legs, and all three-phase legs [1] producing zero voltage across the load. This shoot-through state has the same effect, i.e. producing zero voltage across the load as the traditional zero states. However, these shoot-through states can boost the output voltage. The shoot-through state is forbidden in the traditional VSI, because it would cause a short circuit across the dc-link and damage the converter. The Z-source network and the shoot-through zero state provide a unique buck–boost capability for the inverter by varying the shoot-through time period and modulation index (M) of the inverter. Theoretically, the output voltage of the inverter can be written as

$$\hat{v}_{ac} = \frac{MB}{2} \tag{7.3}$$

$$= M[1 - 2d_{ST}]^{-1}(V_{in}/2) \tag{7.4}$$

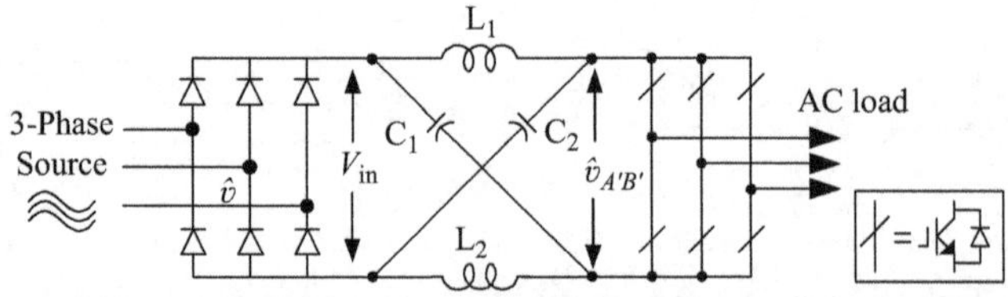

Figure 7.4 Voltage-fed Z-source three-phase inverter – an example of a Z-source inverter

The output voltage in (7.4) can be set to any value between 0 and ∞ by varying the value of M and d_{ST}. However, some practical aspects and performance of the converter need to be considered for large voltage buck or boost operation in order to avoid exceeding device limitations. Some of the waveforms during boost operation are illustrated in Figure 7.5.

All traditional pulse width modulation (PWM) schemes can be used to control the impedance-source converter, and their theoretical input–output relationships still hold true. However, in addition to all states in the traditional modulation techniques, a new state called a "shoot-through state" is introduced and embedded in the modulation strategy for the impedance network–based power converter without violating the volt-sec balance in the operating principle. With the unique feature of these shoot-through states, several new PWM methods modified from the traditional techniques are prevalent in the literature. Figure 7.6 shows a broad categorization of the impedance-network–based power converter using different switch configurations [15].

7.1.3 Modeling and control

A *Z*-source network shows non-minimum phase behavior due to the presence of zero in the right-half-plane (RHP) which could impose a limitation on the controller design. In order to implement a good control strategy, it is imperative to have a good dynamic model of the converter. Various small-signal analysis and mathematical models are presented in the literature to study the dynamic behavior of the system, which then can be implemented in different closed-loop control strategies with different complexities based on various applications [16–24].

To derive an accurate small-signal model, various state variables are selected, such as the input current $(i_{\text{in}}(t))$, inductor currents $(i_{\text{L1}}(t), i_{\text{L2}}(t), \ldots)$, capacitor voltages $(v_{\text{C1}}(t), v_{\text{C2}}(t), \ldots)$, and load currents $\left(i_{\text{L}}(t), i_{\text{d}}(t), i_{\text{q}}(t)\right)$. The small-signal model provides the required transfer function for the controller design and provides a detailed view of the system dynamics, which helps to understand the system

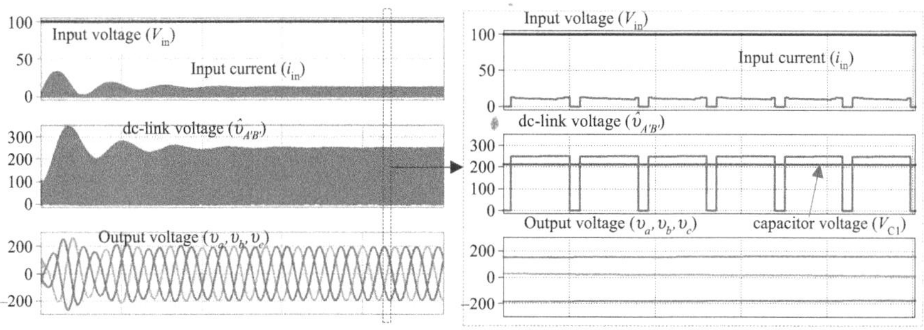

Figure 7.5 Illustration of some of the voltage and current waveforms in voltage-fed Z-source inverter

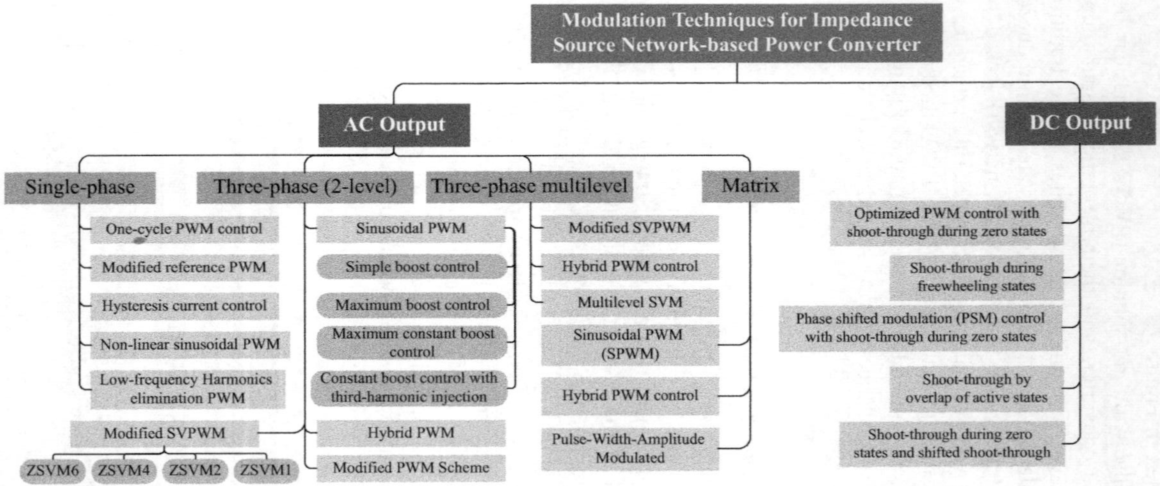

Figure 7.6 Categorization of modulation techniques for impedance source network–based power converter

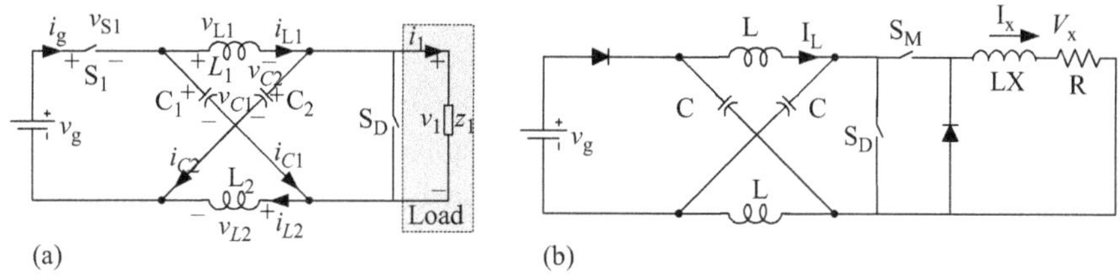

Figure 7.7 Simplified equivalent circuit of Z-source converter for small-signal modeling: (a) D_{st} as control variable and (b) D_{st} and M as control variable

limits, and provides guidelines for system controller design. In general, M and D_{st} are considered as control variables and the capacitor voltage ($v_{\mathrm{C}}(t)$) or the dc-link voltage ($v_{\mathrm{PN}}(t)$) and the load voltage ($v_{\mathrm{x}}(t)$) as variables to be controlled. Figure 7.7(a) shows the simplified Z-source converter model for small-signal analysis, where $v_{\mathrm{C}}(t)$ is controlled using D_{st} as a control variable (control switch $\mathrm{S_D}$). This is the most simplified model, however it does not guarantee tight control of $v_{\mathrm{x}}(t)$, which requires an additional control variable M (control switch $\mathrm{S_M}$) as shown in Figure 7.7(b) [16]. In addition to the state variables, the parasitic resistance of the inductor (r) and the equivalent series resistance (ESR) of the capacitor (R) also influence the dynamics of the impedance source networks and hence are also considered during modeling of the converter to analyze the sensitivity of the circuit under parameter variations [17–19].

Based on some of the above state variables, several small-signal models have been proposed for symmetrical or asymmetrical ZSI [16,20,21] and qZSI [17]. Considering the symmetry of the network (using $v_{\mathrm{C1}}(t) = v_{\mathrm{C2}}(t) = v_{\mathrm{C}}(t)$ and $i_{\mathrm{L1}}(t) = i_{\mathrm{L2}}(t) = i_{\mathrm{L}}(t)$), a simplified small-signal model is presented in [18,20] for ZSI, where the load current is represented by a constant current source. However, such a model describes only the dynamics of the impedance network and fails to describe the dynamics of the ac load. To overcome this disadvantage, a third-order model is presented in [16] using $v_{\mathrm{C1}}(t) = v_{\mathrm{C2}}(t) = v_{\mathrm{C}}(t)$, $i_{\mathrm{L1}}(t) = i_{\mathrm{L2}}(t) = i_{\mathrm{L}}(t)$, and $i_{\mathrm{l}}(t)$ as state variables. In this model, the ac side of the inverter is referred to the dc side with R_{L} load and taking its current as a third state variable. A similar third-order small-signal model is presented in [22], which considers the dynamics of the input-side current. In this, the current-fed qZSI is analyzed using $v_{\mathrm{C}}(t)$, $i_{\mathrm{L}}(t)$, and $i_{\mathrm{in}}(t)$ as state variables to demonstrate the transient response of the inverter during the motoring and regeneration modes of operation for application in electric vehicles. Subsequent fourth- and higher-order small-signal models are also presented for inverters [23] and rectifiers [24] to better understand the dynamics of the input/output (load/source) and the impedance network. However, the complexity in formulating the small-signal model and the control-loop design increases with the increase in state variables. To simplify this, various assumptions (symmetry in impedance network, balanced load) and simplifications (representation of ac load/source by its equivalent dc load/source) are prevalent in the literature without loss of generality and changes to dynamic performance.

The state-space-averaged small-signal modeling provides a derivation of various control-to-output ($G_{\hat{d}_0}^{\hat{v}_c}(s)$) and disturbance-to-output ($G_{\hat{v}_{in}}^{\hat{v}_c}(s)$, $G_{\hat{i}_{load}}^{\hat{v}_c}(s)$) transfer functions, which helps to predict the system dynamics under the influence of various parameter changes. The root-locus of the control-to-output transfer function in the s-domain gives a clear map of the converter dynamics. In addition, predicting a RHP zero in the control-to-output transfer function is a major advantage of small-signal modeling. The presence of RHP zeros indicates that the non-minimum phase undershoot (the controlled capacitor voltage dips before it rises in response to a D_{st} increase), generally tends to destabilize the wide-band feedback loops, implying high gain instability and imposing control limitations. This means that the design of a feedback loop with an adequate phase margin becomes critical when RHP zeros appear in the transfer function. Various analyses of the pole-zero location and the impact of parameter variations on the converter dynamics are studied considering the wide operating ranges of different sources, e.g. fuel cells and photovoltaics. Figure 7.8 shows the locus of the poles and zeros with changes in various parameters such as L, C, D_{st}, R, and r. The impact of these parameter variations on the converter dynamics is summarized in Table 7.1.

The impact of parameter variations on the system dynamics as discussed above can provide direction to designers beforehand to choose component values while

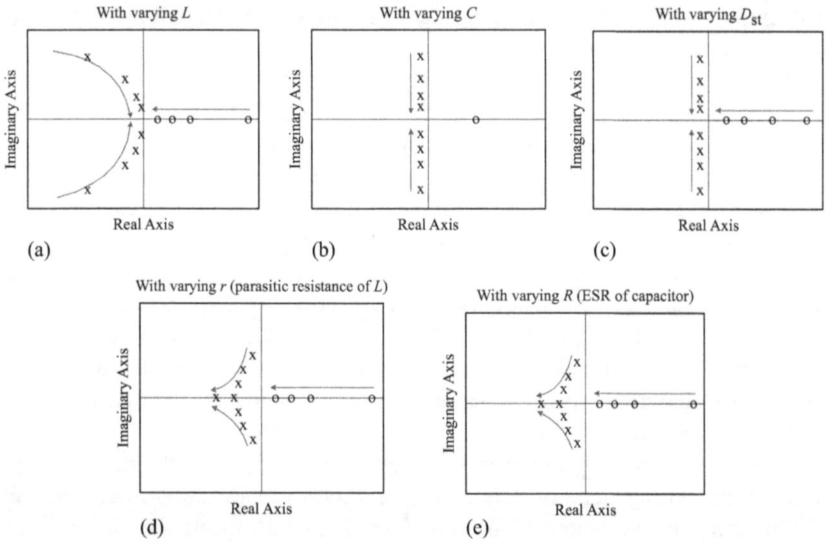

Figure 7.8 Locus of poles and zeros of control-to-output transfer function with varying (a) inductance (L), (b) capacitance (C), (c) shoot-through duty cycle (D_{st}), (d) parasitic resistance of inductor (r), and (e) equivalent series resistance (ESR) of capacitor (R). The direction of the arrow indicates an increasing value of the parameter

Table 7.1 Summary of impact of parameter variations on the Z-source converter dynamics

Parameter	Change	Effect on position of		Impact on system dynamics
		Conjugate poles	RHP zeros	
Inductance (L)	Increasing	Move towards the imaginary axis	Move towards the imaginary axis	• Increase non-minimum phase undershoot • Increase settling time • Increase oscillatory response • Decrease damping ratio • Decrease natural frequency
Capacitance (C)	Increasing	Move towards the real axis	Constant	• Increase system damping • Increase rise time • Increase system settling time • Decrease natural frequency
Shoot-through duty cycle (d_{ST})	Increasing	Move towards the real axis	Move towards the imaginary axis	• Increase non-minimum phase undershoot • Increase system settling time • Decrease natural frequency
Equivalent series resistance (ESR) of capacitor (R)	Increasing	Move towards the real axis	Move towards the imaginary axis	• Increase system damping • Increase non-minimum phase undershoot • Increase current ripple through C
Parasitic resistance of inductor (r)	Increasing	Move towards the real axis	Move towards the imaginary axis	• Increase system damping • Increase non-minimum phase undershoot • Increase voltage ripple across L

considering the design constraints, such as feedback control bandwidth, ripple content, size and cost of components, damping factor, resonant frequency, and overshoot/undershoot in the desired output.

When considering the effect of parameter variations and the effect of poles and RHP zero, several closed-loop control methods are proposed in the literature to achieve a desired performance and to control the dc-link voltage and the ac output voltage of the impedance-source converter [25–29]. In all these control methods, there are two control degrees of freedom (D_{st} and M). The dc-link voltage is controlled by the shoot-through duty ratio (D_{st}) and the output voltage is controlled by the modulation index M.

The dc-link voltage across the inverter bridge can be controlled both directly and indirectly. In the direct dc-link voltage control method [25], the voltage across the inverter bridge is sensed directly by special sensing and scaling circuits as shown in Figure 7.9(b). This improves the transient response, enhances disturbance rejection capability and simplifies the controller design process. In the indirect method, the capacitor voltage at the impedance network is typically sensed and compared with the desired voltage as shown in Figure 7.9(a), (c), and (d). There are two control methods in this category: (1) measurement of capacitor voltage (V_C) in the impedance source network as shown in Figure 7.9(a) [23,26] and (2) measurement of capacitor voltage (V_C) and input voltage (V_{in}) to estimate the peak dc-link voltage as shown in Figure 7.9(c) and (d) [27].

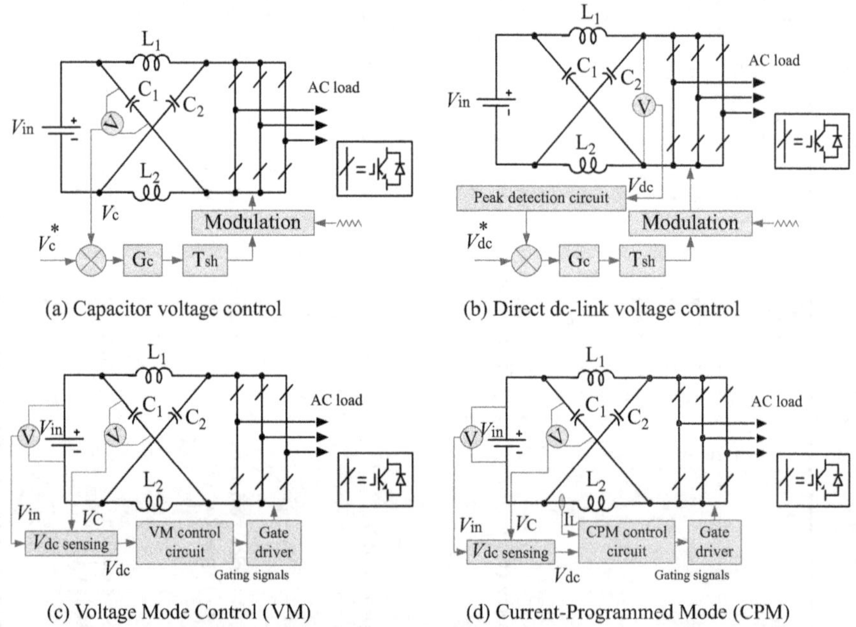

Figure 7.9 Control of dc-link voltage by indirect ((a), (c), and (d)) and direct methods (b) for Z-source inverters

In the indirect control methods, the peak dc-link voltage becomes uncontrollable, while regulating the shoot-through duty cycle for fast changing input voltages. This effect is unacceptable as it affects the output voltage, which forces a change in the modulation index. This may result in higher semiconductor stress and increases the harmonic distortion [26,27] in the output waveforms. The peak dc-link voltage is kept constant in the direct measurement technique; however, the control scheme becomes more complex, with additional circuitry. To resolve this limitation, the peak dc-link voltage is estimated by measuring the input voltage and capacitor voltage as shown in Figure 7.9(c) (voltage mode (VM)) and Figure 7.9(d) (current-programmed mode (CPM)), however additional voltage/current sensors are required.

Figure 7.10 shows a block diagram of two popular closed-loop voltage control methods of ZSI which consist of both dc-link control and ac output control [28,29]. In both control methods, the capacitor voltage is controlled by regulating the shoot-through duty cycle (D_{st}) and the output voltage is controlled by regulating the modulation index (M), by using separate control loops with propotional (P) and/or Proportional & Integral (PI) controllers. However, both control parameters are dependent on each other, as a change in one parameter imposes a limitation of the changeability of the other due to the insertion of a shoot-through time inside the null period. Putting a maximum limit on the control variable could mitigate this limitation.

7.2 Categories of impedance source power converters based on conversion functionality

Figure 7.11 shows a broad categorization of the impedance source power converters using different switching configurations, which will be discussed in the following subsections in details.

7.2.1 DC–DC converter topologies

Various isolated/non-isolated dc–dc converters are proposed in the literature with different control and modulation techniques. For example, a dc–dc converter realized using a qZSI with two- or three-leg H-bridge switching topology is proposed for distributed generation [30,31] as shown in Figure 7.12(a). A new quasi-*Z*-source push–pull converter topology with a reduced number of switching device is also proposed in [32] as shown in Figure 7.12(b). It has the same gain as in [30,31], however, with reduced complexity in gate circuit design.

A family of four-quadrant dc–dc converters using a *Z*-source or a quasi-*Z*-source network with a minimal number of switches and passive devices is presented in [33]. The converter provides four-quadrant operation using four-quadrant switches. Two basic converter topologies derived from ZSI and qZSI are shown in Figure 7.13. The converter has both buck/boost characteristics in the 0–1 range of the duty cycle. This feature along with changing the polarity of the load voltage by just controlling

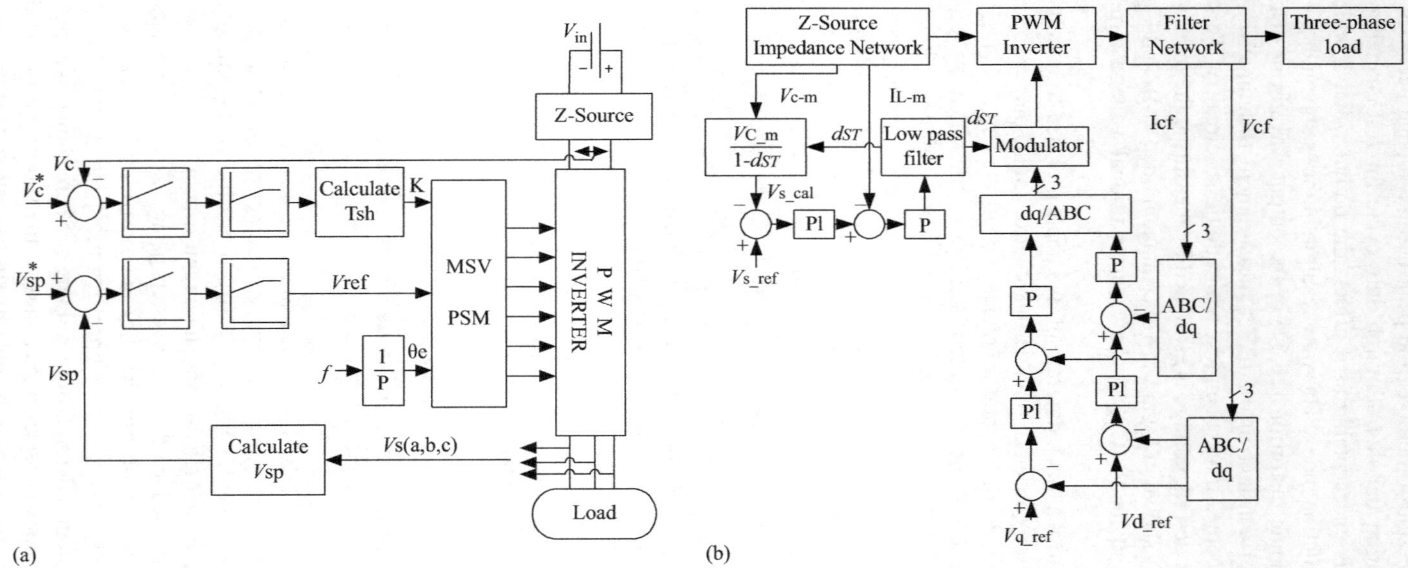

Figure 7.10 *Indirect dc-link voltage controller modules with two degrees of freedom: (a) controller for both dc boost and ac output voltage of Z-source inverter and (b) multi-loop closed-loop controller*

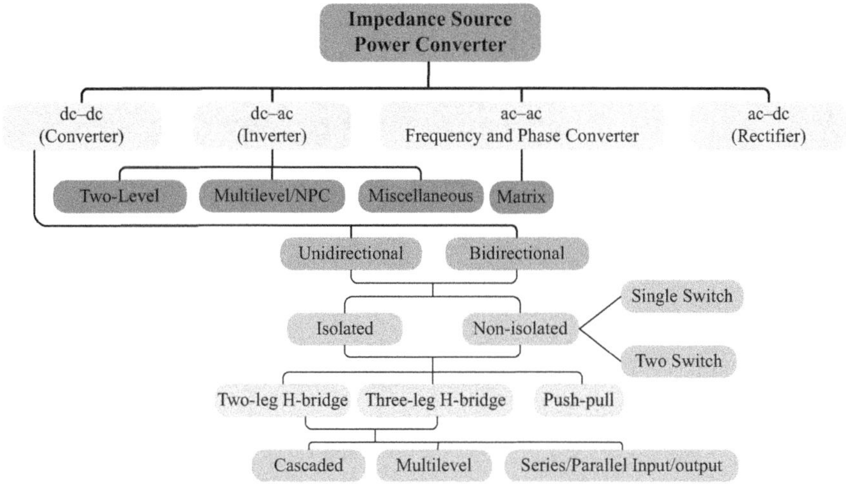

Figure 7.11 Categorization of impedance source network–based power converters

the duty cycle of the switch makes the converter very simple and more economical for many applications, e.g. dc drives and other renewable energy systems.

A new boost dc–dc converter topology was proposed in [34] with three coupled inductors and called a *Y*-source converter as shown in Figure 7.14. It can achieve a very high boost at a lower shoot-through duty cycle of the switch and has one more degree of freedom to choose the voltage boost. The converter can be designed compact with integrated magnetics, and fewer components are required to attain a high-voltage boost. An isolated high-voltage boost converter using a *Y*-source impedance source network is also proposed in [35] which reduced the number of switching devices and the corresponding switching complexities.

Modulation of single-switch [34] and two-switch [35] dc–dc converter topologies is fairly simple and can be achieved by controlling the duty cycle of the switch depending on the dc-link voltage. However, a dc–dc converter using an intermediate H-bridge as shown in Figure 7.15 involves more complex control, as four switches are required to be switched optimally to get the desired output and performance.

Various modulation techniques are proposed in the literature, including shoot-through during freewheeling states, shoot-through during zero states, phase-shift modulation (PSM) control with shoot-through during zero states [31,36] shoot-through by overlap of the active states [31,37] and shifted shoot-through [38], as shown in Figure 7.16. Due to insertion of shoot-through in PWM, the switches in the H-bridge are compelled to commute at 2–3 times higher than the switching frequency. In shoot-through during freewheeling states, the top-side and bottom-side switches of the inverter bridge operate at three times the switching frequency $\left(f_{\mathrm{sw,top}} = 3f_{\mathrm{s}} \text{ and } f_{\mathrm{sw,bottom}} = 3f_{\mathrm{s}}\right)$ as shown in Figure 7.16(a). This leads to high switching loss. Similarly, the number of commutations of the bottom-side switches remains the same while reducing the switching frequency of the top-side switch $\left(f_{\mathrm{sw,top}} = f_{\mathrm{s}} \text{ and } f_{\mathrm{sw,bottom}} = 3f_{\mathrm{s}}\right)$ in the shoot-through during zero states modulation

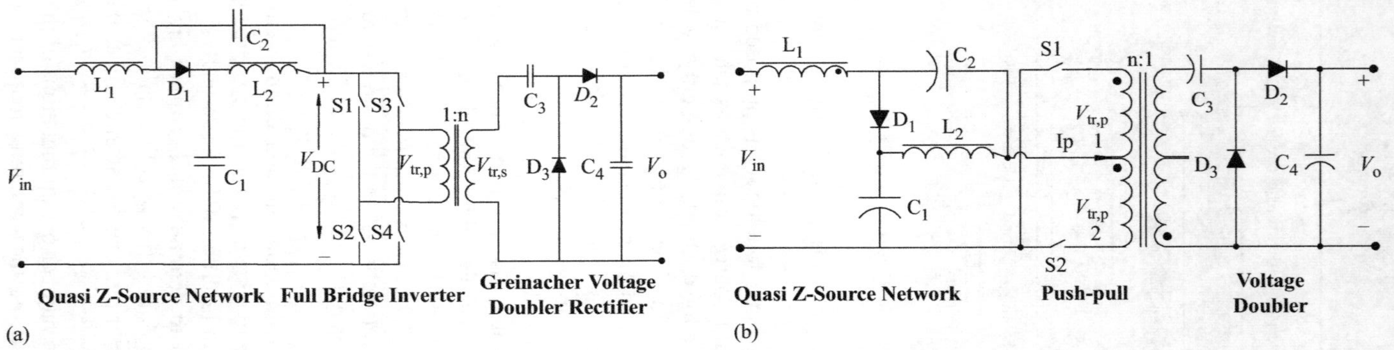

Figure 7.12 qZSI-based isolated dc–dc converter with (a) intermediate H-bridge switching topology and (b) push–pull topology

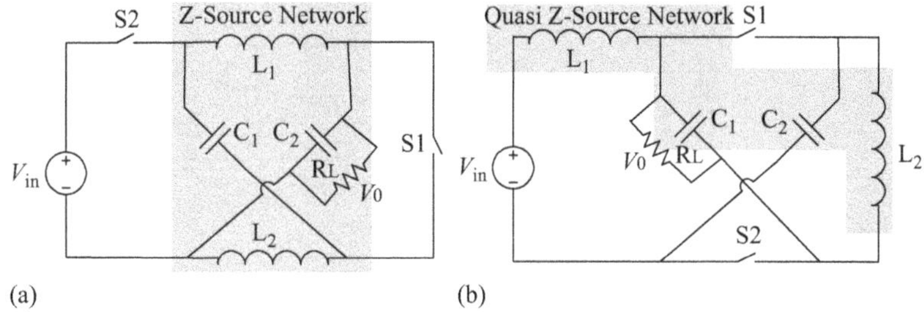

Figure 7.13 Family of four-quadrant dc–dc converters using (a) Z-source and (b) quasi-Z-source network

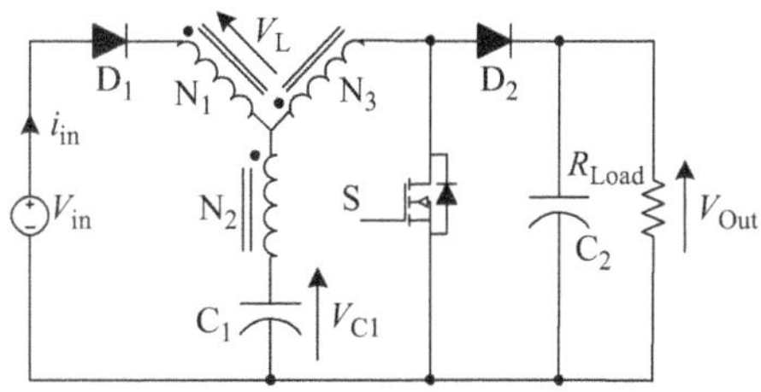

Figure 7.14 Y-source boost dc–dc converter

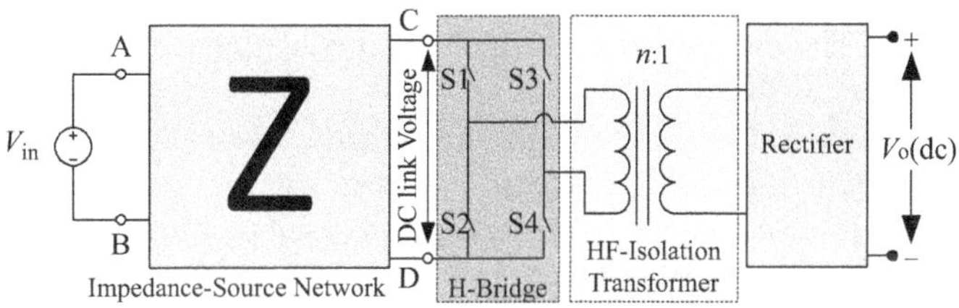

Figure 7.15 A general circuit configuration of impedance source network for dc–dc power conversion with intermediate H-bridge topology using various impedance source networks

techniques (see Figure 7.16(b)). PSM control with shoot-through during zero states (see Figure 7.16(c)) equalizes the switching losses of the top-side and bottom-side switches $(f_{sw,top} = 2f_s$ and $f_{sw,bottom} = 2f_s)$, but this method is not effective in reducing the commutation time of the switches. The switching loss is minimized in the shoot-through by the overlap of active states $(f_{sw,top} = f_s$ and $f_{sw,bottom} = f_s)$ as shown in Figure 7.16(d); however, the active state and shoot-through state duty cycles are not independently controllable. The inter-dependency of the active state and shoot-through state duty cycle could cause problems in the output-voltage compensation and also for systems which require independent control of active and shoot-through state. A shifted shoot-through modulation technique (see Figure 7.16(e)) could reduce the switching frequency of the switches but it is complex and difficult

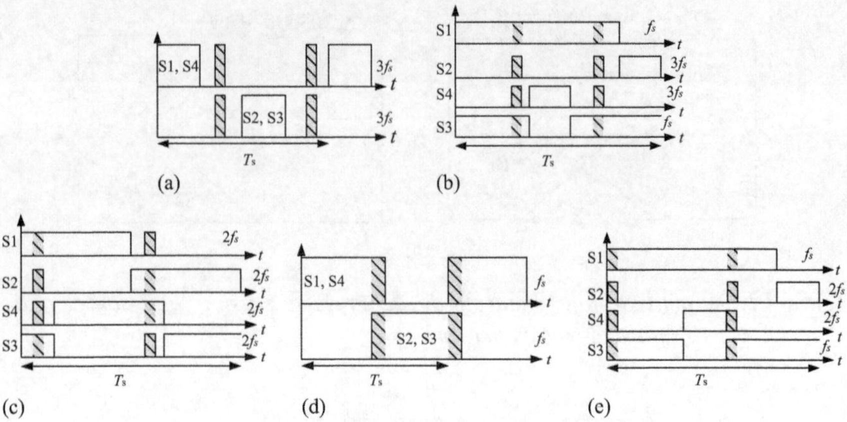

Figure 7.16 Some of the modulation techniques for a dc–dc converter with intermediate H-bridge shown in Figure 7.15. (a) Shoot-through during freewheeling states, (b) shoot-through during zero states, (c) phase-shift modulation (PSM) control with shoot-through during zero states, (d) shoot-through by the overlap of active states, and (e) shifted shoot-through (////→ shoot-through)

to implement particularly in the microcontroller due to large number of comparator requirement. It also requires additional external components (logic gates, etc.) for implementation. The major disadvantage of this modulation technique is the loss of full soft-switching properties.

Every addition of the shoot-through state increases the commutation time of the semiconductor switches and so increases the switching loss in the system. Hence, minimization of the commutation time by optimal placing of the shoot-through state in a switching time period is necessary in order to minimize the switching loss. In addition, the following points should be considered while placing the shoot-through state in the PWM control system for a qZSI-based dc–dc converter:

1. The maximum shoot-through duty cycle should never exceed 0.5; otherwise, the system may become unstable.
2. The minimum number of shoot-through states per switching period is two. One shoot-through state per period causes a discontinuous current and the converter will behave abnormally.
3. The shoot-through state should be in the zero state, i.e. the intact active state.
4. The active state and the shoot-through state should be independently controllable.

A new modulation technique called "optimized PWM control with shoot-through during zero states" incorporating the shoot-through state to minimize the number of commutations in semiconductor switches for any impedance network–based dc–dc converter with intermediate H-bridge switching topology is presented in [30]. Figure 7.17 shows a block diagram of the optimized PWM control with shoot-through during zero states and the corresponding switching sequence.

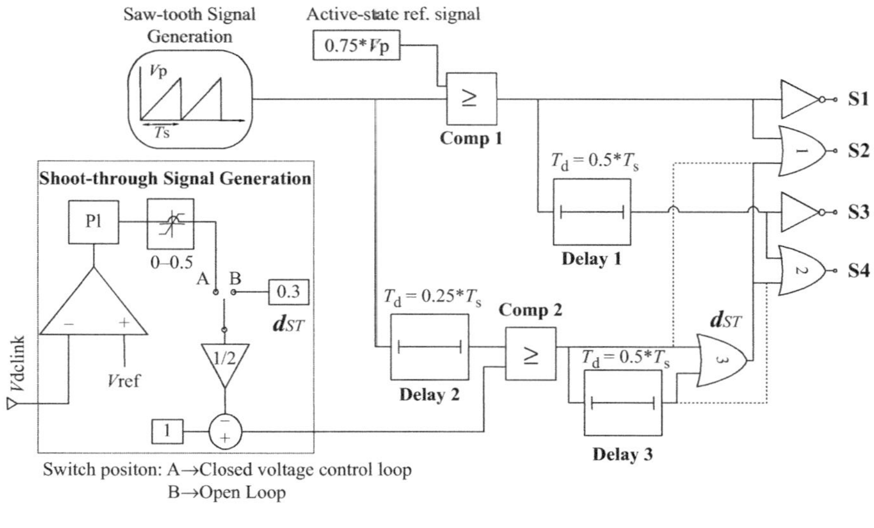

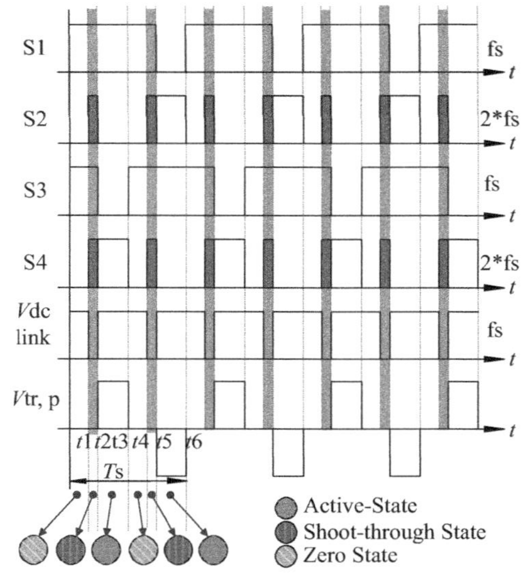

Figure 7.17 Optimized PWM control with shoot-through during zero states illustrating (a) a block diagram and (b) switching sequence for a dc–dc converter

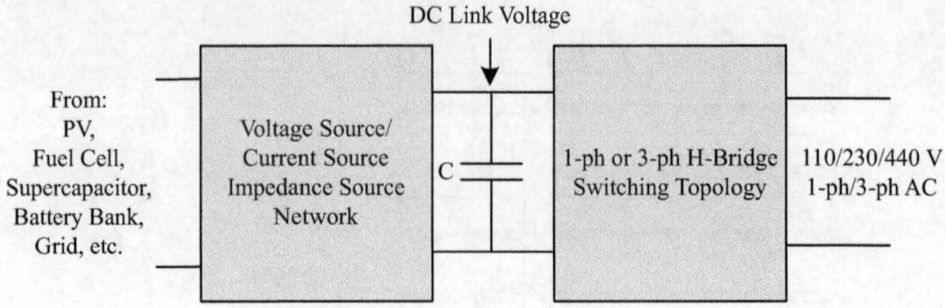

Figure 7.18 General impedance source inverter for one-phase or three-phase

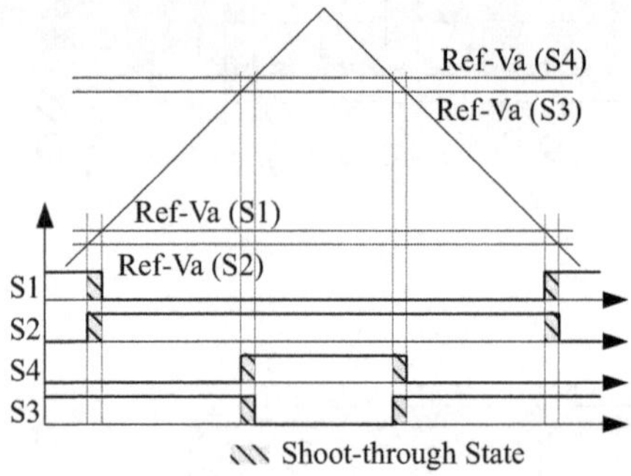

Figure 7.19 Generation of switching signal for one-phase Z-source inverter using (a) non-linear sinusoidal pulse width modulation (SPWM), (b) one cycle control method and (c) modified space vector modulation (SVM)

7.2.2 DC–AC inverter topologies

Different dc–ac topologies inspired from the traditional converter topologies are implemented to utilize the basic properties of the impedance source network, e.g. two-level H-bridge, multilevel/neutral point clamped, as well as the dual bridge. This will be discussed in detail in the following subsections.

7.2.2.1 Two-level H-bridge topologies

The impedance source network was originally implemented with a three-phase H-bridge switching topology as voltage source and a current source to demonstrate its superiority over the traditional VSI and CSI. Both single-phase as well as three-phase inverters can be implemented with an impedance source using the H-bridge circuit. Most of the topologies derived for two-level voltage inversion have the general topology shown in Figure 7.18.

A standard carrier-based PWM as shown in Figure 7.19(c) is used for a single-phase H-bridge topology [39]. A shoot-through state is placed instead of null state without altering the normalized volt-sec average voltage. The duration of each active state in a switching cycle is kept the same as in the traditional sinusoidal pulse width modulation (SPWM). Therefore, the output waveforms are still sinusoidal; however, they are boosted to the desired level by properly controlling the shoot-through time period. The paper extends the modulation concepts to the more complex three-phase H-bridge and four-phase H-bridge topologies for a voltage-fed *Z*-source inverter in both continuous and discontinuous modes.

For three-phase H-bridge topologies (two-level), SPWM and space-vector PWM (SVPWM) are used. SPWMs include simple boost control, maximum-boost control, maximum-constant-boost control, and constant-boost control with third-harmonic injection [40–42]. Simple boost control is the most basic and is derived from the traditional SPWM where a carrier triangular signal is compared to the three-phase reference signal for sinusoidal output voltage and two straight lines (V_p and V_n) to create shoot-through for voltage boost as shown in Figure 7.20(a). The disadvantage of this modulation technique is a decrease of the modulation index with an increase of the shoot-through range. The maximum shoot-through duty ratio of the simple boost control is limited to $D_{st,\max} = (1 - M)$ which limits the boost factor to $B = [2M - 1]^{-1}$. As a result, the device stress increases for the application, which

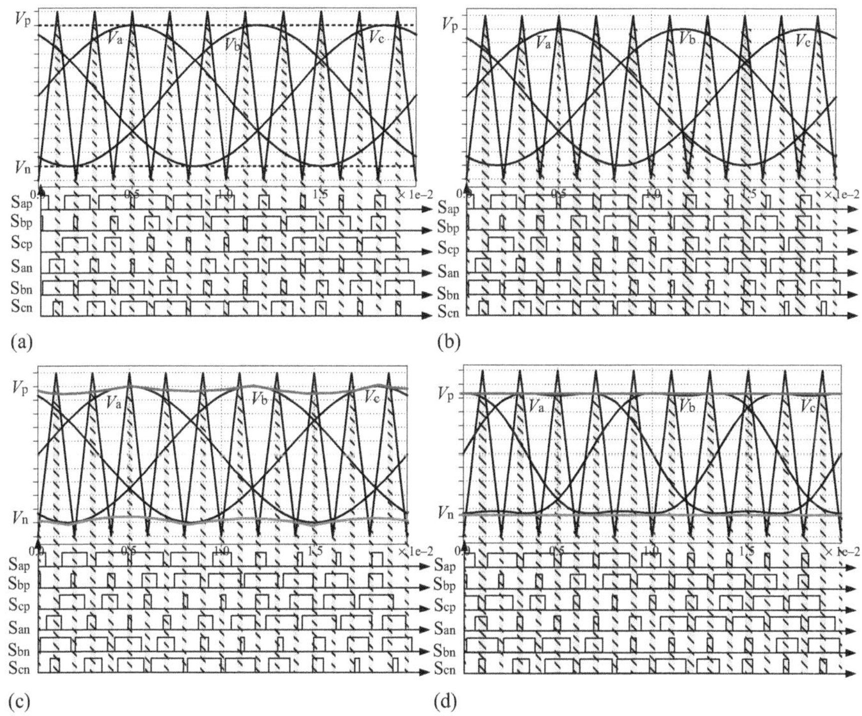

Figure 7.20 Sine wave PWM (a) simple boost control, (b) maximum-boost control, (c) maximum-constant-boost control, and (d) constant-boost control with 1/6 of third harmonic injection (\\\\→ shoot-through period)

requires a higher voltage boost. To address this issue, a maximum-boost PWM control method is presented in [40]. This modulation technique maintains six active states unchanged from those of the traditional carrier-based PWM method; however, it utilizes all zero states to make shoot-through states as shown in Figure 7.20(b). These increase the range of the boost factor $B = \pi[3\sqrt{3}M - \pi]^{-1}$ compared to using simple boost, which reduces the device stress.

However, due to the variable shoot-through time intervals, low-frequency ripple components are present in the capacitor voltage and inductor current, which increase the size and cost of the components in the impedance network. To achieve a constant shoot-through duty ratio and a maximum-boost factor, a maximum-constant-boost PWM control method is proposed in [41,42] which eliminates the low-frequency harmonic component in the impedance source network. Figure 7.20(c) illustrates the switching waveforms of the maximum-constant-boost PWM control. The range of the modulation index is extended from 1 to $2/\sqrt{3}$ by injecting a third-harmonic component with 1/6 of the fundamental component magnitude to the three-phase voltage references. In this modulation technique, two straight lines V_p and V_n are required to generate a shoot-through time period as shown in Figure 7.20(d).

Besides SPWM, space vector pulse width modulation (SVPWM) has similarly been proven to be an effective modulation technique for traditional inverter topologies as it effectively reduces the commutation time of the switches, reduces the harmonic content in the output voltage/current, and better utilizes the dc-link voltage, and consequently reduces the voltage stress and switching loss. This benefit encourages researchers and engineers to retrofit SVPWM for various impedance source inverters. However, proper insertion of the shoot-through state in the switching cycle without altering the volt-sec balance is crucial to reduce additional commutation time of the switches and corresponding switching loss. Various SVPWM techniques are presented in the literature, e.g. ZSVM2 [43], ZSVM4 [44], and ZSVM6 [45]. Modified SVPWM techniques are also being used to reduce the common-mode voltage and leakage currents for photovoltaic systems [46,47] and motor drives [48].

7.2.2.2 Multi-level/neutral point clamped

The concept of a multilevel approach is adopted with an impedance source network to overcome some of the disadvantages of the classical multilevel topologies. Various impedance-sourced multilevel topologies with special modulation techniques are proposed to control the multilevel converter to get reduced harmonic distortion and device commutations. Figure 7.21(a) shows a three-level neutral point clamped (NPC) converter using two ZSIs [49–52]. The neutral point is connected to a common point of the two ZSIs and is grounded. There are other topologies implemented with a reduced number of passive [53] and active devices [54] to reduce the size and cost of the system (see Figure 7.21(b)).

The PWM schemes for *Z*-source NPC inverters are developed from classical three-level SVPWM modulation concepts using two-dimensional vectorial representations with "origin shifting". However, correct integration of the shoot-through state sequence with the classical PWM is essential for proper *Z*-source NPC

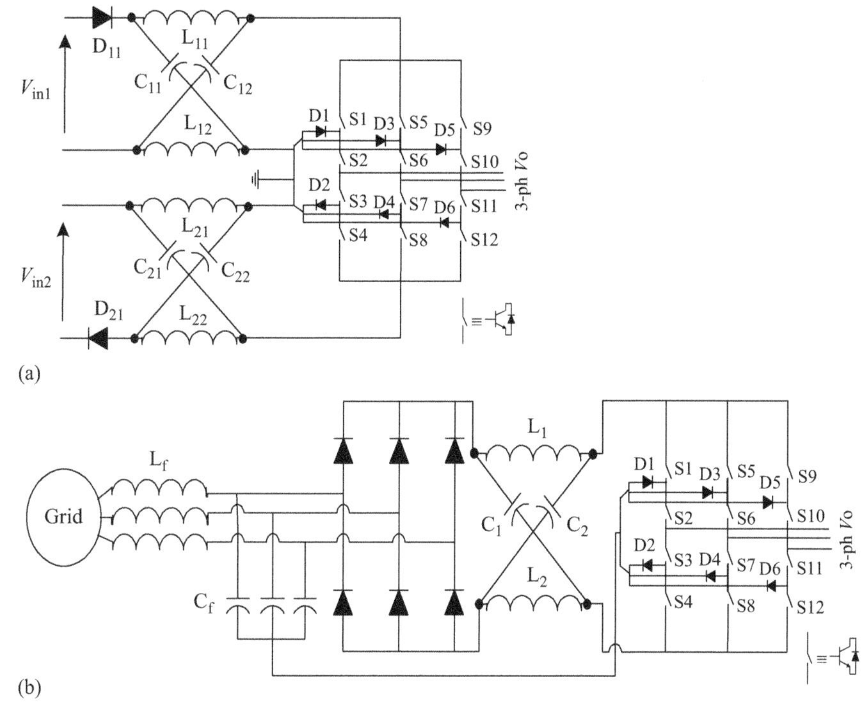

Figure 7.21 Various impedance network source multilevel converter topologies: (a) two-sourced NPC and (b) single-sourced NPC

operation, as some of the vectors can cause a short circuit across the full dc-link, which then results in zero voltage output. It is important to maintain the normalized volt-sec balance while sequencing the shoot-through states to accurately reproduce the desired three-phase sinusoidal voltages. In addition, careful integration of shoot-through with the conventional switching sequences is required to achieve maximal voltage-boost, minimal harmonic distortion, lower semiconductor stress and a minimal number of device commutations per switching cycle. To achieve this, there are various continuous and discontinuous PWM schemes reported in [52] for controlling *Z*-source NPC inverter with two impedance networks at its input side. The former modulation scheme is divided into two, i.e. continuous edge insertion (EI) PWM with symmetrical voltage boost and continuous modified reference (MR) PWM with minimal device commutation, and the latter modulation scheme is also divided into two, i.e. conventional 60 °-discontinuous PWM and origin-shifted 60 °-discontinuous PWM with reduced common-mode (RCM) switching. The device commutation count with EI PWM is the maximum of eight, which reduces to six with the continuous MR technique. The number further reduces to a minimum of four with the discontinuous PWM technique is opted for. However, with the same shoot-through duty ratio, the reduced number of shoot-through per switching cycle in the discontinuous scheme will produce lower common-mode voltage but higher inductor current ripples which significantly increase the size of the passive component in the impedance network. A detailed comparison of continuous and discontinuous PWM schemes is provided in [52].

A nearest-three-vector modulation principle and RCM switching is proposed in [49] to minimize harmonic distortion, device commutation and common-mode voltage of the inverter. A hybrid PWM strategy similar to [55] for a two-level *Z*-source inverter is implemented for the *Z*-source NPC topology [56] to reduce its algorithm calculation by combining the theory of SVPWM and triangular-comparison PWM. A reduced component count *Z*-source NPC converter with modified modulation technique is reported in [53] using a single *Z*-source network. This topology reduces the requirement of an additional impedance network to create a neutral point as explained in [49,52,56]. The modulation scheme is modified to create a full DC-link and a partial DC-link shoot-through state to boost the output voltage without increasing the commutation time as in the conventional NPC modulation techniques [49,52]. An effective control method for a cascaded quasi-*Z*-source inverter using multilevel space vector modulation is presented in [57] for single phase and in [58] for three phase to generate seven-level voltage. This control scheme achieves independent control of maximum power point technique for each photovoltaic panel and also balances the dc-link voltage across each H-bridge inverter to accomplish premium power quality for grid integration of photovoltaic panels for low switching frequency design.

7.2.3 AC–AC converter topologies

The matrix converter consists of nine bidirectional switches that allow any output phase to be connected to any input phase. It is a direct ac–ac converter with sinusoidal input/output waveforms and a controllable input power factor.

The maximum voltage gain of a traditional matrix converter does not exceed 0.866. In addition, the switches are more vulnerable to shoot-through on the same output phase leg. These limitations are overcome by the impedance source matrix converter [59] as shown in Figure 7.22. A traditional matrix converter with an impedance network can buck–boost the voltage as well as the frequency for the ac

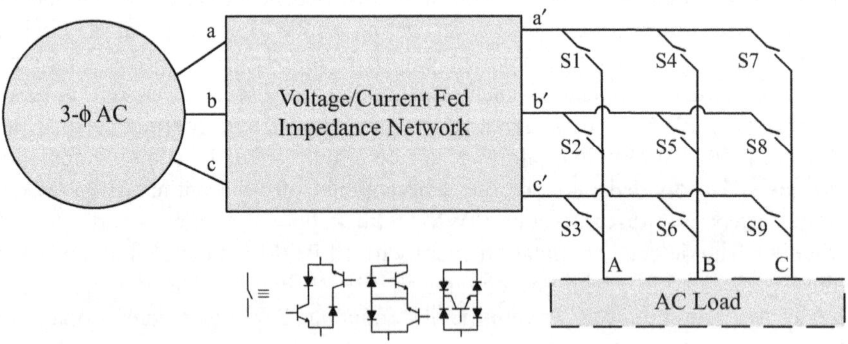

Figure 7.22 Impedance source network–based matrix converter

load requiring adjustable voltage and frequency, e.g. motor drives. Many modulation and control methods, e.g. simple maximum-boost control, maximum-boost control, maximum-gain control, and hybrid minimum-stress control have been proposed to improve the performance and reliability of the converter [60,61].

With some modifications made to the conventional modulation techniques (SVPWM, carrier-based PWM, PWAM, etc.) to incorporate the shoot-through state, various modified modulation techniques are implemented to control direct and indirect impedance-network–based matrix converters. The traditional carrier-based SPWM [40–42] is applied to control and modulate various *Z*-source and quasi-*Z*-source direct matrix converters with a few modifications, e.g. four control strategies: simple maximum-boost control, maximum-boost control, maximum-gain control, and hybrid minimum-stress control are proposed in [59, 61]. In the simple maximum-boost control, the modulation index is limited to $M = 0.5$ which means that the maximum voltage gain can only go up to 0.944. Maximum-boost control utilizes all the zero states as shoot-through states. The range of modulation index is extended to 0.866 by injecting 1/6 of the third harmonic signal in the reference signal. The maximum-gain control method can obtain the maximum gain at the same modulation index among all four techniques, and the hybrid minimum-voltage-stress control can obtain the minimum-voltage stress at the same voltage gain. In terms of the total harmonic distortion at the output, the maximum-constant-boost control method is effective in eliminating low-order harmonics compared to the simple boost and maximum-boost controls. A comparison of various control methods is also presented in [61].

Pulse-width-amplitude-modulation (PWAM) with a maximum-constant-boost shoot-through control strategy is also implemented in [62] to control a voltage-fed quasi-*Z*-source direct-matrix converter. This modulation technique reduces the switching frequency of the matrix converter by 1/3 compared to the SVPWM, which helps to reduce the switching losses by more than 50% compared to the SVPWM and 87% compared to the SPWM.

7.2.4 AC–DC converter topologies

Rectifiers based on impedance source networks have the capability to both buck and boost the input voltage in a single stage compared to a traditional rectifier. In addition, they can provide good input power factor, low line-current distortion, regeneration and improved reliability. Figure 7.23 shows *Z*-source [63,64] and quasi-*Z*-source [65] based rectifier topologies. A new concept of a bidirectional converter based on a matrix converter is also presented in [59]. The advantages of the ac–dc matrix converter are controllable input power factor, tight dc voltage regulation, wide bandwidth with quick response to load variations and single-stage buck voltage ac–dc power conversion.

7.3 Impedance source network topologies

Impedance source networks are mainly derived from the *Z*-source network by modifying the original impedance network, or by rearranging the connections of

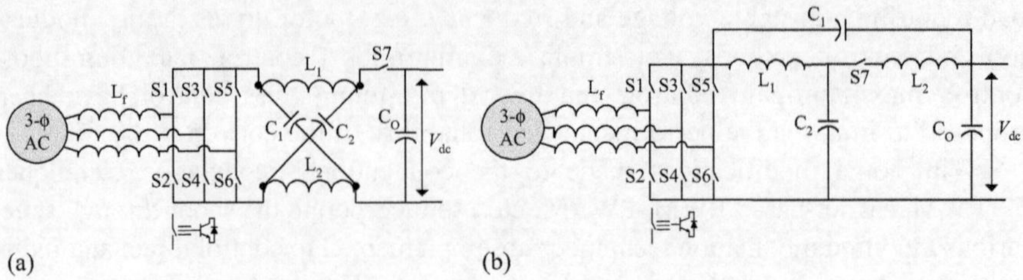

Figure 7.23 Ac–dc rectifier topologies based on (a) Z-source and (b) quasi-Z-source network

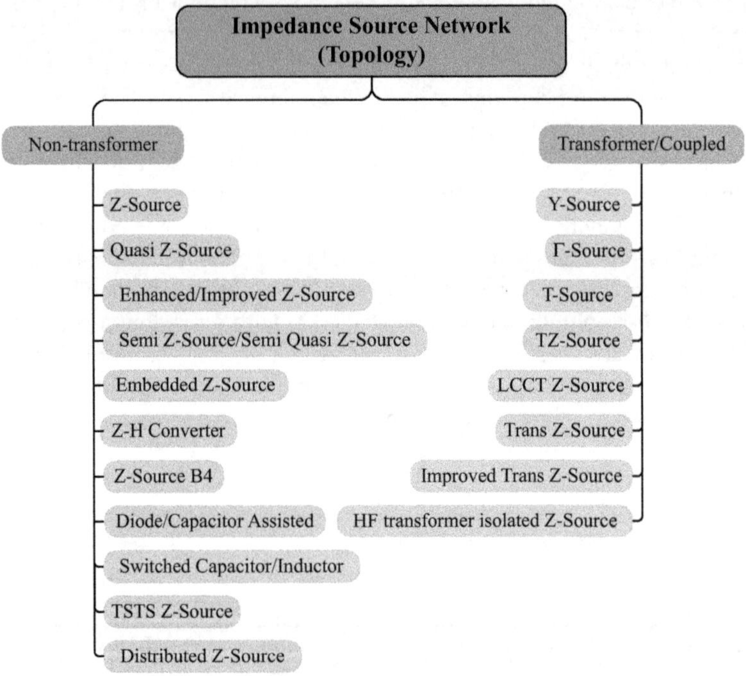

Figure 7.24 Impedance source network topologies

inductors and capacitors [66–87]. New topologies are still being developed, mainly for four reasons: (1) reduction of the *Z*-source network component count and rating; (2) extension of voltage gain range; (3) achieving higher power density; and (4) application oriented optimization and improvement.

The impedance source network is broadly classified into two categories based on magnetics: (a) non-transformer based and (b) coupled or transformer based as shown in Figure 7.24. Each topology has distinct features and advantages and will be discussed briefly in the following subsections.

7.3.1 Non-transformer based

7.3.1.1 Z-source/quasi-Z-source

Z-source converters are broadly classified into two types, voltage-fed and current-fed. However, unlike the traditional voltage-fed/current-fed inverter, the impedance source network provides a buffer between the source and the inverter bridge and facilitates a short- and an open-circuit at any time depending on the mode of operation. Traditional voltage-fed/current-fed Z-source impedance networks suffer from problems like discontinuous input current in the boost mode for the voltage-fed ZSI (Figure 7.25(a)) and high current stress on the inductor in the current-fed ZSI (Figure 7.26(a)). Various voltage- and current-fed topologies derived from ZSI and qZSI (Figures 7.25(b) and (c) and 7.26(b) and (c)) with improved performance were proposed in [66] to solve the problem of ZSIs [1]. It should be noted that all three current-fed ZSIs, Figure 7.26(a)–(c), are capable of bidirectional power flow and buck–boost operation, although the switches have to be reverse-blocking devices. These new benefits extend this topology to applications of renewable energy generation and motor drives. Enhanced-boost Z-source inverter is also proposed in [67] with alternate-cascaded switched and tapped-inductor cells using some lower-rated components. Similarly, an improved Z-source [68] and an improved trans-Z-source [69] are proposed, respectively, to reduce the capacitor voltage stress. However, additional components are required to implement the circuit. This makes the system more costly and also decreases the power density of the converter.

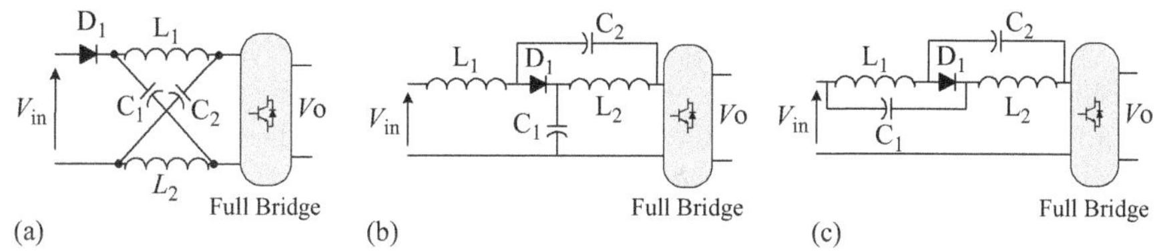

Figure 7.25 Various voltage-fed ZSI topologies: (a) ZSI with discontinuous input current; (b) qZSI with continuous input current; and (c) qZSI with discontinuous input current

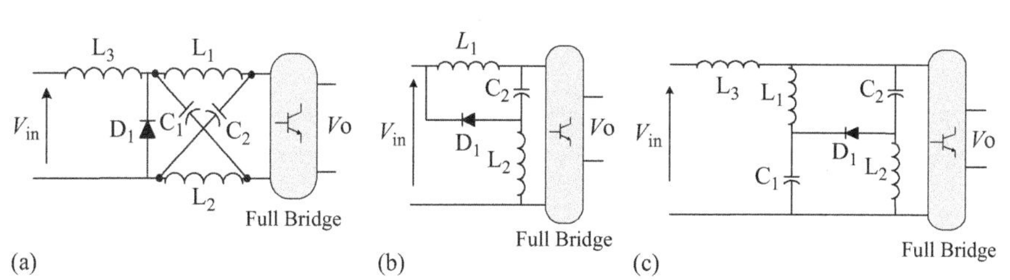

Figure 7.26 Various current-fed ZSI topologies (switches have to be reverse-blocking devices): (a) ZSI with continuous input current; (b) qZSI with discontinuous input current; and (c) qZSI with continuous input current

7.3.1.2 Semi-Z-source/semi-quasi-Z-source

Semi-Z-source inverters were proposed (see Figure 7.27) to achieve low cost and high efficiency in applications such as single-phase grid-tied PV power systems. A semi-Z-source inverter with only two active switches has a voltage boost function and a double-ground feature (both PV panel and ac output can be grounded) that eliminates the need to float/isolate PV panels without leakage current and which improves safety [70]. Unlike the traditional ZSI/qZSI, a shoot-through state is not applicable to a semi-Z-source inverter. An improved non-linear SPWM method is used to get the desired duty cycle to generate a sinusoidal output.

The advantage of the semi-Z-source and semi-quasi-Z-source inverters [71] is that they can be implemented using fewer switches compared to a traditional ZSI and qZSI, but unfortunately the voltage stress on the switching devices is high. This topology is suitable for a grid-connected micro-PV inverter with high-voltage SiC devices.

Two modulation techniques are prevalent in the literature to control and modulate the two switches of a single-phase Z-source/quasi-Z-source to get the desired output voltage, namely one-cycle control [72] and non-linear SPWM [70].

The voltage gain of the semi-Z-source/quasi-Z-source converter is not a straight line as with a full-bridge inverter. So, instead of a sinusoidal reference signal ($v = V\sin\,\omega t$), a non-linear sinusoidal reference signal $v = [2 - M\sin\,\omega t]^{-1}$ is compared with the carrier signal to generate the gate drive signal for the two switches as shown in Figure 7.28(a).

A one-cycle control method is adopted to control a single-phase semi-Z-source topology in [72]. In this control method, two switches work in a complementary fashion where the clock signal (CLK) is used to turn-on any one switch. The turn-on time of the switch is determined by the integrated voltage across the switch, and when it reaches the sinusoidal signal $(v_{\mathrm{i}} - v_{\mathrm{ref}})$, the integrator is reset and the switch turns off as shown in Figure 7.28(b). This control method has the ability to reject input perturbations and is insensitive to the system model, which provides a high-efficiency constant-frequency control.

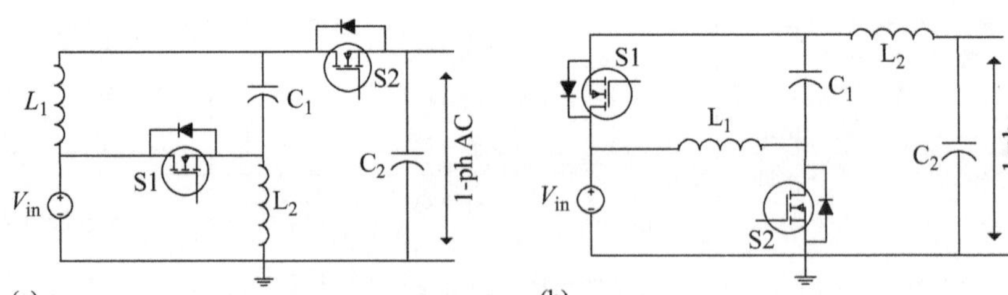

Figure 7.27 Semi-Z-source topologies: (a) semi-Z-source inverter and (b) semi-quasi-Z-source inverter

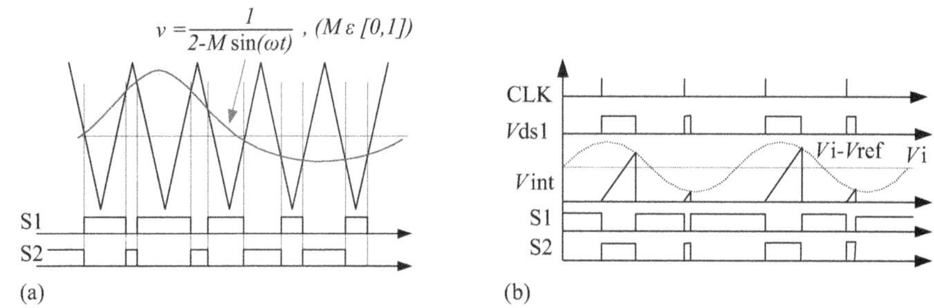

Figure 7.28 Generation of switching signal for one-phase Z-source inverter using (a) non-linear SPWM and (b) one-cycle control method

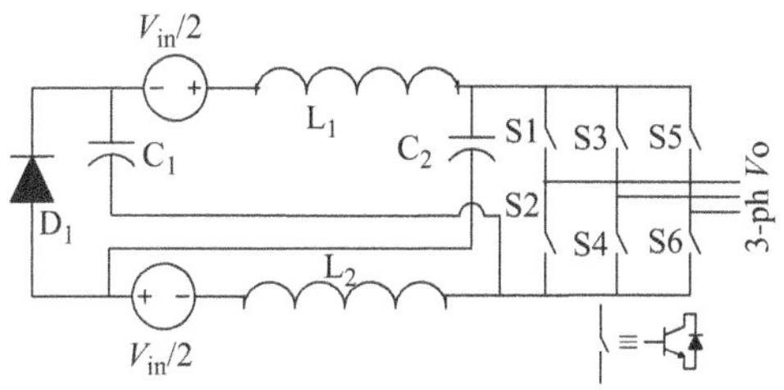

Figure 7.29 Two-level embedded Z-source inverter

7.3.1.3 Embedded *Z*-source

The embedded *Z*-source was proposed to achieve continuous input current and lower capacitor voltage rating, and its multi-source feature is especially suitable for PV power generation [73]. Figure 7.29 shows the circuit topology of a two-level embedded *Z*-source inverter. There are other similar embedded topologies with one or two DC sources suitable for battery storage systems.

7.3.1.4 *Z*-source B4 converter

Inspired by the traditional B4 VSI, a *Z*-source B4 topology is proposed in [74] with a reduced number of active components, enhanced reliability and lower cost. Figure 7.30 shows the *Z*-source B4 converter topology for a three-phase power conversion.

7.3.1.5 Switched inductor/capacitor

Extra inductors and capacitors have been added in the *Z*-source and quasi-*Z*-source impedance network, with the aim of improving the boost capability of the circuit. Many topologies are presented in the literature to reduce the stress on the passive components and also to eliminate the start-up inrush current. A switched inductor/capacitor ZSI/qZSI provides continuous input current and reduced voltage stress on the capacitor [75]. An embedded *Z*-source with a switched inductor combines the advantages of both topologies, e.g. high boost ratio, reduced capacitor voltage stress, and low input ripple current [76]. However, this switched inductor/capacitor

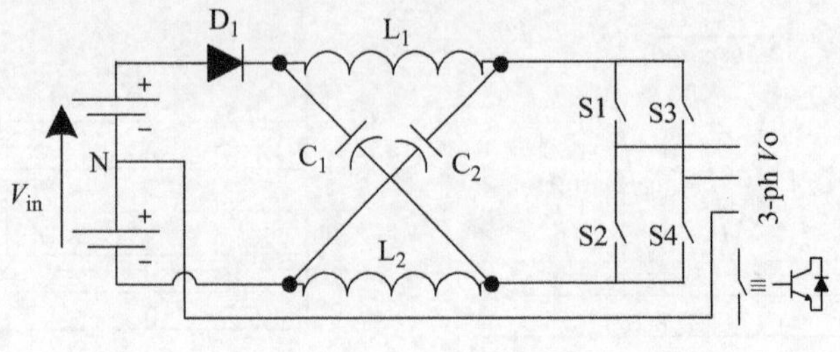

Figure 7.30 B4 Z-source inverter topology

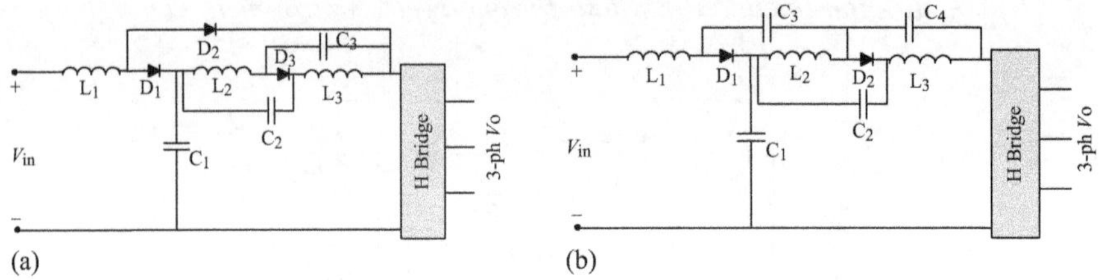

Figure 7.31 Assisted qZSI topologies: (a) diode assisted and (b) capacitor assisted

topology needs a large number of passive devices, which increase cost and volume of the converter.

7.3.1.6 Capacitor/diode assisted

The voltage-boost capability of the *Z*-source and quasi-*Z*-source is extended with the aid of capacitors and diodes in order to meet the needs of applications requiring very high-voltage boost [77]. The impedance network can be extended by using diodes and/or capacitors as shown in Figure 7.31. The advantages of this topology are high voltage gain and reduced capacitor stress; however, this is at the cost of extra passive components.

7.3.1.7 TSTS *Z*-source

Three-switch three-state single-phase *Z*-source inverters (TSTS-ZSIs) were proposed recently in [78], and are classified into two groups, boost-TSTS-ZSI and buck–boost-TSTS-ZSI as shown in Figure 7.32.

This topology can be assembled using fewer switches compared to the traditional impedance source topologies, so higher power density can be achieved. In addition, it has a lower voltage stress and dual grounding, which makes it suitable for PV power generation.

7.3.1.8 Distributed *Z*-source

Distributed impedance networks such as transmission lines and hybrid inductor (LC) components can be used for a *Z*-source network [79] as shown in Figure 7.33. These distributed *Z*-source networks are difficult to implement, but a distributed *Z*-source

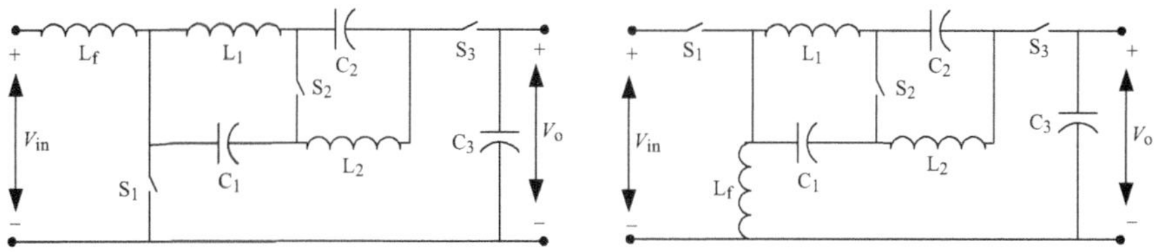

Figure 7.32 Three-switch three-state single-phase Z-source inverters (TSTS-ZSIs): (a) boost topology and (b) buck–boost topology

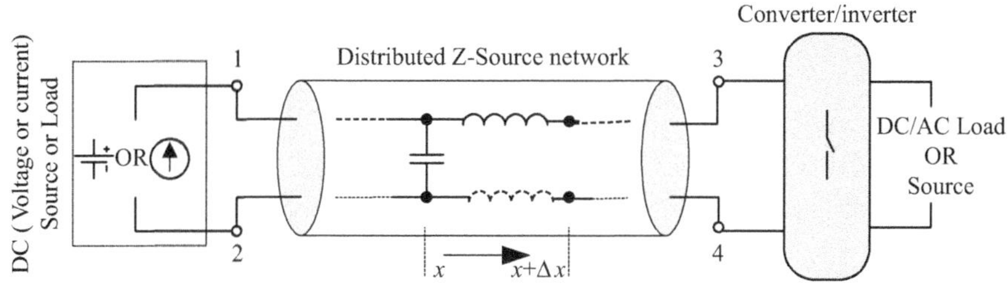

Figure 7.33 A general topology of distributed Z-source converter

inverter does not need any extra diode or switch to achieve the voltage boost function, thus having the minimum component count. This topology could open a door for an impedance source networks to radio-frequency power converter design by utilizing the distributed inductance and capacitance prominent at higher frequencies.

7.3.2 With transformer or magnetic coupling

Magnetically coupled inductors and transformers find a niche in impedance networks to improve the voltage boost capability as well as the modulation index. In addition, they reduce the number of passive components needed in the network, which improves the power density and reducing the cost of the system. A generic method is presented in [80] to illustrate the derivation of some of the magnetically coupled network topologies. The following subsections describe the impedance network topologies based on transformer or coupled inductor.

7.3.2.1 *Y*-source

A unique impedance source network called the "*Y*-source network" topology was proposed in [33] using coupled inductors with three windings (N_1, N_2, and N_3) having versatile characteristics and features (see Figure 7.14). The gain of the converter is presently not matched with existing networks operated at the same duty ratio. The proposed converter has one more degree of freedom (three windings and shoot-through duty cycle of switch (d_{ST})) to choose the voltage boost, as compared to a classical impedance network–based boost converter [81]. Theoretically, any magnitude of voltage boost can be obtained by adjusting the turns ratio and shoot-through duty cycle of the switch.

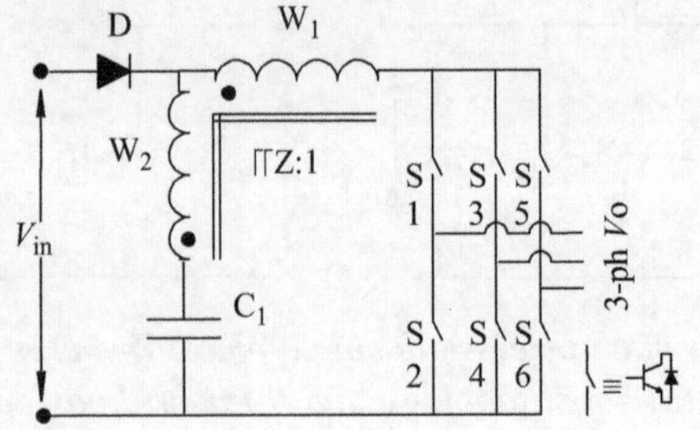

Figure 7.34 *Γ-Z-source network topology*

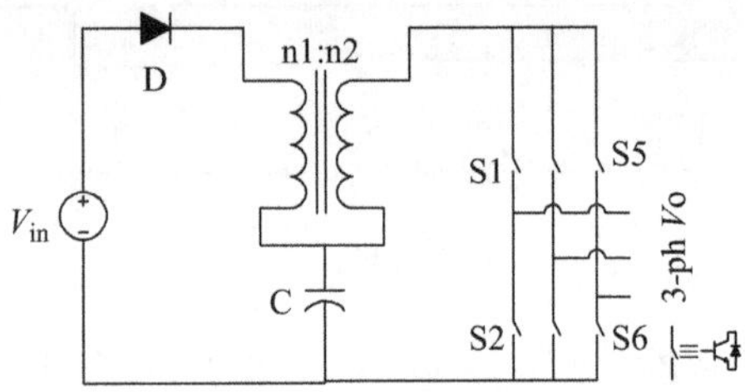

Figure 7.35 *T-source network topology*

7.3.2.2 Γ-*Z*-source

A unique Γ-shape two-winding coupled transformer is also implemented in an impedance network [82] to increase the gain and modulation ratio simultaneously, while reducing the component count. Unlike other transformer-based impedance networks whose gains increases with an increased turns ratio, e.g. T-source, trans-*Z*-source, TZ-source or inductor-capacitor-capacitor-transformer *Z*-source (LCCT *Z*-source), the Γ-*Z*-source gain increases with a decrease in the turns ratio. Figure 7.34 shows the Γ-*Z*-source network topology for an inverter.

7.3.2.3 T-source

The T-source inverter also utilizes a two-winding coupled inductor and one capacitor [83] as shown in Figure 7.35. The gain of the converter can be set higher than the traditional ZSI and qZSI using a transformer turns ratio greater than 1. This topology is suitable for a NPC converter as it shares a common voltage source for both the passive arrangement and the converter circuit.

7.3.2.4 Trans-*Z*-source

Theoretically, the original *Z*-source, quasi-*Z*-source, and embedded *Z*-source all have unlimited voltage gain. Practically, however, a high voltage gain (>2) will result in a high-voltage stress imposed on the switches. Trans-*Z*-source (two voltage-fed and two current-fed) inverters were proposed to have higher voltage gains while keeping voltage stress low and reducing the *Z*-source network to one transformer (or one coupled inductor) and one capacitor [84] as shown in Figure 7.36.

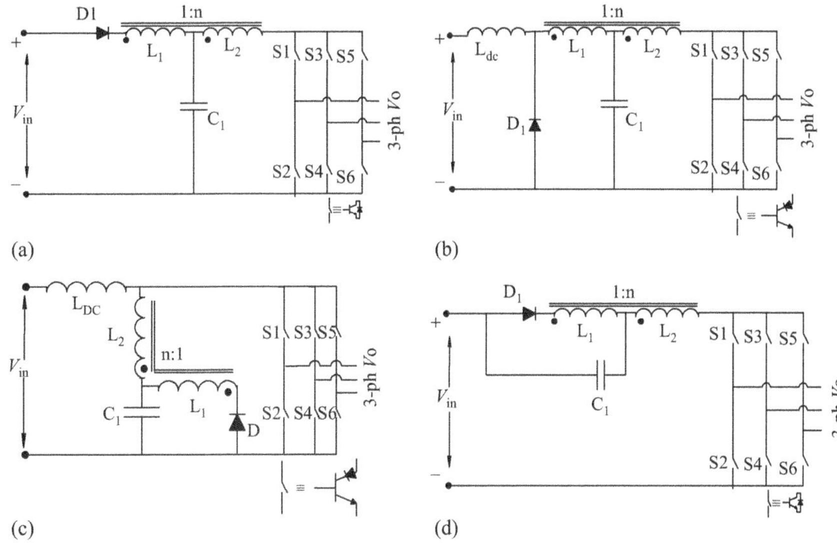

Figure 7.36 *Trans-Z-source inverter. (a) Voltage-fed trans-Z-source; (b) current-fed trans-Z-source; (c) current-fed trans-quasi-Z-source; and (d) voltage-fed trans-quasi-Z-source*

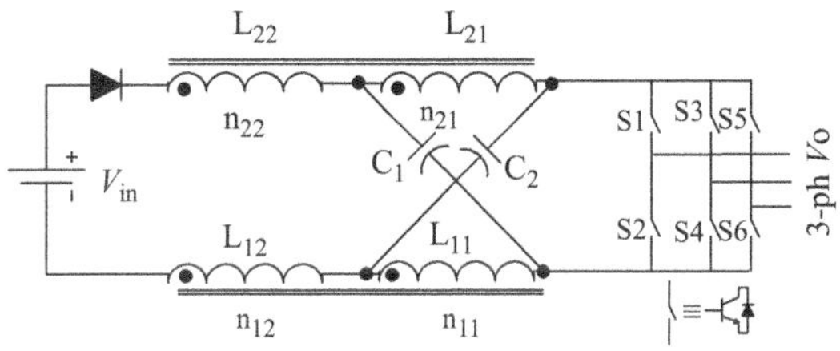

Figure 7.37 *TZ-source network topology*

7.3.2.5 TZ-source

The TZ-source also achieves high voltage gain by setting the turns ratio of the transformer to greater than 1 [85]. Comparatively, it requires a lower transformer turns ratio than the trans-ZSI; however, it requires four coupled inductors as well as the same number of other passive components as the traditional ZSIs. So, this topology is not very effective in reducing the number of components and size. Figure 7.37 shows the circuit topology of the TZ-source network for a three-phase inverter.

7.3.2.6 LCCT Z-source

With the inductor and a transformer integrated into a common core, the LCCT-*Z*-source inverters as shown in Figure 7.38 achieve higher voltage gains and

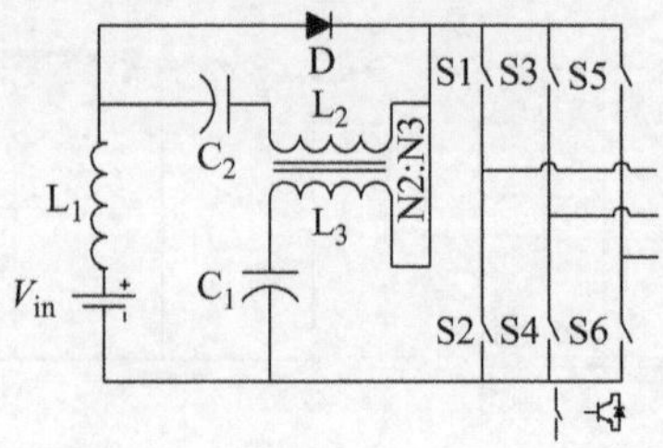

Figure 7.38 LCCT network topology

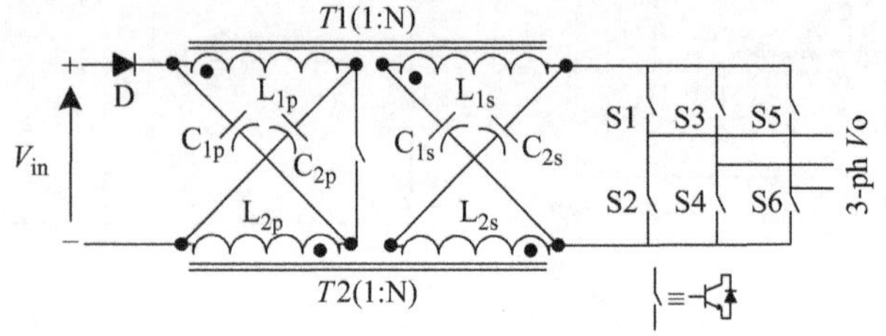

Figure 7.39 HF transformer isolated Z-source inverter

modulation index [86]. This topology maintains a continuous input current even at a light load, and also filters out high-frequency ripples from the input current.

7.3.2.7 HF transformer isolated Z-source/quasi-Z-source/trans-Z-source

A family of impedance source networks with intermediate high frequency (HF) isolation is presented with different topologies for applications requiring isolation for safety reasons [87]. These topologies inherit all the benefits of the Z-source networks along with a higher boost ratio and lower device stress. However, this topology increases the number of active and passive components. In addition, the coupled transformer must be designed properly to minimize the leakage inductance. One example of a voltage-fed HF-isolated ZSI is shown in Figure 7.39.

7.4 Conclusions

Impedance source networks have added a new chapter in the field of power electronic topologies with their unique features and properties that overcome most of the problems faced by traditional converter topologies. Since the publication of the first Z-source network, there have been numerous contributions in the literature modifying the basic topology to suit the needs of many applications. In principle, it overcomes the conceptual and theoretical barriers and limitations of the traditional VSI and CSI and provides a novel power conversion concept. The superior performance of the impedance source network to design more robust and versatile converter topologies for various applications attracts researchers and designers from both academia and industry to explore it in depth. Many topologies were developed to overcome the limitations and disadvantages of the traditional impedance source network. The number of combinations is large and the topologies might in cases be confusing. In this chapter, an attempt to provide a global picture of the impedance source networks proposed in the literature to comprehend and identify their pros and cons. A broad classification of the topologies and modulation techniques in this chapter with further sub-classification aims to provide easy selection of control and modulation techniques for appropriate topologies for particular application. Further, a comparison of various modulation techniques for particular switching topologies based on theoretical complexity and performance informs the selection of the correct control and modulation technique for the respective switching topologies to achieve maximal voltage boost, minimal harmonic distortion, lower semiconductor stress and a minimal number of device commutations per switching cycle.

References

[1] F. Z. Peng, "Z-Source Inverter," *IEEE Trans. Ind. Appl.*, vol. 39, no. 2, pp. 504–510, Mar./Apr. 2003.

[2] F. Z. Peng, X. Yuan, X. Fang and Z. Qian, "Z-Source Inverter for Adjustable Speed Drives," *IEEE Trans. Power Electron. (Letters)*, vol. 1, no. 2, pp. 33–35, Jun. 2003.

[3] Z. J. Zhou, X. Zhang, P. Xu and W. X. Shen, "Single-Phase Uninterruptible Power Supply Based on Z-Source Inverter," *IEEE Trans. Ind. Electron.*, vol. 55, no. 8, pp. 2997–3004, Aug. 2008.

[4] Y. Li, S. Jiang, J. G. Cintron-Rivera and F. Z. Peng, "Modeling and Control of Quasi-Z-Source Inverter for Distributed Generation Applications," *IEEE Trans. Ind. Electron.*, vol. 60, no. 4, pp. 1532–1541, Apr. 2013.

[5] B. Ge, H. Abu-Rub, F. Z. Peng, Q. Li, A. T. de Almeida, F. J. T. E. Ferreira, D. Sun and Y. Liu, "An Energy Stored Quasi-Z-Source Inverter for Application to Photovoltaic Power System," *IEEE Trans. Ind. Electron.*, vol. 60, no. 10, pp. 4468–4481, Oct. 2013.

[6] L. Yushan, G. Baoming, F. Z. Peng, A. R. Haitham, A. T. de Almeida and F. J. T. E. Ferreira, "Quasi-Z-Source Inverter Based PMSG Wind Power Generation Systems," in *Proc. ECCE 2011*, pp. 291–297, 17–22 Sep. 2011.

[7] C. J. Gajanayake, D. M. Vilathgamuwa and P. C. Loh, "Development of a Comprehensive Model and a Multiloop Controller for Z-Source Inverter DG Systems," *IEEE Trans. Power Electron.*, vol. 22, no. 4, pp. 1453–1463, Jul. 2007.

[8] J. Liu, S. Jiang, D. Cao and F. Z. Peng, "A Digital current Control of Quasi-Z-Source Inverter With Battery," *IEEE Trans. Ind. Informatics*, vol. 9, no. 2, pp. 928–936, May 2013.

[9] F. Z. Peng, M. Shen and K. Holland, "Application of Z-Source Inverter for Traction Drive of Fuel Cell-Battery Hybrid Electric Vehicles," *IEEE Trans. Power Electron.*, vol. 22, no. 3, pp. 1054–1061, May 2007.

[10] A. S. Khlebnikov and S. A. Kharitonov, "Application of the Z-Source Converter for Aircraft Power Generation Systems," in *Proc. EDM 2008*, pp. 211–215, Jul. 2008.

[11] S. J. Amodeo, H. G. Chiacchiarini and A. R. Olivia, "High-Performance Control of a DC/DC Z-Source Converter Used for an Excitation Field Driver," *IEEE Trans. Power Electron.*, vol. 27, no. 6, pp. 2947–2957, Jun. 2012.

[12] J. C. R. Caro, F. Z. Peng, H. Cha and C. Rogers, "Z-Source-Converter-Based Energy-Recycling Zero-Voltage Electronic Loads," *IEEE Trans. Ind. Electron.*, vol. 56, no. 12, pp. 4894–4902, Dec. 2009.

[13] K. A. Corzine and R. W. Ashton, "A New Z-Source DC Circuit Breaker," *IEEE Trans. Power Electron.*, vol. 27, no. 6, pp. 2796–2804, Jun. 2012.

[14] Y. P. Siwakoti, F. Z. Peng, F. Blaabjerg, P. C. Loh and G. E. Town, "Impedance Source Network for Electric Power Conversion – Part I: A Topological Review," *IEEE Transaction on Power Electron.*, vol. 30, no. 2, pp. 699–716, Feb. 2015.

[15] Y. P. Siwakoti, F. Z. Peng, F. Blaabjerg, P. C. Loh, G. E. Town and S. Yang, "Impedance Source Network for Electric Power Conversion – Part II: Review of Control and Modulation Techniques," *IEEE Transaction on Power Electron.*, vol. 30, no. 4, pp. 1887–1905, Apr. 2015.

[16] M. Shen, Q. Tang and F. Z. Peng, "Modeling and Controller Design of the Z-Source Inverter with Inductive Load," in *Proc. IEEE PESC 2007*, pp. 1804–1809, Jun. 2007.

[17] Y. Li and F. Z. Peng, "AC Small Signal Modeling, Analysis and Control of Quasi-Z-Source Converter," in *Proc. IEEE ECCE* 2012, pp. 1848–1854, Jun. 2012.

[18] C. J. Gajanayake, D. M. Vilathgamuwa and P. C. Loh, "Small-Signal and Signal-Flow-Graph Modeling of Switched Z-Source Impedance Network," *IEEE Trans. Power Electron. (Letters)*, vol. 3, no. 3, pp. 111–116, Sep. 2005.

[19] C. J. Gajanayake, D. M. Vilathgamuwa and P. C. Loh, "Modelling and Design of Multi-loop Closed Loop Controller for Z-Source Inverter for Distributed Generation," in *Proc. IEEE PESC 2006*, pp. 1–7, Jun. 2006.

[20] P. C. Loh, D. M. Vilathgamuwa, C. J. Gajanayake, Y. R. Lim and C. W. Teo, "Transient Modeling and Analysis of Pulse-Width Modulated Z-Source Inverter," *IEEE Trans. Power Electron.*, vol. 22, no. 2, pp. 498–507, Mar. 2007.

[21] V. P. Galigekere and M. K. Kazimierczuk, "Small-Signal Modeling of PWM Z-Source Converter by Circuit-Averaging Technique," in *Proc. IEEE ISCAS 2011*, pp. 1600–1603, May 2011.

[22] M. Shen, Q. Tang and F. Z. Peng, "Modeling and Controller Design of the Z-Source Inverter with Inductive Load," in *Proc. IEEE PESC* 2007, pp. 1804–1809, Jun. 2007.

[23] F. Guo, L. Fu, C. H. Lin, C. Li and J. Wang, "Small Signal Modeling and Controller Design of a Bidirectional Quasi-Z-Source Inverter for Electric Vehicle Applications," in *Proc. IEEE ECCE 2012*, pp. 2223–2228, Sep. 2012.

[24] Q. Lei, S. Yang, F. Z. Peng and R. Inoshita, "Steady State and Transient Analysis of a Three Phase Current-Fed Z-Source PWM Rectifier," in *Proc. VPPC 2009*, pp. 426–432, Sep. 2009.

[25] X. Ding, Z. Qian, S. Yang, B. Cui and F. Z. Peng, "A Direct Peak DC-Link Boost Voltage Control Strategy in Z-Source Inverter," in *Proc. IEEE APEC 2007*, pp. 648–653, Mar. 2007.

[26] X. Ding, Z. Qian, S. Yang, B. Cui and F. Z. Peng, "A Direct DC-Link Boost Voltage PID-Like Fuzzy Control Strategy in Z-Source Inverter," in *Proc. IEEE PESC 2008*, pp. 405–411, Jun. 2008.

[27] G. Sen and M. E. Elbuluk, "Voltage and Current-Programmed Modes in Control of Z-Source Converter," *IEEE Trans. Ind. Appl.*, vol. 46, no. 2, pp. 680–686, Mar./Apr. 2010.

[28] Q. V. Tran, T. W. Chun, J. R. Ahn and H. H. Lee, "Algorithms for Controlling Both the DC Boost and AC Output Voltage of Z-Source Inverter," *IEEE Trans. Ind. Electron.*, vol. 54, no. 5, pp. 2745–2750, Oct. 2007.

[29] C. J. Gajanayake, D. M. Vilathgamuwa and P. C. Loh, "Modelling and Design of Multi-Loop Closed Loop Controller for Z-Source Inverter for Distributed Generation," in *Proc. IEEE PESC* 2006, pp. 1–7, Jun. 2006.

[30] Y. P. Siwakoti and G. Town, "Improved Modulation Technique for Voltage Fed Quasi-Z-Source DC/DC Converters," in *Proc. APEC 2014*, pp. 1973–1978, Mar. 2014.

[31] D. Vinnikov, I. Roasto, "Quasi-Z-Source-Based Isolated DC/DC Converters for Distributed Power Generation," *IEEE Trans. Ind. Electron.*, vol. 58, no. 1, pp. 192–201, Jan. 2011.

[32] Y. P. Siwakoti, F. Blaabjerg, P. C. Loh and G. E. Town, "High Voltage Boost Quasi-Z-Source Push–Pull Isolated DC/DC Converter," *IET Power Electron.*, vol. 7, no. 9, pp. 2387–2395, 2014.

[33] D. Cao and F. Z. Peng, "A Family of Z-Source and Quasi-Z-Source DC–DC Converters," in *Proc. APEC 2009*, pp. 1097–1101, Feb. 2009.

[34] Y. P. Siwakoti, P. C. Loh, F. Blaabjerg and G. Town, "Y-Source Impedance Network," *IEEE Trans. Power Electron. (Letters)*, vol. 29, no. 7, pp. 3250–3254, Jul. 2014.

[35] Y. P. Siwakoti, P. C. Loh, F. Blaabjerg and G. E. Town, "Magnetically Coupled High Gain Y-Source Isolated DC/DC Converter," *IET Power Electron.*, vol. 7, no. 11, pp. 2817–2824, 2014.

[36] D. Vinnikov, T. Jalakas, I. Roasto, H. Agabus and K. Tammet, "Method of Shoot-Through Generation for Modified Sine Wave Z-Source, Quasi-Z-Source and Trans-Z-Source Inverters," Patent Owner: Tallinn University of Technology, World Intellectual Property Organization, Publication number WO2012/116708 A2, Publication date Sep. 2012.

[37] H. Lee, H. G. Kim and H. Cha, "Parallel Operation of Trans-Z-Source Network Full Bridge DC/DC Converter for Wide Input Voltage Range," in *Proc. IEEE ECCE Asia 2012*, pp. 1707–1712, Jun. 2012.

[38] I. Roasto, D. Vinnikov, J. Zakis and O. Husev, "New Shoot-Through Control Methods for qZSI-Based DC/DC Converters," *IEEE Trans. Ind. Informatics*, vol. 9, no. 2, pp. 640–647, May 2013.

[39] P. C. Loh, D. M. Vilathgamuwa, Y. S. Lai, G. T. Chua and Y. Li, "Pulse-Width Modulation of Z-Source Inverters," *IEEE Trans. Power Electron.*, vol. 20, no. 6, pp. 1346–1355, Nov. 2005.

[40] F. Z. Peng, M. Shen and Z. Qian, "Maximum Boost Control of the Z-Source Inverter," *IEEE Trans. Power Electron.*, vol. 20, no. 4, pp. 833–838, Jul. 2005.

[41] M. Shen, J. Wang, A. Joseph, F. Z. Peng, L. M. Tolbert and D. J Adams, "Constant Boost Control of the Z-Source Inverter to Minimize Current Ripple and Voltage Stress," *IEEE Trans. Ind. Appl.*, vol. 42, no. 3, pp. 770–778, May/Jun. 2003.

[42] M. Shen, J. Wang, A. Joseph, F. Z. Peng, L. M. Tolbert and D. J Adams, "Maximum Constant Boost Control of the Z-Source Inverter," in *Proc. IEEE IAS Annual Meeting 2004*, Oct. 2004.

[43] U. S. Ali and V. Kamaraj, "A Novel Space Vector PWM for Z-Source Inverter," in *Proc. First International Conference on Electronic Energy Systems*, pp. 82–85, Jan. 2011.

[44] O. Ellabban, M. J. Van and P. Lataire, "Experimental Study of the Shoot-Through Boost Control Methods for the Z-Source Inverter," in *Proc. Eur. Power Electron. Drives Assoc.*, vol. 21, no. 2, pp. 18–29, Jun. 2011.

[45] Y. Liu, B. Ge, F. J. T. E. Ferreira, A. T. de Almeida and A. A. Rub, "Modeling and SVPWM Control of Quasi-Z-Source Inverter," in *Proc. IEEE EPQU 2011*, pp. 1–7, Oct. 2011.

[46] Y. Siwakoti and G. Town, "Three-phase Transformerless Grid Connected Quasi-Z-Source Inverter for Solar Photovoltaic Systems with Minimal Leakage Current," in *Proc. IEEE PEDG 2012*, pp. 368–373, Jun. 2012.

[47] F. Bradaschia, M. C. Cavalcanti, P. E. P. Ferraz, F. A. S. Neves, E. C. dos Santos Jr. and J. H. G. M. da Silva, " Modulation for Three-Phase Transformerless Z-Source Inverter to Reduce Leakage Currents in Photovoltaic Systems," *IEEE Trans. Ind. Electron.*, vol. 58, no. 12, pp. 5385–5395, Dec. 2011.

[48] Y. Siwakoti and G. Town, "Common-Mode Voltage Reduction Techniques of Three-Phase Quasi-Z-Source Inverter for AC Drives," in *Proc. IEEE APEC 2013*, pp. 2247–2252, Mar. 2013.

[49] P. C. Loh, F. Gao, F. Blaabjerg, S. Y. C. Feng and N. J. Soon, "Pulsewidth-Modulated Z-Source Neutral-Point-Clamped Inverter," *IEEE Trans. Ind. Appl.*, vol. 43, no. 5, pp. 1295–1308, Sep./Oct. 2007.

[50] P. C. Loh, F. Gao, F. Blaabjerg, "Topological and Modulation Design of Three-Level Z-Source Inverters," *IEEE Trans. Power Electron.*, vol. 23, no. 5, pp. 2268–2277, Sep. 2008.

[51] F. Gao, P. C. Loh, F. Blaabjerg and D. M. Vilathgamuwa, "Performance Evaluation of Three-Level Z-Source Inverters Under Semiconductor-Failure Conditions," *IEEE Trans. Ind. Appl.*, vol. 45, no. 3, pp. 971–981, May/Jun. 2009.

[52] P. C. Loh, F. Blaabjerg and C. P. Wong, "Comparative Evaluation of Pulse-Width Modulation Strategies for Z-Source Neutral-Point-Clamped Inverter," *IEEE Trans. Power Electron.*, vol. 22, no. 3, pp. 1005–1013, May 2007.

[53] P. C. Loh, F. Gao, P. C. Tan and F. Blaabjerg, "Three-Level AC–DC–AC Z-Source Converter Using Reduced Passive Component Count," *IEEE Trans. Power Electron.*, vol. 24, no. 7, pp. 1671–1681, Jul. 2009.

[54] F. B. Effah, P. Wheeler, J. Clare and A. Watson, "Space-Vector-Modulated Three-Level Inverters With a Single Z-Source Network," *IEEE Trans. Power Electron.*, vol. 28, no. 6, pp. 2806–2815, Jun. 2013.

[55] E. C. dos Santos Jr., E. P. X. P. Filho, A. C. Oliveria and E. R. C. da Silva, "Hybrid Pulse Width Modulation for Z-Source Inverters," in *Proc. IEEE ECCE 2010*, pp. 2888–2892, Sep. 2010.

[56] J. H. G. Muniz, E. R. C. da Silva and E. C. dos Santos Jr., "A Hybrid PWM Strategy for Z-Source Neutral-Point-Clamped Inverter," in *Proc. IEEE APEC 2011*, pp. 450–456, Mar. 2011.

[57] Y. Liu, B. Ge, H. A. Rub and F. Z. Peng, "An Effective Control Methods for Quasi-Z-Source Cascade Multilevel Inverter Based Grid-Tie Single Phase Photovoltaic Power System," *IEEE Trans. Ind. Informatics*, vol. 10, no. 1, pp. 399–407, Feb. 2014.

[58] Y. Liu, B. Ge, H. A. Rub and F. Z. Peng, "A Modular Multilevel Space Vector Modulation for Photovoltaic Quasi-Z-Source Cascaded Multilevel Inverter," in *Proc. IEEE APEC 2013*, pp. 714–718, Mar. 2013.

[59] B. Ge, Q. Li, W. Qian and F. Z. Peng, "A Family of Z-Source Matrix Converters," *IEEE Trans. Ind. Electron.*, vol. 59, no. 1, pp. 35–46, Jan. 2012.

[60] X. Liu, P. C. Loh, P. Wang and X. Han, "Improved Modulation Schemes for Indirect Z-Source Matrix Converter with Sinusoidal Input and Output Waveforms," *IEEE Trans. Power Electron.*, vol. 27, no. 9, pp. 4039–4050, Sep. 2012.

[61] Q. Lei, B. Ge, F. Z. Peng, "Hybrid PWM Control for Z-Source Matrix Converter," in *Proc. ECCE 2011*, pp. 246–253, Sep. 2011.

[62] Q. Lei, F. Z. Peng, B. Ge, "Pulse-Width-Amplitude-Modulated Voltage-Fed Quasi-Z-Source Direct Matrix Converter with Maximum Constant Boost," in *Proc. APEC 2012*, pp. 641–646, Feb. 2012.

[63] X. Liu, P. C. Loh, F. Z. Peng and P. Wang, "Optimal Modulation of Indirect Z-Source Matrix Converter," in *Proc. IPEC 2010*, pp. 3049–3056, Jun. 2010.

[64] X. Ding, Z. Qian, Y. Xie, F. Z. Peng, "A Novel ZVS Z-Source Rectifier," in *Proc. APEC 2006*, pp. 951–955, Mar. 2006.

[65] Q. Lei, S. Yang, F. Z. Peng and R. Inoshita, "Steady State and Transient Analysis of a Three Phase Current-Fed Z-Source PWM Rectifier," in *Proc. VPPC 2009*, pp. 426–432, Sep. 2009.

[66] J. Anderson and F. Z. Peng, "Four Quasi-Z-Source Inverters," in *Proc. PESC 2008*, pp. 2743–2749, Jun. 2008.

[67] D. Li, P. C. Loh, M. Zhu, F. Gao and F. Blaabjerg, "Enhanced-Boost Z-Source Inverters with Alternate-Cascaded Switched- and Tapped-Inductor Cells," *IEEE Trans. Ind. Electron.*, vol. 60, no. 9, pp. 3567–3578, Sep. 2013.

[68] Y. Tang, S. Xie, C. Zhang and Z. Xu, "Improved Z-Source Inverter with Reduced Z-Source Capacitor Voltage Stress and Soft-Start Capability," *IEEE Trans. Power Electron*., vol. 24, no. 2, pp. 409–415, Feb. 2009.

[69] M. K. Nguyen, Y. C. Lim and S. J. Park, "Improved Trans-Z-Source Inverter with Continuous Input Current and Boost Inversion Capability," *IEEE Trans. Power Electron.*, vol. 28, no. 10, pp. 4500–4510, Oct. 2013.

[70] D. Cao, S. Jiang, X. Yu and F. Z. Peng, "Low-Cost Semi-Z-Source Inverter for Single-Phase Photovoltaic Systems," *IEEE Trans. Power Electron.*, vol. 26, no. 12, pp. 3514–3523, Dec. 2011.

[71] H. Haimovich, R. H. Middleton and L. De Nicolo, "Large-Signal Stability Conditions for Semi-Quasi-Z-Source Inverters: Switched and Averaged Models," in *Proc. 52nd Conference on Decision and Control 2013*, Dec. 2013.

[72] Y. Tang, S. Xie and C. Zhang, "Single-Phase Z-Source Inverter," *IEEE Trans. Power Electron.*, vol. 26, no. 12, pp. 3869–3873, Dec. 2011.

[73] P. C. loh, F. Gao and F. Blaabjerg, "Embedded EZ-Source Inverter," *IEEE Trans. Ind. Appl.*, vol. 46, no. 1, pp. 256–267, Jan./Feb. 2010.

[74] P. C. Loh, N. Duan, C. Liang, F. Gao and F. Blaabjerg, "Z-Source B4 Inverter," in *Proc. PESC 2007*, pp. 1363–1369, Jun. 2007.

[75] D. Li, P. C. Loh, M. Zhu, F. Gao and F. Blaabjerg, "Generalised Multicell Switched-inductor and Switched-Capacitor Z-Source Inverters," *IEEE Trans. Power Electron.*, vol. 28, no. 2, pp. 837–848, Feb. 2013.

[76] H. Itozakura and H. Koizumi, "Embedded Z-Source Inverter with Switched Inductor," in *Proc. IECON 2011*, pp. 1342–1347, Nov. 2011.

[77] C. J. Gajanayake, F. L. Luo, H. B. Gooi, P. L. So and L. K. Siow, "Extended-Boost Z-Source Inverters," *IEEE Trans. Power Electron.*, vol. 25, no. 10, pp. 2642–2652, Oct. 2010.

[78] L. Huang, M. Zhang, L. Hang, W. Yao and Z. Lu, "A Family of Three-Switch Three-State Single-Phase Z-Source Inverters," *IEEE Trans. Power Electron.*, vol. 28, no. 5, pp. 2317–2329, May 2013.

[79] S. Jiang and F. Z. Peng, "Transmission-line Theory Based Distributed Z-Source Networks for Power Conversion," in *Proc. APEC 2011*, pp. 1138–1145, Mar. 2011.

[80] P. C. Loh and F. Blaabjerg, "Magnetically Coupled Impedance-Source Inverters," *IEEE Trans. Ind. Appl.*, vol. 49, no. 5, pp. 2177–2187, Sep./Oct. 2013.

[81] Y. P. Siwakoti, P. C. Loh, F. Blaabjerg, S. J. Andreasen and G. E. Town, "Y-Source Impedance Network Based Boost DC/DC Converter for Distributed Generation," *IEEE Trans. Ind. Electron.*, vol. 62, no. 2, pp. 1059–1069, Aug. 2014. DOI: 10.1109/TIE.2014.2345336.

[82] P. C. Loh, D. Li and F. Blaabjerg, "Γ-Z-Source Inverters," *IEEE Trans. Power Electron. (Letters)*, vol. 28, no. 11, pp. 4880–4884, Nov. 2013.

[83] R. Strzelecki, M. Adamowicz, N. Strzelecka and W. Bury, "New type T-source inverter," in *Proc. CPE 2009*, pp. 191–195, May 2009.
[84] W. Qian, F. Z. Peng and H. Cha, "Trans-Z-Source Inverters," *IEEE Trans. Power Electron.*, vol. 26, no. 12, pp. 3453–3463, Dec. 2011.
[85] M. K. Nguyen, Y. C. Lim and Y. G. Kim, "TZ-Source Inverters," *IEEE Trans. Ind. Electron.*, vol. 60, no. 12, pp. 5686–5695, Dec. 2013.
[86] M. Adamowicz, R. Strzelecki, F. Z. Peng, J. Guzinski and H. A. Rub, "New Type LCCT-Z-Source Inverters," in *Proc. EPE 2011*, pp. 1–10, Sep. 2011.
[87] S. Jiang, D. Cao and F. Z. Peng, "High Frequency Transformer Isolated Z-Source Inverters," in *Proc. APEC 2011*, pp. 442–449, Mar. 2011.

Chapter 8

Switching power supplies

Dariusz Czarkowski and Mariusz Bojarski**

8.1 Introduction

Power supplies for modern electronic systems should be small, lightweight, reliable, and efficient. Linear power regulators, whose principle of operation is based on a voltage or current divider, are inefficient. They are limited to output voltages smaller than the input voltage. Also, their power density is low because they require low-frequency (50 or 60 Hz) line transformers and filters. Linear regulators can, however, provide a very high-quality output voltage. Their main area of application is at low power levels as low dropout voltage regulators. Semiconductor components in linear regulators operate in their active (linear) modes. At higher power levels, switching regulators are used. Switching regulators use power electronic semiconductor switches in on and off states. Since there is a small power loss in those states (low voltage across a switch in the on state, zero current through a switch in the off state), switching regulators can achieve high-energy conversion efficiencies. Modern power electronic switches can operate at high frequencies. The higher the operating frequency, the smaller and lighter the transformers, filter inductors, and capacitors. In addition, dynamic characteristics of converters improve with increasing operating frequencies. The bandwidth of a control loop is usually determined by the corner frequency of the output filter. Therefore, high operating frequencies allow for achieving a fast dynamic response to rapid changes in the load current and/or the input voltage.

The functions of dc–dc converters are:

- to convert a dc input voltage V_{IN} into a dc output voltage V_{OUT};
- to regulate the dc output voltage against load and line variations;
- to reduce the ac voltage ripple on the dc output voltage below the required level;
- to provide isolation between the input source and the load (isolation is not always required);
- to protect the supplied system and the input source from electromagnetic interference (EMI);
- to satisfy various international and national safety standards.

*New York University, New York, NY, USA

Two main types of dc–dc converters can be distinguished: hard-switching pulse width modulated (PWM) converters, and resonant and soft-switching converters. This chapter describes the PWM converters which have been very popular for the last four decades. They are widely used at all power levels. Topologies and properties of PWM converters are well understood and described in the literature. Advantages of PWM converters include low component count, high efficiency, constant frequency operation, relatively simple control and commercial availability of integrated circuit controllers, and ability to achieve high conversion ratios for both step-down and step-up applications. A disadvantage of PWM dc–dc converters is that PWM rectangular voltage and current waveforms cause turn-on and turn-off losses in semiconductor devices which limit practical operating frequencies to a megahertz range. Rectangular waveforms also inherently generate high levels of EMI.

This chapter starts from a section on non-isolated dc–dc converters. Four basic dc–dc converter topologies are presented in Sections 8.2.1–8.2.4: buck, boost, buck–boost, and integrated buck and boost converters. In Section 8.2.5, the concept of power factor correction is explained. Popular isolated converter topologies are discussed in Sections 8.3.1–8.3.4: flyback, forward, half-bridge, and full-bridge converters. Various rectifier types are presented in Section 8.3.5. Operation of converters is explained under ideal component and semiconductor device assumptions. Section 8.4 discusses effects of non-idealities in PWM converters. Section 8.5 introduces concepts of various types of conduction modes. Section 8.6 presents topologies for increased efficiency at low output voltage. In Section 8.7, examples of bidirectional power converter topologies are presented. Section 8.8 discusses a concept of paralleling multiple converters. Section 8.9 reviews control principles of PWM dc–dc converters. Two main control schemes, voltage-mode control and current-mode control, are described. Finally, a list of modern textbooks on power electronics is provided in Section 8.10. These books are excellent resources for deeper exploration into the area of dc–dc power conversion.

8.2 Non-isolated converters topologies

Converters discussed in this chapter utilize PWM to control the output. The PWM signal is obtained with a switch, for instance a transistor, which is turned-on for time t_{ON} and then turned-off for time t_{OFF}. The process of turning on and off is repeated continuously with a period $T = t_{ON} + t_{OFF}$. Hence, the switch duty ratio D could be defined as

$$D = \frac{t_{on}}{t_{on} + t_{off}} \tag{8.1}$$

It can be seen from the above equation that the theoretical range of duty ratio D is from 0 to 1.

8.2.1 Buck converter

Figure 8.1 shows a step-down dc–dc converter, commonly known as a buck converter. The buck converter consists of dc input voltage source V_{IN}, controlled

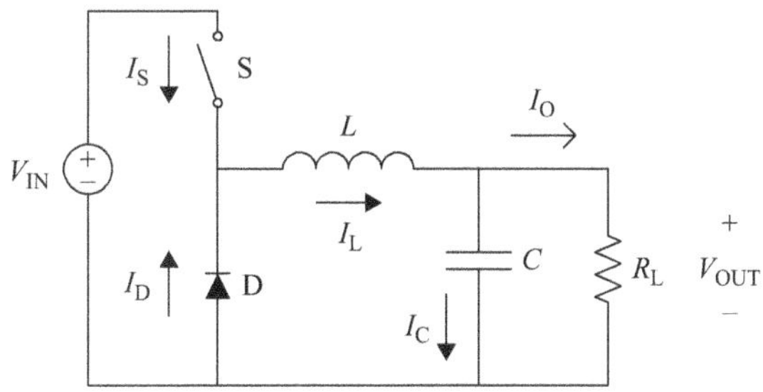

Figure 8.1 The buck converter topology diagram

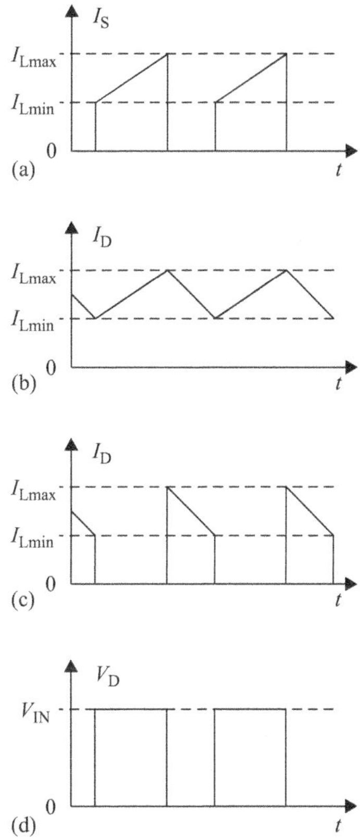

Figure 8.2 Typical voltage and current waveforms in the buck converter: (a) switch current, (b) inductor current, (c) diode current, and (d) diode voltage

switch S, diode D, filter inductor L, filter capacitor C, and load resistance R_L. Typical waveforms in the converter are shown in Figure 8.2 under the assumption that the inductor current is always positive. It can be seen from the circuit that when the switch S is driven to the on state, the diode D is reverse biased.

When the switch S is off, the diode conducts to support an uninterrupted current in the inductor. The relationship among the input voltage, output voltage, and the switch duty ratio D can be derived from, for instance, the waveform of the voltage across the inductor L, which is a difference between the output voltage V_{OUT} and the voltage across the diode D. The voltage across the inductor is equal to $V_{IN} - V_{OUT}$ when the switch S is on and equal to $-V_{OUT}$ when the switch is off. According to Faraday's law, the inductor volt–second product over a period of steady-state operation is zero. For the buck converter

$$(V_{IN} - V_{OUT})DT = V_{OUT}(1 - D)T \tag{8.2}$$

Hence, the dc voltage transfer function, defined as the ratio of the output voltage to the input voltage, is

$$M_V = \frac{V_{OUT}}{V_{IN}} = D \tag{8.3}$$

It can be concluded from the above equation that the output voltage of the buck converter is within the range from 0 to V_{IN}.

The filter inductor current i_L consists of a dc component I_O with a superimposed triangular ac component. This ac component forms the inductor current ripple. Its peak-to-peak value can be calculated as

$$I_{Lr} = \frac{V_{OUT} t_{off}}{L} = \frac{V_{OUT}(1 - D)T}{L} \tag{8.4}$$

The maximum and minimum values of the inductor current i_L are

$$I_{Lmax} = I_O + \frac{V_{OUT}(1 - D)T}{2L} \tag{8.5}$$

$$I_{Lmin} = I_O - \frac{V_{OUT}(1 - D)T}{2L} \tag{8.6}$$

In most design cases, the filtering inductance value L is selected to keep the current ripple below certain limit I_{Lrmax}. Thus, the inductance value can be calculated as

$$L_{min} = \frac{V_{OUT}(1 - D)T}{I_{Lrmax}} \tag{8.7}$$

Almost the entire ac component of the inductor current flows through the filter capacitor as a current i_C. Current i_C causes a small voltage ripple across the dc output voltage V_{OUT}. To limit the peak-to-peak value of the ripple voltage below certain value V_r, the filter capacitance C must be greater than

$$C_{min} = \frac{V_{OUT}(1 - D)}{8V_r L f^2} = \frac{I_{Lrmax}}{8V_r f} \tag{8.8}$$

Equations (8.7) and (8.8) are the key design equations for the buck converter. The input and output dc voltages (hence, the duty ratio D), and the range of load resistance R_L are usually determined by preliminary specifications. The designer needs to determine values of passive components L and C, and of the switching frequency f. The value of the filter inductor L is based on desired level of inductor ripple current using (8.7). The value of the filter capacitor C is obtained from the voltage ripple condition (8.8). For the compactness and low conduction losses of a converter, it is desirable to use small passive components. Equations (8.7) and (8.8) show that it can be accomplished with a high switching frequency f. The switching frequency is limited, however, by the type of semiconductor switches used and by switching losses. It should be also noted that values of L and C may be altered by effects of parasitic components in the converter, especially by the equivalent series resistance (ESR) of the capacitor. The issue of parasitic components in dc–dc converters is discussed in Section 8.4.

8.2.2 Boost converter

Figure 8.3 presents a step-up or a PWM boost converter. It is composed of dc input voltage source V_{IN}, boost inductor L, controlled switch S, diode D, filter capacitor C, and load resistance R_L. Typical waveforms in the converter are shown in Figure 8.4 under the assumption that the inductor current is always positive. When the switch S is in the on state, the current in the boost inductor increases linearly. The diode D is off at that time. When the switch S is turned off, the energy stored in the inductor is released through the diode D to the output RC circuit. The voltage across the inductor is equal to V_{IN} when the switch S is on and equal to $V_{IN} - V_{OUT}$ when the switch is off. Using the Faraday's law for the boost inductor

$$V_{IN}DT = (V_{OUT} - V_{IN})(1 - D)T \tag{8.9}$$

from which the dc voltage transfer function could be derived as

$$M_V = \frac{V_{OUT}}{V_{IN}} = \frac{1}{1 - D} \tag{8.10}$$

It can be concluded from (8.10) that the output voltage of the boost converter is within the range from V_{IN} to infinity. Thus, a duty ratio limitation or the over

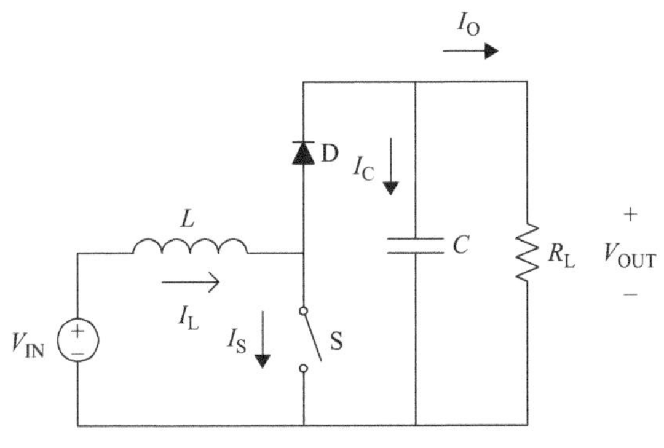

Figure 8.3 The boost converter topology diagram

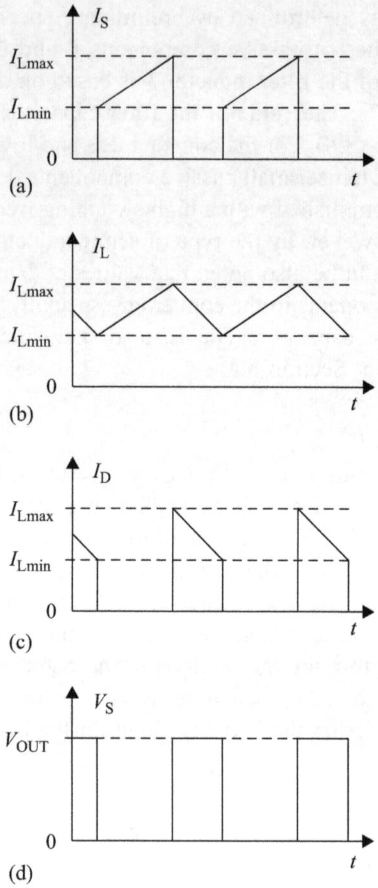

Figure 8.4 Typical voltage and current waveforms for the boost converter: (a) switch current, (b) inductor current, (c) diode current, and (d) switch voltage

voltage protection at the output is strongly recommended as the boost converter can generate very high voltage at the output.

The peak-to-peak value of the inductor current ripple can be calculated as

$$I_{Lr} = \frac{V_{IN}t_{on}}{L} = \frac{V_{IN}DT}{L} \tag{8.11}$$

The maximum and minimum values of the inductor current i_L are

$$I_{Lmax} = I_O + \frac{V_{IN}DT}{2L} \tag{8.12}$$

$$I_{Lmin} = I_O - \frac{V_{IN}DT}{2L} \tag{8.13}$$

In most cases, the filtering inductor value L is selected to keep the current ripple below certain limit I_{Lrmax}. Thus, the inductor value can be calculated as

$$L_{\text{min}} = \frac{V_{\text{IN}}DT}{I_{\text{Lrmax}}} \tag{8.14}$$

In the boost converter, the current is delivered to the output RC circuit only when diode D is conducting. Thus, this current is discontinuous which implies a larger filter capacitor requirement compared to that in the buck-derived converters to limit the output voltage ripple. The filter capacitor must provide the output dc current to the load when the diode D is not conducting. The minimum value of the filter capacitance that results in the voltage ripple V_r is given by

$$C_{\text{min}} = \frac{V_{\text{OUT}}D}{V_r Rf} \tag{8.15}$$

8.2.3 Buck–boost converter

Figure 8.5 depicts a buck–boost converter. It is comprised of dc input voltage source V_{IN}, inductor L, controlled switch S, diode D, filter capacitor C, and load resistance R_L. Typical waveforms in the converter are shown in Figure 8.6 under the assumption that the inductor current is always positive. When the switch S is in the on state, the current in the inductor increases linearly. The diode D is off at the time. When the switch S is turned off, the energy stored in the inductor is released through the diode D to the output RC circuit. The voltage across the inductor is equal to V_{IN} when the switch S is on and equal to $-V_{\text{OUT}}$ when the switch is off. Using the Faraday's law for the inductor

$$V_{\text{IN}}DT = V_{\text{OUT}}(1 - D)T \tag{8.16}$$

from which the dc voltage transfer function turns out to be

$$M_V = \frac{V_{\text{OUT}}}{V_{\text{IN}}} = -\frac{D}{1 - D} \tag{8.17}$$

Equation (8.17) shows that the output voltage of the buck–boost converter is within the range from 0 to negative infinity. Hence, a duty ratio limitation or the over

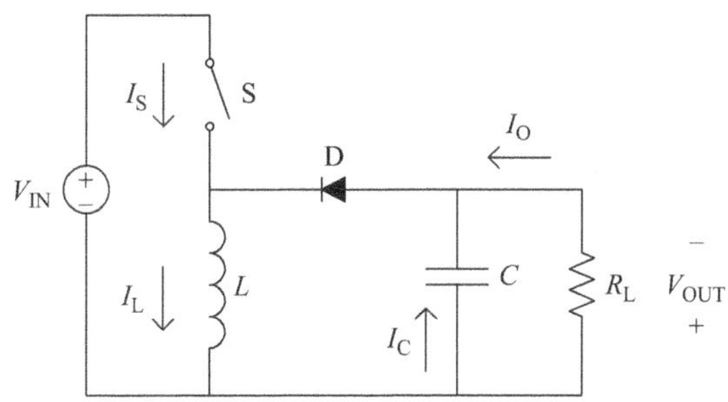

Figure 8.5 The buck–boost converter topology diagram

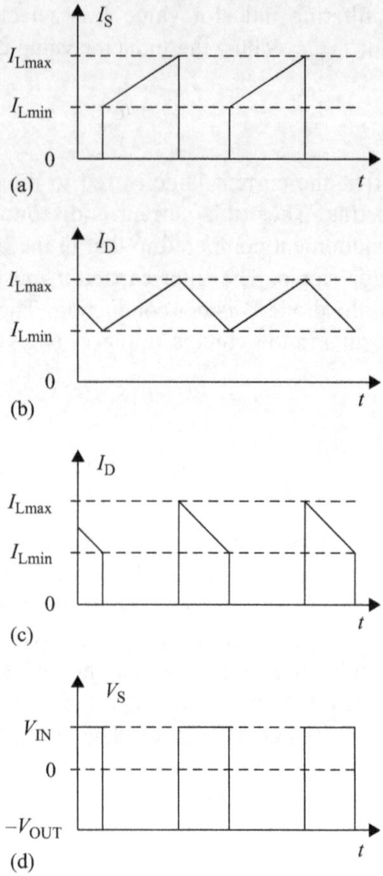

Figure 8.6 Typical voltage and current waveforms for the buck–boost converter: (a) switch current, (b) inductor current, (c) diode current, and (d) switch voltage

voltage protection at the output is strongly recommended since the buck–boost converter can generate very high voltage at the output.

The output voltage V_{OUT} is negative with respect to the ground. Its magnitude can be either greater or smaller (equal at $D = 0.5$) than the input voltage as this converter name implies.

The structure of the output part of the converter is similar to that of the boost converter (reversed polarities being the only difference). Thus, the same equation for current ripples, inductor L value, and filtering capacitor C value can be applied.

8.2.4 Integrated buck and boost converter

Figure 8.7 depicts an integrated buck and boost converter. It is comprised of dc input voltage source V_{IN}, inductor L, two controlled switch S_1 and S_2, two diodes D_1 and

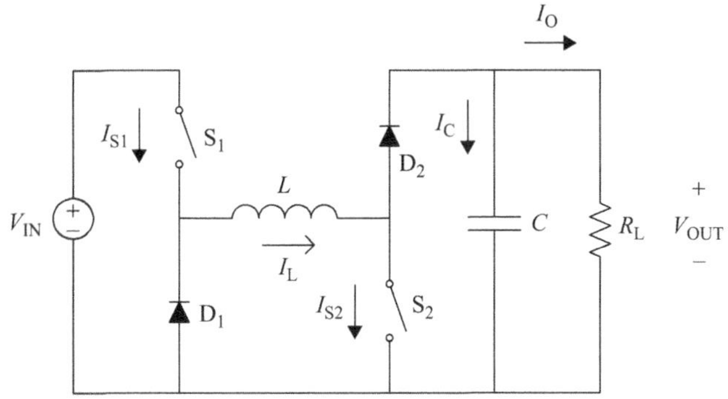

Figure 8.7 The integrated buck and boost converter topology diagram

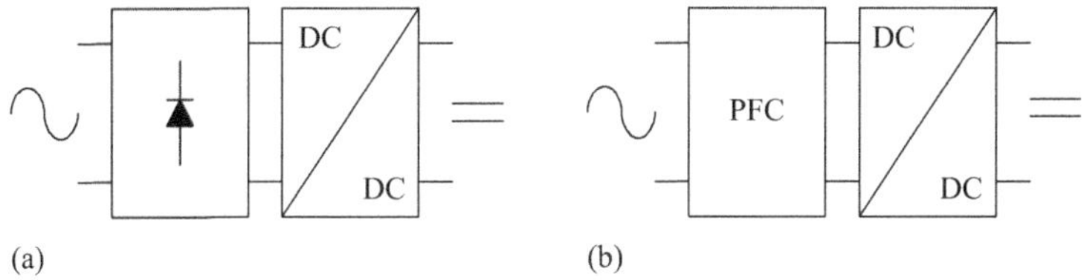

Figure 8.8 Block diagrams of ac–dc converters: (a) with a traditional rectifier and (b) with a power factor corrector

D_2, filter capacitor C, and load resistance R_L. This converter has two operation modes: buck and boost. In the buck mode, the switch S_2 is off and the switch S_1 is being turned on and off with duty ratio D and frequency f. Assuming a positive current in the inductor L, diode D_2 always conducts in this operation mode. Thus, the converter works as a regular buck converter and all related equations are applicable.

In the second operation mode, the converter behaves as a boost converter. In this case, switch S_1 is always on and diode D_1 does not conduct. The switch S_2 is being turned on and off with duty ratio D and frequency f. Thus, the converter works as a regular boost converter and all related equations apply.

The integrated buck and boost converter is used where the output voltage needs to be regulated in a range which extends from below to above the input voltage. Such regulation is not achievable by the buck or boost converter alone. It can be achieved by sepic or Cuk converters which are, however, higher order converters and their analysis and control are more complex. Thus, the integrated buck and boost converter is preferred by the industry in most such applications.

8.2.5 Power factor correction

Typical power supplies take the ac voltage from the grid (50 or 60 Hz) as an input. Thus, the power supply usually consists of a low-frequency rectifier followed by a dc–dc converter as shown in Figure 8.8(a). In recent years, regulations and standards have put tight restrictions on the quality of input current of power supplies. In many cases, a sinusoidal input current with low harmonic distortion level is required. Thus, the power factor corrected pre-regulators are used instead of traditional rectifiers as shown in Figure 8.8(b).

The power factor corrector (PFC) consists usually of a rectifier followed by the boost converter. Other topologies, including bridgeless variants, are also used. The primary objective of the PFC is to provide a sinusoidal input current. Thus, the control is designed accordingly in order to follow the input voltage as a reference for the current. The input current to the power supply with and without PFC is shown for comparison in Figure 8.9, neglecting distortions that typically happen at current values close to zero.

8.3 Isolated converters topologies

8.3.1 Flyback converter

A PWM flyback converter is a very practical isolated version of the buck–boost converter. The circuit of the flyback converter is presented in Figure 8.10. The inductor of the buck–boost converter has been replaced by a flyback transformer.

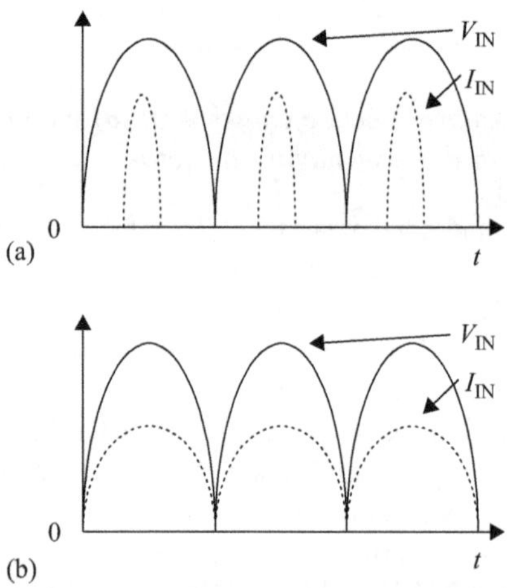

Figure 8.9 Input voltage and current of an ac–dc converter: (a) with a traditional rectifier and (b) with a power factor corrector

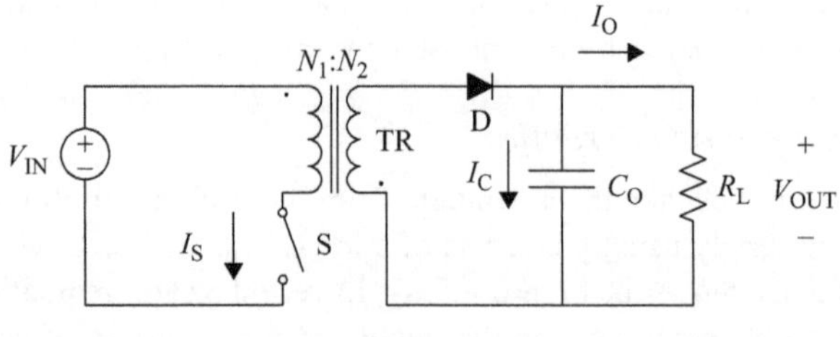

Figure 8.10 The flyback converter topology diagram

The input dc source V_{IN} and switch S are connected in series with the primary transformer. The diode D and the RC output circuit are connected in series with the secondary of the flyback transformer. Refer to Figure 8.10 for the converter operation. When the switch S is on, the current in the magnetizing inductance of the transformer increases linearly. The diode D is off and there is no current in the secondary side winding of the transformer. When the switch is turned off, the magnetizing inductance current is discharged through the diode D, and the transformed magnetizing inductance current is supplied to the RC load. The dc voltage transfer function of the flyback converter is

$$M_V = \frac{V_{OUT}}{V_{IN}} = \frac{D}{n(1-D)} \tag{8.18}$$

It differs from the buck–boost converter voltage transfer function by the turns ratio factor $n = N_1/N_2$. A positive sign has been obtained by an appropriate coupling of the transformer windings.

Unlike in other isolated dc–dc converters, the magnetizing inductance L_m of the flyback transformer is an important design parameter, as it works also as the filtering inductor.

8.3.2 Forward converter

The circuit diagram of a forward converter is depicted in Figure 8.11. When the switch S is on, diode D_1 conducts and diode D_2 is off. The energy is transferred from the input, through the transformer, to the output filter. When the switch is off, the state of diodes D_1 and D_2 is reversed. Thus, the output filter operates exactly like in the buck converter. When the switch S is on, the voltage across the inductor is equal to $V_{IN}/n - V_{OUT}$, and when the switch is off, the inductor voltage is equal to $-V_{OUT}$. The transformer turns ratio $n = N_1/N_2$. Thus, the dc voltage transfer function of the forward converter is

$$M_V = \frac{V_{OUT}}{V_{IN}} = \frac{D}{n} \tag{8.19}$$

In the forward converter, the energy-transfer current flows through the transformer in one direction. Hence, an additional winding with diode D_3 is needed to bring the

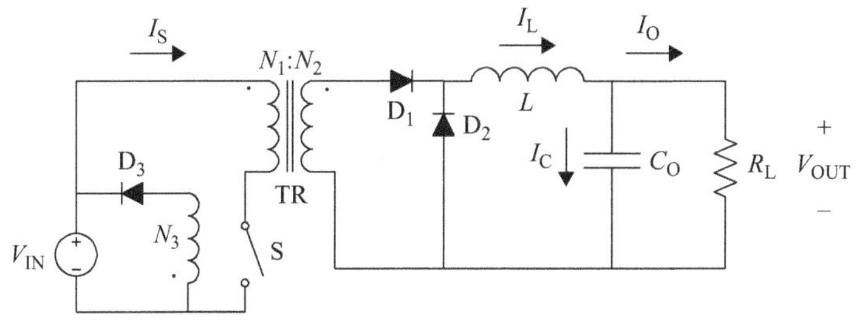

Figure 8.11 The forward converter topology diagram

magnetizing current of the transformer to zero. This prevents transformer saturation. The turns ratio N_1/N_3 should be selected in such a way that the magnetizing current decreases to zero during the time interval when the switch is off. In the most common configuration, the duty ratio D is limited to 0.5 and the mentioned turns ratio $N_1/N_3 = 1$.

A disadvantage of this topology is a high-voltage stress across the switch S. In the most common configuration, the voltage across the switch S when it is open is equal to $2V_{IN}$. This affects the efficiency of the converter and makes it not suitable for high- or medium-power applications. As the structure of the converter is relatively simple and only one transistor is employed, the forward converter is commonly used for low-power applications.

The high-voltage stress problem of the basic forward converter topology could be solved by the two-switch topology variant, which is presented in Figure 8.12. The operation principle is very similar. Both switches S_H and S_L are turned on and off at the same time. When the switches are on, diodes D_H and D_L do not conduct. On the secondary side, diode D_2 is off and diode D_1 conducts delivering current to the output filter. When switches S_H and S_L are off, diodes D_H and D_L conduct allowing transformer to demagnetize before the next cycle. Diode operation at the secondary is reversed, as in the basic forward converter described above. Thus, the output filter design is identical for both of them. In the two-switch forward converter, duty ratio D must be limited to 0.5 to allow demagnetization of the transformer.

8.3.3 *Half-bridge converter*

The PWM dc–dc half-bridge converter is shown in Figure 8.13. In the half-bridge converter, two switches generate a rectangular voltage waveform in a range from 0 to V_{IN}. To avoid saturation of the transformer, a dc offset is cut off by the capacitor divider to which transformer is connected. The switches S_H and S_L operate shifted in phase by $T/2$ with the same duty ratio D. The duty ratio must be smaller than 0.5. When switch S_H is on, diode D_1 conducts and diode D_2 is off. Diode states are

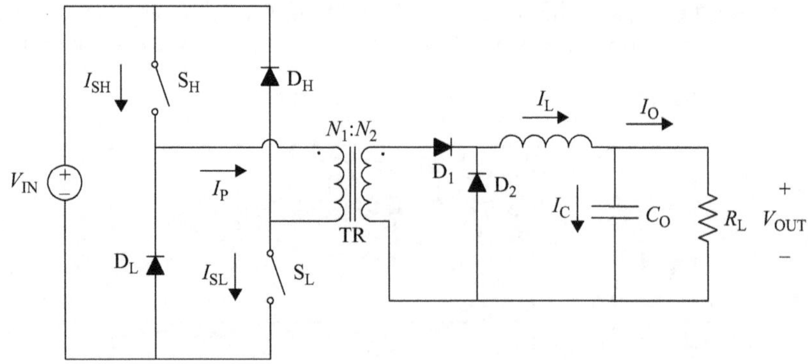

Figure 8.12 The two-switch forward converter topology diagram

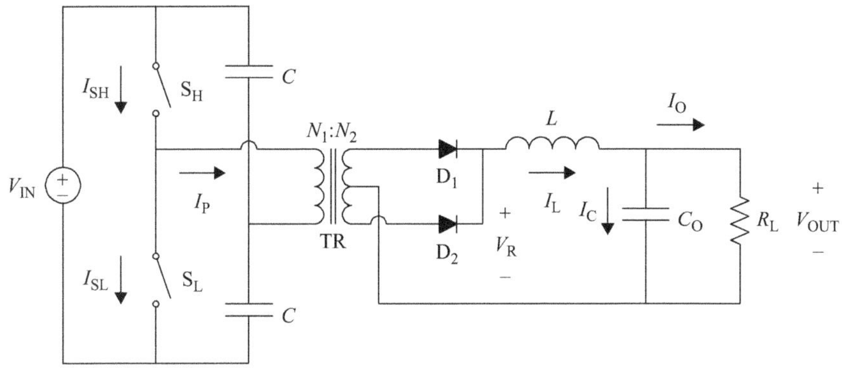

Figure 8.13 The half-bridge converter topology diagram

reversed when switch S_L is on. When both controllable switches are off, the diodes are on and share the filter inductor current equally. Typical waveforms in the converter are shown in Figure 8.14 under the assumption that the inductor current is always positive. The dc voltage transfer function of the half-bridge converter is

$$M_V = \frac{V_{OUT}}{V_{IN}} = \frac{D}{n} \tag{8.20}$$

where $D < 0.5$ and n is the turns ratio equal to N_1/N_2. The inductor current ripple is calculated in a similar manner as in the buck converter. Its peak-to-peak value is

$$I_{Lr} = \frac{V_{OUT}(1-2D)T}{L} \tag{8.21}$$

The maximum and minimum values of the inductor current i_L are

$$I_{L\max} = I_O + \frac{V_{OUT}(1-2D)T}{2L} \tag{8.22}$$

$$I_{L\min} = I_O - \frac{V_{OUT}(1-2D)T}{2L} \tag{8.23}$$

The filtering inductor value L is typically selected to keep the current ripple below certain limit I_{Lrmax}. Thus, the inductor value can be calculated as

$$L_{\min} = \frac{V_{OUT}(1-2D)T}{I_{Lr\max}} \tag{8.24}$$

Almost all of the inductor current ripple flows through the filter capacitor as current i_C. Current i_C causes a small voltage ripple at the dc output voltage V_{OUT}. To limit the peak-to-peak value of the ripple voltage below certain value V_r, the filter capacitance C must be greater than

$$C_{\min} = \frac{V_{OUT}(1-2D)}{32Lf^2} \tag{8.25}$$

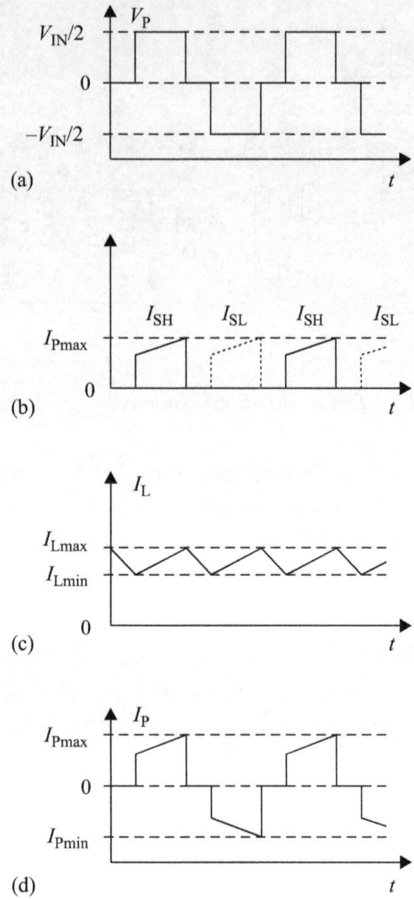

Figure 8.14 Typical voltage and current waveforms for the half-bridge converter: (a) transformer primary voltage, (b) switch currents, (c) inductor current, and (d) transformer primary current

8.3.4 Full-bridge converter

The PWM dc–dc full-bridge converter is shown in Figure 8.15. Its comparison to the half-bridge converter reveals that the capacitive voltage divider has been replaced by two controllable switches. The controllable switches are operated in pairs. When S_{H1} and S_{L2} are on, voltage V_{IN} is applied to the primary of the transformer and diode D_1 conducts. With S_{L1} and S_{H2} on, there is voltage $-V_{IN}$ across the primary transformer and diode D_2 is on. With all controllable switches off, both diodes conduct, similarly as in the half-bridge converter. The dc voltage transfer function of the full-bridge converter is

$$M_V = \frac{V_{OUT}}{V_{IN}} = \frac{2D}{n} \tag{8.26}$$

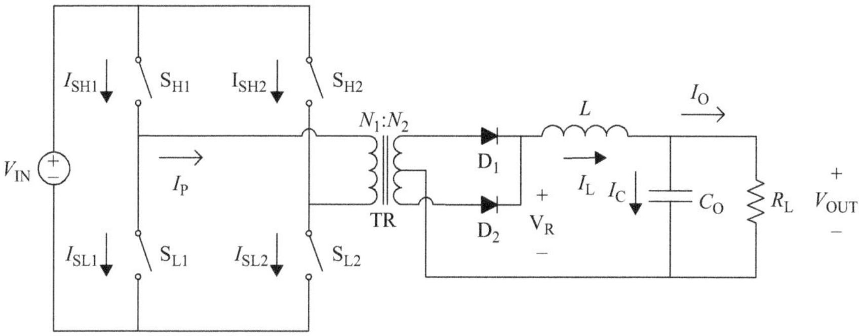

Figure 8.15 The full-bridge converter topology diagram

Filtering components and current ripples are calculated in identical manner as for the half-bridge converter.

The full-bridge topology allows also for another very popular control strategy, called phase-shift control. In this scenario, each half-bridge generates a square wave with the duty cycle equal to 0.5. The regulation is obtained by the phase-shift of those two square waveforms as shown in Figure 8.16. The phase-shift control allows for obtaining zero voltage switching conditions for all switches. At the secondary, this control strategy provides similar waveforms as in a regular half- or full-bridge converter. The phase-shift of 0° is equivalent to duty cycle equal to zero, and the phase-shift of 180° is equivalent to duty cycle of 0.5.

8.3.5 Rectifiers

There are three basic rectifier topologies applicable for half- and full-bridge converters topology: full-bridge rectifier, center-tap rectifier, and current-doubler rectifier. The full-bridge rectifier is presented in Figure 8.17. It consists of four diodes D_1–D_4, inductor L, capacitor C, and load resistance R_L. The rectifier is supplied by the single winding of the transformer TR. The advantage of the full-bridge rectifier is that the voltage stress across diodes is equal to the peak input voltage to the rectifier. It also works with a simple two-winding type of the transformer. A disadvantage of the full-bridge rectifier is that the inductor current always flows through two diodes which increases conduction losses. It makes this type of the rectifier not suitable for low-voltage applications.

The center-tap rectifier is presented in Figure 8.18. It is composed of two diodes D_1 and D_2, inductor L, capacitor C, and load resistance R_L. The rectifier is supplied by the center-tapped secondary winding of the transformer TR. The advantage of this rectifier is that the inductor current flows through only one diode at a time, which makes it more suitable for low-voltage applications than a full-bridge rectifier. The disadvantage of this rectifier is a requirement for more complicated transformer winding and high-voltage stress for the diodes. In this kind of a rectifier, diodes experience two times higher voltage compared to an equivalent case with the full-bridge rectifier topology.

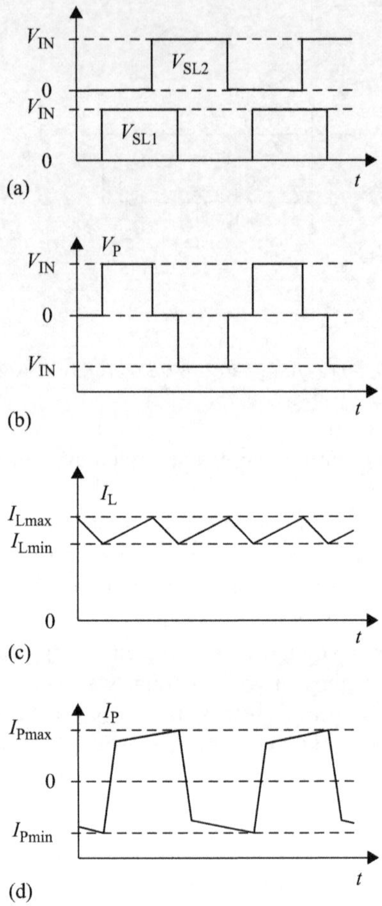

Figure 8.16 Typical voltage and current waveforms for the phase-controlled full-bridge converter: (a) the lower switches voltages, (b) transformer primary voltage, (c) inductor current, and (d) transformer primary current

The current-doubler rectifier is presented in Figure 8.19. It comprises two diodes D_1 and D_2, two inductors L_1 and L_2, capacitor C, and load resistance R_L. The rectifier is supplied by the single winding of the transformer TR. The advantage of this rectifier is that the inductor current flows through only one diode at a time, which makes it more suitable for low-voltage applications than a full-bridge rectifier. The output current is shared by the two inductors which simplifies design for high-current applications. Also, the transformer winding in this case provides two times higher voltage, but there is two times lower current flowing through it. Thus, the transformer winding is greatly simplified, especially for high-current and low-voltage applications. The disadvantage of this rectifier is a high-voltage stress

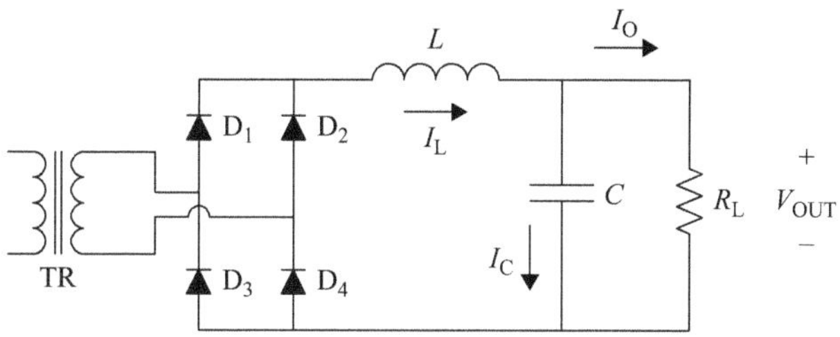

Figure 8.17 The full-bridge rectifier diagram

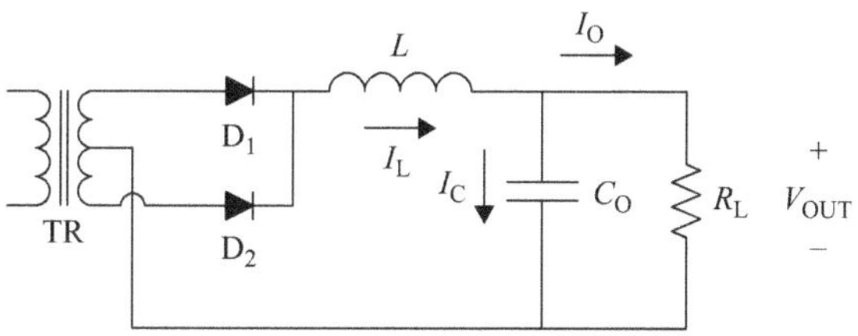

Figure 8.18 The center-tap rectifier diagram

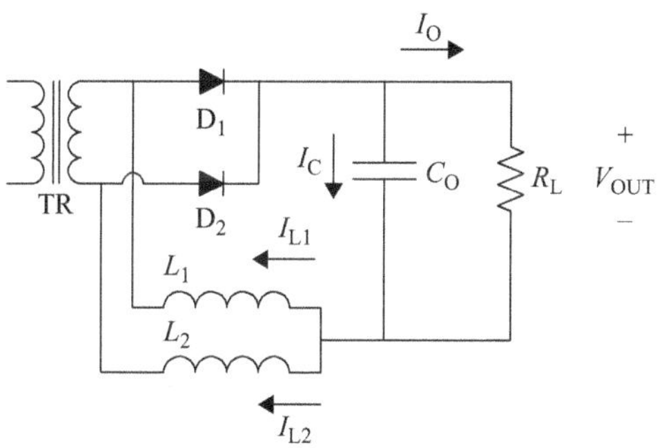

Figure 8.19 The current-doubler rectifier diagram

across the diodes. In the current-doubler rectifier, diodes are subject to two times higher voltage in comparison with an equivalent case of the full-bridge rectifier topology. Typical voltage and current waveforms for this type of rectifier are presented in Figure 8.20. In the current-doubler rectifier, both inductors operate at the same frequency as the transformer but the current in them is shifted in phase by 180°. Thus, after summation of those currents, there is a partial current ripple cancelation at the output.

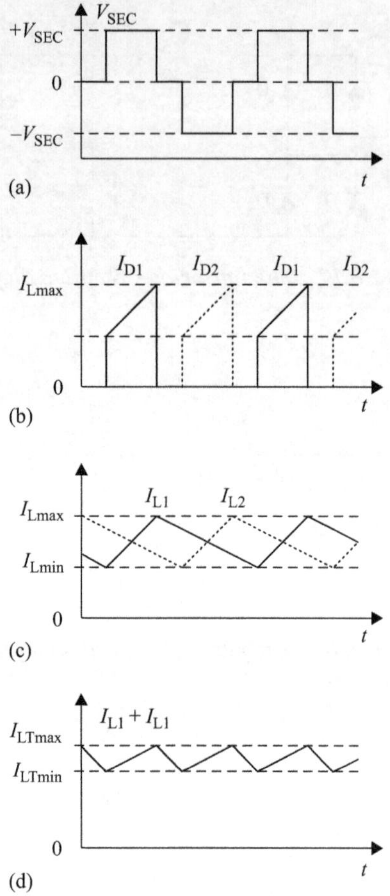

Figure 8.20 Typical voltage and current waveforms for the current-doubler rectifier: (a) transformer primary voltage, (b) diode currents, (c) inductor current, and (d) sum of inductor currents (not in scale)

8.4 Parasitics in DC–DC converters

The analysis of converters in Sections 8.2 and 8.3 has been performed under ideal switch, diode, and passive component assumptions. Non-idealities or parasitics of practical devices and components may, however, greatly affect some performance parameters of dc–dc converters. In this section, modeling and effects of parasitics on output voltage ripple, efficiency, and voltage transfer function of converters are discussed.

An ideal and a more realistic model of a capacitor are shown in Figure 8.21. The realistic model consists of a series connection of capacitance C and resistance r_C. The resistance r_C is called an ESR of the capacitor and is due to losses in the dielectric and physical resistance of leads and connections. Recall (8.8) which

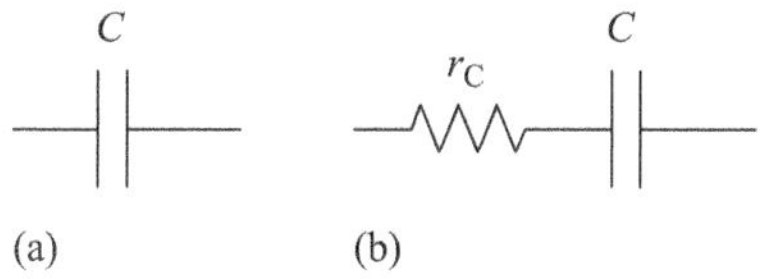

Figure 8.21 Capacitor model: (a) ideal one and (b) with series resistance

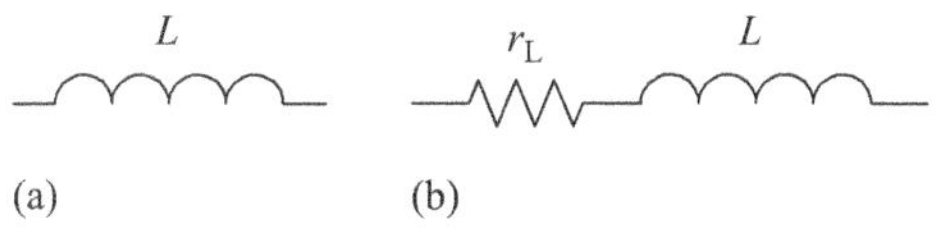

Figure 8.22 Inductor model: (a) ideal one and (b) with series resistance

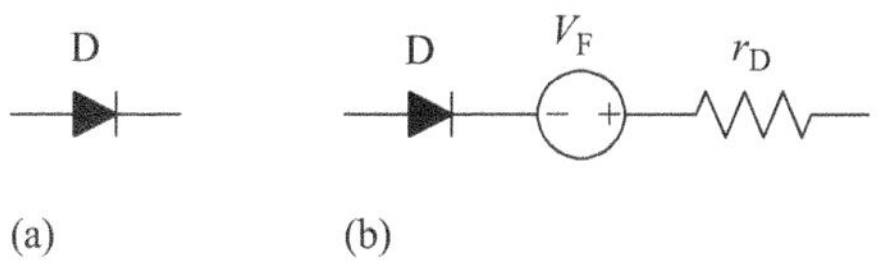

Figure 8.23 Diode model: (a) ideal one and (b) with series resistance and constant forward voltage

provided a value of the filter capacitance in a buck converter that limits the peak-to-peak output voltage ripple to V_r. The equation was derived under an assumption that the entire triangular ac component of the inductor current flows through a capacitance C. It is, however, closer to reality to maintain that this triangular component flows through a series connection of capacitance C and resistance r_C. Thus, the ripples across the resistance r_C adds to the ripples across the capacitor. The peak-to-peak voltage ripple value across the resistance r_C can be calculated as

$$V_{rC} = r_C I_{Lr} = \frac{r_C V_{OUT}(1 - D)T}{L} \tag{8.27}$$

where I_{Lr} is the peak-to-peak current ripple value in the buck converter inductor. Depends on the used capacitor type and size one of the ripples can be dominant. In most cases, especially when electrolytic capacitors are employed, the ripples over resistance r_C are significantly larger than ripples over a capacitance itself. Thus, many times for the ripples requirements, only the series resistance effect can be considered.

Ideal and realistic models of an inductor, a diode, and a transistor switch are presented in Figures 8.22–8.24. An efficiency of a dc–dc converter is defined as the ratio of the output power to input power and is commonly expressed in percent. All the parasitics shown in parts (b) of the figures in this section decrease converter efficiencies. In well-designed practical converters, efficiencies range from high eighties to medium nineties percent. Parasitic components also lower voltage transfer functions of practical converters resulting in maximum values of about ten instead of theoretical limits of infinity for boost or buck–boost converters.

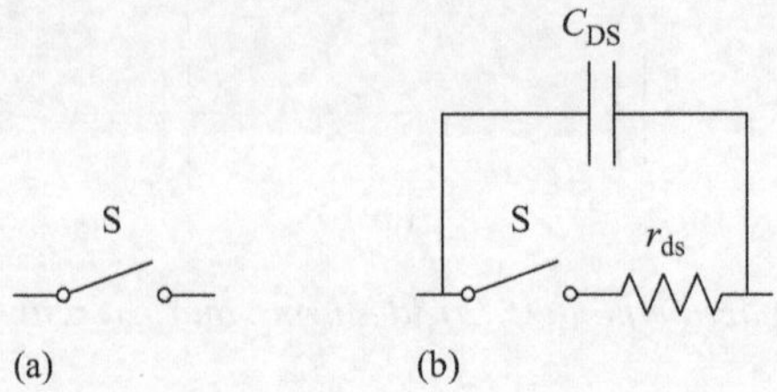

Figure 8.24 Switch model: (a) ideal one and (b) with series resistance and parallel capacitance

8.5 Continuous and discontinuous conduction modes

PWM dc–dc converters may operate in three of conduction modes: continuous conduction mode (CCM), critical conduction mode, and discontinuous conduction mode (DCM). The analysis of the converters in this chapter was done under assumption that current in the inductor is always positive, which is specific to the CCM. The critical conduction mode is a boundary between the other two and is obtained when the current in the inductor drops to zero and then immediately starts rising. This boundary relates the inductor ripple current peak-to-peak value to the output current as

$$I_{Lr} = 2I_O \tag{8.28}$$

Thus, for a converter to operate in CCM, the inductor value needs to be calculated accordingly. For instance, for a buck converter it should be calculated as

$$L_{min} = \frac{V_{OUT}(1 - D)T}{2I_{Omin}} \tag{8.29}$$

where I_{Omin} is the minimum output current for which CCM is desired.

In the DCM, the inductor current drops to zero, remains there for some time, and starts increasing only at the beginning of the next cycle. The disadvantage of this mode is a high-current stress on the converter components. Advantages of DCM are simpler control, as the current in the inductor starts from zero in each cycle, and a lower size of the inductor. Thus, DCM is often used in low-power applications. Typical inductor current waveforms for the three conductions modes are presented in Figure 8.25.

8.6 Synchronous rectification

The idea of synchronous rectification is to replace a diode, which has a constant voltage drop, by a switch with a very low on-resistance, usually a metal–oxide–semiconductor field-effect transistor. It is especially important in a low-voltage applications, where the typical diode forward voltage drop is significant compared to the converter output voltage even when Schottky diodes are used. There are

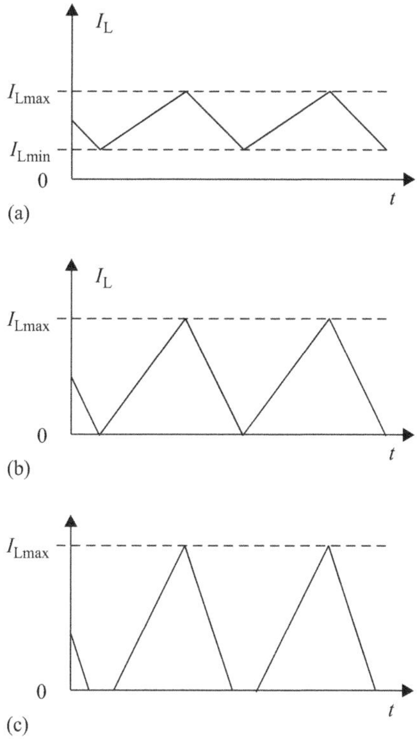

Figure 8.25 Power converter inductor current for various conduction modes: (a) continuous conduction mode, (b) critical conduction mode, and (c) discontinuous conduction mode

many such applications, for instance supplying central processor unit (CPU) cores. The synchronous rectification can greatly increase the efficiency of the power converters in low-output voltage supplies. Nevertheless, a synchronous rectifier needs an additional control circuit which increases the converter complexity.

8.7 Bidirectional converters

Some applications, for instance battery storage converters, require a bidirectional power flow capability of the converter. In non-isolated converters, it can be achieved by applying synchronous rectification and proper control methods. Figure 8.26 shows an example of such a converter. It can be observed that after replacing a diode with a controlled switch, buck and boost converter topologies are identical, only the power flow is reversed. Thus, the converter can operate in one direction as a buck, in reverse as a boost. The integrated buck and boost converter topology gives much more flexibility in bidirectional operation in terms of the input and output voltage ratio.

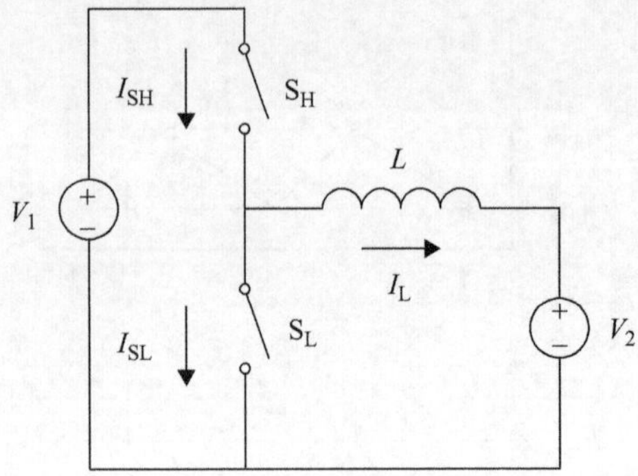

Figure 8.26 The bidirectional non-isolated converter topology diagram

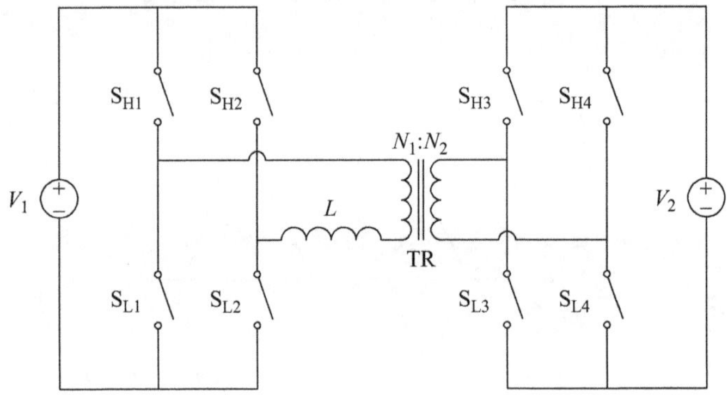

Figure 8.27 The dual active bridge converter topology diagram

For the applications where galvanic isolation between two sides is required, the dual active bridge (DAB) converter can be used. Its block diagram is presented in Figure 8.27. It consists of two full bridges, transformer, and inductor. As an inductor, the leakage inductance of the transformer could be used. Since this converter has a symmetrical structure, it naturally allows for bidirectional operation. The DAB is usually controlled using the phase-shift control approach.

8.8 Interleaving

In the recent years, the concept of interleaving has become very popular. It is related to the growing requirements on the power quality in supplying CPU cores. Modern CPU cores are supplied with very low voltages (below 2 V) and in many cases require significant amount of power (tens to hundreds of watts) which implies very high currents. To meet those requirements, multiple converters are connected in parallel to split the current among them. In addition, a phase-shift between converters is introduced to allow for ripple cancelation. An example of a two-phase interleaved buck topology is presented in Figure 8.28. Typical voltage and currents waveforms are

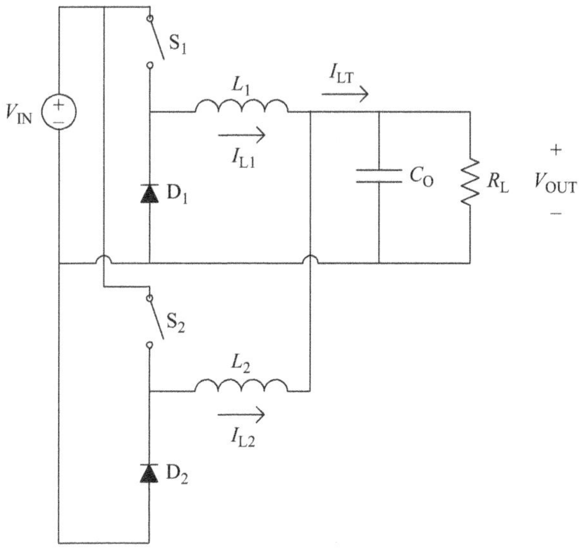

Figure 8.28 The interleaved buck converter topology diagram

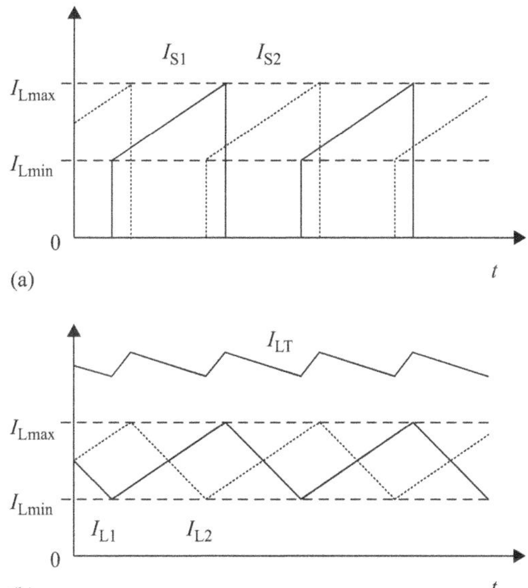

Figure 8.29 Typical voltage and current waveforms for the interleaved buck converter: (a) switch current and (b) inductor current

shown in Figure 8.29. The two paralleled converters are controlled to share the current equally to minimize losses. The introduced phase-shift allows for partial ripples cancelation. Thus, interleaving facilitates an increase in the efficiency of converters, reducing the filtering components size, and improving the output power quality.

8.9 Control principles

A regulated dc output voltage must be provided by a dc–dc converter under varying load and input voltage conditions. The converter component values also change with time, temperature, pressure, etc. Hence, the control of the output voltage should be performed in a closed-loop manner using principles of negative feedback. Two most common closed-loop control methods for PWM dc–dc converters, namely, the voltage-mode control and the current-mode control, are presented schematically in Figure 8.30.

In the voltage-mode control scheme shown in Figure 8.30(a), the converter output voltage is sensed and subtracted from an external reference voltage in an error amplifier (EA). The EA produces a control voltage that is compared to a constant-amplitude sawtooth waveform. The comparator produces a PWM signal which is fed to drivers of controllable switches in the dc–dc converter. The duty ratio of the PWM signal depends on the value of the control voltage. The frequency of the PWM signal is the same as the frequency of the sawtooth waveform. An important advantage of the voltage-mode control is its simple hardware implementation and flexibility.

The EA in Figure 8.30(a) reacts fast to changes in the converter output voltage. Thus, the voltage-mode control provides good load regulation, that is, regulation

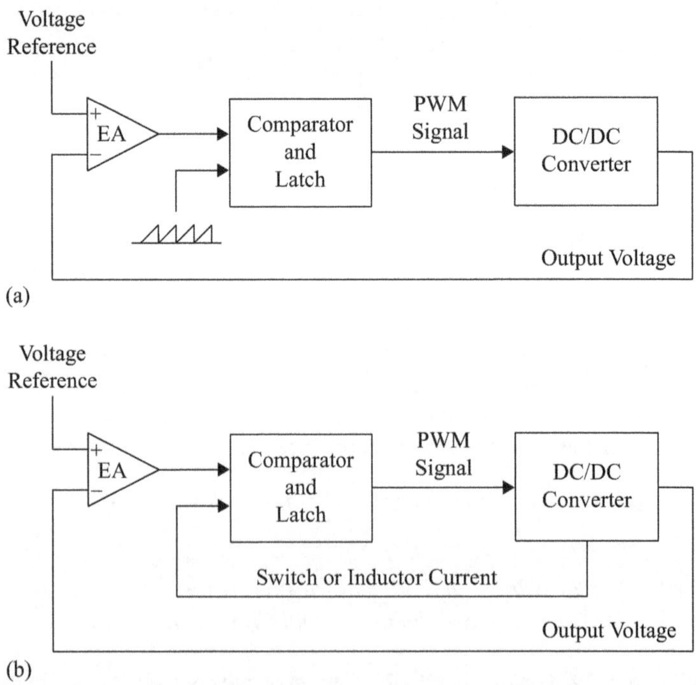

Figure 8.30 Main control schemes for dc–dc converters: (a) voltage-mode control and (b) current-mode control

against variations in the load. Line regulation (regulation against variations in the input voltage) is, however, delayed because changes in the input voltage must first manifest themselves in the converter output before they can be corrected. To alleviate this problem, the voltage-mode control scheme is sometimes augmented by so-called voltage feedforward path. The feedforward path affects directly the PWM duty ratio according to variations in the input voltage. As will be explained below, the input voltage feedforward is an inherent feature of current-mode control schemes.

The current-mode control scheme is presented in Figure 8.30(b). An additional inner control loop feeds back an inductor current signal. This current signal, converted into its voltage analog, is compared to the control voltage. This modification of replacing the sawtooth waveform of the voltage-mode control scheme by a converter current signal significantly alters the dynamic behavior of the converter. The converter takes on some characteristics of a current source. The output current in PWM dc–dc converters is either equal to the average value of the output inductor current (buck-derived and Cuk converters) or is a product of an average inductor current and a function of the duty ratio. In practical implementations of the current-mode control, it is feasible to sense the peak inductor current instead of the average value. Since the peak inductor current is equal to the peak switch current, the latter can be used in the inner loop, which often simplifies the current sensor. Note that the peak inductor (switch) current is proportional to the input voltage. Hence, the inner loop of the current-mode control naturally accomplishes the input voltage feedforward technique. Among several current-mode control versions, the most popular is the constant-frequency one which requires a clock signal. Advantages of the current-mode control include input voltage feedforward, limit on the peak switch current, equal current sharing in modular converters, and reduction in the converter dynamic order. The main disadvantage of the current-mode control is its complicated hardware, which includes a need to compensate the control voltage by ramp signals (to avoid converter instability).

Among other control methods of dc–dc converters, a hysteretic (or bang-bang) control is very simple for hardware implementation. The hysteretic control results, however, in variable frequency operation of semiconductor switches. Generally, a constant switching frequency is preferred in power electronic circuits for easier elimination of EMI and better utilization of magnetic components.

Application specific integrated circuits (ASICs) are commercially available that contain main elements of voltage- or current-mode control schemes. On a single 14- or 16-pin chip, there is EA, comparator, sawtooth generator or sensed current input, latch, and PWM drivers. The switching frequency is usually set by an external RC network and can be varied from tens to hundreds of kilohertz. The controller has an oscillator output for synchronization with other converters in modular power supply systems. A constant voltage reference is generated on the chip as well. Additionally, the ASIC controller may be equipped in various diagnostic and protection features: current limiting, overvoltage and undervoltage protection, soft start, dead time in case of multiple PWM outputs, and duty ratio limiting. In several dc–dc converter topologies, e.g., buck and buck–boost, neither control terminal of semiconductor switches is grounded (so-called high-side switches). ASIC controllers are usually designed for a particular topology and their PWM drivers may be able to drive high-side switches in low-voltage applications. In high-voltage applications, external PWM drivers must be used. External PWM drivers are also used for switches with high input capacitances. To take a full advantage of the input–output isolation in transformer versions of dc–dc converters, such isolation must be also provided in the control loop. Signal transformers or optocouplers are used for isolating feedback signals.

Another approach to dc–dc converter control implementation is digital control. It is usually realized using microcontrollers (MCU), digital signal processors (DSPs), or programmable logic, for instance field programmable gate arrays (FPGAs). In the digital controller, all analog signals, like voltages and currents, are converted to digital signals by analog to digital converters (ADCs). An ADC sampling frequency is usually synchronized with the converter switching frequency. The obtained digital signal corresponding to the output voltage is subtracted from the reference value to obtain an error signal. The error signal is processed by the digital implementation of the controller, for instance, a PID controller. The digital controller is recalculating its output every time when new sample is obtained. In other words, it is synchronized with the ADC sampling frequency. The calculated controller output is then passed to a digital pulse width modulator (DPWM). The DPWM operates in a very similar way to a voltage-mode controller. The input signal is compared with a ramp (sawtooth) and proper pulse width is generated. In digital domain, both signals are digital and have discrete values. The ramp is usually generated by the counting-up counter, which is reset periodically in order to obtain the desired switching frequency. Thus, the resolution of the DPWM is proportional to a ratio of the counter clock frequency to the switching frequency. There are techniques, commonly used in dedicated for power converters DSPs, which allow to increase the resolution of the DPWM in some limited range. The other limitation of digital controllers is computation time. The dynamics of the digital controller is naturally limited by the sampling frequency. The sampling frequency can be easily increased, but it may lead to overloading an MCU or a DSP. It means that the update of the controller output will not be calculated before the next cycle is started. It can be overcome by using FPGA and paralleling calculations, but it is much more complex and expensive solution. The biggest advantage of digital controllers is their flexibility. They allow modification of parameters of the controller by reprogramming, which can be even done on the fly, while the converter is running. Moreover, it allows for implementation of extensive diagnostic and protection algorithms that could be not available or hard to implement in analog controllers. Depending on the application, a low-cost MCU, powerful DSP, or FPGA can be used. All of them offer various performances and costs. In recent years, there have been also available dedicated ASICs with implemented digital controllers together with gate drivers and other necessary circuits for low- and medium-power applications.

Dynamic characteristics of closed-loop controlled dc–dc converters must fulfill certain requirements. To simplify analysis, these requirements are usually translated into desired properties of the open loop. The open loop should provide a sufficient (typically, at least 45°) phase margin for stability, high bandwidth (about one-tenth of the switching frequency) for good transient response, and high gain (several tens of decibels) at low frequencies for small steady-state error.

The open-loop dynamic characteristics are shaped by compensating networks of passive components around the EA. Second- or third-order RC networks are commonly used. Since the converter itself is a part of the control loop, the design of compensating networks requires knowledge of small-signal characteristics of the converter. There are several methods of small-signal characterization of PWM dc–dc converters. The most popular methods provide average models of converters under high switching frequency assumption. The averaged models are then linearized at an operating point to obtain small-signal transfer functions. Among analytical averaging methods, state-space averaging has been popular since late 1970s. Circuit-based averaging is usually performed using PWM switch or direct replacement of semiconductor switches by controlled current and voltage sources. All these methods can take into account converter parasitics.

The most important small-signal characteristic is the control-to-output transfer function. Other converter characteristics that are investigated include the input-to-output (or line-to-output) voltage transfer function, also called the open-loop dynamic line regulation or the audio susceptibility, which describes the input–output disturbance transmission; the open-loop input impedance; and the open-loop dynamic load regulation. Buck-derived, boost, and buck–boost converters are second-order dynamic systems; Cuk and sepic converters are fourth-order systems. Characteristics of buck and buck-derived converters are similar to each other. Another group of converters with similar small-signal characteristics is formed by boost, buck–boost, and flyback converters. Among parasitic components, the ESR of the filter capacitor r_C introduces additional dynamic terms into transfer functions. Other parasitic resistances usually modify slightly the effective value of the load resistance.

Buck-derived converters can be easily compensated for stability with second-order controllers. The boost converter (as well as buck–boost and flyback converters) is a non-minimum phase system. Non-minimum phase dc–dc converters are typically compensated with third-order controllers. Step-by-step procedures for a design of compensating networks are usually given by manufacturers of ASIC controllers in application notes.

Finally, the behavior of dc–dc converters in distributed power supply systems should be considered. An important feature of closed-loop regulated dc–dc converters is that they exhibit a negative input resistance. As the load voltage is kept constant by the controller, the output power changes with the load. With slow load changes, an increase (decrease) in the input voltage results in a decrease (increase) in the input power. This negative resistance property must be carefully examined during the distributed supply system design to avoid conditions that may lead to instabilities.

Further reading

[1] R. Erickson and D. Maksimovic, *Fundamentals of Power Electronics*, 2nd Ed., Norwell, MA: Kluwer Academic, 2001.
[2] N. Mohan, T. M. Undeland, and W. P. Robbins, *Power Electronics: Converters, Applications and Design*, 3rd Ed., New York, NY: John Wiley & Sons, 2002.
[3] M. K. Kazimierczuk, *Pulse-Width Modulated DC–DC Power Converters*, New York, NY: John Wiley & Sons, 2008.
[4] A. I. Pressman, K. Billings, and T. Morey, *Switching Power Supply Design*, 3rd Ed., New York, NY: McGraw-Hill, 2009.
[5] D. W. Hart, *Power Electronics*, New York, NY: McGraw-Hill, 2010.
[6] A. M. Trzynadlowski, *Introduction to Modern Power Electronics*, 2nd Ed., New York, NY: Wiley Interscience, 2010.
[7] A. Ioinovici, *Power Electronics and Energy Conversion Systems: Fundamentals and Hard-Switching Converters* (Volume 1), Chichester: John Wiley & Sons, 2013.
[8] M. H. Rashid, *Power Electronics: Circuits, Devices, and Applications*, 4th Ed., Upper Saddle River, NJ: Pearson Prentice Hall, 2013.
[9] P. T. Krein, *Elements of Power Electronics*, 2nd Ed., New York, NY: Oxford University Press, 2014.

Chapter 9
Smart power electronic modules

*Dorin O. Neacşu**

9.1 History

With the simultaneous advent of power semiconductor devices and microcontroller platforms in late 1980s and early 1990s, the circuit topologies and the principle of power electronics conversion have matured. Certain solutions have emerged as winners of worldwide acceptance. This allowed the transition into the next technological phase with a paradigm shift towards continuous improvement of the way power converters are actually built. It was the time for new packaging methods to emerge for the hardware and novel control architectures and platforms to gain popularity.

Initiated in late 1994, the Power Electronics Building Blocks (PEBB) program, conceived in the Office of Naval Research (ONR, a top research sponsorship institution of the US Department of Defense), Ship Hull, Mechanical and Electrical Systems Science and Technology Division, has sponsored and led a series of electronics, materials, and manufacturing techniques. Another major program of ONR Division 334, the "Ship Hull, Mechanical, and Electrical Systems Science and Technology Division" was the Advanced Electrical Power Systems (AEPS) [1]. Both PEBB and AEPS programs included academia and industry, with notable contributions from CPES at Virginia Tech and SatCon Technology Corporation.

Additional to the actual R&D effort [2,3], these programs set the basis for standardization in power electronics converters of medium power (tens to hundred kilowatt). Similar programs were carried within Oak Ridge National Laboratory [4,5] or major power converter manufacturers. Later on, ABB Corporation has actually extended the PEBB program into the multi-megawatt range. Even more recently, the PEBB became a technology of choice for the development of future *more-electric aircraft systems.*

Springing off this R&D effort, a series of principles responded the modularity demands of the production lines and after-market service expectations. The effort was continued by industry in mid-2000s with the implementation within hybrid integrated circuits, especially for the low-kilowatt range. While it is hard to draw a line and to identify the precise moment the first smart power module has been

*Technical University of Iaşi, Romania

created, it certainly came as a natural technological evolution and it has been adopted by power semiconductor manufacturers and system integrators alike [6–9].

Figure 9.1 illustrates the technology evolution in low-voltage, medium-power range.

An explosive development in all semiconductor device technologies was occurred after 1980 [10–12]. According to the Moore's law, the number of transistors on the same integrated circuit chip has continuously doubled at each two years. This increasing technological capacity combined with the advances in computer-aided design has allowed creation of new very large-scale integrated circuits able to achieve impressive performance in signal processing and size reduction in electronics equipment. Integrated circuits spread into various fields of application including power electronics conversion.

While the main issue against smaller packaging of the power electronics equipment relies on the limited heat extraction capability, progress has been achieved in system level packaging of the gate driver, sensing logic, and power semiconductor (Figure 9.2). A new set of devices has hence emerged and various manufacturers called them either "intelligent power modules", "smart power modules", or "advanced power converters". For the sake of uniformity, we will address herein such devices as *Smart Power Modules* or SPM.

Advantages of the SPM-type devices consist of

- Reduction of system packaging space and weight.
    - The current targets for year 2020, for complete three-phase inverter design (including filters and cooling) in the tens kilowatt range: power density = 14 kW/l and specific power = 14 kW/kg [2]. These milestones decrease with the installed power. In horsepower range, the expectancy is within 8 kW/l range.
    - Easier debugging and field repair of the overall electronics. Less packaging speeds up replacement of damaged components, and smaller package simplifies storage.
    - Simplified power connection ($V_{DC}+$, $V_{DC}-$, A, B, C) and controller connection through logic-level inputs.

- A better thermal design and an optimized layout, both with effects on the system reliability.
    - Using a power module supplied from the manufacture rather than individual components yields hence recommended for the inverter application [6–9].
    - Improved reliability and lifetime [6–9]. This topic will be expanded later on in this chapter.
    - Better power cycling capability than conventional devices [6–9].

- Lower circuit inductance than discrete solutions.
    - Benefits in voltage spike reduction and operation at higher switching frequency with lower switch loss.
    - Improved reliability and lifetime [6–9].

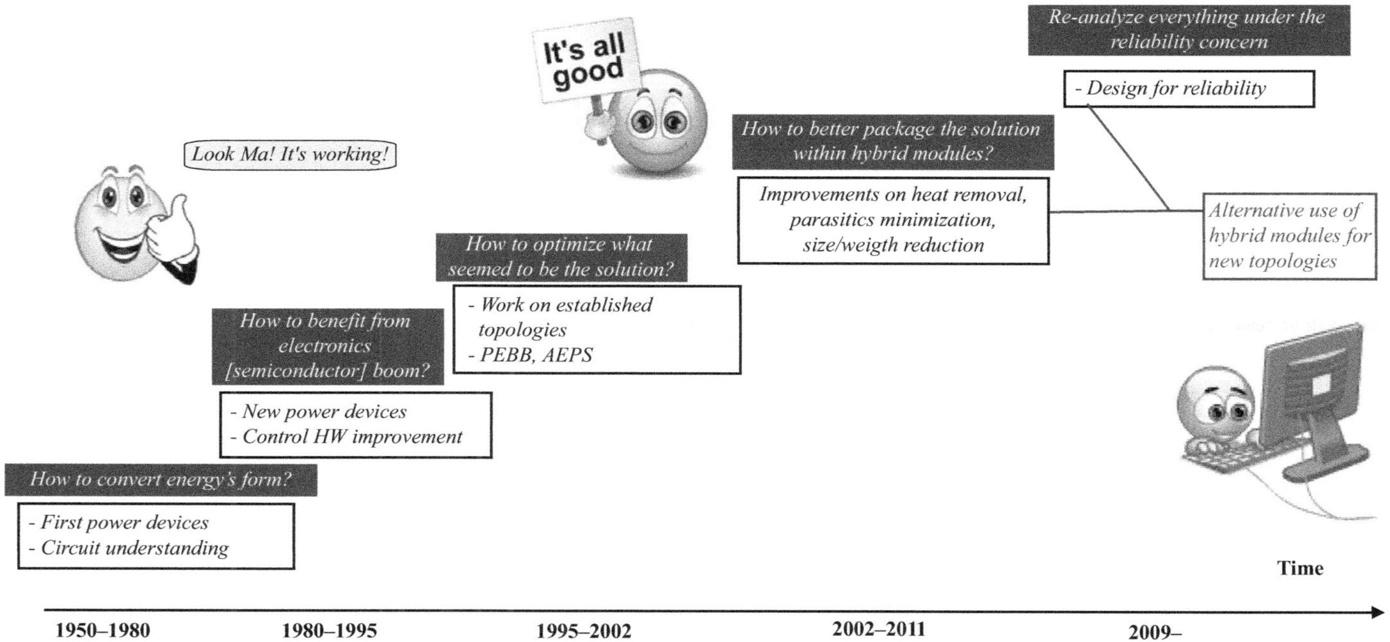

Figure 9.1 Time evolution of technology foci

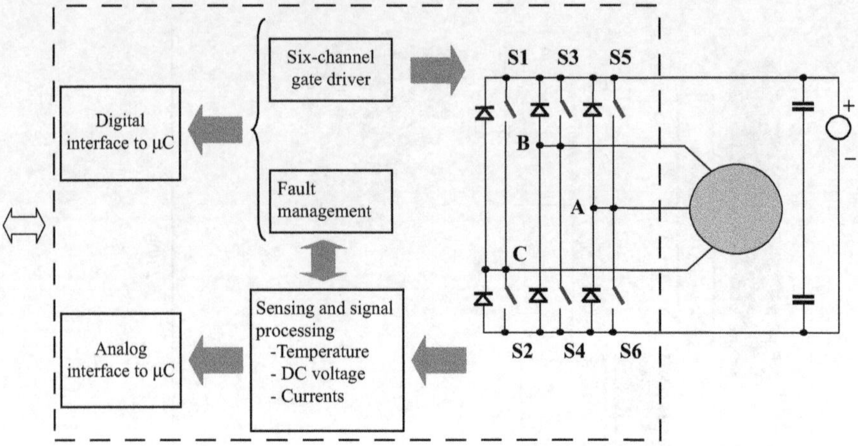

Figure 9.2 Typical architecture of a SPM

- Expanding possibilities within control integrated circuits.
  - Propagation delays for all low-side and high-side insulated gate bipolar transistors (IGBTs) are matched within the same implementation platform.
  - Protection to over-current and over-temperature.
  - More sensing and measurements of temperatures, currents, and voltages.

The initial limitations of using SPM devices are nowadays mitigated [13–17]. The first SPM devices were dedicated strictly to horsepower range applications due to the limited thermal capabilities of the existing packaging. After December 2011, SPM products were extended to currents like 300 A, covering basically the entire range of medium-power applications [18]. Another initial concern related to maintaining the switching frequency under 10 kHz due to the same limited thermal capabilities. Even if the switching frequencies have been increased over years, the use of these SPM devices in motor drive applications justifies maintaining the switching frequency below 10 kHz. Finally, let us mention that all the merits of SPM technology do not eliminate the need for cooling and DC bus filtering. These two components are occupying the largest volume within the power conversion unit.

The success of SPM devices would have not been possible without the technology achievements related to IGBT device technology, packaging materials, and improved gate driver control. Their importance is outlined in the next section.

9.2 Technology background

While the goal of this chapter is not to provide a comprehensive review of power semiconductor technologies, the most important trends that encouraged the development of *Smart Power Modules* are described herein. This is especially important since the richness in advanced technological equipment allowed a wide variety of power semiconductor devices with a clear trend for specialization with application.

The IGBT used within SPM devices are highly specialized for medium-power motor drives [10–12].

The performance of modern power switches is quantified

- with reduction in power loss through lowest $V_{CE(ON)}$, tentatively with zero temperature coefficient,
- with controlled or softer transients able to reduce electromagnetic interference (EMI) [10–12].

9.2.1 IGBT device technologies and their performance

The historical evolution of IGBT technologies after 1990 and their advantages to the creation of a smart or intelligent power module worth are briefly mentioned. Several major device structures marked the evolution of IGBT devices: punch-through, non-punch through and more recently, field-stop. A possible classification can be further extended with the possible gate structures: planar or trench. Without entering technology details, Figure 9.3 illustrates the internal structures for each class of IGBT power devices.

Punch-through planar IGBT devices are mostly recommended for voltages in the range of 250–1,200 V, and represent today a cost effective technology, optimized for speed or short-circuit rating up to 10 μs. The non-punch-through planar IGBT devices are mostly recommended for higher voltages, like 600–1,200 V, for their ruggedness, with short-circuit tolerance up to 10 μs, and optimized for high switching frequency. A more recent option, the field-stop trench IGBT devices are

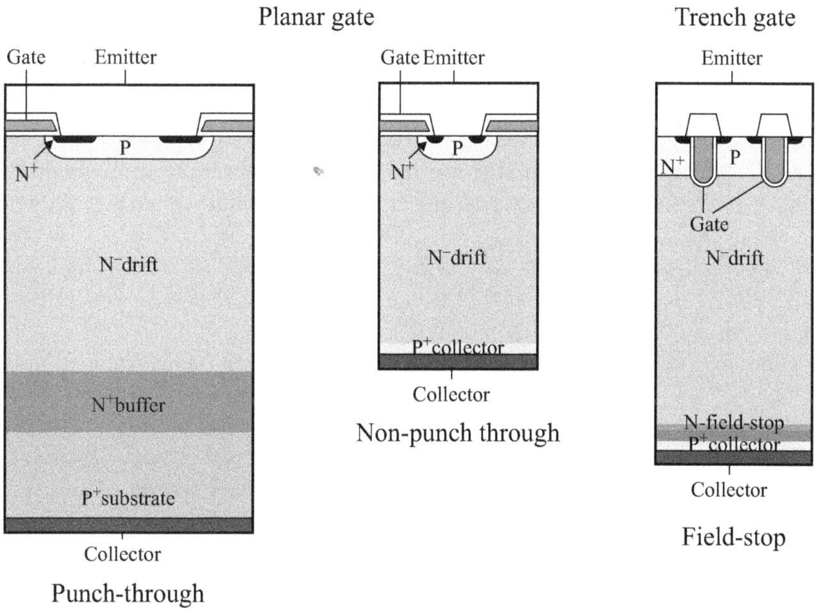

Figure 9.3 Typical IGBT device structures

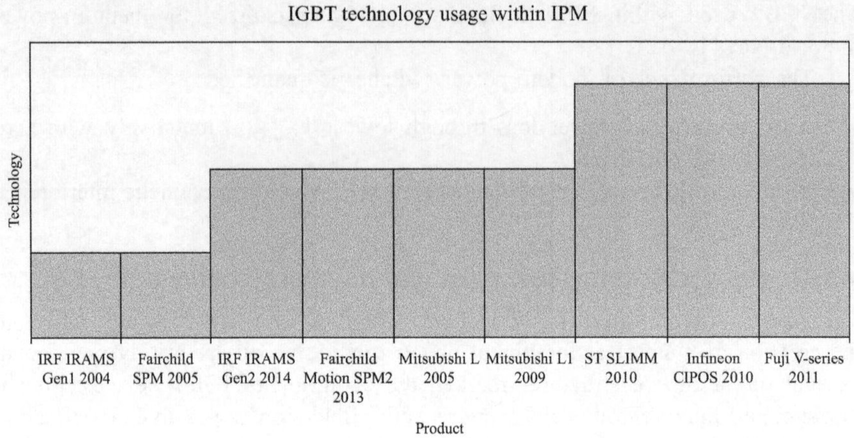

Figure 9.4 Usage of IGBT chip technology within smart power module or intelligent power module (SPM/IPM), with product examples

recommended for lower voltages, like 350–650 V, optimized for both conduction and switching performance, with a rated 5 μs short-circuit capability, and able to allow higher current rating in smaller packages. Alternatively, some rare punch-through trench IGBT devices may be used for 600 V, and they are a cost effective technology at low switching frequencies, below 5 kHz, with a 3 μs short-circuit capability.

For illustration purposes, some actual examples of the usage of these technologies within intelligent or smart power modules (SPM devices) are shown in Figure 9.4 [13–17].

The current trend is the specialization of the IGBT chip depending on application and power level. This leads to products ranging from very low power levels, under 100 W, to tens of kilowatt for a motor drive application. From application point of view, most SPM products are created for appliance market, where the application is a motor drive with a switching frequency below 5 kHz, with transitions generally below 5 kV/μs (or 5 V/ns) to contain EMI emissions, and with a focus of IGBT design on reduction of the conduction loss. This is mostly the case of a *Smart Power Module* detailed within this chapter.

Since products from various historical technological stages (punch-through, non-punch through, or field-stop, associated with either planar or trench gate configurations) are simultaneously present on the market, let us quickly illustrate the performance evolution with the example of a single set of devices to avoid competitive interpretations. We have chosen for this exercise, the Fuji's IGBT products [11]. First, Figure 9.5 illustrates the inverter power loss in watt, when the inverter switch is implemented with 300 A IGBT devices carried out with diverse technologies over the years. Figure 9.6 illustrates the evolution in chip area, while Figure 9.7 expresses the resolution with the design rule used for semiconductor creation.

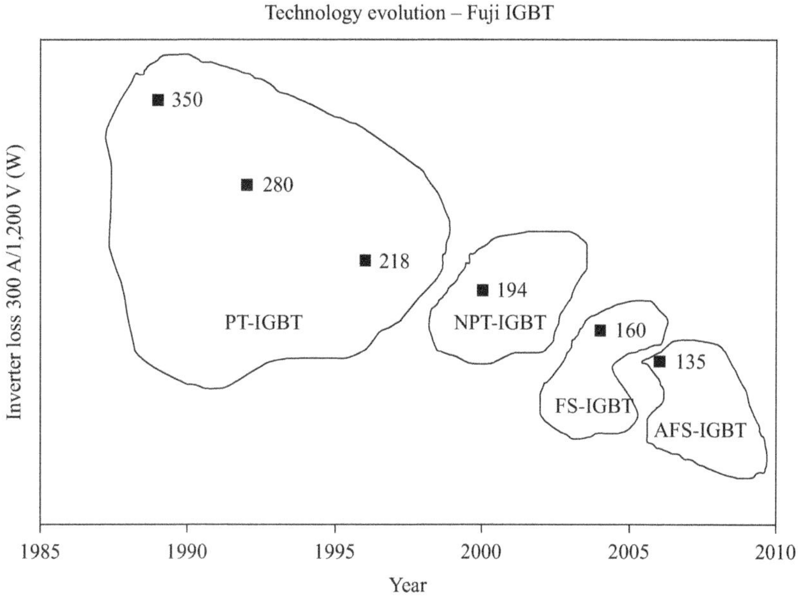

Figure 9.5 Example for evolution of IGBT-based inverter power loss, with time and technology [11]

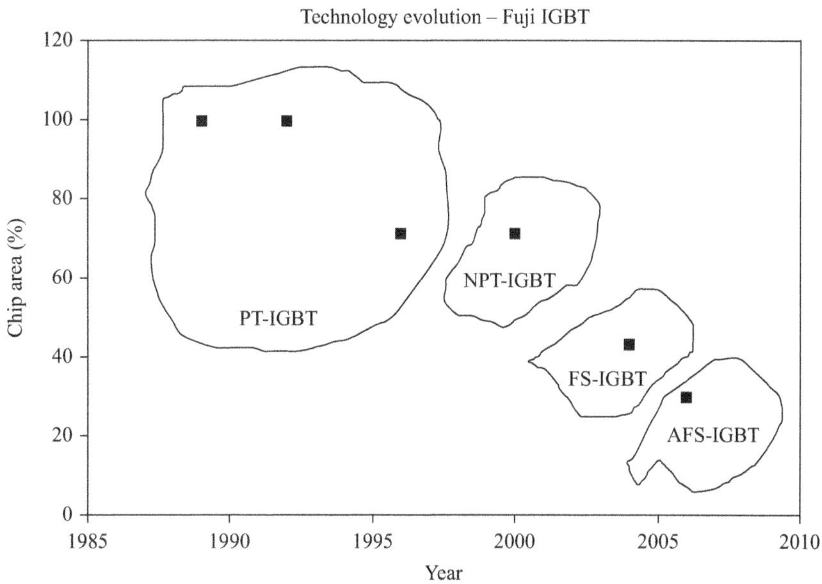

Figure 9.6 Example for relative decrease in IGBT chip size, with time and technology [11]

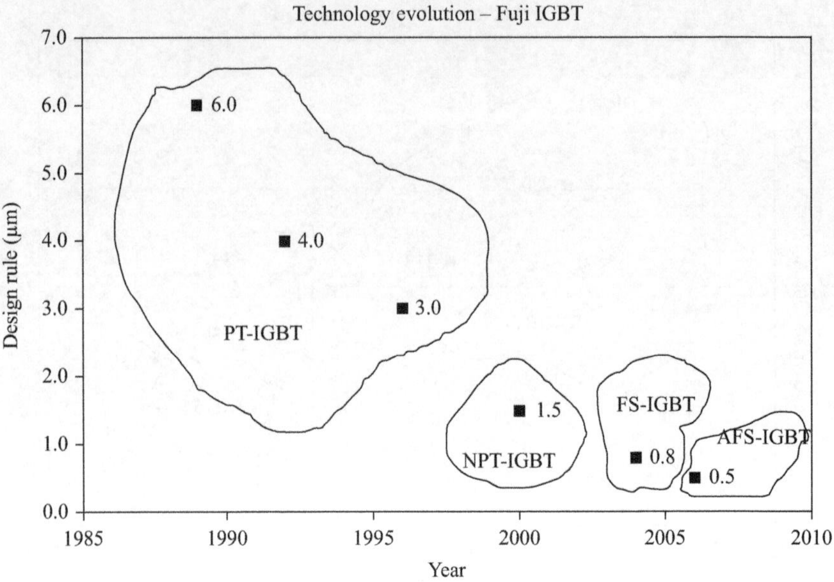

Figure 9.7 Example for semiconductor design rule decrease, with time and technology [11]

9.2.2 Gate driver technology

9.2.2.1 Variable gate resistance

The performance of the IGBT chip is further delineated with the optimization of the gate driver [19,20] through inclusion of all its requirements within high-voltage integrated circuits.

The effect of the two obvious extreme values for the gate resistance (small and large) on various system performance indices is shown in Figure 9.8.

A commonly used solution provided improvement of the gate transition with usage of different turn-on and turn-off resistors, eventually in accordance with the slope of the commutated current. Such solution was further improved with a variable resistance controlled with a feedback from the collector voltage. A step closer to a full semiconductor implementation replaced the variable resistance with a controlled MOS transistor inside the integrated circuit (Figure 9.9) [20].

9.2.2.2 Mitigation of the Miller current

Other advanced gate drivers (like STMicroelectronics' TD351 [21]) mitigate the effect of the Miller current during the turn-off process. At turn-off of any IGBT, the gate-collector capacitance is subject to a large dv/dt able to create a gate current often referred to as Miller current. This parasitic current may in certain circumstances trigger the gate and switch the IGBT into conduction. The conventional solution, mostly used by industry consists in using a negative voltage to bring the IGBT to the OFF state. This is not very easy to implement with integrated circuits

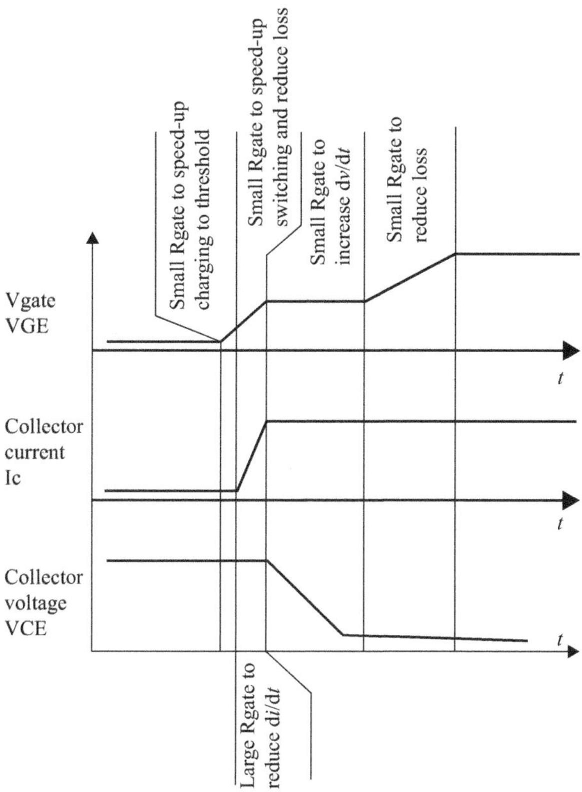

Figure 9.8 Nonlinear gate resistance

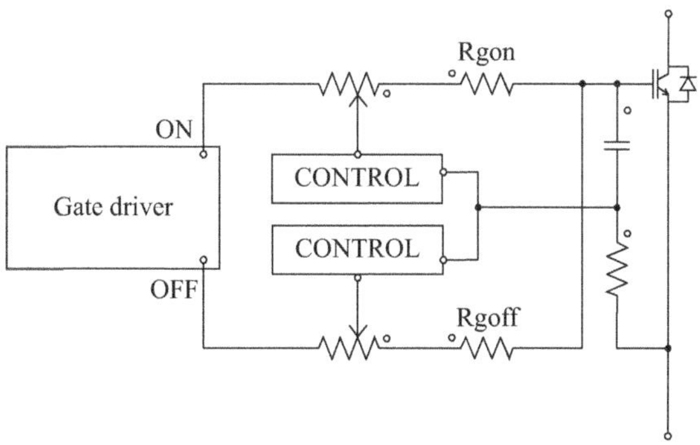

Figure 9.9 Implementation of the nonlinear gate resistance [20] shown with a drawing from the power electronic simulation software PSIM

since it would require a second power supply. Various compromises are herein possible with local power supplies converting the existing positive voltage into a smaller, local negative power supply; or the use of a discharge capacitor in between gate and emitter to provide a friendly path for the Miller current. Unfortunately, the

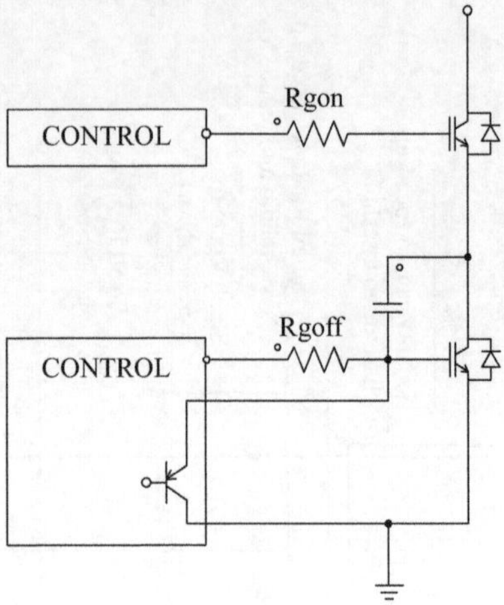

Figure 9.10 Mitigation of the Miller current

use of an additional capacitor may interfere with the ability of the driver to control the gate.

Advanced gate drivers like TD351 [21] use an additional path for the Miller current during the OFF state of the IGBT (Figure 9.10). A low impedance path is established between IGBT's gate and emitter to carry the Miller current. It is important to mention that this solution does not affect the IGBT's turn-off characteristics while it rather keeps the gate at a low level during the entire off-time.

9.2.2.3 Reduction of the IGBT's tail current and minimum dead time

Another important improvement for the gate circuit requirements refers to the reduction of the IGBT's tail current through an optimized design of the power semiconductor. This has effects in reduction of the required dead time from something around 2.4 μs to a value around 1 μs for kilowatt range drives, where the most IPMs are working.

9.2.3 Packaging technologies

Most importantly, the technology of package fabrication has seen several evolutionary steps [22–25]. First, the *transfer mold technology* based on copper lead frames has been a good candidate for power semiconductor devices. For larger power levels, the requirements for heat extraction increase and the heat transfer through the copper leads is not enough. Hence, the integration of some cooling structures like *mold resin and aluminum heat sink* provided a performance leap. Since the SPM devices expanded towards 300 A, even higher power levels require even more heat extraction which is now possible with *integrated ceramic substrates* (Figure 9.11).

The role of the substrate in a power module is to provide the circuit connections as well as cooling. Different from low-power integrated circuit (IC) technology, the substrate of a power module must carry higher currents, provide larger voltage

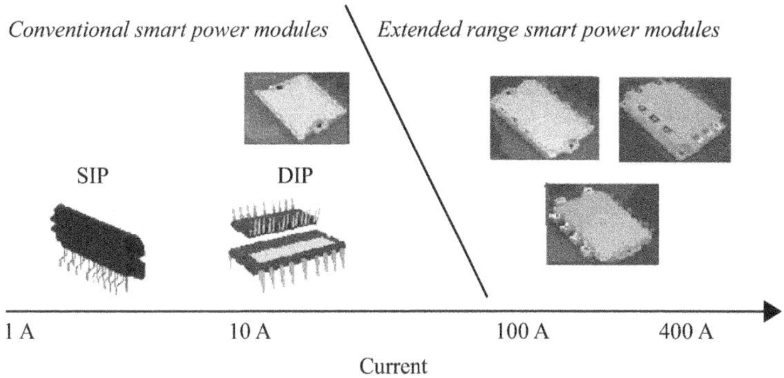

Figure 9.11 Packages for SPM devices using integrated ceramic substrates

insulation, and deal with increased amount of power loss and heat. It is common to face operation up to 150 or 200 °C. The most used substrate technologies are the *Direct Bonded Copper (DBC) substrate* and the *insulated metal substrate (IMS)*.

DBC substrates are composed of a ceramic tile (commonly alumina) and a sheet of copper bonded to one or both sides. The copper and ceramic tile are heated to a controlled temperature, in an atmosphere of nitrogen and oxygen, until a copper–oxygen eutectic forms which bonds successfully both to copper and the oxides used as substrates. The top copper layer can use printed circuit board technology to form an electrical circuit, while the bottom copper layer is usually kept plain for further attachment to a heat spreader.

The *DBC substrates* offer a low coefficient of thermal expansion, capable for good thermal cycling performances (up to 50,000 cycles). Other advantages include excellent electrical insulation and good heat spreading characteristics.

The ceramic material used within *DBC substrates* can be:

- alumina (Al_2O_3)
  - Widely used because of its low cost despite the limited thermal performance (24–28 W/mK) and despite being somewhat brittle.
  - The thickness of this substrate was reduced from 0.63 to 0.38 mm over the past years in order to reduce the thermal resistance (R_{th}) from the chip to the heat sink.
- aluminum nitride (AlN)
  - more expensive, but with better thermal performance (>150 W/mK).
- beryllium oxide (BeO)
  - good thermal performance, but somewhat toxic.

IMS represents a newer technological solution. It starts with a metal baseplate like aluminum, covered by a thin layer of dielectric (usually an epoxy-based layer), followed with a layer of copper (35 μm to more than 200 μm thick). Due to its structure, the *IMS* is a single-sided substrate. Thus, it can accommodate components on the copper side only.

The SPM can be designed with or without a baseplate. If the baseplate is missing, the *DBC substrate* is directly placed on the heat sink. Sometimes the copper baseplate is replaced with composite materials such as AlSiC or Cu–Mo due to their low thermal conductivity.

An important aspect related to packaging concerns the terminal connection. The conventional solution of soldering joints has been proven critical to device lifetime. Modern solutions propose replacement of solders with ultrasound welding [25].

This brief review of technologies incorporated within a SPM allows us to understand the efforts that made possible the integration of the entire power inverter within a single highly integrated module. Let us see how these technologies fit into the complete three-phase inverter.

9.3 Basic usage

SPM have been introduced for a compact solution of large-volume production motor drives. Hence, a series of precautions have to be considered before usage.

9.3.1 Protection

All SPM have internal protection of the IGBT against:

- under-voltage lockout – it triggers when the low voltage (~15 V) was required to supply the gate drivers and miscellaneous electronics falls below a certain threshold when the gate driver IC would cease operation.
- over-current – it triggers when a DC or phase current within the power stage gets above a certain value. The simplest version is a simple *desaturation circuit* which monitors the IGBT's collector–emitter voltage and triggers if such voltage is above a threshold (usually 7 V) for a longer than usual time interval. Most complex solutions include a current sensing device and a comparison with a fixed threshold. Such circuit does not fully sense the current and there may not be any current information available for the microcontroller.
- over-temperature – it triggers when the semiconductor die gets above a certain level. The most used solution implies a thermistor. However, monitoring the voltage across another semiconductor junction is in certain cases used.

The phase or DC bus currents, the DC bus voltage, and semiconductor die's temperature are sometimes measured and reported back to the microcontroller as feedback information. This is not always a feature of a SPM, only the newer devices with advanced mixed-mode ICs have this implemented feature.

Just a few SPM devices allow an isolation barrier for either digital or analog signals sent to the microcontroller. Most SPM devices work directly on the grounding provided by the high-voltage DC bus. This in turn requires that the microcontroller is separated from any other circuit, or the external communication channels (RS232, RS485, SPI, IC2, CAN, etc.) are isolated. Commonly, the isolation requirements for grid-connected low-voltage drives are under 2,000 Vrms.

9.3.2 Bootstrap power supply

Most SPM are designed for low-kilowatt power ranges, with phase currents less than 30 A. The power supply required for the control of each high-side IGBT travels together with the controlled device from ground of the DC bus voltage to the positive DC bus voltage. This means hundreds of volts, and the local power supply needs to be isolated from the low-side electronics. The cheapest solution, possible at such power levels consists in a bootstrap power supply (Figure 9.12).

Obviously, the newer SPM operated at currents towards 300 A require proper power supplies for the high-side IGBTs. They are usually included within the hybrid integrated circuit.

A typical bootstrap circuit is shown in Figure 9.12. When the low-side IGBT turns-on, a current path is created from the power supply V_{DC} (usually 15–18 V_{dc}) through the $R_{bootstrap}$, $D_{bootstrap}$, $C_{bootstrap}$. The capacitor charges towards the V_{dc}. When the low-side IGBT turns-off and the high-side IGBT turns-on, its gate circuitry is supplied from the capacitor $C_{bootstrap}$. The diode needs to be designed to carry the initial charging current and to withstand the entire bus voltage as reverse voltage. The series resistor is used to limit the initial charging current, and it is missing in low-power applications. The capacitor needs to be selected to maintain the voltage decrease within an allowable band while the IGBT is kept ON. The initial charging of the capacitor is usually achieved with a special charging procedure where the microcontroller turns-on the low-side IGBT for a time long enough to allow the voltage rise from 0 to V_{dc}. Alternatively, the initial charging current is limited with a duty cycle control of the low-side IGBT.

In order to avoid the initial struggle with proper microcontroller program, some low-power systems use a low-power resistor (like 100 kΩ, 1–3 W) in parallel with the low-side IGBT, and simply wait for several seconds before starting

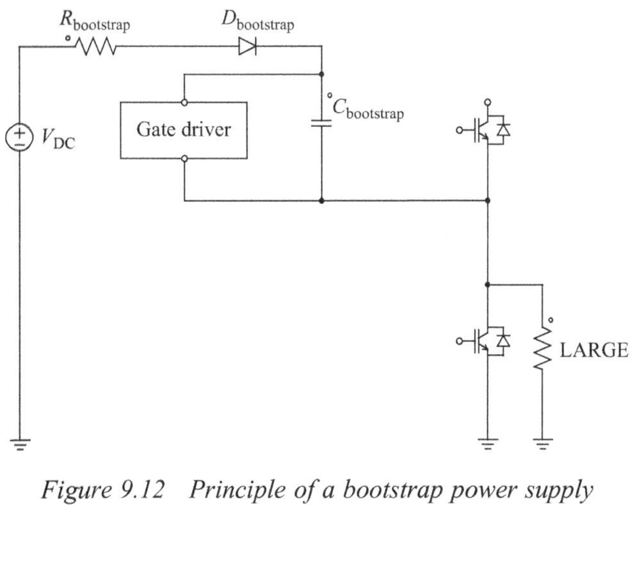

Figure 9.12 Principle of a bootstrap power supply

the actual pulse width modulation (PWM) generation. Unfortunately, it is all a qualitative process since it is virtually impossible to monitor the high-side gate circuit voltage. Ultimately, an *Under Voltage Lock-Out* protection prevents the IGBT against a very low-high side gate power supply, usually around 8 V. Alternatively, the STMicroelectronics products consider an active-low state for the low-side IGBTs and an active-high state for the high-side IGBTs. This allows a quick charging of the bootstrap power supply at initial power-up, when the low-side IGBTs turn-on and the high-side IGBT stay off.

The bootstrap capacitor is usually a multi-layer ceramic capacitor (MLCC) able to behave well in high frequency while storing some energy. While traditionally a combination of ceramic and electrolytic capacitors was of choice, recently the expansion of ceramic capacitors into the large value domain would limit the bootstrap capacitor to a single ceramic capacitor of 1–47 μF, at 25 V.

The bootstrap diode needs to withstand 1.5 times the bus voltage, that is, usually a requirement for a 1,000 V diode for a conventional 600 V IPM. Moreover, it has to be a fast recovery diode, 50 ns or better, at 1–2 A.

An alternative solution would obviously be the independent supply of each gate driver. This is done in industrial systems with special configurations of flyback power supplies, having a large number of secondary circuits and counting on asymmetrical loading of the gate driver circuits. Hence, the specific of such power supplies is that the installed power is considerably higher than the actual power, allowing thus the proper operation of the flyback converter. Interesting advances on the flyback control IC are also seen recently. They include advanced start-stop, provisions for quasi-resonant flyback, valley switching for low EMI, stand-by requirements, meeting the no-load power requirements (usually less than 300 mW), green mode status indicator, burst operation at low load, pulse skipping or pulse density modulation. All of these functions are implemented on the same integrated circuit platform as the high-power gate drivers.

A more modern highly integrated alternative suggests directly using a switch-mode unregulated DC/DC converter, that is, an ideal circuit substitute for the bootstrap diode. The RECOM™ DC/DC switch-mode converter comes in a small SIP4 package, works without any external components and without any heat sink, and offers unregulated 15 V/15 V conversion, 1,000 V isolation, 2 W output (133 mA) at an efficiency of 80%–85% (Figure 9.13). It can accommodate a capacitive load of up to 680 μF, more than enough for a large majority of gate driver applications. For instance, considering a maximum gate current request of 4.5 A for 500 ns, the minimum supply capacitor was required to minimize the voltage drop at 1 V yields 2.25 μF. A ceramic capacitor MLCC of 10 μF at 25 V, in a surface mount 1206 package can be adopted for any IRF IRAMS (*Smart Power Module*). Alternatively, a 22 or 47 μF are also available in ceramic technology.

Obviously the use of a DC/DC hybrid converter is more expensive, but it solves a series of problems with bootstrap power supply and can provide for a great laboratory setup needed to develop new control algorithms, like PWM with 60° no-switching intervals.

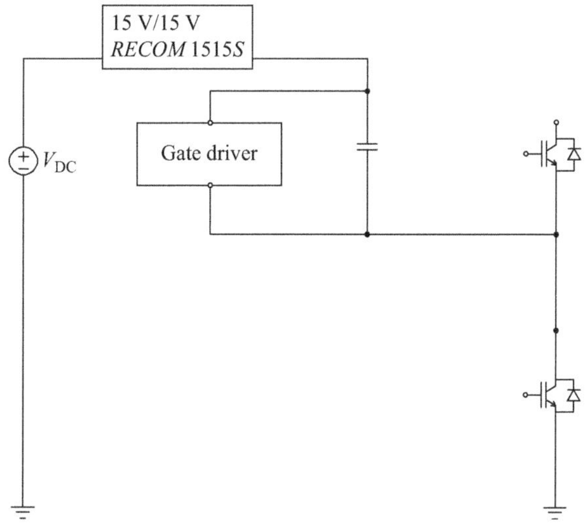

Figure 9.13 Using the unregulated 15 V/15 V power supply, with 1,000 V isolation

9.3.3 Digital interface

Ideally, the SPM device is directly connected with the microcontroller through proper selection of the logic family (like 3.3 or 5 V). Even modern microcontrollers with a core operated at 1.2 or 1.8 V must have a higher voltage interfacing at 3.3 or 5 V. It is important to pay attention to the ON-state logic of these signals, since most European and Japanese companies use an active-low signal for turning ON the IGBTs, while certain American companies use an active-high signal for turning ON the IGBT. As already mentioned, *STMicroelectronics* has a special convention of using active-high signals for control of the high-side IGBT, and active-low signals for turning on the low-side IGBTs. Apparently, this helps with the start-up of the bootstrap power supply before the microcontroller starts the PWM.

Additionally, these logic inputs may benefit from a Schmidt or hysteresis trigger.

9.4 Reliability

Since the power converter structure and the core technologies are well known, contemporary efforts are more and more targeting the lifetime and reliability improvement [22–25].

The advent of SPM subscribe to this desiderate and it can already be concluded that the implementation of a three-phase power inverter with a SPM rather than discrete power devices improves reliability. This is possible through a series of technological improvements, mostly leading to loss reduction, and better thermal aspect.

Obviously, modern research approaches to reliability improvement include online junction temperature monitoring [23], or active gate drivers [19–20].

Even before applying such approaches, the power modules are inherently more reliable since they are built through paralleling of power IGBT devices. It has been demonstrated in [24] that the power loss within a 1,200 A IGBT is higher than when using three parallel 400 A IGBT devices. This observation in addition to the inherent redundancy in operation comes to improve the reliability somewhat against the common wisdom that a circuit with more components is less reliable.

Hence, IGBTs are used with a parallel connection for the large capacity inverters [25]. This translates into a set of requirements suitable for paralleling. The voltage drop during the conduction state of both IGBT and anti-parallel diode must have a positive coefficient to avoid the concentration of current within any of the paralleled chips. Furthermore, it is necessary that the transition times are matched for the switching current balance. A typical figure of merit represents achieving a current balance as close as 2% of each other [25].

Reliability and lifetime of SPM are characteristics strongly depending on thermal performance of the module. Hence, the rating of power converters in the low-kilowatt range is mostly depending on thermal constraints. Given the complex structure of a SPM, the precise thermal modeling is difficult. It is therefore very common for the manufacturer to design and launch preliminary SPM devices ("engineering samples") without complete power or thermal rating. That is a blank space in the rating section, to be filled after more testing or customer assessment.

For the most tested devices, empirical thermal models are made available to account for the internal losses without a physical modeling. An empirical model for an *International Rectifier's* IRAMS SPM device [16] considers each switch individually and calculates the power loss with the following set of equations on dependence with the tested current:

$$E_{\mathrm{ON}} = (h_1 + h_2 \cdot I^x) \cdot I^k = \left[(7.69e-4) + (2.99e-2) \cdot I^{-1.159}\right] \cdot I^2 \tag{9.1}$$

$$E_{\mathrm{OFF}} = (m_1 + m_2 \cdot I^y) \cdot I^n = \left[(1.76e-2) + (4.34e-2) \cdot I^{-0.492}\right] \cdot I^1 \tag{9.2}$$

$$V_{\mathrm{CEON}} = V_{\mathrm{T}} + a \cdot I^b = 0.51 + 0.46 \cdot I^{0.649} \tag{9.3}$$

The results of this method used for a conventional motor drive application built with a 20 A IRAMS IPM (SPM) show 2.3 W power loss per switch, and 14.1 W per entire package, when operated in ambient temperature and trying to prevent a junction temperature close to 125 °C.

Other manufacturers provide loss curves for their SPM.

9.5 Variety of products

All semiconductor manufacturers [13–18] have lines of products dedicated to intelligent or SPM (Table 9.1). Instead of a comprehensive list of products, let us focus herein on performance evolution over the years. This would ultimately allow us to see how far we are from the technology's performance saturation curve.

Table 9.1 Example of SPM devices

	Voltage (V)	Current (A)	UVL	OC	OT	MaxDisip	SC-rated	Bootstrap	TempSns	CrtSns	Isolation	AdvGD
IRF IRAMS Gen1 in 2004	600	20	X	X	X	20 W/phase	10 μs	X	X		2,000	
IRF IRAMS Gen2 in 2014	600	30	X	X	X	32 W/IGBT	5 μs at ambient	X	X	X	2,000	
ST SLIMM 2011	600	30	X	X	X	52 W/IGBT	5 μs	X	X	X	2,500	X
Fairchild Motion SPM 2005	600	30	X	X	X		10 μs				2,500	X
Fairchild Motion SPM 2 2013	600/1,200	75/35	X	X	X	93 W/IGBT	5 μs				2,500	X
Allegro/Sanken SCM 2007	600	15	X	X	X	41.7 W	10 μs	X			2,000	X
Mitsubishi L 2005	600	215	X	X	X	833 W			X		2,500	
Mitsubishi L1 2009	1,200	144	X	X	X	833 W			X		2,500	
Infineon CIPOS 2010	600	22	X	X	X	29 W/mini 59 W/SIL		X	X		2,000	

9.6 Future usage and emerging solutions

In addition to the conventional usage of SPM already discussed in Section 9.3, let us here take a closer look of the futuristic benefits of this class of devices. The inclusion of an entire inverter circuitry within a single hybrid module allows us to compare such trend with the moment of the transition from discrete devices to integrated circuits in analog electronics. An entire inverter becomes a single *Bill-of-Materials'* item, with a cost close to what a single IGBT device used to be. Hence, complex structures can be built with multiple SPM devices, solving some of the limitations of the current implementation.

This concept of building a complex power structure with a multitude of smaller, more intelligent switches was first introduced to the academic community by Professor Ramu Krishnan of Virginia Tech (USA), at a round-table about the future of power electronics, held in 1999 in San Jose, at IECON Conference. While at the time, the concept seemed way ahead of time, it became a favorite for the future usage of SPM. It is one of the cases when power electronics meets high-tech, with high-scale integration of semiconductor-based functions to reduce the usage of passive components.

It is important here to acknowledge the paradigm shift proposed by this idea. The conventional design sequence starts with a topology selection, its simulation-based improvement, and the look for proper parts to build. Now, the design starts from the amazing features already existing on the SPM and – if they are not enough – multiple devices are used. The system designer is thus absolved of a series of concerns that move into the power semiconductor courtyard. A new generation of converters is possible, leading either to system performance able to fulfill the new demanding standards of the European Union or other governments, or to high performance, maximum reliability, and redundancy of custom converters for special applications.

Independent efforts are seen over the last years in studying multi-module converters [1], either as multi-level converter structures, or complex parallel and/or interleaved two-level inverters. A series of conference's special sessions occur on the topic. More interestingly, new topologies are now reported when considering the entire six-switch inverter as a singular integrated circuit, without showing the drawing of the complete schematic. Some results are next briefly reviewed from both industry and academia to illustrate this topic of R&D, even if it is barely starting to show its merits.

9.6.1 Grid interface with multiple power modules

A large number of three-phase power converters are dedicated to setting up grid interfaces with reduced harmonics. This is mostly the result of the multitude of standards and regulation efforts undertaken by most countries in the world. This effort is even more important at higher power levels where the same percentage loss in harmonics represents more power loss in watts.

The three-phase AC/DC conversion employing IGBTs can be achieved with a full six-switch power converter, or with a six-switch power converter used as an active filter in addition to a higher power diode rectifier. For illustration purposes, these two solutions are shown in Figure 9.14.

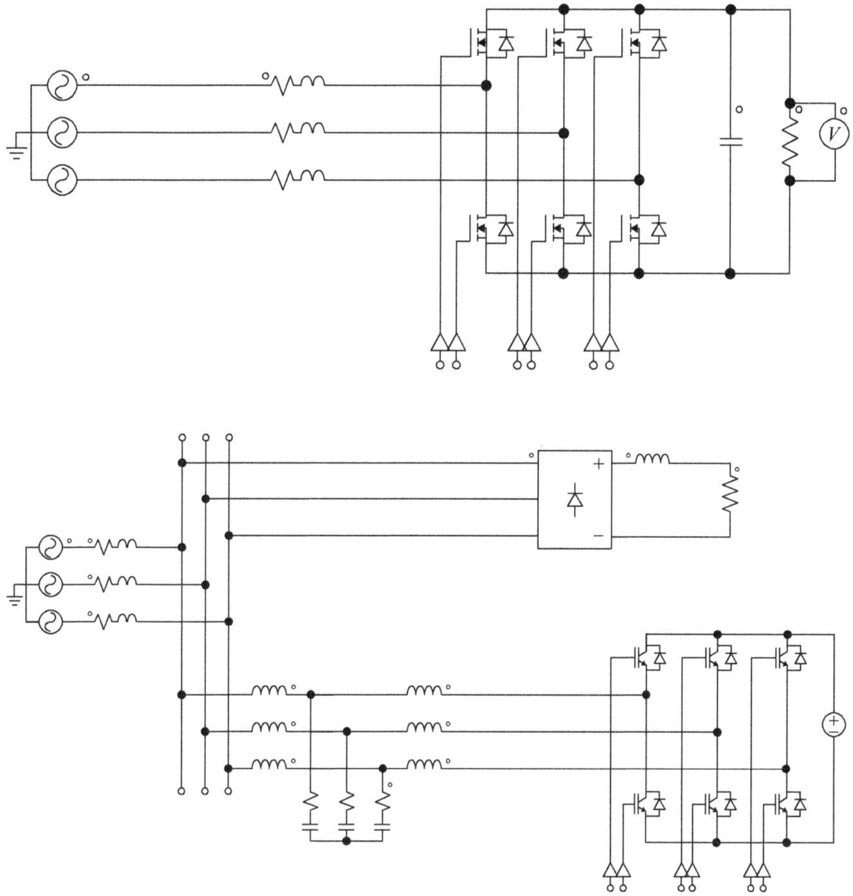

Figure 9.14 Previous solutions for AC/DC interfacing

Obviously, any of these two solutions can benefit from a SPM with a direct implementation of the power stage through a SPM. Unfortunately, despite the tremendous advent of SPM devices, their power levels tend to be limited.

Hence, the opportunity to propose a novel power conversion principle [2,26] able to augment the power capability of a SPM device with a diode rectifier, or reversing the logic, to correct the input harmonics of a diode rectifier with a SPM device. The principle of this new conversion system can be understood from the current waveforms shown in Figure 9.15, with the converter circuit from Figure 9.16.

The conventional operation of the diode rectifier sets the proper current references for the SPM multi-converter module during the intervals with no diode conduction. The current generated by the IGBT compensator converter closes through the two ON-state diodes and back to the power converter. This will overload a little the diode rectifier but this is really not important (cost-wise) in this application. The current circulation though diodes yields an improved grid current shown in Figure 9.17. Additional to the proper synchronization of the waveforms, a closed-loop current controller ensures the current waveforms.

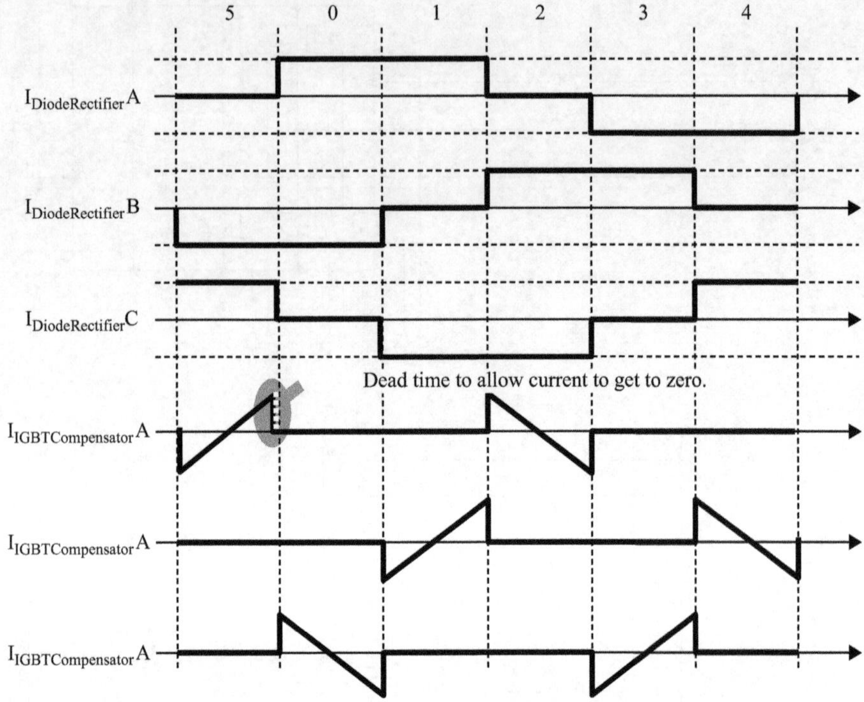

Figure 9.15 Principle of the novel multi-module grid interface

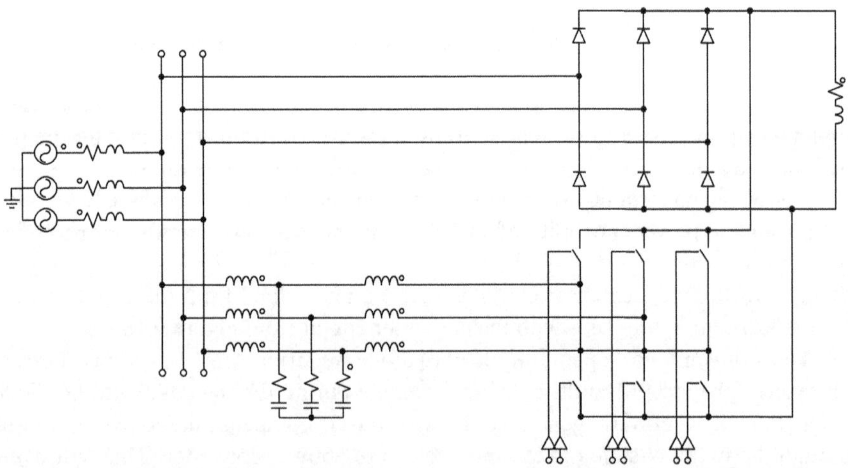

Figure 9.16 Novel multi-module grid interface targeting SPM implementation of the IGBT compensator

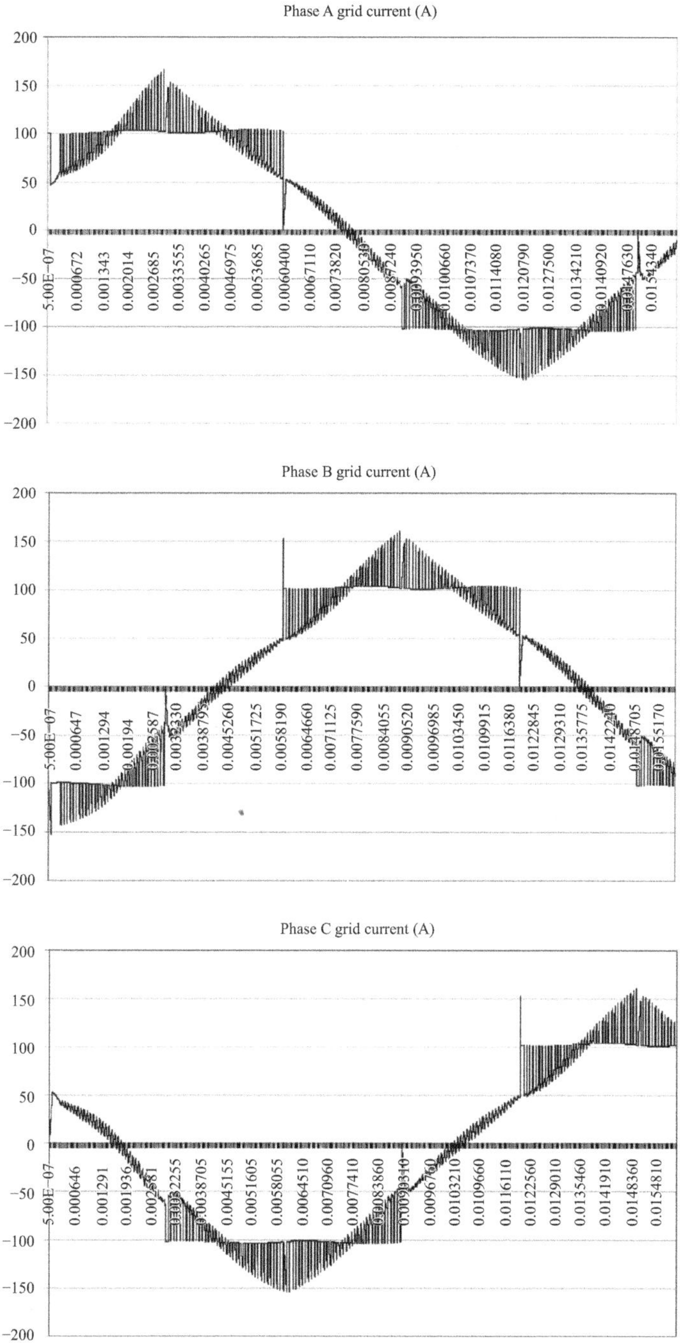

Figure 9.17 Grid phase currents based on this new hardware and current control (data samples shown with Microsoft Office's Excel)

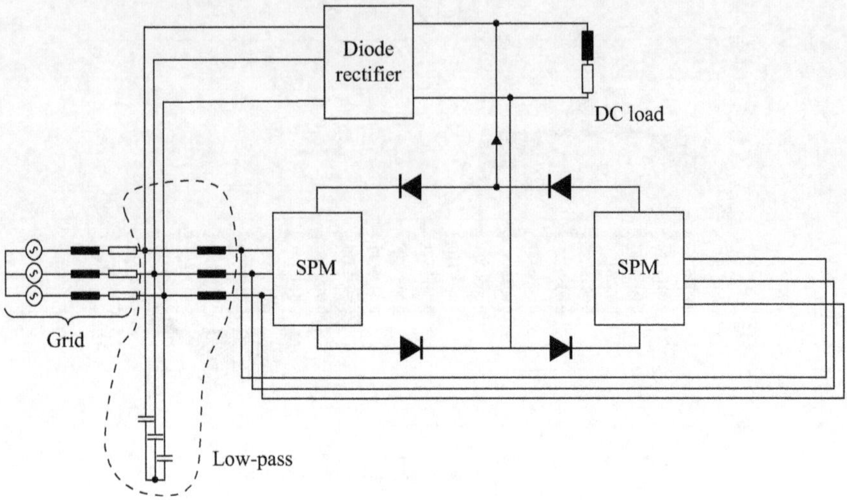

Figure 9.18 Implementation with SPM

Table 9.2 Production of DC current through three different approaches – new method allows more load current under nearly the same power loss stress in the semiconductor devices

Topology	**Direct IGBT bridge**	**Active filter**	**New method**
Maximum load current (DC)	$I_{rectifier} = 6.5$ A DC	$I_{rectifier} = 20$ A DC	$I_{rectifier} = 50$ A DC
IGBT operation	$I_{ph} = 5.0$ Arms, $m = 0.8$ Unity power factor	$I_{ph} = 22$ Arms Waveform = harmonic difference	$I_{ph} = 20$ Arms Power cycle = 0.33
SPM's PWM frequency	$f_{sw} = 6.00$ kHz	$f_{sw} = 12.5$ kHz	$f_{sw} = 12.5$ kHz
$P_{switching}$ (switch + diode)	0.87 W	1.68 W	2.42 W
P_{cond}, switch	3.42 W	2.99 W	1.46 W
P_{cond}, diode	0.80 W	0.35 W	1.06 W
P_{loss}, IPM	30.54 W	30.12 W	29.64 W

The bidirectional switches are implemented with IGBT through two SPMs (Figure 9.18). The advantage of this setup is demonstrated with Table 9.2 where comparison is made while considering the same (numerically close enough) power loss. It is proven that more load current can be processed with the novel solution, which results in a better utilization of both diodes and IGBTs.

The entire design represents a paradigm shift from conventional reasoning of saving or reducing the number of semiconductor components. Using highly integrated power modules provides herein an opportunity for reducing the count of passive components, with advantages in size, weight, efficiency, and reliability. In other words, we trade passive components for installed semiconductor power.

Overall, one achieves a very low-cost solution, with a low component count (three active and three passive components apart from microcontroller setup), without additional power supplies or DC link capacitors, and a fully integrated, fully protected electronics within the newly introduced intelligent power modules.

9.6.2 Matrix converter with SPM devices

A three-phase AC–AC matrix converter [27,28] is made of nine bidirectional switches which allow any output terminal to be connected to any input terminal (Figure 9.19). Any symmetrical three-phase output voltages can be obtained from a set of input three-phase voltages by a suitable switching of the matrix. The PWM operation needs to prevent a short circuit between the input-side terminals, and any output must be connected somewhere to close possible inductive load current. These both constraints require special precautions at switching of the power stage.

The major historical drawback with matrix converters relies in its rather difficult implementation. Various solutions for bidirectional power switches have been presented in the literature. We will consider next an alternative based on the use of SPM devices. First, let us notice that the bidirectional switches can be spread onto Current Source type Converters, as shown in Figure 9.20. Second, let us attempt the implementation of each Current Source Converter with a SPM device, as shown in Figure 9.21. Repeating this for all the three phases leads to the converter architecture from Figure 9.22, where the energy conversion is achieved with an all-semiconductor solution.

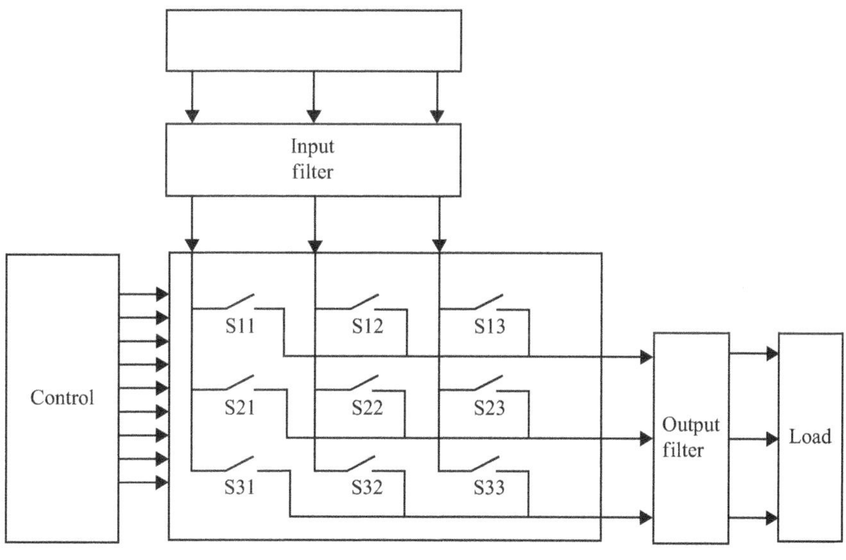

Figure 9.19 Typical configuration for AC/AC matrix converter

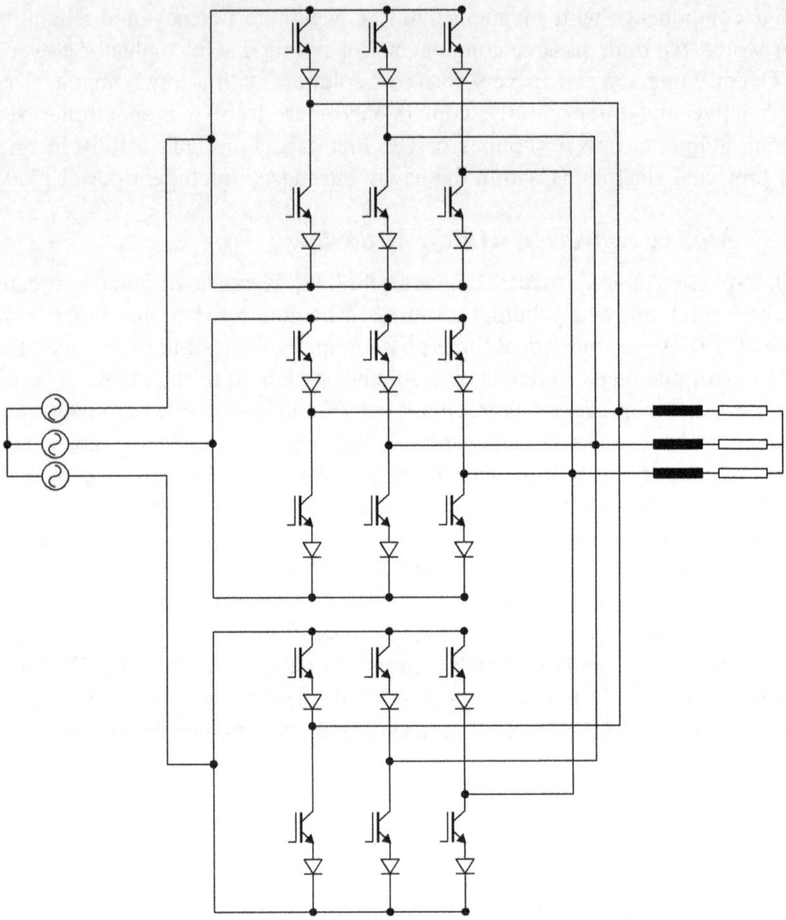

Figure 9.20 Redrawn of the converter from Figure 9.19 with three three-phase current source inverter modules (passive or EMI filters not shown)

9.6.3 *Multilevel converter with SPM devices*

Multilevel converters are seeing an increasing attention in academic environment, mostly because of the challenges and enthusiasm related to the definition of the PWM algorithms. They can make up for an excellent student project. The original use of multilevel converters came from the need to provide electronic solutions to medium and high voltage applications where we did not have semiconductor devices. Multilevel converters allow the use of devices rated at lower voltages within applications where the DC bus voltage is higher. For instance, they can implement grid-level converters with low-voltage IC technologies (under 40 V) [29].

We will continue to expose here the use of SPM devices to building complex power converter structures able to reduce the amount of passive components.

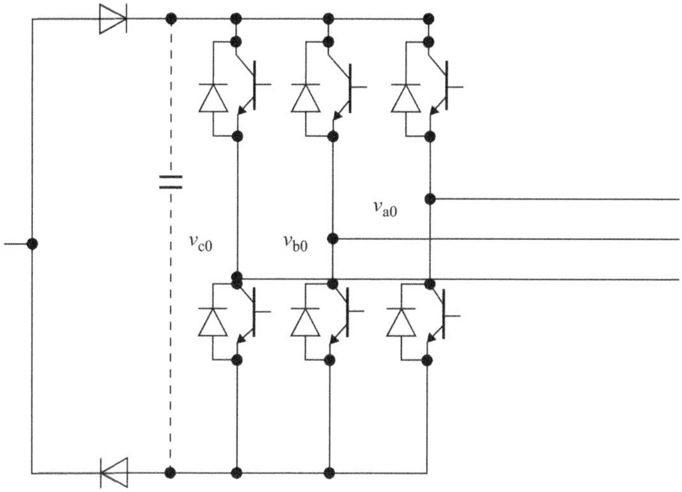

Figure 9.21 Realization of each Current Source Inverter with conventional three-phase power modules

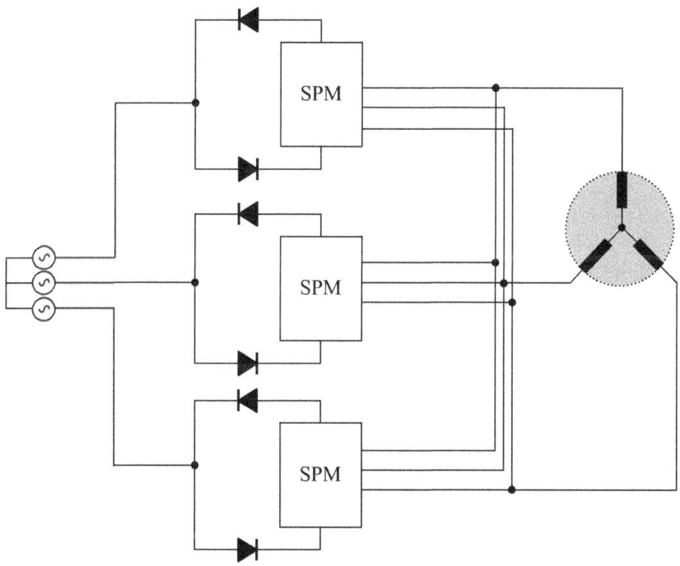

Figure 9.22 Implementation of a matrix converter with SPM devices (passive or EMI filters not shown)

A first solution is shown in Figure 9.23 [30]. A second approach for using SPMs to multilevel converters has been the subject of a PhD thesis at University of Bologna [31]. The base schemes are shown in Figure 9.24.

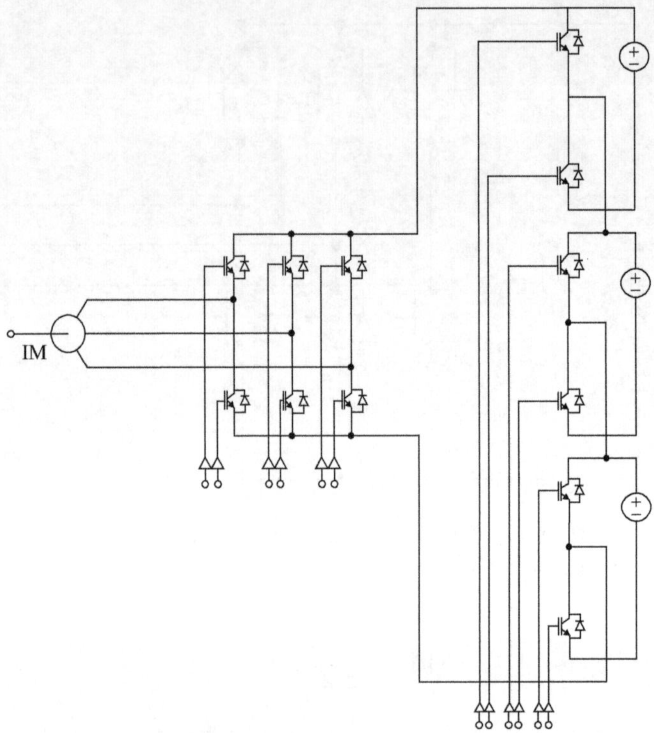

Figure 9.23 Novel multilevel topology built of Fuji IPM [30]

9.6.4 A different direct converter with SPM devices

The modern reliability concerns have pushed for the study of new converter topologies. Among them, the direct conversion from an AC source to an AC load gain the most interest since it does not need the cumbersome DC capacitor bank. We have seen the most common solution as being the nine-switch AC/AC matrix converter in the previous section.

Despite the high academic interest in the nine-switch AC/AC matrix converters, they are facing several major problems: the unconventional converter structure, the somewhat limited output voltage, and the need for additional external components for clamping the reactive energy.

An alternative to the conventional matrix converters is next considered. The idea was first proposed in 1976 [32] and it did not get too much attention for some difficulties of implementation at the time. However, the control challenges were solved many years later with the advent of microcontrollers [33–36], while the hardware implementation became really attractive after the success of SPM [37].

Operation of this converter is derived from the backwards operation of a *Current Source Rectifier* (Figure 9.25). Conventionally, the operation of this power converter assumes a large inductance on the DC side and the IGBT devices are switched after the so-called 120° program (Figure 9.25, like the diodes in Figure 9.14).

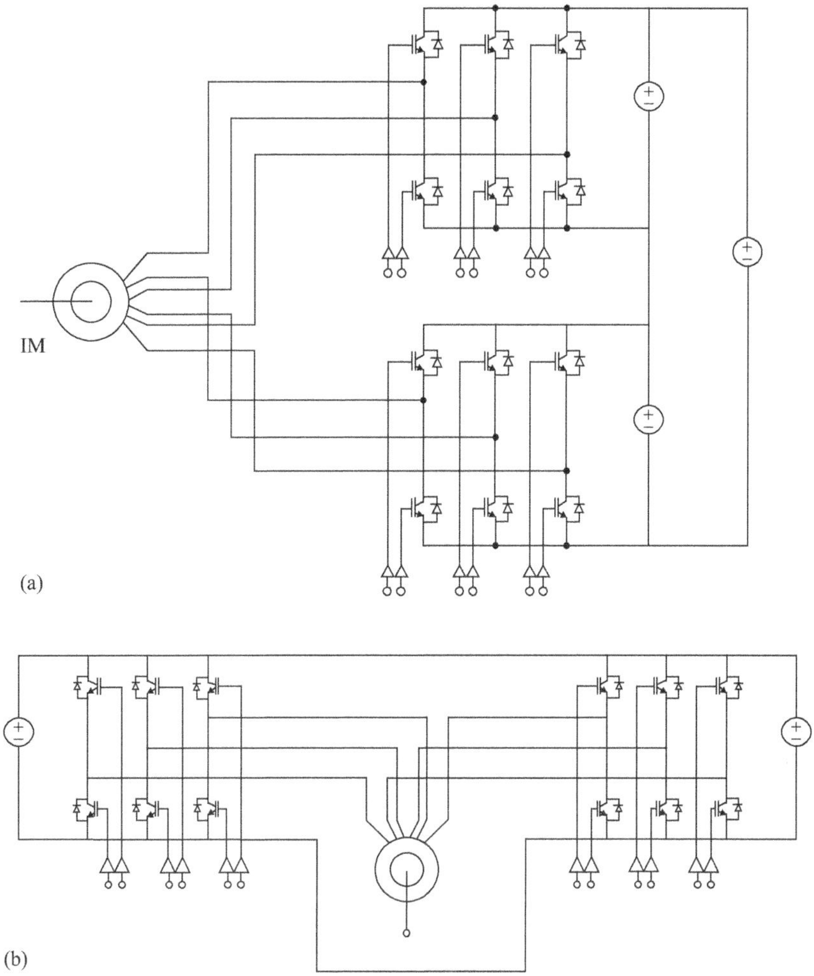

Figure 9.24 Series and parallel dual-SPM-based multilevel converters

The control system is sensing the grid voltage through a low-power diode rectifier. The phase currents are also sensed for closed-loop control. The PWM controller follows the 120° program and it is herein based on the *Space Vector Modulation* principle outlined within Figure 9.26. Designing the PWM algorithm can be made using the current space vectors corresponding to the input phase currents or based on voltage vectors with constant magnitude equaling the DC side voltage. This allows us to convert the conventional current vector control onto a voltage vector control.

The operation of a Current Source Rectifier requires only one high-side switch and one low-side switch must be turned-on at a time. This yields into nine possible combinations for the "ON"-switches. Each pair of two switches turned-on

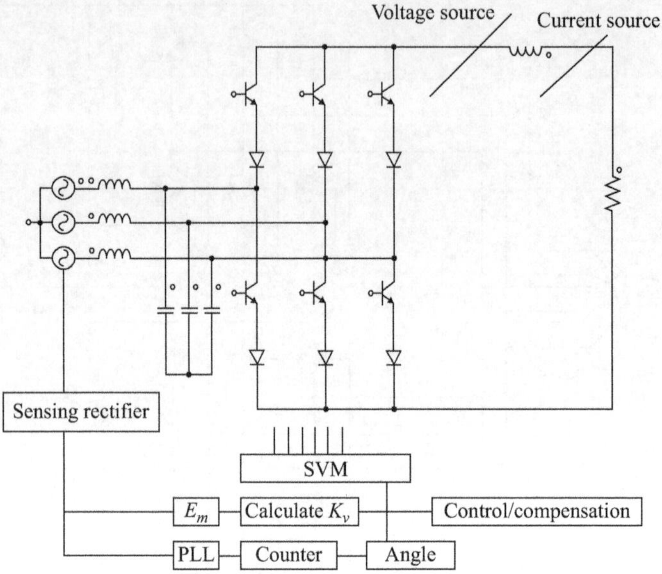

Figure 9.25 Grid interface with current source converter seen as voltage source on the DC side

determines a specific state of the converter. A space vector can be associated with each such state to represent the system of line input currents in the complex plane. During such a conduction state, the output rectified voltage coincides with any of the line-to-line voltages or can be zero. Only seven distinct positions of the space vector can be obtained: I_1–I_6 and zero I_0, and they are shown in Figure 9.26.

A desired position of the current space vector $\boldsymbol{I}$ is always placed between two neighboring vectors $\boldsymbol{I_a}$ and $\boldsymbol{I_b}$, $a,b = 1,\ldots, 6$ which represent the two active states considered within the switching process. The sampling interval is completed with a *zero*-state that can be obtained by turning-on the switches on the same leg so that there is always a current path for the output inductive current.

Writing the appropriate average relationship yields:

$$\bar{I}_a \cdot t_a + \bar{I}_b \cdot t_b + \bar{I}_0 \cdot t_0 = \bar{I} \cdot T \tag{9.4}$$

where T is the sampling interval; t_a is the time assigned for the state I_a; t_b is the time assigned for the state I_b; and t_0 is the time assigned for the state I_0.

Similar reasoning is possible using voltage vectors from Figure 9.26, where each voltage vector is defined with a line-to-line combination from the input voltages. Designing the PWM circuitry based on voltage vectors allows us to neglect the load character and to alleviate the need for a large inductance on the DC side. Moreover, there is no need for a closed-loop control of the output voltage. A phase locked-loop (PLL) loop is used as frequency multiplier locked to the grid frequency in order to produce the desired sampling frequency. The influence of the supply frequency variations can be thus reduced.

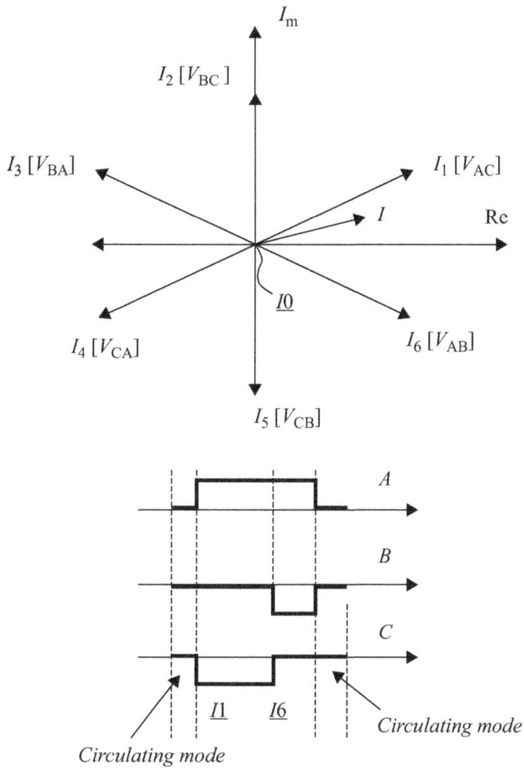

Figure 9.26 Space vector modulation applied to a current source rectifier

Observing the *Space Vector Modulation* algorithm allows us to work with an AC voltage reference for the voltage on the conventional "DC side". A direct AC/AC converter can thus be derived.

A second advantage derived from the application of the *Space Vector Modulation* algorithm consists of the opportunity to compensate for voltage harmonics or other grid-side distortion (Figure 9.27). Considering E_m the envelope of the rectified input voltage [38] and V the desired voltage vector at the output of the converter, the space vector theory can be rewritten in voltage terms:

$$
\begin{aligned}
t_a &= T \cdot k_v \cdot \sin(60 - \alpha) \\
t_b &= T \cdot k_v \cdot \sin \alpha \\
t_0 &= T - t_a - t_b \\
k_v &= \frac{R \cdot I_d}{E_m} = \frac{V}{E_m}
\end{aligned}
\tag{9.5}
$$

where V is one of the desired references for the three phases (v_{1d}, v_{2d}, v_{3d}).

Any unbalance of the grid voltage is reflected in the real-time envelope of the grid voltage E_m. The above equations allow for the proper adjustment of the pulse

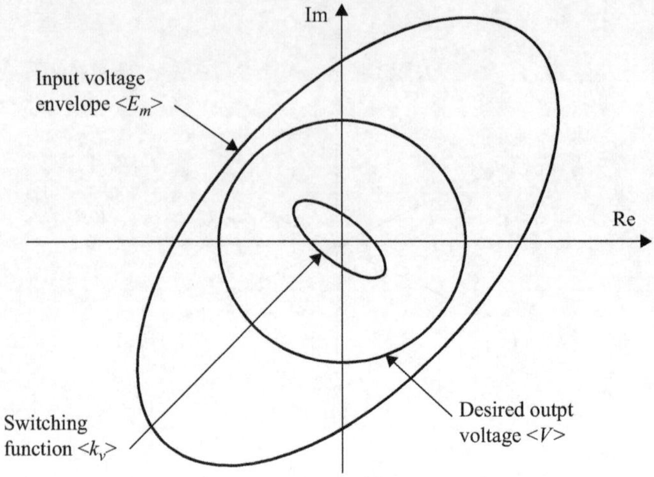

Figure 9.27 Space Vector Modulation for each power converter unit

widths to take into account any distortion or unbalance within this envelope. This method has the drawback of limiting the maximum available voltage to the minimum value of E_m.

Using the conventional Current Source Rectifier as a voltage AC/AC direct converter faces the same drawbacks as any current source converter, that is not yet available as a power module. Attentive to the developments on the market of power semiconductor devices, a solution based on SPM is proposed in Figure 9.28, similar to the solution in Figure 9.21.

The PWM algorithms previously defined for Current Source Rectifiers cannot work for the hardware solution of Figure 9.28(b) since the previous solutions assume shorting the DC bus during the zero-states. Such operation is prevented by the internal operation of the SPM and a short dead time is generally introduced by such module to prevent shoot-through. Additional requirements for the bootstrap power supply should be met, that is, frequent enough low-side switch-mode operation.

A new PWM algorithm is herein considered to use opposite active vectors during the zero-states in order to avoid shoot-through and to comply with the usage of SPMs.

Instead of producing a shoot-through during the zero-states, the new algorithm uses two opposite vectors to compensate each other within the average vector equation used for *Space Vector Modulation* generation (Figure 9.29). The vector applied during the first zero-state is selected to be the same as the closest active vector, and the vector applied during the second zero-state is selected as the opposite of the vector applied during the first zero-state.

Figure 9.30 expands this concept to a three-phase system and illustrates the implementation with SPM. It is important to understand that this converter behaves as a voltage source on the output side (see Figure 9.25). Hence, the overall system

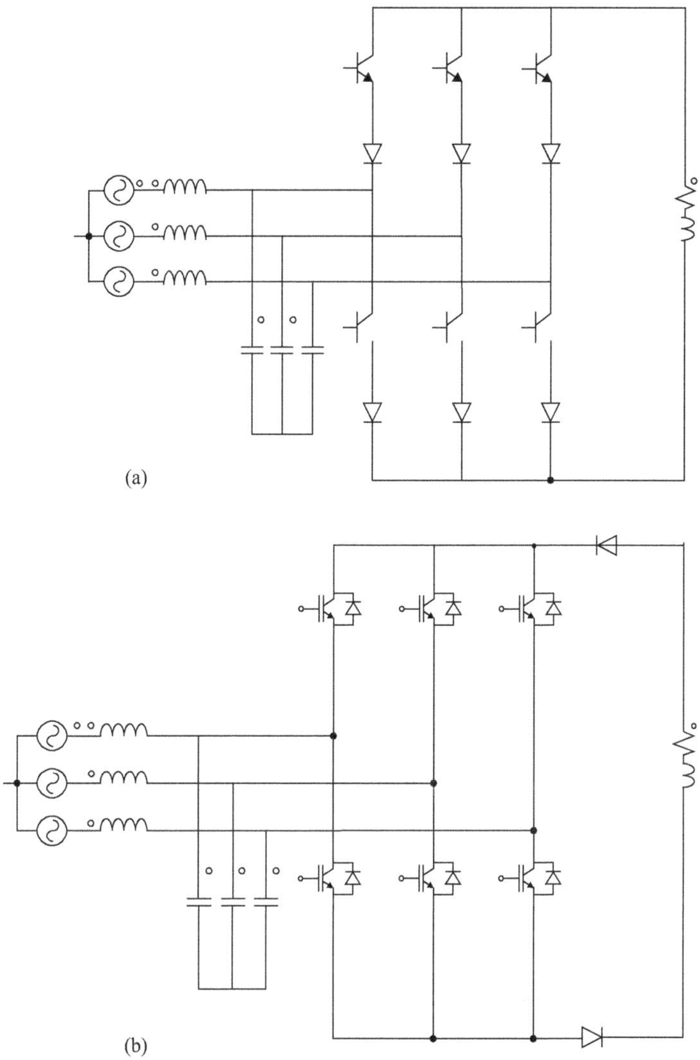

Figure 9.28 Conventional (a) and modified (b) Current Source Rectifiers

represents a three-phase voltage source compatible with the conventional back-to-back converter. This novel converter is able to provide *sqrt(3)* times more load-side phase voltage than a conventional matrix converter without any special over-modulation algorithm, and with considerably less passive components. On the grid or input side, each phase current is a summation of six currents, yielding a high content in fundamental. The grid-side currents can be improved with an interleaved carrier placement of the modules contributing to each phase current.

Harmonic results are shown in Figure 9.31 and they are superior to the conventional back-to-back converter when switched at the same frequency. Among the

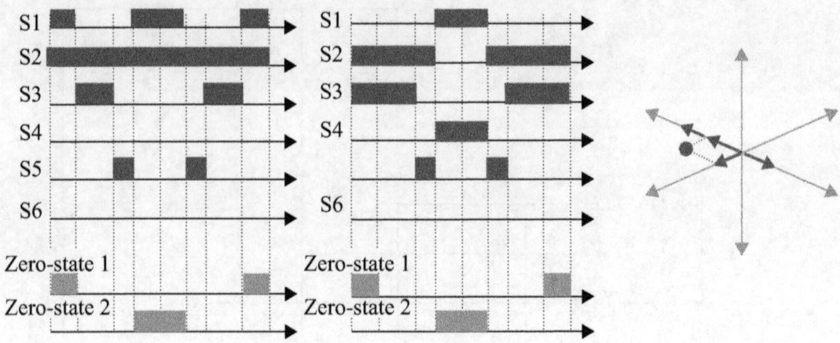

Figure 9.29 Conventional CSI SVM (space vector modulation for current source inverter) algorithm (a) and adaptation for SPM (b)

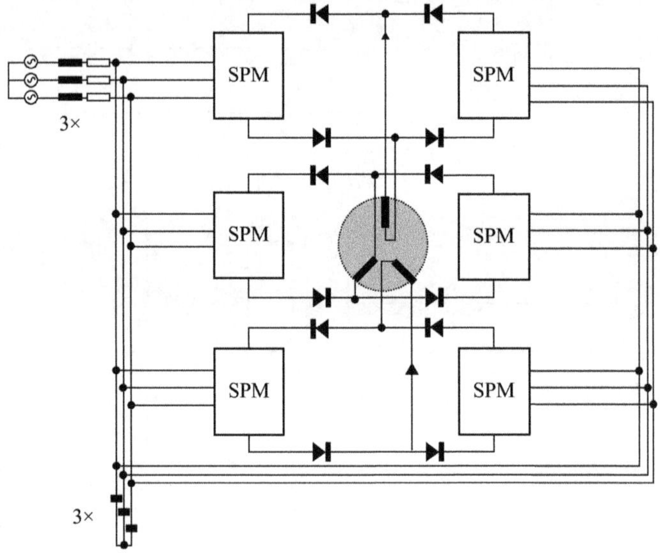

Figure 9.30 Implementation of a matrix converter with SPM devices

multiple coefficients available for harmonic characterization of power converters, we have chosen herein the *total harmonic distortion* (THD) coefficient

$$\mathrm{THD}(\%) = \frac{100}{V_{(1)}} \sqrt{\sum_{n=2}^{\infty} \left[V_{(n)}\right]^2} \tag{9.6}$$

Since the grid-side highly inductive impedance is basically a low-pass filter (LPF) for currents, the higher order current harmonics will be attenuated. The remaining spectrum of the current will be different from one PWM method to another, and from one switching frequency to another. A coefficient regarding current harmonics would better define the performance of a PWM method. Such a

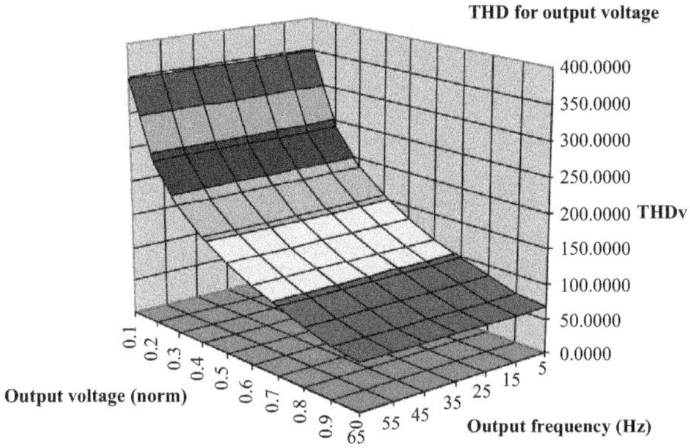

V(RMS)		17.9	35.9	54.0	72.2	90.4	108.6	126.2	145.4	162.2	171.6
m		0.1	0.2	0.3	0.4	0.5	0.6	0.7	0.8	0.9	1.0

(a)

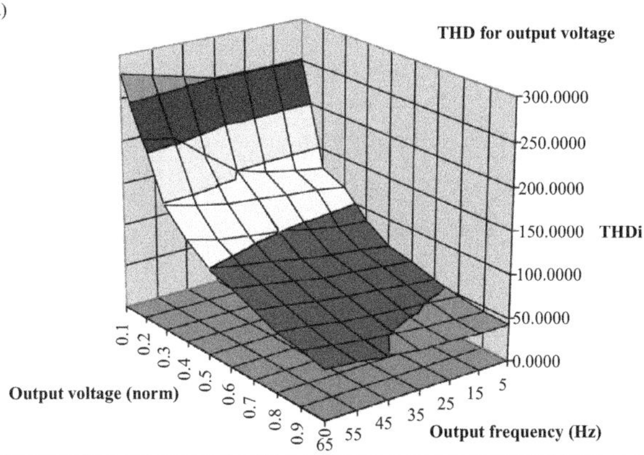

V(RMS)		17.9	35.9	54.0	72.2	90.4	108.6	126.2	145.4	162.2	171.6
m		0.1	0.2	0.3	0.4	0.5	0.6	0.7	0.8	0.9	1.0

(b)

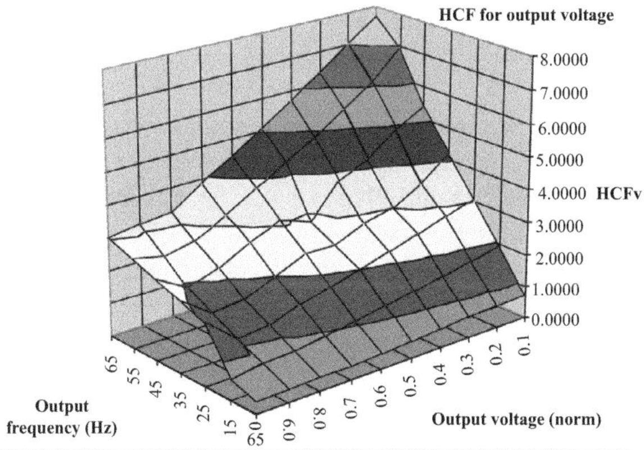

V(RMS)		17.9	35.9	54.0	72.2	90.4	108.6	126.2	145.4	162.2	171.6
m		0.1	0.2	0.3	0.4	0.5	0.6	0.7	0.8	0.9	1.0

(c)

Figure 9.31 Harmonic results

coefficient is called *harmonic current factor* (HCF), and it can be expressed also based on the voltage harmonics at the converter output:

$$\mathrm{HCF}(\%) = \frac{100}{I_{(1)}}\sqrt{\sum_{n=5}^{\infty}\left[I_{(n)}\right]^2} = \frac{100}{(V_{(1)}/\omega L)}\sqrt{\sum_{n=5}^{\infty}\left[\frac{V_{(n)}}{n\omega L}\right]^2}$$

$$= \frac{100}{V_{(1)}}\sqrt{\sum_{n=5}^{\infty}\left[\frac{V_{(n)}}{n}\right]^2} \tag{9.7}$$

This analysis shows that a converter system comprised of six power modules IPM/SPM/IRAMS setup as shown in Figure 9.30 can drive a 2 kW motor, in the similar operation and thermal conditions as a back-to-back dual SPM would drive an 1 HP = 745 W motor.

The power density is also competitive as the entire power stage (with straight-pin mounting, six individual heat sinks, passive LC filtering and power connectors) accounts for (2 in × 2 in × 7.5 in =) 0.49 l for 2 kW delivered power (i.e. 4.1 kW/l). This should compare to contemporary industry goals of 4 kW/l, for this class of converters [36] for low-kilowatt range, required to switch below 10 kHz due to motor requirements.

The advantages over the conventional back-to-back solution consist of:

- longer lifetime, improved reliability;
- considerable smaller package, improved power density;
- fully standard compatible for both grid and EMI.

The drawbacks are:

- slightly higher cost of the system with current costs;
- limited choice of power ratings available for power modules [6–9].

9.7 Conclusion

The advent of power semiconductor technology allowed the integration of semiconductor power devices and their control integrated circuits within the same hybrid device. This chapter has briefly reviewed the advantages of this integration, the technology options, and provided a historical perspective on the topic.

Finally, the conventional usage of SPM has been extended with a new class of multi-module power converters based on a network of switches, implemented with multiple SPM and ready to provide even better performance.

References

[1] Neacşu, D.O., *Current Advanced Power Conversion Concepts with Multiple Power Modules*. Habilitation Thesis, Romanian Ministry of Education, May 2012.

[2] Neacşu, D.O., *Switching Power Converters – Medium and High Power*. Boca Raton, FL: CRC Press; December 2013.
[3] Trzynadlowski, A., *Introduction to Modern Power Electronics*. Hoboken, NJ: Wiley; March 2010.
[4] Tolbert, L.M., 'Smart Integrated Power Module', *2012 U.S. DOE Hydrogen and Fuel Cells Program and Vehicle Technologies Program Annual Merit Review and Peer Evaluation Meeting*, Project ID: APE046, May 15, 2012.
[5] Staunton RH, Ozpineci B, Theiss TJ, Tolbert LM, 'Review of the State-of-the-Art in Power Electronics Suitable for 10 kW Military Power Systems', ORNL/TM-2003/209 Annual Report, October 2003.
[6] Motto, E., 'Application Specific Intelligent Power Modules – A Novel Approach to System Integration in Low Power Drives', *PCIM Conference*, 1998.
[7] Bhalerao, P., Wiatr, R., 'New Intelligent Power Module Series', *PCIM Europe Journal*, vol. 15, no. 2, 2008, pp. 21–22.
[8] Lee, F.C., 'Evolution of Power Electronics Technologies – A CPES's Perspective'. Presented at IEEE *21st International Symposium on Industrial Electronics*, Hangzhou, China, June 2012.
[9] Neeb, C., Boettcher, L., Conrad, M., DeDoncker, R.W., 'Innovative and Reliable Power Modules: A Future Trend and Evolution of Technologies'. *IEEE Industrial Electronics Magazine*, vol. 8, no. 3, 2014, pp. 6–16.
[10] Lutz, J., Schlangenotto, H., Scheuermann, U., DeDoncker, R. *Semiconductor Power Devices*. Berlin: Springer; 2011.
[11] Hosen, T., Yanagisawa, K., 'Fuji Electric's Semiconductors: Current Status and Future Outlook', *Fuji Electric Review*, vol. 57, no. 3, 2011, pp. 68–71.
[12] Gorgerino A, 'New-generation IGBTs increase motor drive efficiency in appliances', *ECN Magazine*, July 2012.
[13] Fairchild Semiconductor International Inc. *FSBS10CH60– Smart Power Module*. Datasheet; 2005.
[14] International Rectifier. *IRAMS10UP60B-iMotion Series, Integrated Power Hybrid IC for Appliance Motor Applications*. Datasheet; 2008.
[15] Wintrich, A., Nicolai, U., Tursky, W., Reimann, T., 'Application Manual – Power Semiconductors', *SEMIKRON Internal Documentation*, 2011.
[16] Wood, P., Battello, M., Keskar, N., Guerra, A., *IPM Application Overview – Integrated Power Module for Appliance Motor Drives*. International Rectifier AN-1044; 2002.
[17] STMicroelectronics. *STGIPS10K60A – IGBT Intelligent Power Module*. Datasheet; 2010.
[18] Shimizu, N., Karasawa, T., Takagiwa, K., 'New Lineup of Large-Capacity "V-Series" Intelligent Power Modules', *Fuji Electric Review*, vol. 58, no. 2, 2012, pp. 65–69.
[19] Motto, E.R., Donlon, J., 'Speed Shifting Gate Drive for Intelligent Power Modules', *IEEE Applied Power Electronics Conference*, Dallas, TX, USA, March 2006.

[20] Neacşu, D.O., Nguyen, H.H. *Gate Drive Circuit with Feedback-Controlled Active Resistance*, US Patent #6,459,324, 2002.

[21] STMicroelectronics, AN2123 Application Note, 'TD351 Advanced IGBT Driver – Principles of Operation and Application', 2005.

[22] Volke, A., Hornkamp, M., *IGBT Modules – Technologies, Driver, Applications, Infineon Documentation*, Munich, German, 2nd ed., 2012.

[23] Baker, N., Liserre, M. , Dupont, L. , Avenas, Y., 'Improved Reliability of Power Modules: A Review of Online Junction Temperature Measurement Methods', *IEEE Industrial Electronics Magazine*, vol. 8, no. 3, 2014, pp. 17–27.

[24] Neacşu, D.O., Wagner, E., Borowy, B., 'A Simulation Benchmark for Selection of the PWM Algorithms for Three-Phase Interleaved Converters', *IEEE Transactions on Industrial Electronics*, vol. 55, April 2008, pp. 1628–1636.

[25] Igarashi, S., Kakiki, H., Nishimura, Y., Goto, T., 'Design of High Reliability Packaging for Fuji High Power Module', *2011 International Conference on Electrical Machines and Systems (ICEMS),* 2011, pp. 1–6.

[26] Neacşu, D.O., 'Principle of a Novel Component Minimized Active Power Filter for High-Power Magnet Supplies', *IEEE Industrial Electronics Conference IECON*, Montreal, Canada, October 2012.

[27] Friedli, T., Kolar, J.W., 'Milestones in Matrix Converters', *IEEJ Journal of Industry Applications*, vol. 1, no. 1, July 2012, pp. 12–14.

[28] Huber, L., Borojevic, D., Burany, N., 'Analysis, Design and Implementation of the Space Vector Modulator for Forced-Commutated Cycloconverters', *IEE Proceedings Series B*, vol. 139, no. 2, March 1992, pp. 103–113.

[29] Neacşu, D.O., 'Fault-Tolerant Isolated Converter in Low-Voltage Technology for Automotive AC Auxiliary Power', *IEEE Industrial Electronics Conference IECON*, Vienna, Austria, November 10–13, 2013, pp. 8184–8189.

[30] Su, G.J., Adams, D., 'Multilevel DC Link Inverter for Brushless Permanent Magnet Motors with Very Low Inductance', *IEEE IAS Annual Meeting 2001*, Chicago, Illinois, USA, September 30–October 5, 2001, vol. 2, pp. 829–834.

[31] Lega, L., 'Multilevel converters: Dual Two-Level Inverter Scheme', *PhD Dissertation*, University of Bologna, 2009.

[32] Jones, J., Bose, B.K., 'A Frequency Step-Up Cycloconverter Using Power Transistors in Inverse Series Mode', *International Journal of Electronics*, vol. 41, no. 6, 1976, pp. 573–587.

[33] Neacşu, D.O., 'Effectiveness of AC/AC Current-Controlled Matrix Converter to Open-Winding Induction Machine Drive' *IEEE-CAS International Symposium SCS99*, Iasi, Romania, July 7–9, 1999, pp. 379–382.

[34] Neacşu, D.O., 'Controlled AC/AC Voltage Source Matrix Converter for Open-Winding Induction Machine Drives', *IEEE Industrial Electronics Conference IECON 1999*, San Jose, CA, USA, Nov. 29–Dec. 3, 1999.

[35] Neacşu, D.O., 'IGBT-Based "Cycloconverters" Built of Conventional Current Source Inverter Modules', *IEEE-CAS International Symposium SCS2003*, Iasi, Romania, July 10–11, 2003, vol. 1, pp. 217–220.
[36] Neacşu, D.O., 'Analysis and Design of IGBT-Based AC/AC Direct Converters Built of Conventional Current Source Inverter Module', *IEEE IAS 2004*, Seattle, WA, October 2004, vol. 3, pp. 1824–1831.
[37] Neacşu, D.O., 'Towards an All-Semiconductor Power Converter Solution for the Appliance Market', *IEEE Industrial Electronics Conference IECON*, Glendale, AZ, USA, November 2010.
[38] Enjeti, P., Xie, B., 'A New Real Time Space Vector PWM Strategy for High Performance Converters', *IEEE/IAS Annual Meeting Conference Record 1992*, Houston, Texas, USA, October 4–9, 1992, *Conference Record 1992*, vol. 1, pp. 1018–1025.

Part II

Applications

Chapter 10

Permanent magnet synchronous motor drives

*Mauro Zigliotto**

10.1 Introduction

As every scientific paradigm, the control of electrical drives is a system in a dynamic equilibrium among innovative push and technological constraints, with some rather influencing parameters, as cost and market demand. Permanent magnet synchronous motor (PMSM) drives play a key role in a productive world that requires energy-awareness, reliability, high-performance and cheapness all together. At the forefront of any challenging application, there are the human mobility in the emerging cultural and social context of smart-cities, the more-electric propulsion systems and the migration in a large share of industrial products from uncontrolled, stand-alone motors to smart and efficient net-connected AC drives. In vehicle propulsion, PMSM are dominant, due to the required peak power rating, that ranges from less than 10 kW in mild hybrid configurations to 175 kW and above in high-performance strong hybrid and battery-electric drive trains. Quite surprisingly, permanent magnet (PM) synchronous motors show a continuous expansion also in harsh environments with important cost limitations such as in-home appliances [1]. The range of application of PM motors is also enhanced by the recent introduction of new constructive topologies, which are essentially the migration, at different degrees, of isotropic PM motors to pure synchronous reluctance (SynR) motors. The addition of a correct amount of PMs to the synchronous reluctance rotor yields a higher power factor and improved efficiency, with limited back-EMF even in the flux-weakening speed range and without a necessary change in the stator design [2]. Figure 10.1 shows a sketch of two structures, among the several possible ones, and the relative axes definition. While it is important to point out that no standards have been developed in terms of control of electric drives [3], nevertheless, some indications come out clearly from the analysis of the most recent and qualified literature, and they will be examined fully hereinafter. The aim of the chapter is to enable the power electronics community to address emerging (but proven) topics in electrical drives, with special emphasis on permanent magnet synchronous motor (PMSM) drives. Among the many alternatives, the collected material was classified according to the type of control, further detailed when

*University of Padova, Padova, Italy

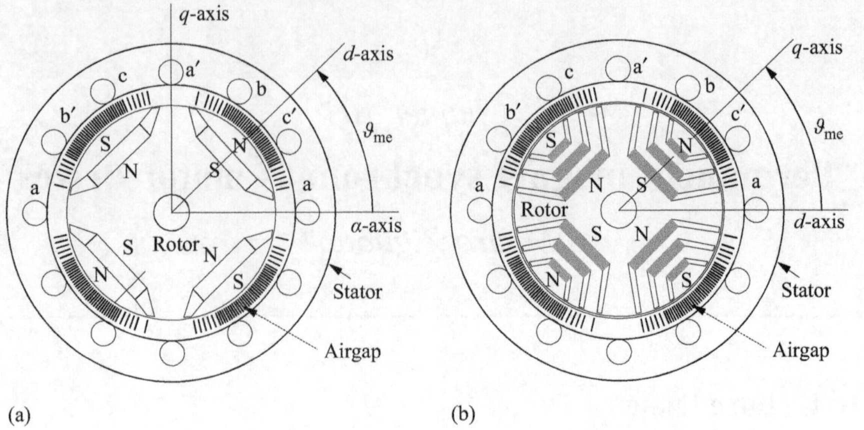

Figure 10.1 Cross-sectional views of four-pole PM motors. (a) Internal PM (IPM) motor. (b) PM-assisted SynR (PMASR) motor

necessary for the specific motor topology. The applications (automotive, industry, mechatronics and so forth) are impressively various, and they will be timely cited as soon as they either justify, enrich or aid the comprehension of a particular control technique. Section 10.2 reports the state of the art in sensorless control for PMSM, with the special target of highlighting the delicate connections between any theoretical algorithm and its implementation in the power electronic converter. With the same critical sensibility, Section 10.3 illustrates the direct torque control (DTC) and the model predictive control (MPC), reporting a selected example of application, as bright and promising trend in multi-object, energy-efficient control techniques for PMSM drives.

10.2 Trends in sensorless control of PMSM

The delicate position sensors, inherent in every PMSM control, increase the cost of the drive while reducing its reliability. A consolidated trend is the reconstruction of the position through the measurement of electrical quantities only. Although sensorless control is arousing the research interest since three decades, it still represents a challenge for scientists. And even when the position sensor cannot be completely removed, e.g. for safety reasons, the sensorless control is used to provide a backup for the rotor position information in case of a sensor failure.

The extensive study over years has almost completely defined the most interesting control techniques. The challenge is now moving on variants that make the sensorless drive economic, efficient and reliable, all weighted by the application needs, of course. Among the few possible hardware improvements, there are:

- the direct measurement of the motor voltages. It will profitably substitute the use of the reference voltages and the associated nonideality compensation techniques [4], increasing the robustness and the accuracy of practically all the sensorless techniques.

- the development of special PM motors, as for example PM-assisted synchronous reluctance (PMASR) motors, designed to account for certain sensorless machine characteristics [1]. The saliency ratio, the level of cross-saturation and eventual presence of secondary saliencies, over a broad load range, are all design aspects that may heavily affect the performance of the sensorless drive, and they still deserve the attention of the researchers. Particular care will be paid to avoid that the saturation of L_q under load causes a zero saliency condition. To this aim, an interesting solution is proposed in [5], where a V-shaped PM array and an eccentric rotor are manufactured on purpose, by deriving a formula for the design of the air gap shape, based on the magnetic circuit model of the interior permanent magnet (IPM) motor. That solution is proved to minimise the effect of loading over the position estimation accuracy. The control technique includes a fast initial position estimation using magnetic hysteresis phenomenon. Equally, it is expected that special winding configurations with inherent self-sensing feature will be further investigated [6].

As regards the software solutions, two distinct techniques are established, depending on the dominant speed range of the application. For operations at standstill and low speed (normally below 20%–30% of the nominal speed), efficient methods are based on tracking the position of asymmetries through the injection of high-frequency (HF) voltage signals, and by measuring and post-processing the related HF currents [7–11].

As soon as the speed is high enough to have a favourable signal-to-noise ratio, observers based on signals at the fundamental frequency, as for example the extended back electromotive force (e-EMF) are exploited [5,12–18]. For applications where low-speed operation is only transient, an e-EMF-based sensorless controller coupled with a simple start-up procedure offers an attractive low-cost alternative [19]. As confirmed by the analysis of the most recent publications, a cutting-edge solution is the merging of the aforementioned techniques into sophisticated hybrid controls [1]. Also, the extension of the speed range is arousing lively interest. For the e-EMF-based technique, the improvement is obtained by an accurate parameter estimation [20,21] or by advanced filtering [9], made possible by the increased availability of computing power.

10.2.1 Key factors for sensorless controls evaluation

For both injection and e-EMF methods the literature presents several alternatives, but an insight of pros and cons suggests that there is still room for improvements. In the following, it is first given a rough description of the different groups in which the schemes can be categorised. Then, the key factors for a correct and technically sound evaluation of each sensorless solution are presented, to give the reader the tools to distinguish between real advances and incremental results. For each class, up-to-date references to recently published papers are provided. The aim is not as much the repetition of the details and the explanation of the theoretical foundations

High-frequency injection, at standstill and low speed

$$\text{Pulsating}\left\{\begin{array}{c}\text{sinusoidal}\\ \text{square wave}\\ \text{arbitrary}\end{array}\right\}\text{excitation in}\left\{\begin{array}{c}\text{stationary}\\ \text{synchronous}\end{array}\right\}\text{reference frame}$$

$$\text{Rotating}\left\{\begin{array}{c}\text{sinusoidal}\\ \text{square wave}\end{array}\right\}\text{excitation in}\left\{\begin{array}{c}\text{stationary}\\ \text{synchronous}\end{array}\right\}\text{reference frame}$$

Fundamental frequency signals, at middle and high speed

$$\text{Closed loop}\left\{\begin{array}{c}\text{reduced}\\ \text{full}\end{array}\right\}\text{order e-EMF observer in}\left\{\begin{array}{c}\text{stationary}\\ \text{synchronous}\end{array}\right\}\text{reference frame}$$

Figure 10.2 Classification of modern sensorless control techniques

of each method, but more the elucidation of the main issues for the acceptance or rejection of each of them. The evaluation of what can be really considered a breakthrough can be performed by weighting the following factors:

- physical principle and dependence on special design of the motor;
- additional hardware and computational requirements;
- dependence on the parameters and sensitivity to their variations;
- complexity of tuning and dependence on pre-measured values;
- completeness of theoretical analysis, e.g. convergence domain and dynamics;
- inclusion of nonlinearity, as additional saliency and saturation.

In the higher power range (medium-voltage drives), additional peculiarities have to be taken into account when proposing a sensorless technique in place of a standard motor drive [13]:

- difficult measurement and conditioning of the DC bus voltage;
- necessarily low switching frequency and consequent high current ripple;
- large temperature range.

A thorough and complete technology status review of sensorless control solutions for PMSM is provided in the excellent papers [13,22,23]. Figure 10.2 recaps the main sensorless methods, as a reference for the discussion of the most promising ones.

10.2.2 A glance to HF injection methods

Figure 10.3 shows a complete general block schematic diagram of the inner current loop of a HF injection-based sensorless drive and how it interacts with the power inverter. The output of the current regulators, issued by two proportional-integral (PI) blocks in the synchronous (*d, q*) reference frame, is summed up to HF voltage signals u_α^{h}, u_β^{h}. The resulting stator voltage references u_α^* and u_β^* constitute the input of a space vector modulation (SVM) inverter that computes the switching pattern for the six switches S_1–S_6 of the voltage inverter. The phase currents (normally two

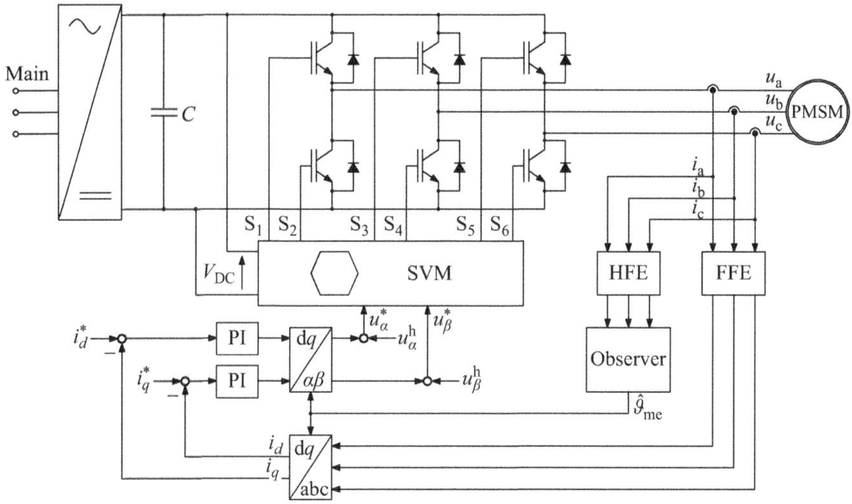

Figure 10.3 General block schematic diagram of a HF injection-based PMSM drive

out of three) are measured and post-processed to get either the fundamental frequency components (FFE block) or the HF components (HFE block). The latter is exploited to estimate the rotor electro-mechanical position ϑ_{me}. Actually, when a rotating carrier-signal voltage at HF $\omega_{\mathrm{h}} = \mathrm{d}\vartheta_{\mathrm{h}}/\mathrm{d}t$ as

$$\mathbf{u}_{\alpha\beta}^{\mathrm{h}} = U_{\mathrm{h}} \begin{bmatrix} \cos(\omega_{\mathrm{h}} t) \\ \sin(\omega_{\mathrm{h}} t) \end{bmatrix} \tag{10.1}$$

is applied to a PMSM, it produces a current vector whose negative sequence component is influenced by the saliency-dependant HF inductances:

$$\mathbf{i}_{\mathrm{n},\alpha\beta}^{\mathrm{h}} = \frac{U_{\mathrm{h}}}{\omega_{\mathrm{h}}} \frac{\left(L_d^{\mathrm{h}} - L_q^{\mathrm{h}}\right)}{2L_d^{\mathrm{h}} L_q^{\mathrm{h}}} \begin{bmatrix} \sin(\vartheta_{\mathrm{h}} + 2\vartheta_{\mathrm{me}}) \\ -\cos(\vartheta_{\mathrm{h}} + 2\vartheta_{\mathrm{me}}) \end{bmatrix} \tag{10.2}$$

where L_d^{h}, L_q^{h} are in general the HF differential inductances of the direct and quadrature axis, evaluated by linearisation around the working point, according to the small signal theory. Further insights can be found in [16,18,24–26].

The negative sequence current vector (10.2) can be profitably used for rotor position detection [23]. As an alternative, a pulsating carrier voltage can be injected in the g_d- or g_q-axis of a tentative synchronous frame (g_d, g_q):

$$\mathbf{u}_{\alpha\beta}^{\mathrm{h}} = U_{\mathrm{h}} \cos(\omega_{\mathrm{h}} t) \begin{bmatrix} \cos(\hat{\vartheta}_{\mathrm{me}}) \\ \sin(\hat{\vartheta}_{\mathrm{me}}) \end{bmatrix} \tag{10.3}$$

where $\hat{\vartheta}_{\mathrm{me}}$ is the estimated position of (g_d, g_q) with respect to the stationary α-axis. It is possible to demonstrate that, similar to (10.2), the resulting carrier current space vector is composed of two components, relative to a positive and negative sequence currents, both of them containing the position information [18].

In general, pulsating carrier observers accomplish an accurate position estimate, but they are less robust and have worse transient response than those based on rotating carrier. Furthermore, the injection of an oscillating signal on the (estimated) flux axis minimises the torque ripple and the related vibration and audible noise [23]. On the other hand, the rotating HF injection is claimed to be effective also at zero speed under load conditions [1]. Another difference is recognisable in that the measured signals provided by the injection in the synchronous reference frame are modulated by the carrier frequency while in the stationary injection methods they are already demodulated around the DC level. Actually, which of the two carrier excitation typologies is preferable has been a matter of discussion for several years [18], even if the HF injection with pulsating voltage vector seems more used than the rotating voltage vector [1]. A detailed comparison of these two approaches has been presented in [24], which analyses the effects of multiple saliencies, cross-saturation, resistive effects and inverter-induced voltage distortions. It is worth to point out that HF injection-based sensorless techniques have reached the scientific maturity, and now the focus is on the best implementation practice.

10.2.3 A HF sensorless technique for IPM and PMASR motors

Besides the injection of a pulsating carrier in the stationary reference frame and of a rotating carrier in the synchronous reference frame, the third way to HF sensorless control is the injection of a HF pulsating carrier into the stationary reference frame, as proposed by Liu and Zhu [14]. The technique is accurate as a rotating carrier method but the position is retrieved from the amplitude-modulated carrier current response, which is simple as the pulsating carrier method. The technique is based on a HF magnetic model of the motor that includes the cross-coupling between the orthogonal axes in the synchronous reference frame, represented by a cross-coupling differential inductance L^{h}_{dq} in the HF inductance matrix $\mathbf{L}^{\mathrm{h}}_{dq}$ [24]. Furthermore, the HF model neglects the speed-related terms, provided that the frequency of the injected voltage is sufficiently higher than the fundamental excitation. Under those hypotheses, a α-axis pulsating voltage excitation

$$\mathbf{u}^{\mathrm{h}}_{\alpha\beta} = U_{\mathrm{h}} \begin{bmatrix} \cos(\omega_{\mathrm{h}} t) \\ 0 \end{bmatrix} \tag{10.4}$$

produces the HF carrier current response

$$\mathbf{i}^{\mathrm{h}}_{\alpha\beta} = \begin{bmatrix} I_{\mathrm{p}} + I_{\mathrm{n}}\cos(2\vartheta_{\mathrm{me}} + \vartheta_{\mathrm{m}}) \\ I_{\mathrm{n}}\sin(2\vartheta_{\mathrm{me}} + \vartheta_{\mathrm{m}}) \end{bmatrix} \sin(\omega_{\mathrm{h}} t) \tag{10.5}$$

where $I_{\mathrm{p}} = U_{\mathrm{h}}/(\omega_{\mathrm{h}} L_{\mathrm{p}})$, $I_{\mathrm{n}} = U_{\mathrm{h}}/(\omega_{\mathrm{h}} L_{\mathrm{n}})$ and ϑ_{m}, L_{p}, L_{n} are scalar functions of the elements of the HF inductance matrix $\mathbf{L}^{\mathrm{h}}_{dq}$. The HF currents are demodulated by a multiplication by 2sin $(\omega_{\mathrm{h}} t)$ followed by a low-pass filter (LPF), so that

$$\mathbf{i}^{\mathrm{h\,LPF}}_{\alpha\beta} = \begin{bmatrix} I_{\mathrm{p}} + I_{\mathrm{n}}\cos(2\vartheta_{\mathrm{me}} + \vartheta_{\mathrm{m}}) \\ I_{\mathrm{n}}\sin(2\vartheta_{\mathrm{me}} + \vartheta_{\mathrm{m}}) \end{bmatrix} \tag{10.6}$$

The retrieval of the position information is relatively simple, once I_p and ϑ_m are pre-detected and eliminated from (10.6). The mechanism of rotor position detection deserves careful attention. The conventional inverse-tangent function is substituted by a two-phase-type phase-locked loop, which is effective at low speed, but that entails the risk of divergence after abrupt speed changes, that must be avoided.

The experimental results reported in Reference 14 show an excellent precision and good dynamic performance under different load conditions. When evaluated by the assessment grid outlined in Section 10.2.1, it is easy to see that the method does not require additional hardware and that the computational requirement is not excessive. Both I_p and ϑ_m depend on motor parameters, but they can be derived from some preliminary measurements, that could be even substituted by a proper finite element analysis (FEA) model of the motor. Also, a complete theoretical analysis is somewhat missing, while the magnetic polarity detection, a classic ambiguity in every reluctance-based technique, is solved by a smart start-up procedure. In synthesis, the technique well represents a good starting point for next development in the specific sector of HF injection methods. Some issues are still open to improvements, but the more and more common availability of FEA models and online parameter estimation techniques [20] should rapidly complete the task. For example, Reference 16 presents a variant of the method, based on injection of square-wave pulsating signal. The bandwidth of the position estimation is significantly improved, and the influence of the rotor resistance, which can cause problems, as shown by Ghazi Moghadam and Tahami [17], is further reduced with respect to the case of sinusoidal HF injection.

10.2.4 A HF injection method for surface-mounted permanent magnet motors

As a matter of fact, isotropic PM motors still represent the largest share in industrial applications. Therefore, even if the trend is towards new PM motor topologies that embed sensorless-oriented features, the study of techniques that extend their validity to surface-mounted permanent magnet (SPM) motors remains a meaningful task. As regards the most suitable method, the HF injection using pulsating voltage vector seems to yield better results for SPM motors, where the machine saliency is given by the magnetic saturation and not by the rotor anisotropy [1].

An interesting example is given by Zaim *et al.* [15]. The estimation of the rotor position is obtained by injecting a voltage pulse u_{gd} along the direct axis of a *tentative* synchronous rotating reference frame (g_d, g_q). It is supposed that the g_d-axis is aligned with the real d-axis at the start-up. The effect of the current adds up to that of the PM, saturating the stator iron and creating a virtual anisotropy. In presence of a mismatch φ between the real and the tentative reference frames, the anisotropy creates a cross-coupling between the tentative axes, so that the injected voltage pulse (along g_d) generates a current i_{gq} along the axis g_q. A simple PI controller is then used to eliminate the mismatch. The domain of convergence and the choice of the PI constants are well discussed in [15]. The study also determines the minimum virtual saliency $(L_q - L_d)/L_q$ required to satisfy the observability conditions on the rotor position. The availability of a FEA model of the motor allowed the computation of the practical values of the amplitude u_{gd} and of the related application time T_{gd} to produce the necessary virtual saliency and the position estimation as well.

According to the check points listed in Section 10.2.1, the technique has light hardware and computational requirements, little tuning complexity, while it reveals a quite heavy dependence on the motor structure, due to need of knowing the behaviour of the virtual anisotropy as function of the injected currents. As regards the completeness of information, the analysis of the dynamics and its link to the frequency of application of the voltage pulse along the g_d-axis would deserve some more deepening.

For the sake of completeness, it is worth to mention the work of Yang and Lorenz [27]. It is an example of a different innovative point of view. It presents a synchronous-frame HF model of the SPM motor in which the eddy currents induced in both the stator and the rotor are represented by means of a couple of additional stator resistances along the (*d*, *q*) axes. The thesis is that the injection of a pulsating HF signal as (10.3) can empower the intrinsic asymmetry of the two resistances, enabling the identification of the rotor position. An in-depth comparison with injection methods, highlighting pros and cons of both, is carried out in [28].

10.2.5 A glance to extended EMF-based methods

The sensorless methods based on spatial inductance variations are suitable at low speed and standstill, while at high speed they generally suffer the effects of back-EMF harmonics and voltage limitation. They also introduce detrimental effects as additional noise and losses. When required by the application, the full operating speed range is obtained by considering the addition of extended EMF-based techniques, which are complementary as regards speed range [29].

In recent times, in correspondence to an increasing interest on alternative PM rotor structures, the concept of back-EMF has been extended to include any possible anisotropy effect. Because the resulting e-EMF model is of general validity for all synchronous motors including SPM, IPM and PMASR types, the study of unified sensorless control techniques valid for all the synchronous motors at high speeds becomes possible [9,10,20].

Adopting the space vector notation, the voltage balance equation can be written highlighting the symmetrical voltage components, and both the terms due to the PM and the anisotropy:

$$\mathbf{u}_{dq} = R\mathbf{i}_{dq} + L_d \frac{\mathrm{d}\mathbf{i}_{dq}}{dt} + \omega_{\mathrm{me}} L_q \begin{bmatrix} 0 & 1 \\ -1 & 0 \end{bmatrix} \mathbf{i}_{dq} + \mathbf{e}_{dq}^{\mathrm{ext}} \tag{10.7}$$

where the e-EMF vector $\mathbf{e}_{dq}^{\mathrm{ext}}$ is given by:

$$\mathbf{e}_{dq}^{\mathrm{ext}} = \mathbf{e}_{dq} + \begin{bmatrix} 0 \\ (L_d - L_q)\left(\omega_{\mathrm{me}} i_d - \frac{\mathrm{d}i_q}{dt}\right) \end{bmatrix} \tag{10.8}$$

The same model can be easily back-transformed into the stationary reference frame, while maintaining the synchronous inductances among the parameters. By applying the inverse Park transformation to (10.7) the voltage balance becomes:

$$\mathbf{u}_{\alpha\beta} = R\mathbf{i}_{\alpha\beta} + L_d \frac{\mathbf{d}\mathbf{i}_{\alpha\beta}}{dt} + \omega_{\mathrm{me}}(L_d - L_q)\begin{bmatrix} 0 & -1 \\ 1 & 0 \end{bmatrix}\mathbf{i}_{\alpha\beta} + \mathbf{e}_{\alpha\beta}^{\mathrm{ext}} \tag{10.9}$$

where

$$\mathbf{e}_{\alpha\beta}^{\mathrm{ext}} = \left((L_d - L_q)\left(\omega_{\mathrm{me}} i_d - \frac{di_q}{dt}\right) + \Lambda_{\mathrm{m}}\omega_{\mathrm{me}}\right)\begin{bmatrix} -\sin(\vartheta_{\mathrm{me}}) \\ \cos(\vartheta_{\mathrm{me}}) \end{bmatrix} \tag{10.10}$$

Similarly to the HF injection methods, the e-EMF-based techniques are well established. Effective implementations of the most conventional schemes are extensively reported in the literature [30], and now the research is focusing on the relevant aspects that still afflict the performances. Signally, some open issues are (a) the presence of LPFs in the estimation chain; (b) the extraction of the position and the speed from the e-EMF quantity and (c) the dependence on the motor parameters.

The e-EMF vector is traditionally obtained by either a full or a reduced-order observer. The observer may include the inverse magnetic model, so that the real PM motor acts as reference model, while the adaptive model consists of the voltage model (that computes the flux linkages) and the inverse magnetic model, which returns the stator currents. The error between the estimated currents and the measured ones is returned to the adaptive model to correct the estimated flux components. At the light of the evaluation scheme of Section 10.2.1, the reduced-order observers are simpler and easier to tune than the full-order observers. On the other hand, some studies indicate that the latter can be made less sensitive to model parameter uncertainties and noise, with good dynamic performance. A comprehensive analysis and comparison were carried out by Hinkkanen *et al.* [7,21].

10.2.6 Sliding mode observers for the extended EMF

A modern trend is the substitution of the conventional observer with a sliding mode observer (SMO) to get the back-EMF signals. This is testified by many recent publications [9,31–33] that highlight the advantages in terms of reduced order, simple implementation and inherent robustness against parameter variations. In a conventional application to SPM motors, the sliding surface is selected as

$$\sigma = \hat{\mathbf{i}}_{\alpha\beta} - \mathbf{i}_{\alpha\beta} = 0 \tag{10.11}$$

where $\mathbf{i}_{\alpha\beta}$ is the stator current vector in the stationary reference frame and $\hat{\mathbf{i}}_{\alpha\beta}$ is its estimated value. The SMO vector equation is

$$L_s \frac{\mathbf{d}\mathbf{i}_{\alpha\beta}}{dt} = \mathbf{u}_{\alpha\beta} - R_s\hat{\mathbf{i}}_{\alpha\beta} - k\Gamma(\hat{\mathbf{i}}_{\alpha\beta} - \mathbf{i}_{\alpha\beta}) \tag{10.12}$$

where k is the observer gain. In principle, $\Gamma(\mathbf{x})$ should be the sign function, but the trend is to use a sigmoid function to reduce the chattering phenomenon:

$$\Gamma(\mathbf{x}) = \frac{2}{1 + e^{-a\mathbf{x}}} - 1 \tag{10.13}$$

and a is a variable parameter that is used to change the sigmoid profile, for example according to speed [32,33]. Based on the *equivalent control* method, when the system on the sliding surface (10.11) comes true, and an estimation of the e-EMF is obtained as

$$\hat{\mathbf{e}}^{\text{ext}}_{\alpha\beta} = k\Gamma(\hat{\mathbf{i}}_{\alpha\beta} - \mathbf{i}_{\alpha\beta}) \tag{10.14}$$

that contains the information about the rotor position ϑ_{me}, as shown by (10.10). The continuous sigmoid function (10.13) has a more favourable harmonic content with respect to the sharp sign function, and this eases the extraction of the low-frequency EMF signal from (10.14). As an example of contribution in this topic, Zhao *et al.* [9] proposed an adaptive line enhancer to filter the estimated e-EMF without introducing any phase delay between the original and filtered e-EMF components.

In case of IPM or PMASR motors, the expression of the e-EMF is much more complex with respect to a SPM motor, since the magnitude of the e-EMF is a function of rotor speed, stator current and derivative of stator current, which means that the vector is load dependent, as shown by (10.10). High-speed and heavy loads applications can exhibit a relevant distortion of the e-EMF. On the other hand, the use of IPM and PMASR motors is destined to grow, and very likely, this will encourage the study of suitable EMF observer structures and of lag-less speed and position estimators. A first good example is given by Zhao *et al.* [34] that develop an e-EMF-based quasi-SMO in the discrete-time domain, proposing an online parameters adaptation mechanism. The result is a rotor position observer that is highly robust to load variations. The paper accomplishes most of the requirements listed in Section 10.2.1 as regards completeness of analysis, innovation and inclusion of several technical aspects. The computational requirement is not a hindrance, especially in view of the probable availability of low-cost field programmable gate array (FPGA) and floating point micro-controllers in the near future.

10.3 Trends in MPC of PMSM

In order to meet the continuous demand for higher performance drives with increasingly faster responses, the trend has been to progressively evolve towards more centralised control structures, with either fewer nested control loops, as in the case of the direct current control and DTC schemes, or even a single feedback with a fully centralised multi-variable controller. The adoption of more advanced control techniques has eased this process, and in this sense, the MPC approach has revealed to be the most promising candidate for current and future developments.

10.3.1 Key factors for MPC

In MPC, a mathematical model of the controlled system is used to predict its future evolution over a receding horizon of finite length. The control sequence to apply at any instant is determined according to optimal criteria, formulated as the maximisation or minimisation of a certain cost function that depends on the system evolution over the prediction horizon. Constraints on the state and input variables can be included in the optimisation problem, so that the feasibility of the control action is automatically guaranteed by design. This is one of salient features that differentiate MPC from other control techniques, where the constraints satisfaction is typically managed after the completion of the control design. Moreover, MPC is inherently oriented to the design of multi-variable control laws with a centralised architecture. Unfortunately, these advantages come at the price of an increased computational effort [35,36], since in principle the controller output must be obtained by solving online, at every sampling instant, the constrained optimisation problem underlying the MPC formulation. This fact imposes some limitations on both the model complexity and the prediction horizon length that can be effectively considered in a practical MPC implementation. Anyway, the advent of faster microcontrollers and FPGA [37] will undoubtedly improve the situation. The research in MPC is very active, and new fundamental results are expected from the advancement of both the theoretical foundations, and the methods used for its design, analysis and implementation. The potentialities of a new contribution should be evaluated according to the following aspects, in addition to those listed in Section 10.2.1:

- easy formulation and management of soft and hard constraints;
- possibility of inclusion of limits and constraints in the design phase;
- limited use of approximated linear models;
- direct control of the inverter switching states, bypassing the modulator.

10.3.2 Direct torque and flux control

Although there is still no common agreement in the terminology, the DTC conventionally identifies any control scheme where the torque and flux are controlled in closed loop, by acting directly on the inverter switching commands, and without resorting to any intermediate current control loop in order to speed up the torque response.

In *nonlinear* DTC schemes, the first to be introduced in the mid-1980s, bang–bang controllers based on hysteresis comparators are used to keep the torque and flux regulation errors within specified error bands. Several efforts have been made over time to improve the overall performances of the basic scheme, trying to overcome its intrinsic issues, such as the increased torque and current ripples, the variable frequency of the switching commands and the difficulty to operate at low-speed regimes. Research is still active nowadays, especially in the context of brushless DC motors, where the major concern still remains the reduction of the torque ripple [38,39].

In *linear* DTC schemes, the torque and possibly also the stator flux are controlled by conventional PI controllers that calculate, at any instant, the required average voltage vector to be applied over a fixed sampling period. The voltage vector is then synthesised by a pulse-width modulator (PWM), which in most cases exploits the SVM technique. A recent contribution to the design of linear DTCs for IPM motors is reported in [40]. The solution comprises both torque and flux limiters, so that the operation in the flux-weakening region can be easily established, as demonstrated in the complementary paper [41].

10.3.3 An MPC method for IPM motors

Future advancements in the field of DTC are expected by the gradual replacement of conventional linear and nonlinear schemes with more advanced control techniques, among which MPC represents the most promising candidate. A recent contribution in this sense is given by Preindl and Bolognani [42,43], and sketched in Figure 10.4.

It consists of a DTC for IPM motors based on a finite control set MPC, which is designed for the optimal torque generation within the current and voltage limitations imposed by the drive. The control input set is constituted by all the admissible switching states of a two-level, three-phase voltage source inverter (operating in three-phase conduction mode). Accordingly, the cost function of the MPC problem is formulated as the sum of a term $c_T(i_d, i_q)$ that penalises any deviation from the specified torque reference τ_{m}^*, i.e.

$$c_T(i_d, i_q) = \left[\tau_{\mathrm{m}}(i_d, i_q) - \tau_{\mathrm{m}}^*\right]^2 \tag{10.15}$$

and a term $c_A(i_d, i_q)$ that penalises any solution violating the optimal operating condition for the given motor speed and torque request. Below the motor base speed, when operations are only limited by the drive rated current I_N, i.e.

$$g_i(i_d, i_q) = \sqrt{i_d^2 + i_q^2} - I_N \leq 0 \tag{10.16}$$

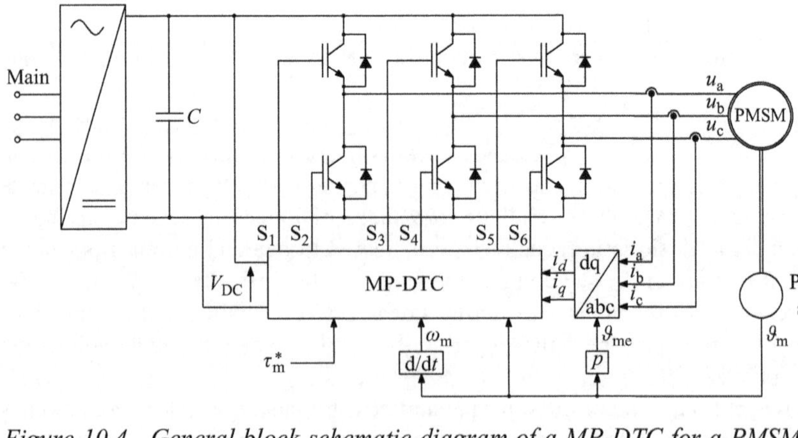

Figure 10.4 General block schematic diagram of a MP-DTC for a PMSM drive

a maximum-torque-per-ampere (MTPA) criterion is adopted to get minimum copper losses. Accordingly, the cost term $c_{A,1}$ is chosen equal to

$$c_{A,1}(i_d, i_q) = \left(i_d + \frac{L_d - L_q}{\Lambda_{\mathrm{mg}}} \left(i_d^2 - i_q^2 \right) \right)^2 \tag{10.17}$$

At higher speeds, the MTPA criteria can be satisfied up to the torque request that violates the constraint imposed by the drive rated voltage U_N, i.e.

$$g_v(i_d, i_q) = \sqrt{\left(\frac{L_q}{L_d} i_q \right)^2 + \left(i_d + \frac{\Lambda_{\mathrm{mg}}}{L_d} \right)^2} - \frac{U_N}{L_d \omega_{\mathrm{me}}} \leq 0 \tag{10.18}$$

To allow operation in such flux-weakening regime, the extra cost

$$c_{A,2}(i_d, i_q) = g_v^2(i_d, i_q) \tag{10.19}$$

is introduced to penalise the solutions off the boundary of (10.18). For sufficiently large motor speeds, the MTPA operation becomes incompatible with the voltage limitation, and therefore the optimal operating points, where a certain torque request is satisfied within the voltage constraints with the minimum amount of current, are always selected on the maximum voltage contour line, up to the point where the maximum torque is delivered, on the maximum-torque-per-voltage (MTPV) characteristic.

The operational constraints (10.16) and (10.18) are included in the MPC formulation by introducing two barrier functions (different from the simple *cost* functions) $c_{L,1}(i_d, i_q)$ and $c_{L,2}(i_d, i_q)$. An extra barrier function $c_{L,3}(i_d, i_q)$ is introduced to avoid solutions located beyond the MTPV curve in flux-weakening regime [42,43]. Finally, an overall cost function c is defined as the weighted sum of all the cost and barrier functions cited above.

Since the control set is finite, the resulting unconstrained optimisation problem is solved by evaluating the cost c for all the admissible control sequences over the prediction horizon, and then by selecting the one yielding the minimum cost.

The above analysis highlights the great potentiality of MPC control, in terms of management of constraints and multi-target optimisation. These factors appoint MPC as one of the most promising techniques for the near future, as further detailed in Section 10.3.4.

10.3.4 A perspective on centralised MPC-based structures

The on-board computational power in most of the current electric drives is still insufficient for running a full-order MPC at the sampling rates (order of tens of kilohertz) usually required to get satisfactory control performances. The trend that can be noticed in recent contributions is either to find a closed form solution of the MPC problem [44,45], or to reduce the number of admissible control values involved in the online optimisation process [46,47].

In [44], it is proposed a continuous control set nonlinear predictive control scheme for the simultaneous speed and d-current regulation in PMSMs. The design is based on a novel predictive output tracking control technique for nonlinear systems developed by the authors. The technique consists of approximating the evolution of the system output over the prediction horizon with its Taylor's expansion with respect to the time variable. The approximation is used to reformulate the optimal output-tracking problem, specified over the prediction horizon, as a parametric optimisation problem for the control input and its derivatives. Since the solution of the problem can be specified in closed form, it follows that no online optimisation is indeed required, and an explicit analytical expression (depending on the system state and input disturbance) of the optimal nonlinear predictive control law can be provided. With respect to the general evaluation criteria outlined in Section 10.3.1, a point that surely needs attention is the inclusion of the constraints, which at present is missing.

A promising approach seems to be that proposed in [46,47], which is based on a finite control set MPC similar to that adopted in [42,43]. The two contributions differ mainly on how the constrained are managed: while in [46] the constraints are converted into penalty terms to be added to the cost function, in [47] the constraints are used to select beforehand the feasible control sequences that will be taken into account by the optimisation process. Moreover, the cost function in [47] has a hierarchical structure, which allows to set priorities in the control goals. Some efforts are still required to solve the two main problems that undermine the applicability of the MPC, namely

- the increased torque ripple compared to the use of a voltage modulator;
- the limited prediction horizon achievable in practice with current technology.

The first issue is fixed in [46], by adding an extra term to the cost function, that weights a high-pass filtered version of the torque command, while in [47] it is suggested to increase the number of voltage vectors in the control set. The second problem, instead, still requires further investigation, even though it is believed that it will be easily overcome with the advent of more powerful computational units.

10.4 Some hints about energy efficiency in PMSM drives

With the ever increasing costs of electrical energy, the aspects related to the energy conversion efficiency are crucial in the design of modern high-performance drives. The improvement of energy utilisation has also a significant impact on reducing global pollution and warming, and thus plays an important role in the emerging field of green economy. Especially for small and medium power variable-speed applications, PMSMs represent the ideal choice, because of their inherent high efficiency over the full speed and load torque ranges. Predicted values of efficiency are about 95%–97% for a 10–100 kW PMSM. Over the past years, thanks to the ever decreasing cost of variable frequency AC power supplies enabled by the remarkable advances in power switching devices and digital control units, this improvement has been mostly achieved by the introduction of adjustable speed drives in applications where constant speed drives were generally used, e.g. pumps, fans, air conditioners, heat pumps, etc.

The major losses in PMSMs occur on the stator, either as joule losses on the stator windings or as core losses in the stator teeth and yoke. Apart from machine design considerations, the stator losses, being controllable, can be reduced by a proper loss minimisation control strategy. This is especially true for IPM and PMASR motors that present a degree of freedom in the composition of the electromagnetic torque. These motors are becoming increasingly popular in traction drives for electric and hybrid vehicles [48], owing especially to their capability of operating over a wide speed range, with a constant power speed range up to five times the base speed, and under intermittent overloading conditions, with up to almost twice the rated torque, at least for short durations [49]. Below the base speed, copper losses can be effectively reduced by a MTPA control strategy, which is a common trend. Available methods are either model-based, comprising parameter adaptation for enhanced robustness and capability to cope with magnetic saturation effects [50], or model-free, mostly perturbation-and-observe methods, based on the extremum seeking control working principle [51].

From a close examination of the solutions proposed so far in the literature concerning the efficient control of PMSMs, it can be noticed that most of them are still focused exclusively on optimising the operating condition of the machine, without concerning about the additional switching and harmonic losses produced by the almost ever-present PWM voltage source inverter. This habit is likely to change, since it may be a source of misunderstanding, for example in the new European Union regulation.

It is worth to note that, in general, the optimal operating conditions of the motor and the inverter do not coincide, and a trade-off must be found. In this sense, better results are expected if the efficiency optimisation is performed at a "system level", in which the tight interaction between the motor and inverter is explicitly taken into account. By proceeding in this way, it becomes also easier to consider additional optimisation objectives (multi-objective optimisation), e.g. the minimisation of the number of inverter switching [52], and operational constraints, e.g. the thermal limits of the power components (power switches and freewheeling diodes) in the inverter [53]. It is believed that even under the energy efficiency perspective, the advent of smarter drives, with more computational power on-board, will help to popularise an integrated approach to optimisation, especially when embedded within a more general centralised MPC-based architecture.

10.5 Final considerations

The use of PMSMs is widespread and the related power electronic control has reached a good maturity. New hybrid (reluctance and PM) motor topologies are studied to provide motors with improved efficiency and self-sensing features. The use of these motors complicates the control algorithms and makes the availability of complete and reliable magnetic model unavoidable. Therefore, the modelling of complex system will probably take a large part of research activity in the field. Further technology steps will be the use of more powerful computing devices, as FPGAs, that in turn will smooth the way for both the direct voltage measurement and a more sophisticated signals filtering. Finally, the emerging energy awareness is going to draw new paradigms, based on multi-variable control, for the contemporary optimisation of different power electronics and drive aspects. In perspective, the best candidate is the MPC, as soon as its online implementation will become technically feasible.

Acknowledgements

The author gratefully acknowledges the support of Dr Riccardo Antonello, who generously helped me in the literature review and in the preparation of some parts of the work.

References

[1] R. Bojoi, M. Pastorelli, J. Bottomley, P. Giangrande, and C. Gerada, "Sensorless control of PM motor drives: A technology status review," in *Electrical Machines Design Control and Diagnosis (WEMDCD), 2013 IEEE Workshop on*, March 2013, pp. 168–182.

[2] K. Rajashekara, "Present status and future trends in electric vehicle propulsion technologies," *Emerging and Selected Topics in Power Electronics, IEEE Journal of*, vol. 1, no. 1, pp. 3–10, March 2013.

[3] F. Betin, G.-A. Capolino, D. Casadei, B. Kawkabani, R. Bojoi, L. Harnefors, E. Levi, L. Parsa, and B. Fahimi, "Trends in electrical machines control: Samples for classical, sensorless, and fault-tolerant techniques," *Industrial Electronics Magazine, IEEE*, vol. 8, no. 2, pp. 43–55, June 2014.

[4] L. Peretti and M. Zigliotto, "FPGA-based voltage measurements in ac drives," in *Electrical Machines (ICEM), 2010 XIX International Conference on*, September 2010, pp. 1–6.

[5] S. Murakami, T. Shiota, M. Ohto, K. Ide, and M. Hisatsune, "Encoderless servo drive with adequately designed IPMSM for pulse-voltage-injection-based position detection," *Industry Applications, IEEE Transactions on*, vol. 48, no. 6, pp. 1922–1930, November 2012.

[6] A. Faggion, E. Fornasiero, N. Bianchi, and S. Bolognani, "Sensorless capability of fractional-slot surface-mounted pm motors," *Industry Applications, IEEE Transactions on*, vol. 49, no. 3, pp. 1325–1332, May 2013.

[7] T. Tuovinen, M. Hinkkanen, L. Harnefors, and J. Luomi, "Comparison of a reduced-order observer and a full-order observer for sensorless synchronous motor drives," *Industry Applications, IEEE Transactions on*, vol. 48, no. 6, pp. 1959–1967, November 2012.

[8] R. Leidhold, "Position sensorless control of PM synchronous motors based on zero-sequence carrier injection," *Industrial Electronics, IEEE Transactions on*, vol. 58, no. 12, pp. 5371–5379, December 2011.

[9] Y. Zhao, W. Qiao, and L. Wu, "Improved rotor position and speed estimators for sensorless control of interior permanent-magnet synchronous machines," *Emerging and Selected Topics in Power Electronics, IEEE Journal of*, vol. 2, no. 3, pp. 627–639, September 2014.

[10] K. Lu, X. Lei, and F. Blaabjerg, "Artificial inductance concept to compensate nonlinear inductance effects in the back EMF-based sensorless control method for PMSM," *Energy Conversion, IEEE Transactions on*, vol. 28, no. 3, pp. 593–600, September 2013.

[11] F. Genduso, R. Miceli, C. Rando, and G. Galluzzo, "Back EMF sensorless-control algorithm for high-dynamic performance PMSM," *Industrial Electronics, IEEE Transactions on*, vol. 57, no. 6, pp. 2092–2100, June 2010.

[12] S. Morimoto, K. Kawamoto, M. Sanada, and Y. Takeda, "Sensorless control strategy for salient-pole PMSM based on extended EMF in rotating reference frame," in *Industry Applications Conference, 2001. Thirty-Sixth IAS Annual Meeting. Conference Record of the 2001 IEEE*, vol. 4, September 2001, pp. 2637–2644.

[13] M. Pacas, "Sensorless drives in industrial applications," *Industrial Electronics Magazine, IEEE*, vol. 5, no. 2, pp. 16–23, June 2011.

[14] J. Liu and Z. Zhu, "Novel sensorless control strategy with injection of high-frequency pulsating carrier signal into stationary reference frame," *Industry Applications, IEEE Transactions on*, vol. 50, no. 4, pp. 2574–2583, July 2014.

[15] S. Zaim, B. Nahid-Mobarakeh, and F. Meibody-Tabar, "Robust position sensorless control of nonsalient PMSM at standstill and low speeds," *Emerging and Selected Topics in Power Electronics, IEEE Journal of*, vol. 2, no. 3, pp. 640–650, September 2014.

[16] J. Liu and Z. Zhu, "Sensorless control strategy by square-waveform high-frequency pulsating signal injection into stationary reference frame," *Emerging and Selected Topics in Power Electronics, IEEE Journal of*, vol. 2, no. 2, pp. 171–180, June 2014.

[17] M. Ghazi Moghadam and F. Tahami, "Sensorless control of PMSMs with tolerance for delays and stator resistance uncertainties," *Power Electronics, IEEE Transactions on*, vol. 28, no. 3, pp. 1391–1399, March 2013.

[18] A. Accetta, M. Cirrincione, M. Pucci, and G. Vitale, "Sensorless control of PMSM fractional horsepower drives by signal injection and neural adaptive-band filtering," *Industrial Electronics, IEEE Transactions on*, vol. 59, no. 3, pp. 1355–1366, March 2012.

[19] Z. Wang, K. Lu, and F. Blaabjerg, "A simple startup strategy based on current regulation for back-EMF-based sensorless control of PMSM," *Power Electronics, IEEE Transactions on*, vol. 27, no. 8, pp. 3817–3825, August 2012.

[20] Y. Inoue, Y. Kawaguchi, S. Morimoto, and M. Sanada, "Performance improvement of sensorless IPMSM drives in a low-speed region using online parameter identification," *Industry Applications, IEEE Transactions on*, vol. 47, no. 2, pp. 798–804, March 2011.

[21] M. Hinkkanen, T. Tuovinen, L. Harnefors, and J. Luomi, "A combined position and stator-resistance observer for salient PMSM drives: Design and stability analysis," *Power Electronics, IEEE Transactions on*, vol. 27, no. 2, pp. 601–609, February 2012.

[22] I. Boldea, "Control issues in adjustable speed drives," *Industrial Electronics Magazine, IEEE*, vol. 2, no. 3, pp. 32–50, September 2008.

[23] F. Briz and M. Degner, "Rotor position estimation," *Industrial Electronics Magazine, IEEE*, vol. 5, no. 2, pp. 24–36, June 2011.

[24] D. Raca, P. Garcia, D. Reigosa, F. Briz, and R. Lorenz, "Carrier-signal selection for sensorless control of PM synchronous machines at zero and very low speeds," *Industry Applications, IEEE Transactions on*, vol. 46, no. 1, pp. 167–178, January 2010.

[25] P. Guglielmi, M. Pastorelli, and A. Vagati, "Cross-saturation effects in IPM motors and related impact on sensorless control," *Industry Applications, IEEE Transactions on*, vol. 42, no. 6, pp. 1516–1522, November 2006.

[26] N. Bianchi and S. Bolognani, "Influence of rotor geometry of an IPM motor on sensorless control feasibility," *Industry Applications, IEEE Transactions on*, vol. 43, no. 1, pp. 87–96, January 2007.

[27] S.-C. Yang and R. Lorenz, "Surface permanent magnet synchronous machine position estimation at low speed using eddy-current-reflected asymmetric resistance," *Power Electronics, IEEE Transactions on*, vol. 27, no. 5, pp. 2595–2604, May 2012.

[28] S.-C. Yang and R. Lorenz, "Comparison of resistance-based and inductance-based self-sensing controls for surface permanent-magnet machines using high-frequency signal injection," *Industry Applications, IEEE Transactions on*, vol. 48, no. 3, pp. 977–986, May 2012.

[29] M. Corley and R. Lorenz, "Rotor position and velocity estimation for a salient-pole permanent magnet synchronous machine at standstill and high speeds," *Industry Applications, IEEE Transactions on*, vol. 34, no. 4, pp. 784–789, July 1998.

[30] S. Bolognani, S. Calligaro, and R. Petrella, "Design issues and estimation errors analysis of back-EMF-based position and speed observer for SPM synchronous motors," *Emerging and Selected Topics in Power Electronics, IEEE Journal of*, vol. 2, no. 2, pp. 159–170, June 2014.

[31] T. Bernardes, V. Foletto Montagner, H. Grundling, and H. Pinheiro, "Discrete-time sliding mode observer for sensorless vector control of permanent magnet synchronous machine," *Industrial Electronics, IEEE Transactions on*, vol. 61, no. 4, pp. 1679–1691, April 2014.

[32] Z. Qiao, T. Shi, Y. Wang, Y. Yan, C. Xia, and X. He, "New sliding-mode observer for position sensorless control of permanent-magnet synchronous motor," *Industrial Electronics, IEEE Transactions on*, vol. 60, no. 2, pp. 710–719, February 2013.

[33] H. Kim, J. Son, and J. Lee, "A high-speed sliding-mode observer for the sensorless speed control of a PMSM," *Industrial Electronics, IEEE Transactions on*, vol. 58, no. 9, pp. 4069–4077, September 2011.

[34] Y. Zhao, W. Qiao, and L. Wu, "An adaptive quasi-sliding-mode rotor position observer-based sensorless control for interior permanent magnet synchronous machines," *Power Electronics, IEEE Transactions on*, vol. 28, no. 12, pp. 5618–5629, December 2013.

[35] Z. Zhu, Y. Li, D. Howe, C. Bingham, and D. Stone, "Influence of machine topology and cross-coupling magnetic saturation on rotor position estimation accuracy in extended back-EMF-based sensorless PM brushless ac drives," in *Industry Applications Conference, 2007. 42nd IAS Annual Meeting. Conference Record of the 2007 IEEE*, September 2007, pp. 2378–2385.

[36] P. Cortes, M. Kazmierkowski, R. Kennel, D. Quevedo, and J. Rodriguez, "Predictive control in power electronics and drives," *Industrial Electronics, IEEE Transactions on*, vol. 55, no. 12, pp. 4312–4324, December 2008.

[37] M. W. Naouar, E. Monmasson, A. Naassani, I. Slama-Belkhodja, and N. Patin, "FPGA-based current controllers for ac machine drives – A review," *Industrial Electronics, IEEE Transactions on*, vol. 54, no. 4, pp. 1907–1925, August 2007.

[38] M. Masmoudi, B. El Badsi, and A. Masmoudi, "Direct torque control of brushless DC motor drives with improved reliability," *Industry Applications, IEEE Transactions on*, vol. 50, no. 6, pp. 3744–3753, November 2014.

[39] M. Masmoudi, B. El Badsi, and A. Masmoudi, "DTC of B4-inverter-fed BLDC motor drives with reduced torque ripple during sector-to-sector commutations," *Power Electronics, IEEE Transactions on*, vol. 29, no. 9, pp. 4855–4865, September 2014.

[40] Y. Inoue, S. Morimoto, and M. Sanada, "Control method suitable for direct-torque-control-based motor drive system satisfying voltage and current limitations," *Industry Applications, IEEE Transactions on*, vol. 48, no. 3, pp. 970–976, May 2012.

[41] Y. Inoue, S. Morimoto, and M. Sanada, "Comparative study of PMSM drive systems based on current control and direct torque control in flux-weakening control region," *Industry Applications, IEEE Transactions on*, vol. 48, no. 6, pp. 2382–2389, November 2012.

[42] M. Preindl and S. Bolognani, "Model predictive direct torque control with finite control set for PMSM drive systems. Part 1: Maximum torque per ampere operation," *Industrial Informatics, IEEE Transactions on*, vol. 9, no. 4, pp. 1912–1921, November 2013.

[43] M. Preindl and S. Bolognani, "Model predictive direct torque control with finite control set for PMSM drive systems. Part 2: Field weakening operation," *Industrial Informatics, IEEE Transactions on*, vol. 9, no. 2, pp. 648–657, May 2013.

[44] R. Errouissi, M. Ouhrouche, W.-H. Chen, and A. Trzynadlowski, "Robust nonlinear predictive controller for permanent-magnet synchronous motors with an optimized cost function," *Industrial Electronics, IEEE Transactions on*, vol. 59, no. 7, pp. 2849–2858, July 2012.

[45] R. Errouissi, M. Ouhrouche, W.-H. Chen, and A. Trzynadlowski, "Robust cascaded nonlinear predictive control of a permanent magnet synchronous motor with anti-windup compensator," *Industrial Electronics, IEEE Transactions on*, vol. 59, no. 8, pp. 3078–3088, August 2012.

[46] E. Fuentes, C. Silva, and J. Yuz, "Predictive speed control of a two-mass system driven by a permanent magnet synchronous motor," *Industrial Electronics, IEEE Transactions on*, vol. 59, no. 7, pp. 2840–2848, July 2012.

[47] M. Carraro, M. Zigliotto, and L. Peretti, "FPGA-based hierarchical finite-states predictive control for PMSM drives," in *Power Electronics, Machines and Drives (PEMD 2014), Seventh IET International Conference on*, April 2014, pp. 1–6.

[48] G. Pellegrino, A. Vagati, P. Guglielmi, and B. Boazzo, "Performance comparison between surface-mounted and interior PM motor drives for electric vehicle application," *Industrial Electronics, IEEE Transactions on*, vol. 59, no. 2, pp. 803–811, February 2012.

[49] G. Pellegrino, A. Vagati, and P. Guglielmi, "Design tradeoffs between constant power speed range, uncontrolled generator operation, and rated current of IPM motor drives," *Industry Applications, IEEE Transactions on*, vol. 47, no. 5, pp. 1995–2003, September 2011.

[50] Q. K. Nguyen, M. Petrich, and J. Roth-Stielow, "Implementation of the MTPA and MTPV control with on-line parameter identification for a high speed IPMSM used as traction drive," in *Power Electronics Conference (IPEC-Hiroshima 2014 – ECCE-ASIA), 2014 International*, May 2014, pp. 318–323.

[51] R. Antonello, M. Carraro, and M. Zigliotto, "Maximum-torque-per-ampere operation of anisotropic synchronous permanent-magnet motors based on extremum seeking control," *Industrial Electronics, IEEE Transactions on*, vol. 61, no. 9, pp. 5086–5093, September 2014.

[52] G. Prior and M. Krstic, "Quantized-input control Lyapunov approach for permanent magnet synchronous motor drives," *Control Systems Technology, IEEE Transactions on*, vol. 21, no. 5, pp. 1784–1794, September 2013.

[53] J. Lemmens, P. Vanassche, and J. Driesen, "Optimal control of traction motor drives under electrothermal constraints," *Emerging and Selected Topics in Power Electronics, IEEE Journal of*, vol. 2, no. 2, pp. 249–263, June 2014.

Chapter 11
Induction motor drives

*Saeed Jafarzadeh**

11.1 Induction motors

With more than 85% of electrical motors, induction motors (IMs) dominate the market for electrical motors and consume more than 60% of total industrial electricity across the globe [1]. Their extensive use is a result of their strong low-cost design along with their reliability and cheap maintenance. IM popularity is not limited to certain applications and IMs are used in a variety of applications with different requirements and constraints. Therefore, control of IMs is regarded as an important field of study ever since introduced. IM drives aim for better efficiency and reliability for IM to achieve more economical operation and to help in energy saving.

When operating at variable speed, IM drives are great assets for energy saving purposes [2]. To achieve this, variable frequency drives are developed. According to the affinity laws, torque in centrifugal load applications changes relative to the square of speed. Power changes in proportion with the cube of speed and therefore, energy uses vary at different speeds. Variable frequency drives can manage energy consumption by wisely changing the speed. Besides, their continuous speed control is desirable for a variety of applications. In short, these drives provide energy efficiency, reliable dynamics, and cost efficiency for a variety of IM applications.

Since variable frequency drives are fundamentally operate by speed feedback, precision in speed measurement is a key in quality of their operation. Speed sensors are often used to measure speed and address such need. However, the IM speed can also be estimated by available estimation methodologies. Such estimation is highly desirable due to the drawbacks of speed sensors. These drawbacks affect the whole IM drive system by adding hardware complexity, requiring additional maintenance, and increasing cost. As a result, estimation of IM speed received considerable attention, in particular, where speed sensors perform poorly or fail.

Most IM speed estimation methodologies fall into one of the two categories. Some estimation techniques estimate IM speed by means of its model. These techniques use voltage and current feedback to estimate IM speed. These feedback signals are typically accurate enough for the estimation purposes. However, the

*California State University Bakersfield, Bakersfield, CA, USA

accuracy in IM speed estimation is highly affected by the accuracy of the model parameters. Therefore, to obtain accurate speed estimate, precise measurement on model parameters is a key. But even with the most accurate measurements, the estimation quality can be decreased by the changes in parameters due to the operation conditions and aging. Such deviations can even put the stability of the IM drive at risk. Therefore, fine parameter tuning must be included in the estimation algorithm to secure reliable speed estimation. The other category of IM speed estimation consists of techniques that use signal injection via exploiting the anisotropy of IM. These techniques address the problems associated with the changes in parameters. They also offer reliable operation as a result of improving the stability of the drive system. They use a carrier signal superimposed on the pulse width modulation (PWM) waveform to estimate rotor speed. The challenges with these techniques are frequency tracking, poor signal-to-noise ratio, and low spectral classification. Therefore, these techniques require advanced signal processing algorithms.

11.2 IM model

The mathematical description of IM is typically in terms of first-order differential equations. For simplicity, we assume stator reference frame to obtain the following differential equations:

$$\frac{\mathrm{d}\underline{\psi}_{\mathrm{s}}}{\mathrm{d}t} = -R_{\mathrm{s}}\underline{i}_{\mathrm{s}} + \underline{u}_{\mathrm{s}}$$

$$\frac{\mathrm{d}\underline{i}_{\mathrm{s}}}{\mathrm{d}t} = -\left(\frac{1}{T_{\mathrm{r}}\sigma} + \frac{1}{T_{\mathrm{s}}\sigma} - j\omega_{\mathrm{r}}\right)\underline{i}_{\mathrm{s}} + \frac{1}{L_{\mathrm{s}}\sigma}\left(\frac{1}{T_{\mathrm{r}}} - j\omega_{\mathrm{r}}\right)\underline{\psi}_{\mathrm{s}} + \frac{1}{L_{\mathrm{s}}\sigma}\underline{\psi}_{\mathrm{s}}$$

where $\underline{\psi}_{\mathrm{s}}$ is a stator flux space vector, $\underline{i}_{\mathrm{s}}$ is a stator current vector, R_{s} and R_{r} are the stator and rotor resistances, L_{s} and L_{r} are the stator and rotor inductances, and L_{m} is the magnetizing inductance. $T_{\mathrm{s}} = \frac{L_{\mathrm{s}}}{R_{\mathrm{s}}}$, $T_{\mathrm{r}} = \frac{L_{\mathrm{r}}}{R_{\mathrm{r}}}$, $\sigma = \frac{L_{\mathrm{s}}L_{\mathrm{r}} - L_{\mathrm{m}}{}^{2}}{L_{\mathrm{s}}L_{\mathrm{r}}}$, ω_{r} is the rotor electrical speed, and $\underline{u}_{\mathrm{s}} = u_{\mathrm{s}\alpha} + ju_{\mathrm{s}\beta}$ is the stator voltage vector, which is the system input. Each space vector represents two states for the IM. In addition to the four resulting states, the rotor electrical speed is another changing state of the IM. Therefore, another differential equation is needed to describe the dynamics of the speed and how it interacts with other states. Researchers employed two main approaches to model the rotor electrical speed. A simplistic approach will assume steady speed and therefore neglect its changes in the mathematical model, which results in a simple differential equation in the following form:

$$\frac{\mathrm{d}\omega_{\mathrm{r}}}{\mathrm{d}t} = 0$$

Another approach uses speed dynamics to obtain

$$\frac{\mathrm{d}\omega_{\mathrm{r}}}{\mathrm{d}t} = \frac{3}{2J}p^{2}\left(\psi_{\mathrm{s}\alpha}i_{\mathrm{s}\beta} - \psi_{\mathrm{s}\beta}i_{\mathrm{s}\alpha}\right) - \frac{T_{\mathrm{L}}}{J}p$$

$$\frac{\mathrm{d}T_{\mathrm{L}}}{\mathrm{d}t} = 0$$

where p is the pole pairs, J is the inertia, and T_L is the load torque. In this case, the load torque is used as another state and assumed constant [3].

These differential equations can be restated as the state space model and utilized for sensorless speed control of IM drives. The difficulties in designing such control arise due to the complexities in the system. From its mathematical model, it can be seen that IM is nonlinear. In addition, changes in its parameters make it a time-varying system. Finally, load torque can be seen as perturbation, which makes the system more complicated. High performance control and estimation techniques for IM drives received much attention due to such complexities. In terms of feedback signals, input currents and voltages are measured for use in control loop. Speed is another measurement quantity, which is skipped in sensorless IM drives.

11.3 Variable frequency drives

The variable frequency drives for IM are classified into three main categories, namely the scalar control, field-oriented control, and direct torque control (DTC). Here, we give a general discussion on each category and then discuss the latter in further detail.

11.3.1 Scalar control

As a simple control technique, scalar control is used for IM drives. It controls the speed in IM by changing the magnitude and frequency of the input voltage [4]. This approach is developed using the steady-state equivalent circuit of the IM. The idea in developing such control is to keep the magnetizing current unchanged by means of keeping the ratio of the voltage magnitude to the frequency unchanged. One popular example of such control is known as the constant volts per hertz. Speed is often measured as a feedback in scalar control to create a closed-loop control system. Without the speed feedback, the resulting open-loop control will have limited performance, which only suits certain applications where load torque is almost constant and quality performance is not required.

11.3.2 Field-oriented control

Unlike scalar control, field-oriented control uses dynamic model of the IM where the voltages, currents, and fluxes are expressed in space vector forms [5]. Since the IM is described by differential equations, the model accounts for both steady state and transient dynamics of the IM. Therefore, field-oriented control can achieve excellent performance in transient and steady-state conditions.

In the rotating rotor flux frame, quantities rotating at synchronous speed appear as DC quantities. With the flux aligned to the d axis of the reference frame, d component of the stator current represents the flux and q component of the stator current represents the torque. Therefore, the control of IM is reduced to a simple control scheme. The simplicity is resulted from the fact that torque and flux components are decoupled in the adopted reference frame. Two types of field-oriented control are defined based on the position of the rotor flux: indirect and direct. By adding the slip position to the measured rotor position, the flux position in indirect field-oriented control is obtained. The flux position in direct field-oriented control is calculated based on the terminal variables and rotor speed.

Since field-oriented control stems from a frame transformation that requires rotor speed, the knowledge of rotor position needs to be acquired accurately in order to perform such transformation. The accuracy of the rotor position estimate has a significant impact on the performance of field-oriented control. If such estimate is not accurate enough, the satisfactory level of decoupling of the torque and flux will not be achieved. Therefore, field-oriented control is desirable only if a good estimate of this quantity is available.

11.3.3 Direct torque control

DTC IM drives are extensively used due to their fast dynamic response and robustness [6]. Besides, they do not need current controllers and coordinate transformation, which make their implementation simple. In DTC scheme, torque and stator flux errors are calculated and used to generate the torque and stator flux control signals. Despite faster torque response, a ripple in torque is inevitable in DTC. Therefore, several variations of DTC are proposed to reduce such ripple to a minimal level while keeping reasonable IM drive performance. Due to the significance of DTC drives, further discussions on IM DTC drives are given in the following section.

11.4 DTC schemes

11.4.1 Space vector modulation DTC

To use space vector modulation (SVM) in DTC, flux speed will be adjusted by zero voltage vector insertion to control the generated electromagnetic torque by IM. With SVM DTC, flux and torque ripple can be avoided to a significant level. Several SVM DTC drives can be formulated and are briefly reviewed here.

One approach in DTC SVM is to use stator flux as the control variable. This approach uses the error between the reference and estimated stator flux to calculate the reference voltage vector. The advantages of this scheme are constant inverter switching frequency and regulated magnitude of the rotor flux. Moreover, additional torque overload is also possible. The main disadvantage of this scheme is its dependence on system parameters, which makes it sensitive to the variations in system parameters.

Another DTC SVM approach is based on control of the load angle. This type of control method was primarily proposed for permanent magnet synchronous motors. However, due to similarities between these motors and IM, this method can be applied for IM too. A proportional-integral (PI) controller is typically used to regulate the torque, where the torque is regulated by changing the angle between the rotor and stator fluxes. As a result of using a single PI controller, tuning task for this control scheme is reduced to two control parameters, which makes the control scheme simple and easy to tune.

Another control scheme combines the two earlier control schemes. In other words, in such scheme both the flux and torque are being controlled. Therefore, the scheme results in improvements that each scheme has to offer and hence, better IM drive performance. Another control scheme is discrete SVM. This scheme takes advantage of predefined time intervals in a period of cycles to reduce the switching frequency. With this approach, further voltage vectors can be synthesized. Therefore, an accurate switching table can be constructed in which voltage vector selection is accomplished according to rotor speed, flux, and torque error.

Discontinuous PWM techniques are based upon power electronic schemes. These methods are called discontinuous due to the use of zero sequence signal [7]. In each sampling time, one phase halts modulation and accompanied phase clamps to negative or positive DC bus. Therefore, the switching losses of the inverter leg involved are removed. The performance of PWM schemes depends on the modulation index. These control schemes perform better in the high modulation range. They have lower overall switching losses, when compared to other DTC control schemes.

11.4.2 Feedback linearization and sliding mode DTC

Feedback linearization (FBL) is a control systems engineering approach used for nonlinear systems. The main idea of FBL is to transform the nonlinear system into an equivalent linear system, design a simple linear controller for the linear system, and then use the inverse transformation to obtain the desired controller for the original nonlinear system [8]. However, because the method is sensitive to modeling errors and disturbances, it has been rarely applied to IM drives. The variable structure control (VSC) is a robust control technique well suited for control systems with uncertainties or perturbations. It has been successfully applied to IM drives and provides excellent performance for a wide speed range operation [9]. A new DTC controller can be designed using FBL together with VSC. The main advantage of FBL is that, one can apply the extensive knowledge of control systems engineering to design a controller for the linear model.

Conventional linearization of a nonlinear system is based on a first-order approximation of the system dynamics at an operating point, while neglecting higher order dynamics. This linearization is satisfactory in many applications where normal system operation remains in the vicinity of a fixed or slowly varying equilibrium, but it is otherwise inappropriate. Linearization is not adequate for variable speed drives, where the IM model is inherently nonlinear. The central idea of FBL is to algebraically transform a nonlinear system model into a linear one, so that linear control can be applied. In this approach, the linearization and the linear behavior are valid in a large subspace or globally, rather than in the vicinity of an equilibrium point. FBL of an IM is achievable by an intuitive transformation of the state variables and a redefinition of the inputs.

Consider the IM state space model in stator reference frame as given earlier. In order to linearize the model, we define the following new states:

$$M = \psi_{sq}\psi_{rd} - \psi_{sd}\psi_{rq}$$

$$R = \psi_{sd}\psi_{rd} + \psi_{sq}\psi_{rq}$$

$$F_s = \psi_{sd}^2 + \psi_{sq}^2$$

$$F_r = \psi_{rd}^2 + \psi_{rq}^2$$

where M is the scaled torque, F_s and F_r are the squared magnitudes of the stator and rotor flux, respectively. For simplicity, we refer M as the torque and F_s as the flux magnitude. We are primarily interested in controlling the torque M and the stator flux magnitude F_s. We must also insure that the remaining state variables, F_r and R, are bounded. We redefine the inputs as

$$w_q = -\omega_r R - \psi_{rq}u_{sd} + \psi_{rd}u_{sq}$$

$$w_d = \frac{2L_m}{L_r T_s \sigma} R + 2\left(\psi_{sd}u_{sd} + \psi_{sq}u_{sq}\right)$$

The IM model after variable transformation and FBL is given as

$$\frac{dM}{dt} = -\left(\frac{1}{T_r\sigma} + \frac{1}{T_s\sigma}\right)M + w_q$$

$$\frac{dF_s}{dt} = -\frac{2}{T_s\sigma}F_s + w_d$$

$$\frac{dF_r}{dt} = -\frac{2}{T_r\sigma}F_r + \frac{2L_m}{L_s T_r \sigma}R$$

$$\frac{dR}{dt} = -\left(\frac{1}{T_r\sigma} + \frac{1}{T_s\sigma}\right)R + \frac{L_m}{L_s T_r \sigma}F_s + \frac{F_r}{2R}w_d - \frac{M}{R}w_q$$

To achieve this linearized system, the following control signals should be implemented:

$$u_{sd} = \frac{\psi_{rd}}{2R}\left(w_d - \frac{2L_m}{L_r T_s \sigma}R\right) - \frac{\psi_{sq}}{R}\left(w_q + \omega_r R\right)$$

$$u_{sq} = \frac{\psi_{rq}}{2R}\left(w_d - \frac{2L_m}{L_r T_s \sigma}R\right) + \frac{\psi_{sd}}{R}\left(w_q + \omega_r R\right)$$

Since the M, F_s, and F_r have dynamics with poles located within the left-hand half-plane, the input–output stability can be easily guaranteed provided that R remains bounded and nonzero. This condition is always true except for the trivial situation when the stator or rotor flux is zero, i.e. at startup. In the drive implementation, provisions should be taken to ensure a smooth flux startup.

The control objective is to control the torque and stator flux magnitude in the machine, i.e. to realize a DTC type controller. To this end, we design controllers for the torque M and the stator flux F_s in the linearized model. Notice that the state equations governing M and F_s are decoupled, and the design of their controllers to obtain the inputs w_d and w_q is simple. Once the controller outputs w_d and w_q found, they will be used to obtain the physical inputs u_{sd} and u_{sq}. Errors in the calculation of the physical inputs are inevitable and must be accounted for and corrected for a robust performance.

Consider the torque dynamics in the linearized system as shown below where g_M represents the uncertain dynamics of the feedback linearized torque equation.

$$\frac{\mathrm{d}M}{\mathrm{d}t} = -\left(\frac{1}{T_r\sigma} + \frac{1}{T_s\sigma}\right)M + w_q = g_M + w_q$$

We assume that the estimation error for g_M is bounded. To design the VSC for the linearized system, we define the sliding surface as the torque error $S_M = M - M_d$. For this choice of the sliding surface, we use the control

$$w_q = -\hat{g}_M - k_M \mathrm{sgn}(S_M), \quad k_M > 0$$

Next, we use the Lyapunov stability theory to prove the stability and to choose the control gain in the form $k_M = G_M + \eta_M$, where G_M is the maximum value of g_M and η_M is a positive constant. The gain k_M includes the term G_M to ensure robust stability and the term η_M to control the speed of convergence of the sliding controller. To avoid chattering, we define a boundary layer $B(t)$ in the vicinity of the sliding surface $B_M(t) = \{x, |S_M(x)| \leq \Phi_M\}$, where Φ_M is a positive constant, the boundary layer thickness. Inside the boundary layer, a proportional control term is added to the control signal w_q. Outside the boundary layer ($|S_M(x)| > \Phi_M$), we use the corrective control $\mathrm{sgn}(S_M)$, to drive the system to the sliding surface. The dynamics of the stator flux F_s is similar to M and with similar analysis its control signal can be found as

$$w_d = -\hat{g}_{F_s} - k_{F_s} \mathrm{sgn}(S_{F_s}), \quad k_{F_s} > 0$$

As for torque, we select the gain $k_{Fs} = G_{Fs} + \eta_{Fs}$ and use a boundary layer with proportional control to avoid chattering.

The controller block diagram is shown in Figure 11.1. Figure 11.2 shows the torque response to a step command and the stator flux. The stator flux was preinstalled and its magnitude shows that FBL is very effective in flux and torque decoupling. Figure 11.3 shows the stator and rotor flux magnitude response to a step command. The response for torque and flux is fast and chattering free and the overall drive system robustly control torque and stator flux. The drive has the same fast and robust response as a conventional DTC drive, while the torque and flux ripple has been eliminated. This investigation proves that FBL is a useful tool for IM drive control. It allows the design engineer to employ a simple design approach and facilitates the integration of linear and nonlinear controllers.

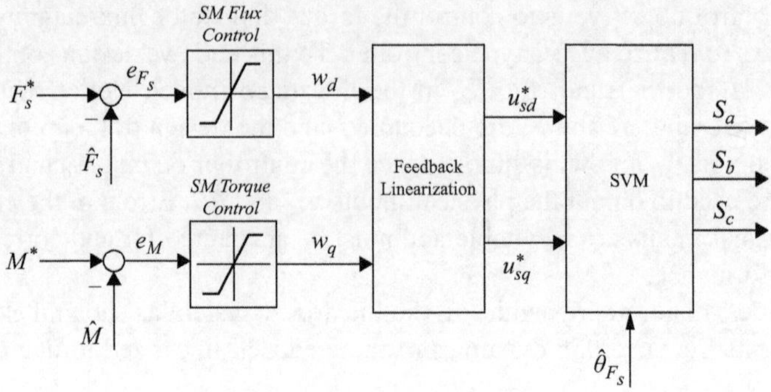

Figure 11.1 Variable structure torque and flux control with FBL

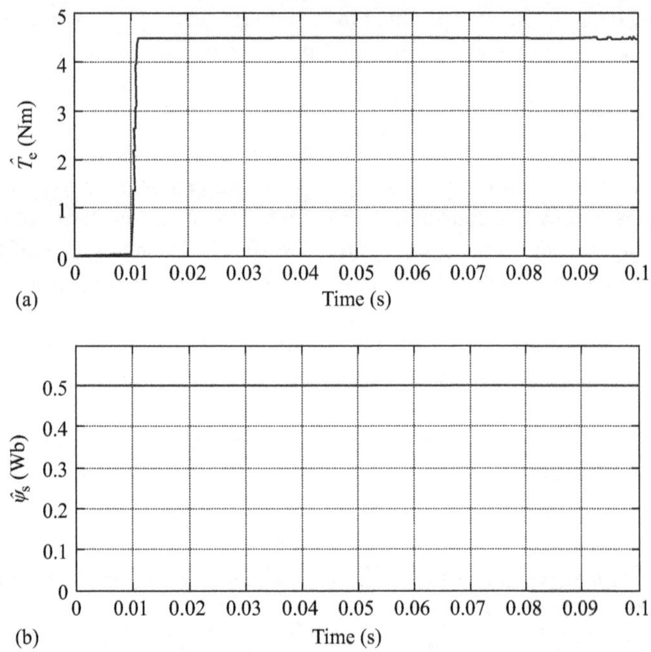

Figure 11.2 Torque transients with feedback linearization and VSC (a) torque and (b) stator flux magnitude

11.4.3 Intelligent DTC schemes

Fuzzy controllers are typically used in DTC schemes to replace the hysteresis controllers and switching table for selecting the space vector [10]. In these cases, the torque and flux errors, and stator flux position are the input to the fuzzy controller. The fuzzy inference engine uses these inputs to select its output using predefined fuzzy rules. The fuzzy rules are either Mamdani or Takagi-Sugeno-Kang (TSK)

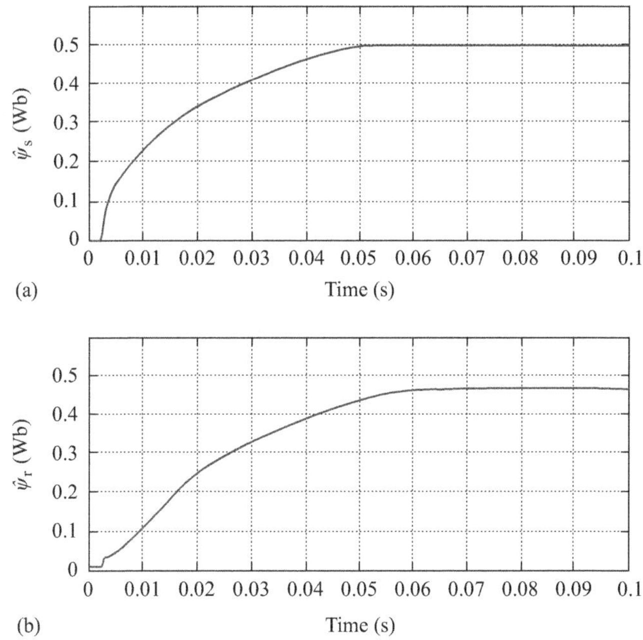

Figure 11.3 Flux response to step command. (a) Stator flux magnitude and (b) rotor flux magnitude

rules. In Mamdani rules, both antecedent and consequent are fuzzy. In TSK rules, only antecedent is fuzzy and the consequent is a mathematical description. In either case, use of fuzzy sets to condition the controller inputs for IM drives is not well justified. Fuzzy controllers are well suited for cases where signals are described in linguistic forms and no clear quantitative measure is available. Since this is not the case for IM drives, as popular they are, fuzzy controllers are not used to tackle linguistic uncertainties and therefore they can typically be replaced with a non-fuzzy controller with similar performance. This issue is overcome in research studies where the linguistic nature exists in some aspect of the IM drive. For instance, to remove chattering from sliding mode IM drives, fuzzy control is used in the vicinity of the sliding surface. Since the distance of the IM trajectory to the sliding surface is described with linguistic terms, fuzzy control can be successfully justified and used in such applications. Even in these cases, the complexity of the fuzzy controllers is of main concern. The number of fuzzy rules in fuzzy controllers should be kept limited to reach a drive system with reasonable implementation. In short, fuzzy controllers are useful in IM drives where they are well justified and simple to use.

Another class of IM DTC drives is neural networks. Neural networks have simple architecture and are easy to train. They provide excellent approximation for nonlinear functions and they are robust to disturbances. Neural networks are used in DTC for control design, observer design, and identification of the IM parameters [11]. In particular, when they are used as observers for speed estimation their performance

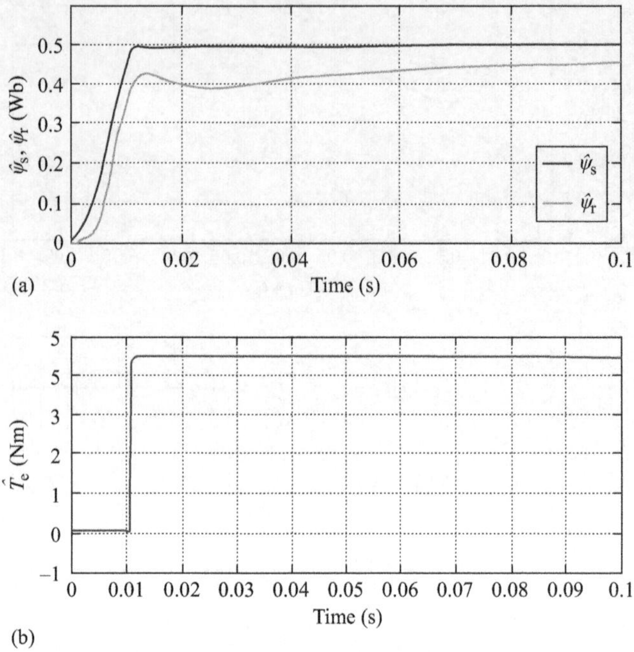

Figure 11.4 Stator and rotor flux dynamics and response of the torque controller

result in quality sensorless DTC IM drives. Neural networks are used in several structures to improve DTC IM drives. These include feed-forward, multilayer, and recursive.

Motivated by the success in functional modeling of emotions in control engineering applications, another intelligent DTC IM drive is successfully designed and implemented using the computational model of emotional learning in the amygdala [12]. Emotional controllers are nonlinear controllers, which have been successfully applied to various systems and have resulted in significant performance improvement. Two independent emotional controllers are implemented for an IM drive, one for stator flux control and the other one for torque control. This configuration simplifies the design and implementation. The dynamic response of the flux controller is shown in Figure 11.4 which depicts the rotor and stator flux magnitudes at startup. The stator flux installation time is about 11 ms. Figure 11.4 also shows the performance of the torque controller, i.e. the torque response to a step command. The torque is ripple free and increases to its rated value (3 Nm) in about 2 ms, which is similar to other modern DTC techniques.

11.5 IM speed estimation with Kalman filtering

The quality of a sensorless IM drive is mostly determined by the quality of its state observer. The low-speed problems reflect the difficulties encountered by observers at very low speeds where the fundamental excitation levels and the signal-to-noise ratio are both very low. The Kalman filter (KF) was originally developed for linear systems but later applied to nonlinear systems using the linearized or extended KF (EKF). EKF is based on a first-order approximation of the system dynamics (linearization) and is only accurate if the error is approximately linear. Although the performance of the EKF is poor in some situations, its performance is acceptable if the system nonlinearity is not severe. Its simplicity, together with the popularity of the KF, makes it the most widely applied nonlinear state estimator.

Nonlinearly mapping an input random variable typically results in a complex distribution with a large number of associated parameters. Hence, optimal nonlinear state estimation requires knowledge of the higher order statistics of this complex density function. Consequently, the exact estimation of the states of a nonlinear system is often impossible in practice. The unscented transform (UT) is a nonlinear transformation which propagates the mean and covariance through a nonlinear function. The UT is based on a set of chosen sample points, known as sigma points, and preserves the nonlinear nature of the system. One way to handle nonlinear models and transformations is to combine the KF with the UT to obtain the UKF. This approach is very promising in IM drives because the IM nonlinear model is known with sufficient accuracy. Three main UTs are known as general UT, simplex UT, and spherical UT. Combination of each UT with KF will results in a UKF. This way, general UKF, simplex UKF, and spherical UKF can be built.

To evaluate the performance of different UKFs, the implementation results for each UKF observer for IM sensorless operation is illustrated here. A direct torque controlled sensorless IM drive is used [13].

11.5.1 High-speed operation

A startup followed by a reversal is illustrated to show the UKFs performance in high-speed operation with fast dynamics. The motor runs at zero speed, accelerates to 300 rad/s, and then reverses to −300 rad/s (electrical speed), all in a time span of less than 1 s. The experimental results for general UKF are shown in Figure 11.5. The speed estimation errors are low during both transients and steady-state operation, and its overall performance in terms of torque response and noise is excellent.

The simplex and spherical UKFs use the least number of sigma points and therefore have simple structure but the latter is more numerically stable. The simplex UKF may lead to problems caused by overflow or underflow, and in the worst case, it may even result in divergence. The experimental results for the simplex UKF are shown in Figure 11.6. The simplex UKF gives poor speed estimation compared to the general UKFs. If the computational load is very critical, the simplex UKF may be selected over the general UKF, at the cost of a modest estimation performance. The spherical UKF is another computationally efficient filter but has better numerical accuracy because its weights are of the same order of magnitude. The experimental results with spherical UKF are shown in Figure 11.7. In high-speed operation, the spherical UKF has similar performance as the general UKF. However, some tracking errors are observed during speed startup in Figure 11.7.

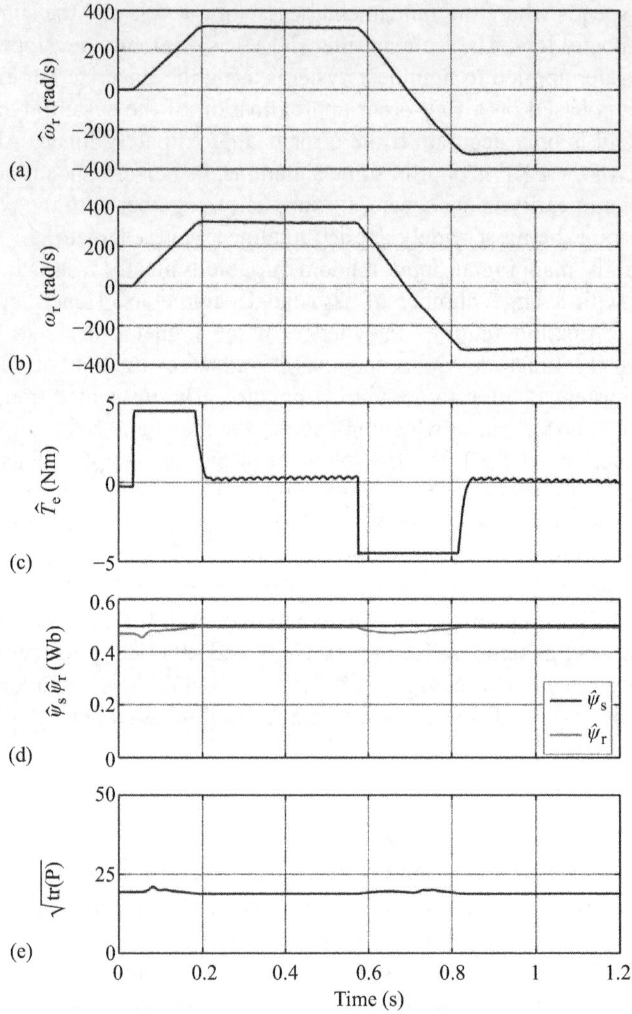

Figure 11.5 Experimental results for high-speed operation of IM with general UKF: (a) estimated speed, (b) measured speed, (c) estimated torque, (d) estimated stator and rotor flux magnitudes, and (e) root mean square of error

11.5.2 Low-speed operation

The observer's low-speed performance was tested with square wave speed reference. Square wave operation with the general UKF is shown in Figure 11.8 (left). The estimation is accurate, with low errors and fast response. Experimental results with the spherical UKF are shown in Figure 11.8 (right). The spherical UKF gives

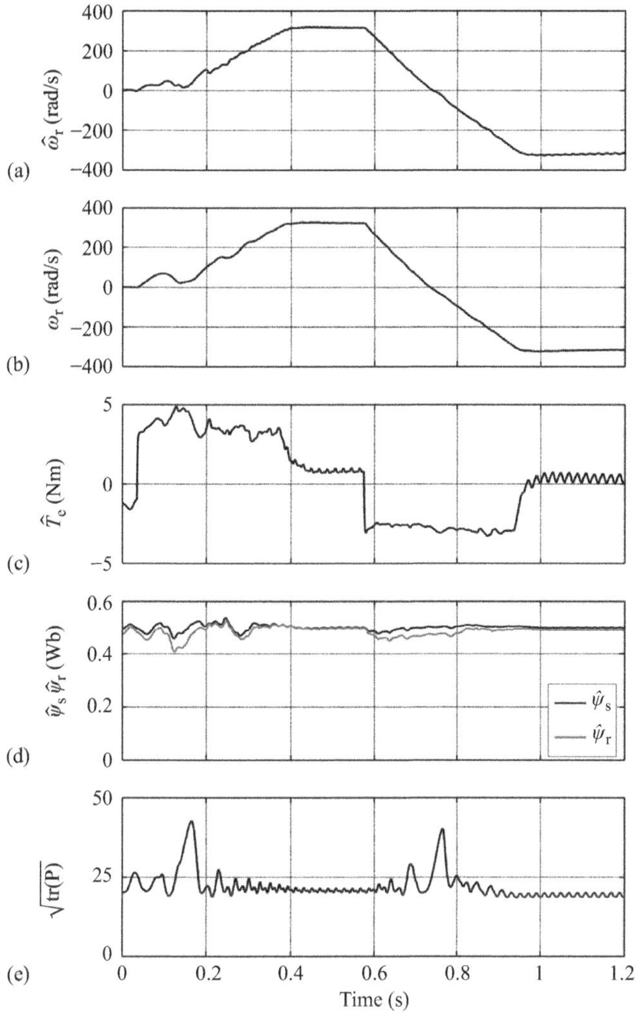

Figure 11.6 Experimental results for high-speed operation of IM with simplex UKF: (a) estimated speed, (b) measured speed, (c) estimated torque, (d) estimated stator and rotor flux magnitudes, and (e) root mean square of error

less accurate estimation than the general UKF, which is reflected in larger speed oscillations and errors. This is not surprising because the spherical UKF uses fewer sigma points and therefore provides less accurate mean and covariance approximation than the general UKF. It is obvious in Figure 11.6 that the simplex UKF has significant problems at low speeds. No experimental results are provided for the simplex UKF at 1 Hz because it cannot run the drive at 1 Hz.

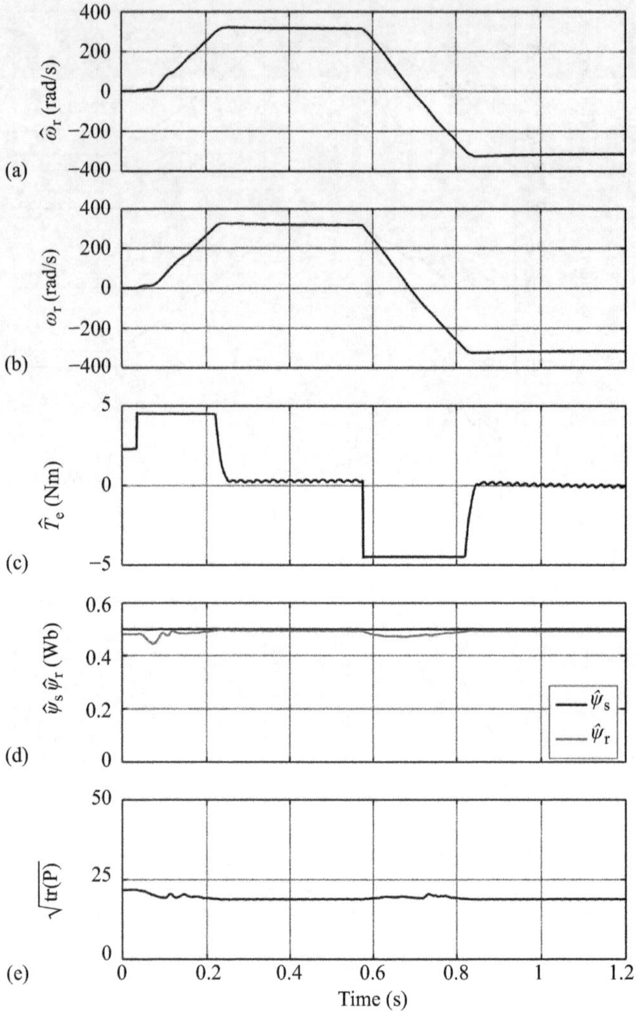

Figure 11.7 Experimental results for high-speed operation of IM with spherical UKF: (a) estimated speed, (b) measured speed, (c) estimated torque, (d) estimated stator and rotor flux magnitudes, and (e) root mean square of error

11.5.3 Implementation considerations

To compare their computational load, the running time and object code size (a measure of memory requirement) for each observer is listed in Table 11.1.

The general UKF provides the shortest computational time and consumes the least memory among UKFs. Although the simplex and spherical UKFs use fewer sigma points and are expected to require less computation than the general UKF, the results unexpectedly show that both require more time and more memory.

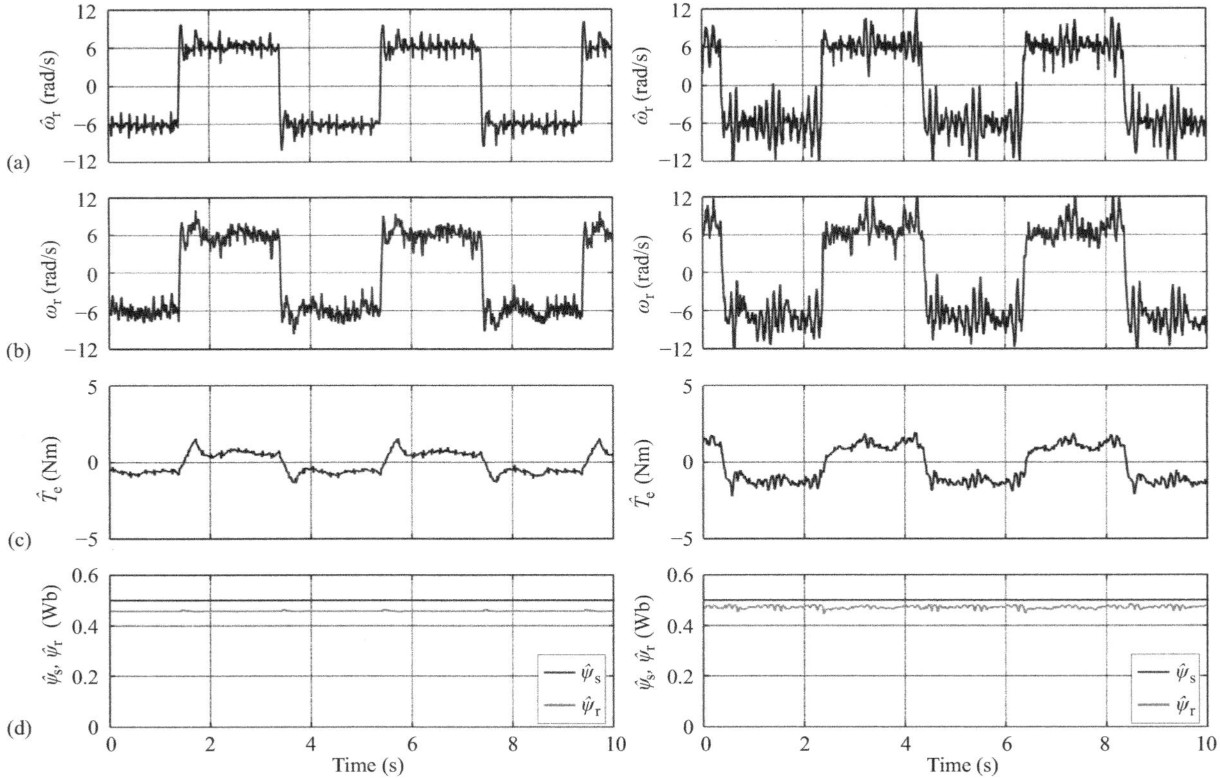

Figure 11.8 Experimental results for low-speed operation of IM with square speed reference of 6.283 rad/s and general (left) and spherical (right) UKFs: (a) estimated speed, (b) measured speed, (c) estimated torque, and (d) estimated stator and rotor flux magnitudes

Table 11.1 The running time and code size for the UKF observers

Kalman filtering	*Running time (μs)*	*Code size (bytes)*
General UKF	68.9	23,869
Simplex UKF	78.2	25,649
Spherical UKF	78.2	25,423

This result is due to the fact that the calculation of the sigma points in the simplex and spherical UKFs needs multiplication of a matrix by a vector which is not needed in general UKF. The matrix computations are costly to perform in microprocessors. The main computationally expensive calculations in UT (square root of a matrix and outer products in obtaining the covariance) are common to all UKFs. Since the simplex and spherical only differ in the set of sigma points, their runtime is the same. Although the UKF is computationally costly, its computational load is acceptable for modern microprocessors. The most costly operations in UT are the Cholesky factorization and outer products in obtaining the covariance of the projected sigma points. While the latter is an inevitable costly operation, Cholesky factorization can be simplified when the covariance matrix is sparse. For application to the IM, symbolic manipulation can be used to simplify the expressions off-line and thereby significantly reduce the computational load.

11.6 Switched reluctance sensorless drives

Switched reluctance motors (SRMs) are suitable for several applications if properly controlled. These applications include automotive, household, electric and hybrid electric vehicles, and compressors. Specific advantages of SRMs include rugged motor structure and absence of magnetic sources in the rotor. SRM drives utilize these advantages to operate the motor at very high speed [14]. Rotor position sensing is a fundamental part of SRM drives due to the nature of reluctance torque production. A position sensor is typically used to provide the rotor position feedback. However, such position sensors add complexity and cost to the overall system and create reliability problems for the SRM drive system. Moreover, the use of position sensors might not be an option due to the restrictions in certain applications. These reasons motivate sensorless schemes for SRM drives. In such schemes, terminal measurements are used for control purposes without the need for additional position feedback.

Since the mechanical time constant of SRM drive is much larger than its electrical time constant, the position information can be recovered from other quantities. With available digital signal processors, estimating the rotor position with desirable accuracy can be integrated in SRM drives. Consequently, model-based sensorless schemes can be implemented in SRM drives to avoid the position sensor.

The fundamental principle of operation of a SRM is based on the variation in flux linkage with the change in the angular position of the rotor. Knowledge of the magnetic characteristics of an SRM plays an important role in determining the rotor position indirectly for sensorless control operation. Invariably, all the existing sensorless techniques use such knowledge to obtain position information. Depending on the geometry and operating point of the motor, a suitable method is chosen such that a very good resolution is obtained. The existing sensorless techniques can be perfected by making use of new advancements made in power electronics and digital signal processing. It is found that significant SRM parameter variations occur in mass production or with motor aging. This necessitates sensorless techniques with self-tuning in the event of parameter variations. Self-tuning control methods with sensor and without position sensor have already been developed and it is found that self-tuning control techniques are essential to squeeze out the best performance from SRMs in the presence of parameter variations. If the control is based on the rotor position obtained from a conventional sensor, which is insensitive to these variations, optimal performance cannot be extracted from the machine. Since the control of SRM is essentially based on the inductance profile, which changes with parameter variations, it necessitates an online self-tuning control strategy for optimum performance. The sensorless techniques based on inductance estimation will have a better performance than the conventional position sensors. Since sensorless techniques are based on the electrical variables, which alter due to parameter variations, the control can be made to adapt to these variations to give better performance. The state of the art in sensorless technology and the recent research advances in sensorless SRh4 drives were discussed. It is found that SRM drives are ready to use in many manufacturing, aerospace, and consumer applications. A range of topics, including the need for sensorless control, classification of existing sensorless techniques, their merits and demerits, and future trends were presented. From the above discussions, it can be seen that the sensorless technology in SRMs is a practical technology. The recent advances in power electronics, digital signal processing, and control systems have already made sensorless SRM a commercially acceptable drive. Further research is necessary in some important issues like the self-tuning to make sensorless SRM drives that give optimal performance in low-cost mass production.

References

[1] Zhu, Z. Q., and Jiabing Hu. "Electrical machines and power-electronic systems for high-power wind energy generation applications: Part I. Market penetration, current technology and advanced machine systems." *COMPEL: The International Journal for Computation and Mathematics in Electrical and Electronic Engineering* 32(1) (2012): 7–33.

[2] Komeza, Krzysztof, and Maria Dems. "Finite-element and analytical calculations of no-load core losses in energy-saving induction motors." *Industrial Electronics, IEEE Transactions on* 59(7) (2012): 2934–2946.

[3] Barut, Murat, Seta Bogosyan, and Metin Gokasan. "Experimental evaluation of braided EKF for sensorless control of induction motors." *Industrial Electronics, IEEE Transactions on* 55(2) (2008): 620–632.

[4] Smith, Andrew, Shady Gadoue, Matthew Armstrong, and John Finch. "Improved method for the scalar control of induction motor drives." *IET Electric Power Applications* 7(6) (2013): 487–498.

[5] Sira-Ramírez, Hebertt, Felipe Gonzalez-Montanez, John Alexander Cortés-Romero, and Alberto Luviano-Juárez. "A robust linear field-oriented voltage control for the induction motor: Experimental results." *Industrial Electronics, IEEE Transactions on* 60(8) (2013): 3025–3033.

[6] Zhang, Yongchang, Jianguo Zhu, Zhengming Zhao, Wei Xu, and David G. Dorrell. "An improved direct torque control for three-level inverter-fed induction motor sensorless drive." *Power Electronics, IEEE Transactions on* 27(3) (2012): 1502–1513.

[7] Wu, Yunxiang, Mohsin Shafi, Andrew M. Knight, and Richard McMahon. "Comparison of the effects of continuous and discontinuous PWM schemes on power losses of voltage-sourced inverters for induction motor drives." *Power Electronics, IEEE Transactions on* 26(1) (2011): 182–191.

[8] Lascu, Cristian, Ion Boldea, and Frede Blaabjerg. "Direct torque control via feedback linearization for permanent magnet synchronous motor drives." *Optimization of Electrical and Electronic Equipment (OPTIM), 2012 13th International Conference on.* IEEE, Brasov, Romania, 2012.

[9] Chen, Shun Zhong, Norbert C. Cheung, K. C. Wong, and Junyong Wu. "Integral variable structure direct torque control of doubly fed induction generator." *IET Renewable Power Generation* 5(1) (2011): 18–25.

[10] Suetake, Marcelo, Ivan N. da Silva, and Alessandro Goedtel. "Embedded DSP-based compact fuzzy system and its application for induction-motor speed control." *Industrial Electronics, IEEE Transactions on* 58(3) (2011): 750–760.

[11] Ghate, Vilas N., and Sanjay V. Dudul. "Cascade neural-network-based fault classifier for three-phase induction motor." *Industrial Electronics, IEEE Transactions on* 58(5) (2011): 1555–1563.

[12] Markadeh, G. R., Ehsan Daryabeigi, Caro Lucas, and M. Azizur Rahman. "Speed and flux control of induction motors using emotional intelligent controller." *Industry Applications, IEEE Transactions on* 47(3) (2011): 1126–1135.

[13] Jafarzadeh, Saeed, Cristian Lascu, and M. Sami Fadali. "State estimation of induction motor drives using the unscented Kalman filter." *Industrial Electronics, IEEE Transactions on* 59(11) (2012): 4207–4216.

[14] de Araujo Porto Henriques, Luis Oscar, Luís Guilherme Barbosa Rolim, Walter Issamu Suemitsu, J. A. Dente, and P. J. Costa Branco. "Development and experimental tests of a simple neurofuzzy learning sensorless approach for switched reluctance motors." *Power Electronics, IEEE Transactions on* 26(11) (2011): 3330–3344.

Chapter 12
Wind energy systems

Mariusz Malinowski[a], Adam Milczarek[a], Dmitri Vinnikov[b] and Andrii Chub[b]

12.1 Introduction

12.1.1 Overview of the wind energy technology

Wind turbines (WTs) are of high potential and most promising among renewable energy sources (RESs), largely contributing to world's energy production (Figure 12.1). Moreover, within renewable energy technologies, wind energy technology is most advanced and the cost of wind power is close to that of fossil fuel power. Recent WT market trends have shown a 12%–13% annual increase in cumulative installed capacity. Such growth rate is anticipated to continue up to 2018, to reaching 596 GW of WT installations [1, 2]. Global leaders in the overall installed capacity of WTs are China, the USA, and Germany; Europe is on the second place behind Asia in the total installed capacity. Most of Europe has good conditions for energy production, hence very fast growing European market holds large producers of WTs, e.g. Enercon, ABB, Siemens, Vestas, Alstom, Gamesa, etc. Further, European wind power industry has formulated generation targets of 180 and 300 GW in 2020 and 2030, respectively [3, 4].

A generalized configuration of a typical wind energy conversion system (WECS) consists of a WT (with vertical or horizontal shaft axis), a generator (with/without a gearbox), and power electronic converters (PECs) [5–7].

Due to huge technological advancements in high power WTs, small wind turbines (SWTs) are also available for individual customers. According to the IEC 61400-02 standard [8], a SWT is characterized by a rotor swept area from 2 to 200 m^2, thus it can be installed in rural or other low density population areas. In the recent years, the SWT market trend has shown an aggressive 35% annual increase in cumulative installed capacity (Figure 12.2). Such growth rate is anticipated to continue until 2020 when it reaches 3 GW of cumulative installed capacity of SWTs [9]. According to a conservative assumption, the market could subsequently see a growth rate of 20% in 2015–2020.

[a]Warsaw University of Technology, Warsaw, Poland
[b]Tallinn University of Technology, Tallinn, Estonia

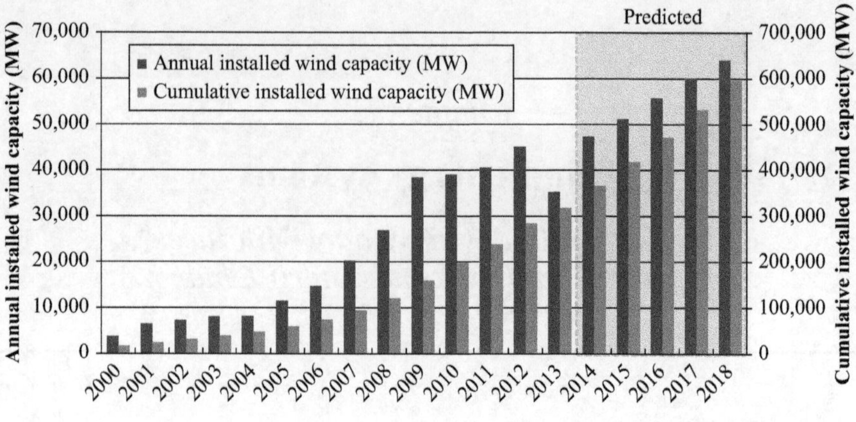

Figure 12.1 Cumulative installed capacity and annual installed capacity (in MW) of high power WTs around the world from 2000 to 2018 [1, 2]. Data used from Global Wind Report 2013 – Market Forecast for 2014–2018, GWEC with permission

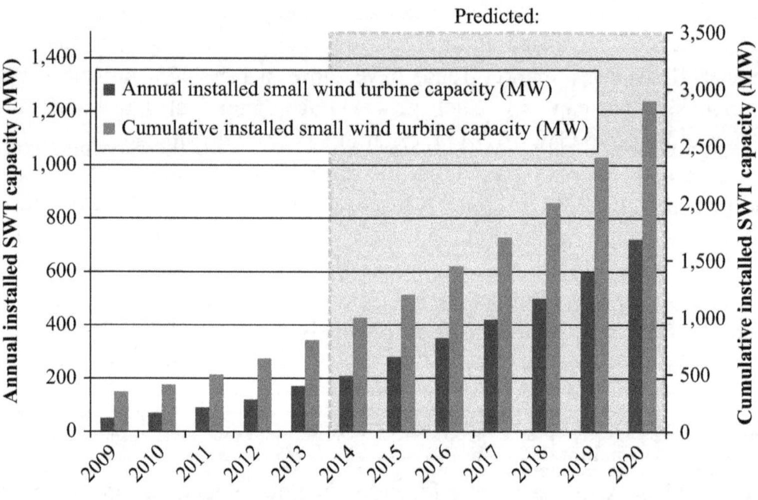

Figure 12.2 Cumulative installed capacity and annual installed capacity (in MW) of small power WTs around the world from 2009 to 2020 [9]. Data used from WWEA Small Wind World Report Summary 2014 with permission

12.1.2 Types of WT rotors

WTs can be defined by the basic parameters presented in Table 12.1. The relations in Table 12.1 are called the rotor's aerodynamic characteristic. It is an individual feature for each turbine and only slightly dependent on the size.

Generally, WTs are divided into two groups:

- horizontal axis wind turbines (HAWTs)
- vertical axis wind turbines (VAWTs)

Table 12.1 Basic parameters (formulas) describing the wind turbine

Feature	Formula	Comment
Generated power	$P_m = \frac{1}{2} A\rho V^3 C_p$	A – rotor swept area (m^2), expressed accordingly for horizontal axis wind turbine (HAWT) as $A = \pi R^2$ and vertical axis wind turbine (VAWT) as $A = 2RH$ R – rotor radius (m) H – rotor height (m) ρ – air density (kg/m^3) C_p – rotor power coefficient V – wind speed (m/s)
Power coefficient	$C_p = \frac{P_m}{P_w}$ $C_{pmax} = \frac{16}{27} \approx 0.593$	P_m – mechanical power (power on the rotor shaft) P_w – power of wind content in the virtual stream tube containing the wind turbine
Tip-speed ratio	$\lambda = \frac{\omega R}{V}$	ω – rotor speed

Commonly, HAWTs with three blades are characterized by low cost and good dynamic properties.

The market share of VAWTs is quite small; however, it is growing obviously due to so-called "urban wind turbines", a relatively new category of small turbines. Two design concepts of VAWTs are usually considered: the Savonius rotor and numerous varieties of the Darrieus rotor. The first is an old solution of a slow speed rotor with large starting torque and its power coefficient halved, as different from any HAWT. Highly variable driving torque with the rotor azimuth has led to the design of a multi-segment structure or rotors twisted along their vertical axis. Modern VAWT is based on the Darrieus type rotor with a self-start capability, which was impossible in older designs. In order to ensure the self-start in VAWTs, the H-shaped rotor (also called a giromill) with a variable pitch angle β was developed. However, the best implementation is based on helix-like shaped rotor blades and its performance is comparable to that of HAWT.

Assuming that all types of turbines can be designed at the same wind speed and nominal power, the rotor dimension, area of blades, and weight of the classical HAWT solution are lower than those of other types. Moreover, it is distinguished by simple control, very high reliability and durability. The disadvantage of this turbine is that its technology is more complex than that of the Savonius VAWT. Among the VAWTs, the Giromill Darrieus WT has the best performance because of its lowest dimensions, weight, and blades area. Similar to the classical HAWT, the control for that turbine is simple, but the technology is complex. This turbine is characterized by good reliability and durability, however, still not as good as those of HAWTs.

Design of wind rotor geometry is a multi-objective process. Therefore, the optimization process has to fulfill some requirements [5, 6]:

- highest overall efficiency of WT for a chosen range of mean long-term wind speed,

- high energy yield,
- reasonable load factor.

The most important factor is the energy yield in the long time scale, i.e. how much energy can potentially be captured for a given turbine design and location.

12.1.3 Generators for WTs and standard configurations of WECSs

Generator is the second most important part of the WECS after the wind rotor. Any WECS could be separated into mechanical and electrical parts. A generator is an element that couples both parts and defines the energy conversion efficiency from the mechanical to the electrical form. Rated power of the generator defines the operation of the whole WECS. WECS performs maximum power point tracking (MPPT) in the range between the cut-in and rated wind speed. It performs stall or pitch control to limit power on the shaft at the rated power of the generator in the range from the rated to the cut-off speed. It is necessary in order to avoid overloading and overheating of the generator. There are different types of generators used in WECSs. Each of them has its own pros and cons. Their performance also depends on the type of the WECS: fixed or variable speed. Combination of the generator and the electric configuration of the WECS defines the overall performance of the WECS.

12.1.3.1 Types of generators used in WECS

In the wind energy industry, two types of electrical machines are commonly used: asynchronous (induction) and synchronous. These two types have been widely used in the industry for years, and their design, manufacturing, and reliability are well known. Asynchronous generators are simple in production, robust, and cheap. Their performance is limited by the consumption of considerable reactive current for magnetizing. This current is required for the generator excitation since it has no permanent magnets or excitation winding. Operation mode of induction generators is defined by slip (relative motion) between the rotor and the stator fields. At start-up induction generators have high slip and consume substantial inrush current. They are usually high speed generators with a complicated gearbox. Synchronous generators contain either permanent magnets or excitation winding and thus require no reactive power for magnetizing. They are expensive and much more complicated in production than the induction machines. Multipole synchronous generators are most widely used in high power variable speed WTs due to the gearless connection with the wind rotor. Direct drive synchronous generators are widely used in multi-MW variable speed WTs [10] because of their high inherent reliability.

Squirrel cage induction generator (SCIG) is one of the simplest and highly reliable solutions [11]. They have penetrated to the fixed and variable speed WECSs due to their low price and considerably high efficiency. Usually, SCIG is designed to operate at frequencies much higher than those of the wind rotor. Complicated gearbox with up to three stages is used to couple a SCIG with the wind rotor and match their operation frequency ranges. Relatively high reactive power has to be compensated in fixed speed grid connected WECSs, or supplied by a converter in variable speed WECSs. Attention should be paid to the start-up of the SCIG, especially in the direct grid connection, when high inrush current may occur. Torque and level of reactive energy drawn by the SCIG depends on the slip, which has to be kept within a very narrow range during normal operation. A grid connected SCIG could be a substantial source of reactive power at low voltage ride through (LVRT) event.

Wound rotor induction generator (WRIG) is a more complicated and expensive asynchronous electrical machine than the SCIG [12]. Its rotor contains windings to enhance the slip operation range. Slip rings with brushes could be used for connection of the external controlled variable resistance. Reliability of this system is lower than that of the SCIG. Slip rings and brushes are the elements that require additional maintenance. Direct connection between the rotor and the controlled variable resistance is preferable due to the higher reliability, but such WECSs are more complicated to design. WRIGs have been used in variable speed WTs with limited speed regulation.

Doubly fed induction generator (DFIG) construction is basically similar to that of the WRIG [11]. The main difference lies outside the generator. WRIGs use external resistance in the rotor circuit to control the slip, while DFIGs are coupled with the grid through a bidirectional AC/AC converter, for example, the back-to-back voltage source converter (VSC). PECs allow bidirectional energy flow, and a DFIG could be magnetized either from the rotor or from the stator side. The term "doubly fed" indicates that both the stator and the rotor are fed from the grid. Such configuration has several times wider regulation range than that of the WRIG. Moreover, control of the DFIG is better at grid faults since the converter can control the frequency and the magnitude of the rotor current. In contrast to the WRIG based WECs, this generator type is intended for variable speed WTs with improved operation range.

Wound rotor synchronous generator (WRSG) as a concept known for more than one hundred years is extensively used in the electric power industry as a reliable and proven solution [11]. Its construction is complicated, it has high weight and quite a high price. Excitation is performed using an additional winding, which is fed by the external converter by means of slip rings and brushes. This eliminates reactive power generation since the magnetizing current is not required. WRSGs need to handle only active power that increases energy yield. The low speed multipole WRSGs are utilized in direct drive variable speed WTs with the full power converter. A multipole WRSG has the same range of rotational frequency that is optimal for bulky high power wind rotors. Because of high weight, the installation of the WRSG is complicated, especially in the case of offshore multi-MW WTs. Its weight could be more than eight times higher than that of the DFIG with equivalent power [10]. Still, this concept is used within the most powerful commercially available WECSs – E-126 from Enercon rated at 7.5 MW [13].

Permanent magnet synchronous generator (PMSG) is the further improvement of the synchronous generator concept [13]. Permanent magnets make these generators self-excited, which leads to high efficiency. As compared to the WRSG, PMSGs are compact, lightweight, and reliable, but complicated in manufacturing and thus expensive. The main concern is the fast growing price of the magnetic materials. PMSGs have proven their performance in numerous commercially available WECSs. They are particularly suitable for the direct drive variable speed WTs. This concept is used within high power commercially available WECSs at the MW level. It is highly challenging from the power electronics point of view, because the output voltage magnitude and frequency of the PMSG vary in a wide range. Most of the PMSG based gearless WTs have low voltage output, which leads to high rated output current that would require the full power converter.

In the wind energy conversion, fast rise of average power per installed turbine, especially in the offshore wind farms, is the current trend [14]. Another trend is the growing number of small residential WT installations. The number of items approaches 1 million, but cumulative installed energy is less than 1 GW. Small WECSs utilize conventional generator types like PMSGs. High power WECSs allow reduction of the energy cost and wind farm capital costs. Currently, a wide commercialization of 6 MW WTs is typical of the wind energy area. The latest development is the 10 MW WT generator by the company AMSC, the design of which incorporates a direct drive high temperature superconductor generator. It could be the most promising emerging technology for the high power WECSs, considering the fast growing prices of the permanent magnet materials. Currently, the industry is ready for producing generators up to 15 MW.

12.1.3.2 Current standard WECS designs

Wind energy conversion industry has been advancing fast starting from the 1980s. It has changed the paradigm of system design for several times. Efforts have been directed to the power level growth, higher energy conversion efficiency, and lower cost of energy. Recently, many countries have imposed grid codes to improve energy supply security. Wide penetration of WECSs in the energy production has raised concerns of grid stability. Grid code directs the behavior of the energy generation unit or plant at grid faults or disturbances. Usually, it regulates the output voltage value, power factor boundaries, reactive power generation, response upon grid frequency variations, LVRT events, short circuits, etc. Each country has its own grid code that complicates the design of the WECS for the worldwide market. This restricts practical use of some WT concepts and imposes numerous requirements for the controllability of WECSs.

Fixed speed WECS concept is based on the direct connection of the SCIG to the grid. High speed generator is driven by the wind rotor through the three-stage gearbox with a high conversion ratio, up to 80–100 [10]. In the steady state, SCIGs operate at nearly fixed frequency at speed variations within 1%–2%. Rotation frequency is defined by the grid frequency: 1,500 rpm for 50 Hz grid and 1,800 rpm for 60 Hz grid. Typical configuration of that WECS is shown in Figure 12.3. This concept, widely used by Danish manufacturers of WTs, is called the "Danish concept", which has been substantially improved during the 1980s and 1990s. Capacitor bank has been added to compensate reactive power from the SCIG. High inrush current has been eliminated with the implementation of the soft starter. The soft starter controls current during grid connection transients, but it is bypassed in a

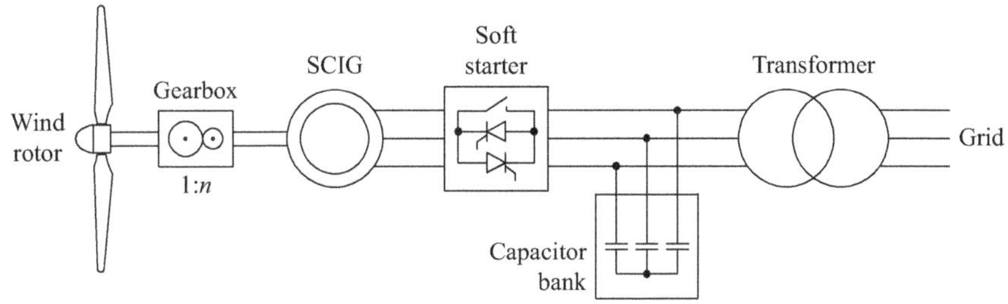

Figure 12.3 Fixed speed WECS

steady state operation. This concept is simple, cheap, and reliable, but suffers from the fixed speed operation. Wind energy conversion is poor over the operation range of wind speeds and reaches its high values only near a given wind speed. Regulation is usually done with stall or pitch control, but output power fluctuations will still follow the wind speed profile. This type of WECSs has almost left the worldwide market, because they cannot satisfy most of the grid codes. In recent years, total share in annual installations has dropped to less than 5% [15].

Variable speed WECS can control the tip-speed ratio by means of rotation speed adjustment. This increases the efficiency of wind energy conversion and abates mechanical stresses in the WECS caused by the wind gusts. Improved mechanical loading allows the implementation of high power WTs at multi-MW level that cannot be achieved with fixed speed WTs. Better operation conditions of the bearings, and especially the gearbox, could substantially increase reliability and therefore reduce the maintenance cost of the WECS. In DFIG, the speed control could be provided with a power converter in the wounded rotor circuit. This approach requires power converters with rated power, which is several times lower than the rated power of the generator. Better performance could be obtained by the full decoupling of the generator from the grid through the high power converter. Rated power of the converter should be equal or higher than the rated power of the generator. Full power converter could handle both induction and synchronous generators.

Variable speed WECS with limited speed control appeared on the market in the mid-1990s as a logical improvement of the fixed speed systems. This concept had a considerable market share for a short period of time and lost its popularity in the early 2000s. It utilizes the WRIG with variable resistance controlled by the converter in the wound rotor circuit, as shown in Figure 12.4. It still requires a soft starter to control the inrush transients. High speed nature of the WRIG deteriorates the reliability of slip rings and brushes, which are commonly used for connection of the external variable resistance. Several companies have proposed novel approaches for higher reliability. The OptiSlip® concept was introduced by the Vestas to improve the reliability of WRIG based WECSs [11]. In this case, resistors and the converter are placed directly on the shaft without slip rings. Control of the converter is implemented without direct connection by use of optical coupling. OptiSlip® WECS allows slip, i.e. rotation speed control in the range of 10%.

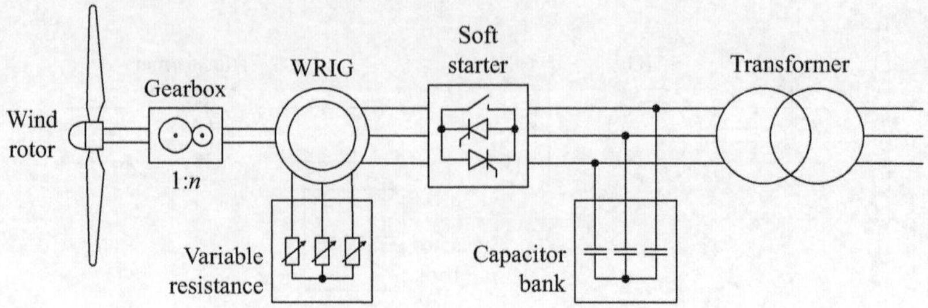

Figure 12.4 Variable speed WECS with limited speed control

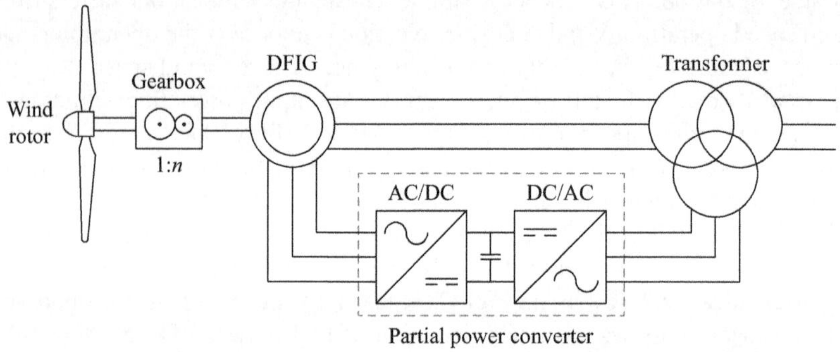

Figure 12.5 Variable speed WECS with a reduced power converter

In another concept, called FlexiSlip® from Suzlon, flexible construction of adjustable slip rings for connection of external resistors is used. WECSs based on that concept are still available on the market up to the MW level. They allow mounting of the resistors outside the generator cage and provide slip regulation up to 17% [16]. WRIG based WECS technology is outdated and its share in annual installations has shrunk considerably during recent years.

Variable speed WECS with reduced power converter appeared on the market in the late 1990s. This concept reached nearly 30% share in annual installations during 1998 and is still very popular. The DFIG allows further improvements in the WRIG based systems. The four-quadrant partial power back-to-back converter is connected to the rotor circuit, as shown in Figure 12.5. There are different WECSs based on that concept with slip regulation up to 30%. Since rotor power is directly related to the slip, partial power converter should be rated at 30% of the generator rated power. It means that care is needed in the selection of the regulation range of the rotation speed to achieve the trade-off between the system cost and the wind power conversion performance. At 30% slip generation, the speed could be regulated in the range ±30% around the synchronous speed, i.e. DFIGs could operate in the sub-synchronous or super-synchronous mode. Bidirectional power flow is required in the rotor side converter: it supplies power to the rotor in the sub-synchronous mode, and consumes power from the rotor in the super-synchronous mode. Also, the converter ensures magnetizing current control, while the stator feeds only active power to the grid. Such systems usually utilize a three-level gearbox with the high speed DFIG, which is equipped with the slip rings for converter connection. Relatively low price and a wide speed regulation range make these WECSs very attractive for the industry. This concept remains the main WECS type with a dominant share in the installed and new WECSs by the power level since 2001 [11]. Its disadvantages are: poor performance under LVRT conditions that leads to additional efforts to satisfy grid codes, and a complicated gearbox that requires additional maintenance. Operation under the LVRT requires additional efforts to be improved. Brushless DFIG is one of the most promising solutions because its behavior is better under the LVRT and it has a simpler gearbox with a lower transmission ratio [17]. In the next years, with constant updates of the grid codes, some countries may lose interest to this concept because of strict regulations, while it may continue market domination worldwide.

Variable speed WECS with full power converter is totally different from the concepts mentioned above. It provides full decoupling between the generator and the power distribution grid. Full power converter adds some extra losses to the system, but this could be compensated by additional improvements in the overall WECS performance. Full decoupling leads to full controllability of the WECS on both the generator and the grid side. It achieves the widest possible range of operation of wind speeds. It also provides superior performance on the grid side: grid support with reactive power, operation during fault conditions within the grid code, etc. Indirect AC/AC converters with intermediate DC link are widely used in this application. It allows easy integration of energy storages and breaking choppers. There are three types of generators that could be used in these WECSs: WRSG, SCIG, and PMSG. In the case of synchronous electrical machine, the generator side converter is usually simple and uncontrolled. For induction machines, the generator side converter has to be fully controlled to supply magnetizing current, which requires from the generator side converter to be rated at the power level higher than that of the induction generator. This type of WECSs has been on the market for more than 15 years. Its share of annual installation has changed from the fourth to the second place during that time. Direct drive implementation of this concept has been widely accepted in offshore wind farms due to the high reliability of the gearless implementation. Such solution is relatively expensive and usually requires a multipole heavy synchronous generator. Gearless construction leads to higher reliability at a considerable price increase. Mechanical design of such WECSs could be complicated due to high torque on the shaft.

Figure 12.6(a) shows the WRSG based WECS with the full power converter. WRSGs are easily acceptable by the wind energy industry due to their well-known performance and reliability proven by decades of operation in energy generation. This has led to the development of the first multi-MW WT E-126 by Enercon that was regarded the most powerful WECS for more than seven years. Low speed direct driven generators are dominating in this field. Their disadvantages are also well proven: they require a second converter for the control of excitation; slip rings are needed for the connection of the excitation converter; weight, production

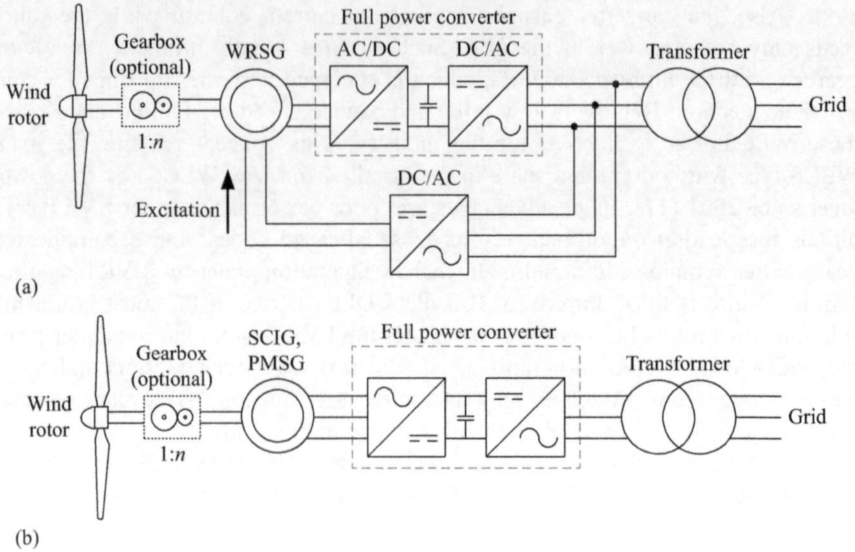

Figure 12.6 Variable speed WECS with the full power converter

complexity and the large diameter complicate their installation. The excitation converter needs to be carefully controlled because this defines the performance of wind energy conversion. Due to numerous disadvantages, this concept is not favored for future designs. However, a combination of a simple single-stage gearbox and a medium speed WRSG could diminish the shortcomings above.

Figure 12.6(b) shows a WECS suitable for PMSG utilization. This concept is close to the previous one based on the WRSG. The PMSG has considerably reduced weight, lower losses, and slightly lower price as compared to the WRSG rated at the same power level [10]. It also provides higher reliability and thus low maintenance since the excitation converter and slip rings are avoided as the permanent magnets provide the excitation field. Due to its improved reliability, the WECS with direct driven multipole PMSG is suitable for the offshore wind farms, especially in the northern regions with harsh environment. It still has price penalty that limits a wide use, and a more complicated installation process than with DFIG based WECSs. This concept shows slow but stable rise of annual installation share, regardless of constant price rise for magnetic materials. It has probably the most enhanced efficiency in the wind energy conversion, but the total system cost has always been a concern. Still, numerous WECSs available on the market at the MW power levels are based on the direct driven PMSG. System cost could be further reduced by use of a small single-stage gearbox and a medium speed PMSG.

The system shown in Figure 12.6(b) could also utilize a SCIG where a complicated gearbox is needed, but the additional cost will be overcome by the low price of the generator. Such solution requires the fully controlled AC/DC converter on the generator side for the control of magnetizing current. Rated power of this converter should be higher than that of the generator, since it needs to handle active and reactive powers. This solution could be a competitor for the PMSG based counterpart, but a complicated gearbox will always be a concern in the operational cost due to the additional maintenance need.

Emerging configurations of the WECS are also available on the market. They are still new and their long-term reliability and operation advantages need to be proven. *Clipper* concept is based on the idea where several generators are placed on the common shaft. It is applicable for variable speed WECSs with the full power converter. In this case, failure of a single generator only decreases the output power, while the WECS remains in operation. *WinDrive* concept from Voith is basically a combination of the planetary gearbox with a hydrodynamic torque converter. This transmission is used for indirect coupling of the variable speed WT and high speed synchronous generator, which is directly connected to the grid. This flexible connection can transfer torque from the variable speed shaft of the WT to the constant speed shaft of the synchronous generator. Hybrid PMSG based *Multibrid* concept looks the most promising. It is based on the combination of a single-stage planetary gearbox and a medium speed PMSG. This approach allows substantial reductions in the weight and dimensions at the same price, but the efficiency of wind energy conversion will be lower. Two high power WTs rated at 8 MW based on this concept will be soon available from Areva and Vestas. They will be the highest power WTs available in the nearest future.

Before 2013 the development trends in the wind energy were quite clear. DFIG based systems reached their maximum and rise of WECSs with the full power converter started. During recent years, governmental support in many countries has been reduced considerably, which in turn has resulted in changes on the wind energy market. This has slowed down installations of offshore wind farms. Along with the fast rise of magnetic material prices, it has substantially limited the predicted penetration of PMSG based variable speed WECSs. These WECSs still have high impact on the offshore wind farms [18]. Another factor is the redistribution of the world market between the countries. For example, China acquires more than 25% of converter sales for onshore WTs. Resulting from high penetration of DFIG based WECSs in China and in other developing countries, they have regained their popularity. Nevertheless, direct drive WTs have 28% share of the global market [19]. Taking into account all these facts, experts predict a delay of shift from DFIG based WECSs to WECSs with full power converters. Finally, it could be stated that WECSs with the full power converter will become a dominant technology in the next 20 years. This time could be further reduced if the prices of magnetic materials decrease or new cheaper generator technologies, like high temperature superconductor generators, appear on the market.

12.2 Power electronic interfaces for variable speed WTs

WRSG, SCIG, or PMSG based variable speed WTs are a popular technology in the wind energy sector, ranging from small power up to multi-MW applications. They require the full power converter for grid integration. This application is highly demanding from the point of view of the PEC since it requires high reliability, efficiency, power density, and minimized maintenance, especially for offshore wind farms. PECs for WTs have to match the output voltage of the generator with the varying frequency and magnitude with the power grid with the constant frequency and magnitude. The AC/AC power conversion could be implemented in two general ways: direct AC/AC conversion (matrix converters) and indirect AC/AC conversion with intermediate DC link. Matrix converters are out of scope of this chapter. The reason is their numerous disadvantages, like large numbers of components, complicated control and design, bulky input filter and concerns regarding long-term reliability [20].

Indirect AC/AC converter performs AC/DC conversion on the generator side and DC/AC conversion on the grid side, whereas DC link with the storage element is placed in the middle. This approach allows grid decoupling from the generator. DC link allows the integration of the energy storages. In addition, a grid side converter can perform additional functions like reactive power compensation, grid support with reactive power, withstand LVRT conditions, delay in generated power transfer to the grid, etc. Indirect AC/AC power conversion is widely used in the renewable energy field because of increased concerns about on-grid operation within grid code boundaries. It is a versatile and proven solution, which is easier to design and control. This chapter deals with different PEC topologies that provide indirect AC/AC power conversion.

Functional diagram of the WECS comprising the indirect AC/AC converter is shown in Figure 12.7. Typically, it contains filters on the generator and grid side to improve current shapes and suppress electromagnetic interference. Decoupled control of the AC/DC and DC/AC converters is commonly used in these converters. Usually, the generator side converter provides MPPT and transfers energy to the DC link, while the grid side converter controls the DC link voltage/current and injects energy to the grid. In some cases, the grid side converter could handle all these functions. Since many WTs today have low voltage output, the line frequency transformer is commonly used to connect the WECS to the medium voltage (MV) power distribution grid. Voltage matching and isolation could also be realized by the high frequency isolated DC/DC converter as a part of the generator side AC/DC converter. In such cases, the line frequency transformer could be avoided. The DC link typically contains the energy storage component – either the capacitor for the

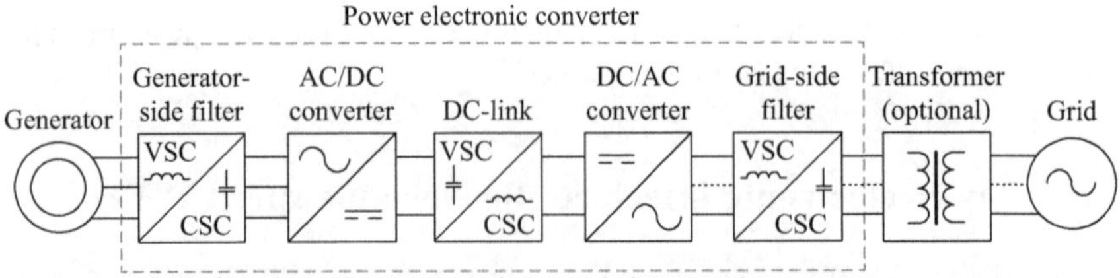

Figure 12.7 Functional diagram of the WECS with an indirect AC/AC power electronic converter

VSCs or the inductor for the current source converters (CSCs). Also, DC link is essential for the integration of braking choppers and energy storages. The converter could be connected to the single-phase or three-phase grid depending on the power rating of the generator. Single-phase grid integration is common for low power residential WECSs. Three-phase grid connection is mainly used in systems with the output power higher than 5 kW. The following sections will describe features of the topologies commonly used for the grid side and generator side converters and their combinations within the WECS.

12.2.1 Conventional power electronic building blocks

WECSs utilize both of the main classes of converters: uncontrolled and controlled. Uncontrolled topologies are typically used in the generator side converter for the PMSG or WRSG based WECSs. Full-controlled converter is essential on the generator side of the SCIG based variable speed WECS to provide magnetizing current, handle reactive power, and to control the torque. Full-controlled converters can work in inverter or rectifier modes depending on the power flow direction. The full-controlled converters could be classified as the phase controlled or the PWM controlled. The switching frequency of the phase controlled VSC is equal to the input/output voltage frequency. The VSCs are the most widespread among the PWM controlled converters. Their switching frequency is much higher than the input/output voltage frequency. This leads to high power density and better dynamic performance.

VSCs have a drawback typical of WECSs – they are placed in the nacelle to minimize stray inductance between the power electronic blocks. It means that the least reliable component of the WECS is hard to maintain. CSCs do not have that problem, since the grid side CSC could be placed in the WT base and utilize interconnection wires as a part of the DC link inductor [20]. It is also possible for VSCs if the boost converter is used as a part of the generator side converter. Moreover, CSCs require two-quadrant switches with reverse voltage blocking possibility, which is often more complicated to drive and associated with either low switching frequency (GTO thyristors) or high conduction losses (switch in series with a diode). CSCs have not been widely accepted in the WT industry and thus are out of scope of this chapter. It is worth mentioning that the CSC could be a suitable solution for PMSG based WTs of medium and high power levels [21, 22].

The VSC is the most commonly used type of converters in the renewable energy field and numerous other applications. The full-bridge topology (Figure 12.8(a)) has a bulky capacitor in the DC link that feeds the converter. VSCs utilize inductive elements for filtering. L-, LC-, or LCL-filters are used for coupling of the VSC and the grid or the generator. The filter removes high order harmonic components of the phase current caused by the inverter switched at high frequency. Filters add some losses and cost to the system, and could raise stability concerns. In PMSGs and WRSGs, filter inductance could be avoided or substantially reduced due to the high phase inductance of the generator. VSCs use two-quadrant switches with reverse current flow possibility that is normally provided by an antiparallel

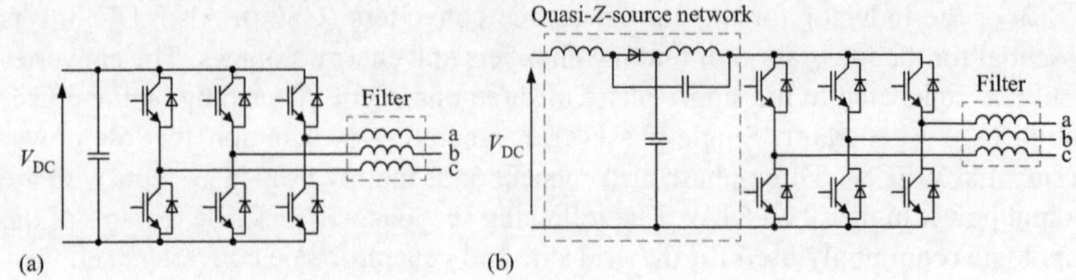

Figure 12.8 Topologies of three-phase full-bridge converters

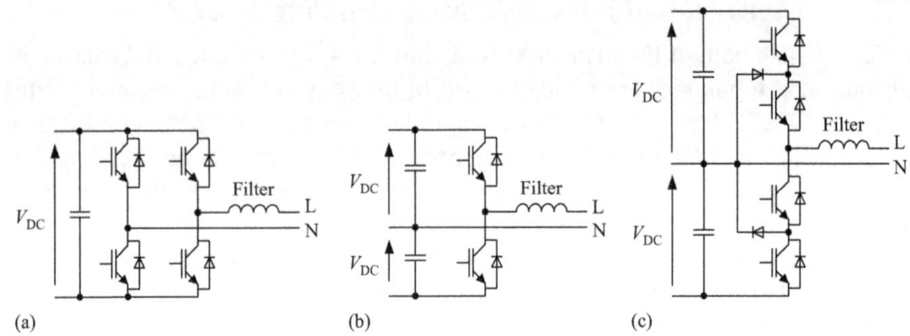

Figure 12.9 Topologies of single-phase voltage source converters

diode. DC link voltage V_{DC} defines the operation of the VSC because it is step down by nature in the inverter mode, and step-up in the rectifier mode. Thanks to its constant DC link voltage, the VSC allows easy integration of the energy storages and braking choppers demanded in several applications. Single-phase systems require much larger DC link capacitance to decouple high instantaneous power ripple from the converter on the other side of the DC link, therefore they are typically used in low power systems.

Recent growing trends in the wind energy sector are directed to the penetration of SWTs in the residential energy systems. Small scale WECSs are usually integrated into the single-phase grid of the household. The PEC for such systems should have a single-phase converter on the grid side. There are three single-phase VSC topologies that could be beneficial in residential WECSs: full-bridge VSC (Figure 12.9(a)), half-bridge VSC (Figure 12.9(b)), and neutral-point clamped (NPC) half-bridge VSC (Figure 12.9(c)). The single-phase VSC requires high capacitance in the DC link to withstand instantaneous power ripple, the frequency of which is two times higher than the grid frequency. The full-bridge VSC has higher DC link voltage utilization, because half-bridge VSCs require two times higher voltage to provide the same output voltage magnitude. On the other hand, the half-bridge VSC requires only two switches with the blocking voltage two times higher than that of the switches of the full-bridge VSC. The NPC half-bridge VSC has the same number of switches with similar blocking voltage as in full-bridge VSC, but requires two additional diodes and one more DC link capacitor;

the NPC half-bridge VSC has the same output voltage as the full-bridge VSC. Despite these disadvantages, the NPC half-bridge VSC could be beneficial when high DC link voltage is imposed by the generator construction. Also, it could be utilized when the generator side VSC comprises the three-level NPC topology with reduced number of switches and the whole three-phase to single-phase PEC could be implemented with a single NPC three-phase power transistor module.

Many existing WTs have low voltage output even at high power levels, which has contributed to a wide use of two-level VSCs. This technology is also widely adopted by industry in adjustable speed drives. It is simple and highly reliable. In high power applications, power rating of a single converter is limited by existing components. Currently, the limit is around 1 MW and rising with semiconductor technology improvement. Parallel connection of converters is used for higher power levels [23]. Full-controlled PEC is a solution preferred because it can provide high power quality, while the price of implementation could be a concern.

The VSCs lack the capability of withstanding the shoot-circuits across the transistor bridge, which limits their reliability in harsh electromagnetic interference environment. That disadvantage is not characteristic of impedance source (IS) inverters, which have recently become popular for grid side converters. Typically, these inverters have buck–boost voltage conversion possibilities: shoot-through states are utilized for voltage step-up, while it could operate as an ordinary VSC when shoot-through is not used. Voltage fed quasi-*Z*-source inverters (qZSIs) shown in Figure 12.8(b) have been verified as a suitable solution for renewable energy applications [24]. The topology contains a typical two-level VSC and a quasi-*Z*-source network on the DC side that provides continuous input current. It means that functions like MPPT could be moved from the generator side converter to the grid side converter.

The PMSG based WECSs accept uncontrolled or half-controlled rectifiers on the generator side [25]. They are simple, cheap, and reliable. The simplest solution for the three-phase systems is the six-pulse diode rectifier (Figure 12.10(a)). It has no control, thus functions, like MPPT, should be performed by the grid side converter. A diode rectifier can be combined with the qZSI for better performance. Also, the grid side converter with a diode bridge could include a DC/DC converter

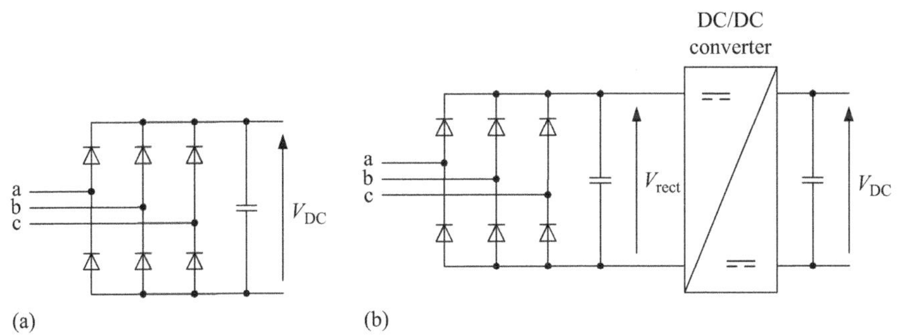

Figure 12.10 Topologies of the generator side VSC with a diode bridge rectifier

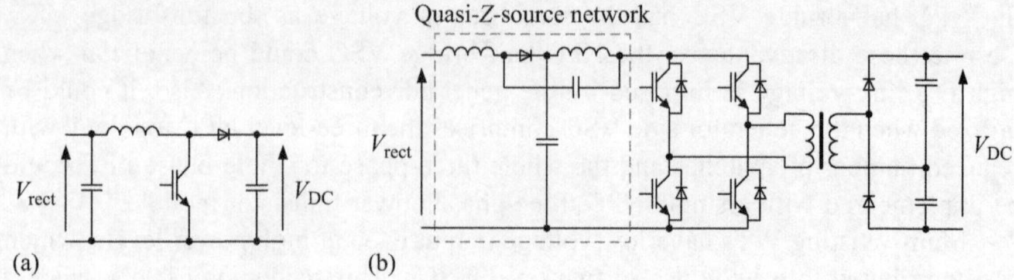

Figure 12.11 Topologies of the DC/DC converter

for MPPT and voltage matching, as shown in Figure 12.10(b). It contains two DC links: an uncontrolled DC link (V_{rect}) with varying voltage that depends on the generator output voltage and a controlled DC link (V_{dc}) with constant voltage that is defined by the grid voltage. The DC/DC converter provides the MPPT control, voltage matching between the varying generator side DC link and the constant grid side DC link. It could contain a galvanic isolation when high voltage step-up is required, or additional safety is a concern, like in low power low voltage residential applications.

The boost converter shown in Figure 12.11(a) is widely used in configurations with a diode rectifier. It has no isolation, but provides high reliability. It is a perfect fit for WT applications because of high efficiency and unity DC voltage gain at the highest power and high DC voltage gain with lower efficiency at the lower power levels. This makes the design of the converter easier, especially in terms of heat extraction. Isolated topology could be a better solution when low voltage WT is integrated into the grid, in which case the bulky line frequency transformer could be avoided owing to the high frequency or medium frequency transformer used for isolation within the DC/DC converter. Among numerous conventional and emerging topologies, the qZSI based DC/DC converter shown in Figure 12.11(b) seems promising for the low power WTs. It has been adopted for fuel cells, solar PV and other modern applications [26]. Experimentally verified for low voltage PMSG based SWTs, it is recommended in [27]. The converter contains the qZSI at the input side, a voltage doubler rectifier at the output side, and an isolation transformer in the middle. The voltage doubler rectifier and the transformer define the main DC voltage gain, while the qZSI is used to adjust the transformer input voltage through the utilization of the bridge shoot-through states.

12.2.2 Common PEC topologies for WTs

The simplest possible unidirectional PEC topology is shown in Figure 12.12. It fits small power PMSG based WTs mostly because of its simplicity and robustness [25]. The DC link voltage is varying in a wide range, which strictly depends on the wind speed. The grid side two-level voltage source inverter injects power to the grid starting from a certain DC link voltage, which is sufficient to start energy transfer using the highest value of the modulation index. Such a system cannot utilize wind power at low wind speeds. Since high wind speeds have low probability, most of the

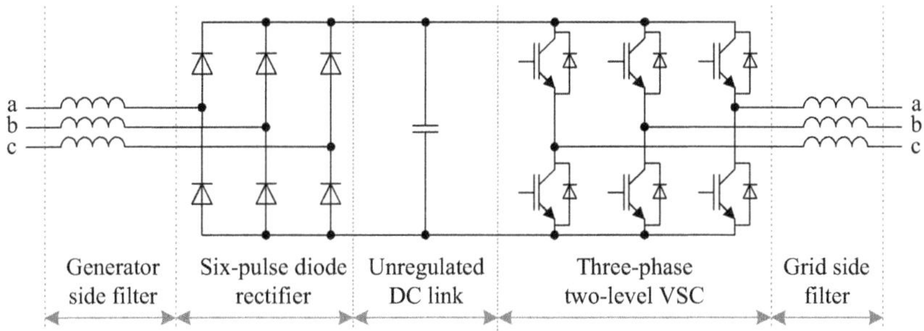

Figure 12.12 Simple power electronic converter for small power PMSG based wind turbines

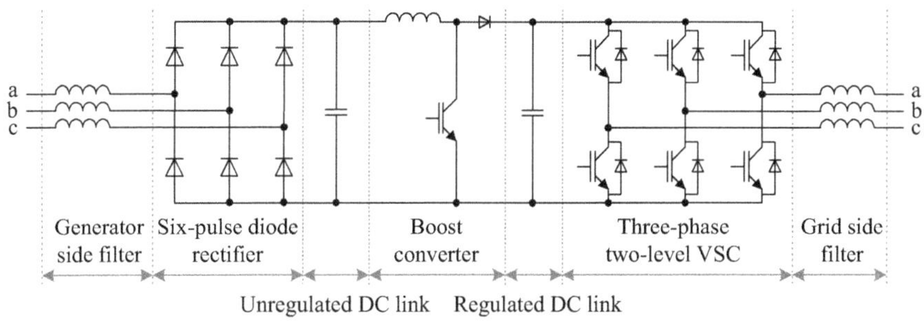

Figure 12.13 Power electronic converter with an intermediate boost converter for PMSG based wind turbines

wind energy is unused by this solution [28]. Another disadvantage is that a diode rectifier consumes non-sinusoidal current from the generator. This technology is one of the first used in wind energy generation. It has very limited operation range and should be avoided in medium and high power applications. Currently, it is replaced by more enhanced topologies, which have attracted interest along with the rapid development of power semiconductor components.

Boost converters can substantially improve the performance of the simplest unidirectional PECs. Figure 12.13 shows the PEC with two DC links. The first one is unregulated and its voltage varies depending on the wind speed. The boost converter provides the MPPT function and transfers energy from the first DC link to the second with the constant voltage. The grid side three-phase two-level inverter injects power to the grid and maintains power balance using the outer control loop with the stabilization of the second DC link voltage. As typical of any VSC, it has inductive filters on both sides. The filter on the generator side could be avoided because of high phase inductance of the PMSG. Also, the boost converter decouples DC links that have different ripple frequency. Thanks to its simplicity and reliability, this topology has been widely adopted by industry at different power levels [29]. Its main advantage over the topology shown in Figure 12.12 is the better performance over all operation ranges of the wind speeds. In this way, much larger quantities of power could be obtained throughout the year, especially in regions with low average wind speed. There are power modules that contain all necessary power electronic devices for the implementation of this PEC topology using a single semiconductor module up to medium power levels. The single-module solution improves the reliability and simplifies the manufacturing process substantially. At higher power levels, the boost converter could be built as a separate block that contains single or several interleaved blocks. Multichannel boost converters with interleaved control are used in industry for WECSs with rated power higher than a few hundred kilowatts [30].

Topologies presented in Figures 12.12 and 12.13 are unidirectional and could be mostly considered for the WRSG or PMSG based WTs. In SCIGs the generator side converter has to be bidirectional and full controlled to process the reactive power and provide a high power factor. Active topologies are also applicable for the PMSG based WTs from the power quality point of view, but the price of such solutions is typically higher. The combination of two bridge VSCs is often referred to as the back-to-back converter. Two-level version of the back-to-back converter is shown in Figure 12.14. This converter supports bidirectional power flow. The generator side VSC works in the active rectifier mode and consumes sinusoidal current from the generator. It provides DC link voltage higher than the generator line voltage amplitude. Control of both VSCs could be decoupled. The control system of the generator side converter provides MPPT and reactive power, i.e. magnetizing current control. The grid side VSC maintains power balance through power injection to the grid and thus controls the DC link voltage. Back-to-back converters have been adopted by the industry for adjustable speed drives for many years. This solution is proven; therefore, it is trusted by manufacturers of power electronic equipment. Modern market of power semiconductor components has a huge variety of power electronic modules that incorporate the single three-phase two-level inverter in a single case. Some manufacturers, like Infineon, have placed ready to use stacks on the market that contain all semiconductors and DC link capacitors specifically for wind energy applications. This topology is most widely used for the WTs even at multi-MW levels when they are connected in parallel [10, 31].

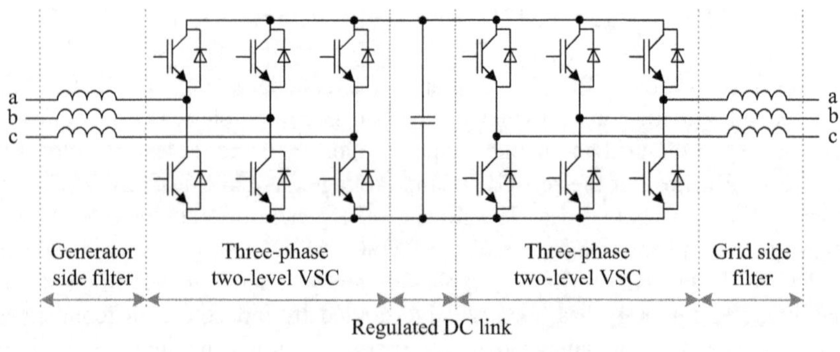

Figure 12.14 Back-to-back voltage source power electronic converter

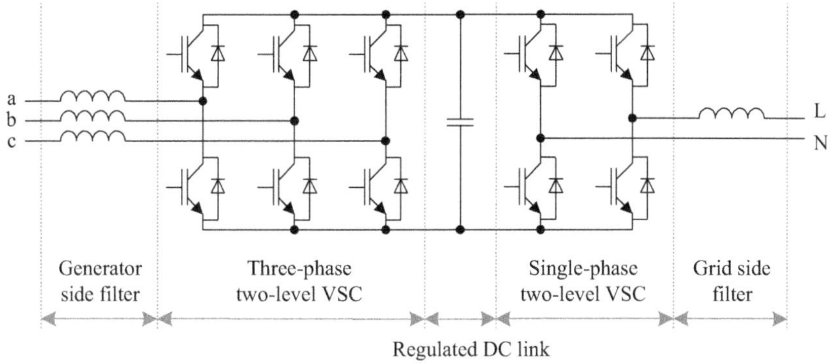

Figure 12.15 Back-to-back voltage source power electronic converter for small residential wind turbines

The back-to-back PEC with three-phase to single-phase energy conversion (Figure 12.15) could be used for grid integration of small residential WTs into a single-phase grid of the household. Three-phase VSCs are used on the generator side, since generators usually have three-phase output. The main disadvantage is the use of single-phase connection to the grid, which is restricted by the application field. It requires a bulky DC link capacitor that decouples high instantaneous power ripple on the grid side from the generator side converter. Despite this, the converter has all the other advantages of the back-to-back converter shown before. Its operation principle is also quite similar, allowing the decoupled converter control. Design of PECs for residential WECSs requires much effort because residential systems impose part of requirements common for consumer electronics field, like low price and high reliability.

12.2.3 Emerging PEC topologies for WTs

Rapid penetration of renewable and alternative energy sources has urged development of new power electronic topologies. New concepts are needed to optimize size, cost, and performance. Multilevel topologies have high performance, but they are costly. Simplified multilevel topologies are interesting in low and medium power applications. Another interesting approach is the IS technology. For example, qZSIs can be used with simple diode rectifiers. These two modern promising technologies are described in this section.

Three-level NPC VSCs that have gained wide acceptance in industry during the last decade are a three-level variation of the diode clamped multilevel VSC. Three-level NPC VSCs are already used for high power MV WTs, but the technology is still expensive for low power WTs. This could be improved through a combination of the multilevel technology and the technology of the four-switch three-phase inverter called the B4 inverter. In this case, we can decrease the number of semiconductor components and the price. The PEC in Figure 12.16 that is proposed in [32] suits for low power applications where the three-phase

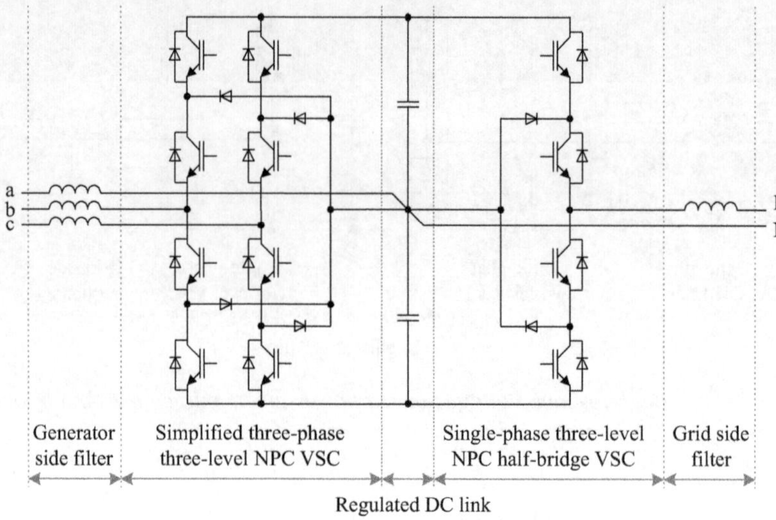

Figure 12.16 Three-phase to single-phase simplified three-level NPC power electronic converter

generator is integrated into a single-phase grid. It could be applied to both induction and synchronous generators. Multilevel approach improves the performance of the B4 inverter considerably because the DC link voltage must be higher than that of the conventional three-phase VSC. The main advantage of this topology is in the use of existing power semiconductor components: the whole PEC could be implemented with a single power electronic module of a three-phase three-level NPC bridge available off-the-shelf. This reduces the price and improves the reliability of the industrial product.

Another promising technology that addresses PMSG based WTs is the qZSI. It overcomes the limitations of VSCs and CSCs since it has the buck–boost operation possibilities. The qZSI can handle all the control functions on the grid side. It allows using the simplest diode bridge three-phase rectifier on the generator side, as shown in Figure 12.17 [33]. DC link is unregulated and depends on the wind speed. It does not require high capacitance for energy storage, because the quasi-*Z*-source network also stores energy. The quasi-*Z*-source network is also used for voltage step-up. It serves as an imaginary DC link for the inverter bridge during the active state, and steps up voltage during the shoot-through state. To improve the efficiency, it is recommended that qZSI works at the unity gain factor or in the buck mode at the rated input power. Like in CSCs, the generator side converter and the grid side converter can be placed far from each other. Then the inductance of the wires will just contribute to the input inductance of the qZSI. Therefore, the highly reliable diode bridge can be placed in nacelle, while the qZSI could be placed in the WT base. This technology is new and requires verification of long-term reliability in the field.

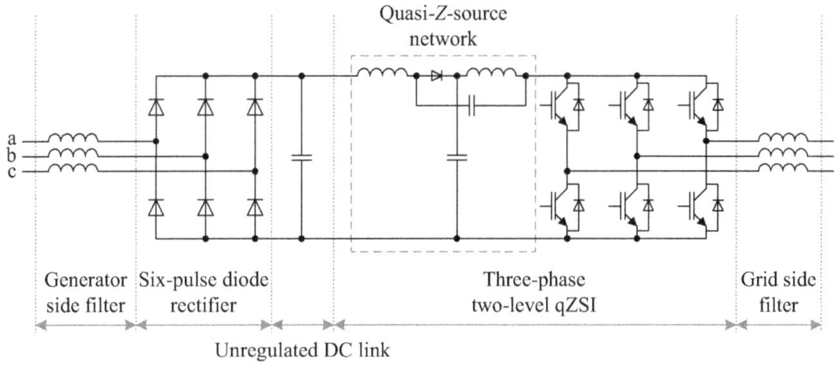

Figure 12.17 Quasi-Z-source inverter based power electronic converter

12.2.4 PEC topologies for high power WTs

Major part of wind energy is generated using high power WTs. Currently, this industry is in the stage of wide commercialization of the variable speed 6 MW WTs with the full power converter. Power rating of WTs on the market is growing to the multi-MW level. High power WTs allow the capital costs and wind energy prices to be reduced. Multi-MW power level is highly challenging because most of the high power WTs are low voltage [10]. To handle high currents, two-level back-to-back VSCs are usually connected in parallel using the same three-phase winding, or single three-phase winding per converter. This approach could be implemented using $n + 1$ redundancy since reliability of the system is considerably lower than that for the single converter. For example, the PMSG based WECS from Gamesa rated at 4.5 MW contains six two-level back-to-back VSCs rated at 950 kVA each [34]. Power density of this solution is 0.58 MW/m^3 and it utilizes 1,700 V insulated gate bipolar transistor (IGBT) to operate with 690 V PMSG.

It is obvious that manufacturers of high power WTs have to switch to the MV range to improve the overall performance of the WECS and decrease energy prices. In this case, the line frequency transformer could be avoided. Multilevel topologies are the most promising for MV applications. There are several possible solutions: diode clamped VSCs, flying capacitor VSCs, multilevel modular converters, etc. Flying capacitor VSCs could provide low price at MW power levels, but their reliability is assumed to be too low [31]. Diode clamped multilevel converters, especially NPC VSCs, have been widely adopted by industry. This urges their utilization in the wind energy field. The back-to-back converter could also be adopted for MV applications when each switch is composed of series connected switches. The number of switches in series depends on the blocking voltage of a single switch and the level of the DC link voltage. This section will review several available and emerging multilevel converters for high power applications.

The three-level NPC back-to-back converter shown in Figure 12.18 has been investigated by ABB [35]. ABB has adopted the topology within their PCS6000 MV WT converter with a rated power up to 8 MW [10]. It utilizes integrated

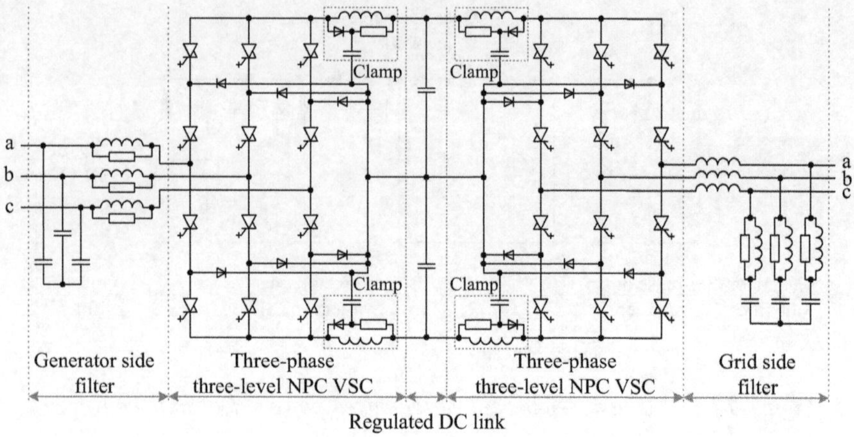

Figure 12.18 Three-level NPC back-to-back power electronic converter

gate-commutated thyristors (IGCTs) with a reverse conduction possibility. The wind energy conversion performance of this converter is similar to that of the two-level version. Increased number of the output voltage levels leads to a smaller grid and generator side filters. Also, a filter would be smaller than in the two-level VSC because the inverter switching frequency is four times higher than the switching frequency of a single device.

The PCS6000 converter is rated for a generator with the output voltage up to 4 kV. Implementation of the converter with the IGCT has some features related to the operation of the high voltage switches. Clamp circuits are used to limit the voltage overshoots across the switch after its turnoff. Clamp circuit also contains an inductor that limits the di/dt when the current is switched from the freewheeling diode to a turning-on the IGCT. This di/dt constraint is imposed by the co-packed diode. Also, the generator side filter includes circuits that limit the dv/dt over the windings. The grid filter may also require a similar dv/dt limitation if the converter is connected to the grid through the transformer that cannot withstand the rapid voltage rise caused by the switching inverter.

The previous topology could be simplified on the generator side for the purposes of the synchronous generator based WTs. In this case, a six-pulse diode rectifier with the three-level boost converter could be used with the three-level NPC VSC [36]. The output voltage of the high power WTs will apparently rise to the 10 kV level in the next years [37]. Four-level VSCs could be more appropriate for the MW-level WTs with higher MV output. The topology shown in Figure 12.19 and described in [37, 38] is intended for future high power WTs in the range of 10–15 MW. It is a good example how to combine the two-level and multilevel converters in a single system. The generator side converter utilizes the two-level diode bridge rectifier, where three diodes are connected in series to withstand high operation voltage. The rectifier feeds unregulated DC link with voltage that depends on the wind speed. The second regulated DC link is coupled with the first one through the four-level boost converter. The number of levels was defined by

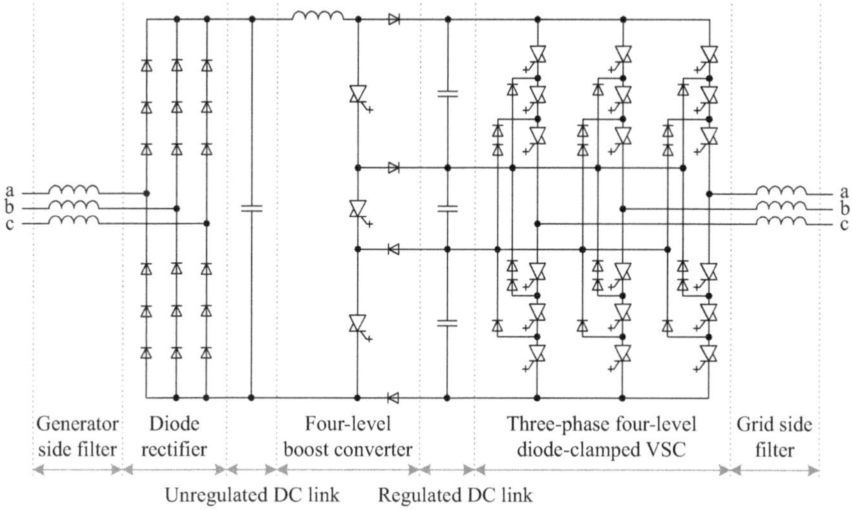

Figure 12.19 Four-level unidirectional power electronic converter

the blocking voltage of a single device and the required voltage of the second DC link. Regulated DC link has three capacitors charged to the equal voltage. They feed the four-level diode clamped inverter. This topology is shown with reverse conducting IGCTs because these switches have the highest blocking voltage among the available high performance full-controlled alternatives. Four voltage levels per phase lead to a lower dv/dt stress, as well as to a smaller grid side filter. Also, the four-level inverter has a higher equivalent switching frequency than the commutation of switches in the inverter.

Past designs of power converters were oriented on the low voltage WTs with the output voltage up to 690 V. The 1,700 V IGBTs have been extensively used for that voltage range. Similar power devices could be used to integrate the low voltage WT into the MV grid without using of the step-up transformer. In this case, several two-level back-to-back VSCs fed by the different isolated windings could be connected in series on the grid side [39, 40]. Modern trends in high power energy conversion systems are within the multilevel modular topologies. Cascaded H-bridge and modular multilevel converter (MMC) topologies are well known. Cascaded topologies seem to have higher reliability than that of the cascaded connection of the discrete converters. The MMC topology seems to be the most promising for the high-voltage direct current (HVDC) systems. However, it has not been widely investigated for WTs, but it could be a high performance solution due to the superior scalability of the voltage and power levels [41]. Due to the high number of the voltage levels, this topology can operate without the grid and generator side filters [23]. Utilization of this topology can be limited by available rating of the generator isolation voltage and overrated capacitors in each cell that is caused by the low frequency input voltage from the generator. This solution needs additional consideration taking into account the constraints of the WT technology.

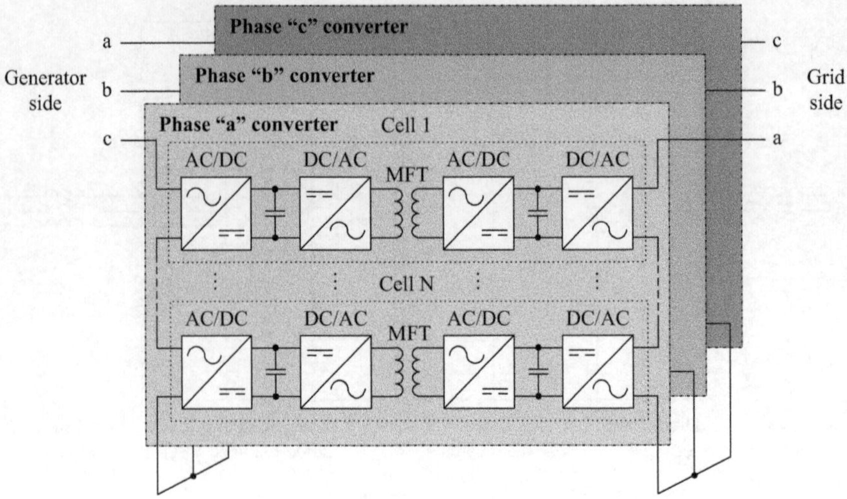

Figure 12.20 Cascaded H-bridge high power converter with medium frequency isolation

There are several solutions for emerging high power WTs that are derived from PEC for electric drives. The cascaded H-bridge converter that utilizes a medium frequency transformer for isolation (Figure 12.20) was first proposed for high power traction applications [42]. Performance of that converter has been tested under the framework of the European UNIFLEX-PM Project [43]. This cascaded converter contains several cells connected in series on the input and output sides. Each cell is based on the back-to-back VSC with a medium frequency isolation transformer. The transformer operates at the frequency from several up to dozen kilohertz. Its structure is modular and $n + 1$ redundancy can be implemented, which is important in WT applications. Filters on both sides could be avoided if the number of levels is high. Implementation of the medium frequency isolation transformer leads to high power density since the line frequency transformer on the grid side could be avoided. Also, this topology can be designed for the direct integration of WTs into the MV power distribution grid with voltages of 10 kV and higher. Since this topology is redundant and versatile, it could be the most realistic candidate for the adoption by industry in the nearest years.

12.3 Control algorithms for PECs

There are many control methods that can be applied to WECSs. Assuming that the energy is transferred from the WT to the electrical system, they can be divided into two groups: AC/DC converter control (also called generator side control) and DC/AC converter control (grid side control). Depending on the converter topologies, the appropriate modulator is applied in the control algorithm [44–48], but other control loops remain the same for a specific algorithm. The most important part of the control system for WTs is the MPPT algorithm, which provides an efficient use of RESs and is implemented usually in the generator side control. Out of many different MPPT types, two basic groups are most common. The first type requires prior knowledge of turbine parameters to calculate the operating point. The second group iteratively searches the optimum with the use of power and rotational speed increments.

12.3.1 Maximum power point tracking

One of the well-known and frequently used methods to extract maximum power from a WT is an algorithm based on the knowledge of the $P(\omega)$ characteristic (Figure 12.21(a)) [49–51]. For each designed WT, the distinctive characteristic of the generated power versus the angular rotor speed can be determined. Furthermore, the maximum power point on these characteristic will be changed with the wind speed. As a result, there is a possibility to extract the curve of the maximum power related to the rotor angular speed (Figure 12.21(b)). There are two methods to implement this characteristic in the control system. First, the curve can be stored in a pre-programmed 2D lookup table and the second is the curve approximation by the third-order polynomial [6]. Therefore, the reference value of the rotational speed is determined by the power calculation as shown in Figure 12.21(a). The main advantage of this method is simplicity and fast response, hence the algorithm immediately finds an optimum operation point. Unfortunately, accurate determination of an optimum $P(\omega)$ curve is unavoidably connected with aerodynamic tests using the wind tunnel, which in practice is omitted due to the high cost. Moreover, it may vary significantly with time and atmospheric conditions. Factors like icing, dirt, and aging are the causes that lower the efficiency of the turbine rotor and cannot be compensated by the control algorithm.

Another well-known solution is the MPPT algorithm based on the measured wind speed [52, 53] shown in Figure 12.22. It requires additional wind

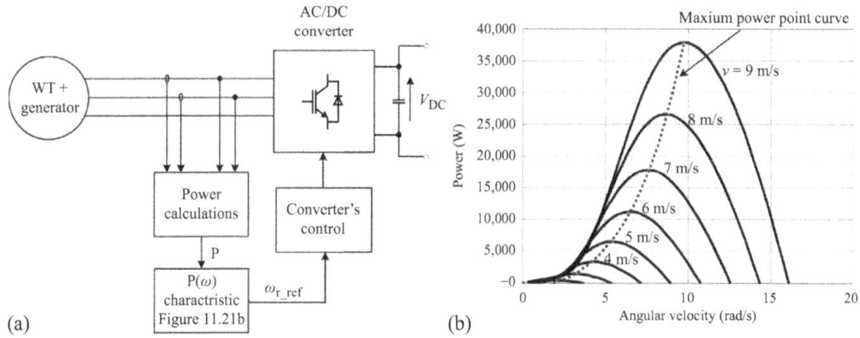

Figure 12.21 (a) Block diagram of the MPPT algorithm with the P(ω) characteristic, (b) example of the P(ω) characteristic

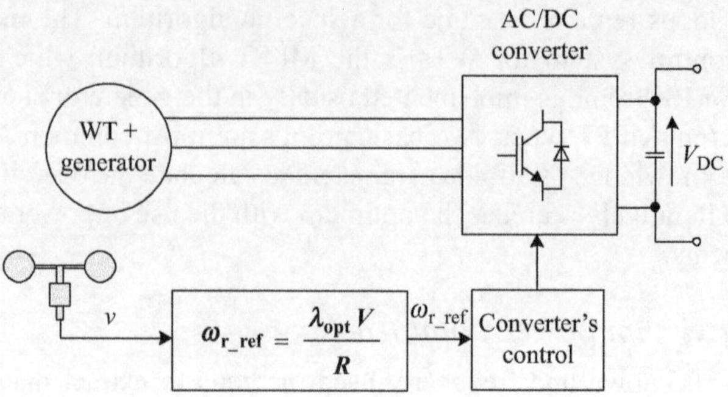

Figure 12.22 Block diagram of the MPPT algorithm with wind speed measurement

measurements by the anemometer that allow the calculation of the referenced generator rotational speed:

$$\omega_{r\_ref} = \frac{\lambda_{opt} V}{R} \tag{12.1}$$

where V is the actual wind speed.

In this method, the knowledge of optimum tip-speed ratio λ_{opt} and the rotor radius R is required, but the measurements of voltages and currents used in the previous method are eliminated. For an optimal λ_{opt}, the power will be always maximized, which corresponds to the maximum power point curve presented in Figure 12.21(b). This control method is characterized by some important features. First, wind speed information used as an input variable allows for the determination of the fast operating point and the dynamic response. On the other hand, data acquisition from the anemometer has to be insensitive to rapid wind gusts to eliminate unwanted changes in the operating point set by the control algorithm, because it may cause high torque differences resulting in the mechanical stress on the turbine rotor blades. As in the case of the previous MPPT, the algorithm λ_{opt} cannot be determined with a high degree of accuracy and varies with time and atmospheric conditions. However, λ_{opt} is a single parameter and it can be adjusted much easier than the polynomial coefficients in the optimum power curve approximation [6].

There is a possibility to reach the maximum power point for WT without any information of WT parameters. It can be realized by iteratively changing the control variable [49, 54–58], which results in rotor rotational speed adjustment (Figure 12.23). Although the steepest ascent method is generally used for that purpose, all algorithms of this type are based on the assumption that for a given wind speed, maximum power is produced when:

$$\frac{\Delta P}{\Delta \omega} = 0 \tag{12.2}$$

where ΔP is the power increase in the next step and $\Delta \omega$ is the rotational speed increment.

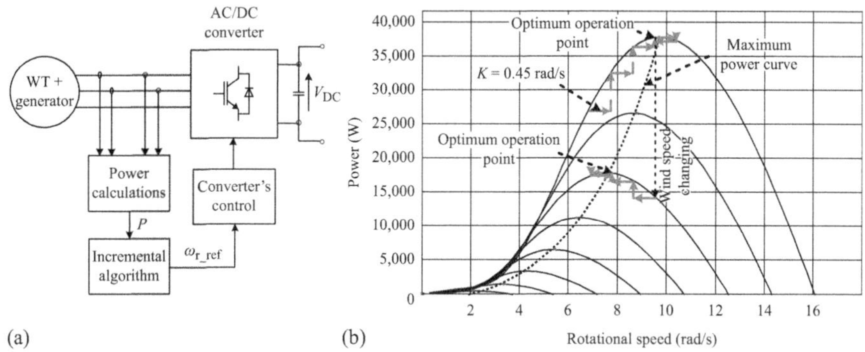

Figure 12.23 (a) Block diagram of MPPT with the incremental algorithm and (b) example of the incremental MPPT process

The incremental algorithm implemented in this solution usually operates at the low sampling frequency f_T because of high wind rotor inertia [6]. It requires voltage and current measurements for the power calculations. The reference rotational speed is incremented in each cycle and the power derivative is calculated. In order to reach condition (12.2), it is necessary to sign the assignation of the result. If $\frac{\Delta P}{\Delta \omega} > 0$, then the reference rotational speed ω_{r_ref} has to be increased in the next step, otherwise if $\frac{\Delta P}{\Delta \omega} < 0$, the value of ω_{r_ref} has to be decreased.

The process of the incremental MPPT is presented in Figure 12.23(b). The most important control parameters in this method are the step size K and the sampling frequency f_T. Step size is usually constant, but there are solutions with a variable step in order to improve the dynamic and minimize the oscillations around the optimum operating point [6]. Proper selection of the sampling frequency f_T is also important. Too high value may effect in false response of the algorithm, but with a very low value, the algorithm will not be able to track the wind changes with adequate precision, decreasing the efficiency of the system.

The main advantage of this method is low complexity and high flexibility, resulting from self-adjusting operation point without knowledge of turbine parameters. These features allow the method to be implemented for each type of WTs. Unfortunately, this algorithm is slower than those described above and the optimum operation point can be determined with an error depending on the selected K and f_T.

12.3.2 Control for DC/DC boost converters

DC/DC boost converter in the WECS is one of the simplest from the control point of view. The block diagram of the control algorithm is presented in Figure 12.24. This method is composed of two control loops with proportional-integral (PI) controllers.

The outer loop corresponds to the rotational speed regulation of the generator in order to achieve the reference value ω_{r_ref} from the MPPT algorithm. As an input for the controller, the speed error between the reference value and the measured or estimated value (sensorless method) ω_r is used. The output of that loop is the referenced value of the inductor current in the boost converter. It is compared with

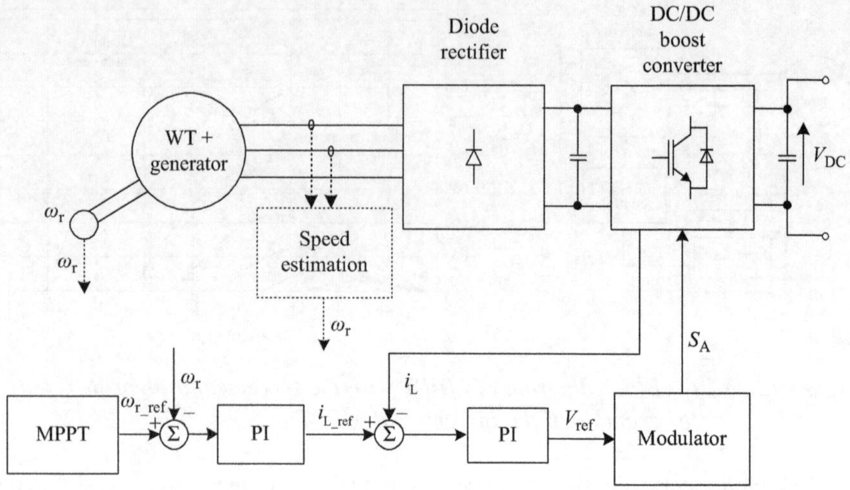

Figure 12.24 Block diagram of the control algorithm for the DC/DC boost converter

the measured value of that current i_L and the error is eliminated by the inner control loop. The signal V_{ref} from the inner loop is used in the modulator to the gate signal generation.

12.3.3 Control for IS converters

IS technology is an emerging trend in electric energy conversion. This technology was applied also for WECSs. An inverter can perform shoot-through and open states without damage and thus achieve a wide input voltage regulation range. Voltage fed configurations of quasi-*Z*-source and *Z*-source inverters have been adopted for grid integration of the PMSG based WTs [59, 60]. In combination with the input diode bridge rectifier, they can perform MPPT and power injection into the grid within the single inverter stage. Example of such a PEC based on the qZSI is shown in Figure 12.17. IS inverters differ from each other in the type of the IS network utilized, and in most cases they could be controlled in the same way. They use the same principle for control as VSCs with modifications related to the voltage control within the inverter stage. It is usually performed with inclusion of shoot-through states into the inverter modulation. This leads to the reduction of the modulation index range, in the case of sinusoidal carrier based PWM. The MPPT for PMSG based WTs could be performed using the dependence between the optimal rotation speed ω_r and the rectifier output voltage V_{rect} that is easily approximated [59, 60]. A typical control system that could be applied to at least *Z*-source and quasi-*Z*-source based PECs is shown in Figure 12.25(a).

The control system is based on the rotating reference frame with modifications. Transformations from the abc to *dq* basis are conventional. The input MPPT block defines the reference output voltage of the diode rectifier. This value could be used for two typical control types that are selected with the control mode selection

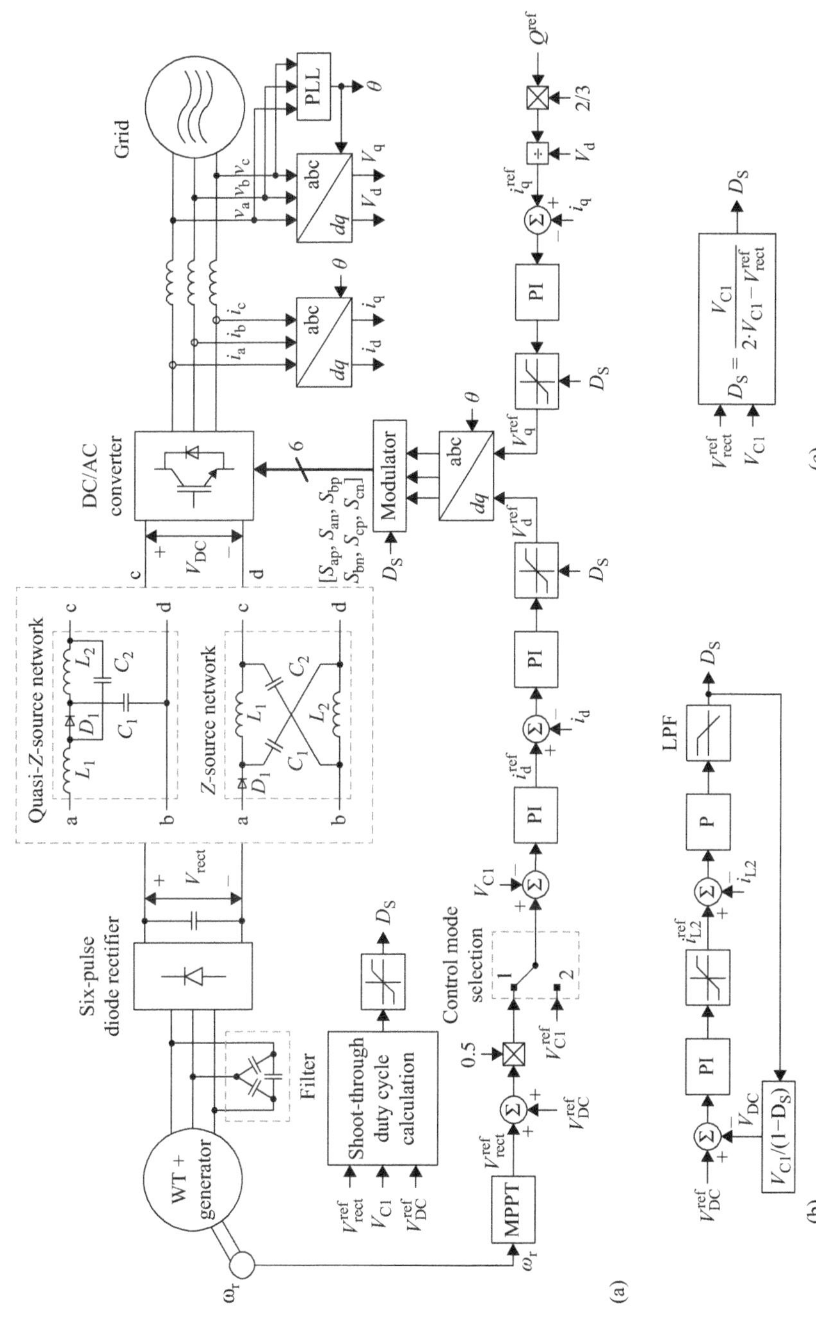

Figure 12.25 Control system of the impedance source converter for PMSG based wind turbines

switch. In position "1" the control system operates in the indirect DC link control mode. Also, the control system can perform direct voltage control of the capacitor C_1 in the IS network when the selector switch is in position "2". In both cases, the capacitor C_1 voltage error defines the input of the PI controller that adjusts the reference value of the grid current direct component i_d. The error of that component defines the direct component of the reference grid voltage, which is applied to the modulator. The grid current quadrature component could be set equal to zero, but in a general case, it is recalculated from the reference reactive power Q_{ref}.

In general, the control system shown in Figure 12.25(a) looks quite similar to that of VSC. The main difference is within the block that calculates the shoot-through duty cycle D_S. It could be calculated in different ways. This value should be adjusted in order to provide the required voltage at the input of the inverter. This could be done by simple recalculation from the equations of the IS network in steady state, as shown in Figure 12.25(c) [60]. The *Z*-source and quasi-*Z*-source networks add right half plane zeros to the dynamic of the converter. To achieve better dynamic performance and improved stable operation range, the double-loop controller could be used as shown in Figure 12.25(b) [59]. The shoot-through duty cycle value has direct influence on the direct and quadrature components of the reference grid voltage through adjusting upper and lower limits of the saturation blocks at the output of the corresponding PI controllers. Also, D_S serves as a control input of the modulator block. This block can contain any of various modulation methods developed for IS inverters. Basic modifications of the sinusoidal carrier based PWM for the IS inverter as well as fundamentals of the IS technology are described in Chapter 7. Modified space vector PWM control methods for IS inverters are also covered there [61].

Control of the voltage fed IS converter is quite similar to that of VSCs. It uses many basic blocks, like Clarke and Park transforms in the direct and inverse form, phase locked loop (PLL), etc. The main difference is within the implementation of the shoot-through states for the input voltage control in a wide range. It imposes limitations on modulation, since the maximum active state duty cycle is limited by the shoot-through state inclusion into the switching pattern.

12.3.4 Field oriented control

The field oriented control (FOC) is most frequently used for the generator side full-controlled VSC. The space vector of the stator current is oriented to the rotor flux vector Ψ_r in the synchronous rotating coordinate system that splits it into the two rectangular components i_d and i_q [62]. The component i_d is proportional to the amplitude of the rotor flux, whereas the component i_q is proportional to the electromagnetic torque of the generator. The advantage of this solution is the possibility to control both components independently.

This method is also known in the literature as the rotor field oriented control (RFOC). Two types of RFOC are known: direct (Figure 12.26) and indirect (Figure 12.27) [62, 63]. In the direct RFOC, the flux vector position θ_r and the speed ω_r are measured or estimated from the generator voltages and currents [63]. The speed ω_r is compared with the reference signal ω_{r_ref} in the outer speed

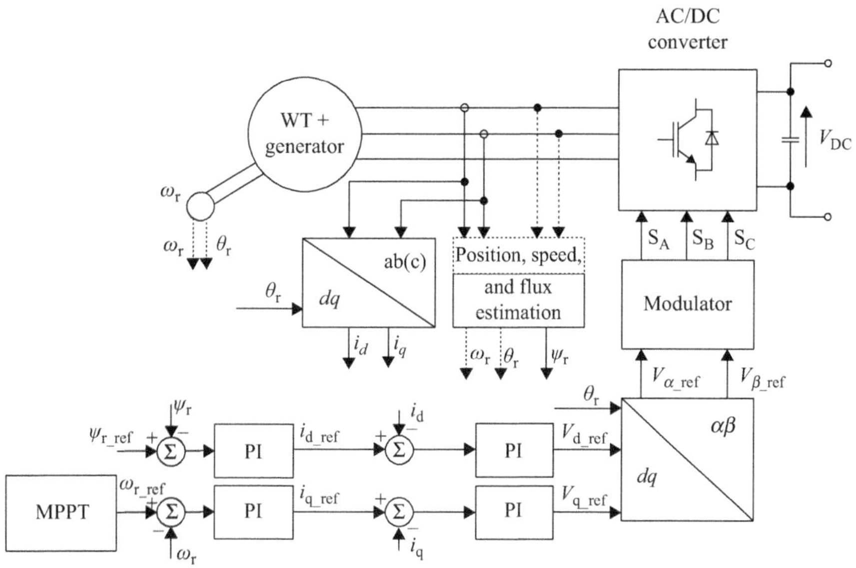

Figure 12.26 Block diagram of the direct RFOC

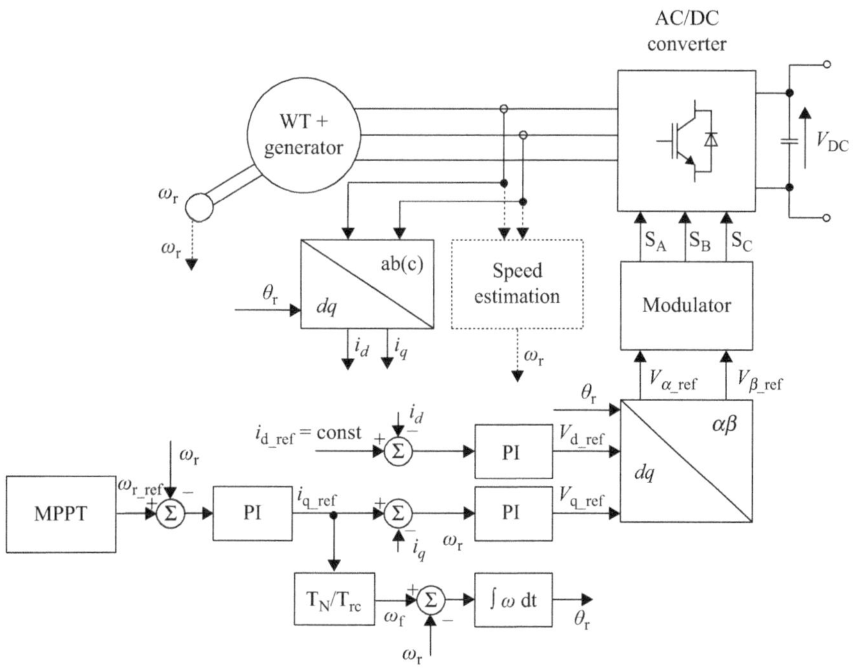

Figure 12.27 Block diagram of the indirect RFOC

controller. Similarly, the estimated rotor flux Ψ_r is compared with the reference rotor flux Ψ_{r_ref} in the flux controller [66]. The outputs of both controllers generate the reference signals of the current i_{d_ref} and i_{q_ref} in the synchronous rotating coordinate system [64]. These signals with measured and transformed current components i_d and i_q [63] create errors for inner control loops. At the output of the inner controllers, the voltage vector components v_{d_ref}, v_{q_ref} are created and used with the flux vector position θ_r in transformation to the stationary coordinate system [63]. After transformation, the voltages v_{α_ref} and v_{β_ref} are used as input for the modulator.

In the indirect RFOC, the flux vector position θ is calculated from the mechanical speed (measurement or estimation) and the current reference value in q-axis (feed-forward control) multiplied by the nominal time constant T_N and divided by the rotor electromagnetic time constant T_{rc}. Indirect RFOC is extraordinarily simple for the PMSG. For that application, the flux vector position θ_r can be calculated only from the speed measurement (estimation). These properties make the method very attractive for practical implementation [65].

12.3.5 Direct torque control-space vector modulated

Another solution for the generator side full-controlled VSC is the direct torque control-space vector modulated (DTC-SVM) [66, 67]. In contrast to the FOC where flux and torque are "indirectly" controlled by a "direct" control of the field oriented current vector components, the DTC-SVM regulates the stator flux Ψ_s and the torque m directly, while the currents are controlled indirectly. In this method, the stator flux Ψ_s and the torque m can be estimated by simple equations [68]:

$$\Psi_S = \frac{1}{T_N}\int_0^t u_k dt \tag{12.3}$$

where u_k is stator voltage, T_N is stator voltage period,

$$m = \frac{3p}{4}(\Psi_{s\alpha} i_{s\beta} - \Psi_{s\beta} i_{s\alpha}) \tag{12.4}$$

where p is the number of poles of the machine.

The block diagram of DTC-SVM is shown in Figure 12.28. The outer speed controller generates the torque referenced value for the inner torque control loop. The second inner control loop is responsible for the flux control. The outputs of the torque and flux controllers define the reference voltages v_{d_ref} and v_{q_ref} in the synchronous rotating coordinate system, which is oriented to the vector of the stator flux Ψ_s. After transformation to the stationary coordinate system [63], the voltage components v_{α_ref} and v_{β_ref} are input signals for the modulator, where the switching signals S_A, S_B, S_C are calculated.

12.3.6 Voltage oriented control

The conventional control method for the grid connected full-controlled VSC is the voltage oriented control (VOC) presented in Figure 12.29. Like the FOC, it is also based on the dq synchronous rotational coordinate system. The referenced DC voltage

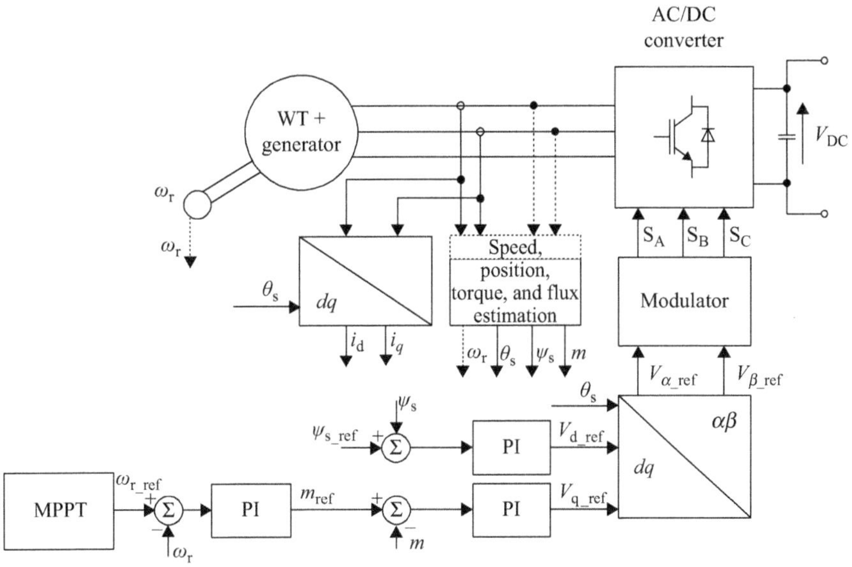

Figure 12.28 Block diagram of DTC-SVM

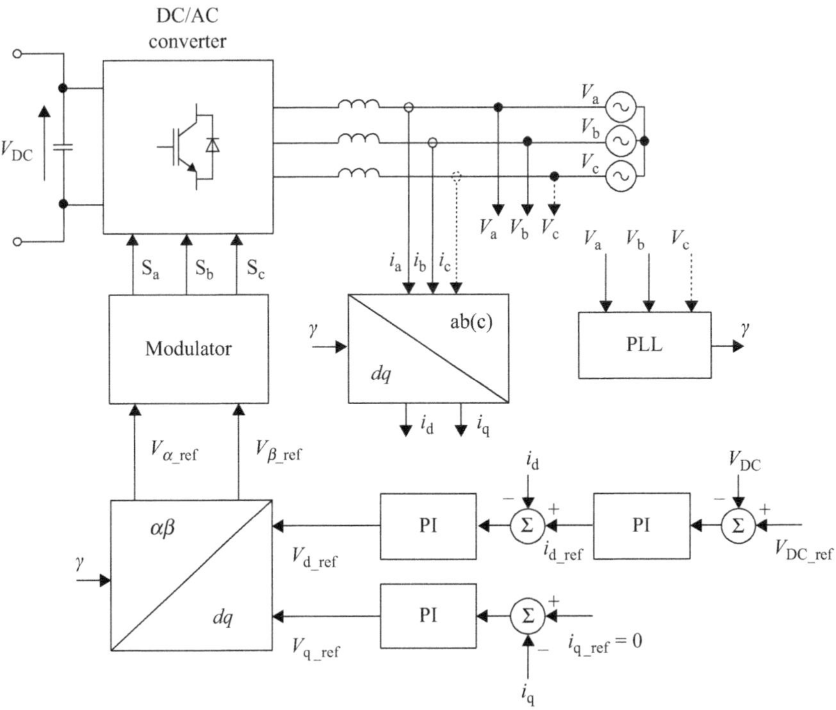

Figure 12.29 Block diagram of the VOC

is maintained by the outer controller, which generates the *d*-axis reference current i_{d_ref}. The *q*-axis current component is zero to obtain the unity power factor. The important feature of this control method is the possibility of the independent control of active and reactive powers by *d*- and *q*-axis current components, respectively. The errors of the measured (i_d, i_q) and the referenced (i_{d_ref}, i_{q_ref}) current values are eliminated by the inner PI controllers. Finally, the referenced converter voltages v_{d_ref} and v_{q_ref} are obtained, which after transformation to the $\alpha\beta$ stationary coordinate system are used in the modulation. In the used transformation to/from *dq* synchronous rotating coordinate system, the grid voltage angle is necessary. Therefore, the simple PLL or second order generalized integrator – phase locked loop (SOGI-PLL) for a more advanced system has to be applied [69].

12.3.7 Direct power control-space vector modulated

Another control method for the grid connected full-controlled VSC is direct power control-space vector modulated (DPC-SVM) [70]. The block diagram of the DPC-SVM is presented in Figure 12.30. In order to obtain direct power control, the instantaneous active and reactive powers are calculated based on the measured and transformed to $\alpha\beta$ stationary coordinate system voltages (v_α, v_β) and currents (i_α, i_β):

$$p = \frac{3}{2} \cdot (i_\alpha v_\alpha + i_\beta v_\beta) \tag{12.5a}$$

$$q = \frac{3}{2} \cdot (i_\alpha v_\beta - i_\beta v_\alpha) \tag{12.5b}$$

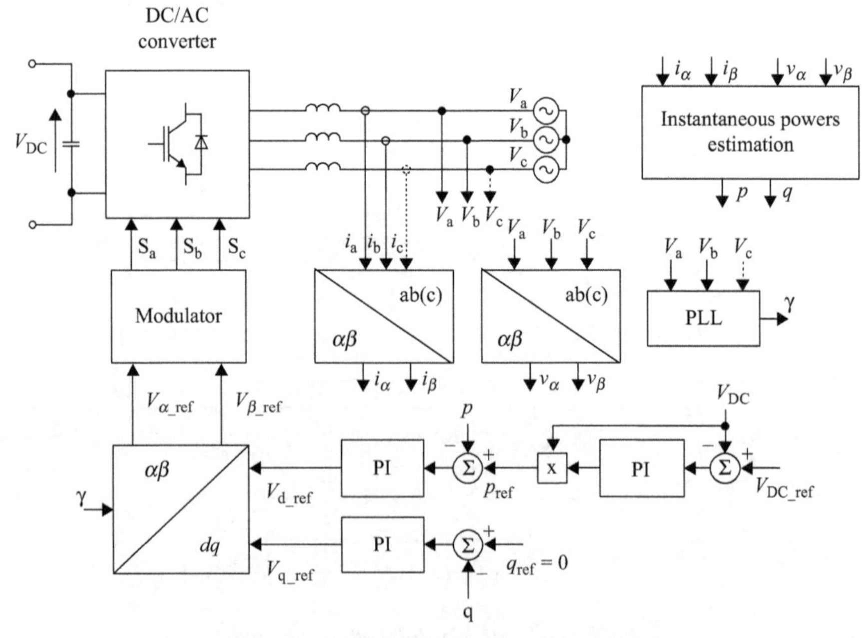

Figure 12.30 Block diagram of the DPC-SVM

Similarly to the VOC, the DPC-SVM allows DC-voltage control at the specified level by using an outer PI controller. The output of the outer control loop is the referenced active power, while the referenced reactive power is set to zero. That solution provides the unity power factor operation. In the inner control loop, two PI controllers are implemented for the active p and the reactive power q independently. Finally, the referenced converter voltages in the synchronous rotating coordinate system v_{d_ref} and v_{q_ref} are generated, which after transformation are input signals for the modulator. For grid connected converters, synchronization is required; hence, the PLL has to be implemented. It generates the voltage vector angle γ, which is used in $dq/\alpha\beta$ transformations.

12.3.8 Single-phase grid converter control

For the grid side single-phase VSC, the simplest control method is created in the stationary coordinate system shown in Figure 12.31 [65]. It consists of an outer DC-voltage control loop and an inner current control loop.

In the inner control loop, the proportional-resonant (PR) controller can be used. The PR controller contains a resonant term with a double imaginary pole adjusted to the fundamental line voltage frequency in order to achieve an infinite gain at the selected frequency. In contrast to the PI controller, PR is able to eliminate not only the amplitude error of the current but also the phase error and it provides a zero static error at the selected frequency.

Another control method for the single-phase VSC is based on the virtual dq synchronous rotating coordinate systems, which is copied from the three-phase system (Figure 12.32). The most complicated issue in this method is the virtual coordinate, since it contains only an one-phase system. The simplest method is to calculate current's $\alpha\beta$ coordinates based on delaying the measured current i_{grid} by ¼ of line voltage period or notch filters with narrow stop-band (e.g. second-order Butterworth). As a result, the i_β is received and i_α is equal to the grid current [71, 72].

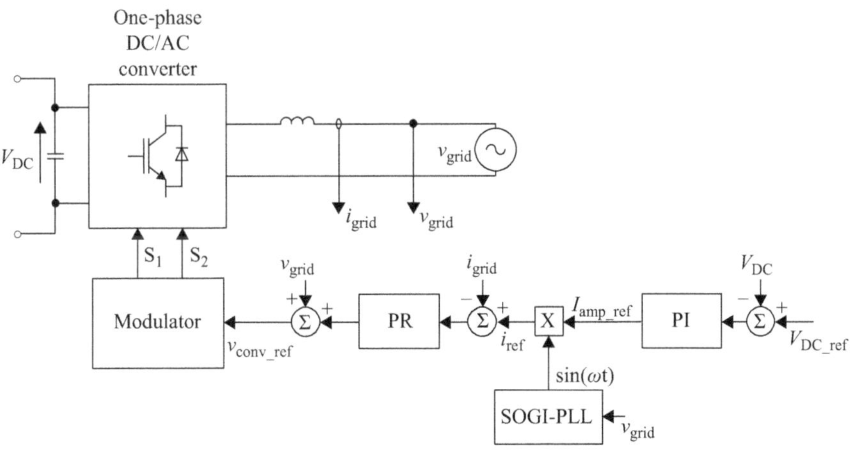

Figure 12.31 Block diagram of one-phase converter control with the proportional-resonant controller

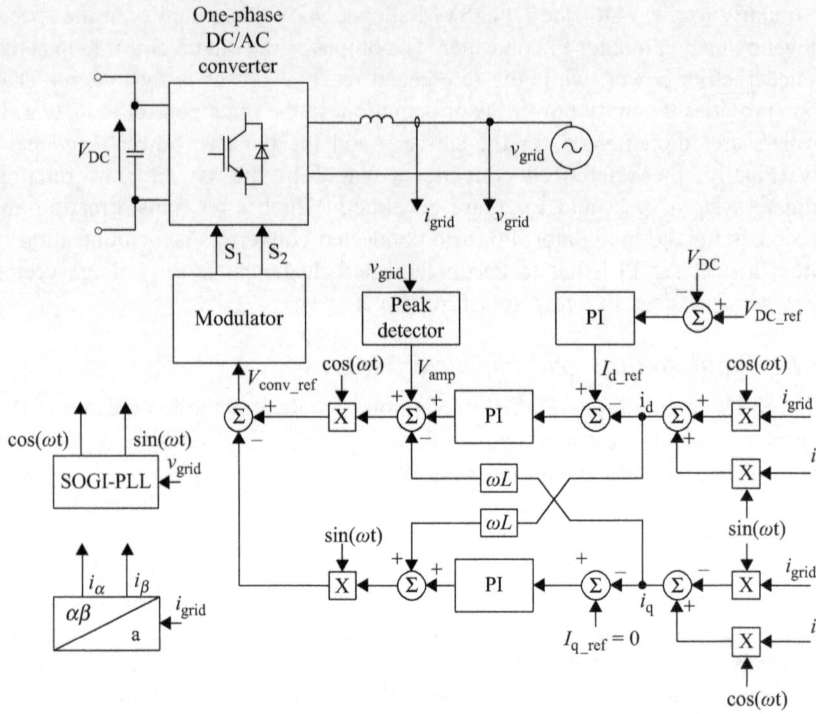

Figure 12.32 Block diagram of the single-phase VSC control with the virtual dq synchronous rotating coordinate system

12.3.9 Control for stand-alone mode of operation

Since WTs are popular in rural places, the stand-alone mode of operation can provide uninterruptible power supply for local loads during the grid faults. Moreover, the stand-alone control methods can be used in places without the grid (e.g. small islands).

The simplest control method is based on the closed voltage control loop (Figure 12.33). The synchronous rotating coordinate system is used to determine the voltage vector aligned with *d*-axis v_{d_ref} and rotated with angle γ calculated from the referenced voltage frequency ω_{ref}. The voltage control loop is keeping correct voltage parameters in conformity with the local grid code.

Another popular solution for an autonomous WT application is the droop control method. Assuming that converter output is inductive, the active and reactive power can be calculated as [73]:

$$P = \frac{EV}{X}\sin\varphi \tag{12.6a}$$

$$Q = \frac{EV\cos\varphi - V^2}{X} \tag{12.6b}$$

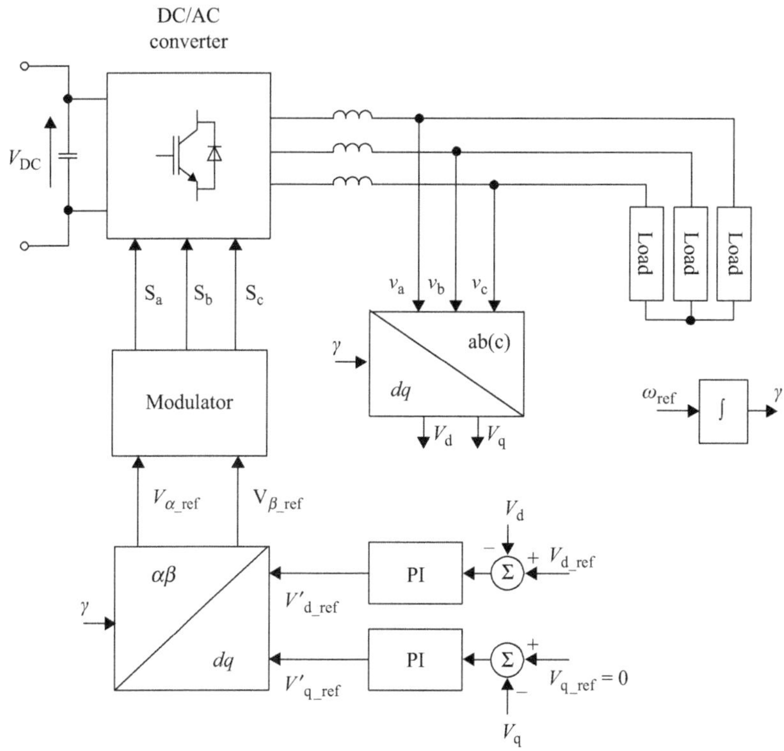

Figure 12.33 Block diagram of the VOC in the stand-alone mode of operation

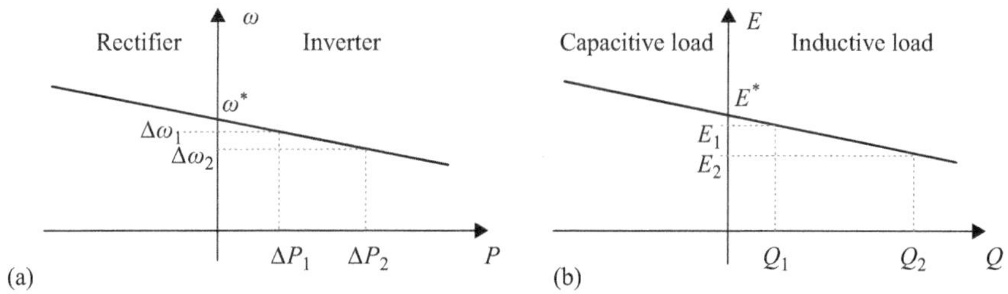

Figure 12.34 Droop characteristics

where P is active power, E is converter voltage amplitude, V is output voltage amplitude (after filter), X is coupling impedance, and φ is angle between the converter voltage E and the output voltage V.

Based on (12.6a) and (12.6b), it can be assumed that active power depends mainly on the angle φ (changing indirectly by the converter voltage pulsation error $\Delta\omega$) and the reactive power depends on the converter voltage amplitude E (Figure 12.34).

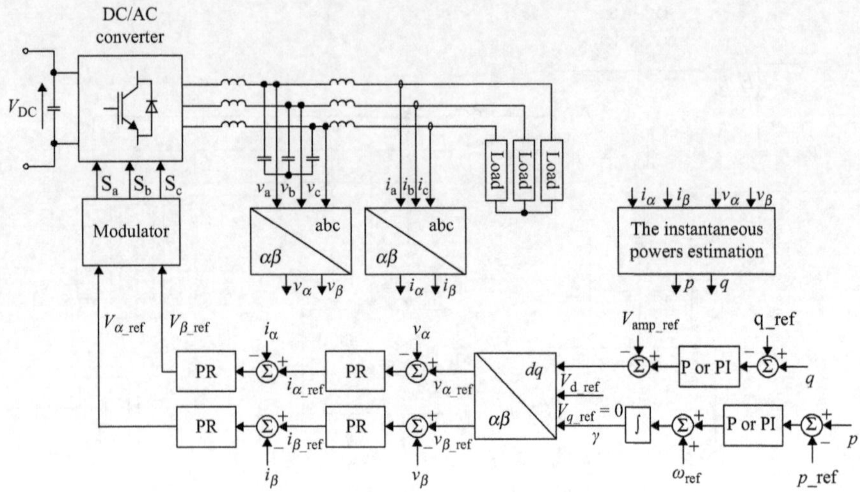

Figure 12.35 Block diagram of droop control in the islanded mode of operation

The above relations are implemented in the control algorithm (Figure 12.35) and they determine the reference converter voltage vector. After transformation to the stationary coordinate system, the voltage inner control loop and the current inner control loop with PR controllers are used. If the microgrid includes no energy storage and the total load cannot always absorb the total injected power, simple proportional controllers are implemented in the droop control loops [73, 74]. In other cases, the PI controllers are used and they allow avoiding power errors and working with MPPT [74–76]. The droop control method is often dedicated to the microgrid system, since many converters can work parallel in the stand-alone mode of operation as well as grid connected.

Acknowledgements

This work was supported by the Estonian Ministry of Education and Research (Project SF0140016s11) and by the Estonian Research Council under Grant PUT744.

This work was supported by Polish National Science Center (project number 6511/B/T01/2011/40).

References

[1] Global Wind Energy Council, "Global Wind Statistics – 2013," February 2014.

[2] Global Wind Energy Council, "Global Wind 2013 Report – Market Forecast for 2014–2018," 2014.

[3] G. Spagnuolo, G. Petrone, S. V. Araujo, C. Cecati, E. Friis-Madsen, E. Gubia, *et al.*, "Renewable energy operation and conversion schemes," *IEEE Industrial Electronics Magazine*, vol. 4, no. 1, pp. 38–51, 2010.

[4] A. Luna, F. K. de Araujo Lima, D. Santos, P. Rodriguez, E. H. Watanabe, S. Arnaltes, "Simplified modeling of a DFIG for transient studies in wind power applications," *IEEE Transactions on Industrial Electronics*, vol. 58, no. 1, pp. 9–20, January 2011.

[5] M. Rolak, R. Kot, M. Malinowski, Z. Goryca, J. T. Szuster, "AC/DC converter with maximum power point tracking algorithm for complex solution of small wind turbine," *Przegląd Elektrotechniczny*, ISSN 0033-2097, R. 87 NR 6/2011, June 2011.

[6] R. Kot, M. Rolak, M. Malinowski, "Comparison of maximum peak power tracking algorithms for a small wind turbine," *Transaction on Mathematics and Computers in Simulation*, ISSN 0378-4754, pp. 29–40, May 2013.

[7] A. Milczarek, M. Malinowski, "Monitoring and control algorithms applied to small wind turbine with grid-connected/stand-alone mode of operation," *Przegląd Elektrotechniczny*, ISSN 0033-2097, R. 88 NR 12a/2012, December 2012.

[8] IEC 61400-02 Design requirements for small wind turbines, IEC, 2007.

[9] World Wind Energy Association, "Small wind world report summary 2014 update," New Energy Husum, March 2014.

[10] M. Liserre, R. Cardenas, M. Molinas, J. Rodriguez, "Overview of multi-MW wind turbines and wind parks," *IEEE Transactions on Industrial Electronics*, vol. 58, no. 4, pp. 1081–1095, 2011.

[11] T. Ackermann (ed.), *Wind Power in Power Systems*, 2nd ed. New York, NY: John Wiley & Sons; 2012.

[12] M. Eremia, M. Shahidehpour (eds.), *Handbook of Electrical Power System Dynamics: Modeling, Stability, and Control*. New York, NY: John Wiley & Sons; 2013.

[13] E. Hau, *Wind Turbines – Fundamentals, Technologies, Application, Economics*, Third, translated edition. Berlin: Springer-Verlag; 2013.

[14] Z. Zhang, A. Chen, A. Matveev, R. Nilssen, A. Nysveen, "High-power generators for offshore wind turbines," *Energy Procedia*, vol. 35, pp. 52–61, 2013.

[15] World Wind Energy Association (Germany), *Small Wind World Report 2014*. Bonn: WWEA; 2014.

[16] P. Jamieson, *Innovation in Wind Turbine Design*. New York, NY: John Wiley & Sons; August 2011, p. 204.

[17] S. Tohidi, P. Tavner, R. McMahon, H. Oraee, M. R. Zolghadri, S. Shao, *et al.*, "Low voltage ride-through of DFIG and brushless DFIG: Similarities and differences," *Electric Power Systems Research*, vol. 110, pp. 64–72, 2014.

[18] IHS Inc. (USA), *The World Market for Wind Converters – 2013 Edition*. Englewood, CO: IHS; 2013.

[19] Navigant Research (USA), *World Market Update 2013. International Wind Energy Development Forecast 2014–2018*. Englewood, CO: Navigant Research; 2014.

[20] R. Teodorescu, M. Liserre, P. Rodriguez, *Grid Converters for Photovoltaic and Wind Power Systems*. New York, NY: John Wiley & Sons; 2011.

[21] J. Dai, D. Xu, B. Wu, "A novel control scheme for current-source-converter-based PMSG wind energy conversion systems," *IEEE Transactions on Power Electronics*, vol. 24, no. 4, pp. 963–972, 2009.

[22] P. Tenca, A. A. Rockhill, T. A. Lipo, P. Tricoli, "Current source topology for wind turbines with decreased mains current harmonics, further reducible via functional minimization," *IEEE Transactions on Power Electronics*, vol. 23, no. 3, pp. 1143–1155, 2008.

[23] F. Blaabjerg, K. Ma, "Future on power electronics for wind turbine systems," *IEEE Journal of Emerging and Selected Topics in Power Electronics*, vol. 1, no. 3, pp. 139–152, 2013.

[24] Y. P. Siwakoti, F. Z. Peng, F. Blaabjerg, P. C. Loh, G. E. Town, "Impedance-source networks for electric power conversion. Part I: a topological review," *IEEE Transactions on Power Electronics*, vol. 30, no. 2, pp. 699–716, 2015.

[25] L. Chen, W. L. Soong, M. Pathmanathan, N. Ertugrul, "Comparison of AC/DC converters and the principles of a new control strategy in small-scale wind turbine systems," *Proceedings of the 22nd Australasian Universities Power Engineering Conference;* Bali, Indonesia, September 2012. New York, NY: IEEE; 2012. pp. 1–6.

[26] D. Vinnikov, I. Roasto, "Quasi-*Z*-source-based isolated DC/DC converters for distributed power generation," *IEEE Transactions on Industrial Electronics*, vol. 58, no. 1, pp. 192–201, 2011.

[27] L. Bisenieks, D. Vinnikov, J. Zakis, "Analysis of operating modes of the novel isolated interface converter for PMSG based wind turbines," *Proceedings of the International Conference on Power Engineering, Energy and Electrical Drives;* Malaga, Spain, May 2011. New York, NY: IEEE; 2011. pp. 1–8.

[28] A. Chub, T. Jalakas, A. Milczarek, A. Kallaste, M. Malinowski, "Grid integration issues of PMSG-based residential wind turbines," *Proceedings of the Electric Power Quality and Supply Reliability Conference;* Rakvere, Estonia, June 2014. New York, NY: IEEE; 2014. pp. 147–154.

[29] F. Blaabjerg, M. Liserre, K. Ma, "Power electronics converters for wind turbine systems," *IEEE Transactions on Industry Applications*, vol. 48, no. 2, pp. 708–719, 2012.

[30] B. Wu, Y. Lang, N. Zargari, S. Kouro, *Power Conversion and Control of Wind Energy Systems*. New York, NY: John Wiley & Sons; 2011.

[31] X. Zeng, Z. Chen, F. Blaabjerg, "Design and comparison of full-size converters for large variable-speed wind turbines," *Proceedings of the 12th European Conference on Power Electronics and Applications;* Aalborg, Denmark, September 2007. New York, NY: IEEE; 2007. pp. 1–10.

[32] P. Mlodzikowski, A. Milczarek, S. Stynski, M. Malinowski, S. Kouro, "Control of simplified multilevel AC–DC–AC converter for small power generation systems," *Proceedings of the 39th Annual Conference of the IEEE Industrial Electronics Society;* Vienna, Austria, November 2013. New York, NY: IEEE; 2013. pp. 5951–5956.

[33] L. Yushan, B. Ge, F. Z. Peng, H. Abu-Rub, A. T. de Almeida, F. J. T. E. Ferreira, "Quasi-Z-source inverter based PMSG wind power generation system," *Proceedings of the Third IEEE Energy Conversion Congress and Exposition;* Phoenix, USA, September 2011. New York, NY: IEEE; 2011. pp. 291–297.

[34] B. Andresen, J. Birk, "A high power density converter system for the Gamesa G10x 4,5 MW wind turbine," *Proceedings of the 12th European Conference on Power Electronics and Applications;* Aalborg, Denmark, September 2007. New York, NY: IEEE; 2007. pp. 1–8.

[35] A. Faulstich, J. K. Stinke, F. Wittwer, "Medium voltage converter for permanent magnet wind power generators up to 5 MW," *Proceedings of the 11th European Conference on Power Electronics and Applications;* Dresden, Germany, September 2005. New York, NY: IEEE; 2005. pp. 1–9.

[36] V. Yaramasu, B. Wu, "Three-level boost converter based medium voltage megawatt PMSG wind energy conversion systems," *Proceedings of the Third IEEE Energy Conversion Congress and Exposition;* Phoenix, USA, September 2011. New York, NY: IEEE; 2011. pp. 561–567.

[37] V. Yaramasu, B. Wu, M. Rivera, J. Rodriguez, "A new power conversion system for megawatt PMSG wind turbines using four-level converters and a simple control scheme based on two-step model predictive strategy. Part I: modeling and theoretical analysis," *IEEE Journal of Emerging and Selected Topics in Power Electronics*, vol. 2, no. 1, pp. 3–13, 2014.

[38] V. Yaramasu, B. Wu, M. Rivera, J. Rodriguez, "A new power conversion system for megawatt PMSG wind turbines using four-level converters and a simple control scheme based on two-step model predictive strategy. Part II: simulation and experimental analysis," *IEEE Journal of Emerging and Selected Topics in Power Electronics*, vol. 2, no. 1, pp. 14–25, 2014.

[39] C. H. Ng, M. A. Parker, L. Ran, P. J. Tavner, J. R. Bumby, E. Spooner, "A multilevel modular converter for a large, light weight wind turbine generator," *IEEE Transactions on Power Electronics*, vol. 23, no. 3, pp. 1062–1074, 2008.

[40] X. Yuan, "A set of multilevel modular medium-voltage high power converters for 10-MW wind turbines," *IEEE Transactions on Sustainable Energy*, vol. 5, no. 2, pp. 524–534, 2014.

[41] L. Popova, J. Pyrhonen, K. Ma, F. Blaabjerg, "Device loading of modular multilevel converter (MMC) in wind power application," *Proceedings of the International Power Electronics Conference (IPEC-Hiroshima 2014 – ECCE-Asia);* Hiroshima, Japan, May 2014. New York, NY: IEEE; 2014. pp. 548–554.

[42] S. Inoue, H. Akagi, "A bi-directional isolated DC/DC converter as a core circuit of the next-generation medium-voltage power conversion system," *Proceedings of the 37th IEEE Power Electronics Specialists Conference;* Jeju, Republic of Korea, June 2006. New York, NY: IEEE; 2006. pp. 1–7.

[43] J. Clare, "Modular high power converter topologies," *IET Seminar on Power Electronics 2010: Improving the Efficiency of the Power Grid;* Birmingham, UK, June 2010. London: IET; 2010. pp. 1–57.

[44] P. R. Mohan, A. Pradesh, T. B. Reddy, M. V. Kumar, "A simple generalized pulse width modulation algorithm for vector control based voltage source inverter fed induction motor drives," *2014 Annual International Conference on Emerging Research Areas: Magnetics, Machines and Drives (AICERA/iCMMD)*; Kottayam, India, 24–26 July 2014. pp. 1–5.

[45] P. Palanivel, S. S. Dash, "Analysis of THD and output voltage performance for cascaded multilevel inverter using carrier pulse width modulation techniques," *IET Power Electronics*, vol. 4, no. 8, p. 951, 2011.

[46] S. Stynski, "Space vector PWM modulator reducing switching losses for three-level flying-capacitor inverters," *2010 IEEE International Symposium on Industrial Electronics*, 4–7 July 2010, Bari, Italy, pp. 3912–3917.

[47] M. Malinowski, M. Jasinski, M. P. Kazmierkowski, "Simple direct power control of three-phase PWM rectifier using space-vector modulation (DPC-SVM)," *IEEE Transactions on Industrial Electronics*, vol. 51, no. 2, pp. 447–454, April 2004.

[48] H. Abu-Rub, M. Malinowski, K. Al-Haddad, Power Electronics for Renewable Energy Sources, *Transportation and Industrial Applications*. Chichester, West Sussex, UK: John Wiley & Sons, Chichester, West Sussex, UK, 2014.

[49] E. Koutroulis, K. Kalaitzakis, "Design of a maximum power tracking system for wind-energy-conversion applications," *IEEE Transactions on Industrial Electronics*, vol. 53, no. 2, pp. 486–494, April 2006.

[50] M. Singh, A. Chandra, "Power maximization and voltage sag/swell ride-through capability of PMSG based variable speed wind energy conversion system," in *2008 34th Annual Conference of IEEE Industrial Electronics*, 2008, pp. 2206–2211.

[51] A. M. De Broe, S. Drouilhet, V. Gevorgian, "A peak power tracker for small wind turbines in battery charging applications," *IEEE Transactions on Energy Conversion*, vol. 14, no. 4, pp. 1630–1635, 1999.

[52] M. A. Abdullah, A. H. M. Yatim, C. W. Tan, R. Saidur, "A review of maximum power point tracking algorithms for wind energy systems," *Renewable & Sustainable Energy Reviews*, vol. 16, no. 5, pp. 3220–3227, June 2012.

[53] M. Druga, C. Nichita, G. Barakat, B. Dakyo, E. Ceanga, "A peak power tracking wind system operating with a controlled load structure for stand-alone applications," *13th European Conference on Power Electronics and*

Applications, 2009. EPE '09. Barcelona, Spain, 8–10 September 2009, pp. 1–9.

[54] K. S. Muhammad Raza, H. Goto, H.-J. Guo, O. Ichinokura, "A novel speed-sensorless adaptive hill climbing algorithm for fast and efficient maximum power point tracking of wind energy conversion systems," in *2008 IEEE International Conference on Sustainable Energy Technologies*, 2008, pp. 628–633.

[55] J. Hui, A. Bakhshai, "A new adaptive control algorithm for maximum power point tracking for wind energy conversion systems," in *2008 IEEE Power Electronics Specialists Conference*, 2008, pp. 4003–4007.

[56] R. Esmaili, L. Xu, D. K. Nichols, "A new control method of permanent magnet generator for maximum power tracking in wind turbine application," in *2005 IEEE Power Engineering Society General Meeting*, 2005, vol. 3, pp. 2090–2095.

[57] R. Datta, V. T. Ranganathan, "A method of tracking the peak power points for a variable speed wind energy conversion system," *IEEE Transactions on Energy Conversion*, vol. 18, no. 1, pp. 163–168, March 2003.

[58] Q. Wang, L. Chang, "An intelligent maximum power extraction algorithm for inverter-based variable speed wind turbine systems," *IEEE Transactions of Power Electronics*, vol. 19, no. 5, pp. 1242–1249, September 2004.

[59] L. Yushan, B. Ge, F. Z. Peng, H. Abu-Rub, A. T. de Almeida, F. J. T. E. Ferreira, "Quasi-*Z*-source inverter based PMSG wind power generation system," *Proceedings of the Third IEEE Energy Conversion Congress and Exposition;* Phoenix, USA, September 2011. New York, NY: IEEE; 2011. pp. 291–297.

[60] S. M. Dehghan, M. Mohamadian, A. Y. Varjani, "A new variable-speed wind energy conversion system using permanent-magnet synchronous generator and *Z*-source inverter," *IEEE Transactions on Energy Conversion*, vol. 24, no. 3, pp. 714–724, 2009.

[61] Y. P. Siwakoti, F. Z. Peng, F. Blaabjerg, P. Loh C., G. E. Town, S. Yang, "Impedance-source networks for electric power conversion part II: review of control and modulation techniques," *IEEE Transactions on Power Electronics*, vol. 30, no. 4, pp. 1887–1906, 2015.

[62] P. Vas, *Vector Control of AC Machines*. Oxford: Clarendon Press, 1990.

[63] M. P. Kazmierkowski, H. Tunia, *Automatic Control of Converter Fed Drives*. London, New York, Tokyo: ELSEVIER Amsterdam; 1994.

[64] S. Stynski, J. San-Sebastian, M. Malinowski, I. Etxeberria-Otadui, "Analysis of multilevel PWM converter based on FLC modules for an AC traction application," in *2009 IEEE International Conference on Industrial Technology*, 2009, pp. 1–6.

[65] P. Mlodzikowski, A. Milczarek, M. Malinowski, "Application of simplified neutral point clamped multilevel converter in a small wind turbine," *Electrical, Control and Communication Engineering*, vol. 5, no. 1, pp. 5–10, January 2014.

[66] M. Zelechowski, M. Malinowski, P. Kaczynski, W. Kolomyjski, M. Twerd, J. Zaleski, "DSP based sensorless direct torque control-space vector modulated (DTC-SVM) for inverter fed induction motor drives," in *Problems of Automated Electrodrives Theory and Practice*, 2003, pp. 90–92.

[67] Y. Xue, X. Xu, T. G. Habetler, D. M. Divan, "A low cost stator flux oriented voltage source variable speed drive," in *Conference Record – IAS Annual Meeting (IEEE Industry Applications Society)*, 1990, pp. 410–415.

[68] G. S. Buja, M. P. Kazmierkowski, "Direct torque control of PWM inverter-fed AC motors – a survey," *IEEE Transactions on Industrial Electronics*, vol. 51, no. 4, pp. 744–757, August 2004.

[69] M. Bobrowska-Rafał, K. Rafał, M. Jasiński, M. P. Kaźmierkowski, "Grid synchronization and symmetrical components extraction with PLL algorithm for grid connected power electronic converters – a review," *Bulletin of the Polish Academy of Sciences: Technical Sciences*, vol. 59, no. 4, pp. 485–497, 2011.

[70] M. Malinowski, M. Jasinski, M. P. Kazmierkowski, "Simple direct power control of three-phase PWM rectifier using space-vector modulation (DPC-SVM)," *IEEE Transactions on Industrial Electronics*, vol. 51, no. 2, pp. 447–454, April 2004.

[71] J. Salaet, S. Alepuz, A. Gilabert, J. Bordonau, "Comparison between two methods of DQ transformation for single phase converters control. Application to a 3-level boost rectifier," in *2004 IEEE 35th Annual Power Electronics Specialists Conference (IEEE Cat. No.04CH37551)*, 2004, vol. 1, pp. 214–220.

[72] U. A. Miranda, L. G. B. Rolim, M. Aredes, "A DQ synchronous reference frame current control for single-phase converters," in *IEEE 36th Conference on Power Electronics Specialists, 2005*, 2005, pp. 1377–1381.

[73] J. M. Guerrero, N. Berbel, J. Matas, J. L. Sosa, L. G. De Vicufia, "Droop control method with virtual output impedance for parallel operation of uninterruptible power supply systems in a microgrid," *Twenty Second Annual IEEE Applied Power Electronics Conference, APEC 2007*. Anaheim, CA, USA, 25 February–1 March 2007. pp. 1126–1132.

[74] J. M. Guerrero, S. Member, J. C. Vasquez, J. Matas, L. G. De Vicuña, M. Castilla, "*Hierarchical control of droop-controlled AC and DC microgrids – a general approach toward standardization*," vol. 58, no. 1, pp. 158–172, 2011.

[75] A. Milczarek, J. C. Vasquez, M. Malinowski, J. M. Guerrero, M. Graells, "Supervisory control for real time reactive power flow optimization in islanded microgrids," in *23rd European Symposium on Computer Aided Process Engineering*, 2013, vol. 32, pp. 325–330.

[76] Q. Shafiee, J. C. Vasquez, and J. M. Guerrero, "Distributed secondary control for islanded MicroGrids – a networked control systems approach," *IECON 2012 – 38th Annual Conference on IEEE Industrial Electronics Society*, pp. 5637–5642, October 2012.

Chapter 13
Photovoltaic energy systems

Eduard Muljadi[a], *Emilio Gomez-Lazaro*[b] *and Antonio Ginart*[c]

13.1 Introduction

13.1.1 Brief overview of photovoltaic generation

The use of photovoltaic (PV) generation was initially started in the aerospace industry to power the satellite, and, outer space sensors, monitors, and International Space Station. In the past few years, PV generation has gained global popularity and acceptance; mostly by the grass root movement from the regular citizens want to do good for the environment. PV power plants (PV plants) have been growing in size, and the installation time is very short. With the cost of PV panels dropping in recent years, it can be predicted that in the next 10 years the contribution of PV plants to the total number of renewable energy power plants will grow significantly.

According to the Solar Energy Industries Association, the PV installation in the US is currently as follows: 2,695 MW operating; 3,608 MW under construction; and 20,155 MW under development, which will bring to a total 26,458 MW PV (http://www.seia.org/research-resources/solar-industry-data).

The advantage of the PV generation especially in the distribution network, it can be placed almost anywhere with a good resource of solar irradiance, it can be placed on the rooftop of buildings, or on the ground without the need to erect tall towers such in wind generation, nor require extensive piping as in the concentrated solar power, nor expensive construction of dams as in hydropower plant.

13.1.2 PV inverter circuit

PV inverter has been developed for many years. It started as off grid applications with earlier power semiconductors switches such as forced commutated silicon controlled rectifiers (SCRs). As the power electronics technologies making progresses via the development in the drives and power supplies industry, the cost of

[a]National Renewable Energy Laboratory, Golden, CO, USA;
[b]Universidad de Castilla-La Mancha, Albacete, Spain;
[c]Sputnik Technology USA, Inc., Norcross, GA, USA

the power switches drop very dramatically at the late eighties, thus, making it possible for widespread applications in the various industries. At the same time, the cost of CPUs also decreases very rapidly, thus, enabling development of smart controls and system protection.

A three-phase PV inverter is usually designed for a three-phase system with a large rating (10 kW–2 MW and above). Most PV inverters are current controlled. To understand a basic PV inverter, it is important to understand the module and PV inverter hardware. In general, a PV inverter consists of a DC bus and three pairs of power semiconductors, also called power electronic switches or power switches. Present-day PV inverter is based on very fast power electronic switches with sophisticated controls, built with high efficiency at low cost. The most common power switches presently used are insulated-gate bipolar transistors (IGBTs). An IGBT can be turned on and off very fast within microseconds (thus, the loss from power switching is low), and it has a low conduction loss.

The component of a PV generation can be simplified as shown in Figure 13.1. It consists of PV panel, maximum peak power tracker, and DC-to-AC inverter converting the power back to the utility grid. The actual implementation of a PV generation is shown in Figure 13.2.

It usually consists of many modules connected in series/parallel to form a string of PV modules with an output voltage and current designed to match the rating of the maximum power point tracker (MPPT). The output of several MPPTs are connected to the DC bus of PV inverter where the DC power is converted to the three phase of single-phase AC distribution network. The PV inverter circuit of the

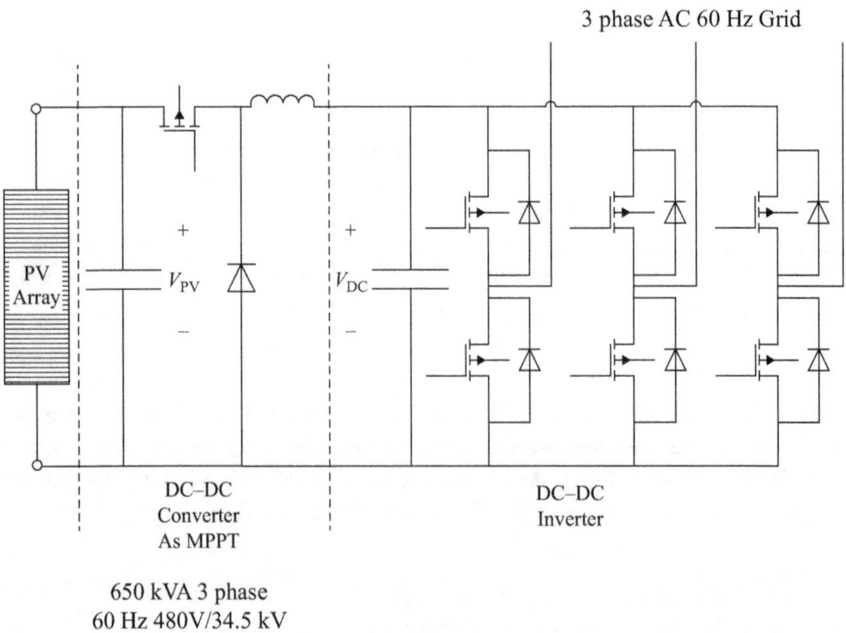

Figure 13.1 A typical three-phase PV inverter with IGBT power switches

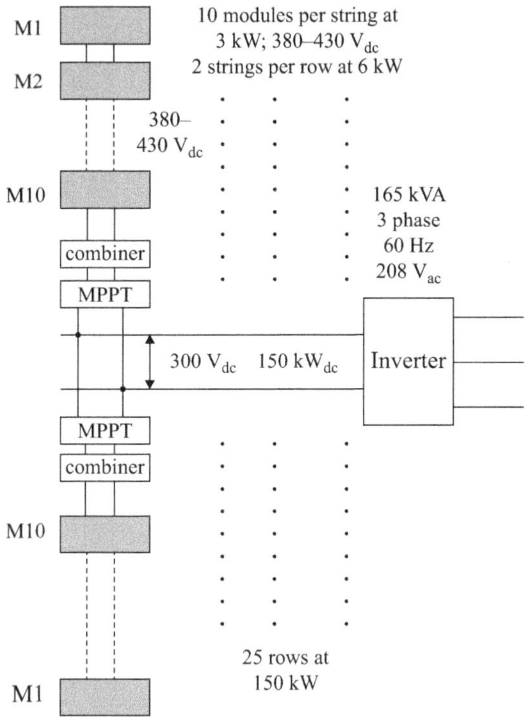

Figure 13.2 Physical implementation of a typical PV installation

size shown in Figure 13.2 is generally installed at the rooftop of commercial buildings.

13.1.3 Centralized PV plant

The centralized PV power plant is a large-scale PV installation consisting of many PV inverters to form a large size 10 MW or higher ratings.

The output of the PV inverter shown in Figure 13.2 has low-voltage output (three phase, 480 V). For a large PV plant, the interconnected underground cables (also called the collector system) carrying power over long distance can lead to very high losses. Thus, a step-up transformer is usually used to step-up the voltage to a medium voltage (three phase, 34.5 kV) commonly used in a PV plant. In this way, the collector system losses can be minimized to 5% or lower of the rated power of the PV plant. For a very large PV plant, and to transmit the output power of the PV plant over long distance the combined output of the PV inverter will be step up further to raise the voltage to the transmission level voltage (say 115 kV) as shown in Figure 13.3.

The Agua Caliente solar project (refer to Figure 13.4), owned by NRG Energy Inc., has come online and is now the world's largest PV plant. This facility has the capacity to generate 290 MW of solar electricity in Yuma County, Arizona.

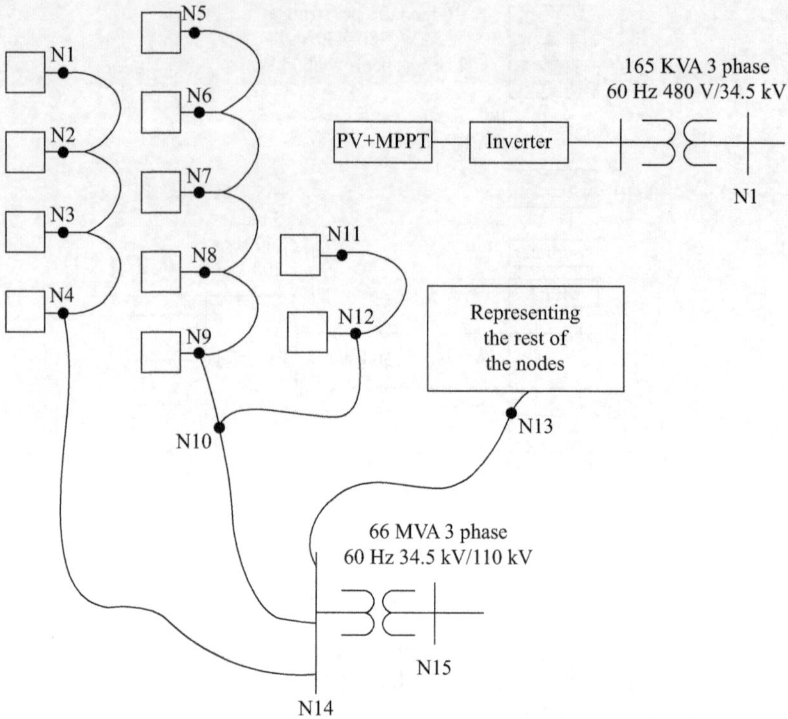

Figure 13.3 Single-line representation of a typical PV power plant

Figure 13.4 Agua Caliente (a 290 MW PV power plant) is located in Yuma County, Arizona (http://energy.gov/articles/agua-caliente-worlds-largest-solar-photovoltaic-plant-helps-advance-americas-solar)

13.2 The technologies

Solar energy is one of the best renewable energy sources available to us. Its success is highly dependent on the ability to deliver an increase in the power generated per dollar. In 2010, the US Department of Energy estimated that installing a one-watt-per-dollar PV energy system would make solar energy without additional subsidies competitive with the wholesale rate of electricity by 2017. To follow this trend, the different components in PV systems, including the inverter, should undergo a cost reduction close to 10% per year. From the inverter's perspective, this goal may be achieved by two main approaches. The first one is to improve the efficiency by taking advantage of more economical inverter topologies and developing more powerful and economic power semiconductor and magnetics components. The second approach is based on increasing the value of the solar proposition by relying on the distributed nature of solar production and the high levels of sophisticated computational and operational capabilities offered by solar inverters. The smart grid will require smart devices with substantial control capabilities and the solar inverter is uniquely qualified to serve this propose.

13.2.1 State-of-the-art technologies

13.2.1.1 Power semiconductors

Due to decreasing cost and other improvement in fabrication technologies of silicon power semiconductors (allowing for higher breakdown voltages, higher current-carrying capability, faster switching, and lower losses), silicon has remained the primary material for power semiconductor, despite competition from emerging materials such as silicon carbide (SiC) and gallium nitride (GaN). Silicon IGBTs remain the preferred semiconductor for inverter bridges. The booster is slowly allowing the incorporation of SiC metal–oxide–semiconductor field-effect transistors (MOSFETs) especially for their ability to deliver very high switching frequencies at relatively high breakdown voltage. SiC diodes have been used in the last four years as integral part of solar inverters due to their low switching and conduction losses.

13.2.1.1.1 Power semiconductor based in silicon

Since the invention of the IGBT and the introduction of the power MOSFET, the progress has been tremendous, largely increasing performance and lowering cost. These performance improvements in the silicon semiconductors have been critical in keeping silicon the semiconductor material of choice, against emerging technologies (Figure 13.5).

13.2.1.1.2 Power semiconductor based in SiC

SiC power semiconductors are leveraging in the industrial development of LEDs, and as a consequence, the price of SiC semiconductors has decreased significantly in the last two years. The first SiC power semiconductor to be developed at a competitive price with better performance than Si was the SiC diode. SiC have a dielectric breakdown field strength that is approximately 10 times higher than that of Si. Consequently, SiC devices can be made to have a much thinner drift layer

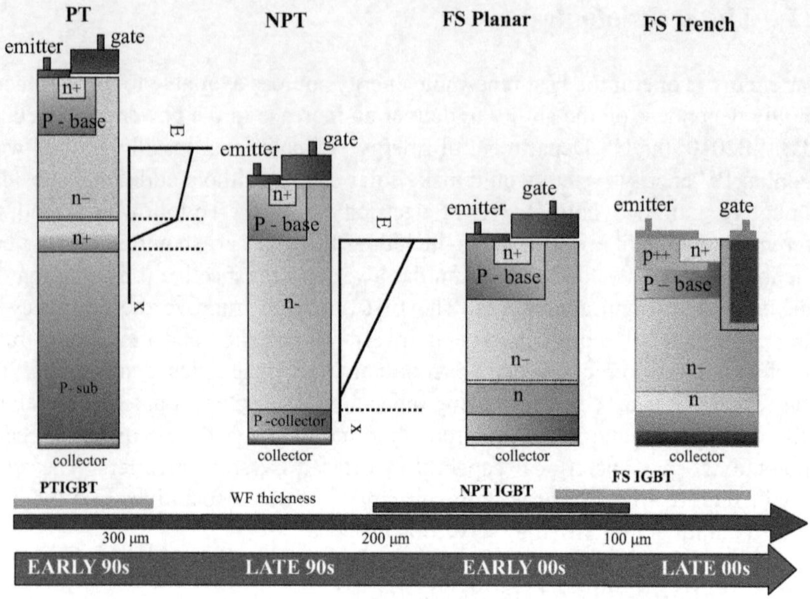

Figure 13.5 Illustration of technology improvements of IGBTS

and/or higher doping concentrations, resulting in power devices that have very high breakdown voltages while exhibiting very low resistance relative to silicon devices. In theory, SiC can reduce the resistance per unit area of the drift layer to 1/300 of Si at the same breakdown voltage.

For the reason explained above, SiC allows the fabrication of MOSFET with lower R_{DS} ON for the same breakdown voltages that silicon. Power MOSFET with breakdown voltages up 1,700 V are manufactured by Cree Inc. Similar devices are now produced by Rohm and Mitsubishi, among others. General Electric is developing a large foundry to deliver a very low cost SiC MOSFET by 2018. Alternatively, SiC-based bipolar junction transistors (BJTs) are developed by GeneSIC and Fairchild Semiconductor International Inc.

13.2.1.1.3 Power semiconductor based in GaN

GaN devices have a theoretical limit for the voltage breakdown with R_{DS} ON ratio that is 20 times greater than that of SiC devices. Until recently, GaN semiconductors were confined to radio frequency (RF) field applications. In the switching power device arena, GaN Systems Inc. and International Rectifier Corporation are developing new power devices, but their solution still remains applicable for voltages under 600 V. Thus, it is mostly viable for the residential solar market. While maturity in fabrication is still very far below the theoretical limits, its intrinsic characteristic allows one to dream of a highly efficient power conversion in the near future.

13.2.1.2 Inverter topology

There are two main complementary hard-switching techniques that produce a sinusoidal wave with low distortion: (1) by generating a high-frequency pulse-width modulation (PWM) and (2) by instating the variation of DC voltage levels produced by the inverter (multilevel inverters). Increasing the switching frequency faces two key obstacles: the need for faster power devices such as wide-band gap devices that can be driven at such high frequencies, as well as for faster and more powerful microcontrollers in order to achieve proper active monitoring of complex systems such as solar inverters. These obstacles lead to higher costs. Driven by the decreasing cost of silicon power semiconductors, the solar industry therefore has moved progressively into the use of multilevel inverters as the more yielding of the two approaches.

In general, a grid-tie PV system is comprised of the devices shown in Figure 13.6. This power converter consists of the PV panel's field arrangement connected in series to provide the input voltage range to the inverter. The inverter block could include a voltage booster stage. A transformer, that may or may not be included in the inverter unit, is used for the final connection to the grid. The booster stages are common in residential and commercial inverters. Residential and commercial inverters have been called string inverters due to the way they are connected to the solar panel. Many of the larger commercial inverters (typically greater than 50 kW) eliminate the booster stage. The same statement holds for utility scale or centralized inverters.

13.2.1.2.1 Multilevel inverter

Over the past 30 years, multilevel inverters have been part of different research topics focused on several areas of interest including new topologies, new controls, and new modulation schemes. One of the main multilevel inverter topologies is the cascade multi-cell inverter, which is based on a series connection of H-bridge

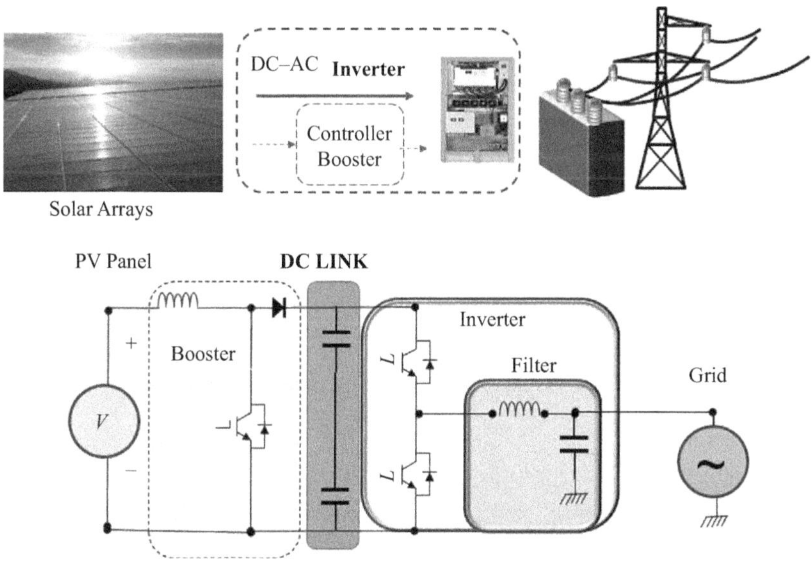

Figure 13.6 PV grid-tie inverters parts and diagram representation

modules with separated DC sources. This topology was first introduced by Baker in 1975 and became widely used until the mid-1990s. The neutral-point diode clamped (NPC) inverter topology was developed as an alternative for the cascade multi-cell concept, and it was described for the first time by Nabae. Since then, the three-level clamped diode inverter has become favorably accepted for the design of high power in medium- and low-voltage applications. Recently, an extension of the concept of multilevel clamping (MNPC) has been introduced. It produces an alternative topology in which a classical three-level NPC can be selectively clamped by a dedicated switching unit in order to increase the number of voltage levels (see Figure 13.7).

The most common multilevel inverters used in solar applications are the NPC (also known as Type I) and the modified NPC, generally known in the industry as MNPC, NPC2, or Type T. Theoretically, the NPC topology requires components that withstand half of the applied voltage, but the MNPC arrangement could deliver a more efficient inverter topology. With the improvement of power semiconductors, the MNPC topology has been increasingly implemented in commercial solar inverters.

13.2.1.2.2 Principle of operation: NPC

In both the MNPC and the NPC, the principle operation can be explained as two buck converters, one each for the positive and negative part of the sinusoidal. The two buck converters disconnect alternatively to avoid a short circuit between each other. Using the NPC structure, Figure 13.8 shows how a three-level inverter can be described as two buck converters. The two IGBTs in the center (SW) select between positive and negative signal generation.

13.2.1.2.3 Topology comparison: 600 V vs. 1,200 V

The comparison among topologies is made assuming IGBT trench stop technology or similar. The 600 V technology is assumed for all switches in the NPC topology.

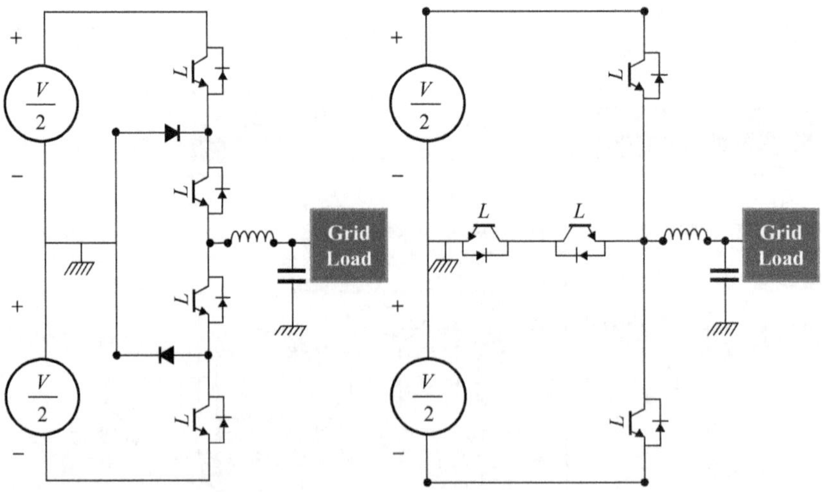

Figure 13.7 Diagram of NPC and MNPC three-level inverter topologies

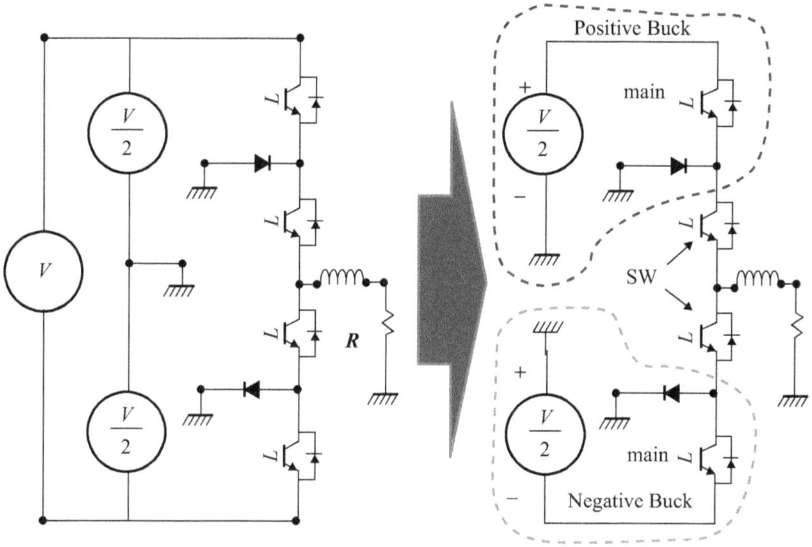

Figure 13.8 Basic operation of the NPC inverter

Table 13.1 Comparison among 600 and 1,200 V IGBTS

	600 V NPC	**1,200 V MNPC**
Conduction (V_{DS})	100%	110%
SW losses E_t	100%	170%

For the MNCP topologies, the main switching device is assumed to be 1,200 V. Table 13.1 compares the average conduction and switching losses between 600 and 1,200 V IGBTs for 40–60 A from two leading IGBT manufacturers Infineon Technologies and Fairchild Semiconductor International Inc. In other moment on time or for other level of current and voltage, the results could be different.

The graphical comparison of topologies, conduction, and switching losses for the different switches in MNPC and NPC configurations are shown in Figures 13.9 and 13.10, respectively. In the NPC topology, for a power factor of approximately one, during the PWM *ON* period, V_{CE1} and V_{CE2} have a voltage drop. In PWM *OFF* period, only V_D and V_{CE2} undergo a voltage drop. In the case of MNPC in the PWM *ON* period, only V_{CE1} presents a voltage drop, and in the PWM *OFF* period, only V_D and V_{CE2} incur a voltage drop. In conclusion, it is clear that in the NPC topology V_{CE2} is on all time, while in the MNPC it is only on during the *OFF* period.

Nowadays, topologies of Type T (MNPC) are dominating the market up to 1,000 V, which is the maximum voltage for certifiable PV systems. If 1,500 V DC is approved in the near future, the topology of Type I may again reemerge as the most economical solution for solar inverters.

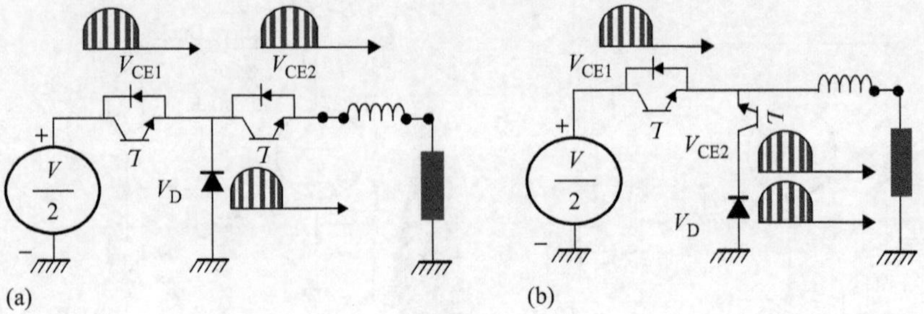

Figure 13.9 Graphical comparison between NPC and MNPC topologies

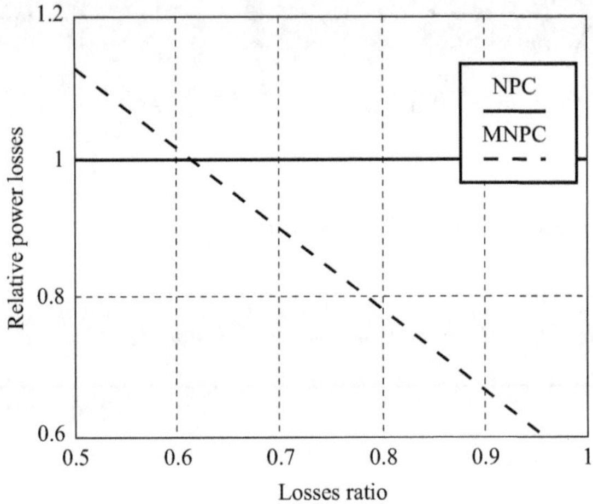

Figure 13.10 Power losses comparison between NPC and MNPC topologies

13.2.1.3 Inverter control

The solar inverter requires a low cost, relatively large, and fast computational system to operate while connected to the grid. The inverter needs to obtain the maximum power available from the PV panels, supervise the grid to make the proper synchronization, and deliver power at the required power factor. The system has to be able to detect independently of any fault (unbalanced voltage, line faults, and short circuits) in the power grid, as well as take the proper action according to the standards and regulations that the specific country establishes.

The block diagram shown in Figure 13.11 summarizes a typical solar grid-tie control system. To have independent control of the reactive and real power, dq transformation is applied. The phase locked loop (PLL) synchronizes the crossing point the grid voltage.

13.2.2 Reliability

Power semiconductors have been proven, to be the weakest link in the reliability of the solar inverters. The graph shown in Figure 13.12 displays the fault distribution

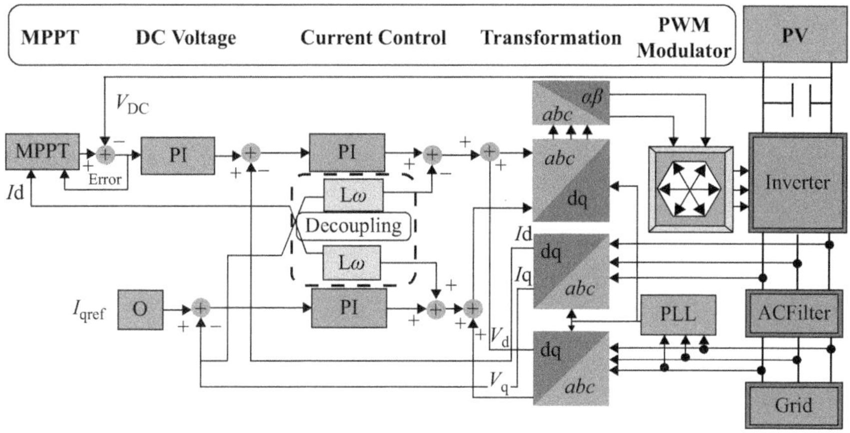

Figure 13.11 Diagram control of a grid-tie solar inverter

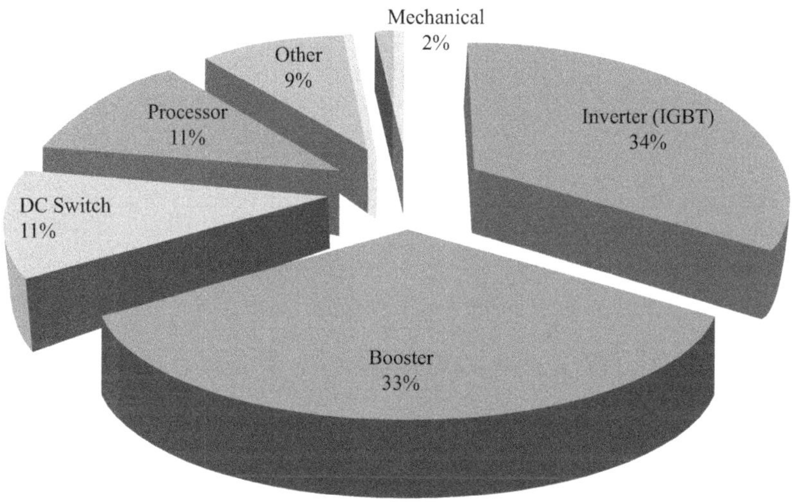

Figure 13.12 Fault distribution for a typical PV inverter

found by an inverter manufacture in one of the commercial inverter series with values from 10 to 20 kW. The historical record of manufacturers' data shows that nearly 70% of the faults are related to the power semiconductor.

In general, the critical parts of a solar inverter are:

- Power semiconductors
- Capacitors: DC capacitor link and AC capacitor filter
- Microcontrollers and peripherals
- Switches, breakers, and sensors

13.2.2.1 Accelerated aging commonality and underlying physics

Aging in semiconductors, like in other type of materials, is determined by the rate of the decay of the component materials. The key point is how long the structure conserves its original properties without being contaminated by the surrounding materials or decomposing into its original elements. For this basic reason, electronic semiconductors present the same type of "progressive degradation" that the materials that constitute them do. The consequence of this assertion is that regardless of the specific aging mechanisms, such as hot carrier injection, electromigration, time-dependent dielectric breakdown, gate sinking (interdiffusion of metal atoms from the gate into semiconductor), Ohmic contact degradation (increase in resistance between metal and semiconductor region), or/and any other mechanism, the aging is affected by the rate of the chemical reaction as described by some form of the Arrhenius equation. This is not the only equation that can model this phenomenon, but also any formula that models movement of particles by diffusion under the presence of a force field can be used for modeling these phenomena.

13.2.2.2 Power device reliability

The previous section highlights some of the common and widely accepted reliability issues presented with the design and operation of MOSFET/IGBT power devices. In this section, traditional semiconductor reliability concepts are related to remaining life estimations. From a reliability engineering point of view, the failure rates caused by factors described in the previous section have to be addressed and taken into consideration.

The mean time to failure (MTTF) is usually defined by the inverse of the failure rate of devices:

$$\mathrm{MTTF} = \frac{1}{\lambda} \tag{13.1}$$

where λ is the failure rate.

A failure rate (λ) could be estimated by

$$\lambda \propto \frac{1}{\mathrm{TDH} * \mathrm{AF}} \tag{13.2}$$

where AF is acceleration factor and TDH is total duration hours.

Semiconductor manufacturers perform millions of hours of testing to determine failure rates. This gives extremely valuable information for the reliability of the product. Additionally, applying this concept with more emphasis on the acceleration factor yields important information for life estimation.

The AF is generally determined from the Arrhenius equation but can also be determined using a power law, as is common industry practice.

13.2.2.3 Field distortion acceleration model

Acceleration models for the reduction of life in equipment have been used extensively. Representative examples include the Arrhenius model for temperature [1]:

$$L(T) = \mathrm{AF} \cdot \mathrm{e}^{\frac{E}{kT}} \tag{13.3}$$

The acceleration factor, AF, is determined generally from the Arrhenius model:

$$\mathrm{AF} = \exp\left(\frac{E_a}{k}\left(\frac{1}{T_1} - \frac{1}{T_2}\right)\right) \tag{13.4}$$

Crook [2] proposed the following expression in the case of two acceleration factors (temperature and voltage):

$$\mathrm{AF} = \exp\left(\frac{E_a}{k}\left(\frac{1}{T_1} - \frac{1}{T_2}\right)\right) \cdot \exp\left(-\frac{E_2 - E_1}{E_c}\right) \tag{13.5}$$

where E_c has been experimentally (Crook) been shown to equal 0.062 MV/cm.

Equation (13.6) shows the power law that is used commonly in determining the life expectancy of power devices is based on the data presented by Pearce *et al.* [3]. Several manufacturers use gate voltages and temperatures for aging estimation. Based on the data [3], the values for the IRF4XX series are computed to be $A = -17.2$, $B = 0.4$, $C = -0.15$ and the respective curves are shown in Figure 13.8.

$$AF = A \cdot 10^{CT} \cdot 10^{B \cdot V_G + a} \tag{13.6}$$

The thermal activation energy of a failure mechanism is determined by testing at a minimum of two different temperature stress levels as is shown in Figure 13.13.

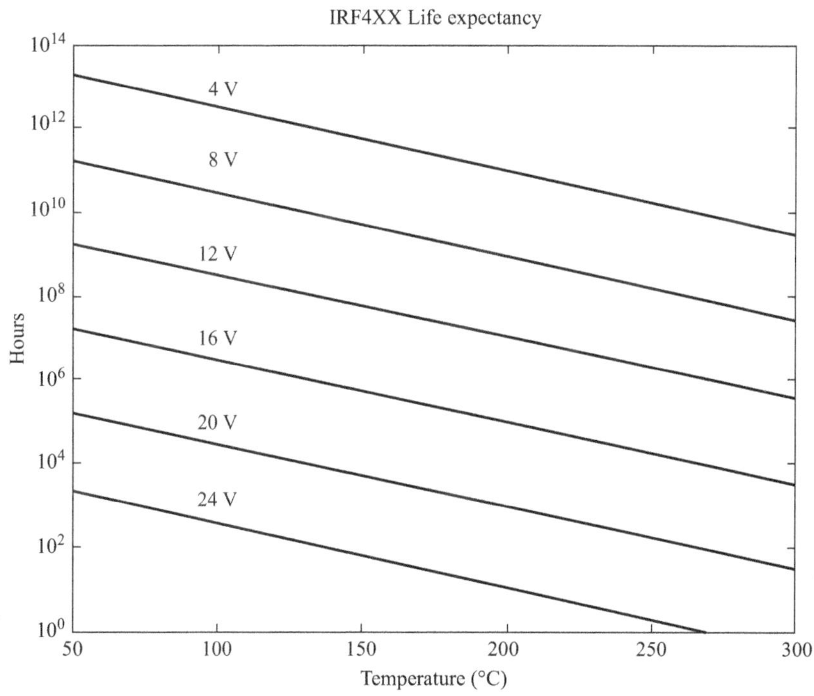

Figure 13.13 Power MOSFET life expectancy under power law

Figure 13.13 shows the power MOSFET life expectancy using power law shown in (13.6). It can be seen that the life expectancy of MOSFET devices changes with the varying bias voltages while uniformly changing as the operating temperature changes. In lab testing and field operations, it has been shown that the device's failure gets accelerated when operating in higher temperature regions.

13.2.2.4 Dominant failure mechanisms

In general, power semiconductors share common failure mechanisms. However, the application and design of the semiconductor often determine the most probable failure event for a specific device or technology. The main semiconductor degradation processes include thermal cycling which affects the packing and the semiconductor's capability to transfer heat. Hot-carrier induced degradation, ION migration, gate sinking degradation, and Ohmic contact degradation are other common semiconductor degradation mechanisms.

13.2.2.4.1 Thermal cycling (void and crack formation)

Thermal cycling is one of the main phenomena that produces semiconductor aging, particularly in power applications. Device degradation occurs because thermal cycling deteriorates the thermal material that allows the device to release generated heat. When a composite of multiple materials is exposed to the stress of thermo-cycling, it deteriorates until a fracture or void space is produced (see Figure 13.14). The semiconductor industry uses different materials to produce the heat transfer path that allows the release of the heat generated. In general, these devices have different coefficients of thermal expansion that make the device more susceptible to cracks or fractures due to the internal stresses originated by thermal expansion and contraction. These fractures among different materials typically do not collapse the process but deteriorate the functionality of the device.

As an example, Figure 13.15 shows the progression of the void formation and detachment of the dice from metalized structure (collector–emitter). The device containing an IGBT and integrated diode was thermo-electrically stressed. The void creation dominated the IGBT region, which was subjected to the thermo-electrical aging. The diode only obtained the heat by thermal conduction. Therefore, less void formation and aging are expected about it. The progression of detachment greatly increases the junction-case thermal resistance. The increase of the junction-case thermal resistance has large consequences in the changes of the parametric characteristics of the IGBT. Under the same operational conditions, the

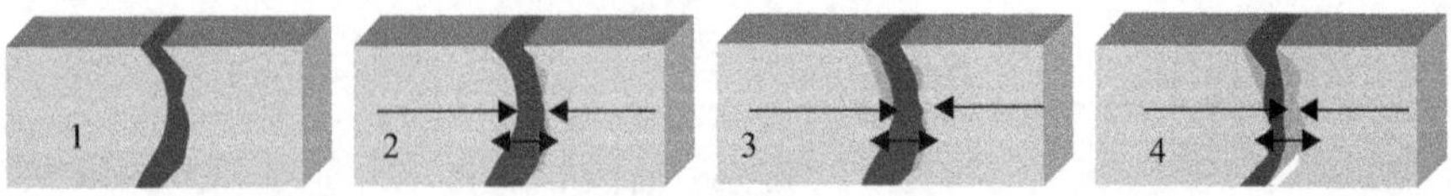

Figure 13.14 Void area creation process due to thermal cycling

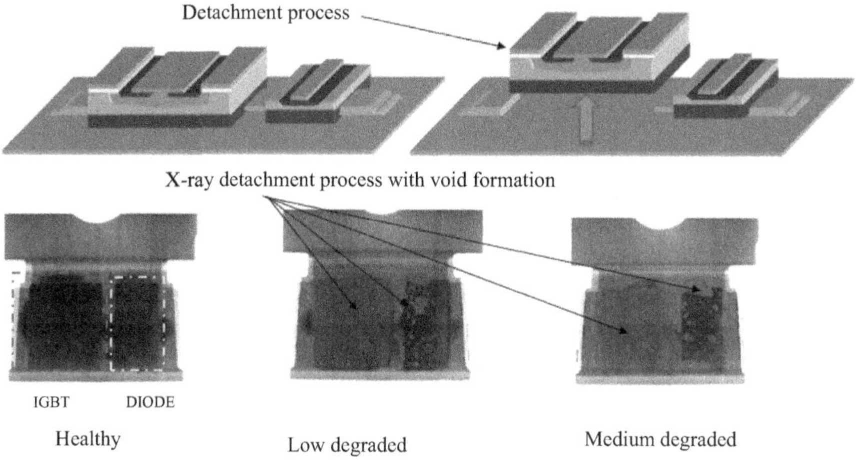

Figure 13.15 Void formation and detachment progression due to thermo-electrical stress

junction temperature increases significantly for the aged device, changing even more the value of the parameter (positive feedback).

13.2.2.4.2 Device aging (voltage gate and breakdown voltage shifting)

When string inverters installed in the field had failures, the standard procedure commonly adopted by most PV inverter manufacturer involved the replacement (swapping) of the inverter for a new one and the used inverter was sent for repairs. During a period of several months, a systematic analysis of the breakdown voltage for the main booster power MOSFET was conducted as follows: every time an inverter was repaired, one of the operational MOSFET was changed for a new one, and the operational MOSFET pulled from the inverter was subject to breakdown voltage measurement. Close to 200 used MOSFET, with one to five years in the field, were removed from returned inverters and had their breakdown voltage measured. In addition, 100 new MOSFET were randomly picked from a purchase and also underwent breakdown voltage measurement. When comparing the measured voltages from the used and the new MOSFET the average difference was 15 V which was negligible with respect to the manufacturer voltage specification of 1,000 V ensuring stability of the breakdown voltage during operation (refer to Figure 13.16).

13.3 The grid interface

13.3.1 Basic control of real and reactive power in a two-bus power system

In this section, we use a two-bus system to illustrate real and reactive power for conventional generation and generation based on power conversion. We assume

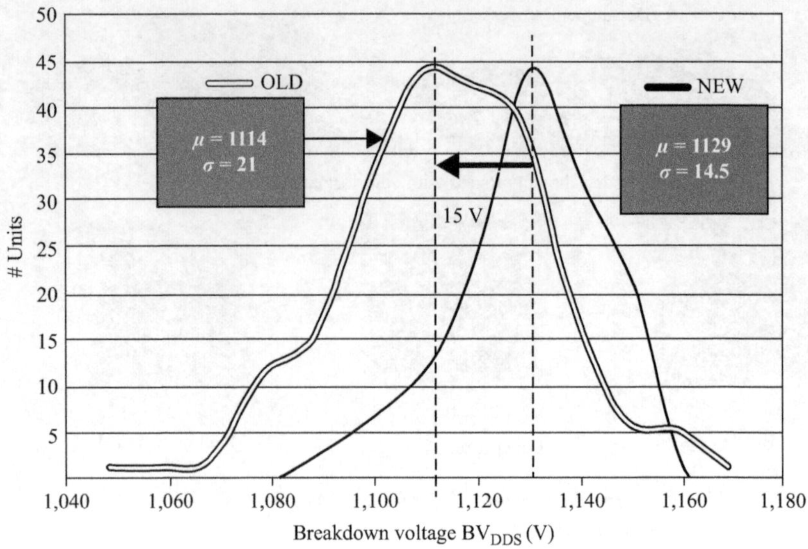

Figure 13.16 Breakdown voltage shift during operation

that the generator, V_{PV} (Bus 1), is connected to an infinite bus, V_S (Bus 2). In a conventional power plant with a synchronous generator, the following quantities are usually used to control the output of the generator:

- The voltage magnitude is used to control the reactive power (or reactive current component, I_q) by controlling the excitation winding, thus increasing the voltage of V_{PV}.
- The mechanical power is used to control the phase angle of the output voltage of the synchronous generator. The power angle, δ, of the output voltage mostly corresponds to the real current component, I_p, of the output current, I_s, of the voltage source.

A similar method can be used for a PV inverter, except we attempt to control the current directly.

13.3.1.1 Reactive power

13.3.1.1.1 Conventional power system

To understand a PV inverter, consider an average model. Because the relationship of the reactive power to the terminal voltage is very tight, either the reactive power control or the voltage control is generally chosen.

To describe the relationship between the reactive power flow and the voltage level, refer to the phasor diagram shown in Figure 13.17. It is assumed that the PV inverter (terminal voltage, V_{PV}) is connected to an infinite bus, V_S, through a reactance, X_S. The PV inverter (V_{PV}) generates an output current, I_S. The equation

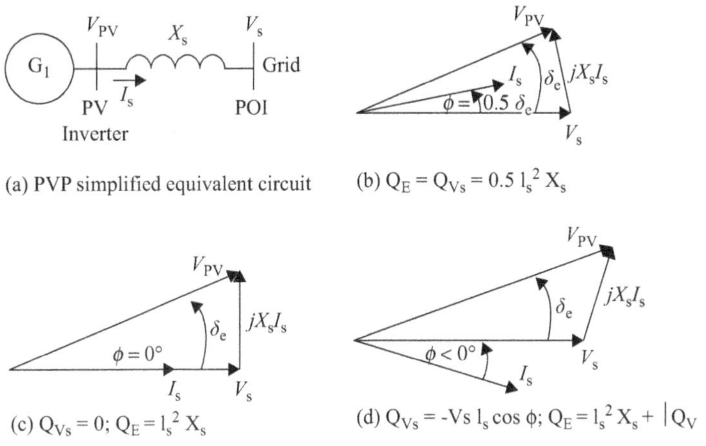

Figure 13.17 Illustration of a two-bus system demonstrating the voltage and reactive power flow in a synchronous generator

describing the relationship of the voltage at the two buses and the corresponding voltage drop across the reactance, X_S, can be written as follows:

$$V_{PV} = V_s + j\,X_s\,I_s$$

The reactive power losses ($Is_2\,X_s$) in the transmission line can be supplied from Bus 1 (V_{PV}) or Bus 2 (V_S), or they can be shared by both sides. Parallel compensation (e.g., capacitor banks) can be implemented on both sides. Reactive compensation can be implemented by controlling the generator itself (internal control), or it can be provided externally, such as from adjustable capacitor banks, synchronous condensers, and static power compensation (a static VAR compensator, or static compensation).

Consider Figure 13.17(a), in which all the reactive power spent in the reactance, X_s, is supplied by Bus 1. As shown, the voltage $V_{PV} > V_s$, to make it possible for the reactive power to flow from Bus 1 to Bus 2. Similarly, as shown in Figure 13.17(b), all the reactive power spent in the reactance, X_s, is supplied by Bus 2. In this case, the relationship between the two voltages is $V_{PV} < V_S$. Now consider Figure 13.17(c), in which the voltages at Bus 1 and Bus 2 are maintained constant, and there is equal magnitude at 1.05 p.u. The reactive power is supplied equally by both Bus 1 and Bus 2. The voltage $V_{PV} = V_S$ requires that the source of the reactive power comes from both sides. Thus, the larger the contribution of the reactive power, the higher the voltage of the source of the reactive power.

13.3.1.1.2 Power-converter-based power system

The impact of reactive power's contribution on voltage regulation can be illustrated by the phasor diagrams. To save some space, we will use the same phasor diagrams shown in Figure 13.20. As shown in Figure 13.20, the phasor current, I_s, is divided into the real current, I_p (in phase with the voltage V_{PV}) and reactive current, I_q (in quadrature with respect to the voltage, V_{PV}) components.

The voltage equation is rewritten as:

$$V_{PV} = V_s + jX_s\,(I_P + I_Q)$$

The real current component, I_p, is proportional to the real power generated, and the reactive current, I_q, is proportional to the reactive power generated. Doubling the reactive current, I_q, will double the voltage drop, $I_q X_s$, and directly increase the terminal voltage, V_{PV}, by an additional $I_q X_s$. Thus, to increase reactive current, we need to increase the voltage, V_{PV}. The larger the value of X_s, the higher the voltage, V_{PV}, will increase as the same amount of the I_q current is increased. As can be expected, increasing the reactive component of the current, I_q, will directly impact the terminal voltage magnitude, V_{PV}, and increasing the real current component, I_p, will more directly impact the power angle, δ between V_{PV} and V.

In PV generation, two limits must be observed:

- The total resultant current, I_s, is limited by the maximum current-carrying capability of the power converter (I_{max}). Thus, the power converter can increase its real or reactive current components (or both) only to a certain level, until the resultant current reaches its maximum ($I_s = I_{max}$ – current-carrying capability of the IGBT switches). At that point, the overcurrent protection will prevent the PV inverter from delivering more current.
- The maximum terminal voltage, V_{PV}, must be limited to the maximum allowable voltage of the power converter. Thus, the power converter can increase its reactive current component, I_q, only so much, until the terminal voltage reaches its upper limit. At that point, the overvoltage protection will prevent the PV inverter unit from delivering more reactive current to the grid.

13.3.1.2 Real power

13.3.1.2.1 Conventional power system

The equation describing the relationship of the voltage at the two buses to the corresponding voltage drop across the reactance, X_s, can be written as follows:

$$P = \frac{V_{PV} V_S}{X_s} \sin\delta$$

Output power can be increased in many ways. For example, increasing the output power while maintaining equal voltage ($V_{PV} = V_S$) can be illustrated by the changes shown in the phasor diagrams in Figure 13.18. As mentioned above, in this equal-voltage condition both sides share the reactive power loss in the line reactance, X_s, equally. As the current, I_s, increases, the power angle, δ_e, also increases while equal voltage is maintained ($V_{PV} = V_S$).

13.3.1.2.2 Power-converter-based power system

Figure 13.19 shows a two-bus system. In a two-bus system, a PV generator is connected to a grid (constant voltage or infinite bus) through a reactance, X_s. The

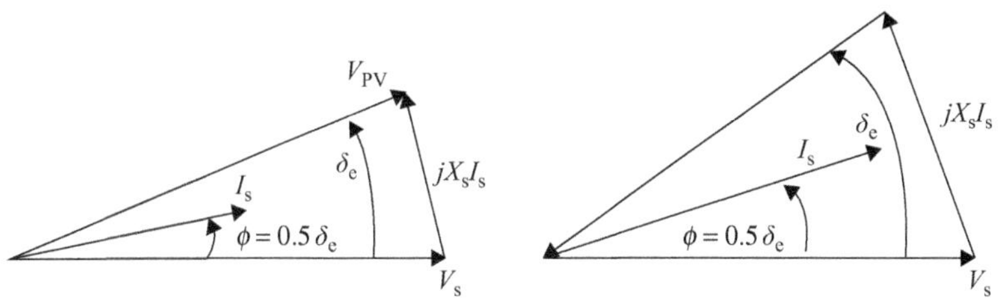

Figure 13.18 Illustration of a two-bus system demonstrating real power control in a conventional synchronous generator

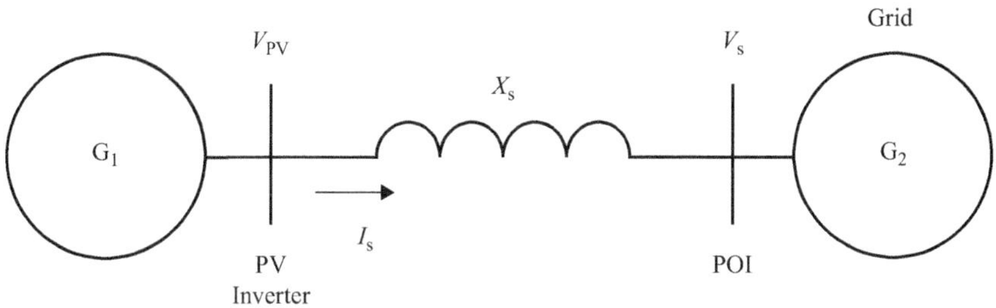

Figure 13.19 Illustration of a two-bus system demonstrating real power control in a PV inverter-based system

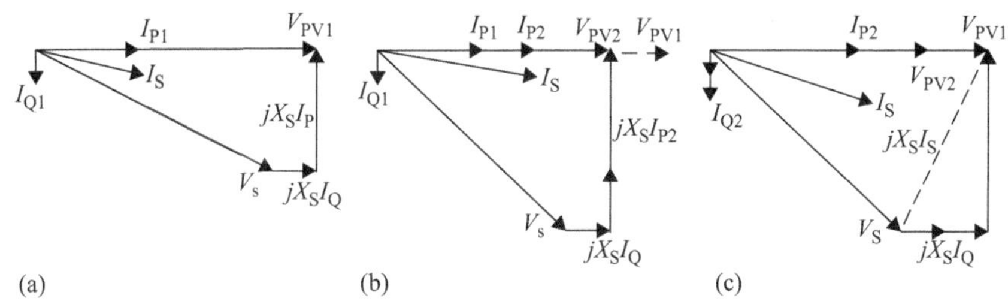

Figure 13.20 Illustration of a phasor diagram demonstrating real and reactive power control in a PV inverter. (a) Baseline: $I_P = I_{P1}$, $I_Q = I_{Q1}$; (b) increased: $I_P = I_{P2}$; and (c) $I_P = I_{P2}$, $I_Q = I_{Q2}$

current, I_S, is decoupled into real and reactive components (I_{P1} and I_{Q1}), as shown in Figure 13.20(a). Assuming that the grid is an infinite bus, V_S is constant. Figure 13.20(b) shows the size of the real current component increased to I_{P2}, while the size of the reactive current component, I_{Q1}, is maintained. The additional voltage drop, $jX_s\,I_{P2}$, is shown to increase by the same proportion. This affects the increase in the angle between V_{PV1} and V_S. Because the size of the infinite bus, V_S, is constant, the increase in I_P increases the voltage drop, but it decreases the voltage, V_{PV2}. The size of the reactive current component stays the same, and the resulting voltage drop, $jX_S I_{Q2}$, also stays the same size. Note that the voltage, V_{PV}, decreases from V_{PV1} to V_{PV2} (indicated by the dashed line in Figure 13.20(b)). To return V_{PV2} to the same level as V_{PV1}, the reactive current component, I_Q, is increased to I_{Q2}. This is shown in Figure 13.20(c), which indicates the size of the voltage, V_{PV1}, and the size of the current, I_{P2}. The increase in the output power is proportional to the change from I_{P1} to I_{P2}, which is equal, and the reactive current contribution to cover $I_S^2 X_S$ is equally distributed between V_{PV} and V_S.

13.4 The standards

Until recently, existing requirements have, for the most part, been centered on conventional utility-scale power plants, such as thermal or hydro units. From past few years, there has been an ongoing effort to define standards and guidelines that allow the accommodation of increasing penetration level of variable generation, in particular wind and PV power plants. Up until a few years ago, only a few countries had developed PV-specific standards but today most of those implementing large-scale PV systems have developed guidelines for the grid interconnection of PV inverter systems.

Therefore, some distribution system operators, as PV penetration increases, are defining their own standards or in some cases adopting the international ones. Besides the available international standards, other requirements may be proposed by national standards, grid codes or combination of regional codes and local codes, or specific utility regulations/guides. These documents, in some cases will be specific to one energy source (e.g., PV) or to a particular generating technology (e.g., the ones using power converters), although they are normally generalized for any distributed/variable generation technologies. On the other hand, these national or local requirements show differences taking into account the actual or expected penetration level of the generation technology, the stiffness or the strength of local grid, and aspects such are the generation mix, grid structure (installed renewable power vs. installed total power), or international interconnections. It is important as well how the PV plants are connected to the power system. For example, in Germany, Japan, or Spain, more than 80% of PV power is connected to the distribution level. Finally, PV systems using static inverters are technically different from rotating electrical machines and this fact has been generally recognized in these new guidelines.

The main standards were introduced by the Institute of Electrical and Electronics Engineers (IEEE) and the International Electrotechnical Commission (IEC). IEEE Std. 929:1988 was the key document for the utility interconnection of PV systems in the US. This document contains recommendations and requirements to ensure compatible operation of the terrestrial PV systems with the electric power system. This standard also addresses safety or equipment protection. IEEE Std. 929:1988 referred to other earlier standards that were developed to ensure that power supplied to the grid by small power producers meets certain standards for frequency, harmonic content, and voltage level. The primary concern of IEEE 929:2000 was to guarantee that the PV inverter is disconnected from the utility grid if the grid loses power lines or operating outside established range of nominal voltage or frequency, even if other PV sources are connected to the grid. IEEE Std. 519:1992 was the standard to specify the limits on the amount of harmonics allowed on the grid. Some aspects and limits from IEEE Std. 519-992 were adopted by IEEE 929:2000. IEEE 929:2000 was replaced by the IEEE Std. 1547:2003, which deals with all types of generation up to 10 MVA. Requirements in IEEE 1547:2003 were based on the assumption that energy production would be predominately consumed on site. The specific requirements were tailored to avoid interference of distributed resources with utility grid operations. Because the relatively low-rated power of PV installations and being connected to the distribution level, the PV industry is most familiar with the IEEE 1547 standard for interconnecting distributed resources. A new revision of this standard in 2008 was adopted by IEC. The IEC/IEEE/PAS 63547:2011 provides the interconnection technical specifications and requirements.

While some countries had specific PV standards, these standards differ from country to country and the harmonization of these standards was intended as the next step, as it is a common practice in the progression of this kind of standards. At the same time, it is clear that due to differing technical boundary conditions in different countries (grid structure, earthing/grounding rules, etc.) it will be difficult to achieve a full harmonization. At the international level, the IEC prepares and publishes international standards for all electrical, electronic, and related technologies.

The IEC Technical Committee 82 (IEC TC82) develops and adopts all PV-related standards. The scope of IEC TC82 is to prepare international standards for PV systems that convert solar energy into electrical energy, as well as for all the elements in the entire PV system. The IEC TC82 is comprised of five working groups, with the working group 3 devoted to the discussion and accumulation of experiences in order to develop guides of practical rules, the connection of dispersed generation on MV/LV grids or the planning and design of MV/LV grids.

In 1995, the IEC published the IEC 61727:1995, specifying the main requirements of a grid interface which ensure that it is both functional and safe for PV connections of 10 kVA or less. The precise values of the parameters are specific to each country, and a copy of the latest national regulations should therefore be obtained through the local electricity company. IEC 61727-2004 edition 2 applies to utility-interconnected PV power systems operating in parallel with the utility and utilizing static (solid-state) non-islanding inverters for the conversion of DC to AC. Another standard, IEC/TR 61000-3-15:2011 is concerned for single- and multi-phase dispersed generation systems up to 75 A per phase, particularly converters connected to the public supply low-voltage network.

At the European level requirements, The European Network of transmission system operators (TSOs) for Electricity (ENTSO-E) is working in a Network Code for requirements for grid connection, applicable to all generators (NC-RfG). This standard will provide legal framework for the definition of local grid codes, focusing on cross border issues. A classification of generators depending on the rated power, voltage level, synchronous zone is realized (Types A, B, C, and D).

According to the four types, requirements to support the grid frequency *P*(*f*), reactive power and fault ride through, regulating reserve, synthetic inertia or black start capability are applied. As well, the European Committee for Electrotechnical Standardization (CENELEC) published the EN 50438:2013 as the European standard for micro generation connected in parallel with the network, requiring similar specification as the IEEE 1547:2008. The EN 50438:2013 together with the technical specifications EN 50549-1 and EN 50549-2 represent standard harmonizing initiatives for the interconnection of distributed generation.

Other countries like Germany have experienced tremendous growth in renewable energy generating sources, like PV and wind systems. To manage better the operations of the electric system as the level of PV and wind generators penetration increases significantly, the grid operators in the German Association of Energy and Water Industries (BDEW), created guidelines for inverter capabilities. While these guidelines are not required by law, electric utilities require interconnecting generators to comply with the BDEW guidelines, in effect turning the guidelines into standards requirements. The BDEW guidelines provide specifications for generator control and communications, frequency control, dynamic reactive support, dynamic grid support including low-voltage ride-through (LVRT), and certification for medium/high voltage. The BDEW guideline refers to other guidelines, such as FGW-TR3, FGW-TR4, or FWG-TR8. Whereas, the VDE-AR-N 4105:2011 (VDE) is applied for low voltage (≤ 16 A per phase) with requirements for power quality, *P*(*f*), VAR, voltage and frequency limits. Therefore, VDE-AR-N 4105 can be considered a "State of Art" guideline, which considers high-level PV interconnections and stipulates grid-supporting features.

Table 13.2 shows the rated power of PV power plants and voltages levels covered by the main standards up to date.

Other countries like Spain, France, Italy, Japan, or China have introduced requirements as well, based on these standards and taking into account their own specific characteristics. Finally in some studies, generic and basic electromagnetic compatibility standards, like those from IEC 61000 series, may also be referred in interconnection standards to define immunity requirements (test levels) or tests procedures.

Table 13.2 Rated power of PV power plants and voltages levels covered by the main standards

Standard	Generation technology	Voltage level	Rated power
IEC 61727:2004	PV	Low voltage	≤ 10 kVA
IEEE 1547:2008	All	Primary/secondary distribution voltages	≤ 10 MVA
VDE-AR-N 4105:2011	All	Low voltage (≤ 1 kV)	≤ 100 kVA
BDEW:2008	All	Medium (1–66 kV)	No limit

The requirements, mainly from IEC 61727:2004, IEEE 1547:2008, and VDE-AR-N 4105:2011, are summarized in three groups: protection, power quality, and ancillary services.

13.4.1 Protection

13.4.1.1 Over/under voltage

Over/under voltage protection is required to ensure appropriate voltage levels. With the proliferation of distributed generation systems, it is becoming difficult and impractical to control dynamic voltage changes using traditional voltage management tools like line regulators and capacitors.

Table 13.3 shows the standards' requirements for the over/under voltage protection taking into account the voltage range in the point of interconnection. The inverter should sense abnormal voltage and respond according to the conditions in that table. In contrast to VDE and IEEE standards, the IEC 61727:2004 states that the PV plants must not disconnect if the time is smaller than the disconnection time. In the IEC 61727:2004, the disconnection time refers to the time between the abnormal condition occurring and the inverter ceasing to energize the utility line.

13.4.2 Islanding

A utility island occurs when a portion of the utility system containing load and operating generators is isolated from the remainder of the utility system. If the island generating source continues to feed the island during the system fault condition, the island may remain energized and several undesirable results may occur. The risk of islanding has been researched in some detail under the International Energy Agency Task 5. Table 13.4 shows the islanding protection together with the disconnection requirement. Existing voltage trip settings prescribed in IEEE 1547 are conservative, forcing generators to trip off line quickly to avoid islanding.

13.4.2.1 Over/short circuit current

In general all forms of DG contribute to the fault, therefore all standards require to assess its impact. Over current refers to the ability of the PV inverter to deliver

Table 13.3 Over/under voltage protection

IEC 61727:2004		**IEEE 1547:2008**		**VDE-AR-N4105:2011**	
Voltage range (%)	***Disconnection time (s)***	***Voltage range (%)***	***Maximum disconnection time (s)***	***Voltage range (%)***	***Maximum disconnection time (s)***
<50	0.10	<50	0.16		
50–85	2.00	50–88	2.00	<80	0.2
85–110	Continue operation	88–110	Continue operation	80–110	Continue operation
110–135	2.00	110–120	1.00	>110	0.2
>135	0.05	>120	0.16		

Table 13.4 Islanding protection

	IEC 61727:2004	IEEE 1547:2008	VDE-AR-N 4105:2011
Method (active or passive)	Active or passive	Active or passive	Active or passive
Disconnection requirement	Cease to energize within 2 s of the formation of the island	Cease to energize within 2 s of the formation of the island	Disconnect in 5 s

Table 13.5 Over/under frequency protection

	IEC 61727:2004 (rated frequency 50 Hz)	VDE-AR-N 4105:2011 (rated frequency 50 Hz)	IEEE 1547:2008 (rated frequency 60 Hz)		
			Rated power ≤30 kW	*Rated power >30 kW*	
Operating range (Hz)	49–51	47.5–51.5 (−5% to +3%)	59.3–60.5 (−2% to +1%)	59.3–60.5 (−2% to +1%)	57–59.3
Disconnection time (s)	0.2	<0.1	0.133	0.133	0.16–300

additional currents (above its rated value) for a specific duration (e.g., 1.1 p.u. for 1 min). The short circuit current is the maximum current that can be delivered by the PV inverter to the grid. Many PV inverters are designed to have the capability to control the reactive power under normal condition and during short circuit to support the grid voltage.

13.4.2.2 Over/under frequency

All the standards explicitly state values for over and under frequency protection are shown in Table 13.5. The purpose of the allowed range and time delay is to ride through short-term disturbances to avoid excessive nuisance tripping, especially in weak-grid situations.

13.4.2.3 Reconnect after grid failure and restoration

The reason for the delay in reconnecting to the PV inverter to the grid is to avoid a recurring connect/disconnect (chattering) condition. Often utilities employ automatic reclosers to restore the service. After a disconnection caused by abnormal voltage or frequency conditions, the inverter can be reconnected only after certain specific conditions on voltage and frequency are fulfilled with the requirements shown in Table 13.6.

Table 13.6 Reconnect after grid failure and restoration protection

Standard	Requirements
IEC 61727:2004	Normal voltage and frequency for 20 s to 5 min
IEEE 1547:2008	Normal voltage and frequency for 5 min
VDE-AR-N 4105:2011	Voltage between 85% and 110% and frequency between 47.5 and 50.05 Hz at least 1 min For short interruptions, reconnection may be immediate

Table 13.7 IEEE 1547:2008 and IEC 61727:2004 harmonic levels

	Odd					Even	THD
Individual harmonic order, *h*	3–9	11–15	17–21	23–35	35+	3–35+	
Allowable limit (%)	<4.0%	<2.0%	<1.5%	<0.6%	<0.3%	25% of odd harmonics	<5%

13.4.3 Power quality

Power converters used in PV plants may introduce harmonics, unbalanced phases, DC injection or network disturbances. Deviation from the limits represents out-of-bound conditions and may require disconnection of the PV system from the grid.

13.4.3.1 Current harmonics and inter-harmonics

The VDE and BDEW refer to the German version of the IEC 61000 electromagnetic compatibility standards and the IEEE 1547 refers to the IEEE 519:2014. The IEEE 1547:2008 standard refers to the values from IEEE 519, which are displayed in Table 13.7. IEC 61727:2004 and IEEE 1547:2008 harmonize very well in this case.

Total harmonic distortion (THD) for the injected grid current should be lower than 5% in normal operation to avoid adverse effects on other equipment connected to the grid.

13.4.3.2 Voltage unbalance

The VDE is the only one that specifies a maximum difference of 4.6 kVA between each phase on the network.

13.4.3.3 Injection of DC into the AC system

The primary reason for keeping DC out of the grid is its effect on inductive loads. With a combination of DC and AC applied to many inductors, the inductor may be driven into saturation, resulting in hysteresis losses beyond the device rating. When an output transformer is not present in the PV system, it is important that the output current is controlled accurately in order to minimize the DC component.

The IEC 61727:2004 assumes <1% of the RMS-rated current as the maximum DC injection allowed. On the other hand, the IEEE 1547:2008 specifies 0.5% of its rated AC output current under any operating conditions.

13.4.3.4 Flicker and fluctuations

The IEEE 1547:2008 and VDE refer to several other standards regarding flicker and fluctuation are shown in Table 13.8.

13.4.4 Ancillary services

New regulations are required gradually due to high penetration of renewable energy in the distribution networks. Inverters must be able to support voltage and frequency to sustain network stability.

13.4.4.1 Network voltage support

The network voltage magnitude can be greatly affected by the distributed generation as they inject power in a power system where previously only loads had been. This is thoroughly considered in every standard.

13.4.4.1.1 Power factor

Keeping the voltage between defined limits and especially avoiding overvoltage is becoming a primary concern of distribution system operators (DSOs) due to the increasing number of generators connected to the distribution network. The power factor is defined at the fundamental frequency for non-sinusoidal current delivered.

According to the IEC61727:2004, the PV system shall have an average lagging power factor greater than 0.9 when the output is greater than 50% of the rated power. On the other hand, on the IEEE 1547:2008, there is not active voltage regulation, although the network voltage of the system should not adversely be affected.

Since 2012, in Germany, PV systems bigger than 3.68 kVA have been providing static grid support by reactive power control. The control strategy depends on the size of the PV system, the voltage level, and the local characteristics of the grid, as shown in Table 13.9. Table 13.9 is considered if the operating voltage is within $\pm 10\%$ of the nominal and the active power output is above 20% of the rated power.

Table 13.8 Flicker and fluctuation levels

Standard	Referred standards
IEEE 1547:2008	IEEE Standard 519-1992, IEEE P1453, IEC/TR361000.3.7, IEC 61000.4.15, IEC 61000.21
VDE-AR-N 4105:2011	DIN EN 610000.3.3, DIN EN 610000.3.11 (up to 75 A)

Table 13.9 VDE-AR-N 4105 reactive power capability limits

$S_{max} \leq 3.68$ kVA	0.95 under-excited to 0.95 over-excited
3.68 kVA $< S_{max} \leq 13.8$ kVA	Characteristic curve from network operator with 0.95 under-excited to 0.95 over-excited
$S_{max} > 13.8$ kVA	Characteristic curve from network operator with 0.90 under-excited to 0.90 over-excited

13.4.4.1.2 Low-voltage ride-through

Fault ride-through (FRT) capability is defined as the PV inverters' capability of remaining connected to the grid in the event of grid failures. A second aspect is also related to FRT, but it refers to the additional capability of injecting reactive power in the grid in case of grid fault, with the purpose to give a voltage support during fault conditions.

The IEC 61727:2004 requires that the PV plant do not have to cease to energize for disturbances shorter than the clearing times whereas the IEEE 1547:2008 requires disconnection from the network in the event of a fault. In the German network code, from January 1, 2011, PV plants have to be capable to participate in full dynamic network support during grid faults: to stay connected to the network and provide voltage support by feeding a reactive current. In fact, Germany has determined that it is necessary to retrofit 315,000 existing inverters to achieve enhanced inverter functions including LVRT capability in order to provide grid support during faults and voltage recovery during post-fault conditions to avoid a system blackout.

13.4.4.2 Frequency support

Where voltage deviations tend to be more localized, frequency deviations will effect an entire interconnection. Generator frequency tolerance is typically coordinated with under-frequency load shedding schemes. To ensure that PV systems and other distributed renewable energy sources provide grid support during system frequency events, it would be desirable that inverters not disconnect too soon and that they disconnect in a random manner. By default, PV inverters designed for distribution interconnection are designed to operate at unity power factor, and are unable to supply reactive power when operating at rated kW output. To maintain a ± 0.95 power factor range at rated voltage and active power, the inverter would need to have a kVA rating at least 5.2% higher than the kW rating.

However, TSOs used to consider distributed generation only as negative load and therefore not relevant for electricity system operation. They required every installation to switch off automatically and instantly as soon as the frequency exceeded a defined range (depending on the country). These obligations are not a consequence of technical limits but are due to grid connection rules set by TSO/DSOs.

During the major system disturbance in Europe on November 4, 2006, part of the decentralized generation (mainly wind power) switched off after the disturbance occurred. Part of Germany belonged to an exporting network region in which the frequency value increased to 50.2 Hz. If the same extreme event was to occur at midday during the summer, more than 20 GW of PV systems would instantaneously disconnect from the grid at the EU level. Such a massive disconnection of PV systems would accelerate the possible cascading breakdown process. National grid codes have been revised in some of the key countries (Germany, Italy, Belgium, and Spain) to integrate new frequency requirements. For the existing PV installed capacity, retrofit programs are under analysis or already under way in countries like Germany and Italy.

Neither IEEE 1547:2008 nor IEC 61727:2004 requires any frequency support. The VDE-AR-N 4105:2011 considers that the inverters will remain active when the frequency is between 47.5 and 51.5 Hz. Above 50.2 Hz, the inverter will start reducing power at a rate of 40% per Hz. The system must disconnect if the frequency is below 47.5 Hz and above 51.5 Hz.

13.5 The field measurements

13.5.1 Intermittence in solar field results

The operation of MT series (15 kW inverter) is shown in Figure 13.21. The measurements were taken in Biel, Switzerland on September 5, 2013. This production is a common example of the intermittence that solar power displays during the fall. The MT series have three trackers that are denoted in colors (dark blue, light blue, and red). The three DC power collected are almost superimposed, showing a good performance of the MPPT trackers and the power AC generated (black) is practically the addictions of the three DC power trackers. Only two times during the day does the inverter deliver close to its nominal power for a few minutes. The DC voltages remain fairly constant, independent of the power generated.

13.5.2 LVRT test results of the 500 kW RX series

In Germany, on June 9, 2013, for the first time ever, solar power accounted for more than 50% of the total power generated, anti-islanding requirements, enforced for years in grid-tie inverters. While anti-islanding is intended for safety concerns,

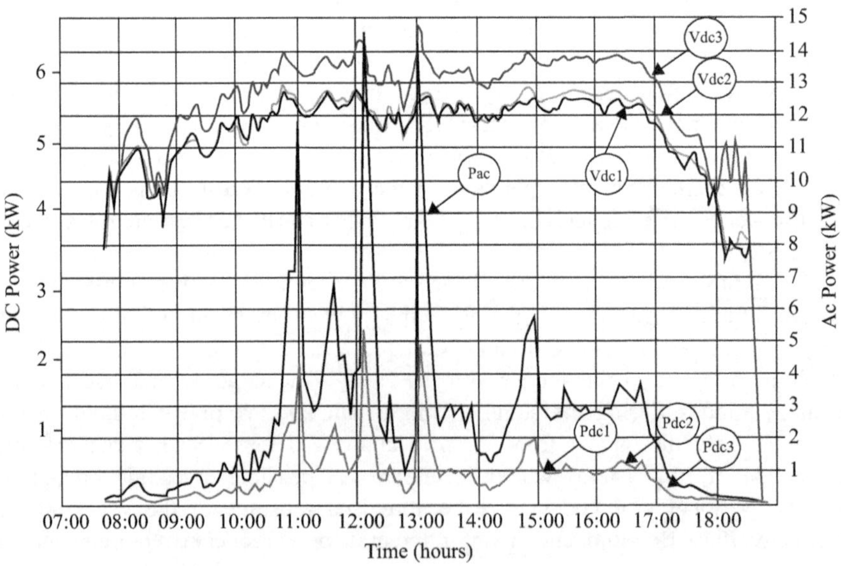

Figure 13.21 Illustration power generated by 15 kW inverter

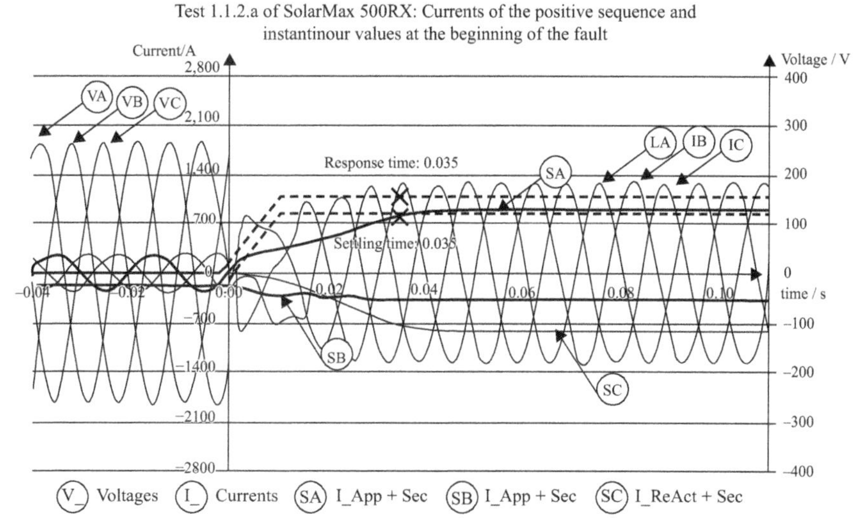

Figure 13.22 LVRT test for 500RX series

i.e., to protect the utility personnel from electric shocks or electrocution, the LVRT is intended to improve the power system reliability. If a fault close to a solar plant surrounded by relatively high-density solar community, the voltage at the local grid may drop significantly. Therefore, if the LVRT is not implemented on the local PV inverters, this situation could generate a cascading effect that could trigger consequence inverters to trip and in seconds the whole German grid could be in jeopardy. This potential risk was clear that the solar system needed to be able to support the grid during fault for power system reliability. The anti-islanding standards should not be in contradiction with the LVRT to support the grid. The latest edition of the IEEE 1547 allows the local utilities modify the range of allowable operating voltage and frequency to ensure grid reliability, thus, avoiding cascading effect. The PV inverters must follow the local utility grid code and LVRT. Figure 13.22 shows the LVRT capabilities of the RX series. When the voltages collapse (in green), the inverters support the grid, injecting nominal current to the system during the fault.

13.6 Summary and conclusions

This chapter summarizes the application of power electronics converters and systems in PV energy system. The chapter is organized with introduction given in Section 13.1 to cover the broad view of the PV generation in the context of power system environment. Section 13.2 is dedicated to the power electronic technologies used in PV generation with topics from the state of the art of technologies and the reliability aspects of the PV inverter. In Section 13.3, the system integration is presented to cover integration of the PV inverter with the power system grid and to understand the type of controls and control actions relevant to power system operation, reliability, and stability. In Section 13.4, the standard commonly adopted for PV generation covering different aspects of acceptable grid integration including protection, islanding, power quality, and ancillary services. Finally, in Section 13.5, the field measurement for PV generation is covered.

The PV generation technology is ever evolving with the technology advancement in power semiconductor (high voltage, high current, high temperature, and high switching frequencies), the future of power system characteristics (less conventional synchronous generators with its associated rotating inertia), the availability of cheap energy storage in the next few years. All of these factors may lead to more affordable PV generation and there will be continuing effort to revise or update the so-called grid codes or standards. It is acceptable that the local and regional conditions are differ from places to places around the world, and it is understandable that the local grid codes may be adapted to local needs and local condition of the power systems.

This chapter is not intended to cover all of the information about PV generation, but, we try to include the most important ones relevant to the present PV generation and power converter technologies.

Acknowledgements

This work was supported by the US Department of Energy under Contract No. DE-AC36-08-GO28308 with the National Renewable Energy Laboratory and "Ministerio de Economia y Competitividad", Spain under No. ENE2012-34603, co-financed with European Union FEDER funds. The authors thank Sputnik Engineering AG for providing data and field results and very especial thanks to Andres Salazar and Carlos Restrepo for their support.

References

[1] Y. Leblebici and S.M. Kang, *Hot-Carrier Reliability of MOS VLSI Circuits*. Boston, MA: Kluwer Academic Publishers, 1993.

[2] D.L. Crook, "Method of Determining Reliability Screen for Time Dependent Dielectric Breakdown," *Proceeding of the International Reliability Physics Symposium*, No. 17, 1979, pp. 1–7.

[3] R. Pearce, S. Brown, and D. Grant, "Measuring HEXFET Characteristics," *HEXFET Power MOSFETs Designer Manual Application Notes and Reliability Data*. International Rectifier, Vol. I, 1993, Application Note 957B.

Chapter 14
Automotive energy systems

Vamsi Krishna Pathipati, Janamejaya Channegowda*, Kunwar Aditya* and Sheldon S. Williamson**

14.1 Electric vehicle batteries

14.1.1 Introduction

Owing to recent well-known trends, renewable resources are becoming increasingly prominent in the complex energy market mosaic. As long as their penetration level is low, they can be handled easily by the current infrastructure, but at present incremental rates, this will not be the case in the future. The intermittent nature of solar and wind generation will require a far more flexible compensation mechanism than is currently available. Because of this, large battery banks that act as buffers between the generator and the grid invariably accompany today's renewable energy installations. Wind power, in particular, is not only intermittent but also it has no day-average predictability, as winds can differ hour-to-hour as easily at night as during the day, adding an extra amount of irregularity to an already varying load. This suggests that plug-in electric vehicles (PEVs) will be called on to perform, not only the more manageable regulation tasks, but also aid in providing peak power. As noted earlier, this might not find approval with PEV owners unless the pricing model is modified. Nevertheless, it is reasonable to ask whether a large PEV contracted fleet could perform this task on a national (US) level. Studies [1] have shown that the answer is yes. With an overconfident 50% estimation for the market penetration of wind energy and 70 million PEVs available, peak power could be provided at the expense of approximately 7 kWh of battery energy per day or about 10%–20% of an average PEV reserve.

14.1.1.1 Important characteristics of battery chemistry

There exist many types of batteries, which can be found in battery reference books such as [2]; however, a large number of them are just produced in laboratory conditions and are still under investigation. They are not commercialized because of many factors, such as non-maturity, low energy density, safety, high rate of toxic

*University of Ontario Institute of Technology, Oshawa, Ontario, Canada

materials, price, etc. Hence, a small group of batteries is available commercially and within this, the most frequently used are lead-acid (PbA), nickel–cadmium (Ni–Cd), nickel-metal hydride (Ni-MH), lithium-ion (Li-ion), and lithium-polymer (Li-Po). Batteries can be divided initially into two categories: primary and secondary. Primary batteries are simply those that can be used only once and after a full discharge cannot be used any more. This is because the chemical reactions happening inside them are irreversible. Secondary batteries, however, can be used many times through recharging. In the case of automotive and traction applications, secondary batteries are of most interest, because utilizing primary batteries in these applications seems unreasonable. Here, we will only consider secondary type batteries and when we mention batteries, we mean secondary batteries unless otherwise stated.

14.1.1.1.1 Lead-acid

For over a century, lead-acid batteries have been utilized for various applications including traction. Their well-improved structure has led to valve regulated lead-acid (VRLA) batteries, which can be considered as maintenance-free batteries, which is a desirable characteristic for plug-in hybrid electric vehicles (PHEVs). In terms of efficiency, they have high efficiency in the range of 95%–99%. The main disadvantage of lead-acid (PbA) batteries is their weight, in other words, they have a low specific energy (30–40 Wh/kg) compared with their counterparts.

14.1.1.1.2 Nickel–cadmium (Ni–Cd)

Considering low power applications, nickel–cadmium (Ni–Cd) batteries also benefit from a mature technology, but considering traction applications, their specific energy is low. The typical specific energy for this type is 45–60 Wh/kg. They are used mainly where long life and price are of high importance. The main applications for this type are portable devices, but in cases that demand high instantaneous currents, there use is desirable. However, considering environmental issues, they do contain toxic metals [3].

14.1.1.1.3 Nickel-metal hydride

Compared with the previous types, these have higher specific energy, but lower cycle life. In general, for the batteries of the same size, Ni-MH batteries can have up to two or three times the energy of a Ni–Cd type. The typical value for the specific energy of the present technology Ni-MH batteries is in the range of 75–100 Wh/kg. This type is widely used in electric vehicles (EVs) and PHEVs.

14.1.1.1.4 Lithium-ion

This type has noticeably high specific energy, specific power and great potential for technological improvements, providing EVs and PHEVs with perfect performance characteristics such as acceleration. Their specific energy is in the range of 100–250 Wh/kg. Because of their nature, Li-ion batteries can be charged and discharged at a faster rate than can Pb-acid and Ni-MH batteries, nominating them as good candidates for EV and PHEV applications. Furthermore, Li-ion batteries have outstanding potential for long life if managed in proper conditions; otherwise, their life cycle can be a disadvantage. One of the main reasons for this is the near absence of memory effect in Li-based batteries. However, safety issues are weak points of Li-based batteries. Overcharge of Li-ion batteries should be carefully prevented, as they have high potential for explosion owing to overheating caused by overcharging. They can easily absorb extra charge and thus could explode. Utilizing advanced battery management systems can ensure a reliable range of operation of Li-ion batteries, even in cases of accidents. In addition, Li-ion batteries contain less environmentally harmful materials compared with nickel-based batteries.

14.1.1.1.5 Lithium-polymer

Li-Po batteries have the same energy density as Li-ion batteries but with lower cost. This specific chemistry is one with the greatest potential for EVs and PHEVs. Recently, there have been significant improvements in this technology. Formerly, the maximum discharge current of Li-Po batteries was limited to about 1 C rate; however, recent enhancements have led to maximum discharge rates of almost 30 times the 1 C rate. This improves and simplifies greatly the storage part of EVs and PHEVs in terms of power density, because in some cases, this can even eliminate the need for ultracapacitors. In addition, there have been outstanding improvements in charging times. Recent advances in this technology have led to some types that can reach over 90% state of charge (SOC) within a couple of minutes, which can increase significantly their attractiveness to EVs and PHEVs because of the noticeable reduction of charging time. Because this type is a solid-state battery, having solid electrolyte, the materials would not leak even in the case of an accident. One of the other advantages of this type is that it can be produced in any size or shape, which offers great flexibility to vehicle manufacturers.

14.1.1.2 Battery parameters

14.1.1.2.1 Battery capacity

This parameter can be assumed simply as the amount of charge that can be drawn from a fully charged battery until it becomes fully discharged. An important effect in batteries is that the higher amount of current drawn from a battery, the lower capacity the battery will have. Hence, theoretically, battery capacity is defined as the amount of current drawn from a battery that discharges it completely in exactly 1 h. For example, a battery capacity of 10 Ah means that if a constant current of 10 A is drawn from the battery, it will become discharged completely after 1 h. However, in practice, battery manufacturers might use other definitions. Usually, tables of different test results are provided showing the amount of time the battery runs under different constant current loads and with different constant power loads. In practice, these tables provide much more practical information rather standard definitions, because after production, different loads with different characteristics might be connected to the battery. Nevertheless, the amount of time that a battery runs is not predictable exactly, because not all the loads are constant current or constant power loads. Even if the loads are one of these types, those tables are valid for new batteries and not for aged ones. Therefore, in many design procedures just rough estimates of battery runtime is calculated. The battery capacity is shown in the literature with letters such as "*C*" or "*Q*" or other notations. The main unit for battery capacity is the ampere-hour (Ah); however, based on the size of the battery, alternative units might be used, such as mAh or even mAs in the case of very small batteries.

14.1.1.2.2 C-*rate*

This parameter is used to show the amount of current used for charging the battery, or that of a load drawn from the battery. We can consider the previous case of a 10 Ah battery as an example. When it is mentioned that the charging process is terminated when the charging current falls below *C*/10 rate (10 h rate), this means the charging should be stopped when the current becomes less than the amount of current with which the battery is discharged after 10 h, in other words, 10 Ah/10 h = 1 A.

14.1.1.2.3 State of charge

In its simplest form, SOC can be visualized as the percentage of the remaining water relative to the entire capacity of a water tank. In terms of charge, this means the percentage of charge available from a battery relative to the entire capacity of the battery. Representing the battery as a water tank gives a good approximation; however, it is not accurate because of some of the effects in the batteries such as relaxation effect, which will be described in the following sections. Furthermore, according to aging, the rated capacity of the battery reduces over time and hence, for determining the SOC, the rated capacity should be measured or calculated regularly.

14.1.1.2.4 Depth of discharge

Again, using the water tank concept, the depth of discharge (DOD) can be assumed as the percentage of water that has been drawn from the water tank relative to the entire capacity of the tank. In terms of charge, the water can be replaced with electric charge. This parameter is usually used in discharge pattern recommendations. For example, the battery manufacturer might recommend the user not to exceed 30% DOD in relation to battery lifetime issues.

14.1.1.2.5 Energy density

Energy density can be defined in two ways. One is "Volumetric Energy Density," which is defined as the amount of available energy from a fully charged battery per unit volume (Wh/liter). The unit liter is used mainly for measuring the volume of liquids. Mostly, batteries have a liquid electrolyte and so in such cases, it easily makes sense; however, even for solid-state electrolytes such as lithium polymer batteries, the same unit is usually used. The other way of defining the Energy Density is "Gravimetric Energy Density," which is usually referred to as "Specific Energy," and is defined as the available energy from a fully charged battery per unit weight (Wh/kg). Based on application and based on the importance of the volume or weight, either definition can be used. In the case of EVs and PHEVs, the weight factor is usually more important than volume; hence, Specific Energy would usually be seen in the literature for this specific application.

14.1.1.2.6 Charging efficiency

The chemical reactions inside the battery during charge and discharge are not ideal and there are always losses involved. In other words, not all the energy used to charge the battery is available during discharge. Some of this energy is wasted in other forms of energy dissipation such as heat energy dissipation. The charging efficiency can be defined as the ratio of available energy from the battery due to a complete discharge, to the amount of energy needed to charge the battery completely. This parameter may be mentioned by other names such as coulombic efficiency or charge acceptance. The types of losses that reduce the coulombic efficiency are mainly losses in the charging process owing to chemical reactions, such as electrolysis of water or other redox reactions in the battery. In general, the coulombic efficiency for a new battery is high; however, it reduces as the battery ages.

Hereafter, this chapter will discuss some aspects of batteries in the specific cases of EVs and PHEVs regarding charging battery packs. This will help greatly in the design of more efficient and flexible chargers based on battery behavior, which will finally lead to the improvement of the battery pack life cycle.

14.1.1.3 Basic requirements of EV/PHEV batteries

The basic preferred characteristics of PHEV batteries can be summarized as follows [4]:

1. high specific energy that results in higher all electric range and fewer recharge cycles
2. high specific power that results in high acceleration characteristics of PHEVs, owing to the high rates of current available from the battery without causing any permanent damage to the battery pack
3. high number of charge/discharge cycles available and high safety mechanisms built into the battery because of high power ratings of battery packs
4. environmentally friendly aspect of the battery, i.e., being recyclable and incorporating low amounts of toxic materials

Cost is also an important concern for commercializing EVs and PHEVs in a large scale.

14.1.1.4 Charging, termination and cell balancing techniques, and SOC estimation

14.1.1.4.1 EV battery charging methods

Generally, charging is the action of putting energy back into the battery in terms of charge or current. Different chemistry needs different charging methods. Other factors that affect the choice of the charging method are capacity, required time and others. The most common techniques are mentioned below.

14.1.1.4.2 Constant voltage

As is clear from the name, "Constant Voltage" or CV is where a constant voltage is applied to the battery pack. This voltage is a value preset by the manufacturer. This method is usually accompanied with a current limiting circuit, especially for the beginning period of charging when the battery can easily accept high rates of current compared with its capacity. The current limitation value depends mainly on the capacity of the battery. Depending on the battery type to be charged, this preset voltage value is chosen. For example, for Li-ion cells, the value of 4.200 $\pm$ 50 mV is desirable. The accurate set point is necessary because overvoltage can damage the cell and undervoltage causes partial charge, which will reduce the battery life cycle over time. Therefore, the circuit used for charging, which can be a simple buck, boost, or buck/boost topology, depending on the voltage ratio of input and output, should be accompanied with a controller to compensate the source and load changes over time. When the cell reaches the preset voltage value, this causes the battery to enter standby mode, ready for later use. However, the duration of this idle time should not be very long and should be limited based on the manufacturer's recommendations. This method is usually used for PbA batteries and Li-ion batteries, using the current limiter to avoid overheating the battery, especially in the initial stages of the charging process [5].

14.1.1.4.3 Constant current

Constant current (CC) charging means simply applying a constant current to the battery with a low percentage of current ripples, regardless of the battery's SOC or temperature. This is achieved by varying the voltage applied to the battery by using control techniques such as current mode control to maintain the constant current. The CC technique can be implemented using a "Single Rate Current" or "Split Rate Current." In single rate, only one preset current value is applied to the battery, which is useful in balancing the cells; however, backup circuits must be used to avoid overcharging. In split rate CC, different rates of current are applied based on the time of the charge, the voltage, or both during different stages of charging. This provides charging that is more accurate and balanced charging; however, backup circuits should still be used to avoid overvoltage of the cells. In some cases, the CC method with high rates and low duration can be utilized to extend the lifetime of batteries. However, this is a very sensitive procedure and it should be performed carefully. Ni–Cd and Ni-MH batteries are charged using this method. Ni-MH batteries can be damaged easily owing to overcharging; therefore, they should be monitored accurately during the charging process [6].

14.1.1.4.4 Taper current

This can be used when the source is a non-regulated DC source. It is usually implemented with a transformer with a high output voltage compared with the battery voltage. A resistance should be used to limit the current flowing to the battery. A diode can also be used to ensure unidirectional power flow to the battery. In this method, the current starts at full rating and gradually decreases as the cell becomes charged. As an example, for a 24 V 12 A battery, the charging begins with 12 A when the battery voltage is 24 V. Then it drops to 6 A when the voltage reaches 25 V and then 3 A for 26 V and finally, it is reduced to 0.5 A for 26.5 V. This is just a hypothetical example and the values are not necessarily valid. This technique is only applicable to sealed lead-acid batteries. Taper charging has other disadvantages. As mentioned before, this technique uses transformers, which add to the weight of charger and generates heat.

14.1.1.4.5 Pulse charge

This technique involves using short current pulses for charging. By changing the width of the pulses, the average of the current can be controlled. Furthermore, the charging provides two significant advantages. One is the noticeably reduced charging time and the other one is the conditioning effect of this technique, which improves greatly the battery life cycle. The intervals between pulses, called rest times, play an important role. They provide some time for the chemical reactions inside the battery to take place and stabilize. In addition, this method can reduce undesirable chemical reactions that might happen at the electrodes. These reactions can be gas formation and crystal growth, which are the most significant causes of life cycle reduction in batteries.

14.1.1.4.6 Reflex charge

During the charging procedure, some gas bubbles appear on the electrodes and this is amplified during fast charging. This phenomenon is called "burping." Applying short discharge pulses or negative pulses, which can be achieved, for example, by short circuiting the battery for very small intervals in a current limited fashion, typically two to three times bigger than the charging pulses during the charging rest period, results in depolarizing the cell, which speeds up the stabilization process and hence, the overall charging process. This technique is called, among others, "Burp Charging" or "Negative Pulse Charging." Different control modes of charging along with waveforms and diagrams can be found in [7]. In addition, there are other charging methods such as current interrupt, which will be explained thoroughly in the charging algorithm section.

14.1.1.4.7 Float charge

For some applications, when the charging process is complete and the battery is fully charged, the batteries should be maintained at 100% SOC for a long time in order to be ready at the time of use. Uninterruptable power supplies are an example of one such application. The batteries should always remain fully charged. However, because of the self-discharge of batteries, they become discharged over time; for example, they may lose 20%–30% of their charge per month. To compensate for self-discharge, a constant is applied, which is determined based on the battery chemistry and ambient temperature. This voltage is called the "Float Voltage." In general, the float voltage should be decreased with the increase of temperature. This causes a very low rate of current, for example, a $C/300$ to $C/100$ rate to the battery, which continuously compensates for the self-discharge rate, also prevents sulfate formation on the plates. This technique is not recommended for Li-ion and Li-Po batteries. Furthermore, this method is not necessary for EV/PHEVs, which are frequently used every day. In addition, float charging involves a protection circuit, which avoids overcharging. This circuit adjusts the float voltage automatically and interrupts charging at some intervals based on battery voltage and temperature.

14.1.1.4.8 Trickle charge

Trickle charging is largely the same as float charging with just small differences. One difference is the usual absence of a protection circuit that avoids overcharging. Hence, it is very important to make sure in the design procedure that the charging current is less than the self-discharge rate. If so, they can be left connected to the battery pack for a long time.

Termination methods. During the charging procedure, it is very important to know when to terminate the charging. This is because of two main reasons. One is to avoid undercharge, i.e., to make sure that the battery is fully charged, not partially, in order to use the full capacity of the battery. The other is to avoid overcharging, which is very dangerous, especially in the case of high energy density lithium-based EV/PHEV battery packs. If not terminated on time, the overcharging of batteries can lead to over gassing of the cells, especially in liquid electrolyte cells, which results in an increase in the volume of individual cells that cannot be tolerated in a battery pack, which is rigidly packed. Another issue is the overheating of the cells, especially in lithium-based batteries, which can lead to an explosion and fire in the entire pack, because lithium is a very reactive material and combines easily with oxygen in the air. The one thing necessary to begin the combination is sufficient heat.

Choosing different termination criteria leads to different termination methods. Selecting the type of termination of charging process depends on different factors, such as the application and the environment in which the battery is used. Listed below are the different termination methods.

14.1.1.4.9 Time

Using time is one of the simplest methods, which is mainly used as a backup for fast charging or normally used for regular charging for specific types of batteries. This method can be implemented cheaply; however, because of diminishing battery capacity over time due to aging, the time should be set for a reduced capacity aged battery to avoid the overcharging of old batteries. Therefore, the charger would not work efficiently for new batteries and this leads to a reduction in battery lifetime.

14.1.1.4.10 Voltage

As mentioned before, voltage can be used as a termination factor. The charging process is stopped when the battery voltage reaches a specific value. However, this method has some inaccuracies, because real open circuit voltage is obtained when the battery is left disconnected for some time after the charging. This is because chemical reactions happening inside the battery need some time to stabilize. Nevertheless, this method is widely used, usually with a constant current technique in order to avoid overheating damage to the battery.

14.1.1.4.11 Voltage drop (dV/dT)

In some chemistry like Ni–Cd, if charged using the constant current method, the voltage increases up to the fully charged state point and then the voltage begins to decrease. This is because of oxygen build up inside the battery. This decrease is significant such that the negative derivative of the voltage versus time can be measured to be a sign of overcharge. When this parameter becomes positive it shows that we are passing the fully charged state and the temperature begins to rise. After this point, the charging method can be switched to trickle or float charge, or terminated completely.

14.1.1.4.12 Current

In the last stages of charging, if the CV method is used, the current begins to decrease as the battery reaches the fully charge state. A preset current value such as the *C*/10 rate can be defined and when the current goes below this value the charging would be terminated.

14.1.1.4.13 Temperature

In general, an increase in temperature is a sign of overvoltage. However, using temperature sensors adds greatly to the cost of the system. Nevertheless, for some chemistry such as Ni-MH, methods such as voltage drop are not recommended, because the voltage drop after the full charge state is not sufficiently significant to be relied on. In this case, the temperature increase is good sign of overvoltage and can be used.

14.1.1.5 Cell balancing

For applications demanding high power and energy such as EV/PHEVs, numerous cells should be connected in series to provide high voltages and connected in parallel to produce high currents. Hence, in general, high power and high energy rates for traction applications are achieved. This seems ideal; however, there are disadvantages involved. It is often claimed that single cells produced by different manufactures are rechargeable hundreds of times over; however, when connected in series their life cycle declines dramatically. This is because of cell imbalances. To get an idea about the significance of this effect, the results of a real experiment from [8] are mentioned here. In an experiment, 12 cells were connected in series. Despite the manufacturer claiming life cycles of 400 cycles, this was reduced to only 25–30 cycles when the cells were arranged in a string. This shows how devastating this effect can be. To deal with this, the reasons for cell imbalance should be understood and managed. Batteries are electro-chemical devices. Even in the case of a simple resistor, there is a percentage of error. In the case of batteries, this is magnified. Two different cells produced in the same factory at the same time will have a slight difference in their parameters. One of these parameters is capacity difference. In the case of a battery pack, there are different reasons leading to cell imbalance. As mentioned in [9], there are four fundamental factors leading to cell imbalance: manufacturing variations, differences in self-discharge rate, differences in cell age, and charge acceptance variance. Similarly, in [10], cell imbalance is classified as Internal Sources, which include "Variations in Charge Storage Volume" and "Variations in Internal Battery Impedance" and External Sources resulting from "Protection Circuits" and "Thermal Differential Across the Battery Pack."

To explain simply what is happening, again we refer to the water tank visualization of cells. Suppose different cells with different capacities are connected in series. It is like assuming different water tanks with different volumes are connected using pipes at the bottom of tanks. If the first tank is supplied with water, the level of water in all the tanks rises evenly. After some time, those tanks with lower capacity become full of water whereas others are only partially filled with water. To fill completely the higher capacity tanks, there is no alternative other than over filling the lower capacity tanks.

Returning to the real situation, it is easy to guess what happens in the case of battery strings. Fully charging the high capacity cells involves overcharging lower capacity cells. This will lead to excessive gassing and premature drying out of lower capacity cells, and at the same time, sulfate formation in the partially charged cells will lead to a reduction in their life cycle. Overcoming this effect is the main task of cell equalization circuits and their control algorithms. It should be noted that in the case of EVs, the batteries are usually completely charged up to 100% SOC; hence, cell balancing is an important issue. However, in PHEVs, batteries are intended be kept in the range of 40%–80% so that they can provide sufficient energy, while being able to absorb regenerative power at the same time. Cell equalization techniques for series strings fall into three main groups: (1) Charging, (2) Passive, and (3) Active.

It is important to note that in cell balancing, the SOC is the key point and not voltage itself, although voltage is a good sign of the SOC. However, if other techniques that can determine the SOC more accurately are available, they may be used. As mentioned in [11], cell balancing in a series string really means equalizing the SOC of the cells, which is equivalent to voltage balancing. Voltage is a useful indicator of the SOC. Different SOC estimation techniques will be studied later.

1. *Charging*: Charging method is simply continuing charging the cells until they are all balanced to some extent. This implies overcharging the cells in a controlled manner, which leads to the full charge of the higher capacity cells. This method is applicable to PbA and nickel-based batteries as they can tolerate overcharge to some degree without significant damage. However, this should be implemented carefully because extra overcharge leads to overheating of the cells and eventually, premature drying of the electrolyte. Despite the simplicity and low cost of this method, there are disadvantages, such as low efficiency and long times required to obtain cell balance. Experimental results from [12] show that for actual cell equalization of 48 V batteries of a specific chemistry, weeks of time are required. Furthermore, results from [8] show that the extra time needed using this method increases with the square ratio of the number of cells added.
2. *Passive*: In this method, the extra energy in the lower capacity cells is dissipated in resistive elements connecting two terminals of the cells. This will provide enough time for the higher capacity cells to become fully charged. This method has also low efficiency because of the energy dissipation; nevertheless, it has a higher speed than the charging method. The passive technique is also cheap and easy to implement and the control algorithm can be easily designed.

3. *Active*: Active cell balancing involves using active electrical elements, such as transistors, op-amps, and diodes to control the power flow between the different cells. This flow can be between groups of cells or single cells. Obviously, extra charge is removed from the lower capacity cells and transferred to the higher capacity cells. This speeds up the charging procedure significantly because no energy is dissipated. Just a small amount of energy is dissipated in the circuitry, which can be minimized using zero voltage or zero current switching techniques if possible.

Let us consider Li-ion batteries, which are one of the most attractive candidates for EV/PHEVs. In this chemistry, the voltage should be carefully monitored and rigorously controlled in the typical range of 4.1–4.3 V/cell because the threshold voltage leading to the breaking down of the cell is very close to the fully charged cell voltage. As mentioned before, lithium batteries cannot tolerate being overcharged. Hence, the charging technique is not applicable to them. According to safety issues related to lithium-based batteries, active balancing is the only reliable cell equalization technique for them.

Various types of cell-balancing techniques can be found in the literature. Hence, there is a need to categorize them based on a certain criterion. Based on energy, flow can be classified into four different groups: (1) dissipative, (2) single cell to pack, (3) pack to single cell, and (4) single cell to single cell. It is easy to imagine the operation of each category based on the name. There are advantages and disadvantages for each group. For instance, the dissipative shunting resistor technique is a low-cost technique. In addition, it is easy to control because of the simple structure leading to simple implementation [13].

In addition to the criterion of energy flow for categorization, cell-balancing techniques can be split into three main groups based on the circuit topology: (1) shunting, (2) shuttling, and (3) energy converter. Non-dissipative techniques such as the pulse-width modulation (PWM)-controlled shunting technique have high efficiency, but need accurate voltage sensing and are somewhat complex to control [14]. In addition, the high number of elements leads to an expensive system. The use of resonant converters increases the efficiency significantly because of the very low switching losses; however, it increases the complexity of the control system [15].

Shuttling techniques work based on transferring the extra charge of a high capacity cell or cells to an energy-storing component, such as a capacitor or a group of capacitors, and then transferring it to the low capacity cell or cells [16]. The system would be cheaper is only one high capacity capacitor were used; however, because of the existence of only one element for charge transfer, the speed of the equalization is lower compared with when a group of capacitors is used. Utilizing a group of low capacity cells instead of one high capacity cell is a good idea, although it increases the complexity of the control system.

Most techniques of energy converter cell equalization utilize transformers. The achieved isolation from transformers is an advantage; however, they suffer from more costly weight. A model and transfer function of the energy converter cell equalization system is derived in [17], which can be used for control design purposes. The above-mentioned cell-balancing techniques are all summarized and explained, together with circuit topologies in [18].

The questions that arise are how much the cells should be balanced and whether the balance range should be in the range of volts or millivolts, etc. As experiments shown in [11] for PbA batteries, cell-to-cell voltage matching should be in the range of 10 mV, which corresponds to the SOC to provide reasonable improvements in life cycle. This is an important factor, because, for example, if the voltage matching should be in the range of 1 mV, this means that the sensors should be ten times more accurate and also that the algorithm might need improvement for this case. This means more cost and might be complex. Therefore, there is a trade-off between expense and life cycle. This parameter should be verified experimentally for different chemistry, environments, and applications.

As EV/PHEV battery packs do not possess a mature technology and because there are few experimental data available, contradictory claims may sometimes be seen in the literature, one of which is mentioned here. As mentioned before, battery packs used in hybrid electric vehicles (HEVs) are usually controlled to remain in the midrange of the SOC. This is in order that the battery has the ability to absorb enough regenerative current, while still being able to provide sufficient power during acceleration. If the battery is in 100% SOC, absorbing regenerative current will lead to the overcharge of the battery. Cell overcharge is usually sensed through measuring the cell voltage. Some researchers believe that switched capacitor cell equalization techniques (Shuttling Method) are suitable candidates for applications with no end-of-charge state like HEVs, because there is no need for intelligent control and it can work in both charge and discharge mode [18]. On the other hand, some others believe that according to the nearly flat shape of the open circuit terminal voltage of Li-ion cells in the range of 40%–80%, the suitability of charge shuttling methods for HEV applications is denied because of the negligible voltage deviation of the cells [13].

An approximate graph of a typical Li-ion cell voltage versus SOC is shown in Figure 14.1. On the contrary to HEVs, EVs and similarly, PHEVs are regularly charged completely. According to Figure 14.1 the slope of the curve below 20% and above 90% is high enough to result in a detectable voltage difference to be relied on by charge balancing control and measurement circuits.

14.1.1.6 SOC estimation

One important piece of information necessary for safe charging is the SOC. Charging algorithms are mainly based directly or indirectly on SOC. Hence, the knowledge of the SOC value is a key parameter in accurate charging. Unfortunately, measuring the SOC directly is impossible, or at least very difficult and expensive to implement. Therefore, generally, the SOC is estimated based on other variables or states of the battery. This involves battery models based on which different estimation methods can be utilized or observers can be designed. Precise estimation of the SOC is not an easy task; however, in usual applications battery voltage, which is a sign of SOC, can be used. In the case of high power/high energy

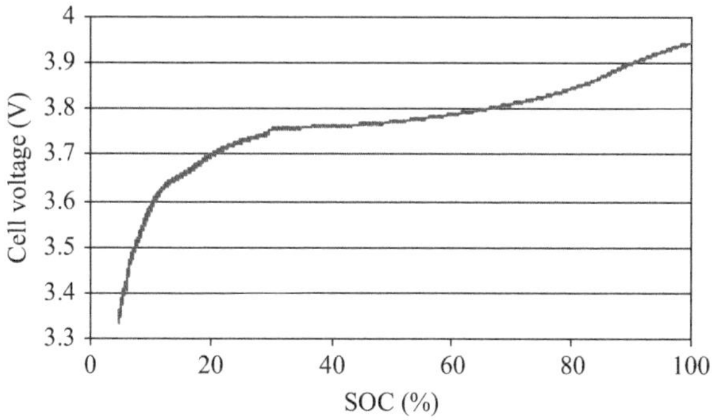

Figure 14.1 Cell voltage in slow discharge (V_{OC} vs. SOC)

EV/PHEV battery packs, methods that are more accurate are advisable, despite being more expensive and complex to implement. The more accurate the SOC estimation, the better the charging algorithms can be implemented, which results in life cycle improvements.

As mentioned before, the SOC is mainly the ratio of available charge to the rated capacity of the cell. One of the important points in SOC estimation is the rated capacity change over time owing to the aging resulting from the degradation of the electrolyte, corrosion of plates and other factors. This issue is in the field of analyzing the state of health of the battery and it is called "State of Health Estimation," which is a field of research not mentioned further here.

However, we will mention some SOC estimation techniques. One of the simplest methods is to *discharge* the battery completely and measure the SOC. Although simple, it is very time consuming and does not seem logical to discharge a battery completely just to measure the SOC. Knowledge of the SOC is useful for assessing the current situation of the battery. Therefore, if the battery is discharged the state of the battery has changed and there is no more use for knowledge of the previous SOC. Furthermore, in the case EVs/PHEVs, this method is not applicable. Although this method is not used in battery packs, it may be used periodically after long intervals to calibrate other SOC methods.

Another method is *Ampere Hour Counting*, which measures and calculates the amount of charge entering the battery or leaving it through integrating the current over time. This is one of the most common methods used; however, there are some deficiencies. There are always inaccuracies in sensors and although very small, because of the integration over time, these errors can accumulate to a considerable value leading to significant errors. In addition, even supposing a very accurate current sensor, because this integration is implemented usually by digital circuits and numerical methods, there are always calculation errors involved and again, these can show up in larger errors over time. Even if assuming both deficiencies could be solved in some way, there is another reason leading to inaccuracy. Even if the amount of charge entering the battery is calculated exactly, because of the coulomb efficiency mentioned before, less amount of charge is available, which is also dependent on the discharge rate when leaving the battery. One way to reduce these inaccuracies is to recalibrate the integration process each time a specific known set point, such as the fully discharged state, is reached.

Another method for SOC estimation is the *Measurement of Physical Characteristics of Electrolyte.* Obviously, this method is applicable mainly to liquid electrolyte batteries, not solid ones like Li-Po. In this method, a chemical fact is used, which is the relation of change of some parameters in the electrolyte with the change of the SOC. One of these parameters is the density of the acid. There is an almost linear relation between change in acid density and SOC. This method is very well known, especially in PbA batteries. The density can be measured directly or indirectly using parameters, such as viscosity, conductivity, ion-concentration, refractive index, ultrasonics, etc.

As discussed before, the *Open Circuit Voltage* of the batteries can be used as an indicator of the SOC. The uncertainty in this method is the fact that batteries under operation need some rest time for their open circuit voltage to become stable. For some cases, this time can be up to a matter of hours. However, this method is also widely used. The key point in this method is the linear relation of open circuit voltage versus SOC in a specific range of SOC. This range and its slope are different in different chemistry, which should be taken into account.

There are other techniques categorized under soft computation techniques, such as fuzzy neural networks [19] or adaptive neuro-fuzzy modeling [20], which can also be utilized for SOC estimation. Other approaches that can be used include the heuristic interpretation of measurement curves mentioned in [21], such as Coup de fouet, Linear Models, Artificial neural networks, Impedance spectroscopy, Internal resistance, and Kalman filters, which are more precise methods but more complicated to implement.

14.1.1.7 Thermal management

The heat generated due to high currents flowing to and from batteries has impacts on batteries characteristics such as capacity, power and self-discharge, leading to cell imbalance and parameters variation which should be taken into account for charging the battery pack weather during the operation of the PHEV on the road or while the car is still and being charged from the charging station. This heat is generated resulting from two actions: (1) electrical resistive heating and (2) electrochemical enthalpy changes. The magnitude of the heat depends on different factors such as chemistry type, SOC, charging or discharging rate and charge/discharge profile. Experiment results, from [22], show that Ni-MH batteries generate more heat than VRLA or Li-ion batteries at temperatures above 40 °C, conversely, at room temperature Ni-MH batteries produce less heat. In terms of performance, Ni-MH batteries are more sensitive to heat than VRLA and Li-ion batteries. Hence, Ni-MH batteries need a higher performance thermal management control. In general, internal resistance of cells increases as temperature decreases. This results in higher rate of heat generation in the cells. Obviously, heat generation increases as the charging rate increases. This is an important issue, specially, in fast charging which involves flowing high amounts of current through the battery pack. High rates of charge will significantly increase the internal temperature of cells which can lead to internal short circuit and thermal runaway. This is a very important safety issue. For instance, in a lithium-based battery if a cell is short-circuited internally a great amount of heat will be generated that may raise the cell temperature to above 180 °C which is the melting point of lithium [23].

In general, battery performance is better in higher temperatures, however, from the life cycle point of view lower temperatures are preferred. There are different reasons contributing to reduction of battery life in high temperatures

Nevertheless, increase of the corrosion is responsible for the most part [24]. Besides, the efficiency of batteries is lower in higher temperatures such as Li/V_6O_{13} mentioned in [23].

Maintaining the battery pack in a specified temperature range can be achieved by passive or active methods. Passive methods only use the ambient environment, however, in active method, heating or cooling is achieved using a built-in source [22]. The heat transfer medium can be air, liquid, phase change material or any combination of these.

Thermal uniformity of the battery pack is a significant point. Thermal unbalance causes parameter deviation in different cells leading to reduced life cycle. Even assuming uniformity of all the cells, the place of different cells causes different temperatures, e.g., the cells near the edges of the pack have lower temperature compared to those near the middle because of better ventilation. The packaging shape also affects the thermal management algorithm, e.g., spiral designs have improved heat transfer characteristics because of the higher surface area to volume ratio. In the case of closed units which contain the power electronics parts such as rectifiers and inverters beside the cells in the pack, high temperature deviations through the pack should be managed by the thermal management system [24].

A team from Optima Battery Inc. in [25] gives the criteria for thermal management of PbA batteries as follows:

- Internal battery temperatures above 60 °C must be inhibited. Otherwise, that will cause battery damage.
- Keep the battery pack in the thermal range of 25–40 °C for all the times (35–40 °C for optimum acceptance of regenerative braking current). There will be noticeable improvement in capacity and power.
- Maintaining the thermal deviation less than 3 °C per module is a highly significant point.

There are different approaches for controlling the pack temperature in the limited range specified by the manufacturer during charging. Some are simpler and cheaper to implement such as current limited recharge schemes which try to maintain the heat generation rate in an acceptable range by controlling the current. According to the fact that heat generation during the battery pack charging is a function of SOC which is an indicator of internal resistance this scheme is not efficient and is only advisable for inexpensive systems. On the other side, using thermal sensors in addition to other sensors for each cell increases the price and complexity of the thermal management system which is a drawback; however, it increases the accuracy and reliability of the system. Recharge current rate can be decided using SOC of the battery.

14.2 EV charging

EVs have rechargeable batteries on-board as a source of energy to propel. Recharging these batteries is similar to refueling the conventional gasoline vehicles. In order to recharge the on-board batteries safe, it is very important to understand the battery characteristics as mentioned in the previous chapter. EV batteries are charged using DC power, as they are DC electric power source.

EV charging can be done in two methods, one is termed as plugged charging and other is wireless charging. We use an electric cable in charging the vehicle in the plugged charging method and in the latter method, charging is done using wireless transmitter and receiver coils avoiding the use of charging cable.

14.2.1 Plugged charging

Plugged charging basically looks like as shown in Figure 14.2. Depending on the charging *c*-rate level, there are two ways to charge the vehicle in the plugged charging method namely normal charging and fast charging.

In normal charging method, AC wall power from grid is rectified using on-board vehicle charger and charge the battery pack. This method charges EV in slower charge rates due to the limitation of on-board vehicle charger hardware and domestic wall socket power. The usual *c*-rates in normal charging is 0.2–0.6 °C.

On the other hand, in fast charging, we use an off-board charger and charge the vehicle battery pack using off-board rectified DC power. Here as the rectifier and charger hardware is off-board, it is possible to charge the vehicle at higher charge rates. The usual *c*-rates for fast charging is >1 °C.

Basic system for the plugged charging is as shown in Figure 14.2. Here, the electric vehicle supply equipment (EVSE) can be either normal charging supply equipment or off-board rectifier charger for fast charging.

The on-board charger power electronic hardware system block diagram is as shown in Figure 14.3. There are numerous on-board charger topologies that are proposed in the literature.

Usually on-board charger power rating depends on the wall socket power availability. There are different charging levels as proposed in various charging standards across the globe.

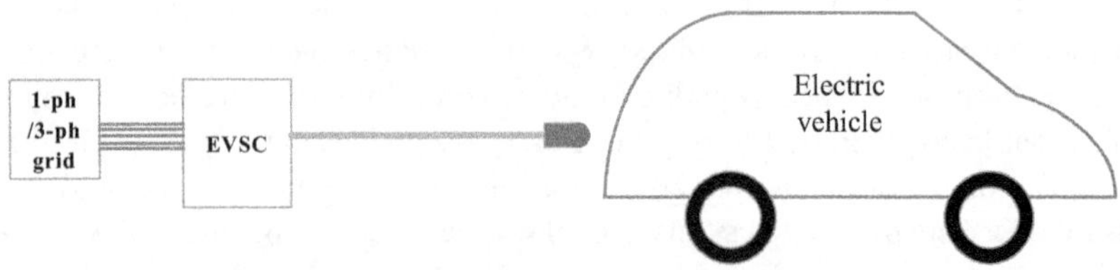

Figure 14.2 Plugged EV charging

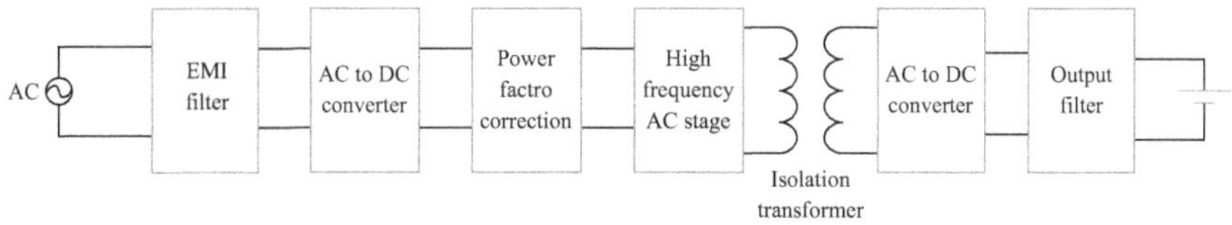

Figure 14.3 On-board EV charger

14.2.1.1 EV normal charging standards

Due to the vehicle battery characteristics, controlled charging of a battery pack is very crucial. There are several EV charging standards and communication protocols developed across the globe for achieving safe and human friendly charging. The EV charging standards across the globe for normal charging are

(a) SAE J1772 (type-1)
(b) IEC 61851-1 (type-2)

14.2.1.1.1 SAE J1772

SAE J1772 is the EV charging standard developed by the Society of Automotive Engineers (SAE). This is adopted mostly in North America and Japan (e.g., Chevy Volt®, Nissan Leaf®, etc.).

Based on the wall power, the standard information for all the levels of charging is provided in Table 14.1. As a test case, the Nissan Leaf® 24 kWh Li-ion battery pack is considered [26].

SAE J1772 interface is developed for use in single-phase system only. The maximum power level for charging in type-1 system is $<$15 kW which essentially falls under level 2 charging.

The type-1 interface is as shown in Figure 14.4. This unique interface is put on the vehicle and an EVSE powers the vehicle with the connector as shown in Figure 14.4.

As shown in Figure 14.5, type-1 interface has a control pilot and proximity pilot connections on both EV and EVSE for ensuring safe charging.

Power circuit consists of Line, Neutral, and Earth as shown. The power circuit connects to the on-board vehicle charger unit which rectifies the AC supply and charges the battery pack.

Control pilot. The control pilot circuit is the primary control means to ensure proper operation when connecting an EV/PHEV to the EVSE. This section

Table 14.1 Standard EV charging levels

Level	Voltage	Phase	Power (kW)	Time (h)
Level 1	120 Vac	One-phase	1.4	17
Level 2	240 Vac	One- and split-phase	4	6
Level 3	208/415 Vac	Three-phase	50	0.5

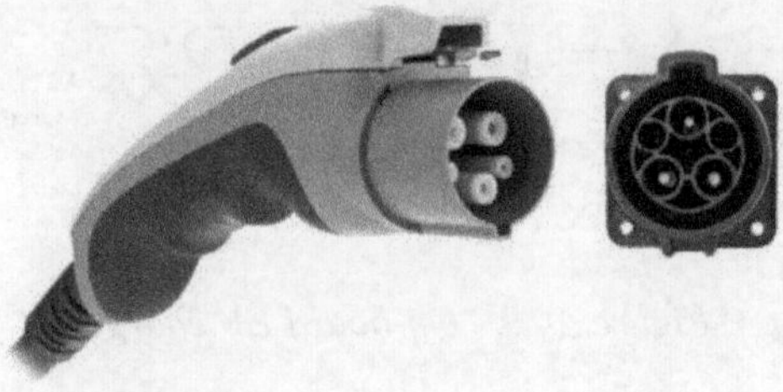

Figure 14.4 Type-1 vehicle and charger Interface

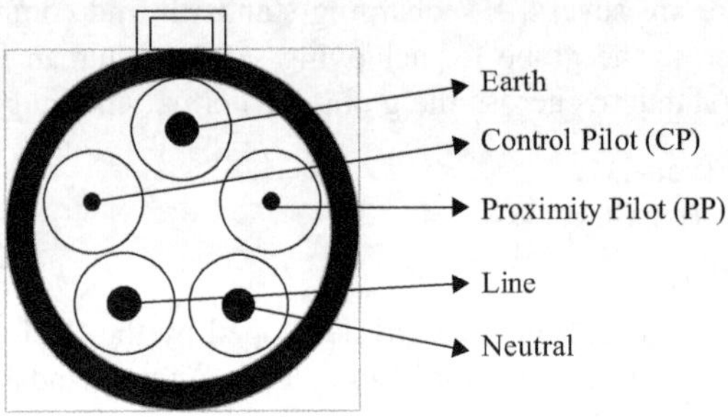

Figure 14.5 Type-1 interface

describes the functions and sequencing of events for this circuit based on the recommended typical implementation or equivalent circuit parameters [27].

The control pilot communication extends from EVSE to vehicle as shown in Figure 14.6 between pin#4 and earth pin#5. It is a one way PWM communication from EVSE to vehicle. The vehicle and EVSE measures the voltage at respective points and identify the states as defined in Table 14.2 [27]. The control flow is as explained below.

1. When no vehicle is connected, the EVSE will measure +12 V between a measurement point after R1 and earth.
2. When an EV is connected to the EVSE for charging, EVSE changes its state from +12 to ±12 V PWM as shown in Figure 14.7.
3. EV starts reading the PWM form the EVSE control pilot. EV gives signal to EVSE for delivering power by turning on the S2 switch as shown in Figure 14.6.
4. The value of R2 resistance depends on the ventilation requirements in the charging area as detailed in Table 14.2
5. Once S2 is turned on, the EVSE sense state change due to difference in the voltage measurement and powers the vehicle to charge.
6. At any point of time if there is any loose connection or connection failure in the control pilot connections, EVSE will immediately stops the power output to the vehicle.

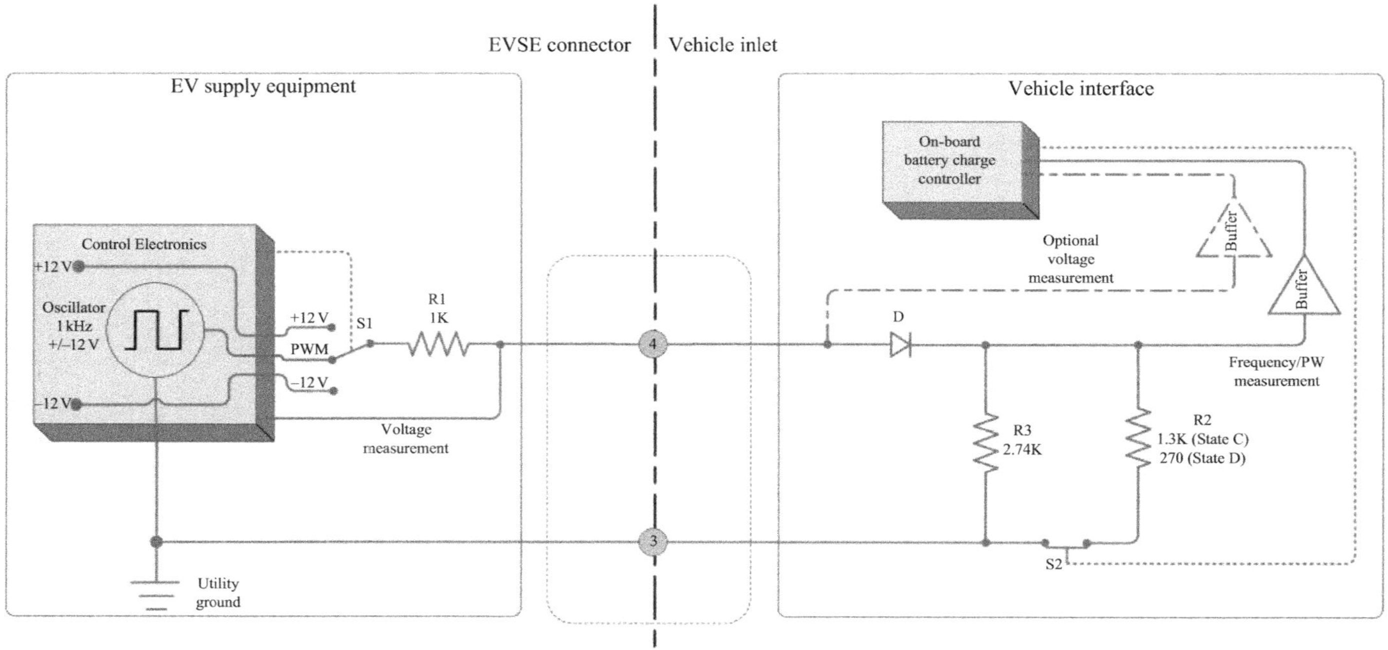

Figure 14.6 Control pilot circuit diagram

Table 14.2 Definition of EV/EVSE states based on voltage drop measurements

State designation	Voltage (Vdc nominal)[(5)]	Description of vehicle/EVSE state
State A	12.0[(1)]	Vehicle not connected
State B1	9.0[(1)]	Vehicle connected/not ready to accept energy EVSE not ready to supply energy,
State B2	9.0[(2)(3)]	Vehicle connected/not ready to accept energy EVSE capable to supply energy
State C	6.0[(2)]	Vehicle connected/ready to accept energy/indoor charging area ventilation not required EVSE capable to supply energy
State D	3.0[(2)]	Vehicle connected/ready to accept energy/indoor charging area ventilation required EVSE capable to supply energy
State E([4])	0	EVSE disconnected from vehicle/EVSE disconnected from utility, EVSE loss of utility power or control pilot short to control pilot reference
State F	−12.0[(1)(2)]	Other EVSE problem

[1]Static voltage.
[2]Positive portion of 1 kHz square wave, measured after transition has fully settled.
[3]The transition from State B1 to State B2 begins as a static DC voltage which transitions to PWM upon the EVSE detection of vehicle connected/not ready to accept energy and EVSE capable to provide energy.
[4]EVSE is not required to actively generate State E.
[5]Voltage measured by EVSE as shown in Figure 4.1.
[6]Optional state. The EVSE may enter State F upon detecting a self-diagnosed fault that prevents the EVSE from delivering power. This option would require user intervention to reset the EVSE to restore normal operation.

Based on the duty cycle of the PWM that EVSE sends, EV can understand the power availability in the EVSE mains. The detailed relation between PWM duty cycle and AC current available is shown in Table 14.3 [27].

Due to the series diode, the EV cannot read the negative half cycles of PWM. So we see only the positive half cycles of the PWM as shown in Figure 14.7(a).

For example, from the duty cycle shown in Figure 14.7, for 50% PWM, the current available in the $50 \times 0.6 = 30$ A. Hence, at 230 Vrms supply, the EVSE can have 6.9 kVA to charge the vehicle.

This PWM helps EV to limit the line current drawn from the source to safe value so as to avoid any overload in the system. The charging current availability in the wall EVSE with respect to the PWM duty cycle shall follow the characteristics as shown in Figure 14.8 [27].

Proximity pilot. Upon insertion of the connector into the vehicle inlet, the coupler shall provide a means to detect the presence of the connector in the vehicle inlet as described as shown in Figure 14.9. The proximity detection is very helpful

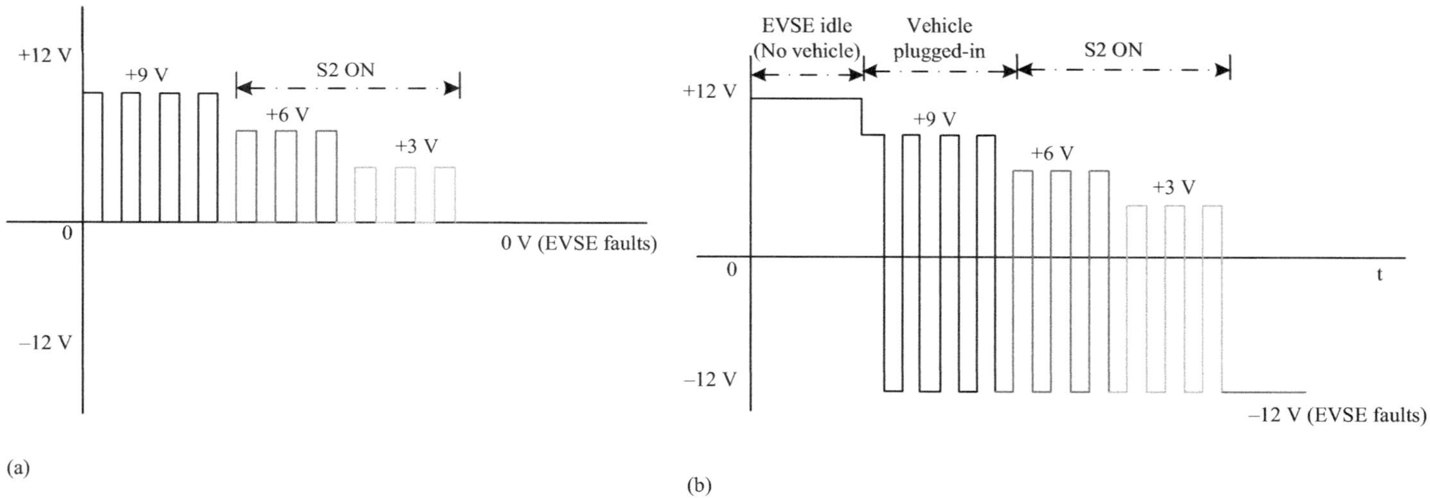

Figure 14.7 (a) PWM read by EV; (b) PWM read by EVSE

Table 14.3 PWM duty cycle interpretation

Vehicle duty cycle interpretation	Maximum current to be drawn by vehicle
Duty cycle < 3%	Error state, no charging allowed
3% ≤ Duty cycle ≤ 7%	Indicates that digital communication is needed
7% < Duty cycle < 8%	Error state, no charging allowed
8% ≤ Duty cycle < 10%	6 A
10% ≤ Duty cycle ≤ 85%	Available current = (duty cycle%) × 0.6
85% < Duty cycle ≤ 96%	Available current = (duty cycle % − 64) × 2.5
96% < Duty cycle ≤ 97%	80 A
Duty cycle > 97%	Error state, no charging allowed

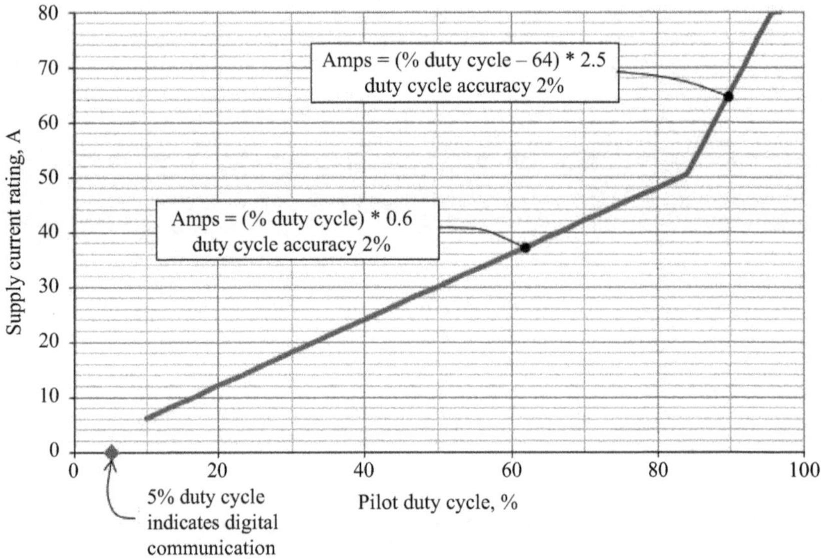

Figure 14.8 Supply current vs. control pilot duty cycle

in ensuring the charging system is making 100% contact with the power conductors. When the vehicle tries to move or when any external user tries to unplug the charging, the EVSE will detect S3 open and shall stop the power to the vehicle avoiding very high spark due to hot plug out.

Proximity detection may also provide a signal to activate the EV/PHEV charge controller and engage the EV/PHEV drive interlock system. EV/PHEV proximity detection may also be used to provide a signal in the vehicle charge control strategy to help reduce electrical arcing of the coupler during disconnect.

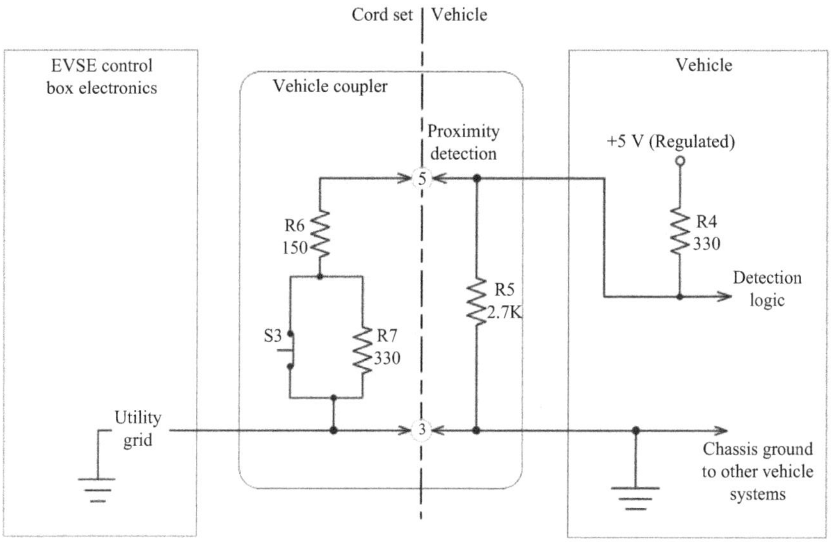

Figure 14.9 Proximity pilot circuit

When the vehicle connector is plugged into the vehicle inlet, switch S3 is closed by a mechanical action. Vehicle internal logic detects the state change and enables charge drive interlock, and ensure vehicle is not drivable as long as it is plugged into charging.

14.2.1.1.2 IEC 61851-1 (type-2 charging)

International Electrotechnical Commission (IEC) developed conductive electric vehicle charging standard IEC 61851-1. The interfaces used in this standard are very different from SAE J1772. However, the communication protocol remains the same as SAE J1772. Commercially available vehicles with type-2 charging system are Renault Zoe® Tesla Model S®, etc.

The type-2 charging system is designed to support three-phase AC system. Which essentially supports charging at semi-fast charging rates. For this to be possible there should be a three-phase charger on-board the vehicle. One can also use the type-2 system with one-phase on-board charger. The maximum power available in type-2 system is 32 A at three-phase 400 V (22 kW) [28].

It is must to use residual current devices in the EVSE to ensure the earth leakage protection.

Type-2 vehicle inlet has a locking actuator as shown in Figure 14.10. This is used to secure the vehicle charging when plugged in at any public EVSE.

Type-2 system has three-phase inputs with one neutral and protective earth (PE) as power conductors (Figure 14.11). The communication is similar to that of SAE J1772 as described earlier.

The control pilot has the same functionality as seen in type-1. In addition to the proximity pilot circuit as seen in type-1 there is one more circuit that is being used widely in type-2 charging systems.

It is as shown in Figure 14.12. This proximity is very important to make the EV and EVSE understand the current carrying capability of the charging cable.

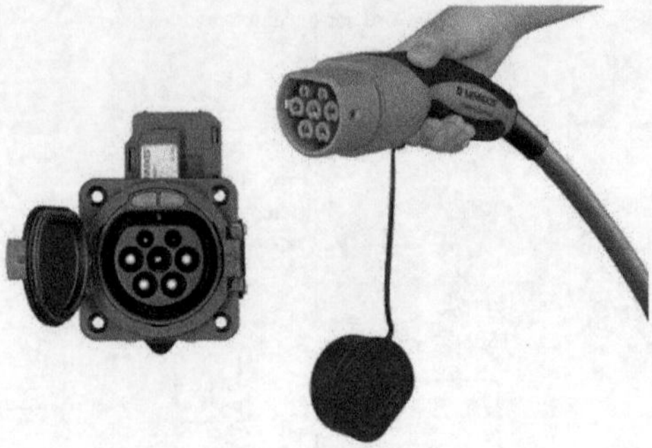

Figure 14.10 IEC type-2 vehicle and charger interface

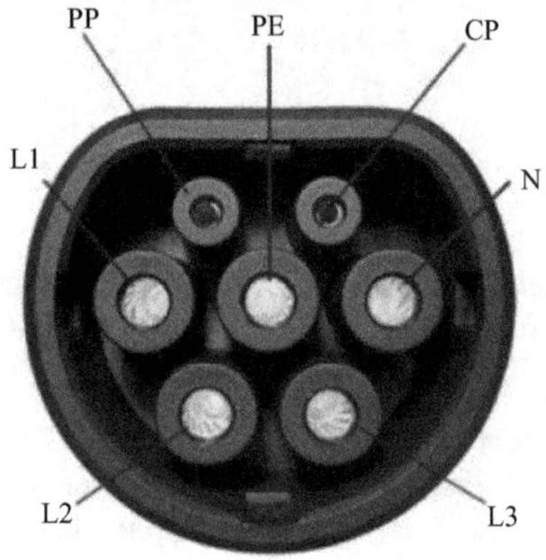

Figure 14.11 Type-2 pin description

As seen in Figure 14.12, the charger cable contains a proximity resistor R_c at the both end couplers of the cable. This R_c value is selected based on the EV charger power and the current carrying capacity of the charger cable. Table 14.4 describes various R_c values and its interpretation on the charge cable current [28].

When an EVSE is connected with a plug for 13 A current, it will not allow any more current than 13 A drawn by EV for charging the car. This means that the EVSE will sense the charging current and if the vehicle demand is more than the cable rating the EVSE will open the power contacts and terminate the charging.

14.2.1.2 EV DC fast charging

EV fast charging is needed to charge the EV very quick. This charging method helps to the range anxiety problem of an EV. One can fast charge the vehicle in less than 1 h and get the vehicle ready to drive closer to full range. This is possible as the vehicle is charged with an off-board power electronic rectifier and charging unit. This does not need any vehicle side power electronics converter on-board.

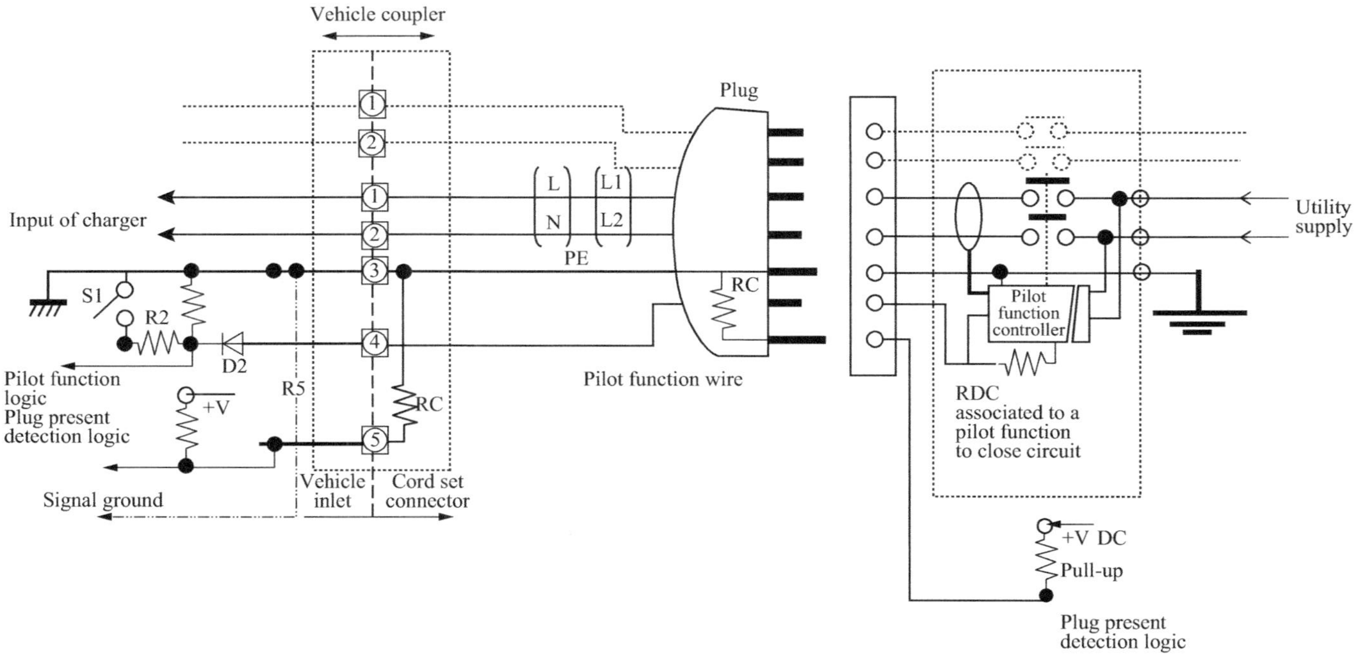

Figure 14.12 Type-2 proximity pilot for charge cable current capacity interpretation

Table 14.4 Proximity resistor coding for charging cable current capacity

Current capability of the cable assembly	**Equivalent resistance of R_c Tolerance $\leq$ 3%**
13 A	1.5 kΩ 0.5 W[a,b]
20 A	680 Ω 0.5 W[a,b]
32 A	220 Ω 0.5 W[a,b]
63 A (three phase)/70 A (one phase)	100 Ω 0.5 W[a,b]

[a]The power dissipation of the resistor caused by the detection circuit shall not exceed the value given above. The value of the pull-up resistor shall be chosen accordingly.
[b]Resistors used should preferably fail open circuit failure mode. Metal film resistors commonly show acceptable properties for this application.
[c]Tolerances to be maintained over the full useful life and under environmental conditions as specified by the manufacturer.

There are two classifications right now based on power flow direction from grid to the load and vice versa, a brief description of each charger type is given below.

14.2.1.2.1 Unidirectional chargers

These chargers can only draw power from the grid but cannot interpose power into the grid. Usually, such converters are designed with only a single stage to reduce its size, weight, and cost. Figure 14.13 depicts a unidirectional full-bridge series resonant converter. There has been much research being done on unidirectional chargers to obtain a reasonable control strategy which help to increase the efficiency of these chargers. These are usually preferred as they do not affect the battery life, as the number of cycles are limited.

14.2.1.2.2 Bidirectional chargers

These chargers typically consist of two stages, namely (1) a grid-connected bidirectional AC–DC converter and (2) a bidirectional DC–DC converter. And they have the two modes, the charge and the discharge mode. Figure 14.14 depicts a bidirectional charger. The presence of two peak-current inductors tends to make the charger a little bulky and expensive. Batteries usually perish faster due to large number of cycles. Table 14.5 gives information in a more organized fashion regarding both the chargers.

Generally, DC fast charging stations for EVs are designed to supply about 50 kW of power [29]. The established trend is to place these chargers off-board. As these stations are bulky, keeping them off-board is convenient. The general block diagram of a DC fast charging station is as shown in Figure 14.15; the charger in the shown scenario is connected to a common AC link.

The output voltage that is fed to the load (either an ultracapacitor or a battery) may be variable or fixed. The filters placed ahead of the DC fast charging station help to maintain a healthy power factor [29–32].

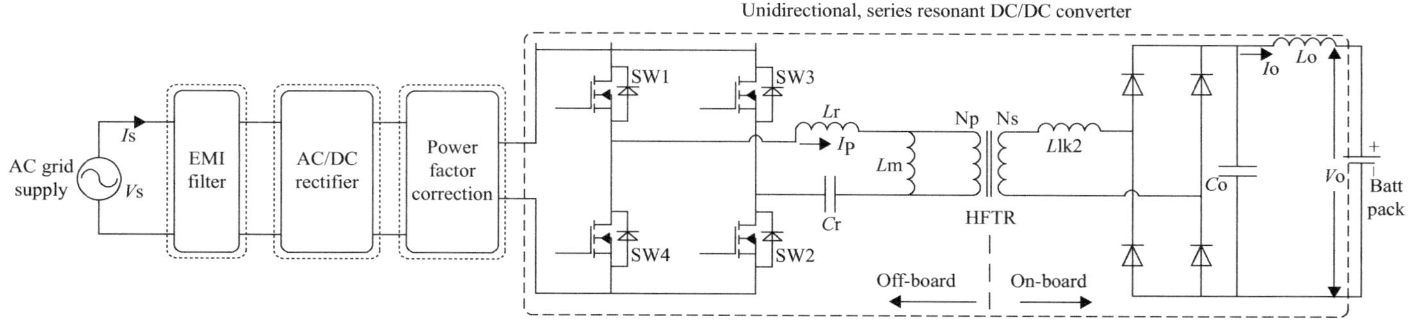

Figure 14.13 Unidirectional full-bridge series resonant converter

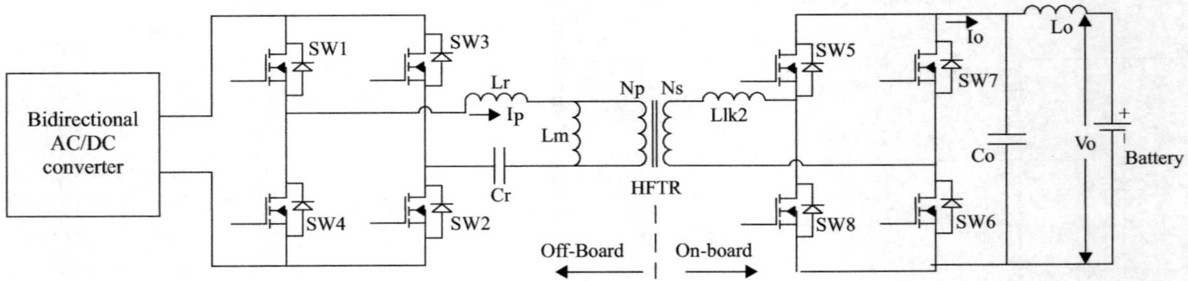

Figure 14.14 Typical bidirectional DC charger for EV

Table 14.5 Comparison of chargers based on power flow direction

Charger type	Present status	Charger level	Power converter	Battery health
Unidirectional	Operational	Levels 1, 2, and 3	Buck and fly back	No degradation
Bidirectional	Unavailable	Only for level 2	Matrix converters	Lifespan reduces

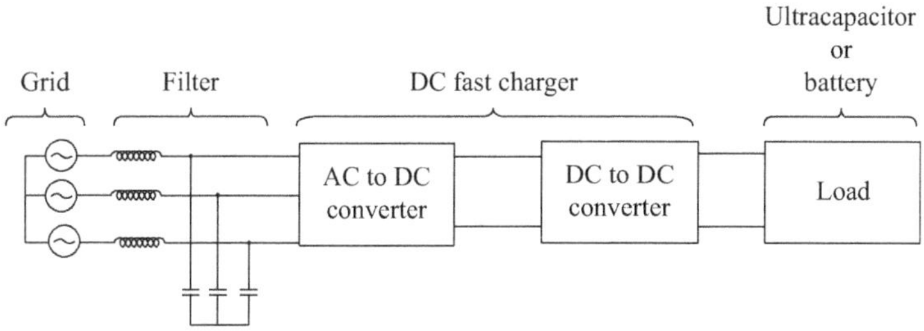

Figure 14.15 General block diagram of a DC fast charging station

14.2.1.3 DC fast charging converter topologies

There are a number of power converter topologies available for the purpose of rapid charging of batteries or ultracapacitors few feasible options are highlighted in this chapter, they are:

1. *Unidirectional boost converters*: The unidirectional boost converter is shown in Figure 14.16; these are employed in situations where the output voltage has to be boosted up for loads, which require higher voltage [33]. The primary goal of using such a boost converter instead of a traditional diode bridge rectifier is to provide better power factor, to remove harmonics at the input end and to have an unvarying DC voltage at the output if unwanted perturbations occur at the AC end.
2. Another popular power converter topology is the Vienna rectifier as shown in Figure 14.17. This too is a popular choice when the aim is to achieve high power factor and to attain lower harmonic distortion. As shown in Figure 14.17 there is only one active switch per phase which makes the Vienna rectifier easier to control and makes it more dependable. This is essentially a PWM converter, the boost inductors present at the input play the role ascertaining power factor correction [18]. Basically, the stored energy acquired by the inductor when the switch is OFF is transmitted to the load via the diodes whenever the switch is turned ON. The advantages of employing this topology include the absence of a neutral point connection and the lack of auxiliary commutation circuits which eliminate dead time problems.
3. *AC–DC reduced-switch buck–boost converter*: The main highlight of this topology is that its inexpensive, has less number of switches and most importantly since this is a buck–boost converter output voltage can be varied over a wide range. The topology is as shown in Figure 14.18. There have been a few

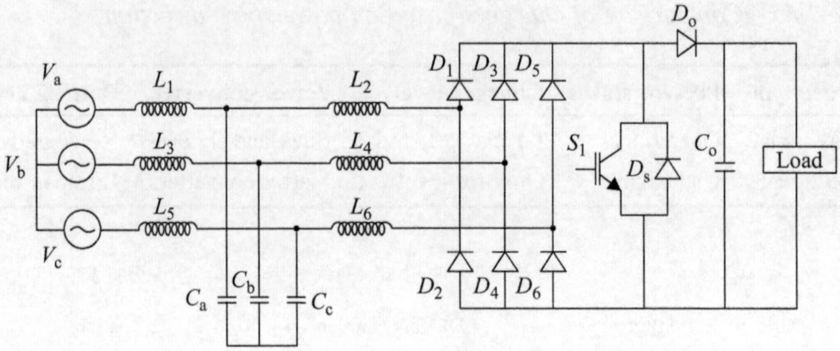

Figure 14.16 Unidirectional boost converter

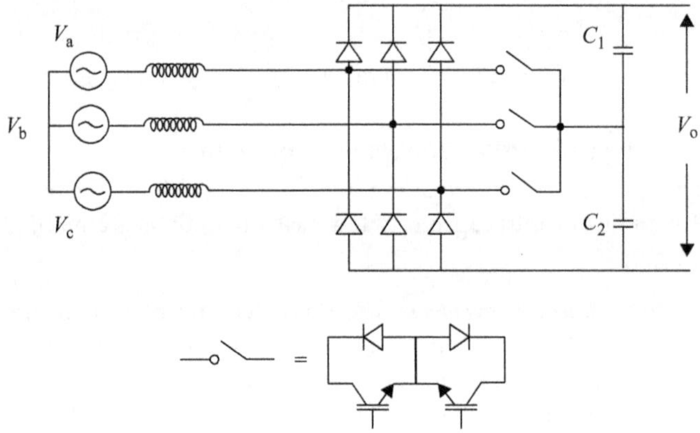

Figure 14.17 Schematic of a Vienna rectifier

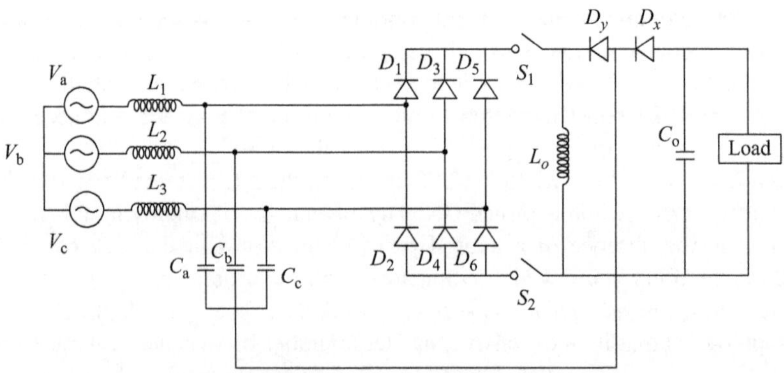

Figure 14.18 Three-phase AC–DC buck–boost converter

three phase front-end rectifiers proposed but they are mostly boost converters which do not allow variation of voltage over wide ranges. This converter topology can operate in buck mode when the duty ratio is below 0.5 and in boost mode when the duty ratio is above 0.5 [34].

Level 3 DC fast charging stations are very demanding in terms of power, the available infrastructure is insufficient to meet those power demands. Usually, simultaneous operation of such charging stations tends to overburden the entire distribution system. Large-scale implementation of such stations requires additional cost which has to be invested in upgrading the transmission cable and transformers. As the fast charging station penetration increases the power demand in the grid also increases proportionally. Apart from increased load demand, level 3 stations also cause drop in voltages and overload of transformers.

14.3.1.1 EV fast charging standards

There are few popular EV DC fast charging standards developed across the globe. Among them Japanese CHAdeMO standard is very popular and is widely used by many EV manufacturers. Chinese GB charging standard is one more popular charging standard. Besides IEC and SAE developed combo charging methods for EV fast charging.

14.3.1.1.1 CHAdeMO DC fast charging

CHAdeMO was formed by the Tokyo Electric Power Company, Nissan, Mitsubishi, and Fuji Heavy Industries (the manufacturer of Subaru vehicles). Toyota later joined as its fifth executive member [35, 36]. Three of these companies have developed electric vehicles that use TEPCO's DC connector for quick charging.

CHAdeMO DC fast charging is developed for a 500 Vdc and 125 A system capable of 62.5 kW charging power. It is a nine-pin interface system as shown in Figure 14.19.

Working of the CHAdeMO DC fast charging system can be understood by the state flow diagram as shown in Figure 14.20.

The interface is mechanically designed in such a way that the hot plug out is not allowed. This can be achieved by making the power contacts open later to the signal contacts opening in the event of the unintentional unplugging. Hot plug out of the high voltage connector is very dangerous as it causes very high arcing in and can lead to human deaths.

As seen in the interface circuit if any of the connection or signal is cut while charging, EVSE will immediately terminate the charging.

Basically, CHAdeMO EV charging works on an 11-bit controller area network (CAN) communication (CAN 2,0). This communication is very helpful in transmitting vehicle battery charging current and charging voltage requirements to the off-board EVSE. The communication also helps to transmit the error messages on the vehicle and the charging unit from the beginning of the charging till the end of charge.

The opto-couplers used in the charging circuit are needed for the sake of isolation from the external EVSE to EV. It is very crucial to protect the EV internal circuits from the external high voltages. The chances of EV components seeing

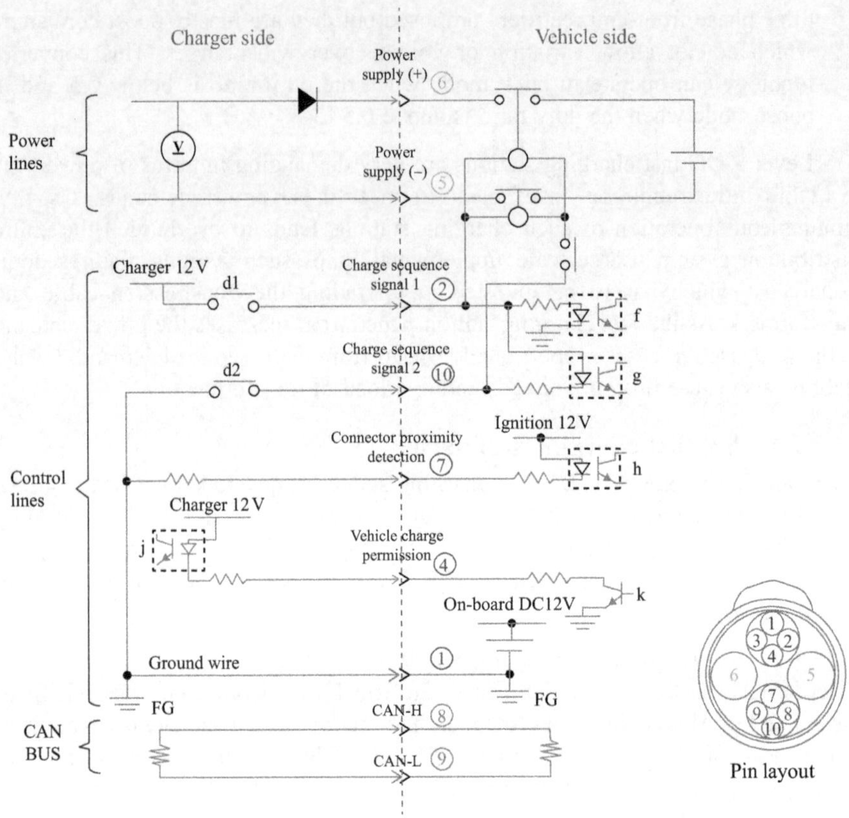

Figure 14.19 CHAdeMO interface and functional circuit. Picture Courtesy: CHAdeMO

high voltages from the EVSE are very high due to the ground being tied between the both units.

Opto-couplers works on the very low drive voltage and current levels from the vehicle engine control units (ECUs) and the EVSE ECUs.

14.3.1.1.2 Chinese GB DC fast charging standard

One other very popular EV DC fast charging standards is Chinese GB 20234-3. This standard is developed for allowing the charging power level higher than the CHAdeMO DC fast charging. This standard can allow user to charge any EV till 750 Vdc and 250 A.

14.3 Wireless charging

14.3.1 Introduction

The wireless charging technology which is based on principle of inductive power transfer (IPT) is not new but was first presented by Nikola Tesla in the 1891 by his

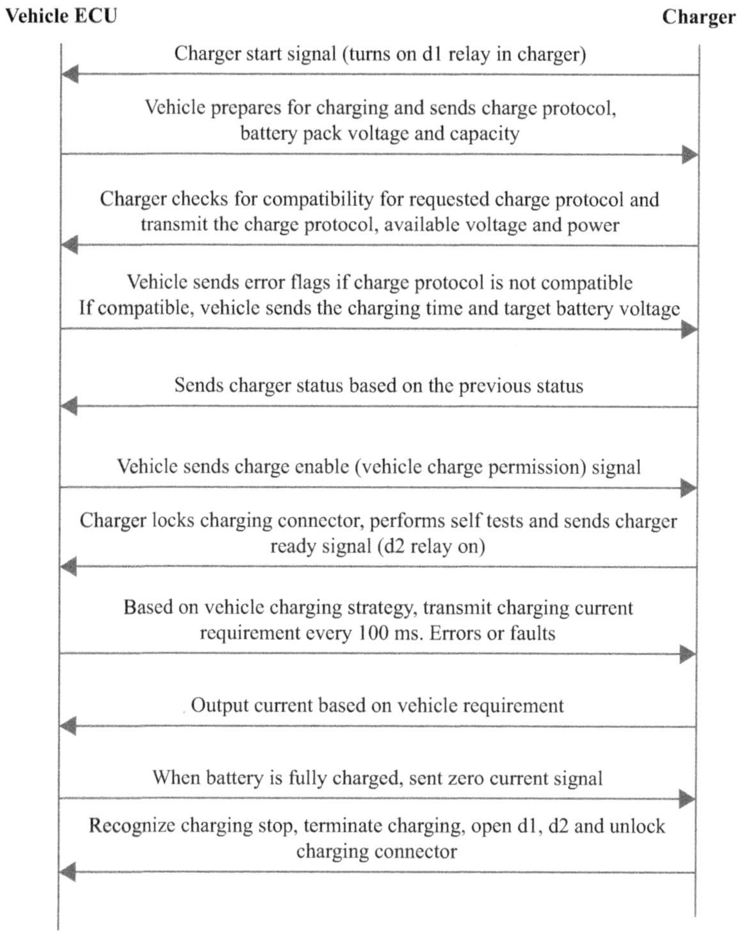

Figure 14.20 CHAdeMO DC fast charging operation flow

well-known Tesla Coil. However, technological limitations at the time eluded the researcher for adopting this means of power transfer for application such as electric vehicle charging [37]. Now, with the advancement of high performance power electronic devices, the possibility of using electromagnetic induction to transfer energy into electric vehicles is now feasible and there have been a number of prototype vehicles produced worldwide [38]. However, major problem associated with IPT system is leakage flux and hence poor efficiency associated with it. To overcome this problem or limitation of IPT system, use of capacitive compensation in both primary and secondary side has been suggested by many authors [38–44]. There exist four basic compensation topologies that are being widely used. These are series–series (SS) topology, series–parallel (SP) topology, parallel–series (PS) topology, and parallel–parallel (PP) topology [45]. Each topology has its own advantages and disadvantages and choice of topology depends upon application for

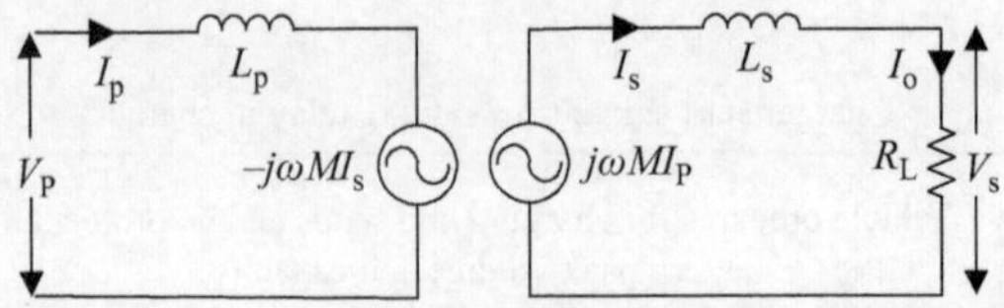

Figure 14.21 Equivalent circuit of an uncompensated loosely coupled IPT system

which it is used. For example, a Li-ion battery has two charging stages, namely: constant current charging and CV charging stage. During the constant current charging stage, the battery is charged at constant current until the specified peak voltage of battery cells is reached. During the CV charging stage, the charging takes place at CV. At the end of constant current charging stage, the battery cell voltage is at the specified peak value and this voltage is maintained across the battery cell throughout the CV charging stage. Hence for charging application the topology which can acts as constant current source as well as CV source will be a good choice [39, 40]. In this chapter, the authors aim to shown that both SS and SP topology can work as constant current source as well as CV source. Following points has been discussed: need of compensation in primary and secondary; analysis of SS topology; analysis of SP topology; efficiency comparison of both topology; control strategies for both topology and finally choice between both topologies.

14.3.2 Necessity of compensation for wireless charging

Wireless charging of electric vehicles, require a significant air gap between primary and secondary winding of IPT system. Large air gap is required to allow for inconsistency in the road surface and better clearance between road and vehicle. Moreover, if its in-motion or dynamic charging, there is always some lateral displacement, i.e., misalignment associated with it. Because of these mechanical constraints mutual coupling between primary and secondary is poor and leakage inductance is much larger compared to mutual inductance [40]. Poor coupling leads to poor transfer of power. To improve coupling and compensate leakage inductance, capacitive compensation in primary and secondary windings is required. However, compensation in primary and secondary serves different purpose. To understand this, consider the equivalent circuit of an uncompensated loosely coupled IPT system shown in Figure 14.21.

Here the subscript "p" and "s" have been used for primary and secondary components, respectively. M, ω, R_L, and I_o represent mutual inductance, operating frequency, load resistance, and load current, respectively.

Maximum power transferred (MPT) to load for circuit shown in Figure 14.21 can be given by:

$$\text{MPT} = \frac{\omega^2 M^2 I_p^2}{2\omega L_s} \tag{14.1}$$

Equation (14.1) can also be verified using Maximum Power Transfer theorem. Equation (14.1) can be seen as:

$$\text{MPT} = \frac{V_{\text{oc}} \times I_{\text{sc}}}{2} \tag{14.2}$$

Here, V_{oc} is open circuit voltage when I_{o} becomes zero and is given by

$$V_{\text{oc}} = \omega M I_{\text{p}} \tag{14.3}$$

I_{sc} is the short circuit current flowing through secondary, when R_{L} is shorted. This is given by:

$$I_{\text{sc}} = \frac{M I_{\text{p}}}{L_{\text{p}}} \tag{14.4}$$

Now, if power transferred to the load needs be more than $\left(\frac{V_{\text{oc}} I_{\text{sc}}}{2}\right)$ then secondary should be compensated [40]. Power transferred for compensated secondary will be the maximum if secondary resonates at frequency, ω_{o}. For this a capacitor (C_{s}) whose value is given by (14.5), can be connected in series or parallel of the secondary as shown in Figure 14.22.

$$\omega_o = \frac{1}{\sqrt{L_s C_s}} \tag{14.5}$$

For parallel compensation, secondary quality factor is given by $\frac{R_{\text{l}}}{\omega_{\text{o}} L_{\text{s}}}$ and for series compensated secondary quality factor is given by: $\frac{\omega_{\text{o}} L_{\text{s}}}{R_{\text{l}}}$. Hence, from Table 14.6 one can say that compensated secondary have a maximum power transfer capability 2**Qs* times greater than uncompensated secondary.

Due to leakage and magnetizing inductance, the IPT is intrinsically inductive in nature. Since high-frequency operation is desirable for effective power transfer, impedance seen by source becomes more and more reactive in nature hence power

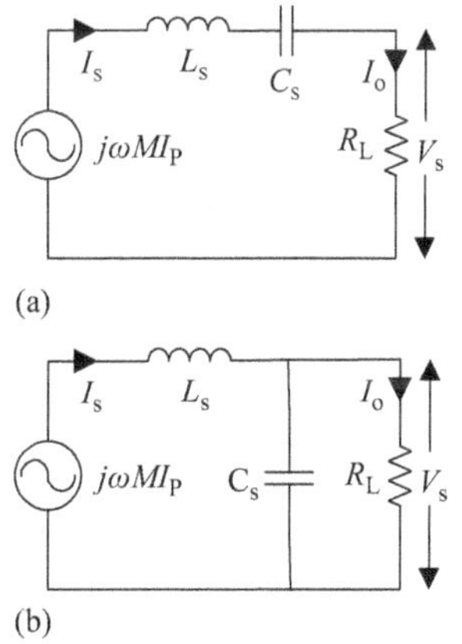

Figure 14.22 (a) Series compensated secondary and (b) parallel compensated secondary

Table 14.6 Value of MPT, C_s and secondary quality factor for uncompensated and compensated secondary

Secondary status	MPT	C_s	Secondary quality factor, Q_s
Uncompensated	$\frac{(\lvert V_{oc}\rvert \times \lvert I_{sc}\rvert)}{2}$	—	—
Parallel compensated	$\lvert V_{oc}\rvert\lvert I_{sc}\rvert\frac{R_l}{\omega_o L_s}$	$\frac{1}{\omega_o^2 L_s}$	$\frac{R_l}{\omega_o L_s}$
Series compensated	$\lvert V_{oc}\rvert\lvert I_{sc}\rvert\frac{\omega_o L_s}{R_l}$	$\frac{1}{\omega_o^2 L_s}$	$\frac{\omega_o L_s}{R_l}$

Table 14.7 Value of primary capacitance, secondary reflected impedance, and total impedance at resonant frequency, ω_o

Topology	C_p	Z_r	Z_T
SS	$\frac{L_s C_s}{L_p}$	$\frac{\omega_o^2 M^2}{R_L}$	$\frac{\omega_o^2 M^2}{R_L}$
SP	$\frac{1}{\omega_o^2\left(L_p-\frac{M^2}{L_s}\right)}$	$\frac{M^2 R_L}{L_s^2}-\frac{j\omega_o M^2}{L_s}$	$\frac{M^2 R_L}{L_s^2}$

factor becomes poorer [39, 40]. As a result of this, for a given amount of output, source needs to have higher volt–ampere (VA) ratings. This is a disadvantage, since greater VA-rating of supply means larger capacity of source side converter, therefore system cost increases. Moreover, circulation of reactive power leads to resistive losses and hence poor system efficiency. To overcome above said shortcomings, it is required to have output voltage and current of the power supply be in phase in order to minimize the VA-rating of the power supply [39, 41].

In the literature, there are many examples in which primary capacitances have been chosen to compensate just the self-inductance of the primary [42–44]. This is only acceptable, if the reflected impedance is negligible in comparison to the primary self-inductance [45]. Therefore, it is better to select primary capacitance which perform compensation for the entire circuit so that the input power factor becomes unity. Similar to secondary, primary capacitor can be connected in series and parallel. Combination of both series and parallel compensation gives total of four topology namely SS, SP, PP, and PS. Table 14.7 gives the value of reflected impedance (Z_r), primary capacitance (C_p) and total impedance seen by source (Z_T), for SS and SP topologies [45, 46].

Primary compensation depends a lot on the application. For long track applications, the impedance is very high. Therefore, the supply voltage needs to be high to drive the current through the primary. By series compensation the voltage across the capacitance compensates the voltage across the primary winding and hence the required voltage rating of the power supply can be reduced to manageable level.

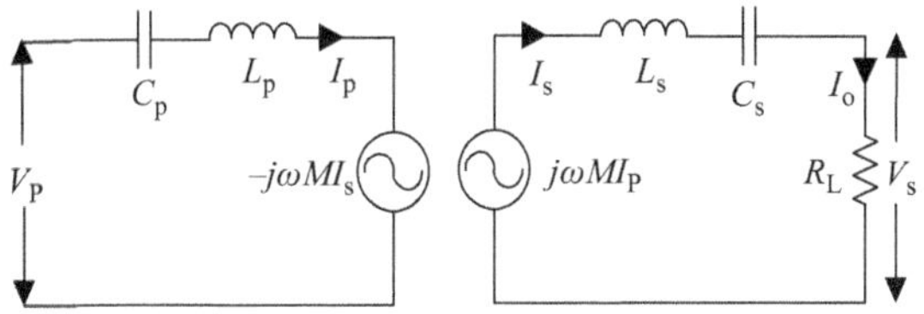

Figure 14.23 SS compensated IPT system

In this paper author consider primary to be series compensated as he intends to present the work for long track applications.

14.3.3 Analysis of series–series topology

Consider the IPT system with SS topology shown below in Figure 14.23. Primary and secondary winding resistance has been neglected in the analysis merely for the purpose of simplicity.

Induced voltage in secondary is given by:

$$E_s = j\omega MI_p \tag{14.6}$$

From (14.6) it can be observed that induced voltage in secondary is directly proportional to current in primary, I_p. However, I_p is determined by primary source and is given by:

$$I_p = \frac{V_p}{Z_T} \tag{14.7}$$

where Z_T is total impedance seen by source. In case of SS topology total impedance, at resonance, seen by primary is given by Table 14.7:

$$Z_T = \frac{w_o^2 M^2}{R_L} \tag{14.8}$$

From (14.7) and (14.8) one can observe if primary current, I_p is maintained constant then secondary induced voltage and therefore load voltage will remain constant. Under this condition, SS topology will acts as CV source. Primary induced voltage, E_P is given by:

$$E_p = j\omega MI_s \tag{14.9}$$

At resonance, E_p is equal to supply voltage, V_p. Hence for a given geometry, if primary is fed from CV source secondary current ($I_o = I_s$) will remain constant and SS topology will acts as constant current source.

Hence, from the above analysis, it can be said that SS topology can behave as constant current source as well as CV depending upon nature of supply.

14.3.4 Analysis of series–parallel topology

Consider the IPT system with SP topology shown below in Figure 14.24.

To understand the behavior of SP topology, consider the Norton's equivalent circuit of its secondary as shown in Figure 14.25.

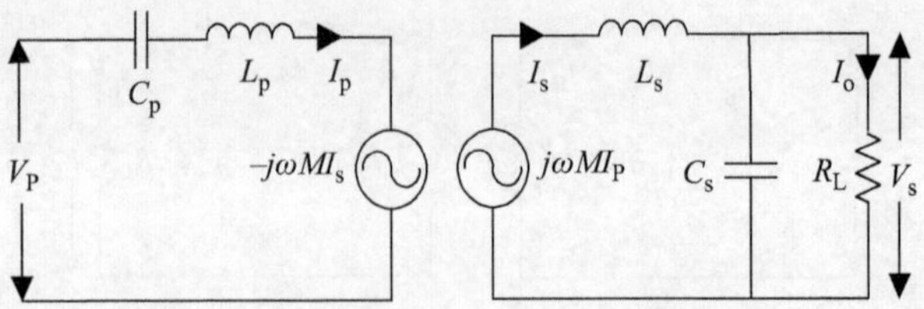

Figure 14.24 SP compensated IPT system

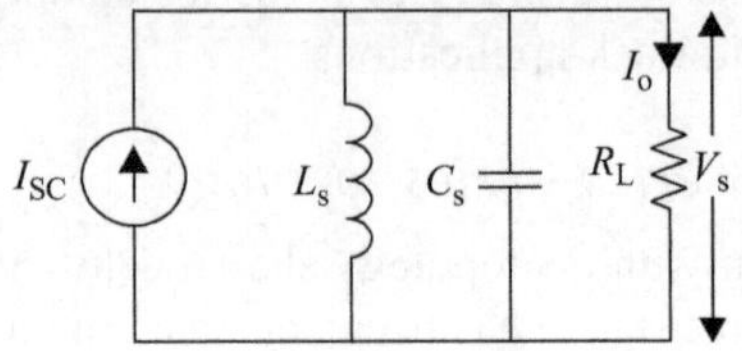

Figure 14.25 Norton's equivalent of secondary of SP topology

Here, $I_{sc}\left(=\frac{MI_p}{L_s}\right)$ is short circuit current given by (14.4). At frequency $\omega_o = \frac{1}{\sqrt{L_s C_s}}$, L_s and C_s behaves as open circuit therefore I_{sc} and I_o becomes equal. From (14.4) it can be observed that expression of I_{sc} is independent of load, R_L. If I_p is maintained constant, i.e., primary is fed from constant current input then I_{sc} and therefore load current I_o will remain constant irrespective of the load. In other words, secondary will behave as constant current source.

Output voltage, V_s of circuit shown in Figure 14.8 is given by:

$$V_s = \frac{MI_p}{L_s} R_L \tag{14.10}$$

Also from circuit shown in Figure 14.7, one can write:

$$I_p = \frac{V_p}{Z_T} \tag{14.11}$$

Using the value of Z_T from Table 14.7 and from (14.10) and (14.11):

$$V_s = \frac{V_p L_s}{M} \tag{14.12}$$

Equation (14.12) states that output voltage is independent of load, R_L and will be constant as long as V_p is maintained constant. In other words, SP topology will behave as CV source if voltage across primary is maintained constant.

14.3.5 Peak efficiency of series–series and series–parallel topology

For any topology average power delivered to the load can be given by:

$$P_o = I_o^2 R_L \tag{14.13}$$

Table 14.8 Efficiency and peak efficiency of SS and SP topology

Topology	Efficiency	Peak efficiency
Series–series	$\frac{R_L}{R_s + R_L + \frac{R_P(R_s+R_L)^2}{\omega_0^2 M^2}}$	$\frac{R_L}{R_s + R_L}$
Series–parallel	$\frac{R_L}{\left(R_S + R_L + \frac{R_P R_S^2}{\omega_0^2 M^2} + \frac{L_S^2 R_P}{M^2} + \frac{R_P R_S^2 R_L^2}{\omega_0^4 L_S^2 M^2} + \frac{2 R_L R_S R_P}{\omega_0^2 M^2} + \frac{R_L^2 R_S}{\omega_0^2 L_S^2}\right)}$	$\frac{R_L}{\frac{R_P L_S^2}{M^2} + R_S + R_L}$

Average power drawn from source can be given by:

$$P_{in} = I_p^2(\text{resistance of primary} + \text{reflected resistance}) \tag{14.14}$$

Therefore efficiency of an IPT system is given by:

$$\eta = \frac{I_o^2}{I_P^2} \frac{R_L}{(\text{resistance of primary} + \text{reflected resistance})} \tag{14.15}$$

Table 14.8 shows the efficiency as well as peak efficiency of SS topology and SP topology at resonance frequency.

Where R_s and R_L are primary coil resistance and secondary coil resistance, respectively. From Table 14.8 one can observe that for a given resonance frequency peak efficiency of series–series topology is always greater than series–parallel topology. Therefore, for fixed frequency systems, it would be better to opt for SS topology.

14.3.6 Control strategies for SS and SP topology

The power supply can be of fixed frequency or variable frequency [46]. For SS topology, primary capacitance is independent of mutual inductance, i.e., there is no volt-ampere reactive (VAR) loading on primary track due to secondary. Hence, it can be controlled by operating at fixed frequency and varying the output voltage of inverter to hold the current over a wide range of loading condition [47]. Also, fixed frequency controller are much simpler and easy to implement. However, in case of SP topology primary capacitance depends upon mutual inductance [48]. Whenever relative orientation between primary and secondary winding changes, tuning is disturbed and significant power transfer will be lost. For this reason, a variable frequency supply is desirable which can detect the change in mutual inductance and tune out the VAR loading on primary [46, 47]. Therefore, system is always operating at unity power factor. However, analysis of an IPT system shows that the system could have as many as three zero-phase angle frequency of which only one is the resonance frequency. This phenomena are called bifurcation [39, 46, 47]. If variable frequency controller cannot deal with uncertainty in bifurcation region the operation frequency of power supply will drift away from the desired operating position and might become unstable. Consequently, power transfer will drop significantly.

Wireless EV charging is the foreseen potential charging option for future EVs in the market. Wireless charging of an EV using IPT has a potential to reduce the EV costs.

14.3.7 Advantages of EV wireless charging

1. no human intervention
2. simple and robust
3. future scope for in motion charging
4. reduced EV weight due to reduced battery capacity in in motion charging
5. over all EV cost reduction due to reduced battery capacity

References

[1] Kempton W, Tomic J (2005) Vehicle-to-grid power implementation: from stabilizing the grid to supporting large-scale renewable energy, *Journal of Power Sources*, vol. 144, no. 1, pp. 280–294.

[2] Crompton TR (2000) *Battery reference book*, 3rd edn, Newnes.

[3] Buchmann I (2001) Batteries in a portable world: a handbook on rechargeable batteries for non-engineers, 2nd edn, Cadex Electronics Inc.

[4] Dhameja S (2002) *Electric vehicle battery systems*, Newnes.

[5] Chen LR (2008) Design of duty-varied voltage pulse charger for improving Li-ion battery-charging response, *IEEE Transactions on Industrial Electronics*, vol. 56, no. 2, pp. 480–487.

[6] Park SY, Miwa H, Clark BT, Ditzler D, Malone G, D'souza NS, Lai JS (2008) A universal battery charging algorithm for Ni–Cd, Ni-MH, SLA, and Li-ion for wide range voltage in portable applications, in *Proceedings of the IEEE Power Electronics Specialists Conference*, Rhodes, Greece, pp. 4689–4694.

[7] Hua CC, Lin MY (2000) A study of charging control of lead acid battery for electric vehicles, in *Proceedings of the IEEE International Symposium on Industrial Electronics*, vol. 1, pp. 135–140.

[8] West S, Krein PT (2000) Equalization of valve-regulated lead acid batteries: issues and life tests, in *Proceedings of the IEEE International Telecommunications Energy Conference*, pp. 439–446.

[9] Brost RD (1998) Performance of valve-regulated lead acid batteries in EV1 extended series strings, in *Proceedings of the IEEE Battery Conference on Applications and Advances*, Long Beach, CA, pp. 25–29.

[10] Bentley WF (1997) Cell balancing considerations for lithium-ion battery systems, in *Proceedings of the IEEE Battery Conference on Applications and Advances*, Long Beach, CA, pp. 223–226.

[11] Krein PT, Balog RS (2002) Life extension through charge equalization of lead-acid batteries, in *Proceedings of the IEEE International Telecommunications Energy Conference*, pp. 516–523.

[12] Lohner A, Karden E, DeDoncker RW (1997) Charge equalizing and lifetime increasing with a new charging method for VRLA batteries, in *Proceedings*

of the IEEE International Telecommunications Energy Conference, pp. 407–411.

[13] Moore SW, Schneider PJ (2001) A review of cell equalization methods for lithium ion and lithium polymer battery systems, in *Proceedings of the SAE 2001 World Congress*, Detroit, MI.

[14] Nishijima K, Sakamoto H, Harada K (2000) A PWM controlled simple and high performance battery balancing system, in *Proceedings of the IEEE 31st Annual Power Electronics Specialists Conference*, vol. 1, Galway, Ireland, pp. 517–520.

[15] Isaacson MJ, Hoolandsworth RP, Giampaoli PJ (2000) Advanced lithium ion battery charger, in *Proceedings of the IEEE Battery Conference on Applications and Advances*, pp. 193–198.

[16] Pascual C, Krein PT (1997) Switched capacitor system for automatic series battery equalization, in *Proceedings of the 12th Annual Applied Power Electronics Conference and Exposition*, Atlanta, GA, vol. 2, pp. 848–854.

[17] Hung ST, Hopkins DC, Mosling CR (1993) Extension of battery life via charge equalization control, *IEEE Transactions on Industrial Electronics*, vol. 40, no. 1, pp. 96–104.

[18] Cao J, Schofield N, Emadi A (2008) Battery balancing methods: a comprehensive review, in *Proceedings of the IEEE Vehicle Power and Propulsion Conference*, Harbin, China, pp. 1–6.

[19] Lee YS, Wang WY, Kuo TY (2008) Soft computing for battery state-of-charge (BSOC) estimation in battery string systems, *IEEE Transactions on Industrial Electronics*, vol. 55, no. 1, pp. 229–239.

[20] Shen WX, Chan CC, Lo EWC, Chau KT (2002) Adaptive neuro-fuzzy modeling of battery residual capacity for electric vehicles, *IEEE Transactions on Industrial Electronics*, vol. 49, no. 3, pp. 677–684.

[21] Piller S, Perrin M, Jossen A (2001) Methods for state-of-charge determination and their applications, *Journal of Power Sources*, vol. 96, no. 1, pp. 113–120.

[22] Pesaran AA (February 2001) Battery thermal management in EVs and HEVs: issues and solutions, in *Proceedings of the Advanced Automotive Battery Conference*, Las Vegas, NV.

[23] Chen Y, Song L, Evans JW (August 1996) Modeling studies on battery thermal behaviour, thermal runaway, thermal management, and energy efficiency, in *Proceedings of the IEEE Intersociety Energy Conversion Engineering Conference*, Washington, DC, vol. 2, pp. 1465–1470.

[24] Jossen A, Spath V, Doring H, Garche J (June 1999) Battery management systems (BMS) for increasing battery life time, in *Proceedings of the IEEE International Telecommunications Energy Conference*, Copenhagen, Denmark.

[25] Olson JB, Sexton ED (January 2000) Operation of lead-acid batteries for HEV applications, in *Proceedings of the IEEE Battery Conference on Applications and Advances*, Long Beach, CA, pp. 205–210.

[26] Hayes John G, Davis K (June 2014) Simplified electric vehicle powertrain model for range and energy consumption based on EPA coast-down parameters and test validation by Argonne National Lab data on the Nissan Leaf, *Transportation Electrification Conference and Expo (ITEC)*, pp. 1–6.
[27] SAE J1772 standard. http://standards.sae.org/wip/j1772/
[28] IEC 61851-1 standard. https://webstore.iec.ch/publication/6029
[29] Hajimiragha A, Canizares CA, Fowler MW, Elkamel A (February 2010) Optimal transition to plug-in hybrid electric vehicles in Ontario, Canada, considering the electricity-grid limitations, *IEEE Transactions on Industrial Electronics*, vol. 57, no. 2, pp. 690–701.
[30] Khaligh A, Dusmez S (October 2012) Comprehensive topological analyses of conductive and inductive charging solutions for plug-in electric vehicles, *IEEE Transactions on Vehicular Technology*, vol. 61, no. 8, pp. 3475–3489.
[31] Veneri O, Ferraro L, Capasso C, Iannuzzi D (October 2012) Charging infrastructures for EV: overview of technologies and issues, *Electrical Systems for Aircraft, Railway and Ship Propulsion (ESARS)*, pp. 1–6.
[32] Dusmez S, Khaligh A (February 5–8, 2012) Cost effective solutions to level-3 on-board battery chargers, in *Proceedings of the Applied Power Electronics Conference*, pp. 2121–2127.
[33] Singh B, Singh BN, Chandra A, Al-Haddad K, Pandey A, Kothari DP (June 2004) A review of three-phase improved power quality AC–DC converters, *IEEE Transactions on Industrial Electronics*, vol. 51, no. 3, pp. 641–660.
[34] Wijeratne DS, Moschopoulos G (March 2014) A novel three-phase buck–boost AC–DC converter, *IEEE Transactions on Power Electronics*, vol. 29, no. 3, pp. 1331–1343.
[35] "Establishment of CHAdeMO Association" (Press release). TEPCO. 2010-03-15. *Retrieved* 2010-05-13.
[36] Chuck Squatriglia (2010-03-16). "Let's Have Tea While Charging Our EV". Autopia. Wired.com. Retrieved 2010-05-13.
[37] Elliott GAJ, Boys JT, Green AW (February 1995) Magnetically coupled systems for power transfer to electric vehicles, in *Proceedings of the IEEE International Conference on Power Electronics and Drive Systems*, Singapore, vol. 2, pp. 797–801.
[38] Wang CS, Stielau OH, Covic GA (December 2000) Load models and their application in the design of loosely coupled inductive power transfer systems, in *Proceedings of the IEEE International Conference on Power System Technology*, Perth, Australia, pp. 1053–1058.
[39] Chopra S, Bauer P (October 2011) Analysis and design considerations for a contactless power transfer system, in *Proceedings of the IEEE International Telecommunications Energy Conference*, Amsterdam, The Netherlands, pp. 1–6.
[40] Stielau OH, Covic GA (December 2000) Design of loosely coupled inductive power transfer systems, in *Proceedings of the IEEE International Conference on Power System Technology*, Perth Australia, pp. 85–90.

[41] Wang CS, Covic GA, Stielau OH (November 2001) General stability criterions for zero phase angle controlled loosely coupled inductive power transfer systems, in *Proceedings of the IEEE Annual Conference of the Industrial Electronics Society*, Denver, CO, vol. 2, pp. 1049–1054.
[42] Esser A, Skudelny HC (September 1991) A new approach to power supplies for robots, *IEEE Transactions on Industry Applications*, vol. 27, no. 5, pp. 872–875.
[43] Green AW, Boys JT (March 1993) An inductively coupled high frequency power system for material handling applications, in *Proceedings of the International Power Engineering Conference*, Singapore, vol. 2, pp. 821–826.
[44] Bieler T, Perrottet M, Nguyen V, Perriard Y (September 2002) Contactless power and information transmission, in *IEEE Transactions on Industry Applications*, vol. 38, no. 5, pp. 1266–1272.
[45] Wang C-S, Covic GA, and Stielau OH (February 2004) Power transfer capability and bifurcation phenomena of loosely coupled inductive power transfer systems, *IEEE Transaction on Industrial Electronics*, vol. 51, no. 1, pp. 148–157.
[46] Wang CS, Stielau OH, Covic GA (October 2005) Design considerations for a contactless electric vehicle battery charger, *IEEE Transactions on Industrial Electronics*, vol. 52, no. 5, pp. 1308–1314.
[47] Covic GA, Boys JT (June 2013) Inductive power transfer, *Proceedings of the IEEE*, vol. 101, no. 6, pp. 1–14.
[48] Green AW, Boys JT (October 1994) 10 kHz inductively coupled power transfer – concept and control, in *Proceedings of the IEEE International Conference on Power Electronics and Variable Speed Drives*, London, UK, pp. 694–699.

Chapter 15
Shipboard power systems

*Herbert L. Ginn III**

Electricity has been utilized in some small fashion onboard ships as early as the 1870s, and by the 1930s, it was in widespread use for various auxiliary machinery and even turbo-electric drives. Turbo-electric drives use electric generators to convert mechanical energy of a turbine into electric energy and electric motors to convert it back into mechanical energy to power the drive shafts. It allows for a decoupling of the prime mover from the propellers so that each can operate at optimum speeds without the need for large mechanical reduction gear sets. Although electric drives have long been in use, modern-day shipboard power systems bear little resemblance to their predecessors of even 30 years ago. Present-day shipboard systems employ power electronics for management of energy in propulsion and elsewhere. Fuel costs constitute a large portion of a ship's life-cycle costs and pressures to improve fuel economy in conjunction with environmental regulations in marine engineering have led to the use of power electronic based drives in ship propulsion and auxiliary equipment similarly as in other industrial applications. Many commercial ships are now built with power electronic drives, including passenger ships, tankers, ice-breakers, cable laying ships, and floating offshore platforms.

15.1 Shipboard power system topologies

Loads in a shipboard system can be categorized as propulsion, ship service, and hotel loads. In all cases, a significant portion of generation is devoted to propulsion. However, the percentage as related to other load categories depends on the type of ship. For example, in passenger cruise ships, the percentage of hotel loads will be much higher than for navy ships. In this chapter, ship distribution systems with total installed generation from several megawatts to many tens of megawatts are considered since ships of various types in that installed power range benefit from electric propulsion utilizing power electronic drives.

Simplified diagrams of a few typical medium-voltage ship distribution architectures with power electronic drives derived from [1] are shown in Figures 15.1–15.3. The electrical power distribution architecture shown in Figure 15.1 features a dual

*University of South Carolina, Columbia, SC, USA

redundant main bus and ship service bus. Two complete propulsion drives are fed from separate buses for good survivability and the ship service buses are fed from two ship service transformers. Prime movers can be either diesel engines for ships such as large cruise ships and tankers or gas turbines such as in some navy ships. Recently some very large cruise ships built after the year 2005 also use gas turbines for the prime movers. In applications where additional redundancy is required, such as ice-breakers, dual winding motors can be supplied from pairs of converters that are cross-fed from opposite main buses. This configuration is shown in Figure 15.2. Finally, a ring bus architecture typical of navy ships is shown in Figure 15.3. Standard voltage levels in ship distribution systems are 13,800, 11,000, 6,600, 4,160, and 440 V. Virtually all US naval ships generate and distribute three-phase 60 Hz electrical power at 440 or 4,160 V.

As evidenced by these distribution system topologies, shipboard power systems are isolated microgrids and as such are presented with the same operational challenges as any isolated microgrid, most of which are due to small supply system inertia. Those challenges are not only present in shipboard power systems but also are compounded by the small physical size required relative to the installed power

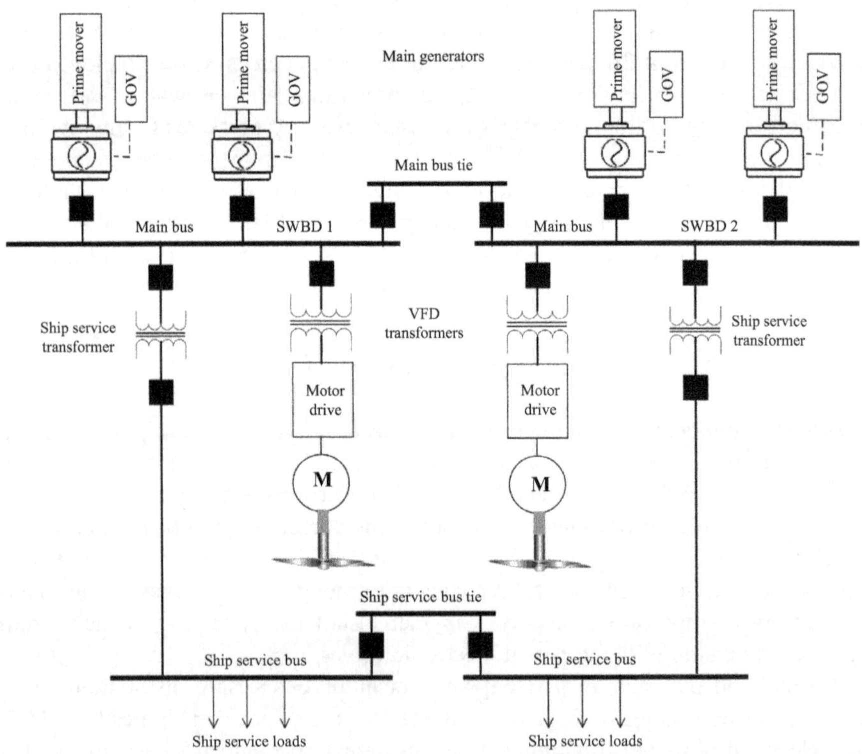

Figure 15.1 Shipboard power system with two separately supplied propulsion drives

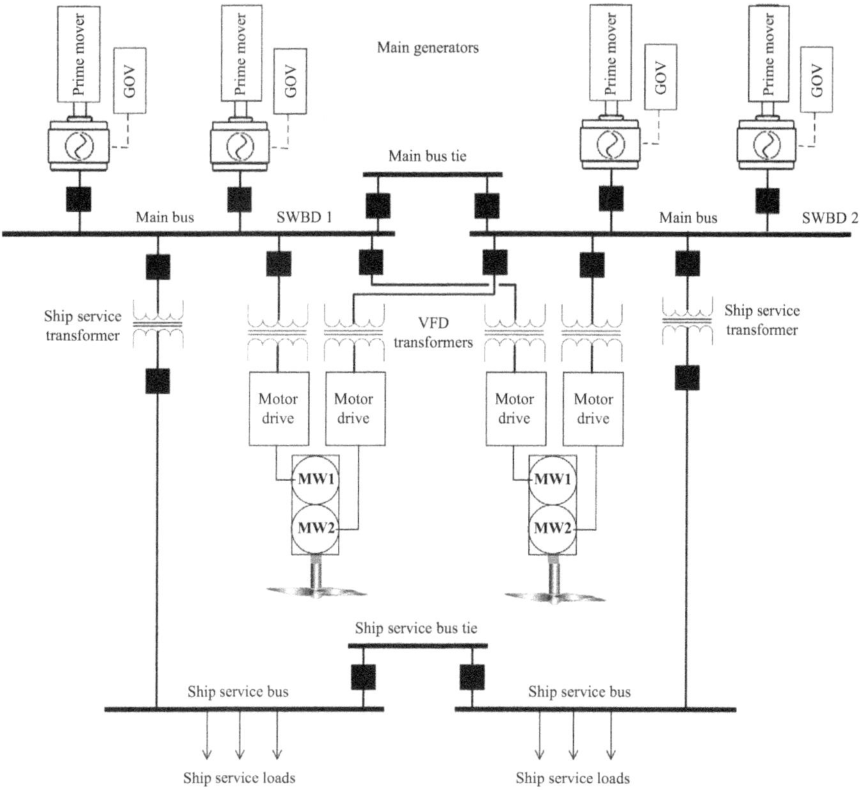

Figure 15.2 Shipboard power system with two cross-fed propulsion drives

of the shipboard distribution system. Some of the key features of a shipboard microgrid are as follows:

- *Non-linear load effects*: Tolerance of poor power quality and electromagnetic interference, the by-products of large-scale connection of power electronics to a grid, is much lower in the shipboard electrical system than in most other applications. These effects must be mitigated through additional filters and enclosures. This results in increased size, weight, and heat in an environment where those attributes often come at a premium.
- *Variable frequency*: Frequency is not constant onboard ships. The low rotational inertia of the prime movers and generators relative to single large loads such as the propulsion drives allows for rapid accelerations and decelerations of the shaft and corresponding frequency fluctuations in response to load changes.
- *Load dynamics*: Shipboard systems must account for dynamics of loads such as propulsion motors, propeller dynamics, large pumps, and ship dynamics. In the case of navy ships, there are also pulsed type loads. Thus, a ships energy resources must be managed in an integrated fashion in order to ensure that the

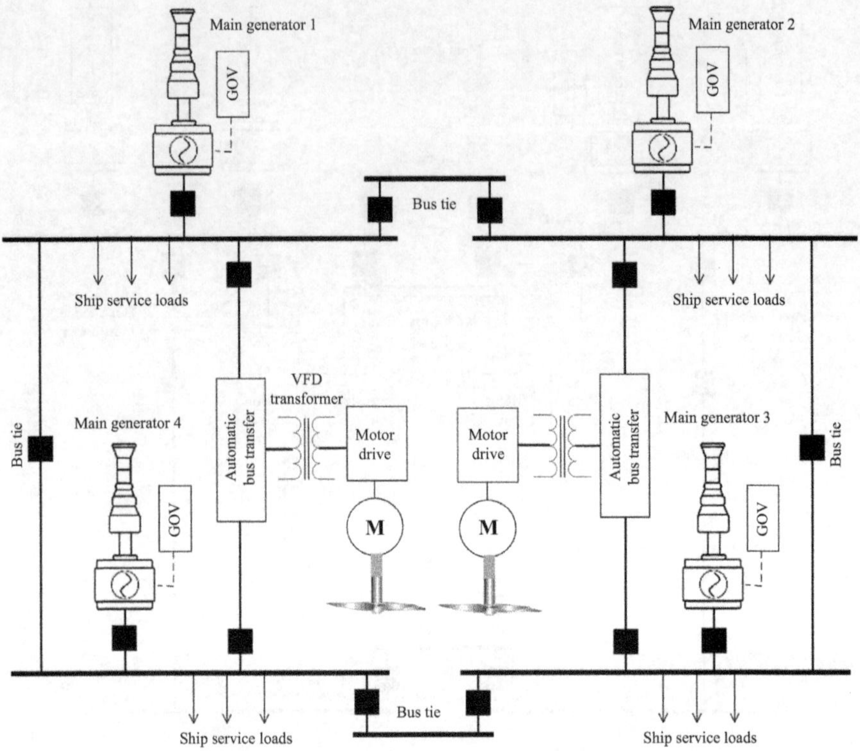

Figure 15.3 Ring bus configuration for a navy ship distribution system

ship accomplishes its mission or reaches its destination under the myriad of scenarios that can occur.

- *Grounding*: The density of electrical equipment on a ship and the proximity to ground potential at all times is more severe in a ship than in most other applications. As a result, the vast majority of the electrical distribution on ships is ungrounded.
- *Harsh environment*: Equipment for shipboard power systems must be able to operate in a pitching, and rolling ship. Vibration, shock, humidity, and salinity must be accounted for in their design.

15.2 Shipboard propulsion drives

Propulsion systems contain the primary power electronics equipment installed in shipboard distribution systems. These are medium voltage industrial power electronic motor drives with the basic architecture shown in Figure 15.4. As with any motor drive system, the power processing unit (PPU) consists of a power electronic converter that transforms the fixed voltage and frequency supply into the appropriate form for a given operating state of the motor. For shipboard applications, the

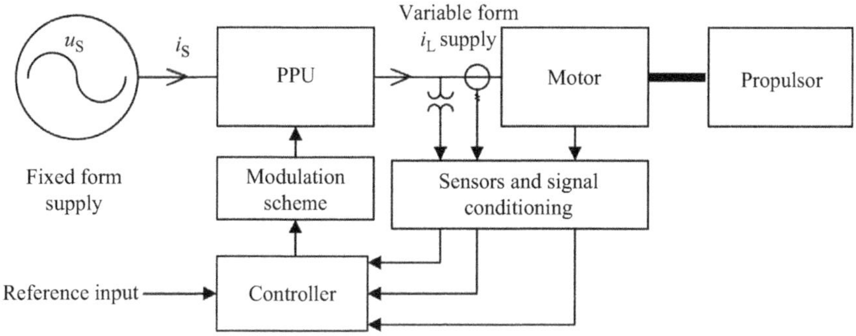

Figure 15.4 Block diagram of the shipboard electric propulsion system

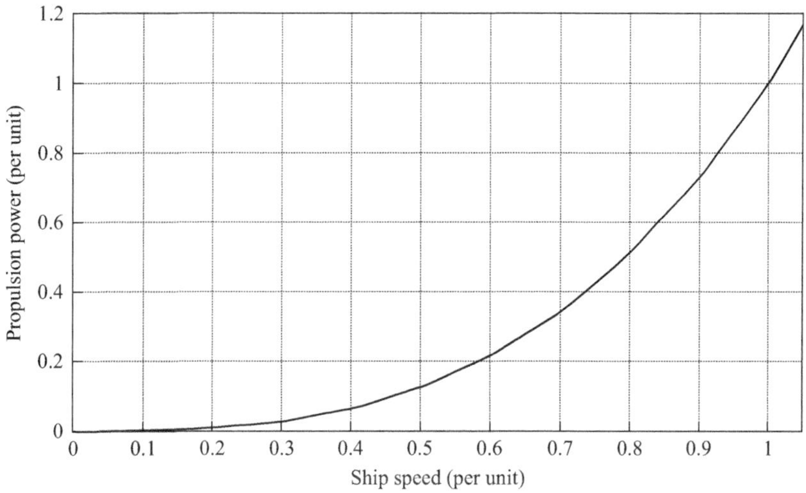

Figure 15.5 Approximate propulsion power curve as a function of speed for a ship

torque ripple to the shaft and input power quality are driving factors in the design. Ship propulsion load profiles can be approximated as a cubic function of power versus ship speed as shown in Figure 15.5.

With the introduction of the thyristor-based rectifier, shipboard propulsion systems originally utilized DC motor drives. In the 1980s, variable frequency drives started to appear in ship propulsion applications. Modern propulsion drives utilize AC frequency converters of some type driving either synchronous or asynchronous AC motors. Converter types for shipboard propulsion drives fall into three broad categories:

1. voltage link systems
2. current link systems
3. direct AC/AC conversion

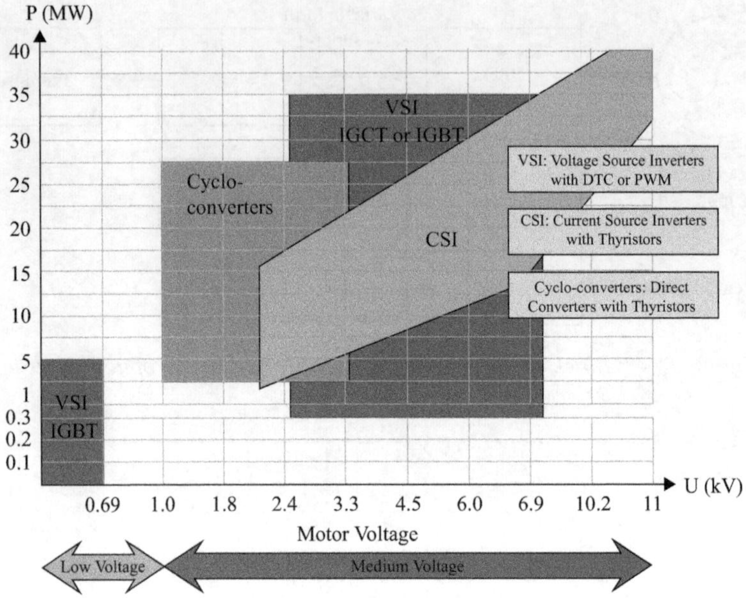

Figure 15.6 Converter topology as an approximate function of output voltage and power in shipboard applications

Motor voltage versus power requirements for these three propulsion system topologies are shown in Figure 15.6.

Other than special shock and vibration mounting requirements, high power medium voltage drives onboard ships are in similar types of enclosures as for any industrial drive application. Two ship drive converters, one for each of two propulsion motors, are shown in Figure 15.7.

15.2.1 Voltage link systems

Voltage link systems provide a stiff DC voltage supply to the converter that supplies the motor. This voltage may be fixed or variable. This motor side converter is often referred to as a voltage source inverter (VSI). The power semiconductors are always forward biased due to the DC supply voltage. Therefore, any fully controllable asymmetric blocking devices are suitable. However, due to the high power requirements of ship propulsion, only gate turn-off thyristors (GTOs) and insulated-gate bipolar transistors (IGBTs) are used. The interface converter between the supply and DC link may be diode rectifiers, phase-commutated thyristor converters, or self-commutated converters similar to the motor side converter. In ship propulsion applications, bidirectional active power flow for converter A is not generally required and therefore self-commutated converters are not typically used (Figure 15.8).

Three-phase bridge inverters are the simplest structure used to supply AC motor loads. For high power applications, GTOs are often used. The switching speed of GTOs is much lower than IGBTs; therefore, for pulse width modulation

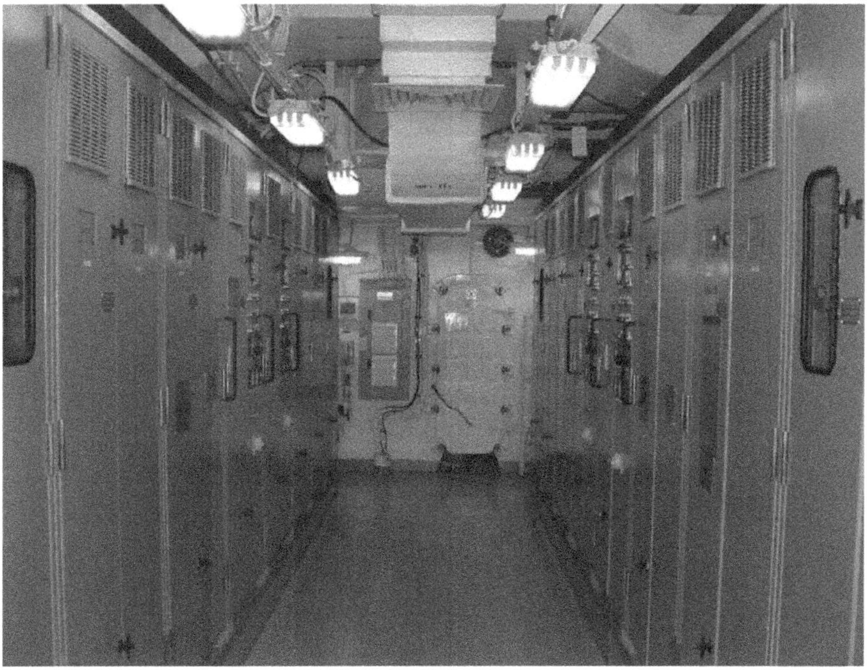

Figure 15.7 *Two medium-voltage cycloconverter propulsion drives onboard a ship*

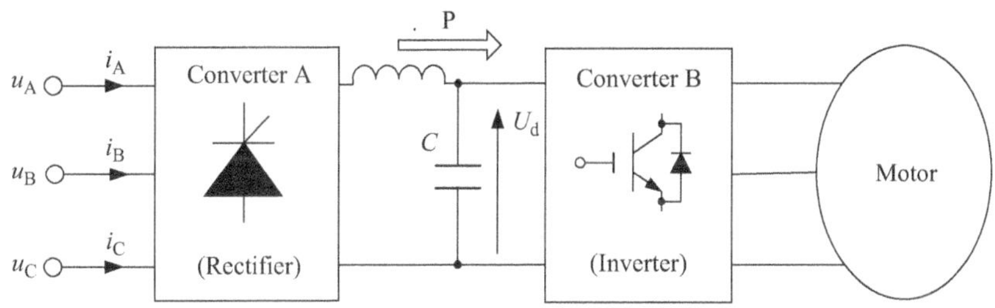

Figure 15.8 *Block diagram of voltage link system with unidirectional active power flow capability*

(PWM) schemes, GTO inverters can only have a carrier frequency of a few hundred hertz. However, GTOs are suitable for square-wave switching. Although it is possible to arrange IGBTs in series strings for higher voltage ratings and in parallel for higher current, this requires difficult device matching and complex gate control algorithms. Multilevel converters are an alternative for high-power applications when fast switching is required.

The most common high-power VSI used in ship propulsion applications is the neutral-point clamped (NPC) inverter shown in Figure 15.9. This inverter has a zero DC voltage center point, which can be switched to the phase outputs creating

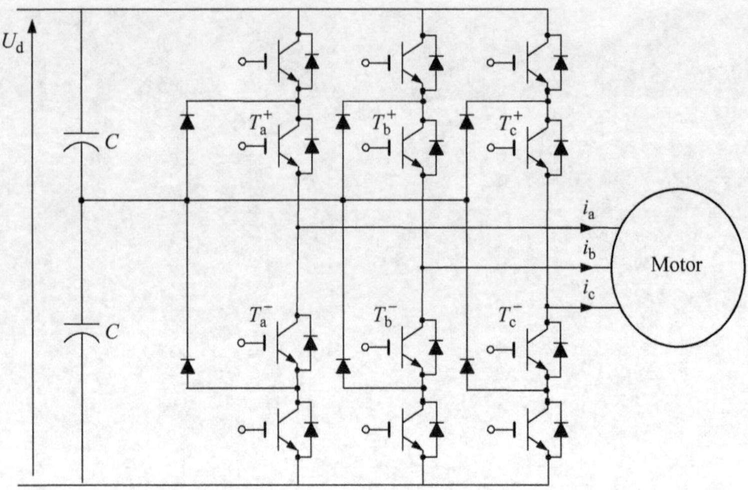

Figure 15.9 Block diagram of three-level NPC VSI

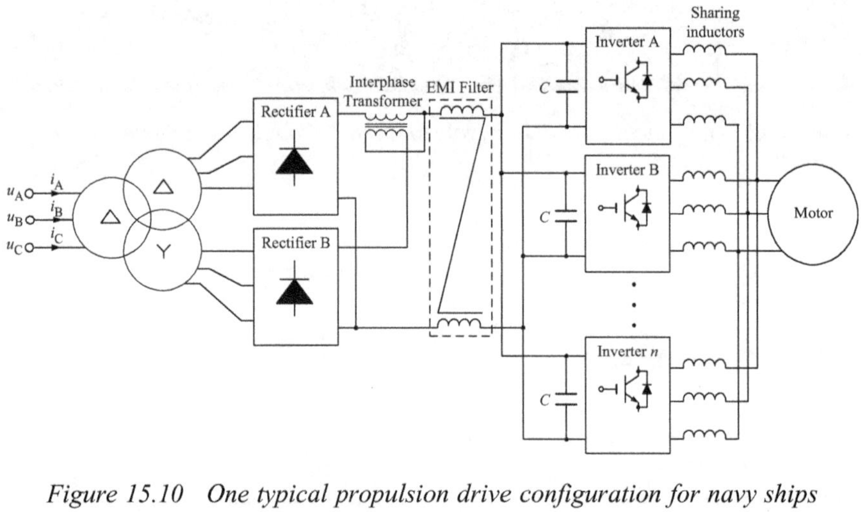

Figure 15.10 One typical propulsion drive configuration for navy ships

the ability to apply one of three voltage levels to each phase leg. Each switch and diode must block only half of the DC link voltage. The voltage across each capacitor can be balanced by utilizing two isolated sources or regulation of the average current into the center point. In this case, feedback control is used to adjust the time each leg dwells on the center point such that the average center point current is zero.

The three-level NPC inverter is often used in naval applications due to high-speed requirements of the drive. In order to increase current capacity for high-power drives, the NPC VSI modules are paralleled with output sharing inductors or interphase transformers. A typical navy propulsion drive configuration is shown in Figure 15.10. In this particular configuration, two six-pulse diode bridges supply the DC bus in parallel and utilize an interphase transformer to ensure equal sharing of DC current. The diode rectifiers are supplied with a phase shifting transformer to cancel characteristic lower order current harmonics. This is a common method to reduce supply distortion caused by the drive. This cancellation method and other harmonic suppression methods used in ship distribution systems are covered in detail later in this chapter.

15.2.2 Current link systems

In current link systems, a stiff DC current source is at the interface between the two converters as shown in Figure 15.11. Therefore, the motor side converter is usually referred to as a current source inverter (CSI). Although it is referred to as an inverter, it may also act as a rectifier (motor in regenerative mode). The semiconductor switches in a CSI must withstand reverse voltage so asymmetric blocking devices cannot be used. Thus, thyristors and symmetric voltage blocking GTOs are used. (Although, some topologies use asymmetric forward blocking devices in series with diodes.) When thyristors are used for the CSI forced commutation topologies are required, with the exception of CSIs supplying over-excited synchronous machines.

Thyristor inverters are the most often used in CSI topologies as shown in Figure 15.12. However, depending on the load, forced commutation may be needed. If leading reactive power is supplied to the load from the CSI then

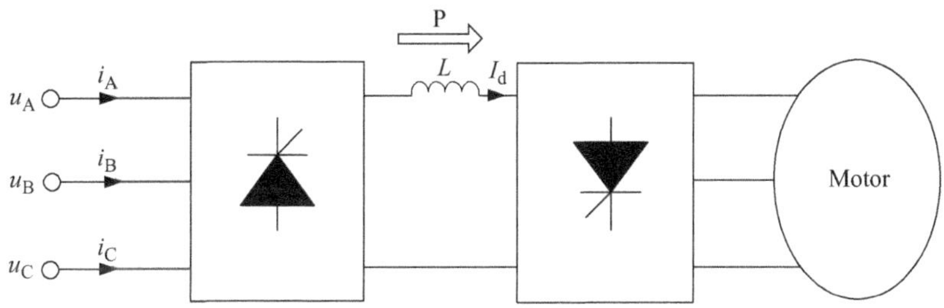

Figure 15.11 Block diagram of current link system with unidirectional active power flow capability

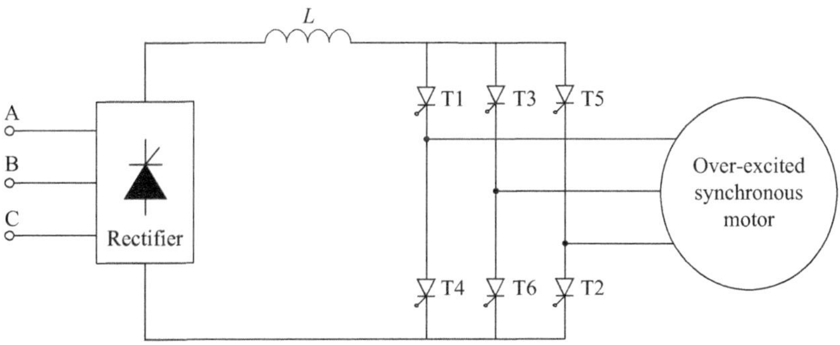

Figure 15.12 Load commutated thyristor CSI

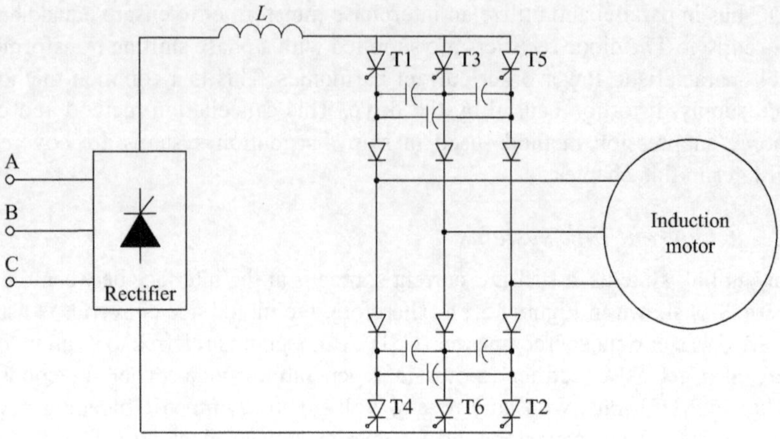

Figure 15.13 Three-phase auto-sequential forced commutation CSI

commutation is provided by the load. This is the case for over-excited synchronous motors. If lagging reactive power is supplied to the load then forced commutation is required. This is the case with induction motors. Figure 15.13 shows the most common type of forced commutation known as the auto-sequential method. Note that induction motor starting for the synchronous machine requires forced commutation, which can be accomplished using the circuit in Figure 15.12 by the pulsed or DC link current interruption method. In shipboard drives for large passenger cruise liners, the load commutated inverter circuit in Figure 15.12 is typically used. It is paired with an over-excited synchronous machine and a thyristor rectifier front end providing DC link interruption. For very low motor speeds, the machine EMF is not high enough to provide correct thyristor commutation of the drive inverter. To overcome this, the DC link current is reduced to zero for a short time by operating the supply side converter in inverter mode. This enables the conducting motor side converter thyristors to return to blocking mode and commutation to occur.

15.2.3 Direct AC–AC conversion – cycloconverters

Cycloconverters are used for very low-speed high-torque applications and provide low-torque pulsations at low speed. They can be used for either synchronous or induction motors, however, in ship propulsion they are usually used to drive synchronous machines. Icebreakers and tanker ships are two classes of ships that commonly use cycloconverter drives. The high torque allows icebreakers to free a propeller from ice or to strike ice without causing a motor stall. Disadvantages are poor supply quality since the modulated supply current is non-periodic with respect to the supply frequency. Therefore, it is a source of "interharmonics" in the supply.

A diagram of a cycloconverter is shown in Figure 15.14. Each phase of the output is supplied by two back-to-back six-pulse thyristor rectifiers in order to allow for bidirectional current flow. The firing angle of each six-pulse rectifier is

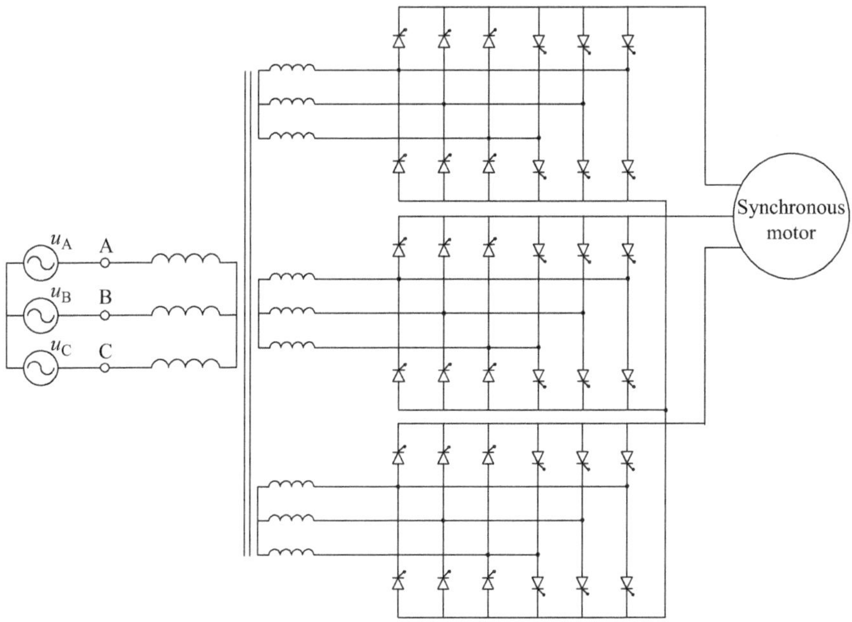

Figure 15.14 Three-phase six-pulse cycloconverter

controlled to yield a low-frequency sinusoidal output. Because the converter is chopping the supply voltage to yield the desired average output value, the output frequency is limited to approximately 30% of the input frequency depending on allowable distortion of the output. In shipboard applications with 60 Hz supply frequencies, the cycloconverter output frequency limit is typically around 15 Hz.

15.3 Power quality requirements in shipboard systems

The presence of non-linear loads in the system cause distorted current. In most cases, these currents are periodic; therefore, using Fourier analysis these distorted voltages and currents can be described in terms of harmonics. Although the range of harmonic frequencies present in power systems is broad, the harmonics in the lower frequency band ranging from the second harmonic to a few kilohertz are the greatest in magnitude, and as harmonic order increases, the magnitude of harmonics declines faster than $1/n$. Therefore, the harmonics in the lower frequency band are the most significant.

A number of harmful effects are known to be caused by harmonic distortion in power systems. A few of the major effects are as follows:

- additional heating and losses in transformers and induction and synchronous machines
- capacitor overloading
- increased probability of relay malfunctions

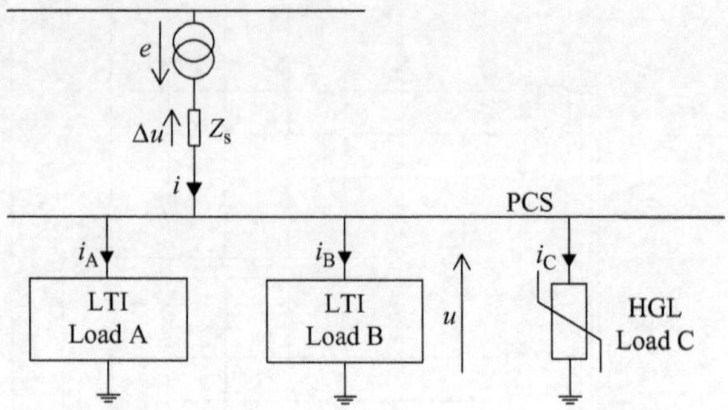

Figure 15.15 Simplified one-line diagram of a distribution system

- disturbances in solid-state and microprocessor-based systems
- interference with communication systems

Because of these harmful effects, the presence of harmonics is considered to be a cause of supply quality degradation in distribution systems. A one-line diagram of a three-phase system with a harmonic generating load (HGL) and other linear time invariant (LTI) loads connected to a common bus is shown in Figure 15.15.

This point where multiple loads are supplied is referred to as the point of common supply (PCS). Harmonic currents generated at the HGL flow through the system impedance Z_s, resulting in a distorted voltage drop Δu across the impedance. The sum of distorted voltage Δu and the distribution voltage e yields a distorted bus voltage u for all loads connected at the bus. Therefore, the supply currents i_A and i_B of LTI loads are also distorted.

In addition to the distortion of the voltage at the PCS, harmonics in the supply current cause power factor to decline. Because power factor is a measure of supply utilization, a low power factor means that the supply current is larger than needed for the transmission of the required energy to the load. The decline of power factor is caused by harmonic as well as by reactive currents. To determine the effect of harmonic currents on the power factor, it is convenient to consider a simplified case where the supply voltage is sinusoidal and the load is a balanced HGL. Then, the current supplied to the load can be decomposed as

$$i = i_l + i_g = i_a + i_r + i_g, \tag{15.1}$$

where the active current, i_a, is the current component responsible for permanent energy transmission; the reactive current, i_r, is the current due to phase shift between the voltage and current; and the load generated current, i_g, is the current component due to non-linear loads. The load-generated current contains all current harmonics generated due to the non-linearity and/or periodic time variance of the load. If all the harmonics orders of the load current except the fundamental harmonic order, $n = 1$, are contained in the set M, then the load-generated harmonic current is equal to

$$i_g = \sqrt{2}\,\mathrm{Re}\sum_{n\in M} \boldsymbol{I}_n \mathrm{e}^{\mathrm{j}n\omega_1 t}. \tag{15.2}$$

All of these components are mutually orthogonal, and therefore, the root mean square (RMS) value of the current is

$$\|i\|^2 = \|i_a\|^2 + \|i_r\|^2 + \|i_g\|^2. \tag{15.3}$$

Then the power factor can be expressed as

$$\lambda = \frac{P}{S} = \frac{\|i_a\|}{\sqrt{\|i_a\|^2 + \|i_r\|^2 + \|i_g\|^2}}, \tag{15.4}$$

which shows that the load-generated current effects the power factor in the same way as reactive current. Therefore, the load-generated harmonics lower the power factor and this requires increased power ratings of power system equipment as well as causes increased active power losses. Increased equipment sizing is very undesirable on ships due to the very limited spaces.

As shown above, harmful effects caused by HGLs are distributed over the power distribution system. Therefore, there are shipboard power system standards that limit the levels of harmonic distortion as well as other power quality related issues like voltage and frequency deviation from the nominal. The IEEE STD-45 [2] provides a number of design and performance criteria for commercial ships including some related to power quality. There is no specific limit on the total harmonic distortion (THD) in the standard but generally acceptable limits are 5% for voltage total harmonic distortion (THD_v) based on the IEEE STD-519 [3] for voltages under 69 kV. For US Navy applications, the governing power quality interface standards are MIL-STD-1399 section 300 and section 680. Some important limits on power quality are summarized in Table 15.1.

In Table 15.1, THD_v is the ratio of the RMS value of the voltage excluding the fundamental to the RMS value of the fundamental as expressed by

$$THD_v = \frac{\sqrt{\sum_{n=2}^{\infty} V_n^2}}{V_1} \tag{15.5}$$

Table 15.1 Summary of some MIL-STD-1399 section 680 limits

Characteristic for all voltage classes	**Value**
Maximum voltage total harmonic distortion (THD_v)	5%
Maximum single-voltage harmonic	3% of V_1
Maximum single-current harmonic above 60–2,000 Hz	3% of I_{1FL}
Current of any frequency above 2,000 Hz–20 kHz	6,000/f% of I_{1FL}
Frequency tolerance	±3%
Frequency transient tolerance	±4%
Frequency modulation	0.5%
Worst-case frequency excursion from nominal	±5%
Nominal frequency	60 Hz

The fundamental of the current at full load, I_{1FL}, is its RMS value with harmonic components removed. Current THD is not limited but is defined as

$$\mathrm{THD_I} = \frac{\sqrt{\sum_{n=2}^{\infty} I_n^2}}{I_1} \tag{15.6}$$

The presence of harmonic distortion in a shipboard power system is primarily due to propulsion drive front ends that are non-linear and/or periodically time variant in nature. However, other sources, such as florescent lamps, rectifiers used in small power supplies, flux distortion in synchronous machines and transformers operated in the non-linear region of their magnetization curve, etc., are present but their contribution is far less. The most widely used of the drive front-end converters, as listed in Section 15.2 are the six-pulse AC/DC converters. The basic circuit configuration of a six-pulse thyristor converter is shown in Figure 15.16. The idealized waveform of the supply current for phase A is shown in Figure 15.17. If the converter's filter inductor has infinite value, the supply has an infinite power, the converter thyristors are perfectly matched, and the supply and control systems are symmetrical then the current waveform will be ideal as shown. The six-pulse diode rectifier supply current will have a similar shape but the pulses cannot be delayed in time as will the thyristor converter as a function of firing angle. In either case, the characteristic harmonics generated by the drive supply side converter will be the same.

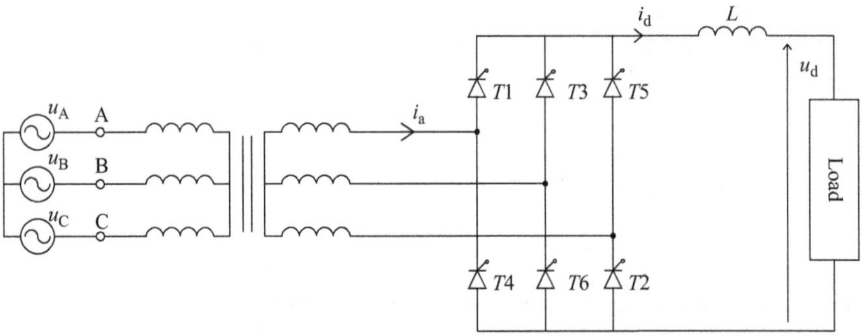

Figure 15.16 Typical configuration of a six-pulse thyristor converter

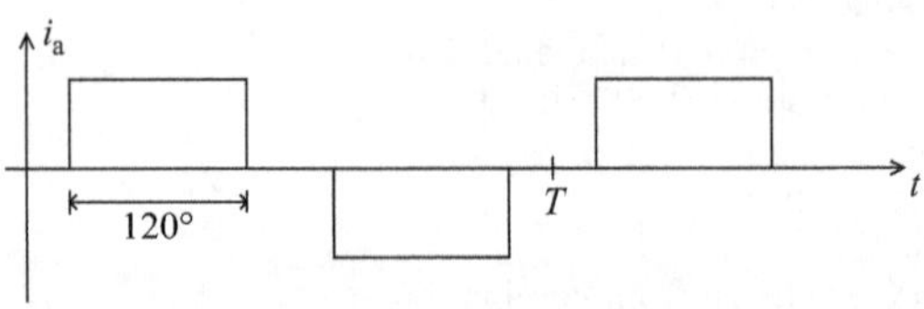

Figure 15.17 Idealized waveform of the converter supply current

Under these idealized conditions and with firing angle equal zero, the phase current has a Fourier series with complex RMS values equal to

$$\boldsymbol{I}_{an} = \frac{2\sqrt{2}I_d}{n\pi} \sin\left(n\frac{\pi}{3}\right) e^{-jn\frac{\pi}{2}} \tag{15.7}$$

for all odd order harmonics n. Therefore, the current only has harmonics of order $n = 6k \pm 1$ where k is a positive integer. These are referred to as the characteristic harmonics of the converter. However, the ideal conditions stated above are never true and, therefore, the supply current always contains some small amount of non-characteristic harmonics. Note that, as harmonic order increases the characteristic harmonics decline in magnitude by $1/n$.

15.4 Harmonic mitigation in shipboard systems

Shipboard distribution systems with high power electric propulsion drives cannot meet harmonic distortion limits if six-pulse thyristor or diode rectifiers are not employed along with a harmonic mitigation method. Harmonic mitigation methods used are harmonic cancellation and/or harmonic suppression. There are several different types of harmonic suppressors that could be used to reduce distortion in power distribution systems. The choice of which harmonic suppressor should be used in a particular case is governed by both technical as well as economic issues. The primary types of suppressors used in shipboard systems belong to one of the following three basic categories:

(i) reactive harmonic suppressors (RHSs)
(ii) switching compensators (SCs)
(iii) hybrid compensators

RHSs are the largest group of suppressors. They modify the frequency properties of the system in order to reduce distortion. Because of this, the design of RHSs is a complex task where the device and system cannot be treated separately. The group of RHSs includes such devices as resonant harmonic filters (RHFs) and low-pass filters. SCs, also referred to as active filters, inject a compensating current which cancels the load-generated harmonics. The compensating current is generated by fast switching of power transistors. The SC is a current or voltage source PWM converter and a signal processing system, and there are several configurations and control strategies that can be used. Finally, hybrid compensators are composed of both a RHS and a SC. The goal of the hybrid compensator is to improve the performance of a RHS by placing a converter in series with the reactive elements. The benefit is that a much lower power rating is required for the converter than for a SC alone.

15.4.1 Harmonic cancellation

Harmonic cancellation by phase shift input transformers is the most commonly employed method of meeting harmonic distortion limits in shipboard systems.

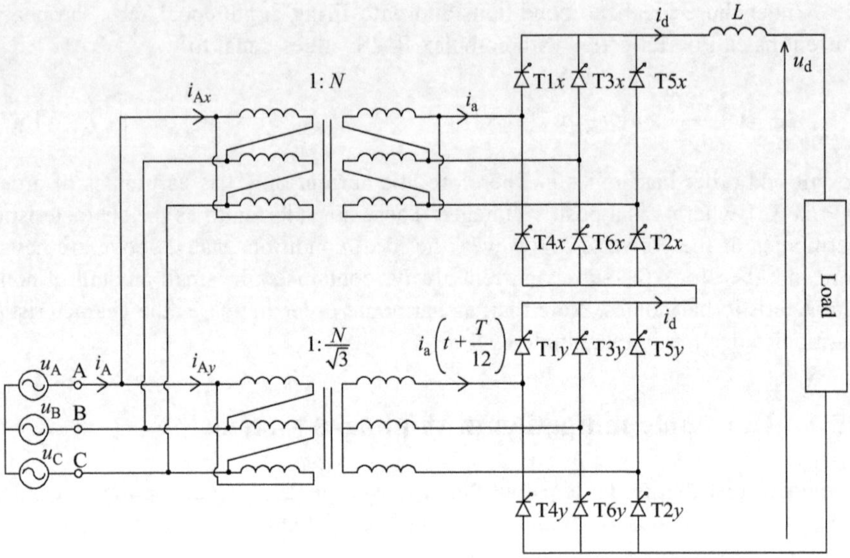

Figure 15.18 Topology of a 12-pulse thyristor rectifier

A 12-pulse thyristor rectifier topology is shown in Figure 15.18. It is comprised of two phase-shifted six-pulse circuits with supply current equal to

$$i_A(t) = i_a(t) + \frac{1}{\sqrt{3}}\left[i_a\left(t + \frac{T}{12}\right) - i_a\left(t + \frac{T}{12} + \frac{T}{3}\right)\right] \tag{15.8}$$

where secondary side phase current $i_a(t)$ is the six-pulse supply waveform shown in Figure 15.17. Then the complex RMS value of the nth-order harmonic is given by

$$\boldsymbol{I}_{An} = \boldsymbol{I}_{an}\left\{1 + \frac{1}{\sqrt{3}}\left[e^{jn\frac{\pi}{6}} + e^{-jn\frac{\pi}{6}}\right]\right\} = \boldsymbol{I}_{an}\left\{1 + \frac{2}{\sqrt{3}}\cos\left(n\frac{\pi}{6}\right)\right\} \tag{15.9}$$

so that the fifth and seventh characteristic harmonics of each six-pulse bridge are cancelled along with other harmonics resulting in a supply current containing characteristic harmonics of order

$$n = 12k \pm 1 \qquad k \in N_1 \tag{15.10}$$

Currents on the primary side of the supply transformers are shown in Figure 15.19. The delta-delta transformer does not modify the shape of the input current waveform while the delta-wye transformer creates a stair-step type waveform. Note that the delta-wye phase shift does not change the characteristic harmonics of that transformer's input current. However, when these two transformer supply currents are summed the result is a much more sinusoidal supply current due to the absence of the fifth- and seventh-order harmonics.

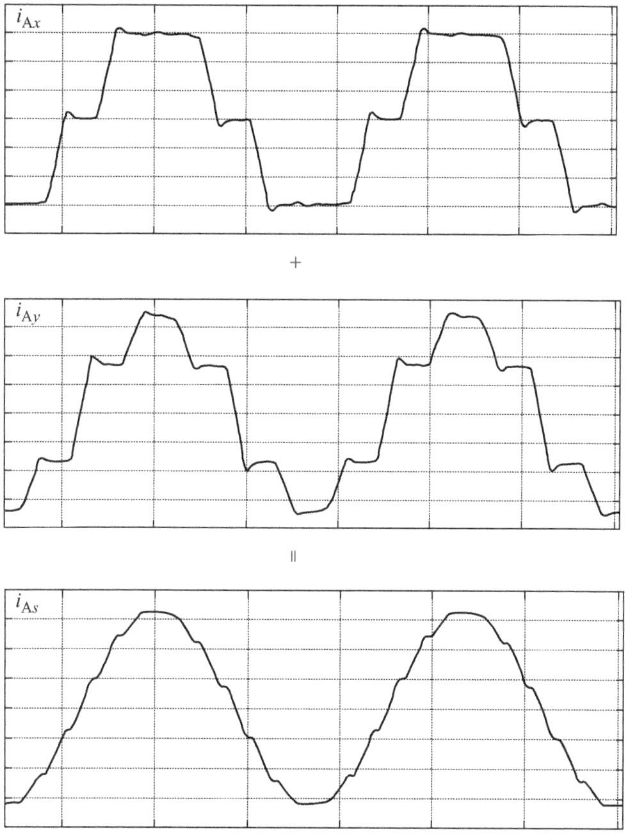

Figure 15.19 Construction of the 12-pulse supply current

It is also possible to further cancel harmonics by increasing the number of transformer secondary windings with appropriate phase shifts. Four transformer windings phase shifted with respect to each other by 15° can be utilized to form a 24-pulse converter. In some ship applications where space is sufficient 24-pulse converters are used. The current harmonic distortion for the three-phase shift cancellation type converters commonly used for shipboard drives are provided in Table 15.2. The table shows that the six-pulse configuration requires additional harmonic mitigation when used for ship propulsion. In some cases, the 12-pulse configuration may also require additional filtering if sensitive loads are present. A case study demonstrating this is provided at the end of the section.

15.4.2 Reactive harmonic suppressors

Conventional RHFs belong to the class of RHSs and are the devices most frequently installed in shipboard distribution systems for reducing distortion caused by propulsion drives. They are reactive devices built of resonant branches, connected in parallel to the drive. Each branch is tuned to a specific harmonic frequency;

Table 15.2 Distortion for each harmonic and THD by pulse number

Characteristic harmonic order	$\frac{I_n}{I_1} \times 100$		
	6-pulse	**12-pulse**	**24-pulse**
$n = 5$	20%	–	–
$n = 7$	14.3%	–	–
$n = 11$	9.1%	9.1%	
$n = 13$	7.5%	7.5%	
$n = 17$	5.8%	–	–
$n = 19$	5.2%	–	–
$n = 23$	4.3%	4.3%	4.3%
$n = 25$	4%	4%	4%
Minimum THD_I	29%	13.2%	5.8%

therefore, it is a notch filter and provides a low impedance path for the load-generated current harmonics for which it is tuned. The most common approach to RHF design is based mainly on [4]. The parameters of individual branches of the filter are calculated based on the chosen value of the reactive power generated by the branch and the chosen resonant frequency of the branch. This frequency, to distinguish it from the frequency of the filter resonance with the distribution system, will be referred to as a tuning frequency. Each branch of a RHF has a capacitive impedance at the fundamental frequency. If a filter has K branches then the reactive power of one branch, denoted Q_{1k}, can be expressed as

$$Q_{1k} = d_k Q_1. \tag{15.11}$$

The coefficient d_k is the *reactive power allocation coefficient*. It has a value between 0 and 1 corresponding to the percentage of reactive power of the entire filter, and it may be chosen at the designer's discretion. The total reactive power generated by all of the filter branches is equal to

$$Q_{\text{tot}} = \sum_{k=1}^{K} d_k Q_1 = Q_1 \sum_{k=1}^{K} d_k. \tag{15.12}$$

Since the reactive power provided by a single branch satisfies (15.11) it is equal to

$$d_k Q_1 = -B_{k1} U^2 \tag{15.13}$$

where B_{k1} is the susceptance of that branch for the fundamental frequency. For a LC branch that has a high quality factor, resistance in the branch can be neglected and the branch susceptance can be approximated as

$$B_{k1} = \mathrm{Im}\left\{\frac{1}{j\omega_1 L_k + \frac{1}{j\omega_1 C_k}}\right\} = \frac{\omega_1 C_k}{1 - \omega_1^2 L_k C_k} \tag{15.14}$$

If the branch is tuned to the frequency $\varsigma\omega_1$ in order to provide a low impedance path for a harmonic of order n, then

$$L_k C_k = \frac{1}{\zeta_k^2 \omega_1^2} \tag{15.15}$$

Consequently, the capacitance and inductance of the branch are equal to

$$C_k = \frac{d_k Q_1 \left(1 - \frac{1}{\zeta_k^2}\right)}{\omega_1 U^2}, \quad L_k = \frac{U^2}{d_k Q_1 \omega_1 (\zeta_k^2 - 1)}. \tag{15.16}$$

Although the process of obtaining the filter parameters is straightforward, the branch tuning frequencies as well as the allocation of the reactive power of the filter among the branches must first be decided. The tuning frequency of each filter branch as well as the number of branches is determined by the harmonic components in the load-generated current that have a significant value. However, observing the impedance magnitude as seen from the load, as shown in Figure 15.20, the addition of a shunt filter branch creates a resonance at a frequency below the tuned frequency of that branch. This is observed as the band of high impedance seen in the plot at a frequency below the branch tuning frequency. The tuning frequency is the point of very low impedance, which located slightly below the fifth-order harmonic in this example.

Changes in the filter parameters due to aging and temperature can cause the tuned frequency and frequency of the resonance to shift. Therefore, filters are often tuned to frequencies slightly lower than the desired harmonic frequency in order to ensure that the resonance does not coincide with a harmonic frequency. This is commonly referred to as de-tuning the filter. A filter might also be de-tuned in order to limit the amount of current carried by the filter branch.

The performance of a RHF is the resultant of the frequency properties of the filter and the system and the harmonic spectrum of the voltage and current. In order to analyze the interaction between the filter and system, a simplified model of a

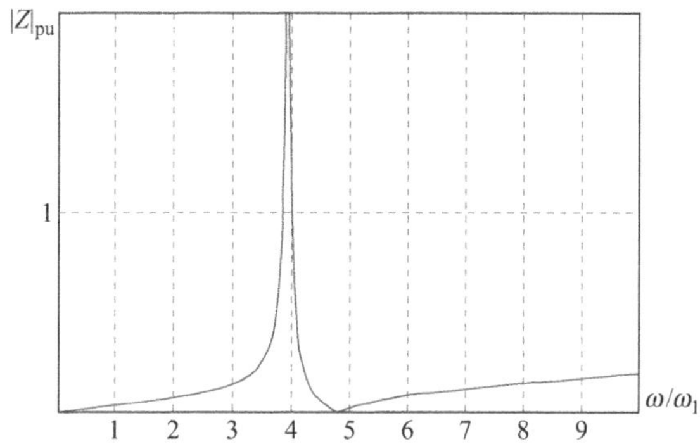

Figure 15.20 Impedance seen from the load

system with an inductive reactance in the supply impedance is shown in Figure 15.21.

A plot of reactance $X_a(\omega)$ is shown in Figure 15.22 along with the negative of the supply system reactance $X_s(\omega)$ for the system shown in Figure 15.21 with a two-branch resonant filter ($K = 2$). For simplicity in this illustration, the reactance, X_s, of the system inductance is assumed to be a linear function of frequency. From this plot, it can be seen that the reactance $X_a(\omega)$ is capacitive in a frequency band below each tuning frequency.

A series resonance occurs at frequency $\omega = \omega_r$ when

$$X_s(\omega) = -X_a(\omega). \tag{15.17}$$

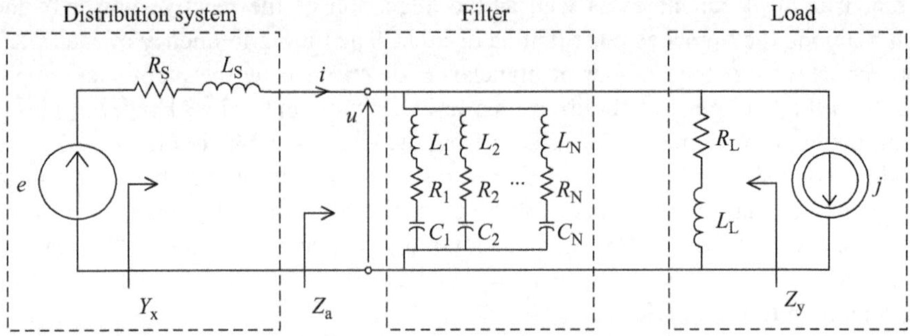

Figure 15.21 Equivalent circuit of a system with a resonant harmonic filter

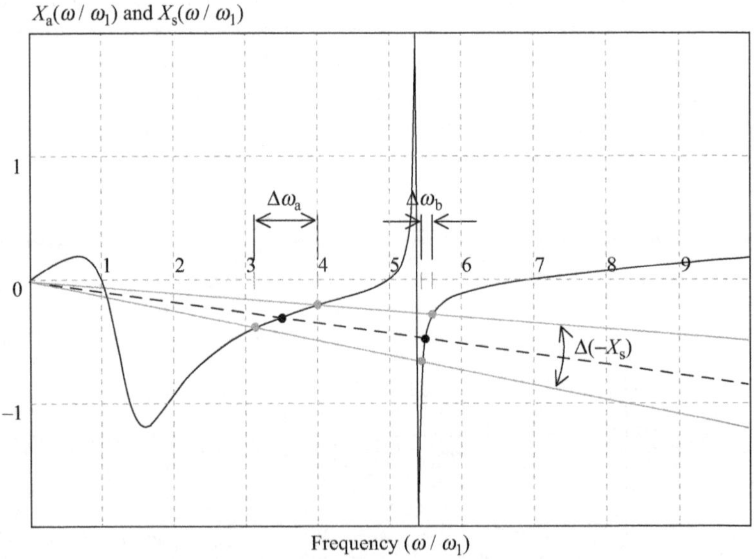

Figure 15.22 Plot of reactance X_a and $-X_s$

Thus, as shown in Figure 15.22, the resonant frequencies are at the points ω_a, ω_b where the plot of minus $X_s(\omega)$ crosses the plot of $X_a(\omega)$. The system inductance shifts the frequency of the zeros of $X_x(\omega)$ with respect to $X_a(\omega)$ to lower frequencies. The transmittance $Y_x(j\omega)$ is the ratio of the spectra of the supply current to the distribution voltage and is simply the admittance as seen from the source of the internal voltage e. Therefore, the ratio of the complex RMS values of the supply current harmonics and the distribution voltage harmonics increases for harmonic frequencies approaching the resonant frequency. In such a case, an increase of current harmonic distortion due to distribution voltage harmonics occurs in the system. The per-unit admittance for most harmonics is much higher than the value of the admittance at the fundamental. Therefore, the performance of tuned branch filters is very sensitive to harmonic distortion in the distribution voltage. Even low levels of distortion of this voltage may cause severe distortion of the supply current and decrease the performance of the filter or even exceed its current rating. Care must be taken to locate the resonances between harmonic frequencies and to consider the range of supply impedance variation, $\Delta X_s(\omega)$, for all operating configurations of the ship distribution system. Note that this is why tuned branch filters cannot be used in conjunction with cycloconverter drives. Cycloconverters produce a broadband of frequencies that shift according to the speed of operation so locating resonances between harmonic frequencies does not avoid amplification of cycloconverter-generated distortion by the filter resonances.

15.4.3 Active filters

The simplified structure of a shunt active filter (SC) comprised of a voltage source converter and its control system is shown in Figure 15.23.

The objective is to adjust the content of current on the supply side of the active filter so that

$$i' = i - j \tag{15.18}$$

It is the reference signal generator that determines the components of the current to be compensated at the cross section of the system where the compensator is connected. The power electronic converter in conjunction with all of the control functions below the reference signal generator simply function as a controlled current source that follows those reference signals. There are numerous strategies for generating reference signals. They separate various components of the voltage or current using either time or frequency domain approaches [5–7]. Additionally hybrid time–frequency methods have been proposed [8]. Thus, component extraction methods can be classified as listed in Table 15.3.

All of the time-domain methods in Table 15.3 are either variants of *p-q* or closely related to it. Several frequency domain methods [9,10] are based directly on Currents' Physical Components (CPC) theory [11]. Whether a reference signal generation method is based on a time-domain or frequency-domain approach, the basic elements are the same. Power or current components present at a measurement point are determined followed by extraction of selected subsets of those components. Finally, a compensation command signal is constructed from the selected subset.

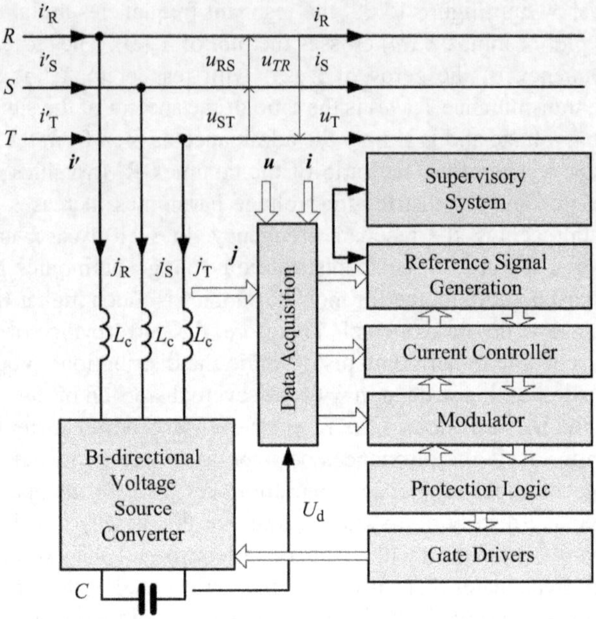

Figure 15.23 Simplified functional diagram of a shunt switching compensator

Table 15.3 Categories of reference signal generators

Category	Component extraction methods
Time-domain methods	1. Instantaneous power *p-q* theory and its variants 2. Synchronous *d-q* frame at the fundamental 3. Synchronous *d-q* frame at individual harmonics 4. Generalized integrators
Frequency-domain methods	1. Recursive discrete Fourier transform (RDFT) 2. Discrete Fourier transform 3. Discrete Kalman filter
Hybrid time–frequency methods	1. Combined *p-q* and RDFT 2. Conservative power theory

Typical reference signal generator structures for shunt compensation in three-wire systems for both *p-q* and *d-q* are shown in Figure 15.24(a) and (b), respectively. Both utilize filters to extract the desired current components. Of course, if reactive power compensation is required the filters for the *q* component will not be present. Also shown is the DC bus voltage controller that injects an appropriate offset into the term associated with the active power at the fundamental as required by the power losses in the SC. In the case of the *p-q* reference generator, the references are converted back into currents by the inverse Clarke transform

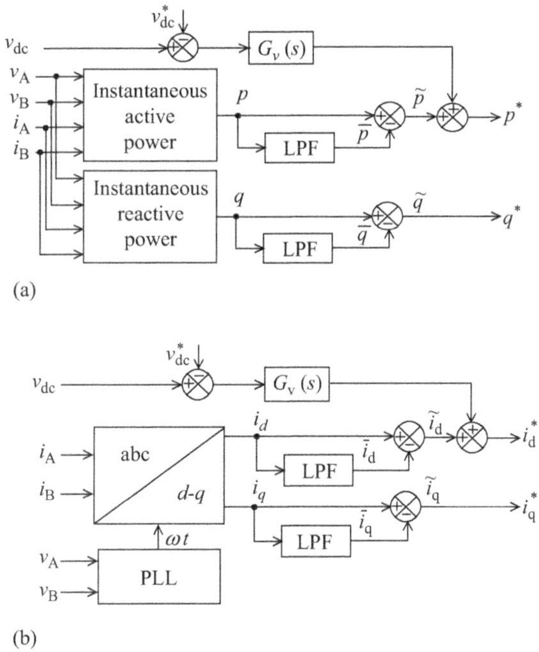

Figure 15.24 (a) General reference generator structure for p-q methods and (b) general reference generator structure for d-q methods

prior to sending to the current control layer where the reference signal alpha components are

$$i_\alpha^* = \frac{u_\alpha}{u_\alpha^2 + u_\beta^2} p^* + \frac{u_\beta}{u_\alpha^2 + u_\beta^2} q^* \tag{15.19}$$

and the reference signal beta components are

$$i_\beta^* = \frac{u_\beta}{u_\alpha^2 + u_\beta^2} p^* - \frac{u_\alpha}{u_\alpha^2 + u_\beta^2} q^* \tag{15.20}$$

For a balanced three-phase system with sinusoidal supply voltage, application of the *p-q* method results in a p term that is associated with active power and a q term that is associated with reactive power. Similarly, for the *d-q* method the d current component is associated with active current and the q current component with reactive current.

Harmonic distortion in the current similarly creates an oscillating component in both p and q or i_d and i_q terms. Thus, distortion will be present in the compensated currents in any case that those pairs are not equal after extraction of the reference signal. These are called "hidden currents" in [12]. Care must be taken when designing filters used in the reference signal generators so that mismatch of the hidden current pairs does not occur. Also, distorted voltages will result in distorted compensating currents.

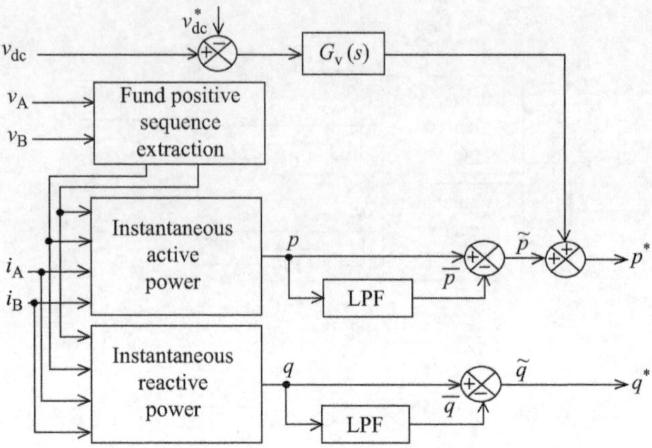

Figure 15.25 General reference generation structure for p-q method for non-sinusoidal and asymmetrical supply

To avoid effects of non-ideal supply voltage, the *d-q* method can be used. However, the impact of non-ideal voltage on the phase-locked loop (PLL) required to synchronize the frame with the fundamental or with a particular harmonic must then be considered. For shipboard power systems in which the voltage may be non-ideal, the positive sequence fundamental of the voltage should be separated out for any of the reference signal generation methods [8]. An example of this for the *p-q* method is shown in Figure 15.25.

15.4.4 Case study

An example of a ship distribution system topology like that shown in Figure 15.2 for an icebreaker is shown in Figure 15.26. The propulsion drive is a 12-pulse cycloconverter comprised of two cross-fed six-pulse cycloconverters each supplying one winding of a doubly wound synchronous motor. The main bus is supplied at 6,600 V and 60 Hz from up to four diesel engine driven generators. Note that in this particular case one of the 450 V ship service buses is designated for sensitive equipment and supplied by motor generator sets. The motor generator sets provide both voltage level adjustment and mechanical isolation of harmonics present in the main bus. This is because although the cycloconverter can be configured to cancel the fifth and seventh harmonics using the same method shown in Figure 15.19, it also modulates the amplitude of the supply current waveforms creating non-periodic effects. The resulting current spectrum is spread out between harmonic frequencies and shifts over time with the converters' operating output frequency. Therefore, resonant type filters cannot be employed since the filter resonances will excite those "interharmonics" as illustrated in Section 15.4.2. The only other option would be an active filter which is not utilized in this particular application since there is space for the motor generator sets supplying the sensitive load bus. Measurements taken of the 6,600 V main bus voltage and of primary side drive

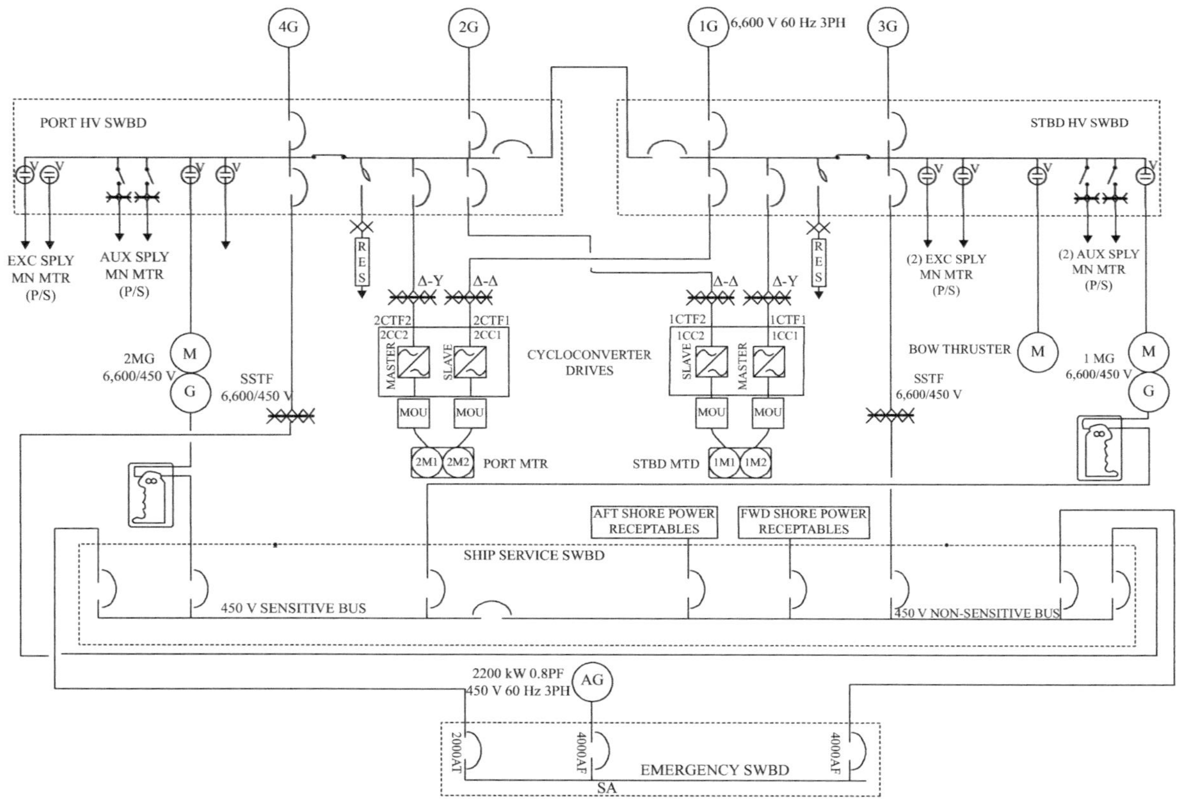

Figure 15.26 One-line diagram of a shipboard power system with 12-pulse cycloconverter drives

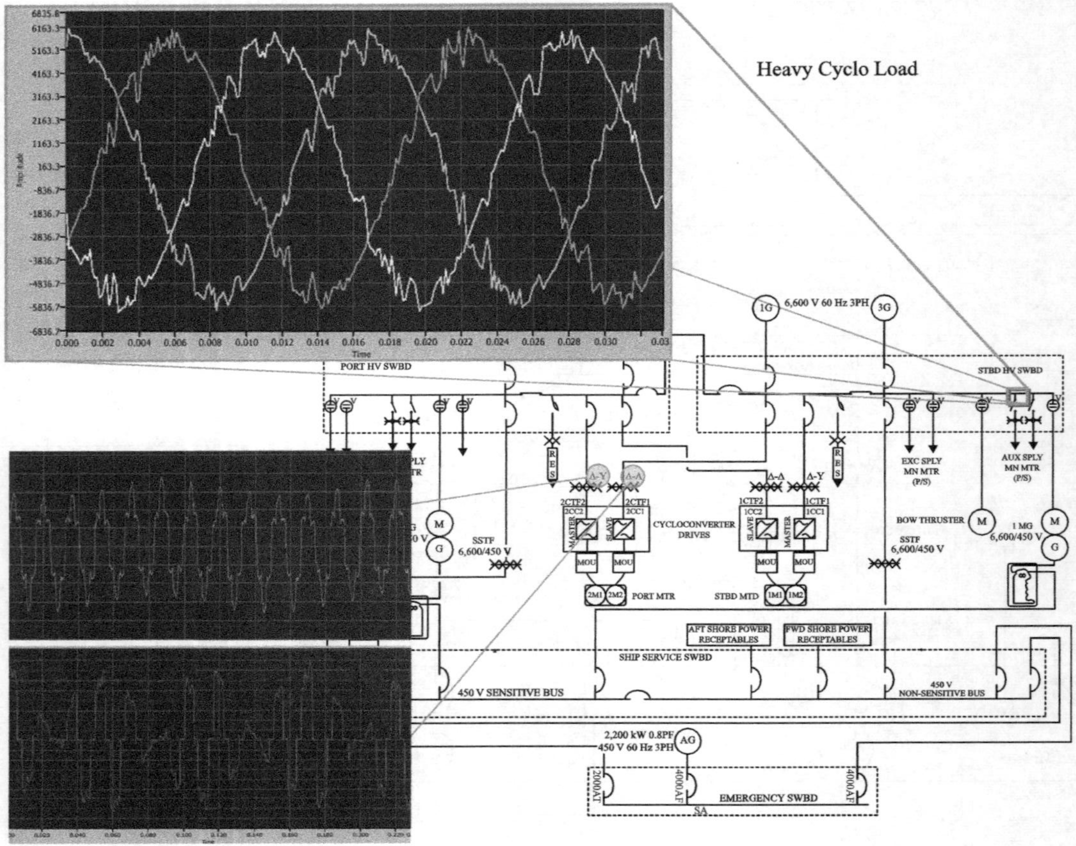

Figure 15.27 Main bus voltage measurements and cycloconverter current measurements during heavy drive loading conditions

transformer currents during heavy propulsion drive load conditions are shown in Figure 15.27. The main bus voltage is heavily distorted despite the 12-pulse configuration of the cycloconverter.

15.5 Frequency variation and converter control

Most digital control architectures for power system applications require synchronization with the distribution system voltage. Therefore, a PLL is generally among the digital control subsystems that make up the overall converter control system. The PLL analyzes the bus voltage and provides power system information for some of the other control subsystems. Thus, the performance of the PLL has a broad impact on the overall converter system performance.

Small-scale power systems, such as naval systems, pose a challenging environment for PLL design due to voltage distortion and variation in the fundamental frequency that is large as compared to large terrestrial systems. The choice of the algorithms should be determined by the actual environment in which the PLL is applied.

15.6 Concepts for future shipboard power systems

The benefits of more efficient and dense energy management systems for ships are driving efforts in standardization of power electronic converter components and interfaces as well as investigation into new distribution system topologies. While there are benefits for any type of ship, the evolution of shipboard power systems is mainly driven by navy ship requirements. Navy ships have an increasing demand to supply high power loads while the space available to install the power system is not increasing and may even decrease in some cases. This leads to highly power electronic converter based microgrid systems.

15.6.1 Power electronics building block

It has long been recognized that innovative and efficient use of electric power requires significant reduction in cost, weight, size, and losses of power electronics. Because of limited space, this need especially applies to ships and other marine applications, and hence reduction of size, and losses, is typically of greater importance for offshore applications than onshore applications.

Power electronics is akin to microelectronics, part of silicon science. Trends in microelectronic applications, i.e. computers, servers, etc., have led to their assembly from functional building blocks with high-volume production and reductions in cost to increase in performance. In contrast, suppliers of high power electronics applications, both civilian and military, designed their application starting from the semiconductor devices all the way to converters control and application. Such end-to-end custom design kept costs high. Following the microelectronics paradigm, power electronics needed a revolution in modularity and commonality. The power electronics building block (PEBB) concept [13] has moved converter power

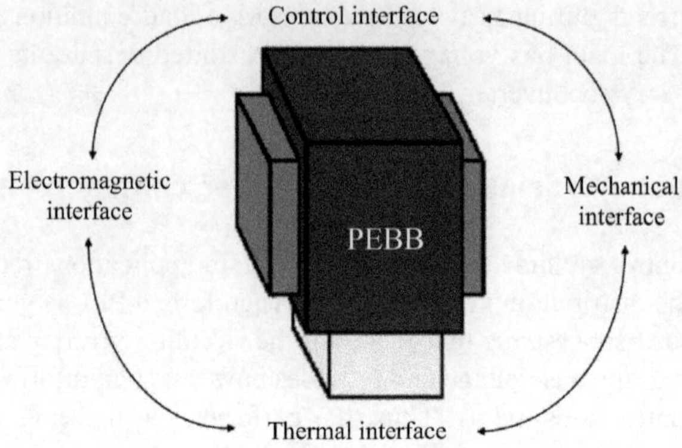

Figure 15.28 Illustration of the power electronic building block (courtesy of Ericsen Innovations)

hardware in that direction. A graphical depiction as a standardized module is shown in Figure 15.28 and a definition of the concept is given in [14] as

> Power Electronics Building Block (PEBB) is a broad concept that incorporates the progressive integration of power devices, gate drives, and other components into building blocks, with clearly defined functionality that provides interface capabilities able to serve multiple applications. This building block approach results in reduced cost, losses, weight, size, and engineering effort for the application and maintenance of power electronics systems. Based on the functional specifications of PEBB and the performance requirements of the intended applications, the PEBB designer addresses the details of device stresses, stray inductances, switching speed, losses, thermal management, protection, measurements of required variables, control interfaces, and potential integration issues at all levels.

Thus, the goal of PEBB-based power systems is to make it possible to integrate identical basic building blocks as major components of the final application. Interconnecting PEBB units creates specific system configurations with various power level ratings. Together, the aggregate of these basic building blocks form a complete power system, including mechanical, cooling, power, auxiliary power, and controls system interconnections. Custom-designed systems and installations, without the benefit of standard interfaces or PEBB concepts, cannot be upgraded without incurring significant design, development, and re-commissioning costs.

15.6.2 Medium voltage DC integrated power system

The medium voltage DC (MVDC) integrated power system (IPS) concept envisions a DC distribution system with power electronic converters between each energy source and the distribution system as well as between the distribution system and all main load centers. This system architecture has been proposed for systems with

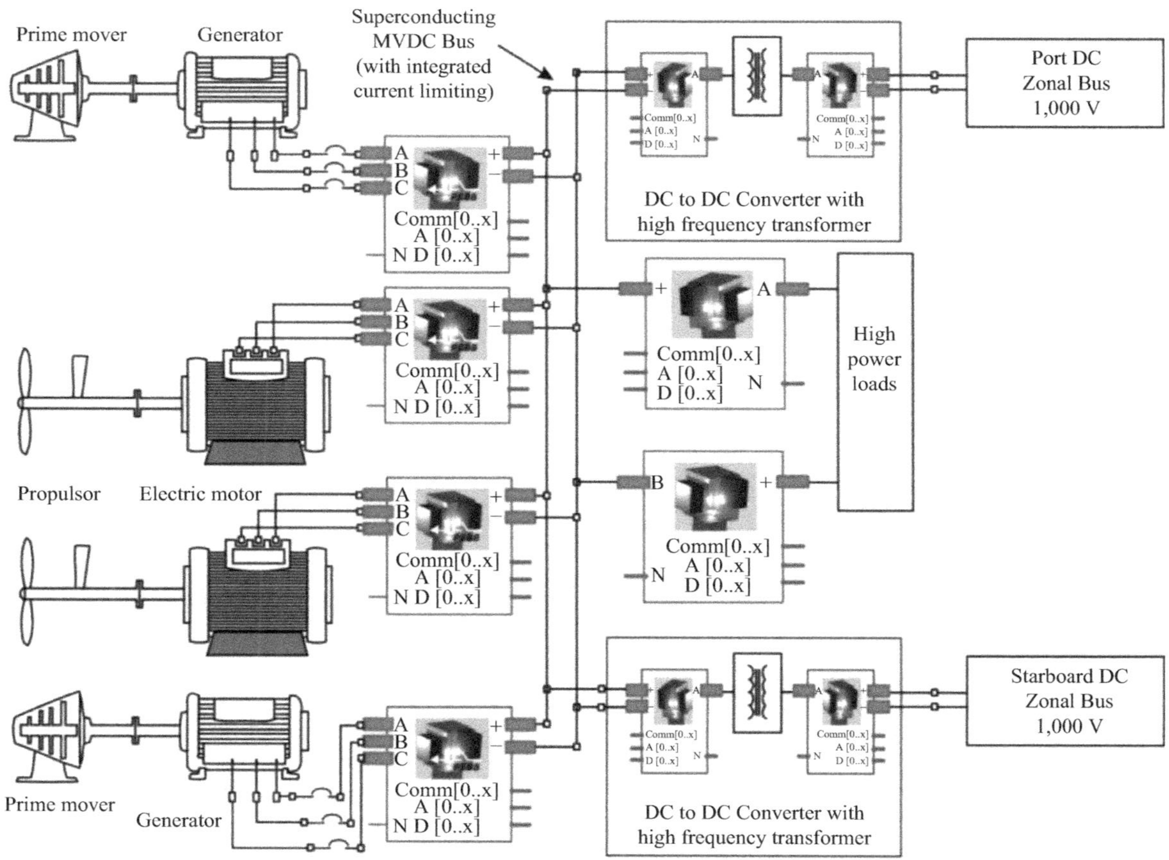

Figure 15.29 Navy MVDC IPS concept (courtesy of Ericsen Innovations)

individual high power loads that are a substantial percentage of the installed generating capacity. Many Naval ships fall into that category. The concept is illustrated in Figure 15.29 with PEBB-based converters at each main interface between supplies and loads. Instead of having power electronics converter equipment with an internal DC bus in each self-contained converter, such as the propulsion drives, the converters are "split" so that a common main DC distribution system is the DC bus between supply and load converter stages. The MVDC IPS architecture reduces power system space requirements by eliminating extra intermediate power conversion stages as well as large low-frequency AC transformers. Although some DC/DC converters may contain high-frequency transformers, this still saves a great deal of size and weight compared to 50 or 60 Hz units. Finally, power electronic converters located at all major interface points enable movement of energy in the system in a very flexible manner. This permits fast real-time scheduling of energy needed to mitigate the small amount of installed generation relative to system loads.

Acknowledgements

I would like to express my gratitude to the Office of Naval Research for support over the past decade of both my research as well as the many others at the University of South Carolina and collaborating universities. That base of support has enabled my research in the area of shipboard power systems resulting in a fascinating journey that also made possible this chapter on the subject.

References

[1] IEEE STD 1662™-2008, IEEE Guide for the Design and Application of Power Electronics in Electrical Power Systems on Ships.
[2] IEEE STD 45-2002™, IEEE Recommended Practice for Electric Installations on Shipboard.
[3] IEEE STD 519™-1992, IEEE Recommended Practices and Requirements for Harmonic Control in Electrical Power Systems.
[4] Steeper, D. E. and Stradford, R. P. "Reactive compensation and harmonic suppression for industrial power systems using thyristor converters," IEEE Trans. IA, Vol. 12, No. 3, May/June 1976, pp. 232–254.
[5] Asiminoaei, L., Blaabjerg, F., and Hansen, S. "Evaluation of harmonic detection methods for active power filter applications," Twentieth Annual IEEE Applied Power Electronics Conference and Exposition, 2005, Vol. 1, pp. 635–641.
[6] Ortega, J. M. M., Esteve, M. P., Burgos Payan, M., Gomez Exposito, A., Garcia Franquelo, L. "Reference current computation methods for active power filters: accuracy assessment in the frequency domain," IEEE Trans. Power Electronics, Vol. 20, No. 2, 2005, pp. 446–456.
[7] Rechka, S., Ngandui, E., Xu, J., and Sicard, P. "A comparative study of harmonic detection algorithms for active filters and hybrid active filters,"

Proceedings of IEEE 33rd Power Electronics Specialist Conference, 23–27 June 2002, Vol. 1, pp. 357–363.

[8] Ginn, H. L. "Comparison of applicability of power theories to switching compensator control," Przeglad Elecktrotechniczny (Electrical Review), Vol. 89, June 2013, pp. 1–10.

[9] Ginn, H. L. and Chen, G. "Flexible active compensator control for variable compensation objectives," IEEE Trans. Power Electronics, Vol. 23, No. 6, Nov. 2008, pp. 2931–2941.

[10] Ginn, H. L. "Control method for grid-connected converters in systems with non-ideal supply voltage," Proceedings of Applied Measurements for Power Systems (AMPS), 2011 IEEE International Workshop on, 2011, pp. 96–101.

[11] Czarnecki, L. S. "Orthogonal decomposition of the current in a three-phase nonlinear asymmetrical circuit with non-sinusoidal voltage," IEEE Trans. Instrum. Meas., Vol. IM-37, 1988, pp. 30–34.

[12] Akagi, H., Watanabe, E. H., and Aredes, M. *Instantaneous Power Theory and Applications to Power Conditioning*. Piscataway, NJ: IEEE Press/ Wiley-Interscience, 2007, ISBN: 978-0-470-10761-4.

[13] Ericsen, T. and Tucker, A. "Power electronics building blocks and potential power modulator applications," IEEE Conference Record of the 23rd International Power Modulator Symposium, New York, 1998, pp. 12–15.

[14] IEEE Power Engineering Society, "Power Electronics Building Block (PEBB) Concepts," IEEE Publication 04TP170, 2004.

Chapter 16
Converters in power grid

Giri Venkataramanan[a] *and Fernando Mancilla-David*[b]

16.1 Introduction

The electrical power system forms the essential backbone of the energy system that enables modern livelihood and supports a comfortable lifestyle for people who have access to it. It is commonly segmented into generation, transmission, distribution, and utilization systems. While electrical power converters have a role to play across all of these segments, their application in generation and utilization systems is rather specific in nature, as has been discussed in various other chapters in this book. The focus of discussion in this chapter is on their application in electrical transmission and distribution system, which are often referred together as the power grid. Since the power grid is largely operated as a three-phase ac system, the discussion in this chapter is mainly focused on three-phase power converters. An outline of different power converters that are used the power grid classified into different categories is presented in Section 16.2. In the discussion, the focus is mainly placed on the circuit configuration of the constituent components, and not on their control, modulation, switching strategy, or design aspects. Such a discussion is beyond the scope of this volume, and may be found in specialized texts, monographs, and references. Various specific applications of these converters in transmission and distribution systems are illustrated in Section 16.3. A brief description of application is restricted to the functional details. Again, detailed discussion of the operation and capabilities of the converters may be found in reference documents. The summary in the concluding section includes a brief discussion of the state of the technology and emerging trends, followed by a list of references.

16.2 Power converter topologies

The feature of power converters used in the power grid that distinguishes them from industrial, residential, and commercial applications, is their power level,

[a]University of Wisconsin-Madison, Madison, WI, USA
[b]University of Colorado Denver, Denver, CO, USA

which may be reached up to thousands of megawatts. This large amount of power throughput places a large demand on the switching power semiconductor devices in terms of voltage and current ratings. This leads to certain topological considerations in the power converters that are often superseded at lower power applications. Various converters that fall under the common ac–dc and ac–ac categories are discussed briefly in this section, followed by a more detailed discussion of two new families of power converters that are emerging to find a unique place in power grid applications.

16.2.1 AC–DC converters

AC–DC converters often form a building block to realize an intermediate dc stage in several power grid applications. They are typically capable of bidirectional power flow, and may form a current stiff (inductive) dc link or voltage stiff (capacitive) dc link, as described further in the following subsections.

16.2.1.1 Current source converters

Current source converter (CSC) technology used in power grids is relatively mature today [1]. The circuit consists of a three-phase bridge network with unidirectional current carrying and bidirectional voltage blocking switching devices connected between each dc terminal and the ac terminal as illustrated in Figure 16.1. The switches used in the converter are silicon control rectifiers (SCRs or thyristors). Since SCRs do not have inherent turn-off capability, these converters employ the voltage reversals of the ac line to commute. Hence, they are termed line-commutated converters. The stiff current requirement on the dc side may be ensured using an explicit inductor, or in some cases part of the dc side circuit inductance. Although the figure indicates a single SCR switching device that interconnects each ac terminal with each dc terminal, a large number of series connected devices are used to meet voltage blocking requirements and also allow the provision of redundancy against device failures in the short circuit mode. Details of such interconnections and snubber circuits to ensure voltage distribution among series connected devices under

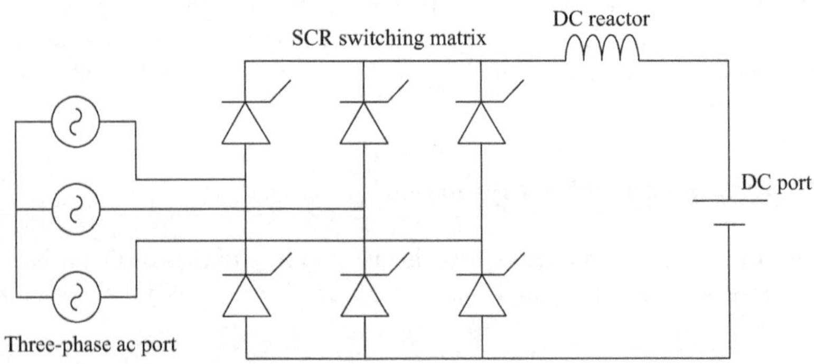

Figure 16.1 Schematic diagram of a current source converter with two dc levels

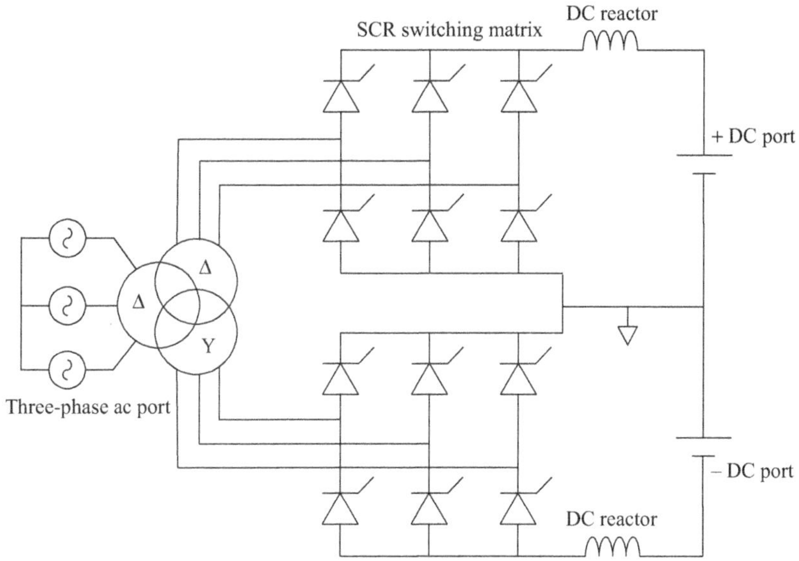

Figure 16.2 Schematic diagram of two current source converters series connected to form a bipolar or three-level dc system

the rubric of "valve design", the application of ac filter devices and dc filter devices are not shown in the figure. Such design details along with a detailed discussion of the operation and control of the converters may be found elsewhere. In most cases, it is common to use two power converters along with an appropriately designed wye-delta transformer to get phase multiplication on the ac side, and series connection on the dc side as shown in Figure 16.2.

While the technology of the line-commutated CSCs is mature, the lack of turn-off controllability of the conventional thyristor results in (i) poor power factor and considerable waveform distortion; (ii) need for a relatively stiff ac voltage supply; and (iii) inability to provide independent control of the active and reactive powers, and often superseded by converters that use gate-turn-off devices [2]. Although it is possible to realize CSCs with gate-turn-off devices, voltage source converters (VSCs) using gate-turn-off devices discussed in the following subsection have enjoyed a more dominant development due to trends in power semiconductor availability and various other application engineering issues [3].

16.2.1.2 Voltage source converters

The technology of VSC that employs a stiff voltage on the dc side realized using a capacitor bank is quite mature, and enjoys a broad and growing installation base for applications in motor drives, uninterruptable power systems, and wind and solar power systems. Figure 16.3 illustrates the power circuit of a two-level dc–three-phase/ac VSC [4]. While the bridge circuit topology of the VSC is similar to that of the CSC, the switching elements in the VSC have unidirectional voltage

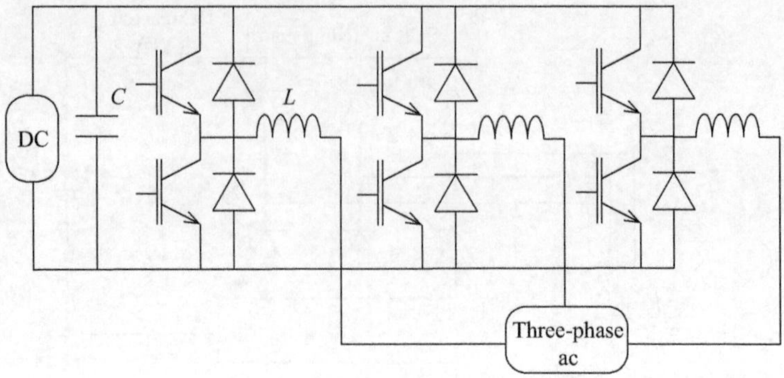

Figure 16.3 Schematic diagram of a VSC with two dc levels

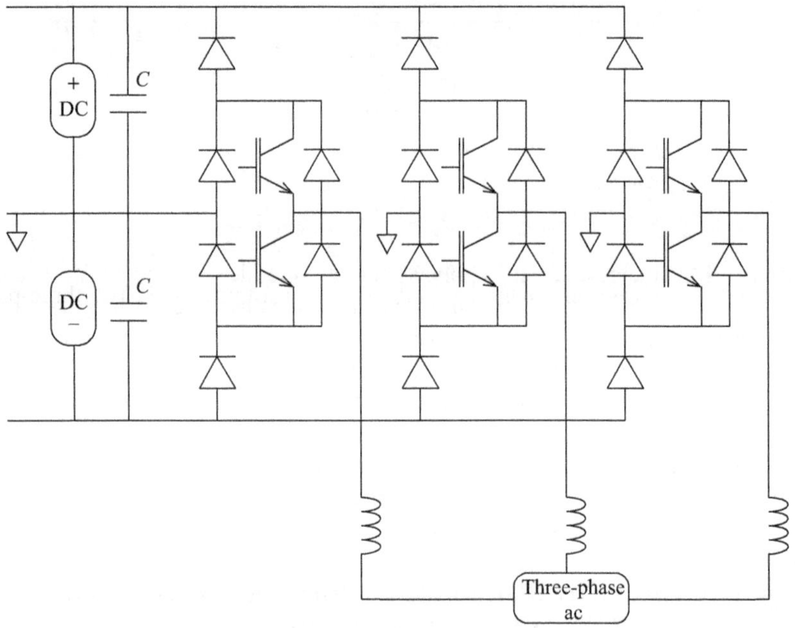

Figure 16.4 Schematic diagram of a VSC with three dc levels

blocking and bidirectional current carrying capability. With the advent of high-power/high-voltage switching devices such as the integrated gate-commutated thyristor (IGCT) and insulated-gate bipolar transistor (IGBT), the realization of VSC for application in the power grid has become practical.

Due to considerations of waveform quality, symmetry and semiconductor switch utilization, often a (diode clamped) three-level converter topology (illustrated in Figure 16.4), that is also called as neutral-point clamped converter is used in power grid applications [5].

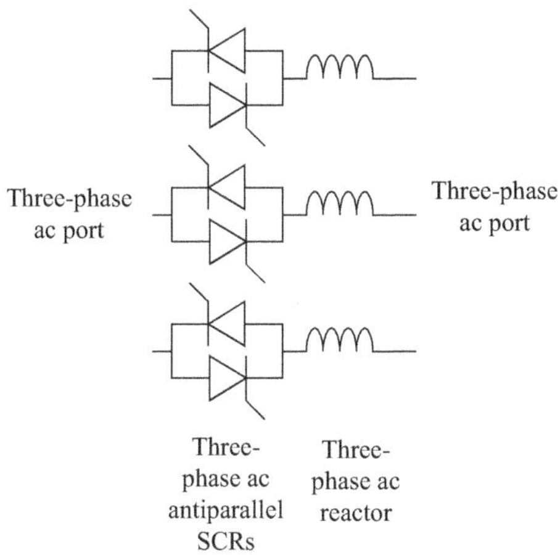

Figure 16.5 Schematic diagram of a thyristor-controlled reactor

16.2.2 AC phase controllers

AC phase-controlled devices correspond to the first generation of solid-state power electronic devices built using SCRs [6,7]. While they are most commonly known for lamp dimming in domestic applications, in power grids, they are applied to realize an electronically regulated inductance, often called the thyristor-controlled reactor (TCR), whose power circuit is illustrated in Figure 16.5. The three-phase TCR consists of three reactors, each of them connected in series with two SCRs that are connected antiparallel, to allow current flow in both directions, as well as voltage standoff in both directions. The firing angle of each SCR, in relation to the zero cross of the line voltage, determines the current flowing through the SCR, and hence provides continuous control of fundamental reactance all the way up to infinity (when the SCRs are not conducting at all), down to the full value of the reactance (when the SCRs are conducting completely). While a variable reactance by itself may not appear to be of such a high value in power grids, they are interconnected appropriately with capacitors to realize various applications, which will be described in Section 16.3.

16.2.3 AC pulse width modulated controllers

The principles of operation of three-phase ac pulse width modulated (PWM) converters have been described in detail [8,9]. Although their realization and applications for ac power flow control can take various forms, herein the focus is on the main features that elucidate their operating features. Since their realizations are not widely known, the operation is developed from primitive principles.

16.2.3.1 Converter circuit

A simplified schematic diagram of an *N*-throw single-pole three-phase ac PWM converter is illustrated in Figure 16.6. The system synthesizes an adjustable

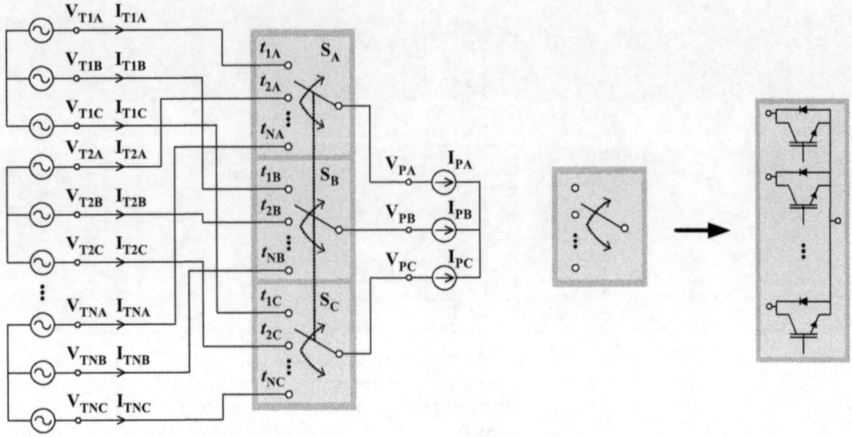

Figure 16.6 Schematic diagram of a three-phase PWM ac converter synthesizing an adjustable pole voltage obtained from N throws

three-phase pole voltage ($V_{P(A–C)}$) by switching among N stiff three-phase voltage sources ($V_{T1(A–C)}, V_{T2(A–C)}, \ldots, V_{TN(A–C)}$). It is important to notice that in ac systems, the designation of the stiff voltage and stiff currents is arbitrary. Stiff voltage sources and stiff current sources can be transformed into either by adding series inductors or shunt capacitors, respectively, in order to fit application considerations [10].

The throws of the switches as well as the sources are assumed ideal for the purpose of the discussion herein, as is common in preliminary functional analysis of switching power converters. Furthermore, the nominal frequency of the three-phase voltages and currents (power frequency) are assumed identical, in the absence of which there could be no net steady-state power transfer among them.

As indicated in Figure 16.6, the three poles are ganged together, i.e., they switch concurrently between the throws to which they are connected. This can be mathematically expressed as $t_{iA} = t_{iB} = t_{iC} = t_i$, for $i = 1, 2, \ldots, N$.

16.2.3.2 Converter model

The transfer properties can be represented using the vector equations consisting of three components for each of the terminal quantities. In this case, the vectors representing the throw voltages, throw currents, pole voltages and pole currents become $\mathbf{V}_{Ti} = [V_{TiA}\ V_{TiB}\ V_{TiC}]^T$, $\mathbf{I}_{Ti} = [I_{TiA}\ I_{TiB}\ I_{TiC}]^T$, $\mathbf{V}_P = [V_{PA}\ V_{PB}\ V_{PC}]^T$, $\mathbf{I}_P = [I_{PA}\ I_{PB}\ I_{PC}]^T$, respectively, for $i = 1, 2 \ldots, N$. Thus, the pole voltage and throw currents can be expressed as

$$\mathbf{V}_P(t) = \sum_{i=1}^{N} H_i(t) \cdot \mathbf{V}_{Si} \tag{16.1}$$

$$\mathbf{I}_{Ti}(t) = H_i(t) \cdot \mathbf{I}_P \quad \text{for } i = 1, 2, \ldots, N \tag{16.2}$$

where

$$H_i(t) = \begin{cases} 1 & \text{if } t_i \text{ is closed} \\ 0 & \text{otherwise} \end{cases}$$

is the switching function of a throw connecting voltage V_{Ti} to the current I_{P}. When the repetition frequency of the switching function (or simply the switching frequency) is large with respect to the power frequency, net power transfer between the voltage ports and the current ports arises from the average value (dc component) of the switching functions. The dc component of the switching functions may be readily represented by the duty ratio of the particular throw. The transfer relationships (16.1) and (16.2) may be simplified as

$$V_{\mathrm{P}}(t) = \sum_{i=1}^{N} \mathrm{d}_i(t) \cdot V_{\mathrm{T}i} \tag{16.3}$$

$$I_{\mathrm{T}i}(t) = d_i(t) \cdot I_{\mathrm{P}} \quad \text{for } i = 1, 2, \ldots, N \tag{16.4}$$

where the duty ratio of the *i*th throw is defined as

$$d_i(\tau) = \frac{1}{T} \int_{\tau}^{\tau+T} H_i(t) \cdot \mathrm{d}t \quad \text{for } i = 1, 2 \ldots, N \tag{16.5}$$

and T is the switching period. As may evident from (16.5),

$$\sum_{i=1}^{N} d_i(t) = 1 \tag{16.6}$$

Since the operating principle of these converters is based on controlling the connectivity between various three-phase ac voltage and/or current vectors by switching among their components *concurrently*, these converters have been termed vector switching converters (VeSCs). The realization of each of the multiple-throw single-pole switch using real semiconductors is shown in Figure 16.6. Each IGBT–diode pair may be replaced by IGCT–diodc or any other device with similar voltage blocking and current conducting capabilities with control in one direction [11].

16.2.3.3 Equivalent circuit

Relationships (16.3) and (16.4) indicate a reciprocal input–output transfer property similar to that of a transformer. Therefore, the fundamental component averaged vector (or single phase) equivalent circuit of the converter system may be represented as shown in Figure 16.7, which describes the input–output reciprocal relationships explicitly as dependent sources. As is evident from Figure 16.7, the net pole voltage, $\mathbf{V}_{\mathrm{R}}$, depends on the value (magnitude and phase) of the throw voltages as well as the corresponding duty ratios. This feature becomes the defining element for application of these converters in power grid as will be described at a later section in this chapter.

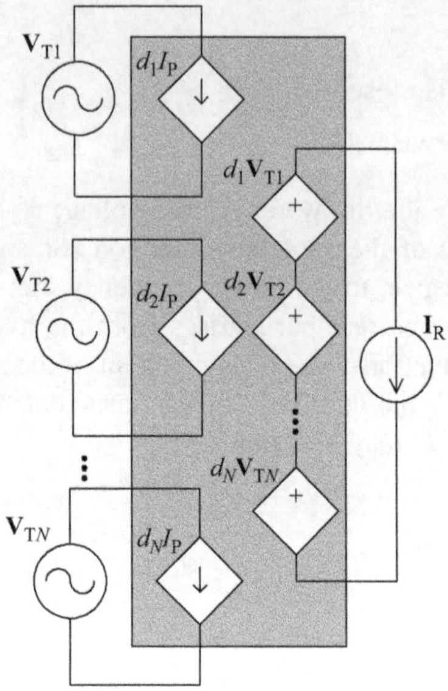

Figure 16.7 PWM ac VeSC equivalent circuit shown using duty ratio controlled coupled current and voltage sources

16.2.4 Modular multilevel converter

The most recent addition to the family of power converters, which has today found a niche in power grid applications, is the modular multilevel converter (MMC) described in this subsection. Several topologies of power converters that use this approach has been presented and demonstrated in recent years. These include three-phase/ac–three-phase/ac matrix converters, single-phase/ac–three-phase/ac dual-bridge converters and dc-three-phase/ac inverters/rectifiers [12–16]. Due to relative novelty of these converters, the topological structure is developed here from the first principles, starting from the description of a particular sub-module (SM) that constitutes the building block of this converter.

16.2.4.1 SM structure

A simplified schematic diagram of a SM in full-bridge configuration is shown in Figure 16.8(a). Bulk energy storage (C_S) in the capacitor is typically in the form of dc, thereby providing an appropriate bias voltage or current for the switching devices that constitute the bridge circuit. The presence of L_B indicates stiffness of the bridge terminal current, which may be provided by an incidental amount of inductance in series with the bridge, with minimal amount of energy storage in it. The voltage and current at the terminals and the dc capacitor along with the constituent switch throws of the SM may be defined as illustrated in Figure 16.8(a).

The realization of the throws of the SM depends on the polarity of terminal currents and the polarity of the bias voltage. Assumption of unipolar bias voltage v_S and bidirectional terminal current i_B leads to a classical H-bridge consisting of

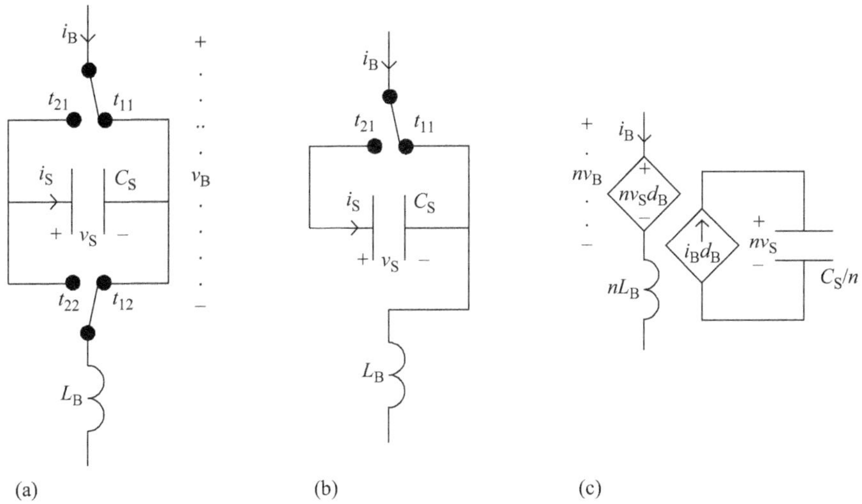

Figure 16.8 Schematic diagram of introduction of sub-modules in MMC as: (a) functional diagram of symmetrical full-bridge configuration, (b) functional diagram of single-ended half-bridge configuration, and (c) averaged model of n series connected sub-modules

gate-turn-off semiconductors such as IGBTs with antiparallel diodes. The full-bridge topology enables bipolar values for v_B and i_B through appropriate control of switch states. On the other hand, a reduced switch topological structure using a single-ended half bridge may be realized as illustrated in Figure 16.8(b), allowing bidirectional values of i_B, and unidirectional values of v_B. The throws constituting the SM may be typically operated under repetitive switching in a duty ratio controlled mode, or generate an ensemble waveform to follow a command. The duty ratio of the throws t_{xy} is denoted as d_{xy}, where x and y represent the throw and pole designations of the switches in the bridge as illustrated. Such a primitive description of the operation of this topology enables the analysis of the power transfer mechanism and convenient application of modeling tools.

16.2.4.2 Averaged model

For sufficiently large switching frequencies and energy storage elements, the terminally stiff electrical variables in switching power converters may be described using their average values. The average transfer properties of the SM may be represented as

$$\begin{bmatrix} v_B \\ i_S \end{bmatrix} = \begin{bmatrix} d_B & 0 \\ 0 & d_B \end{bmatrix} \begin{bmatrix} v_S \\ i_B \end{bmatrix} \tag{16.7}$$

where $d_B = d_{21} - d_{22}$ for the full bridge and $d_B = d_{21}$ for the half bridge. As may be observed from (16.7), the energy storage capacitor current i_S and the SM terminal voltage v_B feature a reciprocity relationship with respect to the SM terminal current i_B and energy storage capacitor voltage v_S, determined by the "bridge duty ratio" d_B, where the bridge duty ratio d_B may take a value between -1 and $+1$ for the full bridge and a value between 0 and $+1$ for the single-ended half bridge.

16.2.4.3 String of SMs

An arbitrary number (n) of SMs either in the full-bridge or in the single-ended half-bridge configuration may be connected in a series string. Typically, these bridges would consist of identical modules for the sake of simplicity and convenience in packaging, design, control, and realization. Since they are connected in a series string, they carry the same current and under most circumstances, their dc bias voltage level will be nominally identical if their respective bridge duty ratios are also identical. Although the duty ratio among these modules may be identical, the switching signals may have phase shifts between them, thereby providing for one of the benefits of multilevel power conversion. Furthermore, imperfections due to parasitic effects may be compensated through second-order perturbations in the control signals. In total, a series connected string of n SMs may be represented as a simplified equivalent circuit as illustrated in Figure 16.8(c).

16.2.4.4 Topology example

Two systems of power sources I and O (and/or loads, e.g. a dc voltage, three-phase ac system, etc.) may be said to be interconnected in a symmetrical bridge configuration if each terminal of System I is connected to each terminal of System O through an identical electrical branch. If each of the interconnection branches itself is a series connected string of SMs, an MMC or bridge-of-bridge (BoB) converter network is formed. BoB converter networks may be configured and operated in a manner to realize particular power conversion functions between different systems of power sources [12]. A dc to single-phase ac power converter using the BoB approach, along with an averaged equivalent circuit is illustrated in Figure 16.9. The dc voltage and the ac voltage at the input and output terminals are defined with respect to their neutral point voltages in Figure 16.9, for the sake of notional convenience. The duty ratio of the different bridges are modulated to follow the relationship

$$D_{Bij} = \frac{V_{Ii} - V_{Oj}}{nV_S} \tag{16.8}$$

where the indices i and j correspond to the location of voltages and the bridge position. Through appropriate duty ratio control of current and voltage using feedback mechanism, the BoB converter may be used to realize various power conversion functions in a versatile manner [15]. More interestingly, the approaches may also be extended to operation and modulation of bridges in staircase or step waveform modulation as opposed to PWM operation, since the power levels in grid applications may preclude PWM operation with high switching frequency.

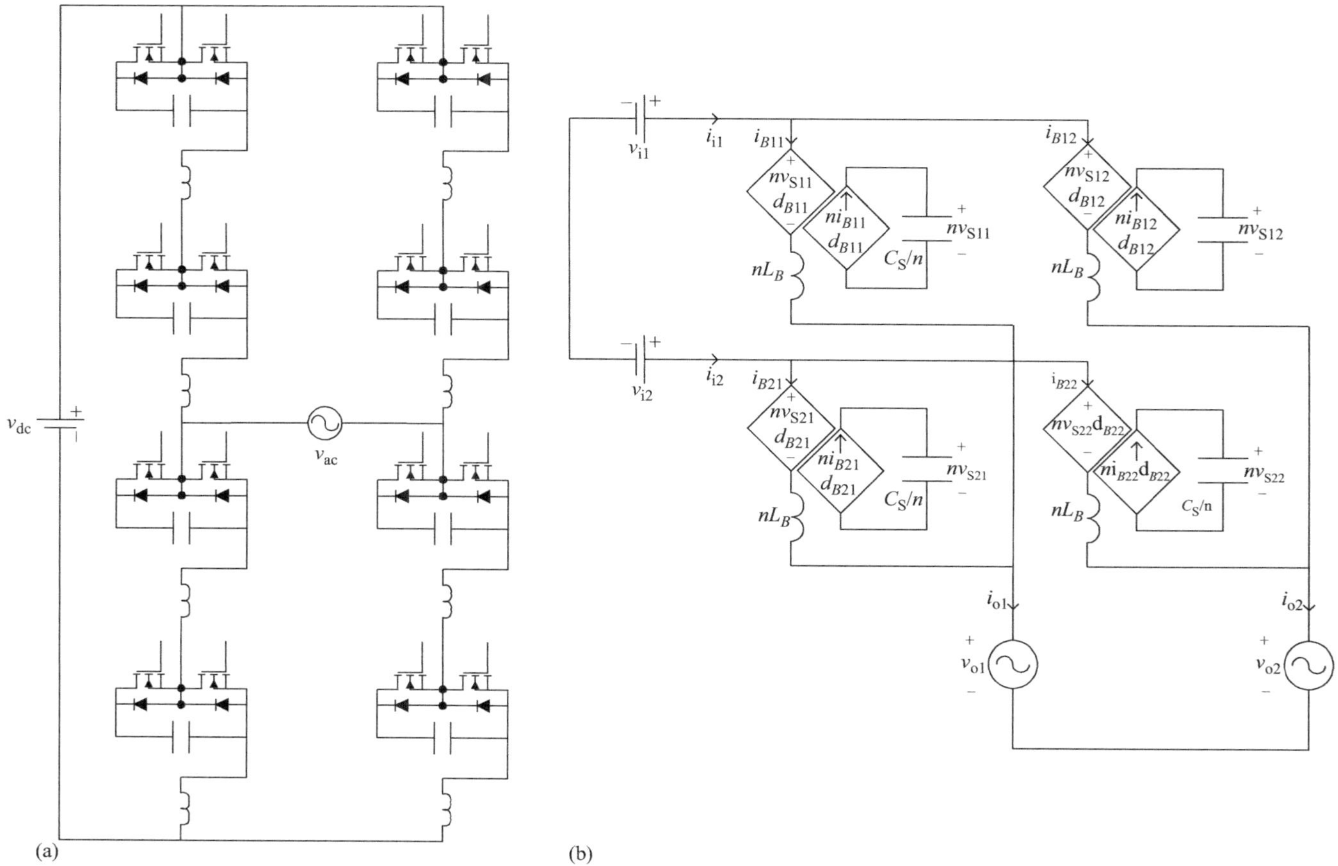

Figure 16.9 (a) Power circuit of a single-phase-dc BoB converter and (b) averaged equivalent circuit

16.3 Application examples of power converters in power grid

16.3.1 Shunt compensation

Shunt compensation is used in power systems to provide voltage support and overcome the impact of lagging power factor loads in transmission systems as well as in distribution systems. While fixed capacitors may be employed for the same purpose, they would not have the required dynamic capacity to react to varying loading conditions. Such applications have become more prevalent with the introduction of wind energy generation systems that have a high degree of variability. Shunt compensation may be use ac phase control, ac–ac power converters, or ac PWM control. Examples of each of these applications are illustrated in this section.

16.3.1.1 Static var compensator

A static var compensator (SVC) is realized using a TCR in parallel with a capacitor as illustrated in Figure 16.10. While the TCR is able to vary the reactance between a fixed value, up to infinity, the parallel connected capacitor provides a negative reactance bias, so that the net reactance of the SVC may be varied between a fixed negative value (X_C) when the SCRs are completely off, and a fixed positive value ($X_L - X_C$), corresponding to the condition when the SCRs are completely conducting. Thus through appropriate choice of C and L, the dynamic range of an SVC may be tailored to meet particular operating conditions. SVCs are utilized in transmission systems for voltage regulation. A detailed presentation of operation of SVCs along with the TCRs, and their controls may be found in [17–20].

An Intellivar is topologically equivalent to the SVC, operated in distribution systems as opposed to the transmission system. The Intellivar has been implemented for flicker mitigation purposes with ratings up to 85 MVA. However, due to its low speed of response, reduction for frequencies above 5 Hz has not been very successful, and a substantial amount of harmonics is introduced in the system.

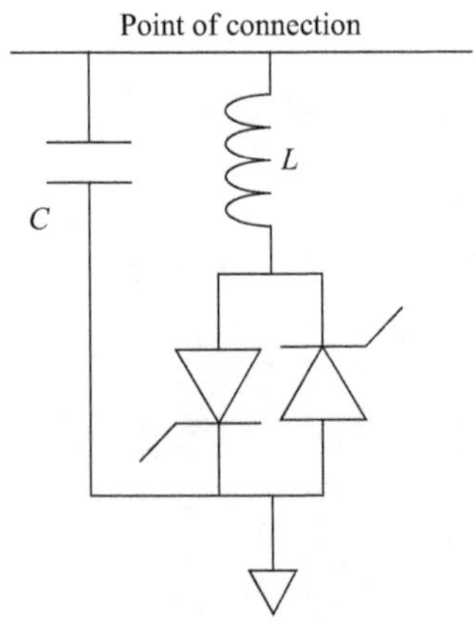

Figure 16.10 Single-phase schematic diagram of a static var compensator

While the use of ac phase control utilizing SVC or Intellivar provides an economic way of realizing dynamic var compensation, it suffers from undesirable terminal current harmonics.

16.3.1.2 STATCOM

STATic COMpensator (STATCOM) is one of the first power electronic devices that use a VSC technology to be introduced under the rubric of flexible ac transmission system (FACTS) in the transmission system [17,21]. It consists of a three-phase VSC that is connected in shunt with the power system through a coupling transformer as shown in Figure 16.11. The transformer provides voltage matching, harmonic cancellation (see Figure 16.16), and a finite amount of inductance between the output of the STATCOM and the point of common coupling (PCC) connection. At steady-state operating conditions, the voltage output of the STATCOM is locked in phase with voltage at PCC. Under this condition, the current injected into the line may be chosen to provide positive or negative vars depending on the magnitude of the STATCOM terminal voltage. When the STATCOM terminal voltage is higher than the PCC voltage, it supplies capactive vars. When the STATCOM terminal voltage is lower than the PCC voltage, it draws inductive vars. Since the current capacity of the system is decided by the converter components and design, the amount of vars in the positive and negative direction is symmetrical. The VSC used for the system may be of two or three levels and may be modulated using PWM or multi-pulse in angle-controlled operation.

The Distribution STATic COMpensator (D-STATCOM) refers to the realization of STATCOM at distribution voltage levels, under various commercial trade names [22].

16.3.1.3 PWM ac VeSC shunt compensator

Voltage fluctuation at a load location becomes a severe power quality problem as it leads to light flicker, fluctuating torque in machinery loads, and the malfunction of protection devices and process control devices. This problem has its origins in loads such as arc furnaces and electric welders, and varying power sources such as certain engine-driven generators and wind turbines. The flicker source draws a current waveform that contains a 60 Hz fundamental component along with random sub-harmonic content in the range of a few Hz to approximately 30 Hz. The sub-harmonic currents propagate into the network and cause voltage fluctuation at the PCC at the corresponding frequencies [23].

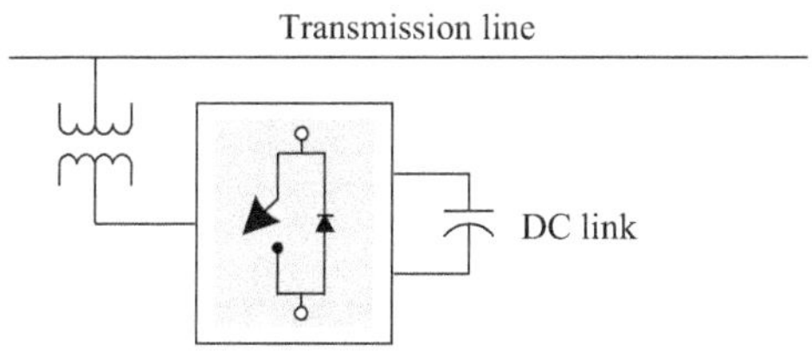

Figure 16.11 Block schematic diagram of a STATCOM

As a result of the voltage fluctuations at PCC, other loads connected to the substation would experience the power quality problems described above. Among them, the light flicker is by far the most serious as it affects humans, producing eye irritations, headaches, and migraines. Ideally, it may be desired that the variations in the voltage at the PCC at a load location (ΔV_{PCC}) to be as close to zero as possible to eliminate flicker.

The classical approach to mitigate these voltage fluctuations has been to increase the size and number of generating units or to make the network more densely interconnected. However, this method has proven to be very expensive and insufficiently effective. More recently, techniques based on reactive power management have become more widespread. The concept of reactive power management is established by sizing the power system according to the maximum demand of real power, and to manage the reactive power by means of compensators and other equipment. This approach may often be more practical and economical.

Within this framework, shunt reactive power compensators can be used to counterbalance the effect of dynamic loads locally. By appropriately injecting a controlled amount of reactive current at the PCC, the effect of active and reactive components of the current in the line impedance can be controlled. In general, the compensation system's most important function is to maintain a substantially steady voltage profile therefore maintaining the relative voltage change (ΔV_{PCC}) within a predetermined margin, over a specified bandwidth [24,25].

The approach of PWM ac VeSC-based controller may be used to inject a fast acting variable amount of capacitance in parallel with the flicker source, as illustrated in Figure 16.12 [23]. The most critical feature of a reactive power compensator to be effective in flicker reduction is the speed response. Typically, if a compensator has a control time delay of 10 ms, no matter what the voltage rating it can reduce flicker, but not by a significant amount. On the other hand, if the delay is greater than 20 ms flicker can become accentuated for several frequencies.

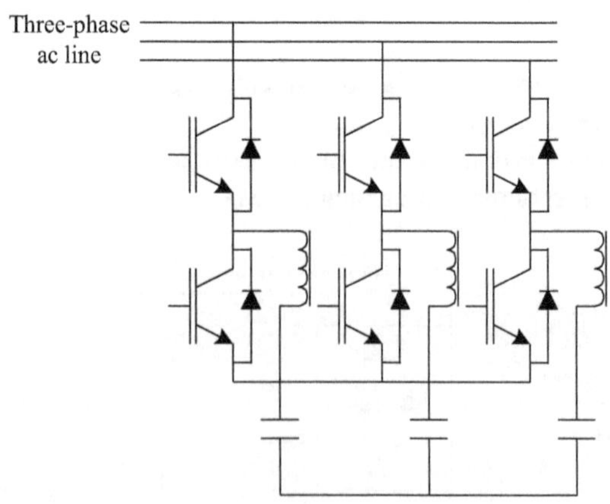

Figure 16.12 PWM ac VeSC shunt controller suitable for the realization of flicker control

Therefore, the application of SVC for this approach is generally limited. On the other hand, the switching times of state-of-the-art semiconductors used in PWM applications have increased dramatically, therefore decreasing the response time, and amount of harmonics. These advances in technology have broadened the possibilities of devices such as the PWM converters in applications such as flicker control, with bandwidths reaching several kilohertz, if necessary.

16.3.2 Series compensation

It has been long recognized that the most effective ways to externally manipulate such a power flow in a transmission line is by injecting some form of additional series reactance with the line, generically known as series compensation. The first attempts at series compensation were carried out by inserting fixed capacitors in series with the line in order to decrease the effective impedance of the line and therefore increase the power flow through it, with the risk of certain undesirable dynamic phenomena such as sub-synchronous resonance. The development of semiconductors with turn-on capability has allowed the utilization of this approach to perform continuous control with faster dynamic properties to realize this without the concomitant dynamic problems. Various approaches for applying power converters to realize controllable series compensation are described in this subsection.

16.3.2.1 Thyristor-controlled series capacitor

The thyristor-controlled series capacitor (TCSC), or sometimes referred as the advanced series compensator (ASC) in the literature has been used to provide precisely controllable series compensation. Similar to an SVC, this device consists of a TCR in parallel with capacitor to provide continuous control of the reactance. The entire device is connected in series with the transmission line as illustrated in Figure 16.13. While the principle of the device is rather simple, based on phase angle control of the reactance, the operational dynamic of the device may become rather complex. This is due to the possible resonances between the reactance of the line, the reactance of the TCR, and the reactance of the parallel capacitor. Since TCR is a variable reactance, as the control angle of the SCRs is varied across its dynamic range, the effective resonant modes of the system also shift. Since the operation of TCR also inherently produces several significant harmonics, the possibility of exciting such harmonics becomes real.

The detailed discussion of such issues including design considerations, dynamic operating range, control approach, etc. may be found in [26–28].

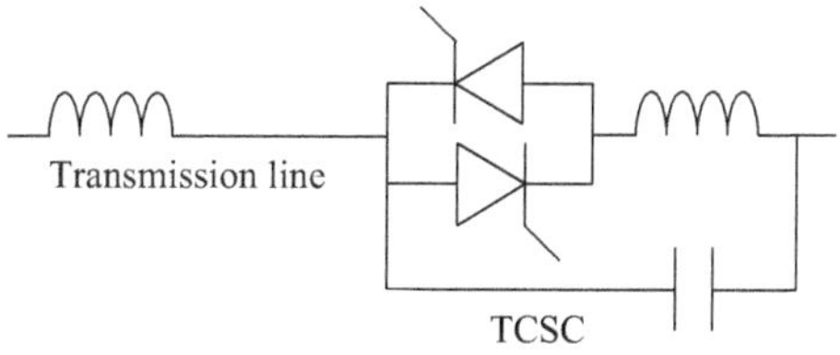

Figure 16.13 Schematic diagram of a three-phase thyristor-controlled series capacitor

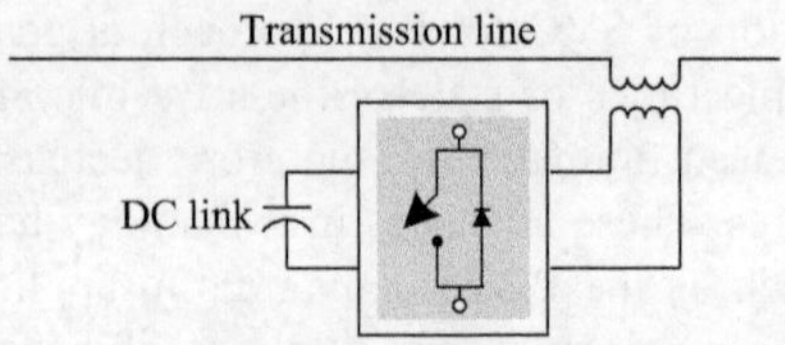

Figure 16.14 Block schematic diagram of a dc link solid-state series capacitor

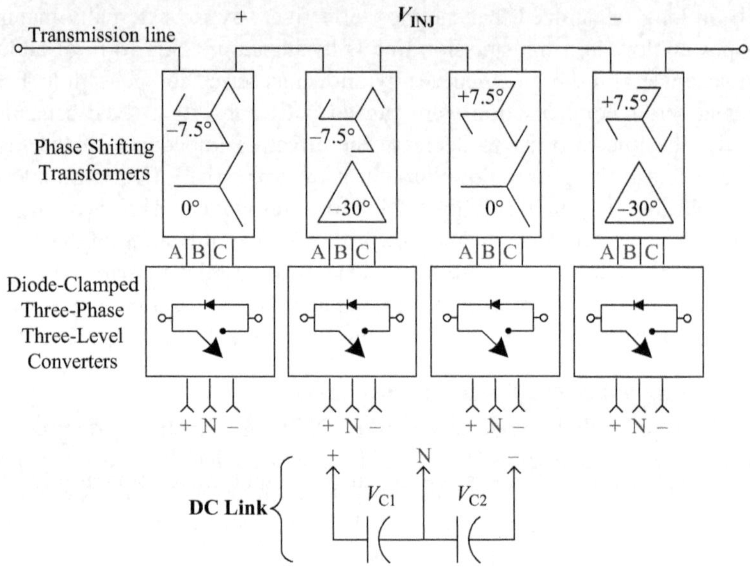

Figure 16.15 Transformer connection details of the 48-pulse SSSC

16.3.2.2 Static synchronous series compensator

Similar to the STATCOM, a VSC may be used to provide var compensation, in series with the line, and is generally termed static synchronous series compensator (SSSC). Since transmission systems operate at voltages beyond the reach of any of the power semiconductor switching devices, coupling of the power converter has to be necessarily made through a series transformer as shown in Figure 16.14 [29,30].

While it may be noted that all three stages, the dc bus, the inverter, and the injection transformer, can be realized in several ways, a realization of a 48-pulse operation is described further. The realization of the VSC-based SSSC consists of three main stages, namely a dc bus, an inverter, and an injection transformer as illustrated in Figure 16.15. As can be seen, the dc bus is realized using two series connected dc capacitors with access to their mid-point N. This allows three voltage levels (V_{C1}, V_{C2}, 0). Four independent diode-clamped three-phase three-level converters are connected to the dc bus. The voltage synthesis for each of the phases has a conduction period of 172.5° for each half cycle. Each three-level converter is then connected to a magnetic structure based on Y-zigzag and Δ-zigzag phase-shifting

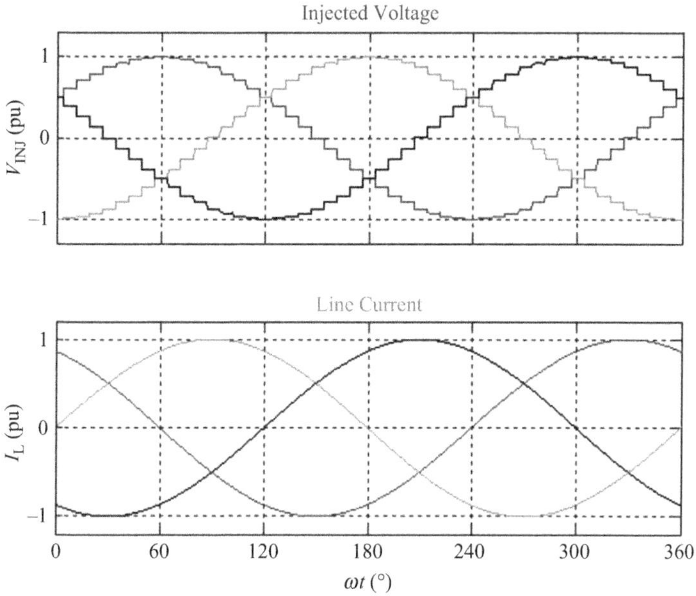

Figure 16.16 Three-phase output voltage waveforms from 48-pulse SSSC

transformers (PSTs) that allows creation of a 48-pulse waveform. Thus, the lowest characteristic harmonic numbers present in the output waveforms are 47 and 49. Figure 16.16 shows typical switched three-phase voltage waveforms along with the corresponding line current. It can be seen that the injected voltage is of high harmonic quality and thus the SSSC realized in this manner can operate with no output filters.

16.3.2.3 PWM ac VeSC series controller

The approach of PWM ac VeSC-based power control may also be used for controlled series compensation [9,11]. The schematic diagram of a power circuit is illustrated in Figure 16.17. The switches S_A, S_B, and S_C that are ganged together form a PWM ac VeSC converter, which operates in a duty ratio controlled mode. The output of the PWM converter is injected in series with the line through a series-coupling transformer. When the switches are in the top position, the capacitors are introduced in series with the line. When the switches are in the bottom position, the coupling transformer is shorted, and the capacitors are isolated from the line. The duty ratio of the switches determines the percentage of the compensation capacitance that is introduced in series with the transmission line.

The equivalent impedance (X_{eq}) between the sending and receiving end of the transmission system may be derived using state-space averaging techniques as

$$X_{eq} = X_L - n^2(1-D)^2 X_C \tag{16.9}$$

where X_C is the reactance of the compensation capacitors, n is the turns ratio of the transformer, X_L is the reactance of the inductors line (which may include the

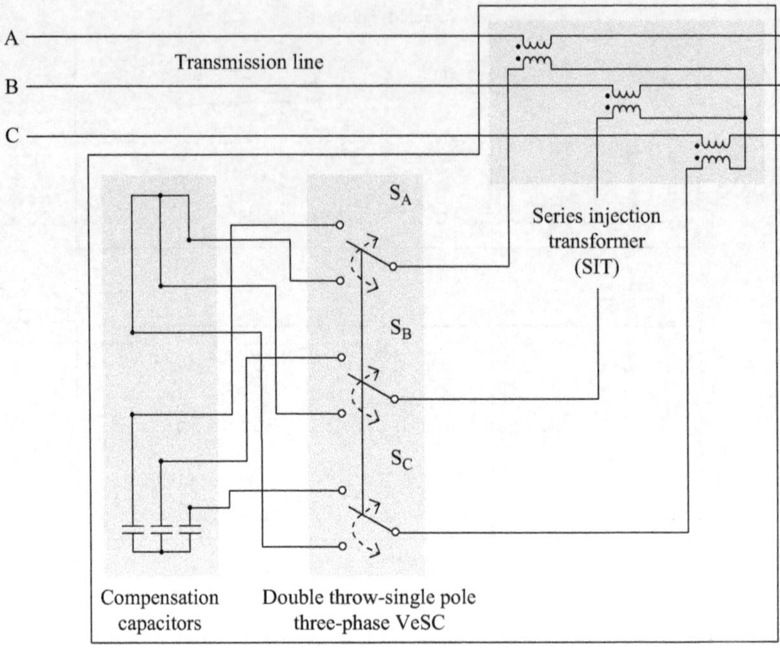

Figure 16.17 Schematic diagram of a three-phase PWM-controlled series capacitor

leakage reactance of the coupling transformers), and D is the duty ratio of the top throws of the switches. The performance of the controller from simulations during a typical case is illustrated in Figure 16.18.

16.3.2.4 Voltage sag correction

Proliferation of electronic equipment results in increasing sensitivity of industry processes to the power quality. On the other hand, the more competitive market environment leads to reduced profit margins, which is the primary reason why the process shutdown caused by the power quality problem is intolerable. The power quality problem of most concern is the voltage quality problem, such as voltage sags. Very commonly voltage sags are due to power system faults, most of which are single line to ground short circuit faults. The short circuit faults can be caused by lighting strokes, storms, animals, and other unpredictable factors. Although the voltage quality can be improved at the power system level, rather often the solution is too costly to implement and it is impossible to eliminate the voltage problem. Alternatively, utilities and customers turn to the practical solution of feeding premium power to sensitive loads, which is realized by the mitigation equipment at the distribution feeder close to the customer [31].

In such cases, if the system is strong (measured by short circuit capacity being much larger than unity) at PCC, the series compensation approach requires less rating size for the power converter. A block schematic diagram of such a system, commonly referred to as a dynamic voltage restorer (DVR) is illustrated in

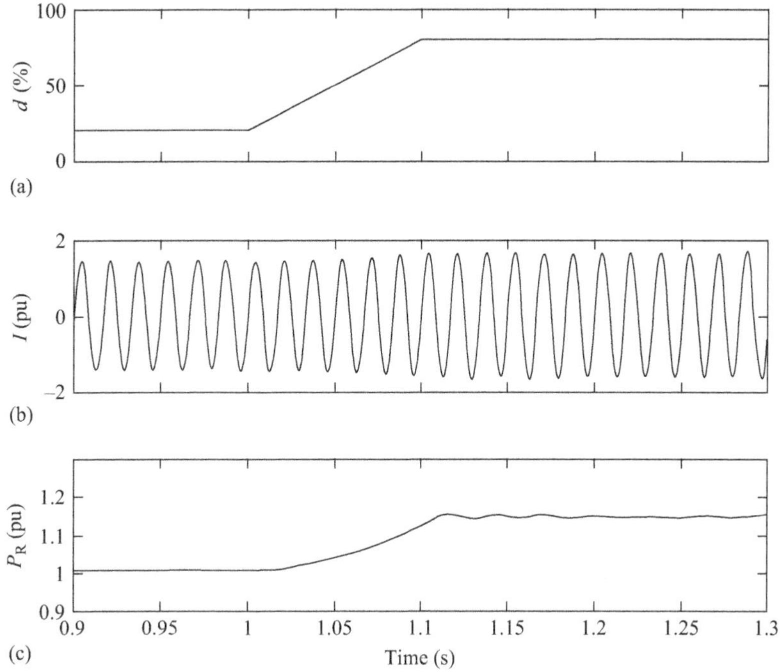

Figure 16.18 A plot of (a) duty ratio, (b) current, and (c) power throughput using a PWM-controlled series compensator obtained from simulation of a transmission system case

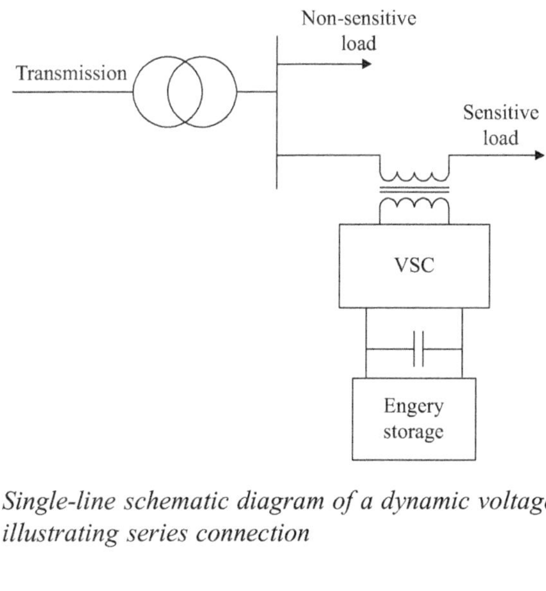

Figure 16.19 Single-line schematic diagram of a dynamic voltage restorer illustrating series connection

Figure 16.19. While VSC-based DVRs can be realized using two-level converters, which are well suited for 480 V systems, for systems operating at distribution power levels a multilevel converter is a more attractive solution. Moreover, since voltage sags occurring in such systems are due to single line to ground faults, it is necessary for a DVR to be able to handle imbalances in sags, as well as zero sequence type of voltages. Therefore, a power converter topology with independent control of each phase as illustrated in Figure 16.20 is suitable for such applications. Moreover, due to the use of transformers with voltage adding capability, the output

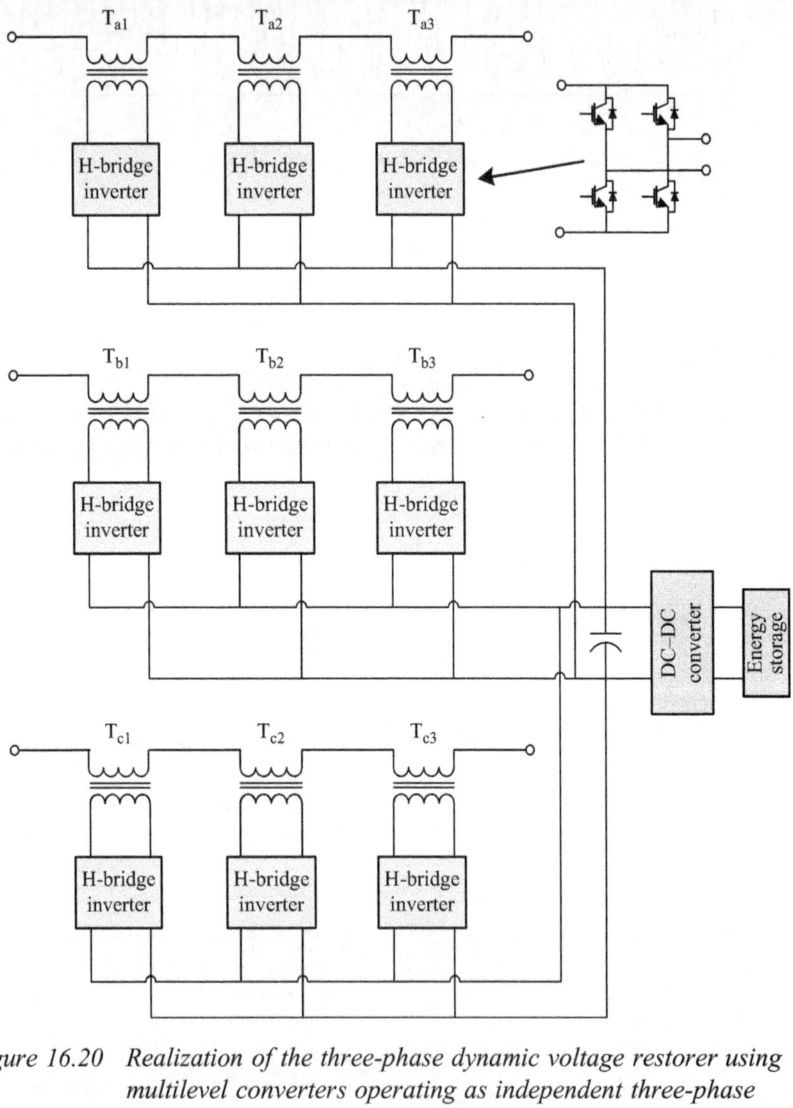

Figure 16.20 Realization of the three-phase dynamic voltage restorer using multilevel converters operating as independent three-phase converters

waveform may be of multilevel nature. In this case, the converters may employ phase-shifted PWM strategy among them to provide excellent quality output waveforms at the output (Figure 16.21), thereby requiring minimal filtering components.

16.3.3 Shunt-series compensation

The shunt-series power flow controller is a device introduced under the rubric of FACTS, that combines the features of shunt compensation such as the STATCOM, and series compensation such as the SSSC together by interconnecting the two converters. VSC-based converters and PWM ac controller-based approaches are discussed in this section.

16.3.3.1 Unified power flow controller

An arrangement of the shunt series power flow controller under title of unified power flow controller (UPFC) provides a means for controlling the active and reactive power flows in a power system by means of the synthesis of a fully adjustable series voltage [32]. This topology, based on a dc link between a shunt

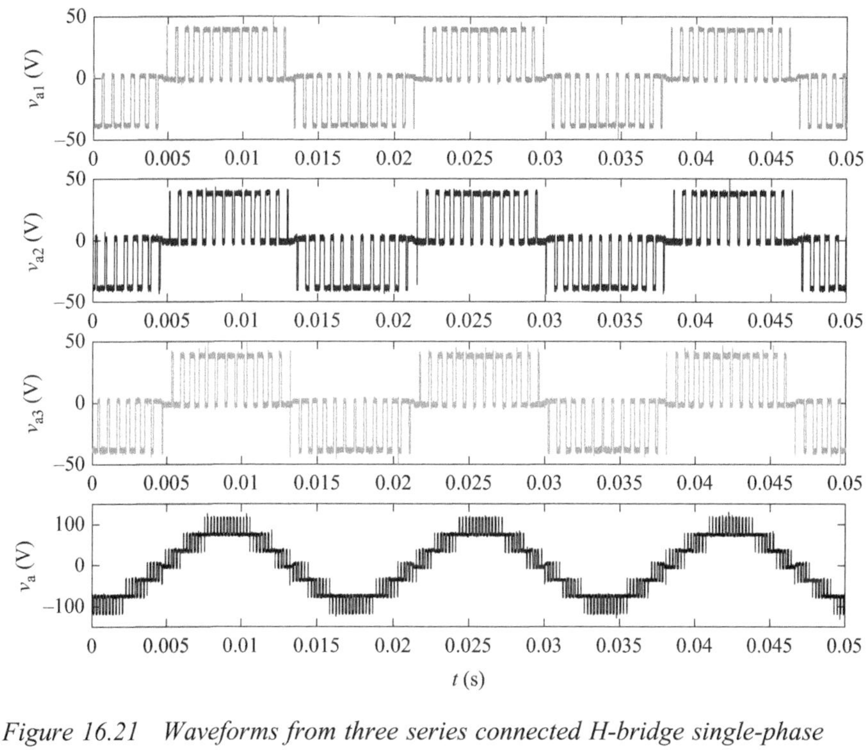

Figure 16.21 Waveforms from three series connected H-bridge single-phase inverters with phase-shifted PWM generating a seven-level output voltage in a DVR

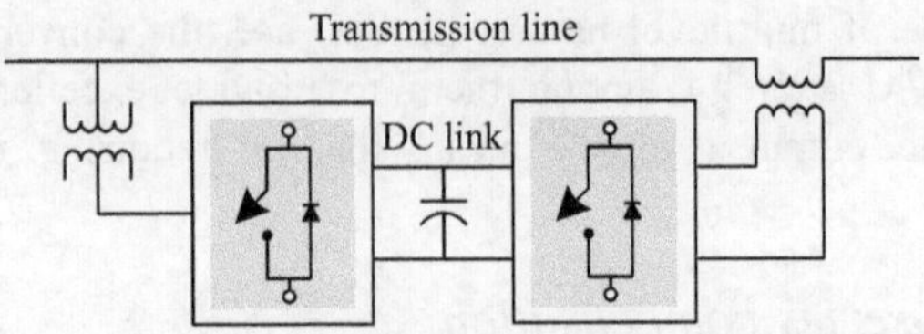

Figure 16.22 Block diagram of a UPFC using shunt and series connected VSCs with a common dc link

VSC (a STATCOM) and a series VSC (a SSSC) connected in a back-to-back configuration, has certain inherent degrees of freedom determining its control capabilities (Figure 16.22). The STATCOM and the SSSC can develop arbitrary terminal voltages and phase angles (four degrees of freedom) within the bounds dictated by their shared dc link voltage. In case of applying a PWM control scheme to these VSCs, the dc link voltage has to be maintained constant. Therefore, the sum of the active power injections of the STATCOM and SSSC must be zero. Taking this into account, the initial four degrees of freedom of the UPFC are reduced to three, being it possible to regulate the active and reactive power flow between the interconnected power systems and the voltage at the PCC of the STATCOM. Each of the STATCOM and SSSC may be realized using multi-pulse converters described in the previous sections.

16.3.3.2 PWM ac VeSC shunt/series power flow controller

The schematic diagram of the PWM ac power converter applied as a shunt/series controller is illustrated in Figure 16.23 [33,34]. The central feature of the approach is to control the power flow along the transmission line by injecting an ac voltage with controllable magnitude and phase angle in series with the line. The system is configured by including a shunt SPT, filter capacitors (FCs), a quadruple-throw single-pole three-phase PWM VeSC, and a series injection transformer (SIT). This system may be located at any point throughout the line.

The SPT (with four secondary winding sets) has the double function of lowering the impressed voltage to match the capability of the semiconductor switches and simultaneously phase shift the impressed voltage by 0°, 90°, 180°, and 270°, respectively, to obtain four sets of three-phase voltage vectors lined up along four cardinal directions. The capacitor bank FC is used to absorb the high-frequency currents introduced due to switching from permeating into the system across the transformer. The VeSC uses a quadruple-throw single-pole three-phase switch to synthesize a controllable voltage at the pole terminals by adjusting the duty ratios d_1, d_2, d_3, and d_4. Given the throw voltages generated by the SPT, the range of values of the pole voltage can be regulated in magnitude and phase angle, both of which can be independently controlled through duty ratios d_1, d_2, d_3, and d_4.

Finally, the SIT accommodates the voltage to an appropriate level to be injected back in series with the transmission line and also provides the necessary electrical isolation. It is important to notice that the stiff voltage and stiff current requirements for the PWM ac converter are given by the capacitor bank and SIT leakage inductance, respectively.

The performance, design, and operational features of this converter may be found in [33,34].

16.3.4 Series-series compensation

The interline power flow controller (IPFC) is another realization of a FACTS device, where two VSC-based SSSC devices in series with two transmission lines share their dc link as illustrated in Figure 16.24. The performance capability and design details of the IPFC may be found in [35,36].

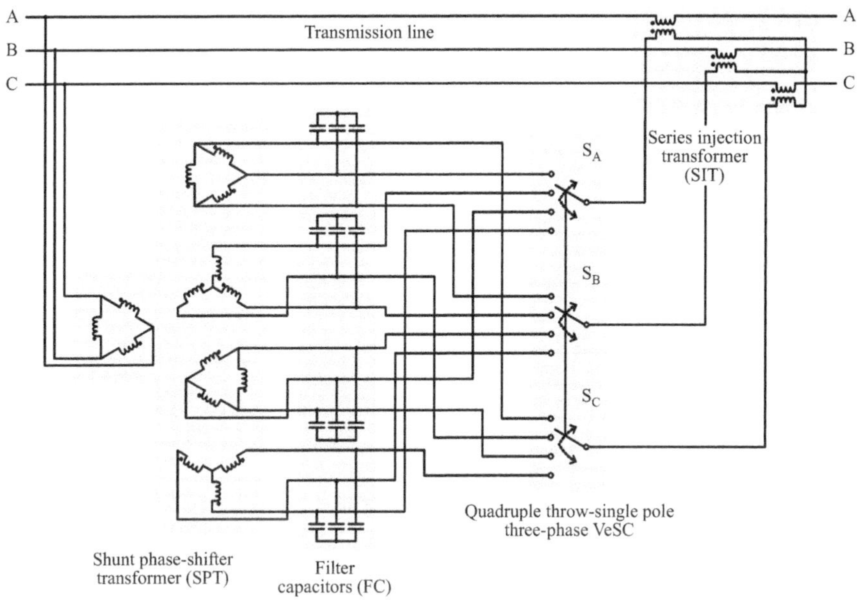

Figure 16.23 Realization of a shunt-series power flow controller using a PWM ac converter

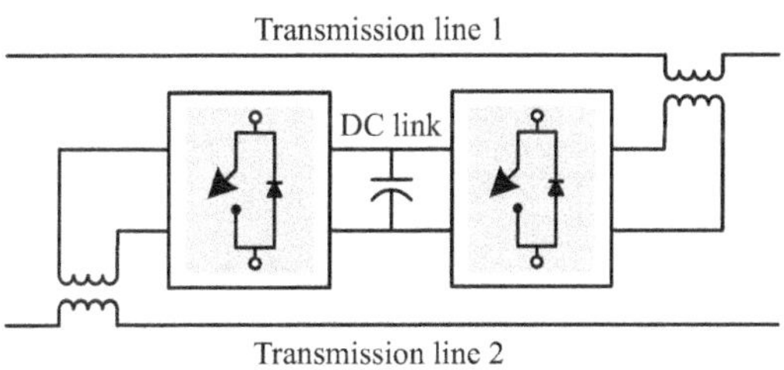

Figure 16.24 Block diagram of an interline power flow controller using interconnected VSCs

16.3.5 High-voltage direct current transmission

16.3.5.1 Conventional CSC

High-voltage direct current (HVDC) based on CSCs has been in use for 50 years, starting evolving from systems using mercury arc valves to modern systems using light fired thyristors. These systems are generally used for bulk power transmission over long distances using overhead lines (500 km or longer) or undersea (50 km or longer in many cases), or to interconnect systems operating out of synchronism or at different frequencies. These systems operate with voltages up to 600 kV to ground and carry currents up to 3,000 A (Figure 16.25).

The lack of turn-off controllability of the thyristors used in the line-commutated converter results in: poor power factors and considerable waveform distortion, need for a relatively stiff ac voltage supply, and inability to provide independent control of the active and reactive powers. Moreover, it requires fast communication channels between the two stations to ensure power balance and appropriate controls [1].

16.3.5.2 VSC-HVDC

Over the last decade, VSC-based HVDC transmission has started to come into use (Figure 16.26) [37]. These systems, referred to as HVDC light by vendors, are based on a modular design, reducing the installation time. VSC transmission is often used in cases where an ac system is not practical, generally due to long underground or undersea cables, and excessive losses. One common application is supplying power to isolated island systems without adequate local generation to provide reactive support for conventional HVDC converters. The converters in VSC transmission are able to supply power to weak or passive ac load systems. Another application where VSC-HVDC transmission is often used is the transmission of power from off-shore wind farms to the mainland. The converters used in this system may be two-level or three-level topologies.

In contrast to CSC systems, the newer VSC-based technology (i) does not require any communications links between the different converter stations; (ii) does not need to rely on the ac networks ability to keep the voltage and frequency stable; and (iii) offers independent control of active and reactive power [16].

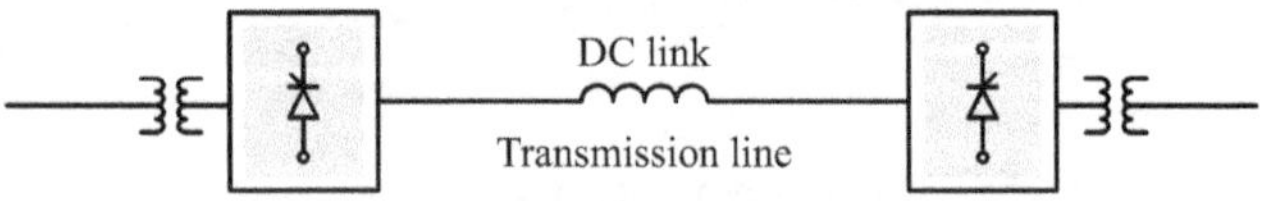

Figure 16.25 Block diagram of an HVDC transmission system

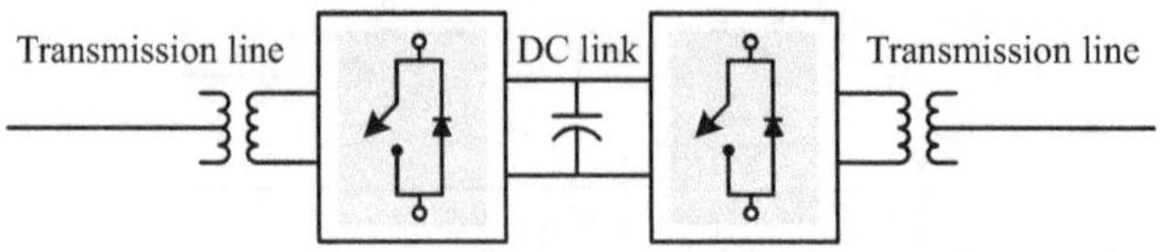

Figure 16.26 Block diagram of VSC-based HVDC system

16.3.5.3 MMC-HVDC

The most recent development in the field of HVDC systems is the application of the MMC or BoB converter to realizing HVDC systems [38–40]. A schematic diagram of the system is illustrated in Figure 16.27. The first field application of the BoB approach is expected to be the Trans Bay Cable Project to be commissioned in 2010 under the product name Siemens HVDC Plus. The SMs are made up of IGBT switches with antiparallel diodes.

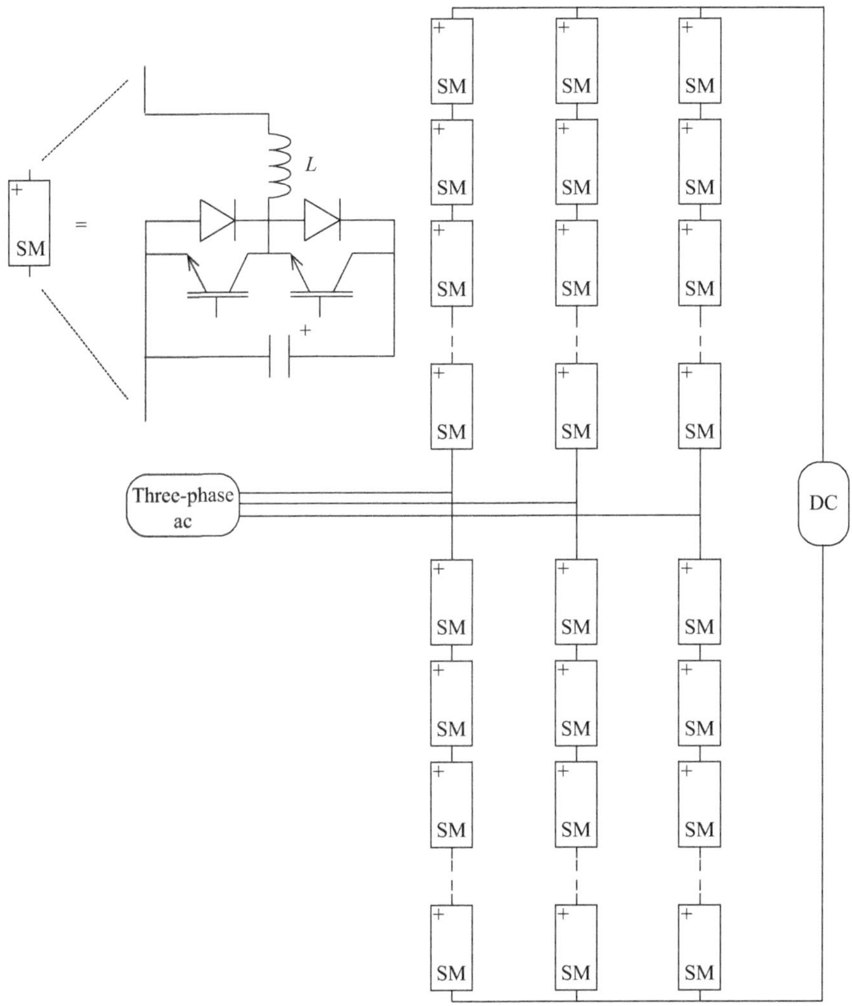

Figure 16.27 Schematic diagram of a three-phase ac-to-dc MMC or BoB converter for HVDC converter stations

16.3.6 Low-frequency high-voltage ac transmission

The low-frequency high-voltage ac (LF-HVAC) transmission approach is a promising alternative to increase power transfer capability and improved voltage profile through the transmission line [41]. Due to reduction in frequency for transmission, the line reactance is also decreased leading to an improve voltage profile along the line and a larger transient stability limit. Past propositions to introduce this technology have been mainly based on cycloconverters which lead to high levels of voltage and current harmonic distortion levels. LF-HVAC systems have been utilized so far for (i) the grid connection of large wind farms at a fractional frequency of 50/3 Hz, and have been termed fractional frequency transmission system [42], and for (ii) cable transmission systems which are limited in transmission length when operated at conventional frequencies [43].

Referring to Figure 16.28(a) the power throughput through a transmission line is roughly proportional to the angle between the voltage at the two ends of the line, and inversely proportional to the impedance of the line. It is well understood in the power industry that viable approaches to control the line throughput power include adjusting the transmission angle and the line impedance. On conventional transmission lines, the line impedance is fixed at $X = 2\pi 60L$ and the transmission angle is always kept below about 40° due to transient stability considerations. The latter is especially critical for long transmission lines, as transient stability considerations limit the throughput power loading. Figure 16.29 illustrates a typical loadability curve for a transmission line, where the limit is expressed as a fraction of the surge impedance loading (SIL). As may be seen, transient stability becomes the predominant limitation for lengths longer than 200 miles (mi). This leads to a very inefficient utilization of the conductors – as suggested in the figure a line of 600 mi can be loaded to less than 30% of its thermal limit. Since the transient stability limit of the line is fixed by the transmission angle, an alternative and convenient means to increase the transient stability limit is to reduce the line's reactance.

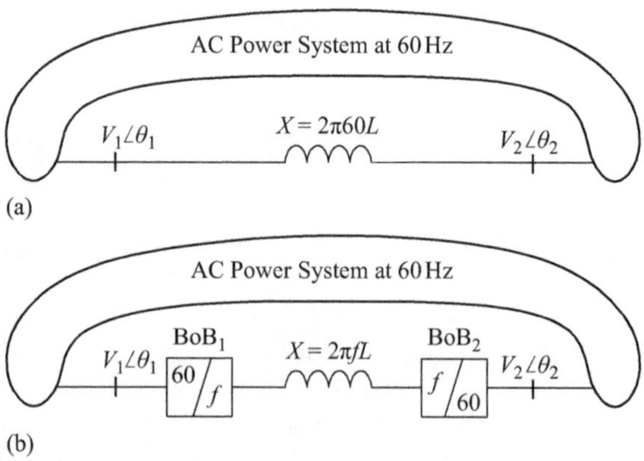

Figure 16.28 Simplified schematic diagram of an ac system illustrating (a) a transmission line and (b) the transmission operating at a lower frequency

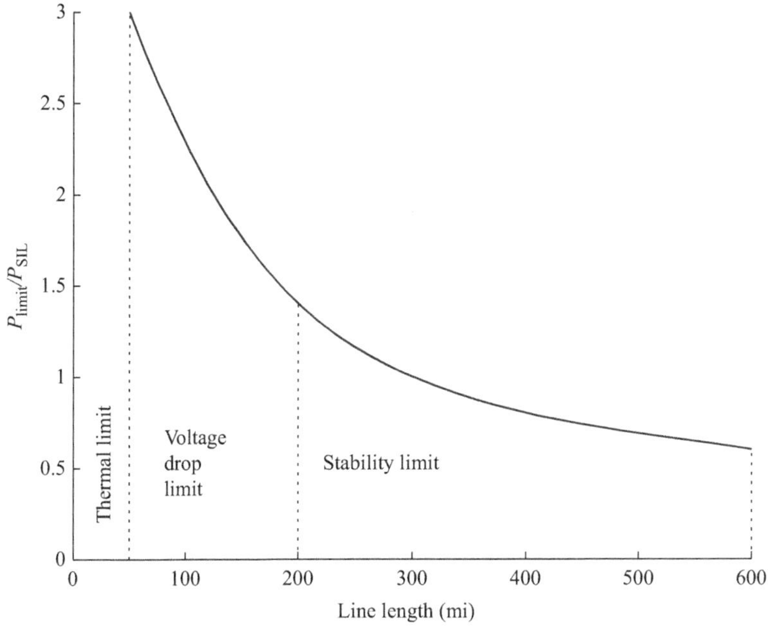

Figure 16.29 Loadability curve of a transmission line as a function of the line length

Figure 16.28(b) shows the implementation of LF-HVAC transmission lines through embedding frequency converters at each end of a number of lines. This would allow the transmission line be operated at a frequency $f < 60$ Hz. Operational benefits of LF-HVAC transmission lines clearly includes the capability for asynchronous operation with full control of the throughput active power and reactive power at each end of the transmission line. Furthermore, since the line at both ends is terminated by power converters interfacing to the rotating machine and load centers, the classical transient stability considerations leading to a nominal limit of 40° is no longer valid. Moreover, power converters have some built-in storage capacity and active control over the power throughput across the line under transient conditions. This may permit the transmission angle to be increased well beyond the nominal limit of 40°, even up to 90°. Thus, by properly selecting the operating frequency, the line reactance can be decreased so that the line capacity can be utilized up to its thermal limit.

Although the previous discussion has focused on overhead transmission lines, similar arguments may be brought forth regarding underground cables. Furthermore, the same ideas may be applied to converting an entire area of a transmission system into LF-HVAC.

Figure 16.30 illustrates a 60 Hz to LF-HVAC converter realized using the BoB converter approach. In this case, all the bridge SMs will be realized using the full-bridge topology.

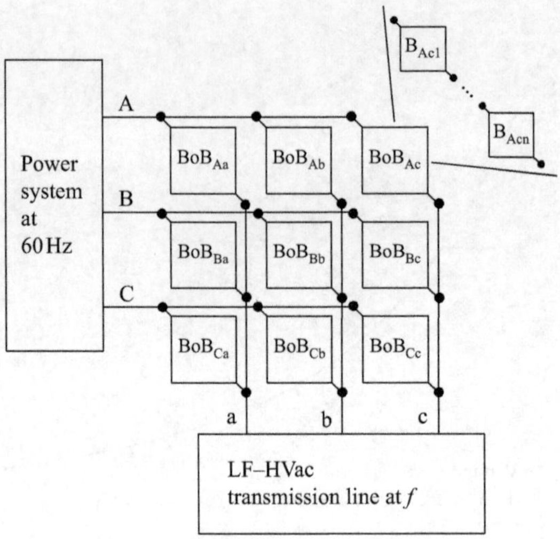

Figure 16.30 Illustration of a BoB converter-based frequency conversion for realizing LF-HVAC systems

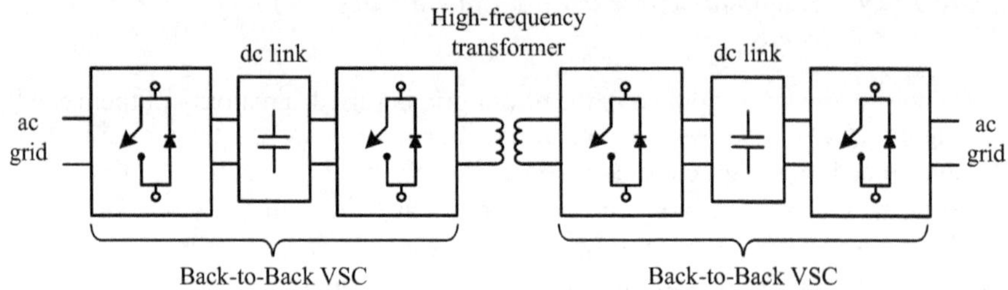

Figure 16.31 Illustration of a solid-state transformer using two dc links and a high-frequency transformer

16.3.7 Solid-state power transformer

The concept of a solid-state power transformer (SSPT) is not recent, but with the advancement of high-power semiconductor, the SSPT may soon become a practical reality [44]. Figure 16.31 shows a possible realization of a SSPT. As the figure suggests, a SSPT is made up of the cascaded connection of a VSC acting as a rectifier; a dc link; a VSC acting as an inverter; a high-frequency ac transformer; a VSC acting as a rectifier; a dc link; and a VSC acting as an inverter. Variations of this topology have been presented in the literature, but all of them follow more or less the same approach. Differences arise from the realization of the various VSC stages.

The main advantage of such configuration is that it provides the benefits of an ordinary transformer – namely, electrical isolation and a potentially large transformation ratio, but, utilizing a reduced amount of magnetic material. The high-frequency transformer processes ac power at a frequency of a few tens of kilohertz, and therefore can be realized utilizing less magnetic material, making it lighter, cheaper, and easier to transport when compared to a standard 50 or 60 Hz standard transformer rated at the same throughput power. Furthermore, due to the presence of VSCs at the input and output, the SSPT can interconnect two ac systems operating at arbitrary frequencies (e.g., 50 and 60 Hz), decoupling the dynamics and providing independent power factor compensation for both systems. The VSCs can also provide active filtering functionality, and the intermediate dc link can serve as a point of interconnection for dc power sources (e.g., photovoltaic arrays). Because of all of these features, SSPTs are considered by some authors as an important building block for the smart grid of the future. Applications of SSPTs have been mostly reported for medium- to low-power systems in the context of ac and dc microgrid networks; for variable speed ac drives; and for the realization of power transformers in distribution systems.

The main drawback of SSPTs stems from the usage of a large number of power semiconductors, leading to a decrease in efficiency, control complexity, and potential failures associated to the semiconductors. Research to improve efficiency has been reported in terms of replacing standard silicon-based switches by silicon carbide and other wide-bandgap semiconductors.

16.4 Summary

While power converters applied in the power grid is varied, a majority of them can be classified to fall into a few categories that are illustrated in Section 16.2. Among these, the SCR-based technologies have in general been the most mature owing to their early date of introduction, and also the availability of power semiconductor technology that reach the power levels warranted by the power grid. While continuing advances in dc link VSC-based technologies they are becoming viable in the power grid, many of the devices introduced under the rubric of FACTS remain dormant due to considerations of economic scale. On the other hand, the recent introduction of MMCs for realizing high-power converters using gate-turn-off switch technologies appears to be a game changer in displacing SCR-based converter technologies. However, the technology of PWM ac power flow control using VeSCs and solid-state transformers continues to stay within the realm of academic research, in the absence of commercial interests required to promote them. Emerging advances in power semiconductors such as silicon carbide and gallium nitride materials may tip the economies of scale to make such newer technologies more attractive in the future.

This chapter has provided a rather concise outline of different power converters used in the power grid, along with selected application examples, chosen to illustrate the diversity of approaches in use. The discussion is necessarily brief due to the nature of this volume and the readers may refer to the primary literature for additional details. The references provided are generally drawn from the more recent literature, as opposed to the classical publications, since they provide a more up to date state of the art.

References

[1] Nozari F., Patel H.S. 'Power electronics in electric utilities: HVDC power transmission systems'. *Proceedings of the IEEE.* 1988;**76**(4):495–506.

[2] Cuzner R.M., Venkataramanan G. 'Current source rectifiers in discontinuous conduction modes of operation'. *IEEE Transactions on Industry Applications.* 2015;**51**(1):470–478.

[3] Cuzner R.M., Drews D., Venkataramanan G. 'Power density and efficiency comparisons of system-compatible drive topologies'. *IEEE Transactions on Industry Applications.* 2015;**51**(1):459–469.

[4] Yazdani A., Iravani R. *Voltage-Sourced Converters in Power Systems: Modeling, Control, and Applications*, Hoboken, NJ: Wiley-IEEE Press; 2010.

[5] Parkhideh B., Bhattacharya S. 'Vector-controlled voltage-source-converter-based transmission under grid disturbances'. *IEEE Transactions on Power Electronics.* 2013;**28**(2):661–672.

[6] Hingorani N., Gyugyi L. *Understanding FACTS: Concepts and Technology of Flexible AC Transmission Systems*, Piscataway, NJ: IEEE Press; 2000.

[7] Hingorani, N.G. 'Flexible AC transmission'. *IEEE Spectrum.* 1993;**30**(4): 40–45.

[8] Venkataramanan G. 'Three-phase vector switching converters for power flow control'. *IEE Proceedings – Electric Power Applications.* 2004;**151**(3): 321–333.

[9] Venkataramanan G., Johnson B.K. 'Pulse width modulated series compensator'. *IEE Proceedings – Generation, Transmission and Distribution.* 2002;**149**(1):71–75.

[10] Garcia-Vite P.M., Mancilla-David F., Ramirez J.M. 'Per-sequence vector-switching matrix converter modules for voltage regulation'. *IEEE Transactions on Industrial Electronics.* 2013;**60**(12):5411–5421.

[11] Mancilla-David F., Bhattacharya S., Venkataramanan G. 'A comparative evaluation of series power-flow controllers using DC- and AC-link converters'. *IEEE Transactions on Power Delivery.* 2008;**23**(2):985–996.

[12] Ludois D.C., Reed J.K., Venkataramanan G. 'Hierarchical control of bridge-of-bridge multilevel power converters'. *IEEE Transactions on Industrial Electronics.* 2010;**57**(8):2679–2690.

[13] Reed J., Venkataramanan G., Martinez F. 'Complex phasor modeling and control of modular multilevel inverters'. *Proceedings of the 2011 IEEE Energy Conversion Congress and Exposition*; Phoenix, AZ, USA, September 2011. pp. 4013–4020.

[14] Ludois D.C., Venkataramanan G. 'Simplified terminal behavioral model for a modular multilevel converter'. *IEEE Transactions on Power Electronics.* 2014;**29**(4):1622–1631.

[15] Akagi H. 'Classification, terminology, and application of the modular multilevel cascade converter (MMCC)'. *IEEE Transactions on Power Electronics.* 2011;**26**(11):3119–3130.

[16] Kawamura W., Hagiwara M., Akagi H. 'Control and experiment of a modular multilevel cascade converter based on triple-star bridge cells'. *IEEE Transactions on Industry Applications.* 2014;**50**(5):3536–3548.
[17] Laszlo G. 'Reactive power generation and control by thyristor circuits'. *IEEE Transactions on Industry Applications.* 1979;**15**(5):521–532.
[18] Manjrekar M., Venkataramanan G. 'Control strategies for a hybrid static reactive compensator'. *Canadian Conference on Electrical and Computer Engineering*; Calgary, AB, Canada, May 1996. pp. 834–837.
[19] Araya P.M., Castro J.M., Nolasco J.C., Behnke R.E.P. 'Lab-scale TCR-based SVC system for educational and DG applications'. *IEEE Transactions on Power Systems.* 2011;**26**(1):3–11.
[20] Sapkota B., Vittal V. 'Dynamic VAr planning in a large power system using trajectory sensitivities'. *IEEE Transactions on Power Systems.* 2010;**25**(1): 461–469.
[21] Al-Gali O.M., Mancilla-David F. 'Realization of an AC-link pulse width modulated shunt converter for STATCOM applications'. *North American Power Symposium*; Champaign, IL, USA, September 2012. pp. 9–11.
[22] Lee T.-L., Hu S.-H., Chan Y.-H. 'D-STATCOM with positive-sequence admittance and negative-sequence conductance to mitigate voltage fluctuations in high-level penetration of distributed-generation systems'. *IEEE Transactions on Industrial Electronics.* 2013;**60**(4):1417–1428.
[23] Gutierrez M., Venkataramanan G., Sundaram A. 'Solid state flicker controller using a pulse width modulated AC–AC converter'. *Conference Record of the 2000 IEEE Industry Applications Conference*; Rome, Italy, October 2000. pp. 3158–3165.
[24] Rahmani S., Hamadi A., Al-Haddad K., Dessaint L.A. 'A combination of shunt hybrid power filter and thyristor-controlled reactor for power quality'. *IEEE Transactions on Industrial Electronics.* 2014;**61**(5):2152–2164.
[25] Bilgin H.F., Ermis M. 'Design and implementation of a current-source converter for use in industry applications of D-STATCOM'. *IEEE Transactions on Power Electronics.* 2010;**25**(8):1943–1957.
[26] Johansson N., Angquist L., Nee H.P. 'An adaptive controller for power system stability improvement and power flow control by means of a thyristor switched series capacitor (TSSC)'. *IEEE Transactions on Power Systems.* 2010;**25**(1):381–391.
[27] Li K., Zhao J., Zhang C., Lee W.J. 'Dynamic simulator for thyristor-controlled series capacitor'. *IEEE Transactions on Industry Applications.* 2010;**46**(3):1096–1102.
[28] Daneshpooy A., Gole A.M. 'Frequency response of the thyristor controlled series capacitor'. *IEEE Transactions on Power Delivery.* 2001;**16**(1):53–58.
[29] Álvarez H.J., Ramírez J.A., Zúñiga P., Mancilla-David F. 'Static synchronous series compensator for active power flow control during unconventional operating conditions'. *Proceedings of the IEEE International Autumn Meeting on Power, Electronics and Computing*; Ixtapa, Mexico, November 2014. pp. 1–6.

[30] Mancilla-David F., Venkataramanan G. 'Modeling and control of the static synchronous series compensator under different operating modes'. *Proceedings of the 38th IEEE Power Electronics Specialists Conference*; Orlando, FL, USA, June 2007. pp. 2443–2449.

[31] Taylor G.A., 'Power quality hardware solutions for distribution systems: Custom Power'. *IEE North Eastern Centre Power Section Symposium on the Reliability, Security and Power Quality of Distribution Systems*. 5 April 1995. pp. 11/1–11/9.

[32] Gyugyi L., Schauder C.D., Williams S.L., Rietman T.R., Torgerson D.R., Edris A. 'The unified power flow controller: a new approach to power transmission control'. *IEEE Transactions on Power Delivery*. 1995;**10**(2): 1085–1097.

[33] Mancilla-David F., Venkataramanan G. 'Realisation of an AC-link unified power flow controller'. *IET Generation, Transmission & Distribution*. 2012;**6**(4):294–302.

[34] Barragán-Villarejo M., Venkataramanan G., Mancilla-David F., Maza-Ortega J.M., Gomez-Exposito A. 'Dynamic modelling and control of a shunt-series power flow controller based on AC-link'. *IET Generation, Transmission & Distribution*. 2012;**6**(8):1754–1763.

[35] Gyugyi L., Sen K.K., Schauder C.D. 'The interline power flow controller concept: a new approach to power flow management in transmission systems'. *IEEE Transactions on Power Delivery*. 1999;**14**(3):1115–1123.

[36] Jiang S., Gole A.M., Annakkage U.D., Jacobson D.A. 'Damping performance analysis of IPFC and UPFC controllers using validated small-signal models'. *IEEE Transactions on Power Delivery*. 2011;**26**(1):446–454.

[37] Venkataramanan G., Johnson B.K. 'A superconducting DC transmission system based on VSC transmission technologies'. *IEEE Transactions on Applied Superconductivity*. 2003;**13**(2):1922–1925.

[38] Saeedifard M., Iravani R. 'Dynamic performance of a modular multilevel back-to-back HVDC system'. *IEEE Transactions on Power Delivery*. 2010; **25**(4):2903–2912.

[39] Qin J., Saeedifard M. 'Predictive control of a modular multilevel converter for a back-to-back HVDC system'. *IEEE Transactions on Power Delivery*. 2012;**27**(3):1538–1547.

[40] Sekiguchi K., Khamphakdi P., Hagiwara M., Akagi H. 'A grid-level high-power BTB (back-to-back) system using modular multilevel cascade converters without common DC-link capacitor'. *IEEE Transactions on Industry Applications*. 2014;**50**(4):2648–2659.

[41] Carrasco M., Mancilla-David F., Venkataramanan G., Reed J. 'Low frequency HVac transmission to increase power transfer capacity'. *Proceedings of the 2014 IEEE PES Transmission and Distribution Conference and Exposition*; Chicago, IL, USA, April 2014. pp. 1–5.

[42] Wang X., Wei X., Meng Y. 'Experiment on grid-connection process of wind turbines in fractional frequency wind power system'. *IEEE Transactions on Energy Conversion*. 2015;**30**(1):22–31.

[43] Miura Y., Mizutani T., Ito M., Ise T. 'Modular multilevel matrix converter for low frequency AC transmission'. *Proceedings of the 2013 IEEE 10th International Conference on Power Electronics and Drive Systems*; Kitakyushu, Japan, April 2013. pp. 1079–1084.

[44] She X., Huang A.Q., Burgos R. 'Review of solid-state transformer technologies and their application in power distribution systems'. *IEEE Journal of Emerging and Selected Topics in Power Electronics*. 2013;**1**(3):186–198.

Chapter 17

Distributed generation and microgrids

Marcelo Godoy Simões[a] *and Tiago Davi Curi Busarello*[b]

17.1 Introduction

The conventional electrical power system is organized by generation, transmission, distribution, and utilization of electrical energy. The main objective of a power system is to transport electricity from the producer (generator) to the consumer (load), while maintaining an acceptable level of reliability and voltage quality [1,2] and a certain degree of reliability for current distortion, or current quality. The generation is normally composed of large power plants, typically a variety of sources based on thermal, hydro, and nuclear energy. They are designed to be very reliable, and almost all produced electricity worldwide comes from these types of power plants. However, the current global scenario is undergoing strong efforts in seeking alternatives in electricity production that minimize the use of power plants based on fossil fuels. The reason for this shift is mainly because thermal plants are directly related to pollution and emissions, and in addition, their primary sources may be depleted within the next hundred years. Nuclear plants present safety concerns and are undesirable by population. Even hydropower plants cause environmental damages and it is not feasible for new installations.

In order to overcome these detriments, combined with increased prices for electricity, the deregulation of the power system industry has been allowing the development and implementation of distributed generators (DGs), particularly with advancements in the current technology.

DGs are located along the electrical power system and often use renewable or alternative energy sources for their prime mover. Some examples are wind, solar, biofuels, with studies of implementation of wave and tidal technologies. Despite the fact that batteries and fuel cells are not renewable, these technologies are also considered DG. DGs may bring potential benefits for both consumers and utilities. The users may reduce their energy bill and could potentially earn a profit when excess energy is sold to the grid. For utilities, DG may increase the hosting capacity and decrease losses when the connecting is close to the loads.

[a]Colorado School of Mines, Department of Electrical Engineering and Computer Science, Golden CO
[b]University of Campinas – UNICAMP, Campinas-SP, Brazil

Nowadays, there is a steady growth of DGs connection to the power grid. Therefore, all agents involved with the electric system are facing challenges in dealing with issues, which have never been problematic in the past. Examples include: reverse power flow in conventional unidirectional distributions system, parameter variation caused by the penetration of DG, likely resonance behavior caused by electronics converter output filters, power fluctuation due to the renewable energy behavior, voltage drops and voltage swells because of DG feeding, reactive power and harmonics, and protection of distribution grids and substations. Microgrids appear to be an attractive solution to deal with the high penetration of DG, because they can properly group a sub-system that is seen as "dispatchable" by the utility and grid operators [3,4].

This chapter presents an overview of DG and microgrids. In Section 17.2, the types of DGs are described with their mathematical models, their technical impacts on the power system and some constraints imposed by standardization. In Section 17.3, microgrids are presented and their features are briefly introduced. Throughout this chapter, a case study is conducted in order to clarify the presented concepts.

17.2 Distribution generators

The use of DG in the United States started in the 1970s; two energy crises on the same decade forced the USA Congress to develop an energy policy, named Public Utility Regulatory Policies Act (PURPA) [5]. The main focus of the PURPA was on small power production as an energy saving methodology. By the 1990s, the electrical system of the United States experienced a meaningful increase in the distribution generation, from 5% of the installed generation capability to 20% by the end of the decade – especially because of such open electricity market.

If the energy demand grows every year, DGs have the potential to manage such growth by, for instance, avoiding construction of new transmission lines by locating DG close to load centers. Additionally, the existent transmission lines benefit from a transmission bottlenecks reduction, allowing the transport of electricity to other regions due the minimization of losses.

DG may also be implemented in stand-alone systems. In this case, remote communities where the electric grid is not available may receive electricity by means of DGs. The use of storage facilities is essential for uninterrupted supply.

Despite many advantages, several technical issues must be taken into account before connecting DGs to the electric system. These issues are necessary in order to guarantee the continuity of energy supply with acceptable quality and they are usually presented in recommendation polices. One of the most common is the IEEE 1547-2003 which is a standard for interconnection of distributed resources with electric power systems (discussed towards the end of this section).

17.2.1 Examples of DGs

The electric sector is very dynamic and constantly improving and getting investments. Some DG technologies are indispensable for a modern grid and this section covers their benefits for the electric power system.

17.2.2.1 Wind energy-based DGs

Wind energy is converted from the kinetic energy of the air movement in order to produce electrical energy in a mechanical shaft, i.e. the process of converting kinetic energy into electrical energy is made by a wind turbine and an AC electrical generator. The wind turns the blades of a turbine, which rotates a shaft connected to the generator rotor, which in turn produces electricity. Wind power is considered to be the most promising new source of electrical energy [6]. However, wind energy-based DGs face unavoidable behaviors. For instance, the amount of energy the wind transfers to the rotor depends on the wind speed and on some machine parameters.

Additionally, the wind power presents intermittent behavior over different timescales, from less than a second to seasonal variations. The wind turbine output power is given by (17.1)

$$P_{\mathrm{wt}} = \frac{1}{2}\rho C_{\mathrm{p}} A v^3 \tag{17.1}$$

where ρ is the specific air density; A is the swept surface by the rotor blades; v is the wind speed; and C_{p} is power coefficient.

The power coefficient as originally discussed by Betz is ideally given by (17.2)

$$C_{\mathrm{p}} = \frac{1}{2}\left(1 - \frac{v_2^2}{v_1^2}\right)\left(1 + \frac{v_2}{v_1}\right) \tag{17.2}$$

where v_1 and v_2 are the airflow speed before and after the rotor blades, respectively.

The power coefficient is the only variable in (17.1) that can be controlled, as the wind speed is a random variable, air density depends on the location, and rotor area is fixed after the turbine is installed. According to Betz's law [7], the theoretical power coefficient presents a maximum value equal to 0.593. Figure 17.1 presents the typical curves of the power coefficient of the speed ratio (λ) and the blade pitch angle (β).

17.2.2.2 Solar energy-based DGs

Photovoltaics (PV) or solar cells are semiconductor devices that convert sunlight into direct current electricity. PV may be combined in modules. An array may be built by series association of modules. A PV array has the advantage of modularity, which allows PV power plants to be easily scaled and facilitates maintenance.

DGs based on PV have great advantages, making them an attractive solution as DGs. In most PV systems, the energy is produced at the same place where it is consumed, such as in the case of a building with PV modules installed on its roof. The energy is consumed in the same building. Moreover, PV systems benefit from the fact that the location of the Sun in the sky is very predictable, even for daily and seasonal timescales. By knowing the location of the Sun in the sky, the PV panels may change their tilt and orientation to track the Sun, resulting in higher electricity production. These tracking PV systems present a higher utilization factor compared to fixed systems, but at an increased cost.

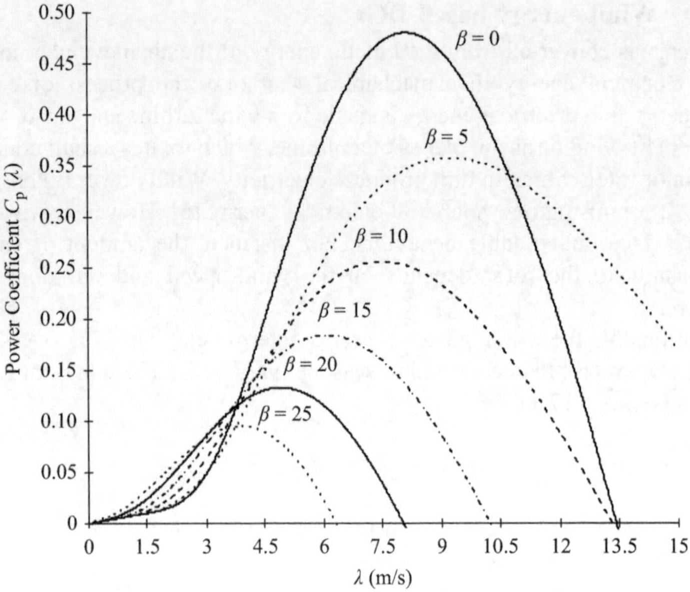

Figure 17.1 Typical curves of the power coefficient of the speed ratio (λ) and the blade pitch angle (β)

PV systems face some disadvantages. The most common are low efficiency and high cost. Typically, commercially available PV modules have efficiency around 6%–15%, but recently researchers have demonstrated modules with efficiencies as high as 30%. The cost per kilowatt is higher than other renewable technologies such as wind power. Additionally, PV faces a potential unpredictability: the clouds. Unlike the position of the Sun in the sky, clouds are hard to predict even for some hours in advance. Certainly, this jeopardizes the PV systems. Figure 17.2 shows how the sunpath, which is dependent on the day of the year and the latitude, will impose a constraint on array power.

Numerous electric circuits describing PV behavior have been presented in the literature [8,9] for modeling and simulation of PV cells and arrays. One of them is presented in Figure 17.3 [10]. The current source corresponds to the generation and the diodes the losses in the photocell. The resistances R_p and R_s are parasitic circuit elements that represent a real solar cell.

The PV cell output current is given by (17.3)

$$i = I_{ph} - I_{s1}\left[e^{\frac{v+iR_s}{kT}} - 1\right] - I_{s2}\left[e^{\frac{v+iR_s}{kT}} - 1\right] - \frac{v + iR_s}{R_p} \tag{17.3}$$

where v is the cell terminal voltage; I_{ph} is the photon current; I_{s1} is the D1 saturation current; I_{s2} is the D2 saturation current; R_s is the cell series resistance; R_p is the cell shunt resistance; A is the diode quality factor; V_g is the band gap voltage; k is Boltzmann's constant; and T is the ambient temperature.

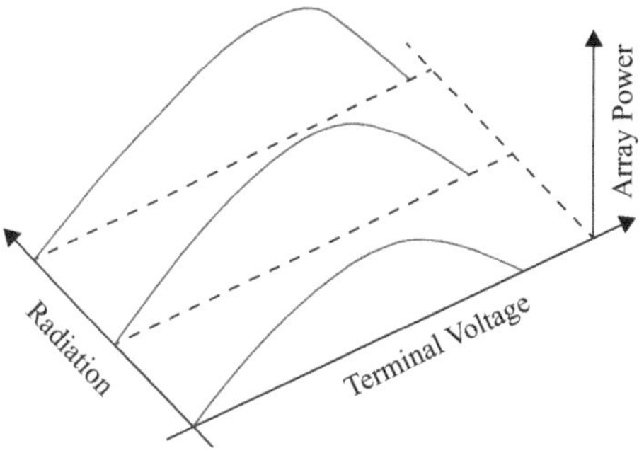

Figure 17.2 Array power variation dependent on the sunpath and radiation

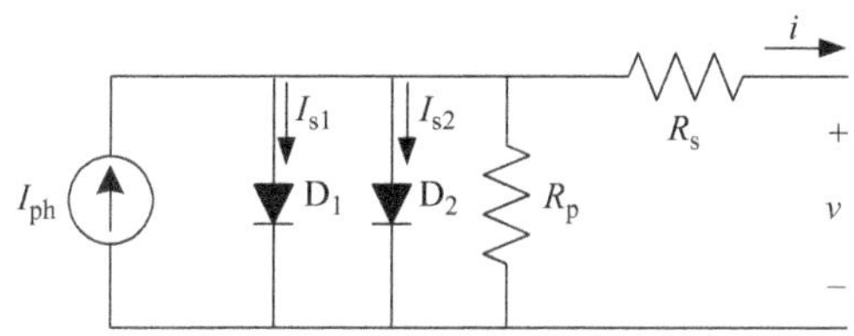

Figure 17.3 Electric circuit that described a PV behavior

Some factors that affect the PV output power include: soiling (dirt and dust at the panels), the wiring losses and conversion efficiency, and the cell temperature (PV power is inversely proportional to temperature).

17.2.2.3 Fuel cell-based DGs

Fuel cells are electrochemical devices that convert chemical energy into electrical energy. They produce electricity and heat when fuel is supplied. Fuel choices include natural gas, propane, methane gas from landfills, anaerobic digester gas, methanol, and hydrogen. Most fuel cells operated with hydrogen, as fossil fuels are finite sources and may be depleted within the next hundred years [10].

Fuel cell power systems are reliable, and a decentralized electricity generation solution for the future. Unlike a battery, which holds a limited fuel supply in a sealed container, a fuel cell requires an ongoing supply of fuel to create a continuous flow of electricity. Fuel cells are based purely on an electrochemical reaction instead of a combustion process and therefore they have no adverse pollutants. Along with that, the byproducts of a fuel cell are water and heat.

Figure 17.4 presents a fuel cell equivalent circuit. The thermodynamic potential, i.e. the reversible voltage (E_{Nernst}) is the fuel cell open-circuit voltage, which is represented by (v_{fc}). The voltage drops v_{act}, v_{con}, and v_{ohmic} are, respectively, due to the activation of the anode and cathode, the resulting from reduction of the reactant gases and the resistances to the conduction of protons through the solid electrolyte, and electrons through the external circuit.

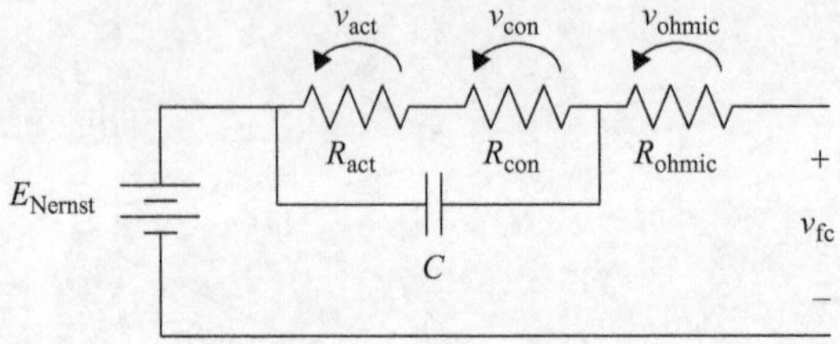

Figure 17.4 A fuel cell equivalent circuit

The reversible voltage is the maximum voltage that the cell can provide for certain operating conditions and it is independent of the load. Its mathematical model is developed from the Nernst equation, which is related to the change in the Gibbs free energy. The reversible voltage is given by (17.4)

$$E_{\text{Nernst}} = \frac{\Delta G}{2F} + \frac{\Delta S}{2F}(T - T_{\text{ref}}) + \frac{RT}{2F}\left[\ln\left(p^*_{\text{H2}}\right) + \frac{1}{2}\ln\left(p^*_{\text{O2}}\right)\right] \tag{17.4}$$

where ΔG is the change in the Gibb free energy (J mol^{-1}); F is the Faraday constant (96,487 C); ΔS is the change of entropy (J mol^{-1}); R is the universal gas constant (8,314 J K^{-1} mol^{-1}); p^*_{H2} and p^*_{O2} are the partial pressures (atm) of the hydrogen and oxygen, respectively; T is the absolute temperature of the operating cell (K); and T_{ref} is the reference absolute temperature in kelvin (298.15 K).

The activation voltage is due to the losses of energy used to break the activation barrier of the chemical reaction and it is given by (17.5)

$$V_{\text{act}} = -\left[\xi_1 + \xi_2 T + \xi_3 T \ln\left(C^*_{\text{O2}}\right) + \xi_4 T \ln(i_{\text{FC}})\right] \tag{17.5}$$

where i_{FC} is the cell operating current (A); ξ_1, ξ_2, ξ_3 and ξ_4 are parametric coefficients for each model. These values are defined based on theoretical equations with kinetic, thermodynamic, and electrochemical foundations. C^*_{O2} is the oxygen concentration on the cathode catalytic interface (mol cm^{-3}).

The concentration voltage drop is due to the losses caused by the changes in the concentration of the reactants in the electrodes and it is given by (17.6)

$$V_{\text{con}} = -B \ln\left(1 - \frac{J}{J_{\text{max}}}\right) \tag{17.6}$$

where J_{max}, J, and B are the maximum current density, the actual current density of the cell, and a constant that depends on which electrode the loss occurs, respectively.

The ohmic voltage drop is due to the resistance of the movement of the electrons through the electrodes and the movement of protons through the membrane.

17.2.2.4 Diesel generators

Generators can run on different types of fuels. Back-up generators usually run on natural gas or propane. A diesel generator is usually used for large commercial applications. Most diesel generators run at 1,800 rpm and are water-cooled, making them last longer and run more quietly than the 3,600 rpm air-cooled gasoline generator. The 3,600 rpm generators are smaller and lighter and are used for portable applications. A diesel generator usually lasts 15,000–30,000 hours before it requires major maintenance [11].

Diesel generators have some disadvantages over some other types of DG. They are noisy, costly, emit pollution, and need fuel storage tanks and transportation. However, they are able to supply constant power and are reliable in offering long-term power supply. When compared to other types of fuels, diesel generators have some advantages. They are cheaper to operate and have better efficiency due to the higher compression rates of the fuel. Diesel fuel has a very high energy density, making it more effective to store and transport. Fuel costs per kWh produced are from 30% to 50% lower than other fuels. Diesel generators range between 8 and 2,000 kW. The most common range is between 20 and 500 kW.

17.2.2.5 Microturbines

Microturbines are generators that produce electricity by burning a fuel. There is a variety of fuels that can be used in microturbines, with natural gas being the most common. Microturbines differ from conventional generation turbines in that they are very small and operate at very high speeds. Microturbines have several advantages, such as low emissions, very fast response to load variation, low weight per horsepower, and a liquid cooling system is not required. Some drawbacks of them are low efficiency (28%–32%), sensitivity to ambient condition, and maintenance requirements.

Microturbines may be found in several applications. One of the most common is the combined heat and power (CHP) in which the produced electricity is used for a specific purpose and the heat generated is used to produce hot water or to heat building space. Other applications of microturbines are in backup/standby power station and microgrids [12].

17.2.2.6 Other types of DG

Other types of DG may be found in prototypes or pilot projects. This is the case of tidal, geothermal, and wave generators; Iceland, for example, gets about 25% of their electricity from geothermal sources. Such new technologies and their installation as commercial plants depend on their proof of technical, economic, and political feasibility studies.

17.2.2 Technical impacts due to DG

The installation of DG into the electric system changes some electrical quantities. Some examples are losses and voltage profile. Figure 17.5 presents an equivalent model for a traditional distribution system. The system contains the main generation, 11 feeders, linear and nonlinear loads, and a DG connected at the feeder 9. This system will be used throughout of this chapter as a case study.

The case study evaluates the impact on the power losses along the system due to the connection of DG at feeder 9. The DG is considered ideal and its nominal

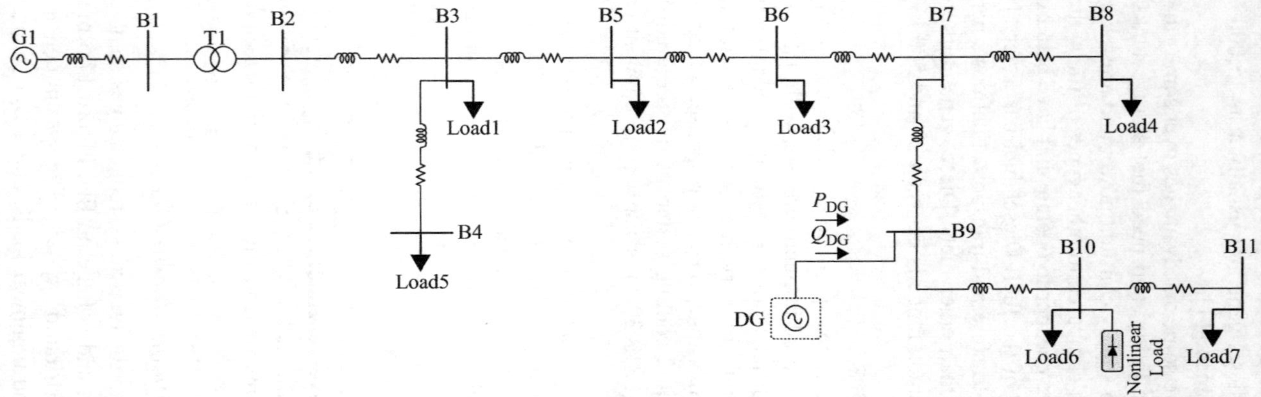

Figure 17.5 Example of distribution system

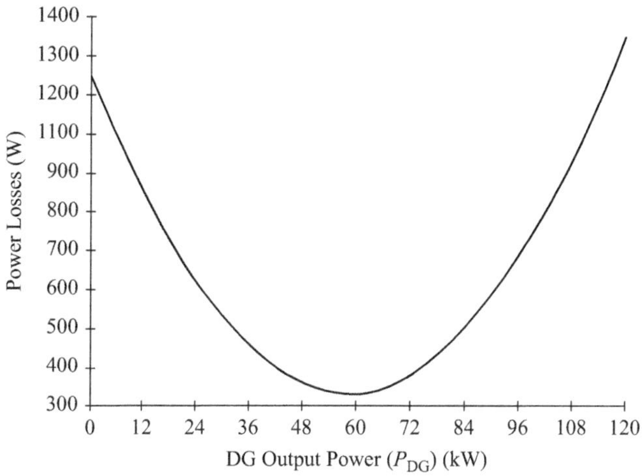

Figure 17.6 Power losses through the feeder for phase A according to the amount of active power the DG is injecting

power is 150 kW. The results were collected in steady-state conditions. The distribution transformer was handling 75 kVA apparent power at a power factor of 0.91, without the DG. Figure 17.6 presents the power losses through the feeder for phase A according to the amount of active power the DG was injecting. The DG was operating with unity power factor, resulting in null reactive power. Initially (at $P_{DG} = 0$), the DG was not injecting any power and the power losses were 1,142 W. As the DG injected more active power, the losses began to reduce. However, for values higher than 110 kW, the losses began to be higher than the case with no DG. This phenomena occur mainly due to the fact the DG is injecting more active power than is required by the system.

The case study leads to the evaluation of the voltage profile. Figure 17.7 presents the voltage profile in all feeders against the injected power. With no DG, it is possible to notice all feeders have voltage below the nominal value. As the amount of injected power rises, the voltage profile also rises. At feeder 9, where the DG is connected there is a noticeably sharper increase.

From the previous case, it can be concluded that the amount of power injected affects the voltage profile. Therefore, it is possible to take advantage of this capability in order to use DG as a voltage regulator. For example, suppose the desire is to raise the voltage at feeder 9 by 25%. To accomplish this, the DG injected power increases. However, the DG may reach its nominal output power without accomplishing this goal. To overcome that, the DG may also inject reactive power concomitantly. The technique to inject or consume reactive power in order to regulate the voltage profile is the principle of operation of a static synchronous condenser (STATCOM). DGs may have a STATCOM function embedded. Figure 17.8 presents the voltage profile in all feeders for some values of reactive power. This study is performed keeping the active power equal to 35 kW.

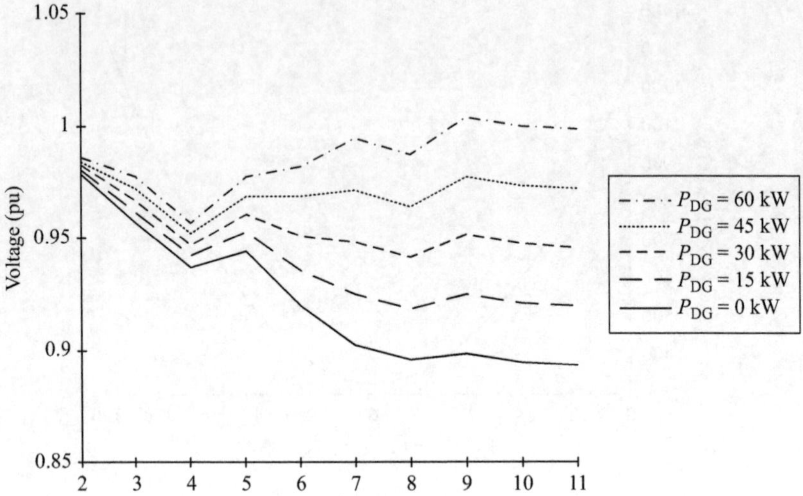

Figure 17.7 Voltage profile in all feeders versus the injected power

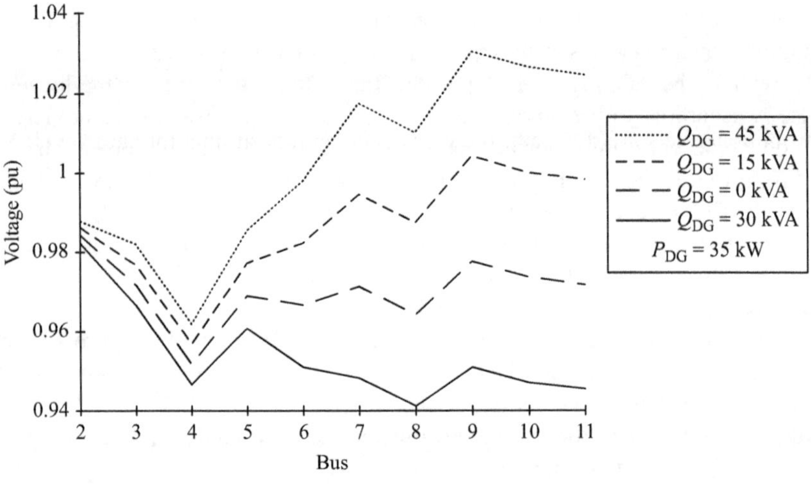

Figure 17.8 Voltage profile in all feeders for some values of reactive power

Likewise, DG can be used to reduce feeder voltage by consuming reactive power. The connection of DG into the system may also change the power factor, the voltage and current Total Harmonic Distortion (THD) the unbalance, and so on. The analysis of such power quality indices is beyond the scope of this chapter [1].

17.2.3 IEEE1547

As could be observed in the previous subsection, the DG penetration affects quantities in the electric system. In order to avoid causing damage or improper

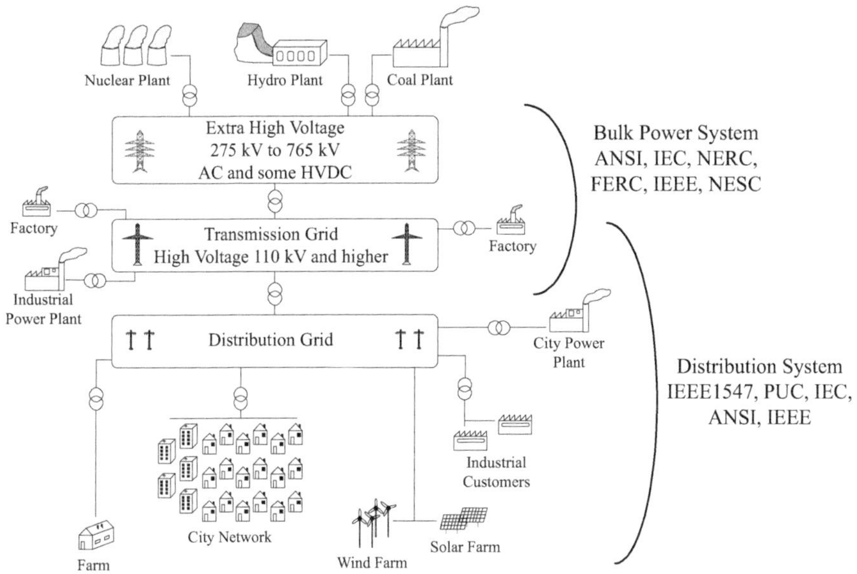

Figure 17.9 A system diagram showing the standards in each level

operation, there are some standards which suggest recommendations to be followed for safe connection of new DGs. As the design and behavior of the transmission and distribution levels are different, there are different constraints for each level. Most of the standardization does not allow DG to perform other tasks more than just injecting active power with unit power factor.

Figure 17.9 presents a simplified diagram showing the system levels and the standardization applied for each one. Bulk power and distribution systems have specific standardization while the transmission faces a conflict. By the moment of writing this chapter, there is no consensus about what is restricted to the transmission system.

One of the most important standards related to the interconnection of DG into the electrical distribution system is IEEE 1547 (parts of which are also incorporated into UL 1741) [13]. IEEE 1547 establishes criteria and requirements that all DG must adhere to in order to connect to the electrical power system. The requirements shall be met at the point of common coupling and they are applicable to all distributed resources technologies with aggregate capacity of 10 MVA or less.

The criteria concern mainly voltage and frequency variation limits, synchronization aspects, intentional and unintentional islanding, response to abnormal conditions, etc. The readers are encouraged to read IEEE 1547 in its entirety.

17.3 Microgrid

A microgrid is a set of DG and loads that are managed by intelligent controls. The installed capacity in a microgrid is capable of supplying the local demand.

Microgrids promise substantial environmental benefits by facilitating the integration of renewables such as PV arrays and small wind turbines. A microgrid can also be reliant on micro-CHP generators that are driven by the production of heat mostly during winter and PV generators that provide electrical energy during summer. Some storage capability can also enhance the microgrid [14,15]. This section presents introductory concepts regarding microgrids.

17.3.1 DC and AC microgrids

Microgrids can be found in DC and AC architecture. In DC microgrids, the distribution of energy is made by direct current and has positive and negative conductors. One of the main advantages of DC microgrids is the absence of frequency and phase dependences among AC generators. Concerning converters, most found in DC microgrids are DC–DC and AC–DC [16].

On the other hand, AC microgrids are very similar to the conventional system. Therefore, it is possible to apply conventional power system operational concepts, such as power flux control, protection schemes, fault detection, and so on [17].

17.3.2 Stand-alone microgrids

A microgrid operating in the absence of a stiff voltage source is considered to be stand-alone. In this scenario, all the power required by the microgrid is supplied by the local DGs. Consequently, sophisticated, coordinated control is used in order to guarantee the reliability, stability, and uninterrupted supply of the micrgrogrid. In AC microgrids, one of the DGs must serve as a voltage and frequency source, to which all the other DGs will then synchronize. Power and frequency droop controls are used in AC microgrids in order to share the supplied power among the DG while voltage droop control is used in DC microgrids. Details about droop control in AC and DC microgrids are found in [18].

17.3.3 Grid-tied microgrids

In this case, the microgrid is attached to one or more stiff generators. The frequency and voltage are imposed by a stiff generator and the DGs must be synchronized to it. Controllers based on droop technique are also applied to grid-tied microgrids.

Grid-tied microgrids must have a feature called "anti-islanding". Islanding is a scenario in which the microgrid was operating with the main grid and suddenly operates alone, usually because a fault in the main grid. As a result, the microgrid alone supplies all the energy for the loads located downstream. But this scenario is not desired by system operators and must be avoided for several reasons. One reason is the uncertainty of when the main grid will return and in what phase angle. Such angle may be different from the current islanded microgrid angle and the reconnection would cause severe transitory or even unpredictable behavior. Therefore, the microgrid must be equipped with anti-islanding detection.

The anti-islanding feature is the capability of the microgrid to switch-off from the main grid in an occurrence of fault. In this case, all loads would be de-energized. Numerous methodologies to implement the anti-islanding feature are presented in the literature [19,20]. There are special cases where some loads are kept energized by the microgrid even during an outage. In this case, the microgrid has a phase-locked loop and feeder switches in order to allow the main grid reconnection with no severe transient. It is important to highlight the fact that an islanded microgrid is different from a stand-alone microgrid. As mentioned before, the first is an undesirable situation right after a fault occurrence and the second is designed for operating without the main grid.

17.3.4 Centralized control

Regardless of whether the microgrid is grid-tied or stand-alone, it is evident that a controller must coordinate the DGs located within the micrgrogrid. Two different control structures are commonly employed: centralized and local. Here, the centralized control is described and the next subsection describes the local control.

In a centralized control architecture, a single controller is intended to manage the entire microgrid; this type of control is advantageous when the whole distribution system has some specific goals, for example, loss reduction. This may be achieved by setting DG injected power set points. The decision about the optimum set point for the DGs is usually based on an optimized algorithm. Voltage and current along the system are the primary inputs for these algorithms. From them, a variety of information can be obtained, such as apparent power, power factor, unbalances, etc. Such variables are good for real-time decisions. But the centralized control may also be capable of daily planning. Input variables about the prediction of load demand, solar irradiation, and wind speed and energy price along one day may be helpful for an optimum planning. Storage devices located along the system may have their contribution to the uninterrupted supply improved by precise scheduling of their charging and discharging.

Figure 17.10 presents the behavior of the solar irradiation, the load demand curve, and the wind speed along one day. The centralized control can take these curves and run an algorithm to best determine how the DG and storage elements should operate along the day, given this date. As a result, the system may offer better reliability, stability, and lower cost, than the traditional distribution system, with fewer service interruptions.

17.3.5 Conventional droop control method

The droop control is beyond the most popular control techniques that has a long history of use for the synchronous generator control in power system. Recently, it has been used for parallel-inverter control, especially in case of inverter-dominated microgrids. This section will provide a brief review on the conventional power droop framework and fundamental concepts.

Nowadays, microgrids are made up of different type of distributed energy production resources, such as solar panels, fuel cells, wind turbines and so on. Every resource needs a power electronic interface to transfer the energy to the common bus. Without loss of generality, we can model a single interface as an inverter connected to the common bus through a decoupling impedance, as shown in Figure 17.11 [21].

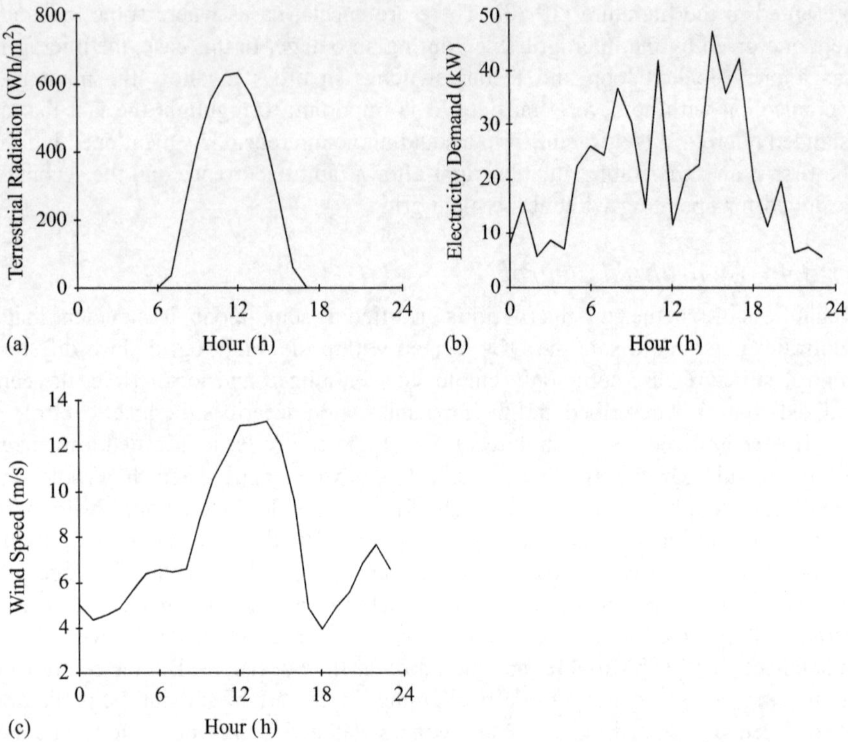

Figure 17.10 Behavior of the terrestrial irradiation (a), the load demand curve (b), and the wind speed (c) along one day

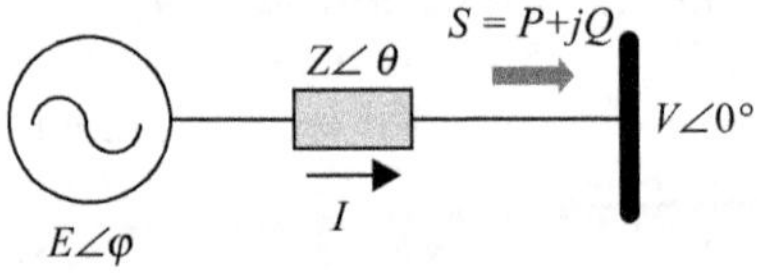

Figure 17.11 Equivalent circuit of an inverter connected to a bus

The active and reactive power transmitted across a lossless line ($\theta = 90$) are:

$$p = \frac{EV}{X} \sin \phi \tag{17.7}$$

$$Q = \frac{EV \cos \phi - V^2}{X} \tag{17.8}$$

Since the power angle ϕ is typically small, we can simplify (17.7) and (17.8) further by using the following approximations ($\sin(\phi) = \phi$, and $\cos\phi = 1$):

$$\phi \approx \frac{PX}{EV} \tag{17.9}$$

$$(E - V) \approx \frac{QX}{E} \tag{17.10}$$

From the above equations, it can be derived that the active power is predominately dependent on the power angle ϕ, while the reactive power mostly depends on the output-voltage amplitude. Following droops are defined for the amplitude and the frequency of the inverter output voltage:

$$\omega = \omega^* - K_p P \tag{17.11}$$

$$E = E^* - K_q Q \tag{17.12}$$

where ω^* and E^* are the output voltage angular frequency and amplitude at no load, respectively, and K_p and K_q are the droop coefficients for the frequency and amplitude, respectively. Equations (17.11) and (17.12) are plotted in the characteristics as shown in Figure 17.12.

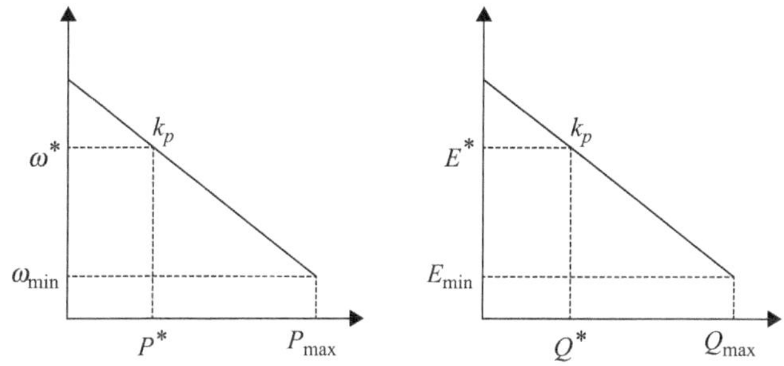

Figure 17.12 Typical frequency and voltage droop characteristics

When frequency falls, the output power of the generating unit is allowed to increase. A falling frequency indicates an increase in loading and a requirement for more active power. Multiple parallel units with the same droop characteristic can respond to the fall in frequency by increasing their output active powers simultaneously. The increase in output active powers will counteract the reduction in frequency and the units will settle at output active powers and frequency at a steady-state point on the droop characteristic. The droop characteristic therefore allows multiple units to share load without the units fighting each other to control the load. The same logic above can be applied to the voltage droop characteristic.

17.3.6 Local control

Local control usually deals with more information than centralized. Several details about the grid may be transferred among DGs connected close to each other details which may be unnecessary or undesirable for the centralized control. An example is the battery temperature information on battery-based storage systems. This information may be useful for local control. A nearby DG may support the storage in order to get the situation normalized. This alleviates the need for that information to travels to the centralized control, and the response to return to the nearby DG.

Local controls may be involved with one or more DGs. They may have goals to improve a local performance in a single feeder, for instance. Consider the same traditional grid depicted in Figure 17.13. A battery-based storage is connected at feeder 9 and the DG already connected is a wind-based generator. Figure 17.13 presents the detail of feeder 9 including the wind- and storage-based DGs. The output power of the wind-based DG is very dependent on climate conditions. Therefore, the storage can support by supplying power in order to minimize the intermittency of the wind-based DG. For that, the storage-based DG needs information about the instantaneous output power of the wind-based DG. This information is transferred locally between them through the local control.

Once the storage has this information, its output power may be set to be the difference between the desired and actual instantaneous power output of the wind-based DG. Figure 17.14 shows the wind-based DG output power, the storage output power, and the total power being injected in feeder 9. The wind-based DG has a variable output power, but when combined with the storage output power, the total power flow into feeder 9 is constant.

The local control allows the DG and the storage to operate in combination for a constant power supply at feeder 9. Other strategy could be applied such as making the DG plus storage to inject active power equals to feeder 9 downstream power. The local control may have its own features for specific applications.

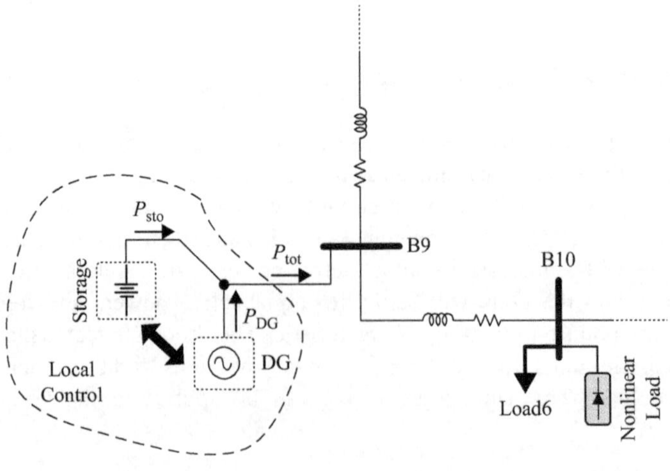

Figure 17.13 Detail of feeder 9 including the wind- and storage-based DGs

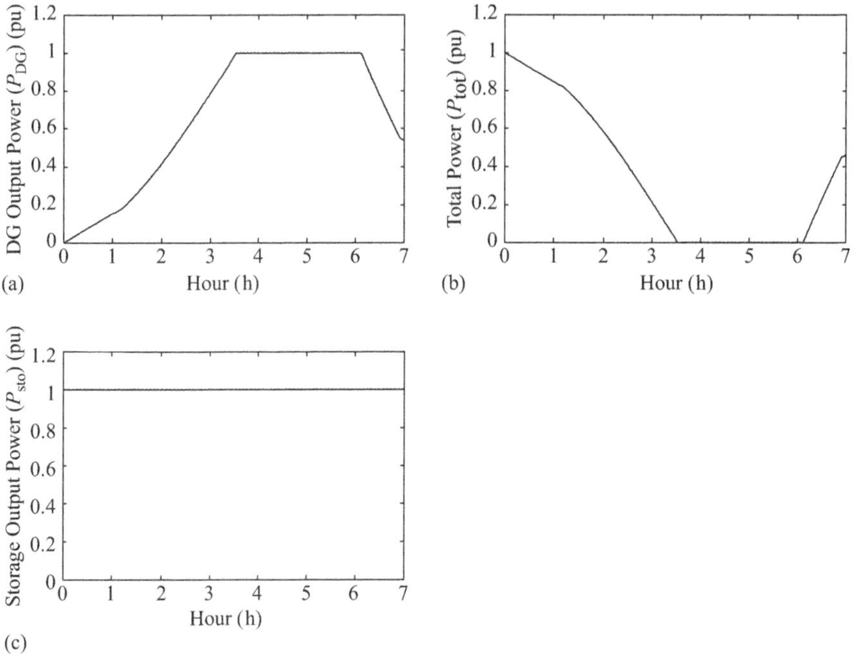

Figure 17.14 The wind-based DG output power (a) the storage output power (b) and the total power (c) being injected in feeder 9

17.3.7 Multifunctionalities

DG may perform different functions other than just injecting power into the grid. Once DGs have power electronics converters with embedded microprocessors, functions like voltage distortion regulation and power quality improvement may be aggregated. Such functions can be performed simultaneously with power injection. Even though almost all system operators do not presently allow DG to perform these ancillary functions, these issues will certainly be under discussion in the coming years, primarily due to the simplicity of employing additional function. Applications of multifunctional compensator can be found in [22].

The case study now presents a possible scenario in which a DG operates not only on its basic function, but also as shunt compensator. Coming back to the system presented in Figure 17.4, the DG is set to operate exclusively as an active filter. The DG compensates all harmonic currents downstream of feeder 9. Figure 17.15 presents the incoming feeder 9 current in the moment where the DG begins to operate exclusively as active filter. Initially, the current is distorted due to the nonlinear load connected at feeder 10. At $t = 2.25/60$ ms, the DG begins to compensate only the harmonic currents. Consequently, the incoming feeder 9 current becomes sinusoidal.

The scenario in which a DG operates exclusively as active filter may happen when the energy price is not economically feasible for the DG to sell it. It is important to highlight that active filtering does not require active power processing, except for the losses.

The case study now leads to a scenario that the DG supplies all the power consumed by the feeder 9 downstream system. All the power means active, reactive, and harmonics quantities. Figure 17.16 presents the DG output current in the

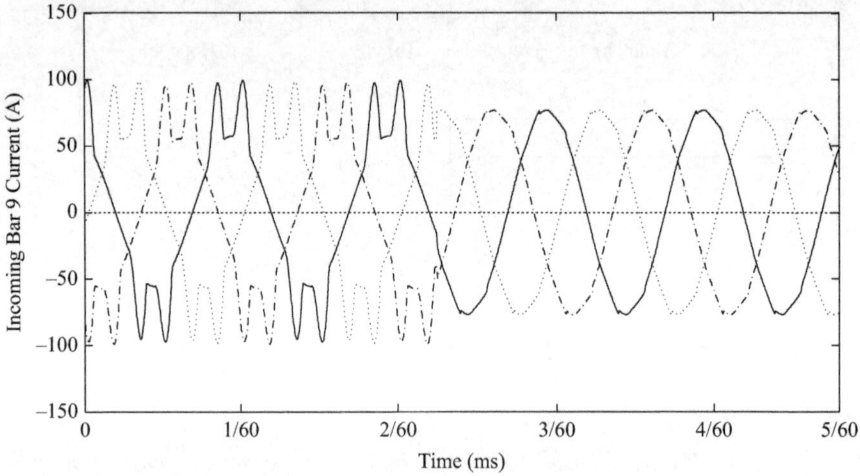

Figure 17.15 Incoming feeder 9 current at the moment the DG begins to operate exclusively as active filter

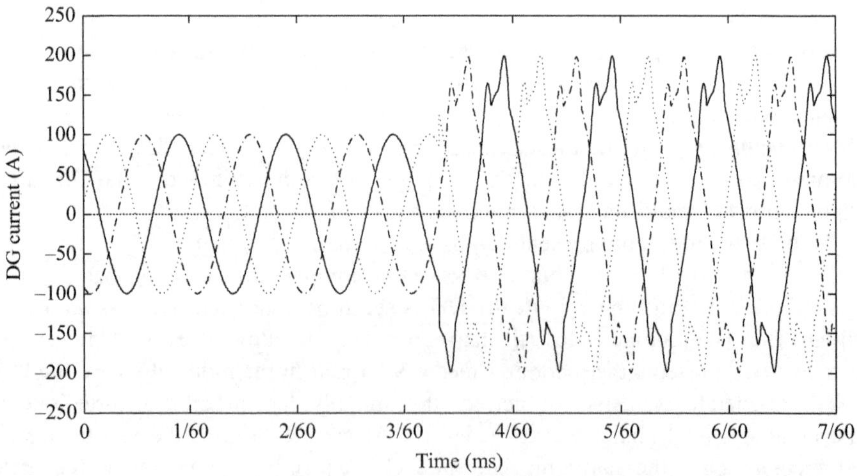

Figure 17.16 The DG output current at the moment the DG begins to supply all the power consumed by feeder 9 downstream system

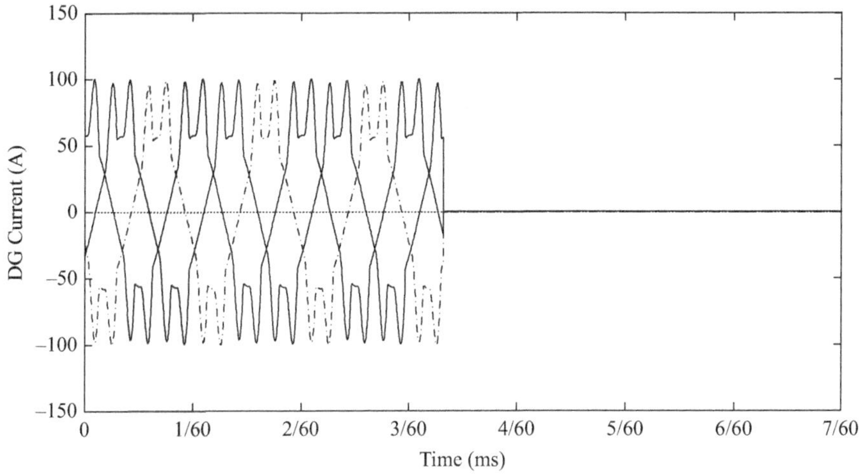

Figure 17.17 The incoming feeder 9 current for the same scenario

moment where the DG begins to supply all the power consumed by feeder 9 downstream system. Initially, the DG is supplying active power with unity power factor. At $t = 3.35/60$ ms, the DG begins to supply all the power. The current now contains distortions due to the nonlinear load.

Figure 17.17 presents the incoming feeder 9 current for the same scenario. The current is zero after $t = 3.35/60$ ms, proving that the DG is supplying all the current at feeder 9 downstream.

References

[1] M. H. Bollen and F. Hassan. *Integration of Distributed Generation in the Power System*. Vol. 1. Wiley-IEEE Press, Piscataway, NJ, 2011.

[2] A. Reznik, M. Godoy Simões, A. Al-Durra, and S. M. Muyeen. LCL filter design and performance analysis for grid interconnected systems. *IEEE Transaction on Industry Applications*. 2014, Vol. 50, No. 2, pp. 1225–1232.

[3] R. Carnieletto, D. I. Brandão, S. Suryanarayanan, F. Farret, and M. Godoy Simões. Smart grid initiative. *IEEE Industry Applications Magazine*. 2011, Vol. 17, No. 5, pp. 27–35.

[4] X. Y. Yu, C. Cecati, T. Dillon, and M. Godoy Simões. The new frontier of smart grids. *IEEE Industrial Electronics Magazine*. 2011, Vol. 5, No. 3, pp. 49–63.

[5] H. L. Harkins. PURPA: New horizons for electric utilities and industry. *IEEE Trans. Power Appar. Syst.* 1981, Vol. 100, No. 6, pp. 2784–2789.

[6] M. Godoy Simões, E. Muljadi, M. Singh, and V. Gevorgian. Measurement-based performance analysis of wind energy systems. *IEEE Instrumentation and Measurement Magazine*. 2014, Vol. 17, No. 2, pp. 15–20.

[7] M. Godoy Simões, B. K. Bose, and R. J. Spiegel. Fuzzy logic based intelligent control of a variable speed cage machine wind generation system. *IEEE Transactions on Power Electronics.* 1997, Vol. 12, pp. 87–95.

[8] M. G. Villalva, J. R. Gazoli, and E. R. Filho. Comprehensive approach to modeling and simulation of photovoltaic arrays. *IEEE Transactions on Power Electronics.* 2009, Vol. 24, No. 5, pp. 1198–1208.

[9] Y. A. Mahmoud, X. Weidong, and H. H. Zeineldin. A parameterization approach for enhancing PV model accuracy. *IEEE Transactions on Industrial Electronics.* 2013, Vol. 60, No. 12, pp. 5708–5716.

[10] C. S. Uriarte. *Electrical Model Development of Distributed Resources and the Integration with the Electric Grid*, Master Thesis, Colorado School of Mines, USA, 2007.

[11] M. Godoy Simões, B. Palle, S. Chakraborty, and C. Uriarte. *Electrical Model Development and Validation for Distributed Resources.* National Renewable Energy Laboratory, 2007.

[12] M. Godoy Simões and F. A. Farret. *Modeling and Analysis with Induction Generators.* 3rd edition. Taylor and Francis/CRC Press, Boca Raton, FL, December 2014.

[13] 1547-2003, IEEE Std. *IEEE Standard for Interconnecting Distributed Resources with Electric Power Systems*. 2003.

[14] M. Godoy Simões, B. Blunier, and A. Miraoui. Fuzzy-based energy management control: Design of a battery auxiliary power unit for remote applications. *IEEE Industry Applications Magazine.* 2014, Vol. 20, No. 4, pp. 41–49.

[15] N. Hatziargyriou. *Microgrids: Architectures and Control.* 1st edition. Wiley-IEEE Press, Chichester, 2014.

[16] L. Che and M. Shahidehpour. DC microgrids: Economic operation and enhancement of resilience by hierarchical control. *IEEE Transactions on Smart Grid*. 2014, Vol. 5, No. 5, pp. 2517–2526.

[17] M. Godoy Simões, R. Roche, E. Kyriakides, S. Suryanarayanan, B. Blunier, K. McBee, P. Nguyen, P. Ribeiro, and A. Miraoui. A comparison of smart grid technologies and progresses in Europe and the US. *IEEE Transactions on Industry Applications.* 2012, Vol. 48, No. 4, pp. 1154–1162.

[18] S. Chakraborty, M. G. Simões, and W. E. Kramer. *Power Electronics for Renewable and Distributed Energy Systems*. 1st edition. Springer, London, 2013.

[19] Zhihong Ye, A. Kolwalkar, Y. Zhang, Pengwei Du, and R. Walling. Evaluation of anti-islanding schemes based on nondetection zone concept. *IEEE Transactions on Power Electronics.* 2004, Vol. 19, No. 5, pp. 1171–1176.

[20] M. Tedde and K. Smedley. Anti-islanding for three-phase one-cycle control grid tied inverter. *IEEE Transactions on Power Electronics.* 2014, Vol. 29, No. 7, pp. 3330–3345.

[21] J. M. Guerrero, L. García de Vicuña, J. Matas, M. Castilla, and J. Miret. A wireless controller to enhance dynamic performance of parallel inverters

in distributed generation systems. *IEEE Transactions on Power Electronics.* 2004, Vol. 19. No. 5, pp. 1205–1213.

[22] T. D. Curi Busarello and J. Antenor Pomilio. Bidirectional multilevel shunt compensator with simultaneous functionalities based on the conservative power theory. *7th IET International Conference on Power Electronics, Machines and Drives (PEMD 2014).* April 8–10, 2014, pp. 1–6.

Chapter 18

Uninterruptible power supplies

Luke G. Weber and Adel Nasiri

18.1 Introduction

Uninterruptible power supply (UPS) systems have been common tools to supply and protect critical loads when the main supply ceases to provide power or the quality of power does not meet load requirements. The need for UPS systems has increased with advancements in information technology, sensitive electronic equipment, and mission critical systems. The UPS concept has moved from rotary [1–5] to off-line [6], to on-line [7], and line-interactive systems [8–13], and evolved into multi-layer, multi-bus systems supporting complex infrastructure such as data centers [14, 15].

In this chapter, various UPS system topologies are described, control methods are explained, and some applications are discussed.

18.2 Topologies

There are typically two modes of operation common to all types of UPS systems: normal mode and stored energy mode. Bypass mode is available in some system configurations. In normal mode, power to the load and energy storage device is supplied by the utility or customer-owned electric power system. In stored energy mode, connection to the electric power system has been severed, and the load is served from the energy storage device. When the system is so equipped, bypass mode is available in the event of UPS system failure, or so that system maintenance may be performed.

18.2.1 On-line UPS systems

In normal mode, on-line UPS systems are designed to serve load isolated from the electric power system through a series rectifier and inverter combination as shown in Figure 18.1. A battery bank is connected to the DC bus between these two converters, and serves the load when the static switch opens. A Normally Open (N.O.) static

Center for Sustainable Electrical Energy Systems, University of Wisconsin-Milwaukee, 115 East Reindl Way, Rm. 201Q, Milwaukee, WI 53212-1255, USA

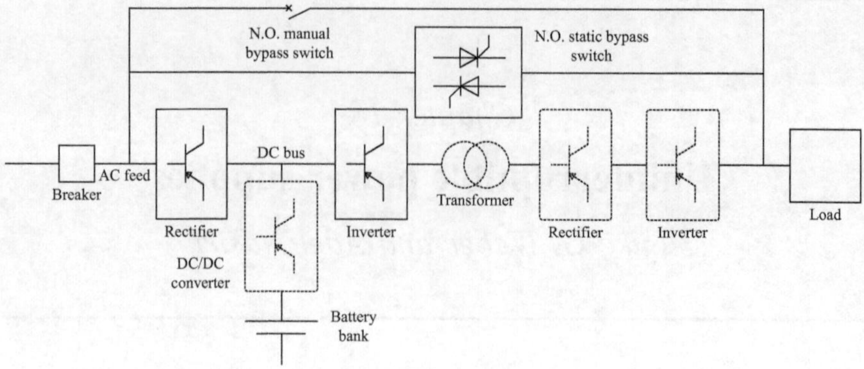

Figure 18.1 An on-line UPS system with optional equipment shown with dashed lines

bypass switch closes upon failure of the UPS system proper. UPS system maintenance is facilitated by the installation of a N.O. manual bypass switch. Battery capacity and connected load determine the duration the battery bank can serve load.

A phase-locked loop (PLL) is used to keep the output voltage in phase with the input waveform to ensure a smooth transition to the electric power system in the event of UPS failure. The transition from energy storage mode to normal mode is also abetted by the PLL, as the output voltage phase is aligned with the input electric power system waveform prior to returning to normal mode.

The advantages of this configuration make it the most common arrangement. The transfer time from normal mode to energy storage mode is zero. The system is highly tolerant to input voltage and frequency variations. The system is capable of providing precise output voltage and frequency regulation except in bypass mode.

Disadvantages of the on-line UPS system include the rectifier capacity and cost, which must be rated to serve 100% of the AC load and DC battery charging load. The system typically has a low power factor and exhibits high current waveform distortion at the AC input terminals. On-line UPS systems have low efficiency in normal mode due to the double conversion topology, with full power flowing through both stages. The system also exhibits poor performance when serving non-linear loads in normal and energy storage modes.

For galvanic isolation, a transformer may be added to the output stage of the system. The secondary terminals of a line frequency transformer connect directly to the load. High-frequency transformation requires another rectifier and inverter. These components are shown in dashed lines in Figure 18.1. A DC–DC converter may also be used to better control battery charging and discharging. These components increase the size, complexity and cost of the UPS, and reduce the system efficiency.

18.2.2 Off-Line UPS

One off-line UPS system configuration is shown in Figure 18.2. Under normal operating conditions, the load is served from the incoming AC line through a normally closed (N.C.) static switch. The rectifier has a lower rating than that of the on-line UPS system because it serves only the DC battery charging load. The inverter must be rated to carry 100% of the connected load as is the case with an on-line UPS system. Transfer time can be as fast as ¼ line cycle, and the inverter may be used for line conditioning or power factor correction during normal operation.

When the static switch opens, load is served from the battery bank through the inverter, and both voltage and frequency regulation are possible.

In Figure 18.3, a second off-line UPS system arrangement is shown. A three-winding transformer is used in this configuration to attain galvanic isolation for the load. To achieve this goal with the design shown in Figure 18.2, a two-winding

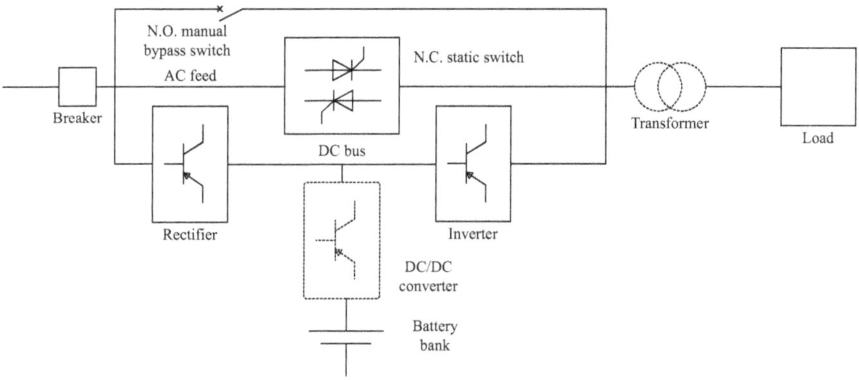

Figure 18.2 An off-line UPS system showing optional components using dashed lines

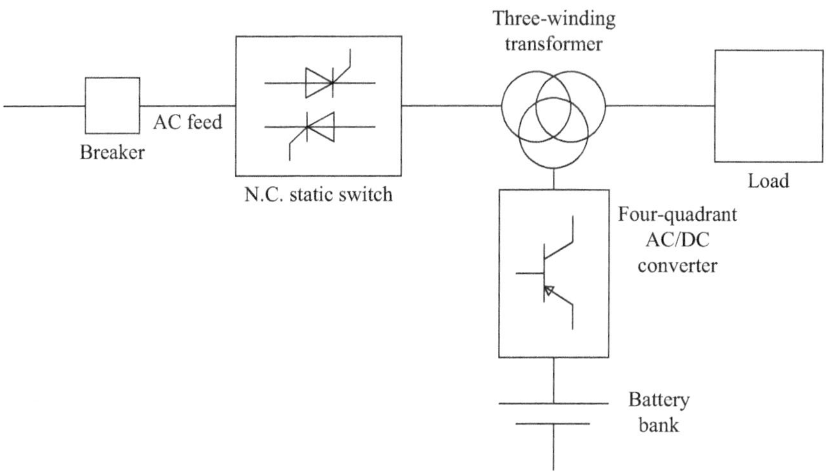

Figure 18.3 An off-line UPS system using a three-winding transformer to attain galvanic isolation for the load

transformer – shown in dashed lines – adds complexity and cost. The four-quadrant converter and controls for Figure 18.3 design are more sophisticated than converters in Figure 18.2. In normal mode, the load is served from the incoming AC line, and the four-quadrant converter is used to charge the battery bank, and can provide line conditioning or power factor correction. Frequency at the load is dictated by the incoming line.

When the static switch opens, the load is fed from the batteries through the four-quadrant inverter, and both voltage and frequency regulation are feasible.

The advantages of off-line UPS systems are lower initial cost, lower losses, and smaller size than on-line systems. Under normal conditions, the off-line arrangement facilitates line conditioning and power factor correction. Disadvantages include lack of isolation from the incoming electric power system and limited output voltage regulation in normal mode, finite transfer time from normal to energy storage mode, poor performance serving non-linear loads in energy storage mode, and a maximum capacity of about 2 kVA [16].

18.2.3 Line-interactive UPS

Line-interactive UPS systems may operate in on-line or off-line configurations, but most operate as on-line systems to improve load power factor or regulate the output voltage [17]. A defining characteristic of this arrangement is the series inductor shown in Figure 18.4. Under normal conditions, power factor control or voltage regulation is achieved by varying the injection of reactive current from the four-quadrant converter. As with the off-line UPS system, load frequency is dictated by the incoming AC electric power system.

When the static switch opens and the UPS is operating in stored energy mode, line-active systems can regulate voltage and frequency.

Using on-line UPS systems as the benchmark, line-interactive UPS systems are simpler, less expensive, and more efficient. Efficiency improvement comes about by avoiding double conversion through a rectifier and inverter in normal mode. Active line current filtering reduces harmonic distortion of the incoming AC line voltage waveform.

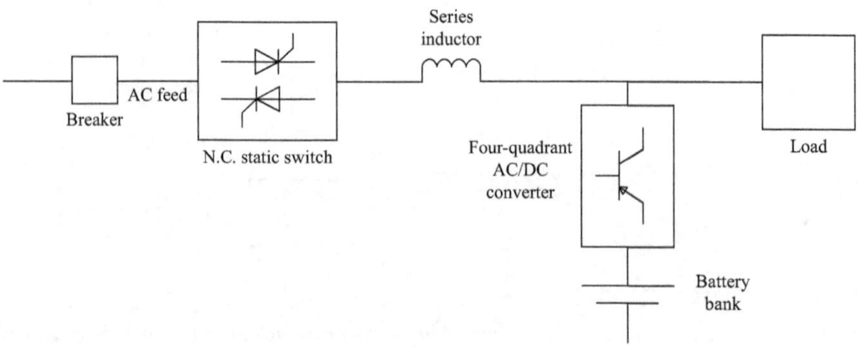

Figure 18.4 A typical line-interactive UPS system configuration

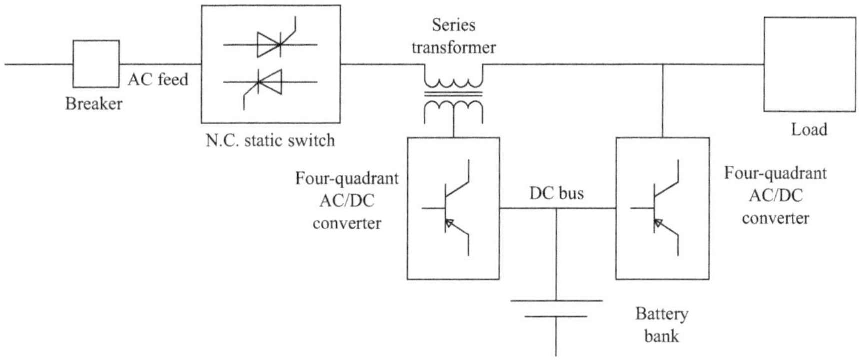

Figure 18.5 Diagram of a series–parallel line-interactive UPS system

The line-interactive UPS system shown in Figure 18.4 provides no galvanic isolation from the incoming AC electric power system. This can be achieved by installation of a transformer at the output, increasing the size, complexity, and cost of the system.

18.2.4 Delta conversion UPS

A variation of the line-interactive topology known as a delta conversion UPS is shown in Figure 18.5. The series transformer permits current waveform shaping using the series converter, and the parallel converter performs voltage waveform shaping and power factor correction [18]. Other control schemes use the series converter to balance and shape the incoming voltage waveform, and the parallel converter to shape the output voltage waveform [8]. The former provides superior performance and a smaller filter package [19].

In normal mode, load is served from the electric power system so energy efficiency is quite high. In energy storage mode, the parallel converter feeds the load, and voltage and frequency regulation is possible. The series converter is rated at about 20% of system capacity, and the parallel converter is rated to carry the full load.

The delta conversion UPS offers high efficiency, unity power factor, low-current waveform distortion at the input, and low-voltage waveform distortion at the output. The system does not provide galvanic isolation for the load and requires sophisticated control topologies.

18.2.5 Tri-mode UPS

The salient feature of the UPS system shown in Figure 18.6 is the ability to capture photovoltaic (PV) solar and wind energy under normal conditions, and to use these resources to lengthen the survival period upon loss of the electric power system in the case of a synchronous generator with limited fuel supply. Under normal conditions, the static switch is closed and load is served from the AC electric power system. The four-quadrant converter connecting the AC system to the DC bus controls the DC bus voltage, and injects energy from the wind and solar into the AC system, while the inverter is idle.

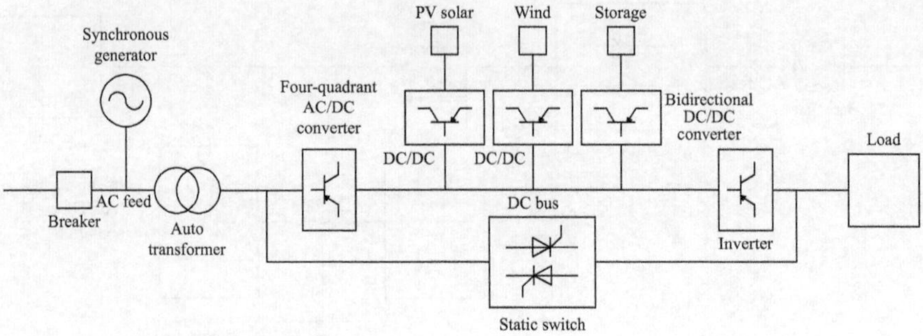

Figure 18.6 Simplified diagram of a tri-mode UPS system incorporating PV solar and wind generating systems

In double conversion mode, the AC supply is available, the static switch is open, and the load energy flows through the inverter. The DC bus voltage is regulated by the four-quadrant converter, and energy from the wind and solar sources feeds into the DC bus. The inverter controls the load AC voltage and energy flow to the load, while the four-quadrant converter either supplies the shortfall from the wind and solar or injects the extra into the AC electric power system.

In case the AC electric power system and synchronous generator are not available, the system operates in backup mode, and the load is served from stored energy via the inverter. The bidirectional DC-to-DC converter for the energy storage device controls the DC bus voltage, and injects the shortfall or absorbs the extra energy supplied by the wind and solar sources. If the system was operating in normal mode immediately prior to loss of the AC electric power system, a momentary interruption might occur. If positioned in the dual conversion mode, transfer to backup mode will be seamless.

The tri-mode system takes a comprehensive approach to resilience by incorporating wind and solar energy sources into the system, extending the duration the system can operate without an AC electric power supply. In normal mode, the system achieves high-energy efficiency. When operating in double conversion mode, energy efficiency drops, but output voltage and frequency regulation are possible. Transition from double conversion to backup mode is seamless. Galvanic isolation from the electric power system and synchronous generator is provided by the autotransformer.

The system requires sophisticated control topologies for all three operating modes as several layers of control network are required to recognize the desired operating mode and alter control objectives. When operating in normal mode there is a risk of a momentary interruption to the load. The autotransformer adds expense and increases the size and weight of the system.

18.2.6 Rotary UPS

A rotary UPS uses three rotating machines located on a common shaft and a battery bank as shown in Figure 18.7. In normal mode, the AC feed provides energy to the motor, and the motor drives the DC machine and the AC generator. The load is

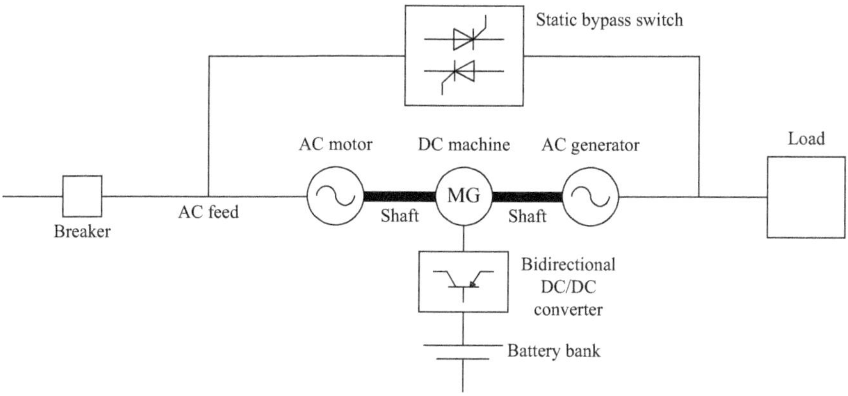

Figure 18.7 Configuration of a rotary UPS system

served from the AC generator, and the DC machine provides energy to restore and maintain charge on the battery through the bidirectional DC-to-DC converter. In energy storage mode, the battery bank provides power to the DC machine which drives the AC generator. Transition from normal to stored energy mode is seamless as inertia of the rotating machines helps to maintain system frequency. A static switch bypasses a failed rotary UPS system.

The rotary UPS provides a low impedance source for the load, exhibits good performance with non-linear loads, and supplies high fault current for protection system coordination. In normal mode, there is excellent electrical isolation between the source AC electric power system and the load. The rotary UPS is capable of a transient overload of 3–6 per unit compared to 1.5 per unit for a static UPS system [17]. Normal-mode system efficiency is determined primarily by the selection of the AC motor and generator. The DC machine efficiency factors into the round trip efficiency of the energy storage components.

The system is heavy, bulky, and expensive compared with static UPS systems. Due to the rotating components, maintenance costs are high.

18.2.7 Hybrid static and rotary UPS

A hybrid between the static and rotary UPS systems is shown in Figure 18.8. In normal mode, the AC motor is connected to the AC electric power system, and drives the AC generator. The bidirectional converter restores and maintains charge on the battery. In energy storage mode, the battery provides power to the AC motor through the bidirectional converter. A static bypass switch is closed in the event the UPS system fails or during scheduled maintenance.

This system provides excellent isolation between the AC electric power system and the load. During momentary interruptions of the AC electric power system, motor and generator inertia is sufficient to avoid disruption to the load. Extended outages of the supply will require use of the stored energy to power the AC motor–generator set. To reduce the capacity requirements of the AC/DC converter, a breaker is installed allowing motor starts motor using the electric power system.

Once up to speed, the breaker is opened, and the motor is served from the converter. This configuration permits seamless transition to battery power in the event of an extended outage. Short duration momentary interruptions rely on the inertia of the motor–generator set.

The hybrid system provides excellent isolation from the electric power system, low impedance for load, seamless transfer to stored energy mode, is simpler than the rotary UPS system, and requires less maintenance due to elimination of the DC machine and intrinsic mechanical commutator [20]. The systems are larger, heavier, and costlier than purely static UPS systems, and require regular maintenance for the rotating components.

18.2.8 Flywheels

Flywheel UPS systems are suited to installations which benefit from a high power density, low energy density supply. Locations with frequent short duration momentary outages, an alternate on-site energy supply such as diesel powered generators, limited space, and life-cycle economic considerations are good candidates [17]. Use of a bidirectional converter is shown in Figure 18.9 to extract maximum energy from the flywheel.

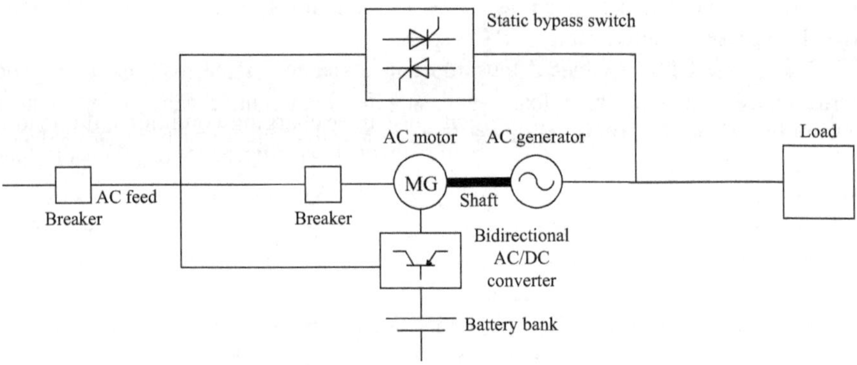

Figure 18.8 A hybrid rotary and static UPS system

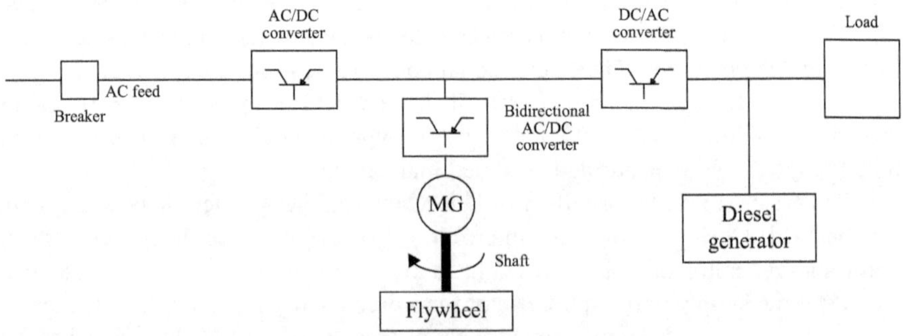

Figure 18.9 A flywheel UPS system with on-site diesel powered generation

Under normal conditions, load is served from the electric power system through the dual conversion topology. The bidirectional converter is used to power the motor–generator set and charge the flywheel by bringing it up to rated speed. Short-term interruptions of the electric power system are mitigated by the flywheel system alone. For long-term loss of the electric power system, the bidirectional converter extracts energy from the flywheel – slowing it down – while the diesel powered generator is starting. The generator synchronizes to the AC system, and serves the load for the duration of the outage. When the electric power system is restored, the DC/AC converter synchronizes with the AC system feeding the load, closed transition switching is possible, and the generator comes off-line.

Unlike battery powered UPS systems, flywheels can be charged and discharged rapidly without degradation to the energy storage capabilities. Flywheel systems are best for high power, low-energy UPS applications where another source with long-term energy delivery capability is available. The system in Figure 18.9 has relatively low efficiency due to dual energy conversion during normal operation, and friction and windage losses required maintaining flywheel velocity. Flywheel systems are more expensive than static UPS equivalents, and require physical containment for flywheel failure.

18.2.9 DC UPS for pulse load with power leveling

The demand reduction UPS system is shown in Figure 18.10. The DC load is characterized as cyclic with short duration. Under charging conditions, the load is less than 5% of full load, and the battery bank is storing energy. When the load demand spikes, a small portion of the energy is transmitted through the rectifier from the AC supply system, and the balance is supplied from the battery bank. This configuration also permits sizing the AC feed and rectifier components at a small percentage of full load. The DC–DC converter is sized at 100% of full load. The configuration operates in four distinct modes: AC on with the battery bank fully charged, AC on with the battery bank charging, AC on with the battery bank discharging into the load, and AC off with load supplied by the battery bank.

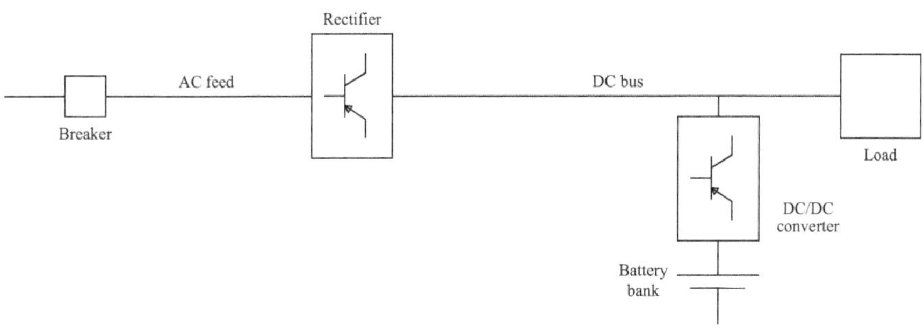

Figure 18.10 The medical equipment demand reduction UPS system

18.2.10 Redundant bus

The redundant bus system using AC distribution is quite common. Because many devices ultimately rectify AC waveforms to DC for end use, DC only UPS systems are gaining traction. An attraction is the potential reduction in the number of conversion stages compared with traditional AC UPS systems, and subsequent reduction in power losses. Another is the simplification of power flows vis-à-vis elimination of reactive power flow within the DC system. The system in Figure 18.11 is a schematic diagram for a redundant bus DC system where hypercritical loads may be served from either of two buses. Each DC bus is equipped with a battery bank and bidirectional DC/DC converter. Assuming proper design, this system has high reliability and resilience.

Downsides to the DC system include sophisticated protection schemes, limited suppliers of DC breakers and DC powered devices, the installation cost for the high reliability and resilience, and the learning curve and comfort curve for designers and installers. As the market for this type of installation grows, it is expected more suppliers will provide useful products.

18.3 Controls for UPS systems

UPS systems necessarily must control input power to loads, and charging of energy storage devices. The former will be referred to as load control, and the latter storage control. Performance specifications are normally stated in terms of output impedance, total harmonic distortion, voltage regulation, transient response to load and mode changes, and ability to serve non-linear loads [21]. Voltage regulation is arguably the most critical performance indicator. The control strategy influences the output impedance of the UPS, with in ideal transfer function of unity (1) at all frequencies [22]. There are typically upper frequency limits as dictated by the system frequency response. This inverse of the output impedance – termed the dynamic stiffness – is often used in place of the output impedance specification.

Ideal control results in fundamental frequency sinusoidal input current and output voltage waveforms, but may result in non-sinusoidal output current waveforms due to non-linear loads. Simple UPS systems with single devices are normally operated in voltage and frequency control mode for load control. This is akin to an isochronous generator with proportional-integral (PI) voltage regulation. Load control might also include line conditioning and power factor regulation. Line conditioning may cause ripple in the DC bus voltage to the detriment of energy storage devices. On-line UPS system load control does not change regardless of AC supply status. The charge control is required only when AC power is available.

Normal conditions for off-line systems require only storage control, but when called on to supply load, a simple off-line system will function in voltage and frequency control mode.

Interactive UPS system control is similar to the on-line system philosophy. Line conditioning, power factor control and ~20% of the load energy is supplied by the device during normal conditions, with one converter used for storage control. For the system depicted in Figure 18.5, the series converter voltage and current is

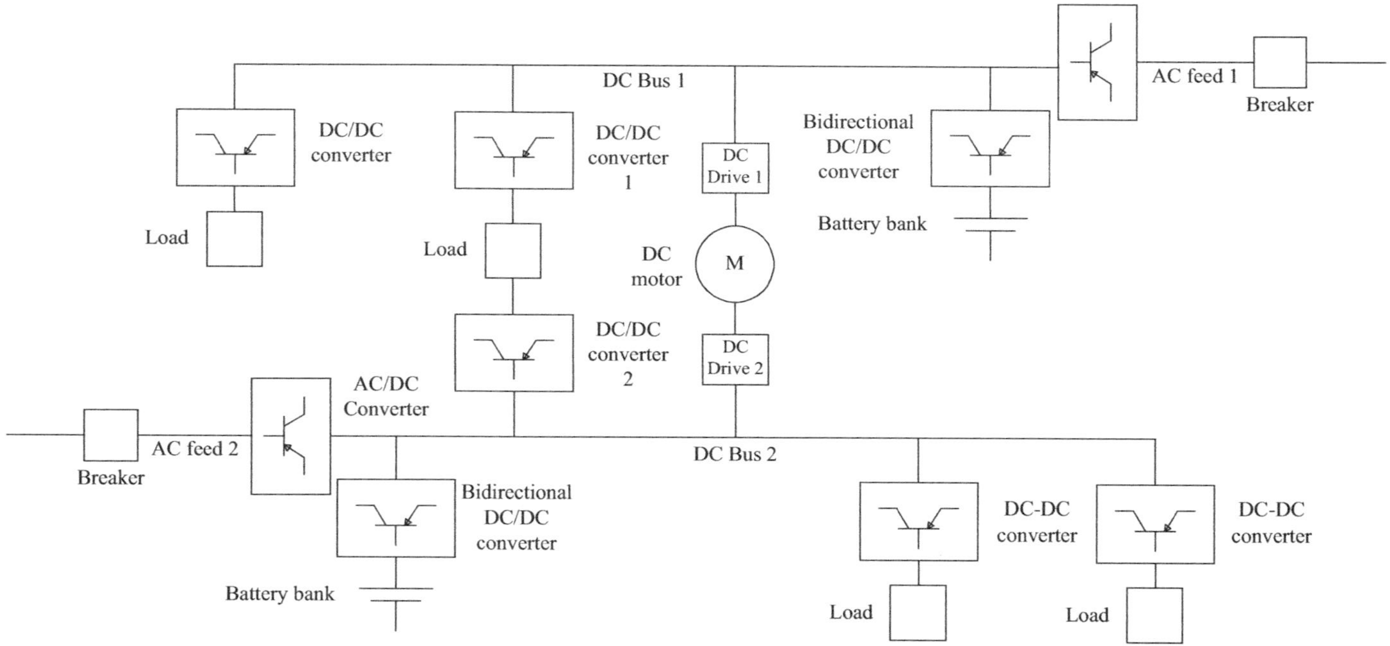

Figure 18.11 Schematic diagram of a redundant bus DC UPS system

controlled to supply nominal fundamental frequency voltage and support the fundamental frequency load current. The parallel inverter is controlled to correct the load power factor to unity and to supply non-sinusoidal current to non-linear loads. When the static switch opens, full load current is supplied by the parallel converter, and the control technique uses voltage and current control. The reference voltage is provided by a fundamental frequency PLL. When the AC power is restored, the PLL synchronizes the UPS output voltage waveform with the incoming AC line before closing the static switch.

As systems become more complex and multiple UPS devices feed into a common supply, only one device will operate in voltage and frequency control mode, and other devices will operate in current control mode. To maintain stability, droop settings for the load current and voltage set points are required. Essentially this means pure proportional control, with the cumulative system error accruing to the voltage control device. For complex, systems with multiple UPS devices operating in parallel, a second control layer for optimal device commitment and loading is warranted.

Because control techniques for single- and three-phase UPS systems are similar, discussion will be limited to three-phase systems. All systems utilize the switching capability of power electronic devices to form the output voltages and currents. Switching is generally slower in high-power devices than for low-power devices to reduce switching losses. Pulse width modulation (PWM) or vector controlled switching techniques are commonly used.

Most UPS control systems rely on a PLL to synchronize the control strategy with either the incoming AC voltage waveform or the UPS output voltage. Many control systems share a common frequency domain model of the output filter as shown in Figure 18.12 [23], where leakage resistances are assumed negligible.

Some control techniques track reference AC waveforms. An example is the controller after Holmes [24] where reference voltages for three phases are modulated with a triangular waveform to establish the voltage pulse width modulation (PWM) signals. The controller fits in the category of deadbeat control techniques where the current or voltage reference is predicted based on sampled existent and previous system states. Ideally, the controlled state will be forced to the command state within a delay stated in terms of sampling periods. This controller is complex, is computationally intense due to the sample rate calculations, and requires knowledge of the load through monitoring and feedback or feedforward.

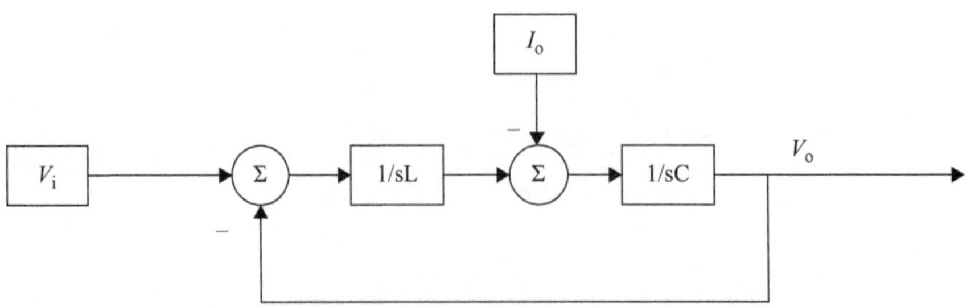

Figure 18.12 Frequency domain model for an LC output filter

Control techniques that use Park's transformation from stationary ABC to synchronous dq frames or a dual transformation from stationary ABC to stationary $\alpha\beta$ to synchronous dq assume balanced three-phase systems, and use the transformation techniques to obtain system states as DC values. The system states are compared to DC reference values and manipulated to arrive at DC command values. The DC command values are then inverse transformed and converted to PWM signals using ramp comparison or vector control commands for the converter. In Figure 18.13, X_{ABC} represents the signal to be transformed, X^*_{dq} is the reference for the state in the dq axis frame, and the output is used to determine the PWM or vector control power electronic device firing. An example deadbeat controller using the synchronous reference frame is found in [25].

The advantages of using synchronous frame control techniques include the use of DC reference values, simple control implementation – especially for PI techniques, elimination of phase errors, and simple techniques for separating fundamental and harmonic frequency components when harmonic filtering is a design consideration. The disadvantage is that all feedback signals must be translated into the synchronous reference frame.

Feedback control comes in many forms. Voltage feedback is common for voltage control strategies, with current feedback common for current control. Sophisticated strategies incorporate several feedback loops [26], with inner loops for current (voltage) control and slower outer loops designed for voltage (current) control. Nested loop strategies generally use on outer voltage feedback loop and for inner current feedback fall into two categories: feedback of filter inductor current versus filter capacitor current [27]. In [28], a single feedback loop is used for voltage control as seen in Figure 18.14, where G_c represents the controller architecture, K_a the PWM or vector control gain, G_p the output filter transfer function, H the sensor dynamics, and G_f the feedback control architecture. G_c and G_f represent series compensation techniques to the inclusion of proportional-integral-derivative, lead, lag, and lead–lag architecture. This topology is relatively simple to deploy, is well understood, easy to troubleshoot and robust to parameter variation. The harmonic performance is acceptable if the sinusoidal PWM is limited to the linear range, permitting good voltage regulation.

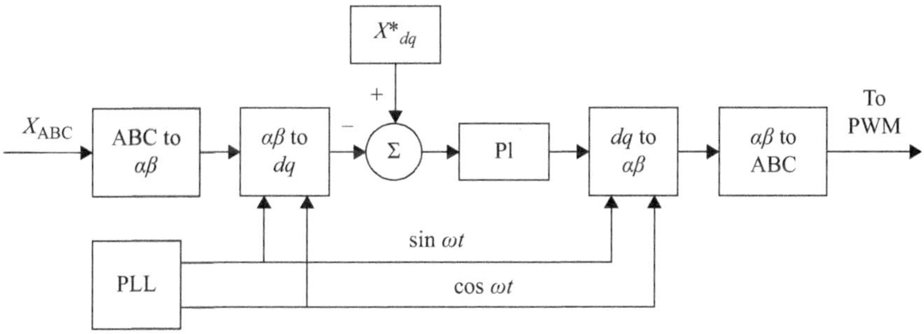

Figure 18.13 Transformation from the stationary ABC frame to the synchronous dq frame

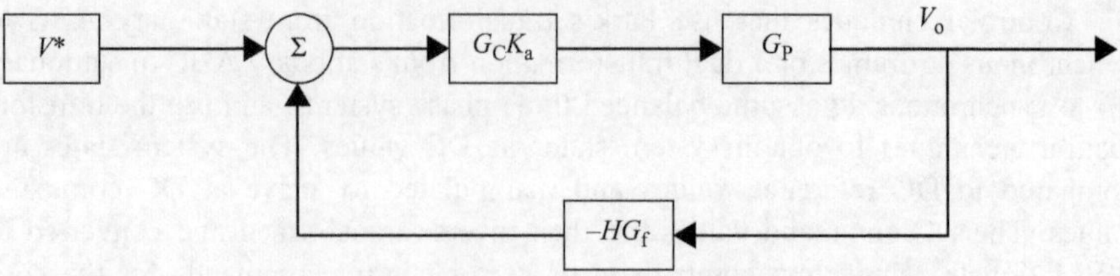

Figure 18.14 Single loop voltage feedback control

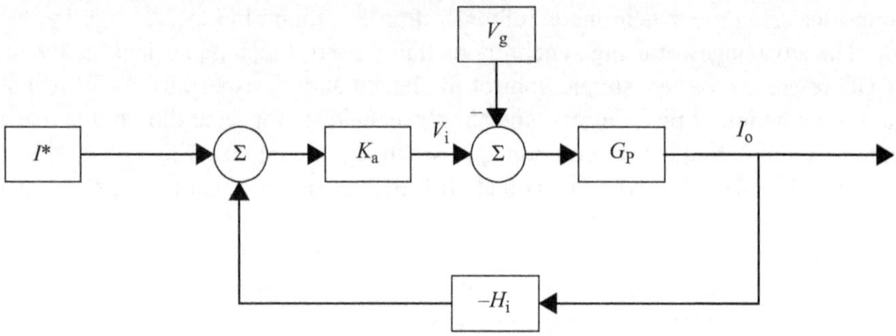

Figure 18.15 Current control with output current and grid voltage feedback

Typically, there is an assumption of linear load, which can lead to degraded performance if most load served is rectifier front end to DC bus capacitor power supplies. The dynamic response is sluggish, steady-state phase errors occur between the reference and output currents, and tuning requirements are load dependent. The voltage set point is normally set to nominal system voltage.

The control techniques for single voltage loop designs range from proportional-integral, digital deadbeat, to sliding mode control.

In Figure 18.15 after Jie [29], the simplest current control feedback loop applies a proportional gain to the current error signal. K_a represents the PWM or vector control modulator gain, G_p represents the interconnection system (output filter) dynamics, and H_i represents current transducer dynamics. Hysteretic current control is simple and straightforward and uses an alternating current for reference. When the output current falls outside a defined band about the command current, the modulator alters the firing pattern such that the current is returned to the hysteresis band. The strengths of this topology are simple implementation, fast dynamic response, and generally good performance. Filtering for hysteresis control can be problematic as switching frequencies vary.

Feedforward may be used in conjunction with feedback to speed the current response of the system [30]. In Figure 18.16, the current input is passed through a model of the plant (output filter) G_p' and then passed to the inverter control input. Under the best conditions, the plant model is an exact representation of the output

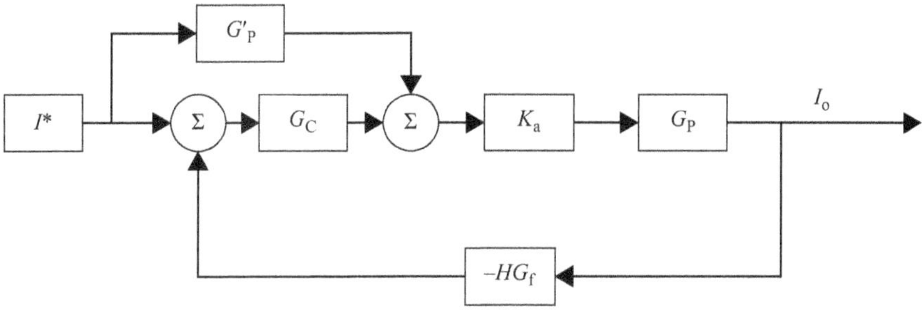

Figure 18.16 UPS current control using feedforward and feedback

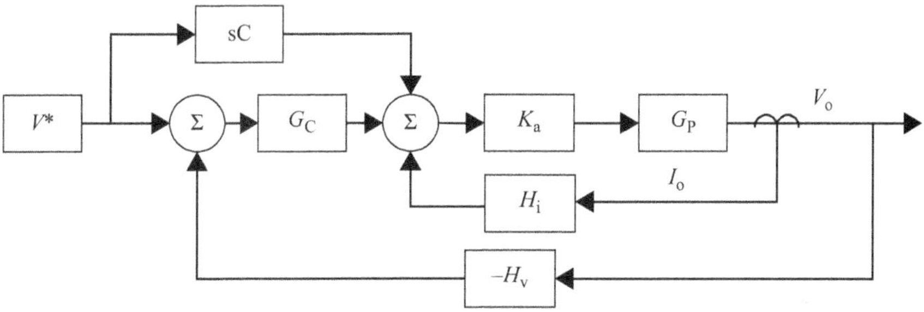

Figure 18.17 Load feedforward control scheme

filter dynamics. The feedback loop compensates for parameter variation and model mismatches.

Current references may be obtained by measuring load current. In case there are parallel UPS systems, the load current reference for each could be obtained based on the individual UPS system capacity compared to the total system capacity.

A variation of the feedforward control with AC reference inputs uses an inner current feedback loop, an outer voltage feedback loop, and a scaled derivative command signal. In Figure 18.17, the voltage reference is compared to the output voltage as altered by the transducer transfer function H_V, and the error is processed through G_C, typically PI architecture. That output is then combined with the feed-forward term – derived by taking the derivative of the input command V^* and scaling it by the output filter capacitor – and the UPS output current as altered by the current transducer transfer function H_i to attain the input to the current regulator K_a.

18.4 Applications

The use and application for UPS systems is quite diverse: small, portable, single-phase UPS systems might be used for a single desk top personal computer, to a system rated several kilowatts including two or more battery banks used for industrial billing and corporate data systems, to a data center with several electrical system power feeds, many battery banks, and on-site generating capability with an electrical demand of several megawatts.

18.4.1 Desktop personal computers

Within the category of small UPS systems there are a variety of sizes and styles. One example is the UPS for a desktop personal computer. A typical unit is rated 410 W, 620 VA, 120 V input, 120 V output and has six (6) 15 A receptacles.

18.4.2 Industrial systems

Industrial manufacturing process and power plant controls are good candidates for using UPS systems. In addition to providing power during electric power system outages, the intent is to provide galvanic isolation from the electric power system, to filter harmonics from the electric power systems, and to protect against low voltage occurring while the electric power system is faulted. These electric power system events may cause control systems to operate incorrectly, with resultant injury, equipment damage, and lost production.

An application that uses the DC system directly is electric power substation systems. Relays and breakers operate on 48 V DC. During normal conditions, the battery chargers apply a float voltage to the battery bank(s). Complete loss of voltage to the battery chargers typically requires two contingencies – there is a normal and reserve alternating current supply. The battery banks are normally sized such that 8 h after complete loss of alternating current power, there is sufficient energy in the batteries to operate each breaker device in the substation.

18.4.3 Data centers

Computing and data centers require continuous power and offer challenges in control design due to the non-linear load characteristics. These are predominantly AC systems, but DC versions are gaining traction for some installations. In addition to the data devices, supporting system devices are loads to be served. In fact, as much as 50% of the total load in some data centers is attributable to heating, ventilating, and air cooling (HVAC), lighting, fire protection, security, losses, etc. The scale of data centers is quite impressive, with the largest demand claim of 250 MW of electrical demand and largest area covering more than 1 million square feet.

Communication systems are often integrated with data centers, but there are sites which are primarily used as switching for landline communication or cellphone repeaters. Landline switching sites typically use 48 V DC, and have several battery banks. Cellular phone towers generally use a small UPS system in conjunction with a natural gas or diesel back-up generator.

18.4.4 Medical equipment

The UPS system shown in Figure 18.10 is intended for use with a computerized tomography (CT) scan device. The CT device has a relatively short on cycle, followed by a longer rest period.

When connected to the electric power system during the CT machine rest cycle, the battery may be charging, or if fully charged simply resting. Power to the CT machine is supplied by the electric power system. If the CT machine is scanning, the battery supplies approximately 80% of the machine power.

If the electric system is not connected, the battery supplies 100% of the machine power during rest and scan periods.

References

[1] Beck, B. J., "The design and performance of a rotary UPS system," *Safeguarding Industrial Plant During Power System Disturbances, IEE Colloquium on*, 1989, pp. 19/1–19/6.

[2] Dolezal, H., "UPS – dynamic-rotary systems with flywheel and diesel engine," *Telecommunications Energy Conference, 1987. INTELEC'87. The Ninth International*, 1987, pp. 187–192.

[3] Iwanski, G. and Koczara, W., "PLL grid synchronization of the standalone DFIG based wind turbine or rotary UPS," *EUROCON, 2007. The International Conference on "Computer as a Tool,"* 2007, pp. 2550–2555.

[4] Kusko, A. and Fairfax, S., "Survey of rotary uninterruptible power supplies," *Telecommunications Energy Conference, 1996. INTELEC'96, 18th International*, 1996, pp. 416–419.

[5] Sachs, K., "New developments in low impedance rotary UPS design," *Telecommunications Energy Conference, 1986. INTELEC'86. International*, 1986, pp. 637–645.

[6] Krishnan, R. and Srinivasan, S., "Topologies for uninterruptible power supplies," *Industrial Electronics, 1993. Conference Proceedings, ISIE'93 – Budapest, IEEE International Symposium on*, 1993, pp. 122–127.

[7] Nasiri, A., Nie, Z., Bekiarov, S. B., and Emadi, A., "An on-line UPS system with power factor correction and electric isolation using BIFRED converter," *Industrial Electronics, IEEE Transactions on*, 2008, pp. 722–730.

[8] da Silva, S. A. O., Donoso-Garcia, P. F., and Cortizo, P. C., "A three-phase series-parallel compensated line-interactive UPS system with sinusoidal input current and sinusoidal output voltage," *Industry Applications Conference, 1999. Thirty-Fourth IAS Annual Meeting. Conference Record of the 1999 IEEE*, Vol. 2, 1999, pp. 826–832.

[9] Bekiarov, S. B. and Emadi, A., "Uninterruptible power supplies classification, operation, dynamics, and control," *Applied Power Electronics Conference and Exposition, 2002. APEC 2002. Seventeenth Annual IEEE*, Vol. 1, 2002, pp. 597–604.

[10] Guerrero, J. M., Berbel, N., Matas, J., Sosa, J. L., and de Vicuña, L. G., "Control of line-interactive UPS connected in parallel forming a microgrid," *Industrial Electronics, 2007. ISIE 2007. IEEE International Symposium on*, 2007, pp. 2667–2672, DOI: 10.1109/ISIE.2007.4375029.

[11] Li, Y. and Hong-Tao, J., "Research of a new type series-parallel converted UPS circuit and its control strategy," *Consumer Electronics, Communications and Networks (CECNet), 2011 International Conference on*, 2011, pp. 1560–1563.
[12] Nasiri, A. and Amadi, A., "Different topologies for single-phase unified power quality conditioners," *Industry Applications Conference, 2003. 38th IAS Annual Meeting. Conference Record of the*, Vol. 2, 2003, pp. 976–981.
[13] Rathmann, S. and Warner, H. A., "New generation UPS technology the delta conversion principle," *Industry Applications Conference, 1996. Thirty-First IAS Annual Meeting, IAS'96, Conference Record of the 1996 IEEE*, Vol. 4, 1996, pp. 2389–2395.
[14] Lacina, R., Besseling, F., and ter Heddle, B., "An integrated data center solution on medium voltage level (UPS)," *Electricity Distribution (CIRED 2013), 22nd International Conference and Exhibition on*, 2013, pp. 1–4.
[15] Li, H., Zhang, W., and Xu, D., "High-reliability long-backup-time super UPS with multiple energy sources," *Energy Conversion Congress and Exposition (ECCE), 2013 IEEE*, 2013, pp. 4926–4933.
[16] Karve, S., "Three of a kind [UPS topologies, IEC standard]," *IEE Review*, Vol. 46, No. 2, 2000, pp. 27–31.
[17] Emadi, A., Nasiri, A., and Bekiarov, S. B., "Uninterruptible Power Supplies and Active Filters," CRC Press LLC, Boca Raton, FL, 2005.
[18] da Silva, S. A. O., Donoso-Garcia, P. F., Cortizo, P. C., and Seixas, P. F., "A three-phase line-interactive UPS system implementation with series-parallel active power-line conditioning capabilities," Industry Applications, *IEEE Transactions on, Vol. 38, No. 6*, 2002, pp. 1581–1590.
[19] Karshenas, H. R. and Niroomand, M., "Analysis and comparison of two control strategies for three-phase series-parallel UPS," *Electrical Machines and Systems, 2008. ICEMS 2008. International Conference on, 2008*, pp. 1726–1731.
[20] Windhorn, A., "A hybrid static/rotary UPS system," *Applied Power Electronics Conference and Exposition*, 1990. APEC'90, Conference Proceedings 1990, Fifth Annual, 1990, pp. 422–427.
[21] Manning, C. D., "Control of UPS inverters," *Uninterruptible Power Supplies, IEE Colloquium on*, 1994, pp. 3/1–3/5.
[22] Ryan, M. J., Brumsickle, W. E., and Lorenz, R. D., "Control topology options for single-phase UPS inverters," *Power Electronics, Drives and Energy Systems for Industrial Growth, 1996, Proceedings of the 1996 International Conference on*, Vol. 1, 1996, pp. 553–558, DOI: 10.1109/PEDES.1996.539673.
[23] Mohamed, I. S., Zaid, S. A., Abu-Elyazeed, M. F., and Elsayed, H. M., "Model predictive control – a simple and powerful method to control UPS inverter applications with output LC filter," *Electronics, Communications and Photonics Conference (SIECPC), 2013 Saudi International*, 2013, pp. 1–6, DOI: 10.1109/SIECPC.2013.6551018.

[24] Holmes, D. G., and Martin, D. A., "Implementation of a direct digital predictive current controller for single and three-phase voltage source inverters," *Industry Applications Conference, 1996. Thirty-First IAS Annual Meeting, IAS'96, Conference Record of the 1996 IEEE*, Vol. 2, 1996, pp. 906–913, DOI: 10.1109/IAS.1996.560191.

[25] Nasiri, A., "Digital control of three-phase series-parallel uninterruptible power supply systems," *Power Electronics, IEEE Transactions on*, Vol. 22, No. 4, 2007, pp. 1116–1127, DOI: 0.1109/TPEL.2007.900546.

[26] Ryan, M. J. and Lorenz, R. D., "A high performance sine wave inverter controller with capacitor current feedback and 'back-EMF' decoupling," *Power Electronics Specialists Conference, 1995. PESC'95 Record, 26th Annual IEEE*, Vol. 1, 1995, pp. 507–513, DOI: 10.1109/PESC.1995.474857.

[27] Escobar, G., Mattavelli, P., Stankovic, A. M., and Leyva, J., "An adaptive control for UPS to compensate unbalance and harmonic distortion using a combined capacitor/load current sensing," *Industrial Electronics Society, 2003. IECON'03. The 29th Annual Conference of the IEEE*, Vol. 1, 2003, pp. 378–382, DOI: 10.1109/IECON.2003.1280010.

[28] Chen, J. and Chu, C., "Combination voltage-controlled and current-controlled PWM inverters for UPS parallel operation," *Power Electronics, IEEE Transactions on*, Vol. 10, No. 5, 1995, pp. 547–558, DOI: 10.1109/63.406842.

[29] Jie, Z., Peng, W., and Jie, H., "Control strategy of microgrid inverter operation in grid-connected and grid-disconnected modes," *Electric Information and Control Engineering (ICEICE), 2011 International Conference on*, 2011, pp. 1257–1260, DOI: 10.1109/ICEICE.2011.5777376.

[30] Phakamach, P., Tiacharoen, S., and Akkaraphong, C., "Control of an uninterruptible power system (UPS) using a feedforward integral variable structure controller," *Industrial Technology, 2002. IEEE ICIT'02. 2002 IEEE International Conference on*, Vol. 1, 2002, pp. 618–623, DOI: 10.1109/ICIT.2002.1189972.

Chapter 19
Wireless power transfer

S.Y. Ron Hui[*#] and Paul D. Mitcheson*[*]*

19.1 Introduction

With the introduction of the Ampere's law and Faraday's law at the end of the nineteenth century, Nikola Tesla [1] pioneered wireless power research and laid down the fundamental principles for wireless power transfer (WPT). WPT can be broadly classified as radiative and non-radiative. Power can be radiated by an antenna and propagates through a medium such as air in the form of a radio frequency (RF) electromagnetic wave. Non-radiative WPT is based on near-field magnetic coupling of magnetic circuits that are generally in the form of conductive loops with a resonant frequency. WPT can be achieved through a range of technologies [2], ranging from near-field magnetic coupling based technologies operating at a relatively low frequency (such as 10 kHz–15.65 MHz) to microwave technologies operating at relatively high frequency (up to a few giga-hertz). This chapter focuses primarily on the former type of research and applications based on near-field magnetic coupling. It covers WPT research and applications from low-power applications. For the descriptions of a full spectrum of WPT technologies, the readers can refer to [2].

Nikola Tesla was a great inventor whose inventions have influenced human society profoundly in the twentieth century [3]. His inventions include ac power transmission (on which modern power transmission systems are based), ac induction machines (which have been the most dominant electric machines used in industry worldwide), radio and tuned circuits (which form the basis for long-distance communication), light sources (including several forms of discharge lamps) and of course WPT.

The WPT principle has in fact been used in the induction machines. Energy is transferred from the stator windings across the air gap to the rotor cage. Energy transfer via magnetically coupled windings has been the basic operating principle in ac electric machines. Tesla's idea of magnetic coupling of coils has been linked to his tuned and radio circuits that emphasize the use of circuit resonance. Whilst the term 'magnetic resonance' has recently been misunderstood by some as a new

*Department of Electrical and Electronic Engineering, Imperial College London, South Kensington, SW7 2AZ, London, United Kingdom
also with the Department of Electrical and Electronic Engineering, The University of Hong Kong, Pokfulam Road, Hong Kong

technology, it must be pointed out that such concept was established by Tesla in his early WPT research.

In a study of Tesla's contribution in wireless transmission of energy [4], some important quotations were cited from a 1943 technical article that:

> Tesla is entitled to either distinct priority or independent discovery of the following:
>
> 1. The idea of inductive coupling between the driving and the working circuits.
> 2. The importance of tuning both circuits, that is, the idea of an 'oscillation transformer'.
> 3. The idea of a capacitance loaded open secondary circuit. [5]

These three aspects of discovery have in fact formed the founding principles of WPT. The use of the tuned circuits as an 'oscillation transformer' means that both of the transmitter and receiver circuits are tuned to operate at a resonant frequency. It has been pointed out [6,7] that the 'oscillation transformer' concept goes beyond pure magnetic induction principle commonly adopted in transformers. It refers more precisely to magnetic resonance between two magnetic coupled coil resonators. The combined use of magnetic induction, tuned circuits and resonance has been a common theme in Tesla's wireless power and radio investigations [8] as well as his low-frequency WPT via the resonant frequency of the Earth [4].

Since Tesla published his work on wireless power in early 1900s, there have not been widespread WPT applications in the first half of last century. The main reason is probably due to the low energy efficiency as the transmission distance increases [9,10]. However, WPT emerged again in the 1960s in transcutaneous energy systems for medical implants [11–15]. The availability of power electronics technology in the 1980 provided the needed technology to generate high-frequency high-power supplies for WPT. In the 1990s, Green and Boys used power inverter as the excitation source for an inductive power transfer system [16] and then for charging electric vehicles [17]. Such WPT initiative has already led to applications in factory automation in clean factories, medical implants and charging of electric vehicles [18]. The dawn of the mobile phone era in 1990s also triggered new WPT research for planar wireless charging pads for portable electronic products including mobile phones in early 2000s [19–23]. The successes in the planar wireless charging technology in meeting the requirements of user friendliness and various safety and electromagnetic compatibility (EMC) standards eventually prompted the formation of the Wireless Power Consortium (WPC) in 2008 [24], which launched the world's first wireless power standard 'Qi' in 2010 [6, 25]. The WPC consists of 210 company members in over 20 countries by 2014.

Qi-compatible charging systems for portable electronics and wireless charging of electric vehicles mentioned so far are considered short-range WPT. Mid-range WPT started to gain lots of attention since mid-2000s. Mid-range WPT refers to a transmission distance being larger that the dimensions of the transmitter and receiver coils [7]. The early forms of mid-range WPT systems involve the use of relay resonators [26,27]. Besides the original two-coil systems, three-coil [28,29], four-coil [30–33] and domino [34–36] WPT systems have consequently emerged for mid-range applications. The following sections will describe the features of these emerging non-radiative WPT systems for both short-range and mid-range power transfer applications.

19.2 Basic principles and two fundamental concepts of WPT

19.2.1 Basic principles

Magnetic (inductive) coupling and magnetic resonance have been the central themes of Tesla's WPT research. Because of the transmission distance, the coupling coefficient between the coupled coils (such as the transmitter and receiver coils) is relatively low. From an electric circuit point of view, a low mutual coupling coefficient means a large leakage inductance. Such leakage inductance has positive impedance that limits the power transfer. Tesla used a capacitance-loaded coil to form a resonator so that the capacitance, which has a negative impedance, can cancel the positive impedance of the leakage inductance. In the two-coil WPT research, Tesla pointed out that optimal WPT could be achieved at the resonance frequency of the WPT systems. In physical terms, Tesla used the resonance of the magnetic field related to the inductance and the electric field related to the capacitance to achieve optimal WPT.

Regardless of the number of coils, a WPT system (Figure 19.1) can be modelled as a mutual magnetically coupled circuit using standard circuit theory as shown in (19.1).

$$\begin{bmatrix} R_1+j(\omega L_1-\frac{1}{\omega C_1}) & j\omega M_{12} & j\omega M_{13} & \cdots & \cdots & j\omega M_{1n} \\ j\omega M_{12} & R_2+j(\omega L_2-\frac{1}{\omega C_2}) & j\omega M_{23} & \cdots & \cdots & j\omega M_{2n} \\ \vdots & \vdots & \vdots & \ddots & \vdots & \vdots \\ j\omega M_{1(n-1)} & \cdots & \cdots & R_{n-1}+j(\omega L_{n-1}-\frac{1}{\omega C_{n-1}}) & & j\omega M_{(n-1)n} \\ j\omega M_{1n} & \cdots & & j\omega M_{(n-1)n} & & R_n+R_L+j(\omega L_n-\frac{1}{\omega C_n}) \end{bmatrix} \cdot \begin{bmatrix} \mathbf{I}_1 \\ \mathbf{I}_2 \\ \vdots \\ \mathbf{I}_{n-1} \\ \mathbf{I}_n \end{bmatrix} = \begin{bmatrix} V_1 \\ 0 \\ \vdots \\ 0 \\ 0 \end{bmatrix} \tag{19.1}$$

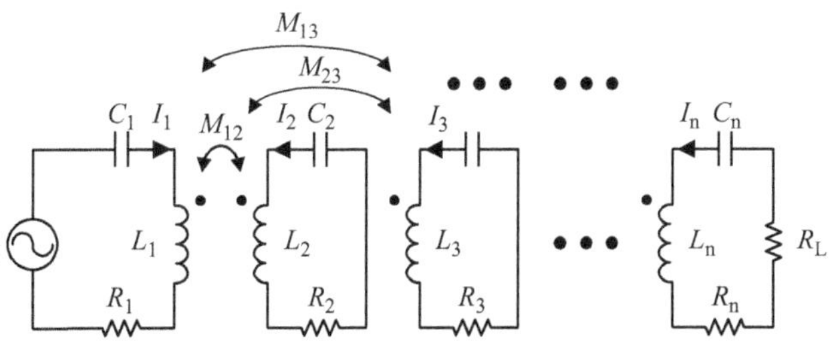

Figure 19.1 Schematic of a general WPT system with n resonators

where $M_{ij} = k_{ij}\sqrt{L_i L_j}(i,j = 1,2,\ldots,n; i \neq j)$ is the mutual inductance between winding-i and winding-j; R_L is the load resistance which is connected to winding-n; I_i is the current in winding-i; L_i is the self-inductance of winding-i; C_i is the compensating capacitance of winding-i; R_i: is the resistance in resonator-i (including the resistance of winding-i and the equivalent series resistance of the capacitor C_i); and ω is the angular frequency.

19.2.2 Two fundamental concepts

The vast number of WPT research published in the last decade can be classified under two fundamental principles, namely (i) maximum power transfer theorem and (ii) maximum energy efficiency principle [7]. It is necessary to understand their differences, advantages and disadvantages.

19.2.2.1 Maximum power transfer theorem (source impedance matching)

Maximum power transfer theorem is a fundamental circuit theory that states that maximum power transfer can be achieved if the source impedance matches the load impedance. Figure 19.2 shows an equivalent circuit consisting of an ac power source, source impedance of $R_s + jX_s$ and load impedance $R_L + jX_L$, where R_s and X_s are the source resistance and source reactance, R_L and X_L are load resistance and load reactance, respectively. Figure 19.3 shows a plot of energy efficiency with the ratio of the load resistance and source resistance. Maximum power transfer can be achieved when $R_s = R_L$ and $X_s = -X_L$. For impedance matching (i.e. $R_S = R_L$), the system energy efficiency η_E that includes the power loss in the power source is

$$\eta_E = \frac{i^2 R_L}{i^2 R_S + i^2 R_L} = \frac{R_L}{R_S + R_L} = 0.5 \qquad (19.2)$$

It is therefore important to note that any WPT system that adopts the maximum power transfer principle via impedance matching has a severe limit on its overall energy efficiency of 50%. Half of the total power consumption occurs in the source resistance (R_s) of the power source. This is an important point that is somehow neglected by some WPT researchers who only consider the power available from the output terminals of the ac power source and forget that there is also power loss in the source resistance of the power source.

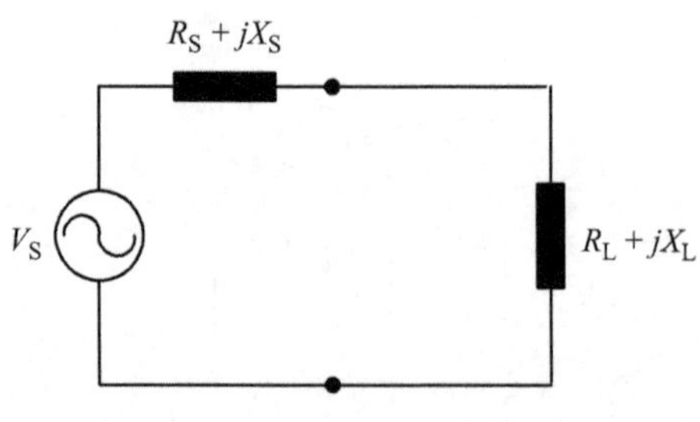

Figure 19.2 Equivalent circuit of an ac power source and an equivalent load

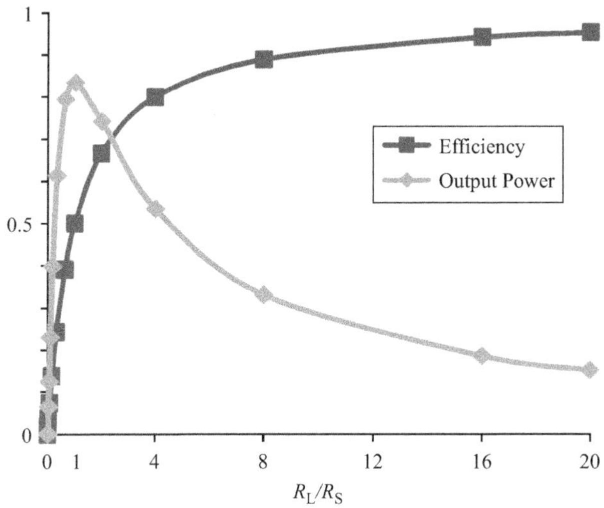

Figure 19.3 Variations of energy efficiency and output power as a ratio of R_L and R_s for the equivalent circuit in Figure 19.2 [Y-axis: per-unit scale]

The impedance matching technique is commonly used by RF researchers in RF and antenna circuits. Most of the four-coil WPT systems reported in recent literature are based on the maximum power transfer theorem. These WPT systems are usually operated at a frequency exceeding several mega-hertz. A good example that highlights the poor energy efficiency is illustrated in [30] which has transmission efficiency (power from the output of the ac power source to the load) of 40% and an overall system energy efficiency (including the power loss in the source resistance) of 15%.

19.2.2.2 Maximum energy efficiency principle

The maximum energy efficiency principle aims at maximizing the energy efficiency in the power transfer process. This principle has commonly been used by power electronics engineers in transformer and switched-mode power supply (SMPS) designs. Figure 19.3 shows that if the source resistance is close to zero, the system energy efficiency will approach 100%. Therefore, a good transformer has small winding resistance and close winding coupling, and a good power supply has a small internal source resistance preferably close to 0 Ω. For a general n-coil WPT system, the system energy efficiency is:

$$\eta_E = \frac{i_N^2 R_L}{i_1^2(R_S + R_1) + i_2^2 R_2 + \cdots \in + i_N^2(R_N + R_L)} \tag{19.3}$$

By using a power source with very low source resistance (R_s) and coils with low winding resistance ($R_1, \ldots, R_N$), most of the input power will be consumed by the load resistance (R_L) according to (19.3). Thus, an overall system energy efficiency higher than 50% is feasible. In fact, most of the power transformers and SMPSs have system efficiency exceeding 90%.

19.2.2.3 Choice of operating principles

The choice of the operating principles depends on the nature of applications. For short-range applications and high-power applications, the maximum energy efficiency principle is the natural source, as reflected by the Qi standard for portable electronic devices [6] and many research reports on wireless charging of mobile robots and electric vehicles [18]. The maximum power theorem can be used to increase the transmission distance, but at the expense of energy efficiency. Therefore, application of the maximum power theorem via impedance matching is more suitable for mid-range and low-power WPT applications.

19.3 Different forms of WPT systems

WPT systems can take various forms and numbers of stages. They can comprise several combinations of series and parallel resonant circuits. The advantages and disadvantages of these combinations will be addressed in Section 19.4. This section highlights the features of two-coil, three-coil, four-coil and other multiple-coil systems.

19.3.1 Two-coil systems

Tesla described the use of near-field inductive coupling and magnetic resonance two-coil systems for WPT a century ago [37]. Two-coil systems (Figure 19.4) have been proven to be an energy-efficient approach to short-range WPT applications and are suitable for both low- and high-power applications. Two-coil systems with fixed-positioning features are presently used for wireless charging of commercial electronic products such as electric toothbrushes and mobile phones. Active research is being conducted in using the two-coil systems for charging electric vehicles, with the coil misalignment being a major research topic. However, the energy efficiency of a two-coil system drops rapidly with increasing transmission distance (Figure 19.5). Therefore, two-coil systems are not suitable for applications in which the transmission distance is comparable to the dimensions of the transmitter and receiver coils. Generally speaking, most of the two-coil systems reported in the literature adopt the maximum energy efficiency principle.

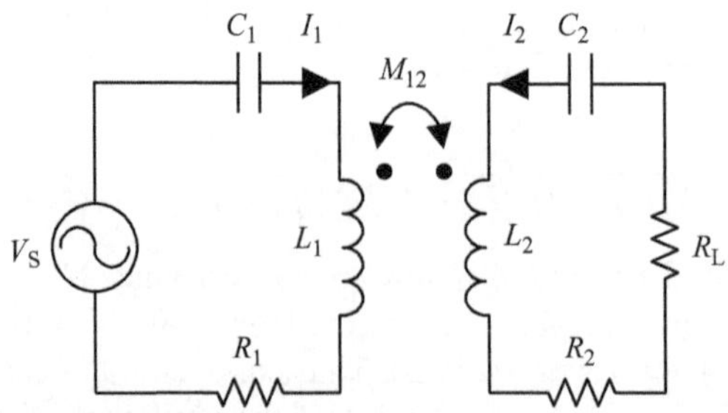

Figure 19.4 An equivalent circuit of a two-coil WPT system

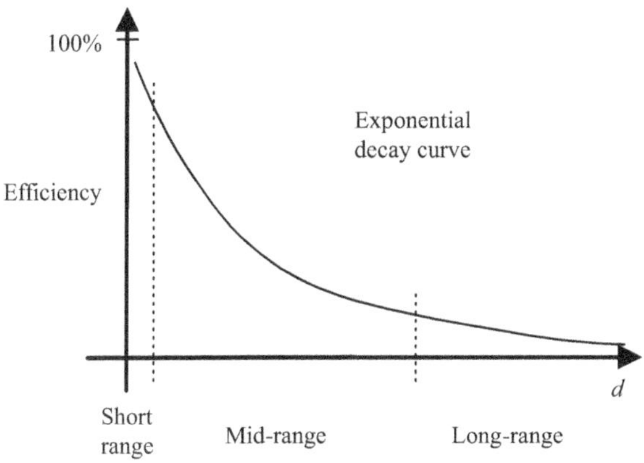

Figure 19.5 Exponential decay of energy efficiency with increasing transmission distance in a two-coil WPT system

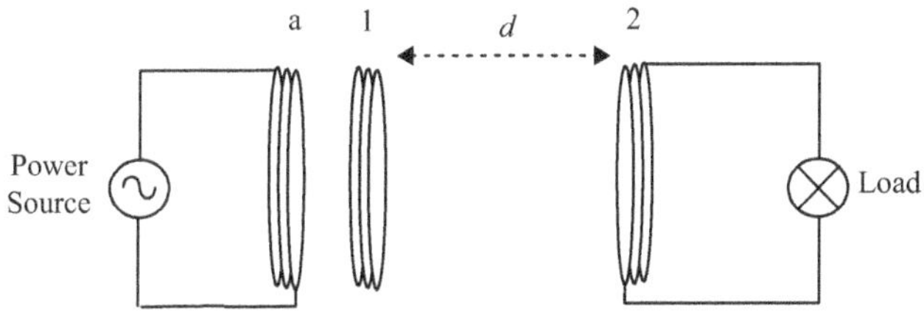

Figure 19.6 A WPT system with an additional resonator at the transmitter side

19.3.2 Three-coil systems

Three-coil systems have attracted attention only recently for both low- and high-power applications. One initiative of the Qi standard is to extend the transmission distance from a typical transmission distance of 5–30 mm [25]. This extension enables new planar wireless charging systems, such as coffee tables and office desks equipped with designated wireless charging areas, to be developed. Figure 19.6 shows the schematic of a three-coil system, in which a relay resonator is included between the transmitter and receiver coils. The relay resonator can be co-planar with the transmitter coil, as described in a three-coil system studied for a coffee table [29,38]. Under certain conditions, it has been demonstrated that a three-coil system can be more energy efficient than a two-coil counterpart. The efficiency improvement is achieved by using the transmitter coil and the relay resonator as a current transformer so that the current stress is shifted from the transmitter circuit and coil to the relay resonator. If the conducting loss of the relay resonator is less than the reduction of the power losses in the transmitter circuit and coil, an efficiency improvement can thus be achieved. It has been pointed out [38] that the three-coil approach may provide a breakthrough in high-power wireless charging of electric vehicles, which currently has a typical system energy efficiency of 92%.

In a study of a three-coil planar WPT system in which the magnetic power loss is ignored, it has been shown that a three-coil system can be more energy efficient than a two-coil one if the following inequality is satisfied:

$$\omega M_{\mathrm{a1}} > (R_1 + R_{\mathrm{E1}})\sqrt{1 + \frac{R_{\mathrm{a}}}{R_{\mathrm{S}}}} \tag{19.4}$$

where ω is the angular frequency of the excitation, M_{a1} is the mutual inductance between the transmitter coil (coil-a) and the relay resonator (coil-1), R_{P1} is the parasitic resistance of coil-1, $R_{\mathrm{E1}} = \frac{\omega^2 M_{12}^2}{R_2}$ is the reflected resistance from the receiver to coil-1, R_{S} is the source resistance and R_{a} is the winding resistance of the new primary coil-a.

19.3.3 Four-coil systems

Recent interests of four-coil systems were sparked off by the report in [30], which describes the use of two resonators, with one magnetic resonator coupled with a driving loop and the other resonator coupled with a load loop (Figure 19.7). The authors of [30] demonstrate that it is possible to power a 60 W light bulb with a transmission distance of 2 m at an overall energy efficiency of 15%. The four-coil system takes advantages of (i) the magnetic coupling between the loops and the resonators and (ii) magnetic resonance between the two resonators. Both techniques were used in Tesla's early WPT research. The operating principle adopts the maximum power transfer theorem that requires impedance matching with the source impedance. The use of four coils introduces three mutual coupling coefficients that can be manipulated to extend the transmission distance. The three mutual coupling coefficients are K_{PS} (between the power driving loop and the sending resonator), K_{SR} (between the sending resonator and the receiving resonator) and K_{RD} (between the receiving resonator and the load driving loop). The condition for impedance matching is to design the input impedance of the entire

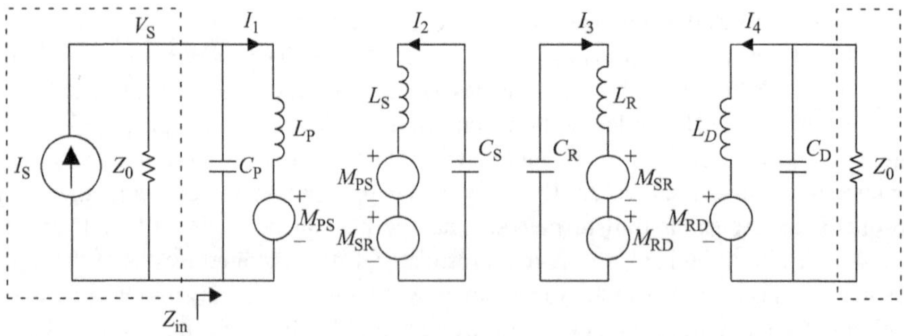

Figure 19.7 An equivalent circuit of a WPT system comprising a power driving coil, a sending resonator, a receiving resonator and a load driving coil [31]

four-coil system (Z_{in}) to be equal to the source impedance (Z_o). This can be achieved by the following equation:

$$\frac{K_{PS}K_{RD}}{K_{SR}} = 1 \tag{19.5}$$

It should be noted that a large transmission distance between the sending resonator and the receiving resonator corresponds to a small K_{SR}. The benefit of the four-coil system is to provide the flexibility of two extra coefficients (i.e. K_{PS} and K_{RD}) to satisfy (19.4) even if K_{SR} is small. For example, for a long transmission distance with a corresponding small K_{SR} (e.g. 0.01), (19.5) can be met by choosing K_{PS} and K_{RD} equal to 0.1. Of course, the use of the maximum power transfer theorem means that the long transmission distance is achieved at the expense of energy efficiency.

19.3.4 WPT systems with relay and domino resonators

WPT systems using relay resonators in the domino arrangements provide a good compromise in terms of the transmission distance and energy efficiency, provided that such resonators are allowed to be placed between the transmitter coil and the receiver coil. These systems adopt the maximum energy efficiency principle. Because of the relatively short distance between adjacent resonators, they enjoy a fairly high energy efficiency in mid-range application. Figure 19.8 shows a straight domino WPT system powering an 18 W compact fluorescent lamp over a distance of 10 feet. Power flow in a domino WPT system can be directed flexibly. Figure 19.9 shows a Y-shape domino WPT system which has the power flow split into two paths driving two 18 W compact fluorescent lamps.

19.4 Power electronics and control

Any WPT system should use high-efficiency circuits for both the driver (generating the sinusoidal currents that create the magnetic field) and for the rectifier on the

Figure 19.8 A straight domino WPT system powering a 18 W compact fluorescent lamp

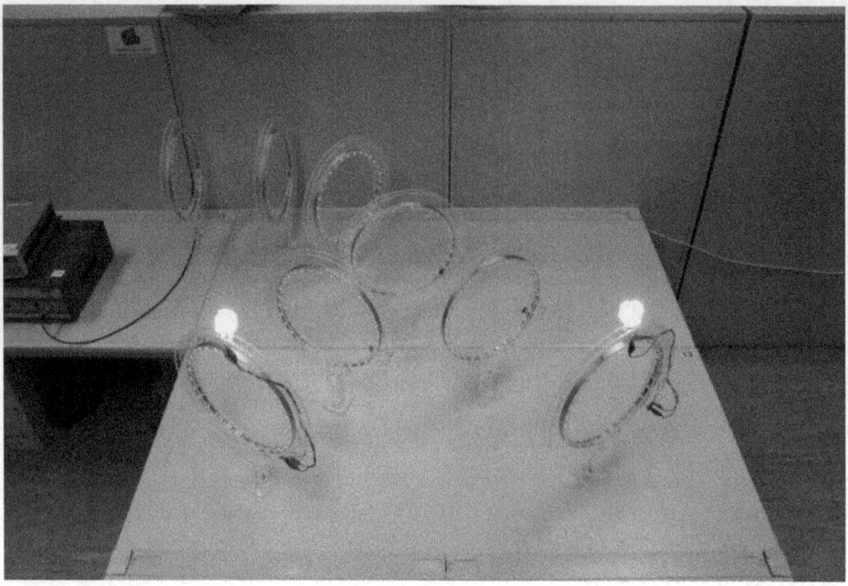

Figure 19.9 A Y-shaped domino WPT system powering two compact fluorescent lamps

secondary side (allowing the systems to provide a dc output). Several difficulties arise with the design and realization of these circuits due to the high frequencies often required, the potential high power requirements and the highly tuned resonant circuits causing high voltages and/or currents. All of these requirements mean that the choice of circuit architecture as well as both active and passive components is critical to realising a high-efficiency end-to-end solution. Voltage and current requirements in the active components can be traded-off by use of series or parallel resonant circuits on the primary and secondary and various circuit topologies exist which aim to reduce component stress and switching loss by exploiting soft switching.

19.4.1 Series and parallel tuned primary and secondary circuits

The basic formula for the maximum link efficiency of a WPT system is well known [39] as:

$$\eta = \frac{k^2 Q_{\mathrm{TX}} Q_{\mathrm{RX}}}{\left(1 + \sqrt{1 + k^2 Q_{\mathrm{TX}} Q_{\mathrm{RX}}}\right)^2} \tag{19.6}$$

where k is the coupling factor and Q_{TX} and Q_{RX} are the Q-factors of the transmitter and receiver, respectively. It is important to note that this equation assumes nothing about the configuration of the link other than that the secondary is set to resonate at the driving frequency of the transmitter and that the receiver's load is optimised to present a specific reflected impedance back to the primary. This means that the efficiency for transferring power over an inductive link is independent of the circuit configuration of the driver side circuit, and hence resonance of the primary is not a fundamental requirement for realising WPT systems at the efficiency given by the fundamental link efficiency formula. In reality, however, the use of primary side resonance allows for greater driving circuit efficiencies to be achieved as it reduces the reactive power capability requirement of the drive circuit. Due to the more fundamental requirements on the secondary circuit to maintain operation according to (19.6), the possible secondary configurations will now be discussed.

As stated, (19.6) holds assuming that the WPT system driven at the resonant frequency of the secondary side circuit and that the secondary reflects back an optimal impedance to the primary side. The choice of series or parallel secondary resonance (Figure 19.10) changes the value of the required optimal load impedance that connects to the secondary circuit.

The optimal resistive load for the link with a parallel secondary resonance (Figure 19.10(a)) is given by:

$$R_{\mathrm{LOAD}} = \frac{\alpha}{\omega C_{\mathrm{RX}}} \tag{19.7}$$

where

$$\alpha = \frac{Q_{\mathrm{RX}}}{\sqrt{1 + k^2 Q_{\mathrm{TX}} Q_{\mathrm{RX}}}} \tag{19.8}$$

whereas for a series secondary resonance (Figure 19.10(b)), the optimal resistance is also given by (19.7), but α is redefined as the reciprocal of (19.8) to:

$$\alpha = \frac{\sqrt{1 + k^2 Q_{\mathrm{TX}} Q_{\mathrm{RX}}}}{Q_{\mathrm{RX}}} \tag{19.9}$$

For inductive links with low coupling values and high-Q coils, this means that, typically, the series resonant circuit requires a low value of optimal load which it feeds at high current, whereas the parallel configuration requires a higher value of optimal load fed at high voltage. In addition, the choice of series and parallel resonance also dictates whether the rectifier that attaches to the system is current driven or voltage driven (Section 19.4.2).

The objective of the primary circuit is to create a sinusoidal current in the transmit coil and, as stated, the link efficiency does not depend on the primary side

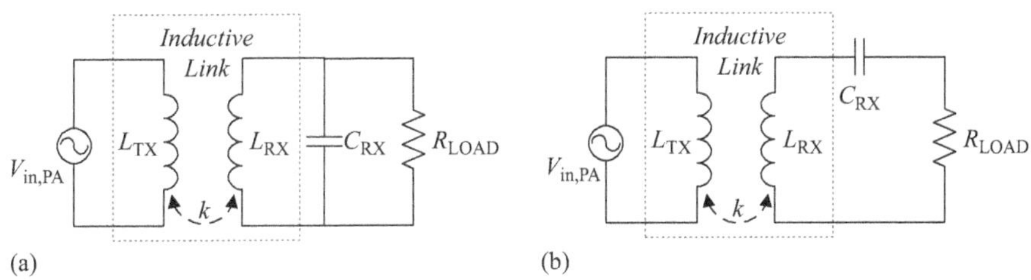

Figure 19.10 An inductive link with (a) parallel secondary resonance and (b) series secondary resonance

circuit topology that is chosen. However, in most practical WPT applications, the primary circuit is made to resonate at or near the driving frequency because this minimises the need for the drive circuit to supply reactive power. A parallel resonant primary will require the drive circuit to supply the resonant tank with low currents, but at high voltage, whereas the series resonant primary requires high currents at low voltage. These differing requirements for series and parallel circuits therefore place different stresses on the drive side components and lead to different circuit topologies and device choices.

One last but important point to consider when choosing between parallel and series resonances on primary or secondary is that the parasitic capacitance of the coils and any connected semiconductor devices are readily absorbed into the tuning capacitors in the parallel resonant circuits, but this is not with series resonance.

19.4.2 Practical circuit topologies

The primary objective of the circuits for driving and receiving power into and out of the inductive link is to do so at the required power level with high efficiency. Component choices in WPT systems can significantly influence the efficiency of the system, but circuit topologies also play a significant role in determining the maximum achievable end-to-end system efficiency. Several topologies for both the inverting transmitter side and the rectifying receiver side are now discussed.

19.4.2.1 Inverters

The most basic way to drive an inductive link primary coil is to use a simple linear analogue power amplifier (e.g. a class AB topology) to provide the excitation to the primary tank. Such circuits can readily be operated without the need to resonate the primary tank and still maintain sinusoidal primary coil currents (the linear mode of the amplifier allows a high fidelity sine wave output to be achieved). However, such topologies suffer from poor efficiency and so practical WPT amplifier circuits tend to operate in switch mode. The most basic switch-mode excitation of a transmitter coil can be achieved using a half-bridge cell operating with class-D (hard) switching. This simple circuit is often used in low frequency inverters (e.g. drive systems) where the devices are driven with a PWM signal that can be low pass filtered to generate a high fidelity sine wave. This typically requires the PWM switching frequency to be greater than the generated sinusoid by a factor of between 10 and 100 times. In a WPT system, where the sinusoidal excitation is rarely below 80 kHz, this places unrealistic requirements on hard switching the devices at a suitable PWM frequency. Consequently, with such half-bridge converters, a series primary resonance is often used. This both reduces the driver's voltage requirements (the voltage across the resonant capacitor and inductor are in anti-phase), whilst also allowing a near sinusoidal current to flow in the coils even when excited at the fundamental link frequency with a square wave from the inverter (because a high-Q resonant tank will filter the harmonic components sufficiently well).

A relatively common extension to the half-bridge topology is to drive the circuit with a full-bridge [18], as shown in Figure 19.11. As this circuit drives both sides of the resonant tank with a non-zero voltage, the full-bridge effectively

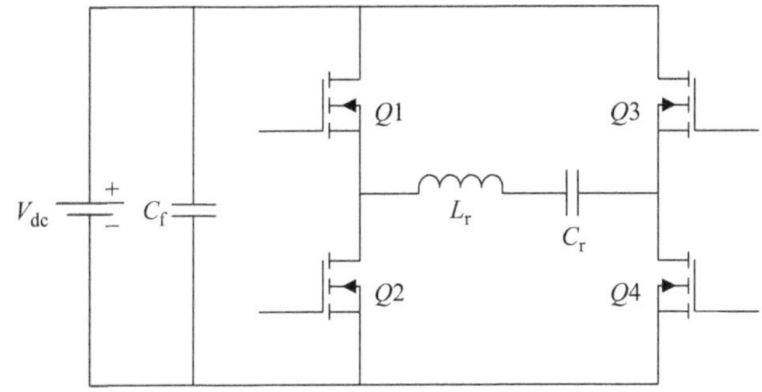

Figure 19.11 Simple H-bridge driver with series resonant primary tank

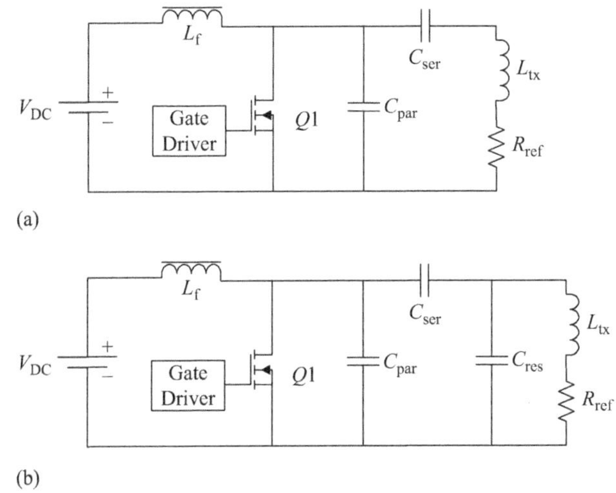

Figure 19.12 Class-E inverter in (a) standard operation and in (b) semi-resonance

doubles the primary circuit voltage capability for the same voltage blocking capability devices as in the half-bridge.

These class-D topologies are relatively simple, can be efficient and have a low component count and are thus common for lower frequency WPT systems, such as for EV charging. However, for systems that require higher frequency operation (such as those working in the 6.78 or 13.56 MHz industrial, scientific and medical (ISM) bands), hard switching becomes undesirable as switching losses can dominate. Consequently, a range of harmonically tuned inverters, which can exhibit (theoretically) zero switching loss, such as the class-E inverter, become desirable. The class-E inverter circuit, first published by Nathan and Alan Sokal in 1975 [40] allows high-fidelity sinusoidal currents to be generated from a simple square wave metal-oxide semiconductor field-effect transistor (MOSFET) gate signal, whilst achieving very low switching losses in the active device. The standard class-E inverter is shown in Figure 19.12(a). A practical improvement to this circuit

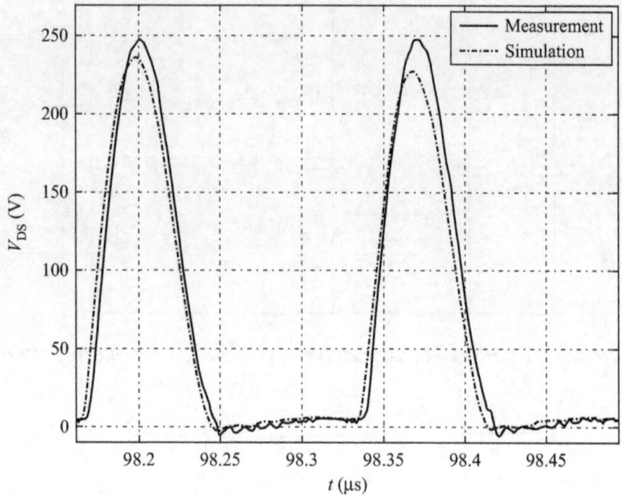

Figure 19.13 Typical drain voltage waveform on class-E inverter (from [41])

topology is to add a capacitor in parallel with the coil (eliminating the requirement for the MOSFET to conduct the full tank current) whilst operating the resonant tank slightly below the tank resonance and thus allowing the series branch inductance of the original class-E circuit to be provided by the semi-resonant tank [39], as shown in Figure 19.12(b). This operation has the advantage of reducing the primary side MOSFET current rating whilst ensuring that the WPT transmit coil provides the required branch inductance to allow true class-E operation. High efficiencies were achieved using this method in a 200 W capable inverter running at 6 MHz [41].

Figure 19.13 shows a measured drain voltage waveform on the class-E inverter: the smooth waveform allows zero switching loss due to both zero voltage and zero rate of change of voltage when the device switches.

19.4.2.2 Rectifiers

As discussed in Section 19.4.1, the receiver circuit must be resonant to maximise link efficiency and the load must be set to an optimal value. In the ideal case, this requires the input impedance of the rectifier to exhibit a constant real input impedance (as if a pure resistance was loading the resonant secondary circuit) whilst providing a dc output voltage. A simple passive rectifier, such as a full-wave passive diode rectifier can achieve the rectification but does not load the resonant tank with a purely resistive load (as the current contains harmonics if driven with a voltage source and the voltage contains harmonics if driven with a current source). In this case, an equivalent resistance at the fundamental frequency can be calculated and if this resistance can be set to that required by (19.7) controllable, the rectifier's loading of the link can be set optimally. If the resonant tank has high Q, then the generation of these harmonics by the rectifier will not cause significant

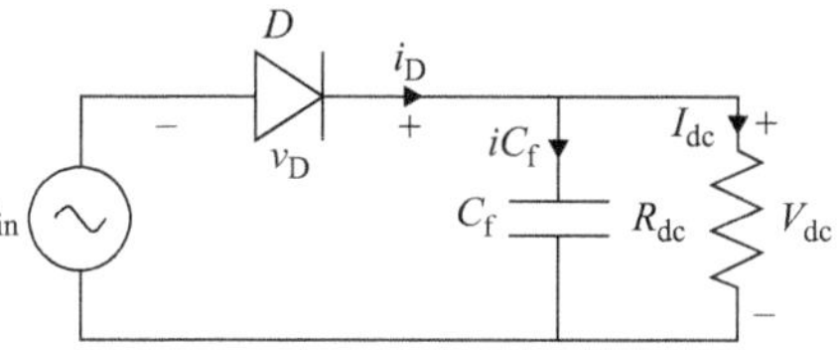

Figure 19.14 Voltage-driven half-wave rectifier

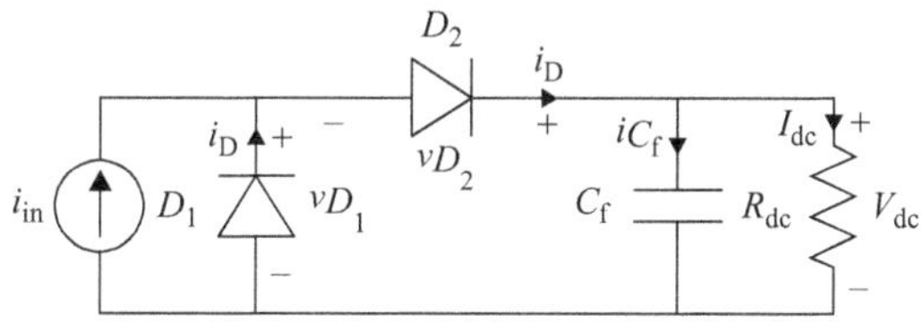

Figure 19.15 Current-driven class-D half-wave rectifier

distortion in the coil current due to the tank presenting a high impedance to these frequencies.

Just as the optimal secondary load value is dependent on whether the secondary tank is a parallel or series resonance, the choice of rectifier topology will also depend on the resonant tank type as rectifiers are designed to be current driven and some voltage driven. A voltage-driven rectifier is attached to a parallel secondary resonant tank, whilst a current-driven rectifier is attached to a series tuned secondary resonant tank [39].

The difference in the two can be seen by considering the simple half-wave voltage-drive rectifier shown in Figure 19.14. This rectifier can be driven from a parallel LC receiver tank, but if driven from a series LC tank, the negative tank current has no path and this result in circuit failure. In order to allow such a rectifier to work from a current source series resonant secondary, an additional diode must be added to provide a path for negative tank current, resulting in a class-D rectifier of Figure 19.15.

The input impedance of this class-D rectifier at the fundamental frequency is resistive with a value of:

$$R_{in} = \frac{2R_{DC}}{\pi^2}$$

which can be set to a value required to optimise the link.

The class-D rectifier is hard switched and at high frequency this can result in excessive losses in the diodes. A possible solution to this is to use a current-driven class-E rectifier topology, which controls the rate of change of voltage across the diode, reducing reverse recovery losses, as shown in Figure 19.16.

Voltage-driven soft-switched rectifier topologies are also possible, and a voltage-driven class-E rectifier topology is shown in Figure 19.17. This topology

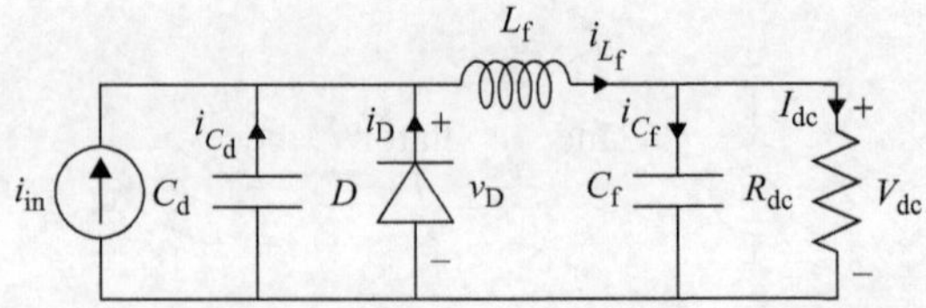

Figure 19.16 Class-E current-driven half-wave rectifier

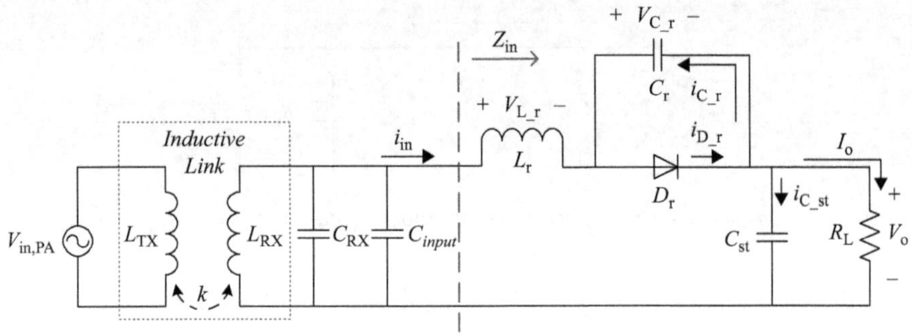

Figure 19.17 Voltage-driven class-E half-wave rectifier

has the additional advantage of inherently drawing a relatively pure sine wave input current from a sine wave voltage source, thus exhibiting very low harmonic content in the input voltage or current waveforms [42].

19.4.2.3 Component choices

Both active and passive components need to be carefully chosen to maximise the performance of a WPT system. High-Q capacitors, such as those available from AVX or Dielectric Laboratories, and fast MOSFETs, such as those from IXYS RF or SiC devices from Cree Inc., can all be used in mega-hertz WPT systems.

19.4.3 Control

Whilst it is possible to operate a WPT system in open loop, most practical WPT systems require some form of closed-loop control for individual constituent blocks, and a global feedback loop to control power throughput in the system. The typical aspects of the WPT system that use some form of closed loop control are:

1. *Optimal loading of the link (as discussed in Section 19.4.2)*: The combination of the rectifier and load must present the optimal impedance to the link, independent of what equipment is actually being powered from the secondary side. In a general WPT system that is designed to power a generic load, the impedance of the load will rarely present an optimal impedance to the link, and in many real-world scenarios the impedance of the load equipment will be time (e.g. when a charging battery). This means that an additional circuit, an impedance emulator,

must generally be inserted into the system between the rectifier and load so that the link is always correctly loaded.

The realisation of the impedance emulator is thus often a switch-mode circuit (which can be a traditional SMPS design) whose input impedance can be controlled. As we have seen, the link efficiency is maximised when the load impedance is set to an optimal value, and this value changes as the coupling between the coils changes, and hence changes as the relative coil positions change. This can require the input impedance of the emulator to be modified ensuring system operation to maintain the optimal load. Impedance emulators themselves fall into two broad categories: those that exhibit a constant input impedance when run in an open loop (i.e. the ratio of the input voltage to current remains constant as both the input voltage and the load on the converter changes) and those that require a closed-loop control to operate with a constant input impedance. Two possible open-loop emulator topologies are the buck–boost converter and the isolated flyback, both operating in discontinuous mode [43] (Figure 19.18).

2. Tuning of the primary driver: As the coupling between primary and secondary coils changes with relative movement, the primary side drive circuit may become detuned and, taking a class-E circuit as an example, no longer exhibit perfect class-E soft switching behaviour. A possible solution to retuning the class-E inverter is to use a saturable reactance [44]. By running a small bias current in the dc feed choke of the class-E inverter, the circuit waveforms can be modified so that the circuit achieves true class-E operation even as the relative coupling between coils changes.

3. Output voltage of the secondary: As is the case in the majority of power supply electronics, the dc output of the system should be controlled. Typically, a suitable energy buffer (e.g. a capacitor) follows the impedance emulation circuit and a regular SMPS topology is attached to this to regulate the output voltage.

4. Power throughput: As the link efficiency is maximised when the load presents an optimal impedance to the link, in order to maintain maximum efficiency at all times, the control of power throughput should not be achieved by modifying the load impedance and detuning the link, but instead by reducing the power input from the transmitter side. This power throughput control requires some form of communications link between the receiver and the transmitter which can be done using the magnetic link with load emulation or via an out of band radio link.

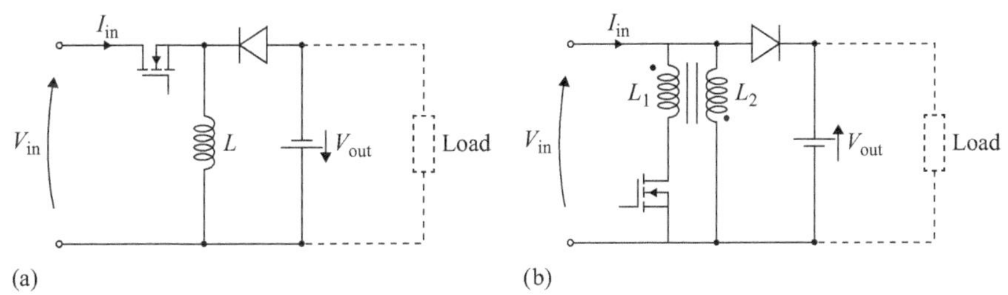

Figure 19.18 Two suitable impedance emulator topologies (a) buck–boost and (b) isolated flyback

5. Safety: In conditions where the WPT system operates where humans can approach the air gap and hence the magnetic field, a foreign object detection system must shut down the system if the magnetic/electric fields would exceed the required exposure limits (discussed in Section 19.5). Foreign Object detection can be done with several means, including detecting a detuning of the link or some form of motion detection and this is a common feature on high-power WPT systems such as the Qualcomm Halo product [45].

19.5 Safety regulations

Safety and regulatory requirements are a significant factor in the design choices for inductive WPT systems because the limits on both radiated emissions and near-field magnetic and electric field strengths are regulated. Due to the recent success of WPT systems, these long-standing regulations are being reviewed and updated with increased interest, with bodies being created to standardise and regulate different use cases. The regulations broadly fall into two categories: EMC for equipment and communications, and exposure limits for humans. The limits are frequency dependent, and the allowed field strengths reduce as the frequency increases. This means that the general trend in WPT design is that lower power WPT systems are able to operate at higher frequency (typically in the low megahertz ISM bands), whilst high power systems, such as those for electric vehicle charging, operate at around 100 kHz.

When considering the safety case for WPT systems, factors other than their electromagnetic output are also important. In many usage scenarios, WPT systems can eliminate the possibility of sparks occurring when physical connections are unplugged and in some medical applications the danger of operating a system above the recommended human exposure limits could be tolerated if the WPT system solves a severe medical issue for the patient where the benefits from using the system are greater than the potential risks associated with using it.

19.5.1 Electromagnetic compatibility

Any WPT system must comply with local radiated and conducted EMC standards for the class of equipment to which it belongs. Basic EMC requirements in different countries are set by local governments and hence differ from region to region. In Europe, an European Union (EU) directive 2004/108/EC [46] exists to ensure that equipment emits levels of electromagnetic energy such that other appliances and telecommunications are not interfered with, and that the equipment will maintain normal operation with acceptable degradation when exposed to an electromagnetic disturbance that might be reasonably be expected during normal use. This EU directive calls on several CENELEC standards for the technical detail, including: EN 55011 which covers household equipment and similar equipment designed to locally generate/use RF energy; EN 55014, which covers emissions from 150 kHz to 30 MHz from household appliances; and EN 61000 which covers conducted emissions of harmonics in the range 100 Hz–2 kHz from household appliances.

The specific use case of the WPT equipment may determine which standard applies, but in Europe the above standards are relevant. There are several frequency bands that are designated for ISM use which are covered by EN 55011. In these bands, equipment is allowed to radiate far-field RF energy provided the levels fall within those allowed by the standards. An efficient WPT system is designed not to radiate but small amounts of radiated RF energy are unavoidable. For this reason, WPT systems are often chosen to operate in the ISM bands, with the slots at 6.78 and 13.56 MHz being useful for WPT systems.

19.5.2 Human exposure

The safety for human exposure for WPT systems is of keen interest to the public and in the authors' experience is one of the first questions asked when demonstrating inductive power transfer to the lay person. The standards for human exposure to RF electromagnetic waves are described by several different bodies worldwide and although based on the same basic principles, the recommendations from each standards association can differ and this leads to different recommendations being made in different countries.

Through the International EMF Project, the World Health Organisation refers [47] to the guidelines set by both the IEEE [48] and ICNIRP [49]. There are two main types of health effect that the standards and directives take into account: tissue heating effects, and non-heating effects such as nerve/muscle stimulation. Both the IEEE and ICNIRP describe what they call 'basic restrictions', i.e. fundamental human-internal quantities of electric field strength and specific absorption rates (SARs) that should not be exceeded. As these levels are difficult to measure, the IEEE and ICNIRP both recommend a set of derived limits for the environment in which the person is situated (called reference levels by ICNIRP) which, if not exceeded, should not cause the internal basic restrictions to be violated. In the inhomogeneous near-field, the spatially varying nature of the E and B fields means that the derived limits/reference levels are specified in terms of independent values of the electric and magnetic field strengths rather than as a simple incident power density. Note that although the basic restrictions specified in IEEE and ICIRP are the same, the derived limits/reference levels differ, with the ICNIRP standards being stricter.

In Europe, the ICNIRP recommendations have themselves been adopted into the new EU Electromagnetic Fields Directive [50], which EU member states must transcribe into law by July 2016. The terminology in the EU directive differs from the ICNIRP, in that the EU calls the basic restrictions 'exposure limit values' (ELVs) and the reference levels are known as 'action levels' (ALs). At the frequencies of interest for WPT systems, the ELVs require that SAR is kept below 10 W/kg in the head and trunk, 20 W/kg in the limbs and below 0.4 W/kg averaged across the whole body. The ELVs also require that internal electric fields be lower than $3.8 \times 10^{-4} f$ V/m (where f is in Hz). A summary of the ICNIRP 1998 reference levels (adopted by the EU as ALs) for the E and B fields is given in Figures 19.19 and 19.20.

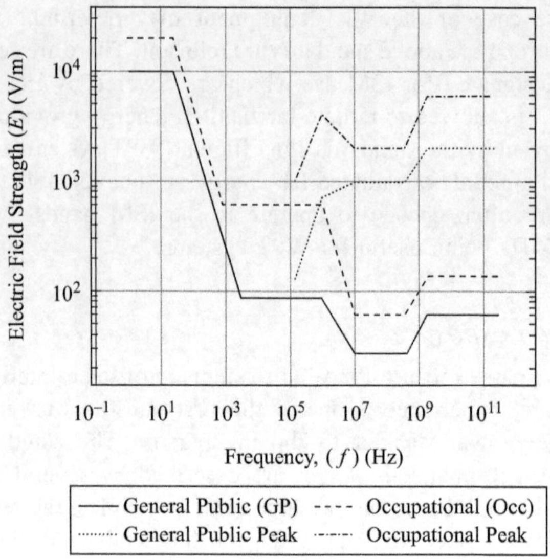

Figure 19.19 1998 ICNIRP E-field reference levels (redrawn from [49])

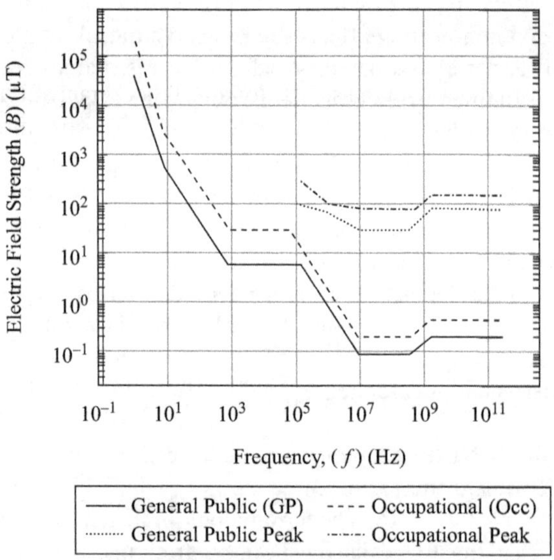

Figure 19.20 1998 ICNIRP B-field reference levels (redrawn from [49])

It should be noted that failure to meet the basic restrictions (ELVs in EU terminology) does not mean a system cannot be compliant with the regulations as it is the basic restrictions (ELVs in EU terminology) which must not be violated. These can be obtained by simulation. Examples of systems that do not necessarily meet the derived values/reference levels (ALs in EU terminology) but do meet the basic restrictions are discussed in [51].

19.6 Conclusion

Active WPT research and developments over the last two decades have enabled such technology to reach commercialization stage for near-field applications. Recent research has been extended to mid-range applications based on multiple coil configurations. This chapter explains the basic principles on which modern WPT is based. The advantages and disadvantages of the maximum energy efficiency principle and maximum power transfer principle are highlighted. Guidelines for choosing these principles and the choice of series and parallel resonant circuits in the receiver modules have been addressed. For practical implementation, relevant international safety regulations cannot be ignored because they set the practical boundaries for the power level and system designs. With the increasing number and variety of modern portable electronic devices and the dawning of the electric vehicle era, it is envisaged that applications of WPT will continue to grow both in power level and in variety in the future.

Acknowledgements

This work was carried out with the support of an EPSRC grant (EP/L00089X/1) and a Hong Kong RGC General Research Fund (17206715). They thank Mr. George Kkelis for his help with the diagrams.

References

[1] N. Tesla, "Apparatus for transmitting electrical energy", US Patent 1,119,732, Dec. 1, 1914.
[2] Special Issue on "Wireless power transmission, technology & applications", *Proceedings of the IEEE*, vol. 101, no. 6, June 2013.
[3] Robert Lomas, *The Man Who Invented the Twentieth Century: Nikola Tesla, Forgotten Genius of Electricity,* Headline Book Publishing, London, 1999, ISBN 0 7472 6265 9, p. 146.
[4] A.S. Marincic, "Nikola Tesla and the wireless transmission of energy", *IEEE Transaction on Power Apparatus and Systems*, vol. PAS-101, no. 10, pp. 4064–4068, Oct. 1982.
[5] L.P. Wheeler, Tesla's contribution to high frequency, Electrical Engineering, New York, Aug. 1943, p. 355.
[6] S.Y.R. Hui, "Planar wireless charging technology for portable electronic products and Qi", *Proceedings of the IEEE*, vol. 101, no. 6, pp. 1290–1301, June 2013.

[7] S.Y.R. Hui, W.X. Zhong and C.K. Lee, "A critical review of recent progress in mid-range wireless power transfer", *IEEE Transactions on Power Electronics*, vol. 29, no. 9, pp. 4500–4511, Sept. 2014.

[8] T.K. Sarkar, R.J. Mailloux, A.A. Oliner, M. Salazar-Palma and D.L. Sengupta, "Chapter 8. Nikola Tesla and His Contributions to Radio Development", in: *History of Wireless,* John Wiley & Sons, New Jersey, USA, 2006.

[9] E. Waffenschmidt and T. Staring, "Limitation of inductive power transfer for consumer applications", in *Proceedings of the EPE*, 2009, pp. 1–10.

[10] J.O. Mur-Miranda, G. Fanti, Y. Feng, K. Omanakuttan, R. Ongie, A. Setjoadi and N. Sharpe, "Wireless power transfer using weakly coupled magnetostatic resonators", in *Proceedings of the ECCE*, 2010, pp. 4179–4186.

[11] J.C. Schuder, H.E. Stephenson and J.F. Townsend, "High level electromagnetic energy transfer through a closed chest wall", *IRE International Convention Record*, vol. 9, no. 9, pp. 119–126, 1961.

[12] W.H. Ko, S.P. Liang and C.D.F. Fung, "Design of rf-powered coils for implant instruments", *Medical & Biological Engineering & Computing*, vol. 15, pp. 634–640, 1977.

[13] E. Hochmair, "System optimization for improved accuracy in transcutaneous signal and power transmission", *IEEE Transaction on Biomedical Engineering*, vol. BME-31, no. 2, pp. 177–186, Feb. 1984.

[14] B. Choi, J. Nho, H. Cha, T. Ahn and S. Choi, "Design and implementation of low-profile contactless battery charger using planar printed circuit board windings as energy transfer device", *IEEE Transaction on Industrial Electronics*, vol. 51, no. 1, pp. 140–147, Feb. 2004.

[15] Y. Jang and M.M. Jovanovic, "A contactless electrical energy transmission system for portable-telephone battery chargers", *IEEE Transaction on Industrial Electronics*, vol. 50, no. 3, pp. 520–527, Jun. 2003.

[16] A.W. Green and J.T. Boys, "10 kHz inductively coupled power transfer-concept and control", *Fifth International Conference on Power Electronics and Variable-Speed Drives*, 1994, pp. 694–699.

[17] G. Elliott, J.T. Boys and A.W. Green, "Magnetically coupled systems for power transfer to electric vehicles", *Proceedings of 1995 International Conference on Power Electronics and Drive Systems*, vol. 2, 1995, pp. 797–801.

[18] G. Covic and J. Boys, "Inductive power transfer", *Proceedings of the IEEE*, vol. 101, no. 6, pp. 1276–1289, Jun. 2013.

[19] S.Y.R. Hui and W.C. Ho, "A new generation of universal contactless battery charging platform for portable consumer electronic equipment", *IEEE Transaction on Power Electronics*, vol. 20, no. 3, pp. 620–627, May 2005.

[20] X. Liu and S.Y.R. Hui, "Simulation study and experimental verification of a contactless battery charging platform with localized charging features", *IEEE Transaction on Power Electronics*, vol. 22, no. 6, pp. 2202–2210, Nov. 2007.

[21] S.Y.R. Hui, "Planar inductive battery charging system", US Patent 7,576,514, 2009.

[22] S.Y.R. Hui, "Planar inductive battery charger", GB Patent 2,389,720, 2005.

[23] S.Y.R. Hui, W.C. Chan, W.C. Ho and X. Liu, "Localized charging, load identification and bi-directional communication methods for a planar inductive battery charging system", *US patent* 7,915,858, 2011.

[24] Wireless Power Consortium Website [online]. Available: http://www.wirelesspowerconsortium.com

[25] Wireless Power Consortium Website [online]. "Qi System Description: Wireless Power Transfer", Volume I: Low Power, Part 1: Interface Definition, Version 1.1. Available: http://www.wirelesspowerconsortium.com/downloads/wireless-power-specification-part-1.html

[26] R. Syms, E. Shamonina and L. Solymar, "Magneto-inductive waveguide devices", in IEE Proc. Microwave, Antennas Propagation, vol. 153, no. 2, pp. 111–121, 2006.

[27] F. Zhang, S. Hackworth, W. Fu and M. Sun, "The relay effect on wireless power transfer using WiTricity", *IEEE Conference on Electromagnetic Field Computation*, Chicago, USA, May 2010.

[28] J.W. Kim, H.C. Son, K.H. Kim and Y.J. Park, "Efficiency analysis of magnetic resonance wireless power transfer with intermediate resonant coil", *IEEE Antennas and Wireless Propagation Letters*, vol. 10, pp. 389–392, 2011.

[29] W.X. Zhong, C. Zhang, X. Liu and S.Y.R. Hui, "A methodology for making a 3-coil wireless power transfer system more energy efficient than a 2-coil counterpart for extended transmission distance", *IEEE Transactions on Power Electronics*, vol. 30, no. 2, pp. 933–942, Feb. 2015.

[30] A. Kurs, A. Karalis, R. Moffatt, J.D. Joannopoulos, P. Fisher and M. Soljacic, "Wireless power transfer via strongly coupled magnetic resonances", *Science*, vol. 317, no. 5834, pp. 83–86, Jul. 2007.

[31] S. Cheon, Y.H. Kim, S.Y. Kang, M.L. Lee, J.M. Lee and T. Zyung, "Circuit-model-based analysis of a wireless energy-transfer system via coupled magnetic resonances", *IEEE Transaction on Industrial Electronics*, vol. 58, no. 7, pp. 2906–2914, Jul. 2011.

[32] M. Kiani and M. Ghovanloo, "The circuit theory behind coupled-mode magnetic resonance-based wireless power transmission", *IEEE Transaction on Circuits Systems – I*, vol. 59, no. 8, pp. 1–10, Aug. 2012.

[33] C.J. Chen, T.H. Chu, C.L. Lin and Z.C. Jou, "A study of loosely coupled coils for wireless power transfer", *IEEE Transaction on Circuits and Systems – II: Express Briefs*, vol. 57, no. 7, pp. 536–540, Jul. 2010.

[34] C.K. Lee, W.X. Zhong and S.Y.R. Hui, "Effects of magnetic coupling of non-adjacent resonators on wireless power domino-resonator systems", *IEEE Transaction on Power Electronics*, vol. 27, no. 4, pp. 1905–1916, Apr. 2012.

[35] W.X. Zhong, C.K. Lee and S.Y.R. Hui, "Wireless power domino-resonator systems with non-coaxial axes and circular structures", *IEEE Transaction on Power Electronics*, vol. 27, no. 11, pp. 4750–4762, Nov. 2012.

[36] W.X. Zhong, C.K. Lee and S.Y.R. Hui, "General analysis on the use of Tesla's resonators in domino forms for wireless power transfer", *IEEE Transaction on Industrial Electronics*, vol. 60, no. 1, pp. 261–270, Jan. 2013.

[37] H. Winfield Secor, "Tesla apparatus and experiments – how to build both large and small Tesla and Oudin coils and how to carry on spectacular experiments with them", *Practical Electrics*, November 1921.

[38] X. Liu, "Inductive power transfer using a relay coil", US patent application 13/907,483, filed on 31-May-2013.

[39] K. Van Schuylenbergh and R. Puers, Inductive Powering, *Basic Theory and Application to Biomedical Systems*, Springer, New York, USA, 2009.

[40] N.O. Sokal and A.D Sokal, "Class E-A new class of high-efficiency tuned single-ended switching power amplifiers", *IEEE Journal of Solid-State Circuits*, vol. 10, no. 3, pp. 168–176, 1975.

[41] G. Kkelis, J. Lawson, D. Yates, M. Pinuela and P. Mitcheson, "Integration of a class-E low dv/dt rectifier in a wireless power transfer system", in *Wireless Power Transfer Conference (WPTC)*, 2014 IEEE, May 2014, pp. 72–75.

[42] M. Kazimierczuk, "Analysis of class-E zero-voltage-switching rectifier", *Circuits and Systems, IEEE Transactions on*, vol. 37, no. 6, pp. 747–755, Jun. 1990.

[43] C.H. Kwan, J. Lawson, D.C Yates and P.D. Mitcheson, "Position-insensitive long range inductive power transfer", *Proceedings of PowerMEMS* 2015, Awaji Island, Japan.

[44] S Aldhaher, P.C.K. Luk and J.F. Whidborne, "Tuning class E inverters applied in inductive links using saturable reactors", *IEEE Transactions on Power Electronics*, vol. 29, no. 6, pp. 2969–2978, June 2014.

[45] http://www.qualcommhalo.com, accessed 9/2/15.

[46] Directive 2004/108/EC Directive of the European Parliament of the Council on the approximation of the laws of the Member States relating to electromagnetic compatibility, December 2014.

[47] World Health Organization, "Electromagnetic fields and public health", Fact Sheet No. 304, May 2006.

[48] IEEE Standard for Safety Levels with Respect to Human Exposure to Radio Frequency Electromagnetic Fields, 3 kHz to 300 GHz, IEEE Std. C95.1-2005.

[49] Guidelines for Limiting Exposure to Time-Varying Electric, Magnetic and Electromagnetic Fields (Up to 300 GHz), ICNIRP Guidelines, International Commission on Non-Ionizing Radiation Protection, *Health Physics*, vol. 74, no. 4, pp. 494–522, 1998.

[50] Directive 2013/35/EU of the European Parliament and of the Council on the minimum health and safety requirements regarding the exposure of workers to the risks arising from physical agents (electromagnetic fields), June 2013.

[51] M. Kesler, Highly Resonant Wireless Power Transfer: Safe, Efficient, and Over Distance, WiTricity whitepaper, 2013.

Chapter 20

Advanced control of power electronic systems

*Sudip K. Mazumder**

20.1 Introduction

Power electronic systems (PESs) are nonlinear hybrid dynamical systems [1]. The instability in such switching systems, owing to their discontinuity, can evolve on slow and on fast scales [2]. Conventional analyses of PESs and their subsystems are based on averaged models, which ignore the fast-scale instability and analyze the stability on a reduced-order manifold [1–3]. As such, validity of the averaged models varies with the switching frequency even for the same topological structure. The prevalent procedure for analyzing the stability of standalone and networked PESs is based on linearized averaged (small-signal) models that require a smooth averaged model. Yet there are systems (in active use) that yield a non-smooth averaged model. Even for systems for which smooth averaged model is realizable, small-signal analyses of the nominal solution/orbit do not provide anything about three important characteristics: region of attraction of the nominal solution, dependence of the converter dynamics on the initial conditions of the states, and the post-instability dynamics. As such, conventional linear controllers for PESs, designed based on small-signal analyses, may be conservative and may not be robust and optimal.

20.2 Brief overview of historic advanced nonlinear controllers for PES applications

The inadequacy of the linear controllers to adequately address most application scenarios of PESs implies that, there is a clear need to design nonlinear controllers for such hybrid (power electronic) systems thereby achieving wider stability margin, improved robustness against parametric variations, feedforward and feedback disturbances, switching nonlinearities and interactions, and enhanced performance.

*IEEE, Piscataway, NJ, USA
*University of Illinois, Chicago, IL, USA

In that regard, historically, there has been work from varied perspective and focus, which fall broadly into the following categories:

- Sliding-mode control (SMC) [4]
- Passivity-based control (PBC) [5]
- Back-stepping control (BSC) [6]
- Feedback linearizing control (FLC) [7]
- Model-predictive control (MPC) [8]

It is worth mentioning and as one can conjecture, there has been several additional works that are based on these foundational control methodologies.

Among all of these historic nonlinear control methodologies for PESs, SMC probably has found greatest acceptance. It yields good transient performance and yields robustness even in the presence of parametric variations. SMC essentially relies on controlling a PES using a twofold strategy: one that guides the system trajectories to a reduced-order manifold (thereby ensuring existence of a hypersurface); and the other that guides the system dynamics on this manifold to the desired equilibrium or orbit. Essentially, this is achieved by controlling the switching states of a PES in such a manner such that the evolving switching sequence leads to the minimization of a (typically) Lyapunov-based (sliding) function. The initial work on SMC, even though yielded great transient results ran into issues with PESs that are susceptible to parasitic dynamics. That led to modified SMC schemes (often referred to as integral variable structure control or IVSC schemes) that incorporate an integral term in the sliding function and mitigate the need for the derivative term leading to robustness against parasitics and paying a small price with regard to the anticipatory response. Notwithstanding the strengths of SMC, it remains a methodology that only ensures sliding-mode convergence. While this acceptable for several applications, it is not the only mechanism or mode via which one can achieve convergence of trajectories or it may not be always desirable to only achieve sliding-mode convergence. An associated difficulty arises with ensuring constant-frequency operation for a PES under steady-state conditions.

Nonlinear PBC algorithms for PESs have proved to be an interesting alternative as well. Passivity is a basic property of physical systems such as a PES that can be defined in terms of energy dissipation and transformation. It is an input–output property in the sense that it quantifies and qualifies the energy balance of a system when stimulated by external inputs to generate some output. This is in contrast to Lyapunov stability which concerns the internal stability of a system. The control objective in PBC is typically realized via an energy-reshaping process and by injecting damping to modify the dissipation structure of the PES. Even though PBC has had some success in the control of PESs, the control formulation, based typically on an averaged sense, requires an external modulator to synthesize the discontinuous signals that are fed to the switching PESs. Thus, unlike the SMC, the time evolution of the switching states is typically dependent on the modulation principle. As such, multi-scale controllability in PBC is not an easy proposition.

BSC outlines and ensures a systematic way of realizing the Lyapunov function for a closed-loop PES by formulating the control input design. The latter in BSC also needs to ensure negative definiteness of the derivative of the control Lyapunov function by typically canceling the indefinite cross-coupling terms. Needless to mention that, such cancellations need to be achieved carefully keeping an eye on the tradeoff among stability, performance, and control-overhead tradeoff. Even though BSC has some issues concerning the fact that either all of the state variables need to be measureable or a subset of the variables need to predictable using an estimator (that typically requires a nonlinear observer) and that BSC is sensitive to parameter variations, work has been conducted in the previous decade that has addressed some of these issues. As such, BSC has found some acceptance for PES applications. Nevertheless, almost all applications of BSC for PES applications require a modulator as the control is primarily average model based. Further, unlike sequence-based control (SBC), evolution of switching sequence in BSC is dependent on the selected modulation strategy.

The conventional FLC has gained some popularity since it enables the utilization of tools for linear system for designing the control of nonlinear systems such as PESs thereby precluding the need relatively more complex design and stability-analysis nonlinear tools. This method is typically restricted to certain classes of nonlinear systems and stable invertibility is required; further special attention may be required regarding the zero dynamics of certain class of PESs. Further, notwithstanding the relative ease of design and analysis using FLC, choice of mechanism for linearization needs careful analysis to prevent elimination of useful nonlinearities or synthesizing overcompensating controls. Yet another issue with a conventional FLC is dealing with parametric uncertainties that affect PES stability and convergence of dynamics. Finally, just like the BSC, majority of the work on FLC has used averaged model of the PES and external modulator to transform the smooth control output to a discontinuous output that can be fed to a switching PES following a predefined switching sequence.

MPC, also known sometimes as receding horizon control, makes use of PES dynamic model to predict the control needs by minimizing a cost function. It has achieved recognition since it provides a systematic methodology to control constrained multivariable dynamical PESs. However, the performance of the controller depends on how well the dynamics of the PES being captured by the model. MPC uses the dynamic model to determine a control sequence that optimizes a desired performance index. A receding horizon strategy is used so that at each instant the horizon is moved towards the future, which involves the application of the control signal of the sequence (that yield the desired behavior) calculated at each step. Even though several of the earlier works on MPC for PESs focused on averaged modeling based approaches along with an external modulator or even discrete modeling based approaches using a predefined modulation scheme, a few recent works have advocated the integration of modulation and control in a monolithic control platform. The latter, like the SBC, precludes the need for transforming the average control following a predefined switching sequence as evident in several of other nonlinear control methodologies discussed earlier including conventional MPC. The synthesis of the sequence is actually an outcome of the evolution of the switching states which is the primary focus of these MPC schemes for PESs.

20.3 Switching SBC [9–11]

Relatively recently, the author has proposed a control methodology (i.e., SBC) that results in control of PESs by controlling the time evolution of stabilizing feasible switching sequences of PESs. Unlike MPC, SBC optimally controls the evolution of the switching sequences that have been determined to be stabilizable. SBC directly controls the switching sequences of the power devices without the need for an intermediate pulse-width modulator as evident in conventional power converter control schemes. SBC involves identifying the set of feasible sequences of switching states using a composite Lyapunov functions for global stability. Subsequently, by solving an optimization problem, an optimal switching sequence (comprising a union of feasible sequences) and corresponding time horizon for each switching state of the sequence is determined. The application of SBC to standalone and homogeneously and heterogeneously networked power converters is outlined briefly in the following subsections.

20.3.1 SBC for standalone PES

Controllers for PESs have to address two key issues related to the PES dynamics: one of them is to ensure steady-state stability, while the other is to ensure orbital existence (i.e., whether the trajectories of the PES converge to an orbit from an arbitrary initial condition). In [10], the authors propose a SBC scheme for PESs that can address both of these issues, i.e., it ensures optimal dynamic performance while ensuring that the PES is globally stable. The control is designed using piecewise linear models of the PES and can directly control the switching actions of the PES power devices, unlike conventional two-stage schemes where the controller (derived based on smooth averaged models) and the modulator (to generate the switching signals) are two separate entities, as illustrated in Figure 20.1. The overall control scheme consists of first determining the set of feasible switching sequences (using a composite Lyapunov function based approach) and then determining the duration of time spent in each switching state (by solving an optimization problem). The SBC control scheme exhibits superior dynamic performance, primarily because it directly controls the switching sequence of the PES. In this section, we illustrate the application of the SBC scheme to a standalone PES and demonstrate how superior control performance can be achieved compared to conventional linear controllers.

20.3.1.1 Description of the SBC scheme

As illustrated in Figure 20.2, the overall control is divided into two parts: (a) a saturated region optimal control (SROC) that controls the transient dynamics of the SPC and (b) an unsaturated region optimal control (USROC) that controls the steady-state dynamics of the SPC. The SBC can be derived using switching models of the converter, which is given by

$$\dot{x}(t) = A_i x(t) + B_i \tag{20.1}$$

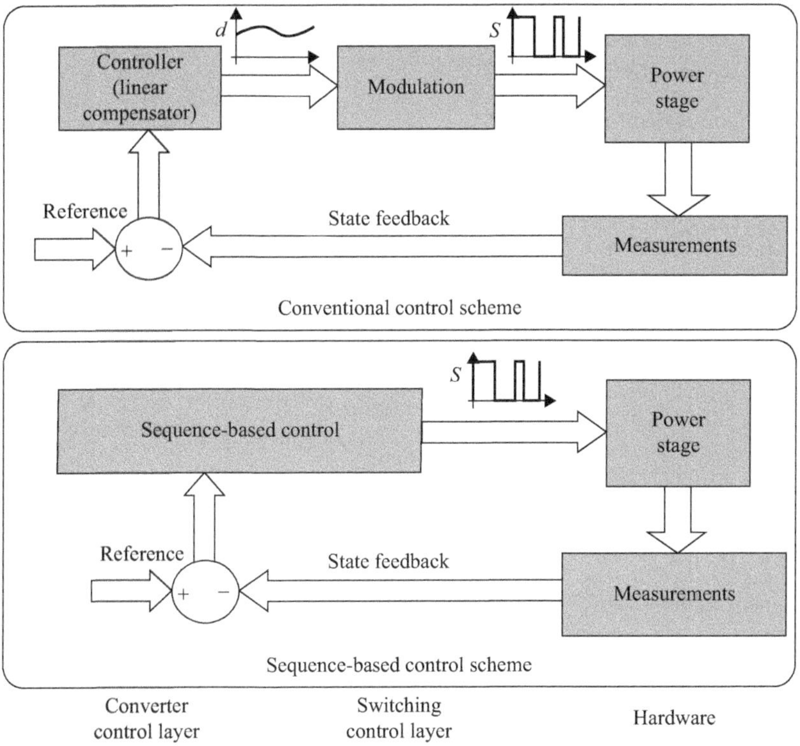

Figure 20.1 Block diagram illustrating a comparison between a conventional control scheme for PESs and SBC scheme

where $x(t)$ represents the states of the converter, i represents the switching states, and A_i and B_i are matrices and vectors of appropriate dimensions. The first step in both these schemes is to choose the feasible switching sequences (among all possible switching sequences) using a composite Lyapunov function based criteria, described in the following equation:

$$\sum_{i=1}^{h} \alpha_{ki} \begin{bmatrix} A_i^T P_{ki} + P_{ki} A_i & P_{ki} \overline{B}_i \\ \overline{B}_i^T P_{ki} & 0 \end{bmatrix} < 0 \tag{20.2}$$

where $P_{ki} = P_{ki}^T > 0$, $\sum_{i=1}^{h} \alpha_{ki} = 1$, $0 \leq \alpha_{ki} \leq 1$, and $\overline{B}_i = -B_i - A_i x^*$. Having determined the set of feasible switching sequences, the second step is to derive a discrete map for the states of the SPC over a time horizon T_w (i.e., $x(t_0 + T_w)$), as illustrated in Figure 20.3, given initial values of the states $x(t_0)$. Note that, for the USROC, the time horizon T_w corresponds to the switching time period, T_s. The overall map can be obtained by patching together the individual maps

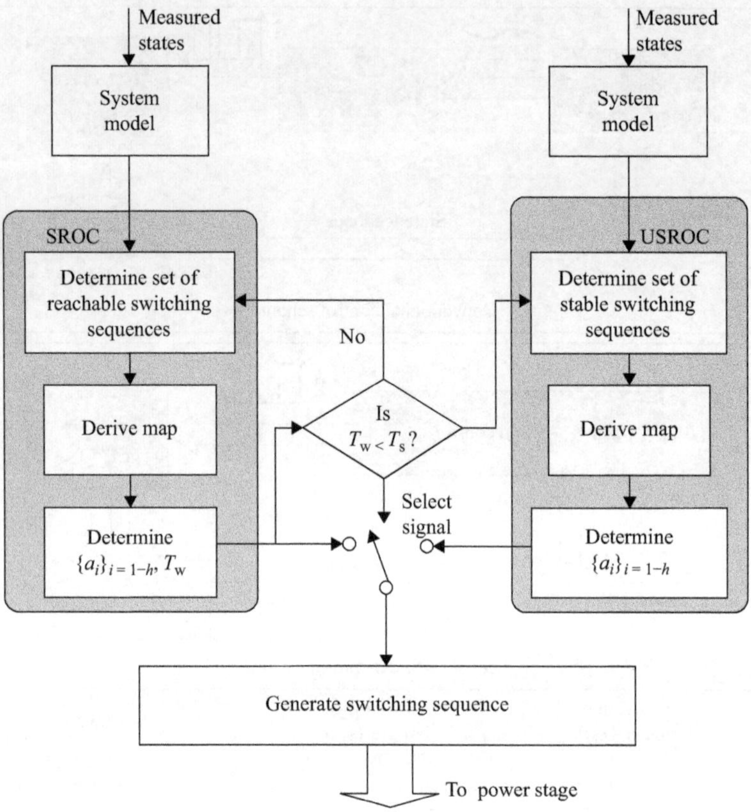

Figure 20.2 Schematic illustrating the functionality of the SBC. The measured states are obtained from the sensors. The "Select Signal" is used to switch from SROC to USROC when $T_w < T_s$

corresponding to each switching state, as described below (I is an identity matrix of appropriate dimension):

$$
\begin{aligned}
x(t_0 + \alpha_1 T_w) &= e^{A_1 \alpha_1 T_w} x(t_0) + \int_{t_0}^{\alpha_1 T_w} (e^{A_1 \tau} - I) B_1 \, d\tau \\
x(t_0 + (\alpha_1 + \alpha_2) T_w) &= \left\{ \begin{array}{l} e^{A_2 \alpha_2 T_w} x(t_0 + \alpha_1 T_w) \\ + \int_{\alpha_1 T_w}^{\alpha_2 T_w} (e^{A_2 \tau} - I) B_2 d\tau \end{array} \right\} \\
&\ \vdots \\
x(t_0 + T_w) &= \left\{ \begin{array}{l} e^{A_h \alpha_h T_w} x(t_0 + (\alpha_1 + \cdots + \alpha_{h-1}) T_w) + \\ \int_{\alpha_1 T_w}^{\alpha_2 T_w} (e^{A_h \tau} - I) B_h d\tau \end{array} \right\}.
\end{aligned} \tag{20.3}
$$

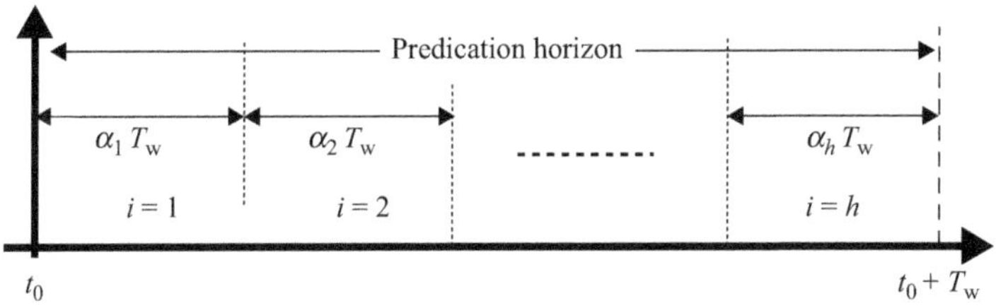

Figure 20.3 Schematic illustrating prediction horizon for computation of the optimal control law

The third and final step is to determine the optimal duration of time the SPC spends in a given switching sequence. The overall control schemes for both the SROC and the USROC are similar except for the presence of the additional optimization variable T_w in the SROC problem that corresponds to the time horizon over which the control is computed, as illustrated in Figure 20.3. Due to the paucity of space, here we describe only the SROC scheme. The SROC scheme is formulated as a cost optimization problem (where the cost is a function of α_i and T_w). Choice of the cost function is user dependent and could also vary with applications and operating conditions. The constraints for the optimal control problem include the map (20.2) and the maximum values that each state can attain (X_{max}). The SROC optimization problem can be summarized as follows.

Determine $\{\alpha_i\}_{i=1-h}$ and T_w that

$$\begin{aligned} \text{minimizes} \quad & J(\{\alpha_i\}_{i=1-h}, T_w) = (x^* - x(t_0 + T_w))^T P (x^* - x(t_0 + T_w)) \\ \text{s.t} \quad & x(t_0 + T_w) = f(x(t_0), \{\alpha_i\}_{i=1-h}, T_w, \{A_i\}_{i=1-h}, \{B_i\}_{i=1-h}) \\ & x(t_0 + T_w) \le X_{max}, \sum_{i=1}^{h} \alpha_i = 1, \text{ and } 0 < \alpha_i < 1; \; i = 1 - h \end{aligned} \tag{20.4}$$

where P is a positive-definite diagonal matrix, and x^* represents the reference value for all of the SPC states. The optimization problem (20.3) is a quadratic programming problem and can be solved numerically using quadratic programming algorithms [12]. The SROC optimization problem is solved repeatedly until $T_w < T_s$, where T_s is the steady-state switching time period. When this condition is satisfied, the SBC transitions to the USROC scheme. This condition ensures that the SPC frequency does not approach infinity.

20.3.1.2 Application of SBC to a standalone PES

In this subsection, we investigate the application of the SBC scheme to a standalone three-phase inverter, as illustrated in Figure 20.4. The switching states of the inverter are described in Table 20.1. Note that the zero states ($S_1 = 0$, $S_2 = 0$, and $S_3 = 0$; $S_1 = 1$, $S_2 = 1$, and $S_3 = 1$) are similar from the control performance point of view and are therefore redundant. The number of possible switching sequences is given by $M = \sum_{l=1}^{(2^N - W)} ({}^{(2l)}C_l)$ [9], where N represents the total number of

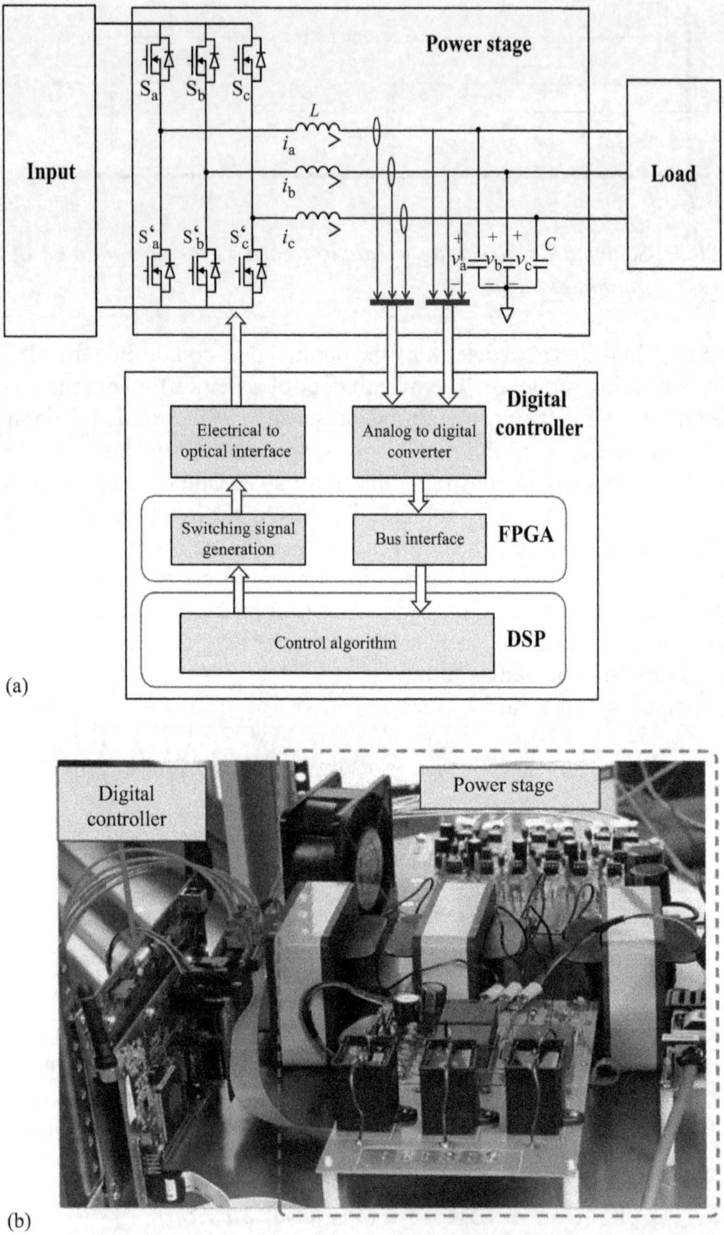

(a)

(b)

Figure 20.4 (a) Architecture and (b) and (c) experimental setup of a digitally controlled three-phase voltage source inverter

switching functions (in this case 3), while W represents the redundant switching states. The control scheme is implemented in the synchronous reference frame [9]. The parameters of the inverter are provided in Table 20.1 while switching models are provided in [11].

Table 20.1 Nominal parameters of the inverter with a linear resistive load

Parameters	**Nominal values**
Input voltage, V_{in}	400 V
Output line–line voltage	208 V (RMS)
Output line frequency	60 Hz
Power	2.5 kVA
Switching frequency, f_{sw}	20 kHz
Line inductors, L_j	1.5 mH
Output filter capacitors, C_j	10 F

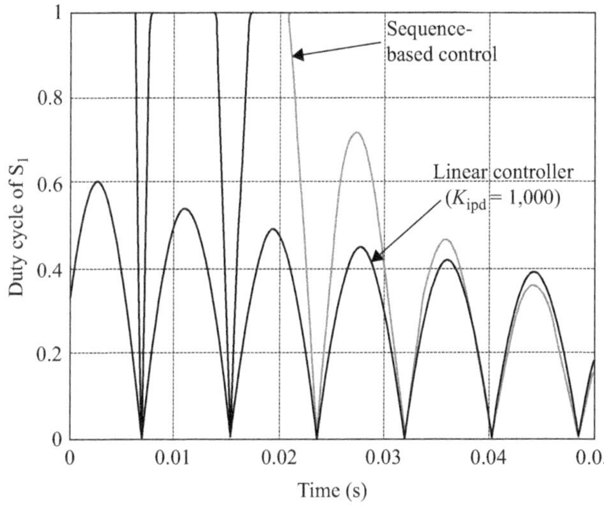

Figure 20.5 Comparison of the duty cycles of a inverter switch generated using the SBC scheme and a conventional linear controller [13]

The quadratic cost function for the SBC scheme is as follows:

$$J(t) = w_1(I_{dref} - i_d(t_0 + T_w))^2 + w_2(I_{qref} - i_q(t_0 + T_w))^2 + w_3(V_{dref} - v_d(t_0 + T_w))^2 \tag{20.5}$$

where the values of the weights are $w_1 = 10$, $w_2 = 10$, and $w_3 = 5$. The reference for the voltage V_{dref} is predefined while the current references I_{dref} and I_{qref} are defined based on the load requirements.

The superior performance of the SBC as compared to state of the linear controllers is demonstrated in Figure 20.5. Superior performance of the SBC scheme can be explained by evaluating the duty ratio of one of the inverter switches, as shown in Figure 20.5. The SBC drives the switches of the inverter into saturation (unlike the linear compensator based scheme); therefore, its response to transients is faster. Further analyses of the switching sequences will be presented in the final paper to demonstrate the superiority of the SBC scheme.

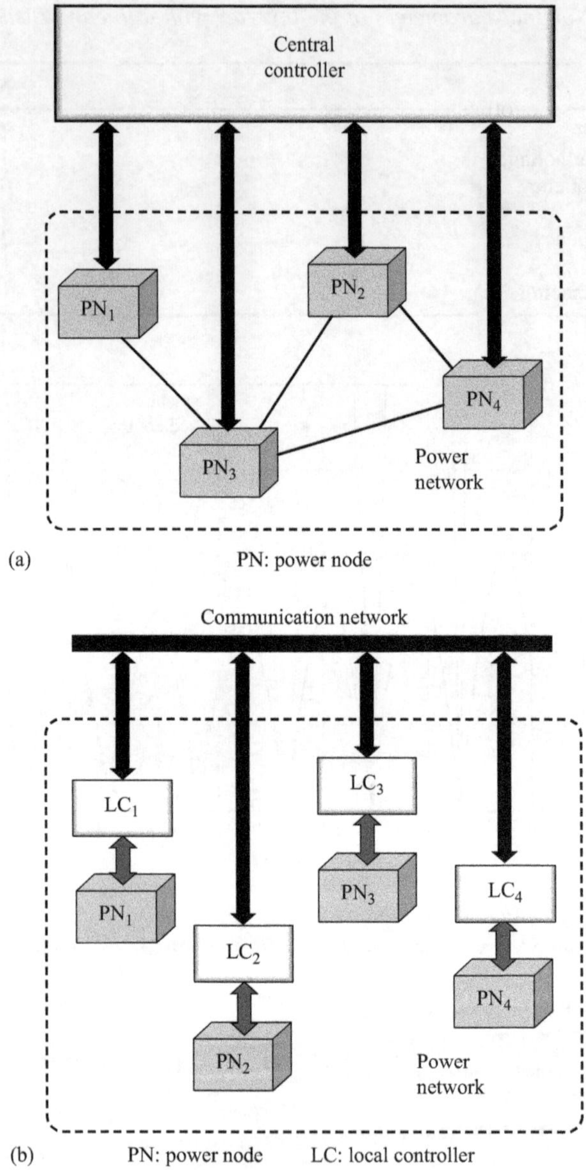

Figure 20.6 Schematics illustrating the (a) centralized and (b) distributed architectures of SBC implementation

20.3.2 SBC for networked PESs

To extend the SBC scheme to networked converters, we consider two implementation architectures, as illustrated in Figure 20.6. In the centralized architecture [4,7], as illustrated in Figure 20.6(a), the entire network is treated as a single entity. State variables of all the modules are transmitted to a central controller. Using this information and knowledge of the overall model, the controller computes the optimal switching sequence and the time spent in each switching state of the sequence. This information is then transmitted to all the modules via a common broadcast. On the other hand, for distributed implementation as illustrated in Figure 20.6(b), the overall control problem is decomposed into multiple local control problems [10,11,14]. However, because each module is affected by interactions with the other converters, inter-module communications are required to solve the local control problems. The clustered scheme is a combination of these two schemes. The SBC control schemes for the centralized and distributed control implementation of the parallel inverter, shown in Figure 20.7, is captured in Figure 20.8, which follows mechanism outlined in Section 20.3.1.1. For the distributed implementation, communication is enabled to handle the uncertainties.

Here, we present experimental results demonstrating the application of the control schemes to a parallel inverter network. Load-sharing errors of the experimental parallel inverter with varying number of modules are shown in Figure 20.9.

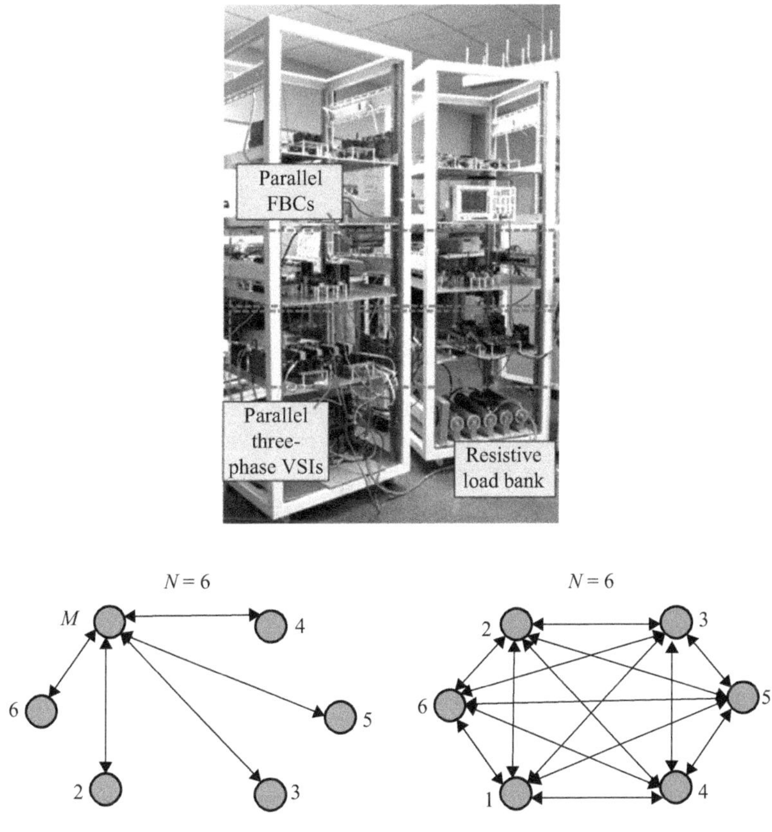

Figure 20.7 (Top) Experimental setup consisting of six inverter modules connected in parallel; and (bottom) schematics illustrating the centralized and distributed architecture for the SBC implementation

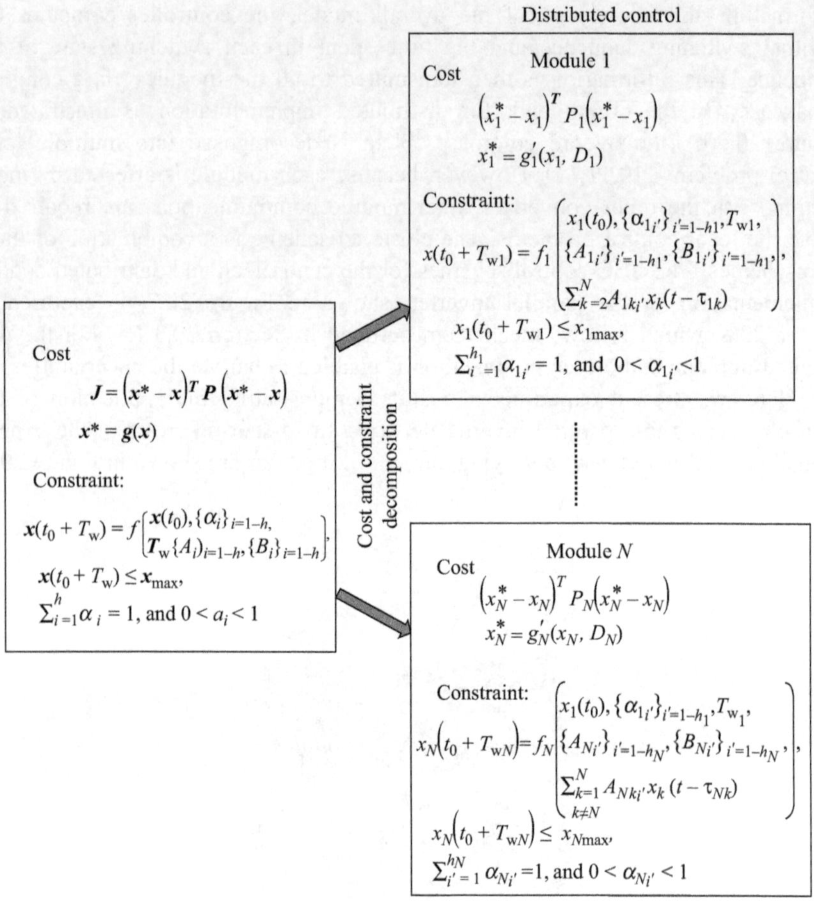

Figure 20.8 Illustration of the decomposition of the centralized SBC problem to a number of smaller (and computationally more tractable) control problems for distributed SBC implementation [15]

The experimental results illustrate that while the steady-state load-sharing error for the centralized and distributed implementations is comparable, the response times for the centralized case increase significantly compared to the distributed case with increase in the number of modules. This can be attributed to the increased computation times for the centralized case. Therefore, while the centralized implementation scheme can be a viable alternative for lower number of modules, distributed/clustered implementation is desirable for higher number of modules. For the distributed implementation also, we observe performance degradation as the number of modules increases due to increased communication requirements.

Next, we consider a microgrid, illustrated in Figure 20.10 [16], as a case illustration for distributed control-communication. We explore communication via differentially encoded data packets to reduce the data packet size since it can

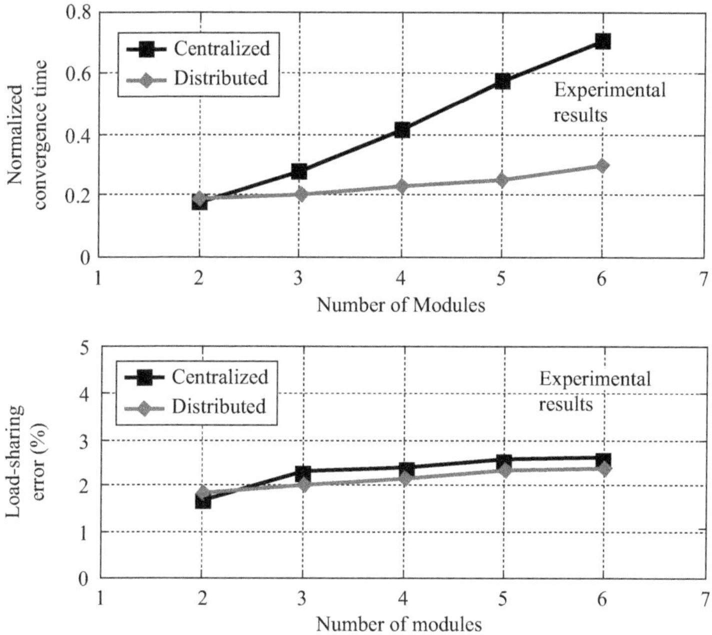

Figure 20.9 Experimental results for the parallel three-phase inverter comparing the centralized and distributed schemes of SBC implementation with the varying number of inverter modules

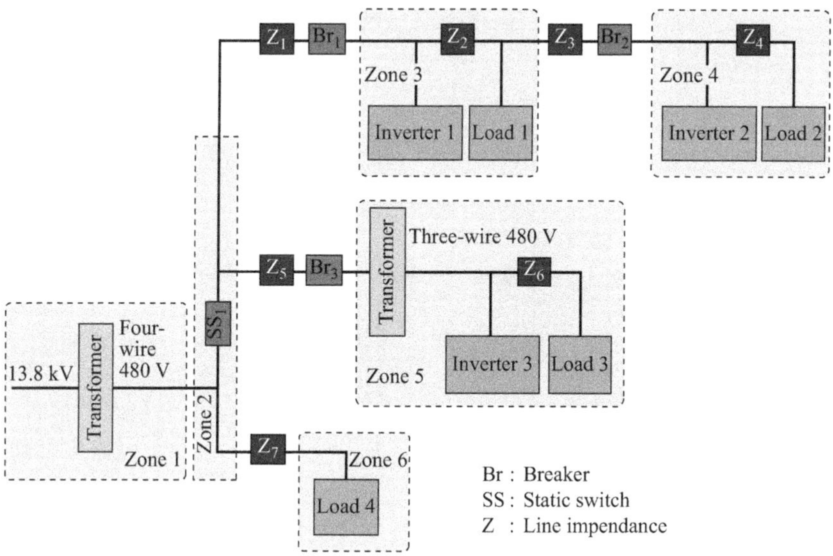

Figure 20.10 Architecture of the CERTS microgrid consisting of six zones [16]. Only the islanded mode of operation is investigated for the results in this paper. Grid connection takes place through Zone 3

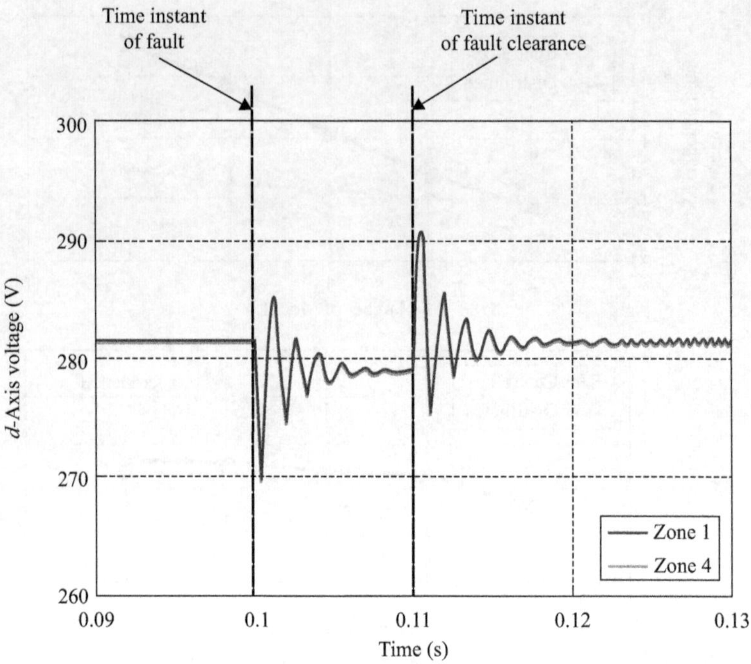

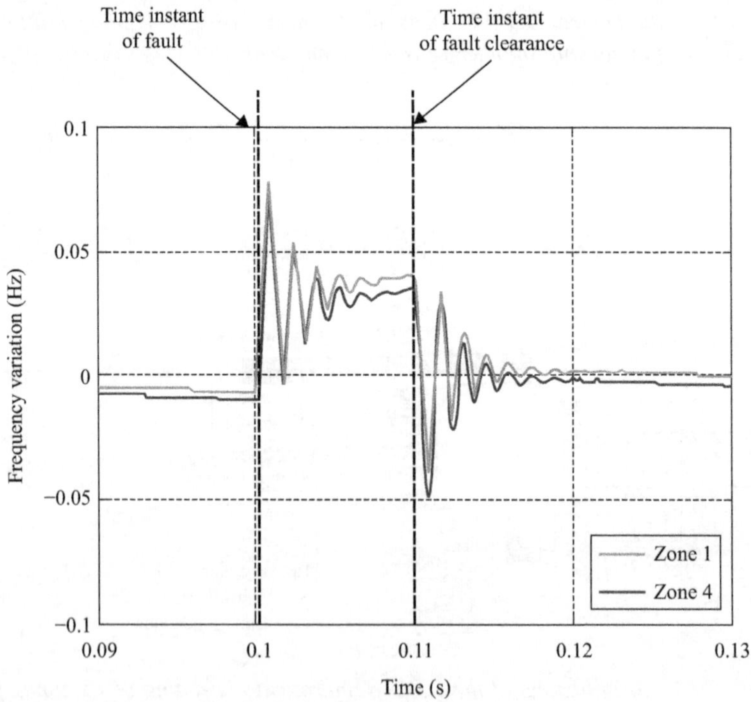

Figure 20.11 (Top) d-axis output voltage and (bottom) frequency of the inverters in Zone 1 and Zone 4, in the presence of a Zone 5 short circuit fault

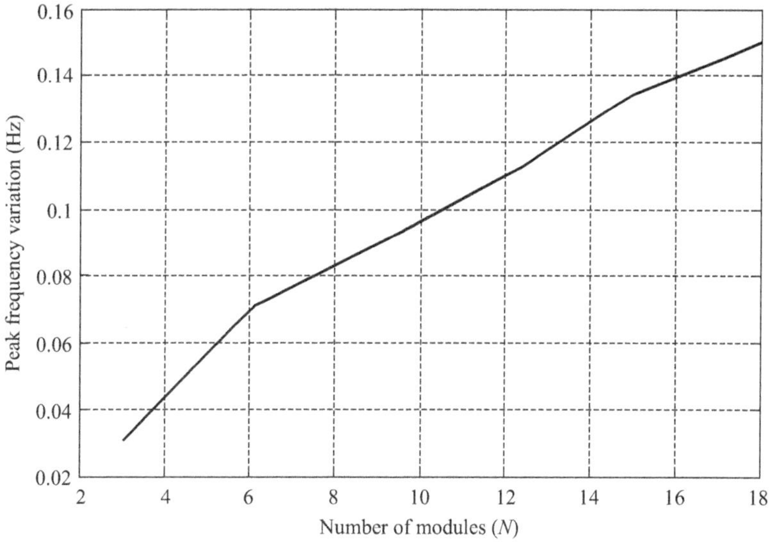

Figure 20.12 Variations of the peak inverter frequency deviations with the number of modules in the presence of the short circuit fault in Zone 5

reduce the information overhead with reference to module scalability. Some have shown that the spatio-temporal state variables of a power network are correlated while other have used this knowledge to demonstrate that, the communication overhead increases sublinearly with increasing modules.

To evaluate the efficacy of this concept for the microgrid under consideration, a three-phase short circuit fault in Zone 5. Figure 20.11 illustrates the effectiveness of the control because the deviation in the *d*-axis voltage and the frequency of the inverter in Zones 1 and 4 during fault and under post-fault condition is found to be small.

Figure 20.12 demonstrates the variation in the difference of peak frequency among the inverters. To reduce these deviations, differentially encoded information exchange is considered. Figure 20.13 clearly shows the resultant improvement in latency and convergence time (i.e., the time required to restore the network to its post-fault steady state). Finally, for the case considered in this chapter, distributed control with actual data transmission is not sustainable for more than 12 modules since the communication delay violates the stability bounds. This is not the case for the differential-data-based control scheme, which shows superior stability margins and control performance.

SBC scheme to heterogeneous power networks, such as microgrid, has also been presented. While the basic mechanism of distributed control for the heterogeneous network is similar, the issue of scalability has to be addressed differently. For instance, while in a homogeneous network, commonality of functionality can be exploited to cluster the control-communication network, in a heterogeneous network, one may need to focus on other approaches for clustering.

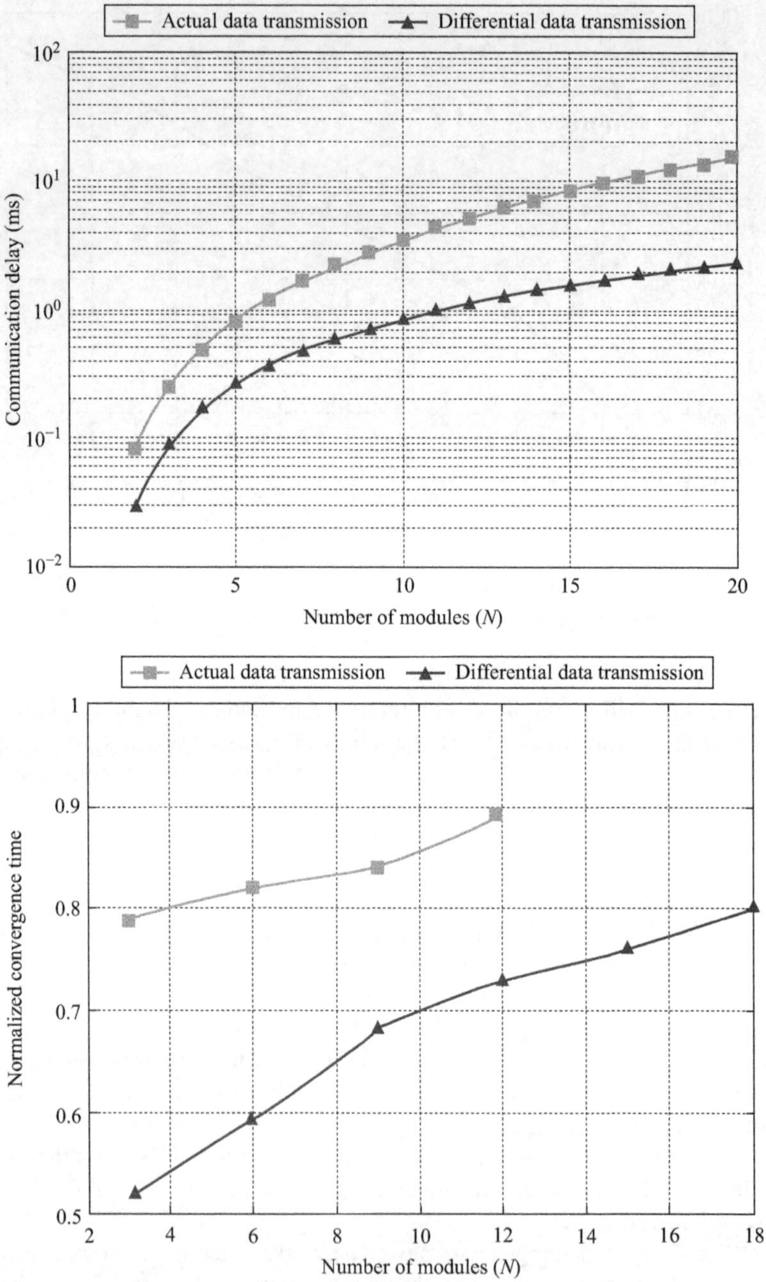

Figure 20.13 Improvement in (top) latency and (bottom) convergence time due to differential information exchange among modules

Another approach is to explore alternate mechanisms for information encoding to reduce communication overhead.

Acknowledgements

This work is supported in part by the National Science Foundation (Award No. 1002369, 0239131) and Office of Naval Research (Award Nos. N000140510594). However, any opinions, findings, conclusions, or recommendations expressed herein are those of the authors and do not necessarily reflect the views of the NSF and ONR.

References

[1] S.K. Mazumder, Nonlinear analysis and control of standalone, parallel DC–DC, and parallel multi-phase PWM converters, Doctoral Dissertation, *Department of Electrical and Computer Engineering*, Virginia Polytechnic Institute and State University, 2001.

[2] S.K. Mazumder, A.H. Nayfeh, and D. Boroyevich, "Theoretical and experimental investigation of the fast- and slow-scale instabilities of a dc/dc converter", *IEEE Transactions on Power Electronics*, vol. 16, no. 2, pp. 201–216, 2001.

[3] R.D. Middlebrook and S. Cuk, "A general unified approach to modeling switching converter power stages", *IEEE Power Electronics Specialists Conference*, pp. 521–550, 1977.

[4] V. Utkin, *Sliding modes in control optimization*, Springer-Verlag, New York, NY, 1992.

[5] H. Rodriguez, R. Ortega, G. Escobar, and N. Barabanov, "A robustly stable output feedback saturated controller for the boost dc-to-dc converter", *Systems & Control Letters*, vol. 40, no. 1, pp. 1–6, 2000.

[6] P.V. Kokotovic, "The joy of feedback: nonlinear and adaptive", *IEEE Control Systems Magazine*, vol. 12, no. 3, pp. 7–17, 1992.

[7] H. Sira-Ramirez and M. Illic-Spong, "Exact linearization in switched-mode dc-to-dc power converters", *International Journal of Control*, vol. 50, no. 2, pp. 511–524, 1989.

[8] T. Geyer, G. Papafotiou, and M. Morari, "Hybrid model predictive control of the step-down dc–dc converter", *IEEE Transactions on Control Systems Technology*, vol. 16, no. 6, pp. 1112–1124, 2008.

[9] S.K. Mazumder and K. Acharya, "Multiple Lyapunov function based reaching condition for orbital existence of switching power converters", *IEEE Transactions on Power Electronics*, vol. 23, no. 3, pp. 1449–1471, 2008.

[10] S.K. Mazumder, Invited tutorials on switching sequence based stability and control, presented or to be presented in various forms at the following conferences: *IEEE PESC* (2007), *IEEE IECON* (2010), *IEEE IECON* (2012),

EPE (2014), *IEEE CDC* (2014), *IEEE PEDES* (2014), *IEEE APEC* (2015), and *IEEE ACC* (2015).

[11] http://ece.uic.edu/~mazumder/Publications_journals_and_conferences.html

[12] E.K.P. Chong and S.H. Zak, *An introduction to optimization*, John Wiley, New York, NY, 2001.

[13] K. Acharya, S.K. Mazumder, and M. Tahir, "Fault-tolerant wireless network control of load-sharing multiphase interactive power network", *IEEE Power Electronics Specialists Conference*, pp. 1167–1174, June 2006.

[14] E. Camponogara, D. Jia, B.H. Krogh, S. Talukdar, "Distributed model predictive control", *IEEE Control Systems Magazine*, vol. 22, no. 1, pp. 44–52, February 2002.

[15] K. Acharya, Global stability analyses and controller design for switching power converters in interactive power-electronics networks, Doctoral Dissertation, *Department of Electrical and Computer Engineering*, University of Illinois, Chicago, 2008.

[16] R.H. Lasseter, "Micro grids", *IEEE Power Engineering Society Winter Meeting*, January 2002, vol. 1, pp. 305–308.

Index

10 kV SiC MOSFETs, characterization of 9–11
11-level inverter 45
12-pulse supply current 485
12-pulse thyristor rectifier 484
15 kV SiC N-IGBTs, characterization of 4–9
1,200 V, 100 A SiC MOSFET 12–14
1,200 V/45 A Infineon SiC JFET module 23
  SiC super-junction transistor characteristics 24–5
1,700 V SiC MOSFET 14
  BiMOSFET switching 16
  SiC MOSFET switching 14–15
  Si IGBT switching 15

ABB 273, 371
ac/ac matrix converter 295
acceleration factor (AF) 406–407
ac/dc converter control 374
ac–dc converters 227, 502
  current source converters (CSC) 502–3
  voltage source converters (VSC) 503–4
ac/dc interfacing 290–1
ac link multi-port power converters 92–7
ac phase controllers 505
ac pulse width modulated controllers 505
  converter circuit 505–6
  converter model 506–7
  equivalent circuit 507
action levels (ALs) 595
active filters 483, 489–92
active front-end converter (AFEC) 34, 36, 38
active harmonic elimination (AHE) technique 133
Advanced Electrical Power Systems (AEPS) 273
advanced power converters 274
advanced series compensator (ASC) 515
Agua Caliente solar project 397–398
Ampere Hour Counting 437
analog to digital converters (ADCs) 270
anti-islanding feature 546–7
application specific integrated circuits (ASICs) 269–70
Arrhenius equation 406
asymmetric cascaded H-bridge (ACHB) converter 49, 52, 65, 113–14
asynchronous generators 354
automotive energy systems 425
  electric vehicle batteries 425
    basic requirements of EV/PHEV batteries 429
    battery parameters 427–9
    cell balancing 433–6
    charging, termination and cell balancing techniques, and SOC estimation 429–33
    important characteristics of battery chemistry 425–7
    SOC estimation 436–8
    thermal management 438–40

EV charging 440
plugged charging 440–56
wireless charging 456
advantages of EV wireless charging 464
analysis of series–parallel topology 461–2
analysis of series–series topology 461
control strategies for SS and SP topology 463–4
necessity of compensation for wireless charging 458–61
peak efficiency of series–series and series–parallel topology 462–3
auto-sequential method 478
averaging voltage control 125, 127

back-stepping control (BSC) 602–3
battery chemistry, important characteristics of 425
lead-acid (PbA) 426
lithium-ion 426–7
lithium-polymer 427
nickel–cadmium (Ni–Cd) 426
nickel-metal hydride 426
battery parameters 427
battery capacity 427–8
charging efficiency 429
C-rate 428
depth of discharge 428
energy density 428
state of charge 428
battery-powered vehicle, multi-port converters for 103–4
Betz's law 537
bidirectional chargers 450–3
bidirectional converters 265–6, 565
bidirectional power flow 87, 358, 368
BiMOSFET switching 16
bipolar junction transistors (BJTs) 400
boost converter 77, 79, 249–51
bootstrap capacitor 286
bootstrap power supply 285–7
bridge duty ratio 510
brushless doubly fed induction generator 359
buck–boost converter 77, 79–80, 251–2
buck–boost QSWC 191, 193
buck converter 77–78, 246–9, 271, 402
buck-derived converters 251, 271
Burp Charging 431
burping 431

cascaded H-bridge (CHB) converter 43, 49–52, 111, 373–4
asymmetric CHB (ACHB) 49, 52
symmetric CHB (SCHB) 49, 52
using three two-level H-bridges 54
center-tap rectifier diagram 259, 261
centralised MPC-based structures, perspective on 325–6
central processor unit (CPU) cores 266
CHAdeMO dc fast charging 455–7
charging efficiency 429
Chinese GB dc fast charging standard 456
circuit-based averaging 271
circulating current control strategies 128–30
clipper concept 361
closed-loop controlled dc–dc converters 270
cluster-balancing control 125
combined heat and power (CHP) 541
constant current (CC) charging 430
constant voltage (CV) 429–30
continuous conduction mode (CCM) 264
control algorithms, for PECs 374
control for stand-alone mode of operation 386–8
dc/dc boost converters, control for 377–8
direct power control-space vector modulated (DPC-SVM) 382, 384–5
field oriented control (FOC) 380–2
IS converters, control for 378–80
maximum power point tracking 375–7

single-phase grid converter control 385
voltage oriented control (VOC) 382–4
control principles 268–71
control strategies, in modular converters 125
circulating current control 128–30
voltage-balancing control 125–8
individual-balancing voltage control 127
SM voltage sorting algorithm 127–8
conventional current source converter 524
conventional matrix converter: *see* direct matrix converter (DMC)
conventional power electronic building blocks 363–6
conventional power system 410–11, 412
converter circuit 505–6
converter control, frequency variation and 495
converter model 506–7
C-rate 428, 440
CSI SVM (space vector modulation for current source inverter) algorithm 304
Čuk converter 78, 196
current-doubler rectifier 260–2
current link systems 477–8
current-mode control 268–9
current source converters (CSC) 295, 298, 300, 363, 502–3
current source inverter (CSI) 205, 477
Current Source Rectifier 298–9, 301–3
Currents' Physical Components (CPC) theory 489
custom power (CP) devices 137
cycloconverters 478–9, 489, 492–3

Darrieus rotor 353
data centers, applications of UPS systems in 572
dc/ac converter control 374
dc and ac microgrids 546
dc/dc boost converters, control for 377–8
dc–dc converters 79, 215–20, 271
capacitor model 263
diode model 263
functions of 245
hard-switching pulse width modulated (PWM) converters 246
inductor model 263
parasitics in 262–4
resonant and soft-switching converters 246
dc/dc hybrid converter 286
dc–dc parallel resonant converters 173–6
dc–dc series resonant converters 170–3
dc fast charging converter topologies 453–5
dc UPS for pulse load with power leveling 565
delta conversion UPS 561
depth of discharge (DOD) 428
desktop personal computers, applications of UPS systems in 572
diesel generators 540–1
digital controller 270
digital interface 287
digital pulse width modulator (DPWM) 270
digital signal processors (DSPs) 270
Direct Bonded Copper (DBC) substrates 283
direct converter with SPM devices 298–306
direct matrix converter (DMC) 147
circuit topology 147
modulation techniques 147
predictive control technique 155–9
scalar technique 148–51
space vector modulation 151–5
switching states of 152
three-phase DMC topology 148

direct power control-space vector modulated (DPC-SVM) 384–5
direct torque control (DTC) schemes 336
  feedback linearization and sliding mode DTC 337–40
  intelligent DTC schemes 340–2
  space vector modulation (SVM) DTC 336–7
direct torque control-space vector modulated (DTC-SVM) 382
discontinuous conduction mode (DCM) 264
distributed impedance networks 232–3
distribution generators (DGs) 536
  diesel generators 540–1
  fuel cell-based 539–40
  IEEE 1547 545–6
  microturbines 541
  solar energy-based 537–9
  technical impacts due to 541
  wind energy-based 537
Distribution STATic COMpensator (D-STATCOM) 137, 513
distribution system operators (DSOs) 414, 420
double-star chopper-cells (DSCCs) 112–13, 121, 123–4, 128–9, 131, 137–8
doubly fed induction generator (DFIG) construction 355, 357–9, 361
drain to source capacitance 13
droop control method 547–9
dual active bridge converter (DABC) 36, 92–3, 266
dual-input single-output converter 80–1
dual-SPM-based multilevel converters 299
dynamic voltage restorer (DVR) 518–20

edge insertion (EI) PWM 225
efficiency of dc–dc converter 263
electric machine drives, special topologies for 70–2
electric vehicle (EV) batteries 425
  battery chemistry, important characteristics of 425
    lead-acid (PbA) 426
    lithium-ion 426–7
    lithium-polymer 427
    nickel–cadmium (Ni–Cd) 426
    nickel-metal hydride 426
  battery parameters 427
    battery capacity 427–8
    charging efficiency 429
    C-rate 428
    depth of discharge 428
    energy density 428
    state of charge 428
  cell balancing 433–6
  constant current (CC) 430
  constant voltage 429–30
  current 433
  EV battery charging methods 429
  EV/PHEV batteries, basic requirements of 429
  float charge 431
  pulse charge 431
  reflex charge 431
  SOC estimation 436–8
  taper current 430–1
  temperature 433
  thermal management 438–40
  time 432
  trickle charge 432
  voltage 432
  voltage drop (dV/dT) 432–3
electric vehicle (EV) charging 440
  plugged charging 440
    dc fast charging converter topologies 453–5
    EV dc fast charging 448–53
    EV fast charging standards 455–6
    EV normal charging standards 441–9
electric vehicle supply equipment (EVSE) 440, 441–8
electromotive force (EMF), extended sliding mode observers for 321–2

electromotive force (EMF)-based methods, extended 320–1
electronic switches to implement converters 47
CHB converter 49–52
combined topologies 53–5
FC converter 49
NPC converter 47–9
emotional controllers 342
energy density 427–8
energy efficiency, in PMSM drives 326–7
equivalent circuit 507
for analyzing the SRC 171
of DAB 92, 93
of LCC resonant converter 179
of PRC 173–4
of two-coil WPT system 582
of *Z*-source converter for small-signal modeling 211
equivalent impedance 517
error amplifier (EA) 268
European Network of Transmission System Operators (TSOs) for Electricity (ENTSO-E) 415
extended electromotive force (e-EMF) 315, 320
-based methods 320–1
e-EMF vector 321
sliding mode observers for 321–2
extended Kalman filter (EKF) 343

fault ride-through (FRT) capability 421
fault-tolerant operation 134–6
external faults 134–5
internal faults 135–6
feedback linearization (FBL) and sliding mode DTC 337–40
feedback linearizing control (FLC) 603
feedforward 269, 570–1
field distortion acceleration model 406–8
field-oriented control (FOC) 335–6, 380–2
field programmable gate arrays (FPGAs) 270, 323
finite element analysis (FEA) model 319
fixed speed WECS 356, 357
flexible ac transmission system (FACTS) 137, 513, 521, 523, 529
FlexiSlip 358
float charge 431
floating dc capacitor pre-charging procedures 136
float voltage 431
flyback converter 93, 254–5
flying capacitors (FC) converter 43, 49, 52–3
flywheel UPS systems 564–5
forward converter 255–6
four-coil systems, for WPT 584–5
four-port converter 88, 101–2
FREEDM Systems Center 4, 36
freewheeling diode (FWD) 7, 12
frequency variation and converter control 495
fuel cell-based DGs 539–40
full-bridge converter 258–9
full-bridge rectifier 259, 261
full-wave ZCS-QRC 185–6
full-wave ZVS-QRC 182, 188
fuzzy controllers 340–1

gallium nitride (GaN), power semiconductor based in 400
GaN Systems Inc. 400
gate driver technology 280
IGBT's tail current and minimum dead time, reduction of 282
Miller current, mitigation of 280–2
variable gate resistance 280
gate turn-off thyristors (GTOs) 474–5
grid converter control, single-phase 385
grid interface with multiple power modules 290–5
grid-tied microgrids 546–7

half-bridge (HB) inverters 111
half-bridge converter 256–8
half-wave ZCS-QRC 183–5
half-wave ZVS-QRC 186–8
harmonic cancellation 483–5
harmonic current factor (HCF) 306
harmonic generating load (HGL) 480–1
harmonic mitigation in shipboard systems 483
  active filters 489–92
  case study 492–5
  harmonic cancellation 483–5
  reactive harmonic suppressors 485–9
H-bridge converter 46–7, 77
  bidirectional 89
hexverter 125
hidden currents 491
high-frequency modulation techniques 130–2
high power WTs, power electronic converter (PEC) topologies for 371–4
high-voltage direct current (HVDC) systems 373
high-voltage direct current (HVDC) transmission 524
  conventional current source converter 524
  MMC-HVDC 525
  VSC-HVDC transmission 524
high-voltage SiC power switches 4
horizontal axis wind turbines (HAWTs) 352, 353
hybrid compensators 483
hybrid electric vehicles (HEVs) 436
hybrid static and rotary UPS 563–4

icebreakers 478
IEC 61727:1995 415
IEC 61727-2004 415
IEC 61851- (type-2 charging) 447–8
IEC Technical Committee 82 (IEC TC82) 415
IEC/TR 61000-61003-15:2011 415
IEEE 1547 544–5
IEEE STD-45 481
impedance source (IS) converters, control for 378–80
impedance source (IS) inverters 365
impedance source network topologies 205, 227
  non-transformer based 227–33
    capacitor/diode assisted 232
    distributed Z-source 232–3
    embedded *Z*-source 231
    semi-*Z*-source/semi-quasi-*Z*-source 230–1
    switched inductor/capacitor 231–2
    TSTS *Z*-source 232
    Z-source B4 converter 231
    Z-source/quasi-*Z*-source 229
  with transformer or magnetic coupling 233–6
    HF transformer isolated *Z*-source/quasi-*Z*-source/trans-*Z*-source 236
    LCCT Z-source 235–6
    trans-*Z*-source 234–5
    T-*Z*-source 235
    Y-source 233
    Γ-*Z*-source 234
impedance source power converters, categories of 215, 217
  ac–ac converter topologies 226–7
  dc–ac topologies 222–6
    multi-level/neutral point clamped 224–6
    two-level H-bridge topologies 222–4
  dc–dc converter topologies 215–21
indirect matrix converter (IMC) 159
  circuit topology 161–2
  modulation techniques 162–5
individual-balancing voltage control 126–7
induction motor (IM) drives 333
  direct torque control (DTC) schemes 336
    feedback linearization and sliding mode DTC 337–40

intelligent DTC schemes 340–2
space vector modulation (SVM) DTC 336–7
IM model 334–5
IM speed estimation with Kalman filtering 342
high-speed operation 343
implementation considerations 346–8
low-speed operation 344–5
switched reluctance sensorless drives 348–9
variable frequency drives 335
direct torque control 336
field-oriented control 335–6
scalar control 335
inductive capacitive (LC) filter cell 81
inductive power transfer (IPT) 121, 456
industrial systems, applications of UPS systems in 572
Institute of Electrical and Electronics Engineers (IEEE) 414
insulated gate bipolar transistors (IGBTs) 165, 277–80, 285–6, 288, 294, 371, 396, 400, 408, 474, 504
insulated metal substrate (IMS) 283
integrated buck and boost converter 252–3, 265
integrated gate-commutated thyristors (IGCTs) 371–2, 504
intelligent DTC schemes 340–2
intelligent power modules 274, 277–8
Intellivar 512–513
interior permanent magnet (IPM) motors 315
HF sensorless technique for 318–19
MPC method for 324–5
interleaving 266–7
interline power flow controller (IPFC) 523
International Electrotechnical Commission (IEC) 414
International Rectifier Corporation 400
inverse-tangent function 319
inverters 588–90
inverter topology 400
multilevel inverter 401–2
neutral-point diode clamped (NPC) 402
topology comparison 402–4
islanding 417, 546
protection 418
isolated converters topologies 254
flyback converter 254–5
forward converter 255–6
full-bridge converter 258–9
half-bridge converter 256–8
rectifiers 259–62

Kalman filtering (KF), induction motor (IM) speed estimation with 342
high-speed operation 343
implementation considerations 346–8
low-speed operation 344–5
Kirchhoff's Current Law (KCL) 77
Kirchhoff's Voltage Law (KVL) 77

LCCT-*Z*-source inverters 235–6
lead-acid (PbA) batteries 426
linear power regulators 245
linear time invariant (LTI) loads 480
line-interactive UPS 560–1
lithium-ion 426–7
lithium-polymer 427
local controls 550
low-frequency high-voltage ac (LF-HVAC) transmission 526–7
low-frequency modulation techniques 132–4
low-pass filter (LPF) 318
low-voltage ride-through (LVRT) 355, 416, 422–3
low-voltage SiC devices 11
common-mode current minimization 12
low-voltage gate drive design 11–12
LSC-PWM technique 131–2

LV side 1,200 V SiC MOSFET-based converters 37–9
Lyapunov function 602–3, 605
Lyapunov stability theory 339

magnetic (inductive) coupling and magnetic resonance 579
Mamdani rules 341
matrix converters (MCs) 147
  direct matrix converter (DMC) 147
    circuit topology 147
    modulation techniques 147–59
  indirect matrix converter (IMC) 159
    circuit topology 161–2
    modulation techniques 162–5
  with SPM devices 295
  technological issues of 165–7
  versus voltage back-to-back converter 167
maximum energy efficiency principle 581–2, 585
maximum power point tracking (MPPT) 354, 366, 368, 377–8, 396
maximum power transfer theorem 458–9, 580–1, 584
maximum-torque-per-ampere (MTPA) criterion 325
mean time to failure (MTTF) 406
medical equipment, applications of UPS systems in 572–3
medium voltage dc (MVDC) integrated power system (IPS) 496–8
metal–oxide–semiconductor (MOS) gate structure 4
metal-oxide semiconductor field-effect transistor (MOSFET) gate signal 589–90
metal–oxide–semiconductor field-effect transistors (MOSFETs) 4, 399
microgrid 388, 545
  centralized control 547
  dc and ac microgrids 546
  droop control method 547–9
  grid-tied microgrids 546–7
  local control 550
  multifunctionalities 551–3
  multi-input converters in 101–3
  shipboard 471–2
  stand-alone microgrids 546
microturbines 541
Miller current, mitigation of 280–2
MMC-HVDC 525
model predictive control (MPC) 322, 603
  centralised MPC-based structures, perspective on 325–6
  direct torque and flux control 323–4
  key factors for 323
  method, for IPM motors 324–5
modified reference (MR) PWM 225
modular converters 111
  asymmetric clamp-double submodule (ACD-SM) 120
  circuit configuration 114
  clamp-double submodule (CD-SM) 120
  control strategies 125
    circulating current control 128–30
    voltage-balancing control 125–8
  double-star bridge-cells (DSBC) topology 124
  double-star chopper-cell (DSCC) topology 123–4
  main applications 136–8
  MMCCs, operational issues in 134
    fault-tolerant operation 134–6
    floating dc capacitor pre-charging procedures 136
  modular multilevel cascade converter (MMCC) topologies 114, 116, 121–2, 133
  modulation techniques: *see* modulation techniques
  submodules (SMs) 111, 112
  SxBC topologies 122–3

topologies and description 115
basic modular multilevel cascade converters 121–5
MMCC arrangements and SM topologies 116–21
triple-star bridge-cells (TSBC) topology 124–5
modular multilevel cascade converter (MMCC) 112–13, 115, 128, 132
arrangements and SM topologies 116–21
operational issues in 134
fault-tolerant operation 134–6
floating dc capacitor pre-charging procedures 136
topologies 121–2
modular multilevel converter (MMC) 111, 373, 508, 529
averaged model 509–10
sub-module (SM) structure 508–9
sub-modules (SMs), string of 510
topology 373
topology example 510
modular multilevel matrix converter (M3C) 124
modulation strategies for multilevel converters 60
space vector–based algorithms 69
nearest vector control (NVC) 69–70
space vector modulation (SVM) 70
special topologies for electric machine drives 70–2
voltage levels–based algorithms 61
multicarrier PWM 61–2
nearest level control (NLC) 63–9
modulation techniques 130
direct matrix converter (DMC) 147–59
predictive control technique 155–9
scalar technique 148–51
space vector modulation 151–5
high switching frequency techniques 130–2
for impedance source network–based power converter 210
indirect matrix converter (IMC) 162–5
predictive control technique 164
space vector modulation (SVM) 162–4
low switching frequency techniques 132–4
Moore's law 274
Multibrid concept 361
multi-input converters 75
applications 97
in micro-grids 101–3
for renewable energy systems 99–101
for vehicular power systems 103–6
current input cells 79
G6 topology 84–5
multi-port converters 87
ac link 92–7
with dc link 89–90
synthesis of, by extending multi-input topologies 87–9
passive elements 76
switching networks 76
topologies, realizing 76–87
topology optimization 87
voltage input cells 79
multi-layer ceramic capacitor (MLCC) 286
multilevel clamping (MNPC) 402
multilevel converters 43
basic concepts 43
one-branch converter 44–5
two branches, "H-bridge" converter 46–7
circuit configuration 114
classification 113
electronic switches to implement converters 47
CHB converter 49–52
combined topologies 53–4
FC converter 49
NPC converter 47–9

modulation strategies for multilevel converters 60
space vector–based algorithms 69–72
voltage levels–based algorithms 61–9
with SPM devices 296–8
three-phase multilevel converters 54
phase-to-phase and phase-to-neutral voltages 55–6
space vector representation 56–60
using sub-cells 54
multilevel inverter 43, 401–2
using sub-cells 54
multi-MW power level 371
multiple-input, multiple-output systems 75
multiple-input, single-output converters 75, 82
multi-port systems 75
multi-resonant converters (MRCs) 190, 191

nearest level control (NLC) 61, 63–9, 133
nearest vector control (NVC) 69–70, 133
Negative Pulse Charging 431
networked PESs, sequence-based control (SBC) for 610–17
neural networks 341
neutral point clamped (NPC) converter 43, 47–9, 224–6, 504
neutral-point clamped (NPC) inverter 402, 475–6
nickel–cadmium (Ni–Cd) batteries 426
nickel-metal hydride 426
non-isolated converters topologies 246
boost converter 249–51
buck–boost converter 251–2
buck converter 246–9
integrated buck and boost converter 252–3
power factor correction 253–4
nonlinear controllers for PES applications 601–3
Normally Open (N.O.) static bypass switch 557–8

Oak Ridge National Laboratory, 273
Office of Naval Research (ONR) 273
off-line UPS systems 558–60
one-branch converter 44–5
on-line UPS systems 557–8
operating principles 62, 207, 209, 507, 577, 582, 584
Optima Battery Inc. 439
OptiSlip concept 357
oscillation transformer 578
output power-switching frequency plane, operating boundary in 18–23
over/short circuit current 417–18
over/under frequency protection 418
over/under voltage protection 417

packaging technologies 282–4
parallel ac-link universal converter 193–4, 196
parallel resonant ac-link converter 177
parallel resonant converters (PRCs) 170, 171
parasitic oscillations in QRCs 189
partial resonant link converter 192
passivity-based control (PBC) 602
PCS6000 converter 371–2
peak-to-peak value, calculating 248
permanent magnet synchronous generator (PMSG) 355–6, 359, 361
permanent magnet synchronous motor (PMSM) 313
energy efficiency in PMSM drives 326–7
final considerations 327–8
trends in model predictive control (MPC) of 322, 323
centralised MPC-based structures, perspective on 325–6

direct torque and flux control 323–4
MPC method for IPM motors 324–5
trends in sensorless control of 314
extended EMF-based methods 320–1
HF injection methods 316–20
HF sensorless technique for IPM and PMASynR motors 318–19
sensorless controls evaluation 315–16
sliding mode observers for extended EMF 321–2
phase locked loop (PLL) 300, 319, 380, 384, 404, 492, 558
phase-shift control 259, 266
phase-shifted carriers (PSCs) 127
phase-shifting transformers (PSTs) 516–17
phase shift modulation 93, 95, 97, 217
phase shift PWM (PSPWM) 61
phase-to-phase and phase-to-neutral voltages 55–6
photovoltaic (PV) energy systems 395
ancillary services 420
frequency support 421–2
network voltage support 420–1
centralized PV plant 397–8
field measurements 422
intermittence in solar field results 422
LVRT test results of the 500 kW RX series 422–3
grid interface 409
reactive power 410–12
real power 412–14
islanding 417
over/short circuit current 417–18
over/under frequency 418
reconnecting after grid failure and restoration 418
power quality 419
current harmonics and inter-harmonics 419
flicker and fluctuations 420
injection of dc into the ac system 419
voltage unbalance 419
protection 417
over/under voltage 417
PV inverter circuit 395–7
reliability 404
accelerated aging commonality and underlying physics 406
dominant failure mechanisms 408–9
field distortion acceleration model 406–8
power device reliability 406
standards 414
state-of-the-art technologies 399
inverter control 404
inverter topology 400–4
power semiconductors 399–400
photovoltaic backup systems 99, 101
photovoltaics (PV) 537–9
plugged charging 440
dc fast charging converter topologies 453–5
EV dc fast charging 448
bidirectional chargers 450–3
unidirectional chargers 450
EV fast charging standards 455
CHAdeMO dc fast charging 455–6
Chinese GB dc fast charging standard 456
EV normal charging standards 441
IEC 61851 (type-2 charging) 447–8
SAE J1772 441–7
plug-in electric vehicles (PEVs) 425
plug-in hybrid electric vehicles (PHEVs) 426–7, 429, 434, 436, 438
PM-assisted synchronous reluctance (PMASynR) motors 315, 322, 327
HF sensorless technique for 318–19
point of common supply (PCS) 480

poly-phase parallel resonant ac-link converters 177
poly-phase parallel resonant dc-link converters 178
poly-phase series resonant ac-link converters 176–7
poly-phase series resonant dc-link converters 177
power-converter-based power system 411–14
power converters, fundamental theory of 77
power device reliability 406
power electronic interfaces for variable speed WTs 361
  common PEC topologies 366–9
  conventional power electronic building blocks 363–6
  emerging PEC topologies 369–70
  PEC topologies for high power WTs 371–4
power electronics and control 585, 592–4
  practical circuit topologies 588
    component choices 592
    inverters 588–90
    rectifiers 590–2
  series and parallel tuned primary and secondary circuits 586–8
power electronics building block (PEBB) 273, 495–6
power electronic switches 396
power factor correction 246, 253–4, 453, 559, 560–1
power factor corrector (PFC) 254
power grid, power converters in 512
  high-voltage direct current (HVDC) transmission 524
    conventional CSC 524
    MMC-HVDC 525
    VSC-HVDC 524
  low-frequency high-voltage ac (LF-HVAC) transmission 526–7
  series compensation 515
    PWM ac VeSC series controller 517–18
    static synchronous series compensator (SSSC) 516–17
    thyristor-controlled series capacitor (TCSC) 515
    voltage sag correction 518–21
  series-series compensation 523
  shunt compensation 512
    PWM ac VeSC shunt compensator 513–14
    STATic COMpensator (STATCOM) 513
    static var compensator (SVC) 512–13
  shunt-series compensation 521
    PWM ac VeSC shunt/series power flow controller 522–3
    unified power flow controller (UPFC) 521–2
  solid-state power transformer (SSPT) 528–9
power processing unit (PPU) 472
power quality requirements in shipboard systems 479–83
power semiconductors 43, 276, 399
  based in GaN 400
  based in SiC 399–400
  based in silicon 399
power switches 396
  semiconductor: *see* semiconductor power switches
practical circuit topologies 588
  component choices 592
  inverters 588–90
  rectifiers 590–2
predictive control technique 148, 155–9
  direct matrix converter (DMC) 155–9
  indirect matrix converter (IMC) 164
progressive degradation 406
proportional-integral (PI) controller 129–30, 336
proportional-integral (PI) voltage regulation 566
proportional-resonant (PR) controller 130, 385
PSC-PWM technique 131–2

Public Utility Regulatory Policies Act (PURPA) 536
pulsating current source cells (PCSCs) 78–9
pulsating voltage source cells (PVSCs) 78–9
pulse charge 431
pulse load, dc UPS for
  with power leveling 565
pulse-width-amplitude-modulation (PWAM) 227
pulse width modulated (PWM) ac vector switching converter (VeSC)
  series controller 517–18
  shunt compensator 513–14
  shunt/series power flow controller 522–3
pulse-width modulation (PWM) 61, 127, 147–8, 209, 302, 324, 334, 401, 474–5, 568
  PWM dc–dc converters 246, 264
    closed-loop control methods for 268
  PWM flyback converter 254–5

Qi-compatible charging systems 578
quasi-multi-input converters 79
quasi-resonant converters (QRCs) 181
  effect of parasitic oscillations in 189
  full-wave ZCS-QRC 185–6
  full-wave ZVS-QRC 188
  half-wave ZCS-QRC 183–5
  half-wave ZVS-QRC 186–8
quasi-square-wave converters (QSWCs) 191–6
quasi-*Z*-source inverter (qZSI) 208, 215, 226, 365, 370
  -based dc–dc converter 220

reactive harmonic suppressors (RHSs) 483, 485–9
reactive power 410
  conventional power system 410–11
  power-converter-based power system 411–12
reactive power allocation coefficient 486
real power 404, 412
  conventional power system 412
  power-converter-based power system 412–14
rectifiers 82, 227, 259–62, 590–2
redundant bus system 566–7
reflex charge 431
relay and domino resonators, WPT systems with 585
renewable energy sources (RESs) 351
renewable energy systems 75
  multi-port power converters for 99–101
resonant converters 169
  multi-resonant converters 190
  quasi-resonant converters (QRCs) 181–9
  second-order 169–81
    dc–dc PRCs 173–6
    dc–dc SRCs 170–3
    poly-phase parallel resonant ac-link converters 177
    poly-phase parallel resonant dc-link converters 178
    poly-phase series resonant ac-link converters 176–7
    poly-phase series resonant dc-link converters 177
  with three or more resonating elements 178–81
reversible voltage 539–40
ring bus configuration, for a navy ship distribution system 472
root mean square (RMS) value 481
rotary UPS 562–4
rotor field oriented control (RFOC) 380–2
rotor's aerodynamic characteristic 352

SAE J1772 441–7
Savonius rotor 353
scalar control 335

second order generalized integrator – phase locked loop (SOGI-PLL) 384
second-order resonant converters 169–81
  dc–dc PRCs 173–6
  dc–dc SRCs 170–3
  poly-phase parallel resonant ac-link converters 177
  poly-phase parallel resonant dc-link converters 178
  poly-phase series resonant ac-link converters 176–7
  poly-phase series resonant dc-link converters 177
selective harmonic elimination (SHE) technique 127, 133
semiconductor power switches 3
  1,200 V, 100 A SiC MOSFET 12
    characterization with complementary device of half-bridge module 12–14
  1,200 V/45 A Infineon SiC JFET module 23
    SiC super-junction transistor characteristics 24–5
  1,700 V SiC MOSFET, hard-switching characterization of 14
    BiMOSFET switching 16
    SiC MOSFET switching 14–15
    Si IGBT switching 15
  10 kV SiC MOSFETs, characterization of 9–11
  12 kV SiC, zero voltage switching characterization of 25
    turn-off characteristics 28–33
    turn-on characteristics 26–8
  15 kV SiC N-IGBTs, characterization of 4–9
  forward characterization 18
  high-voltage SiC power switches 4
  low-voltage SiC devices 11
    common-mode current minimization 12
    low-voltage gate drive design 11–12
  output power-switching frequency plane, operating boundary in 18–23
  performance comparison 16–17
  SiC-based SST 34–9
    LV side 1,200 V SiC MOSFET-based converters 37–9
semi-quasi-*Z*-source inverters 230–1
semi-*Z*-source inverters 230
sequence-based control (SBC) 604
  for networked PESs 610–17
  for standalone PES 604
    application of SBC to standalone PES 607–9
    description of SBC scheme 604–7
series ac-link universal converter 196
series and parallel tuned primary and secondary circuits 586–8
series compensation 515
  PWM ac VeSC series controller 517–18
  static synchronous series compensator (SSSC) 516–17
  thyristor-controlled series capacitor (TCSC) 515
  voltage sag correction 518–21
series injection transformer (SIT) 522
series–parallel topology, analysis of
  analysis of 461–2
  peak efficiency of 462–3
series resonant converters (SRCs) 170
series resonant dc-link converter 177
series-series compensation 523
series–series topology
  analysis of 461
  peak efficiency of 462–3
shipboard electric propulsion system 473
shipboard microgrid 471
  grounding 472
  harsh environment 472
  load dynamics 471–2

non-linear load effects 471
variable frequency 471
shipboard power systems 469
frequency variation and converter control 495
future 495
medium voltage dc integrated power system 496–8
power electronics building block 495–6
harmonic mitigation in 483
active filters 489–92
case study 492–5
harmonic cancellation 483–5
reactive harmonic suppressors 485–9
power quality requirements in 479–83
shipboard propulsion drives 472
current link systems 477–8
direct ac–ac conversion – cycloconverters 478–9
voltage link systems 474–7
topologies 469–72
shunt compensation 512
PWM ac VeSC shunt compensator 513–14
STATic COMpensator (STATCOM) 513
static var compensator (SVC) 512–13
shunt phase-shifter transformer (SPT) 522
shunt-series compensation 521
PWM ac VeSC shunt/series power flow controller 522–3
unified power flow controller (UPFC) 521–2
Shuttling Method 436
SiC-based SST 34–9
LV side 1,200 V SiC MOSFET-based converters 37–9
SiC junction gate field-effect transistor (JFET) 3, 23–5
SiCMOSFET 11, 16, 26
turn-ON loss/turn-OFF of 16–17
SiC MOSFET switching 14–15
Si IGBT switching 15
silicon, power semiconductor based in 399
silicon carbide (SiC) 3, 529
power semiconductor based in 399–400
super-junction transistor characteristics 24–5
silicon controlled rectifiers (SCRs) 395, 502, 505, 512, 515, 529
silicon IGBTs 399
simplex UKF 343, 345, 348
sine wave pulse width modulation 223
single-ended primary inductor converter (SEPIC) 77–9, 84, 86
single-input single-output ac–dc converter 94
single-star bridge-cells (SSBCs) 112, 123, 126, 137
sinusoidal pulse width modulation (SPWM) 222–3
sliding-mode control (SMC) 602
sliding mode observer (SMO) 321–2
slip rings and brushes 355, 357
small wind turbines (SWTs) 351–2, 546
smart power electronic modules (SPM) 273–306
basic usage 284
bootstrap power supply 285–7
digital interface 287
protection 284
example of SPM devices 289
future usage and emerging solutions 290
different direct converter with SPM devices 298–306
grid interface with multiple power modules 290–5
matrix converter with SPM devices 295
multilevel converter with SPM devices 296–8
history 273–6

reliability 287–8
variety of products 288–9
technology background 276
gate driver technology 280–2
IGBT device technologies and their performance 277–80
packaging technologies 282–4
typical architecture of 276
smart power modules 274, 276, 278
soft-switching ac-link buck–boost converter 193
soft-switching converters 169–200
multi-resonant converters (MRCs) 190
quasi-resonant converters (QRCs) 181
effect of parasitic oscillations in 189
full-wave ZCS-QRC 185–6
full-wave ZVS-QRC 188
half-wave ZCS-QRC 183–5
half-wave ZVS-QRC 186–8
quasi-square-wave converters 191–6
resonant converters 169
resonant converters with three or more resonating elements 178–81
second-order resonant converters 169–78
ZVS and ZCS converters 196–200
solar cells 537
solar energy 399, 562
solar energy-based DGs 537–9
solid-state power transformer (SSPT) 528–9
source impedance matching 580–1
space vector–based algorithms 61, 69–72
nearest vector control (NVC) 69–70
space vector modulation (SVM) 70
special topologies for electric machine drives 70–2
space vector modulation (SVM) 69–70, 148, 151–5, 162, 226, 299, 301–2, 316, 336
direct matrix converter (DMC) 151–5
DTC 336–7
indirect matrix converter (IMC) 162–4
space vector pulse width modulation (SVPWM) 223, 224
space vector representation 56–60
space vector theory 301
sparse ac-link buck–boost converter 196
sparse matrix converters (SMCs) 161–2
spherical UKF 343–6, 348
squirrel cage induction generator (SCIG) 354, 359, 361
stand-alone microgrids 546
standalone PES, sequence-based control (SBC) for 604–10
application of SBC to standalone PES 607–9
description of SBC scheme 604–7
state of charge (SOC) 123, 427–8, 436–8
State of Health Estimation 437
static and rotary UPS systems, hybrid between 563–4
STATic COMpensator (STATCOM) 513, 516, 521–2
STATic synchronous COMpensator (STATCOM) 123, 137
static synchronous condenser (STATCOM) 543
static synchronous series compensator (SSSC) 516–17, 522
static var compensator (SVC) 411, 512–13
STMicroelectronics 40, 286–7
storage-based DG 550
submodules (SMs) 111–12
topologies 119
super-junction transistor (SJT) 24–5
surface-mounted permanent magnet (SPM) motors 319
surface-mounted permanent magnet motors, HF injection method for 319–20

SV-PWM technique 131, 132
switch duty ratio 246, 248
switched-mode power supply (SMPS) 581
switched reluctance motors (SRMs) 348–9
switched reluctance sensorless drives 348–9
switching compensators (SCs) 483
switching energy 15–17, 23–5, 37
switching power supplies 245–71
  bidirectional converters 265–6
  continuous and discontinuous conduction modes 264
  control principles 268–71
  dc–dc converters, parasitics in 262–4
  interleaving 266–7
  isolated converters topologies 254
    flyback converter 254–5
    forward converter 255–6
    full-bridge converter 258–9
    half-bridge converter 256–8
    rectifiers 259–62
  non-isolated converters topologies 246
    boost converter 249–51
    buck–boost converter 251–2
    buck converter 246–9
    integrated buck and boost converter 252–3
    power factor correction 253–4
  synchronous rectification 264–5
symmetric cascaded H-bridge (SCHB) converter 49, 52, 54
synchronous-frame HF model 320
synchronous generators 354, 357, 361, 370, 410–11, 561–2
synchronous rectification 264–5
synchronous reluctance (SynR) motors 313

Takagi-Sugeno-Kang (TSK) rules 340–1
tanker ships 478
taper current 430–1
technology evolution, in low-voltage, medium-power range 275
Tesla, Nikola 456, 577–8
thermal cycling 283, 408–9
thermal management 438–40, 496
three-coil systems, for WPT 583–4
three-level neutral point clamped (3L-NPC) converter 34, 36–7, 43, 224
three-phase multilevel converters 54–5
  phase-to-phase and phase-to-neutral voltages 55–6
  space vector representation 56–60
three-phase voltage-fed *Z*-source inverter 208
three-port converter 87, 88
three-port parallel high-frequency ac link converter *98*
thyristor-controlled reactor (TCR) 505
thyristor-controlled series capacitor (TCSC) 515
thyristor inverters 477
thyristors 176, 477, 502, 524
torque and flux control 282, 323–4
total harmonic distortion (THD) 43, 194, 227, 419, 481, 544, 566
total harmonic distortion (THD) coefficient 304
total harmonic distortion (THD) level standards 132
transfer mold technology 282
Transformerless Intelligent Power Substation (TIPS) 34, 36
transmission system operators (TSOs) 421
trans-*Z*-source inverter 234–5
trickle charge 432
tri-mode UPS 561–2
triple-active bridge, MOSFET realization of 97
triple-star bridge-cells (TSBC) topology 124–5, 137
T-source inverter 234
two-bus power system, real and reactive power in 409

reactive power 410
conventional power system 410–11
power-converter-based power system 411–12
real power 412
conventional power system 412
power-converter-based power system 412–14
two-coil systems, for WPT 582

UKF 343–5, 346, 348
ultra-sparse matrix converters (USMCs) 160, 162
unidirectional chargers 450
unified power flow controller (UPFC) 521–2
uninterruptible power supply (UPS) 205, 386, 557
applications 571
data centers 572
desktop personal computers 572
industrial systems 572
medical equipment 572–3
controls for 566–71
dc UPS for pulse load with power leveling 565
delta conversion UPS 561
flywheels 564
hybrid static and rotary UPS 563–4
line-interactive UPS 560–1
off-line UPS systems 558–60
on-line UPS systems 557–8
redundant bus system 566–7
rotary UPS 562–3
tri-mode UPS 561–2
unscented transform (UT) 343
utilization factor (UF) 93, 95–7

valve regulated leadacid (VRLA) batteries 426
variable frequency drives 333, 335–6, 473
for induction motor (IM) 335
direct torque control 336
field-oriented control 335–6
scalar control 335
variable gate resistance 280
variable speed WECS
with full power converter 359
with limited speed control 357–8
with reduced power converter 358–9
variable speed WTs, power electronic interfaces for 361–74
common PEC topologies for WTs 366–9
conventional power electronic building blocks 363–6
emerging PEC topologies for WTs 369–70
PEC topologies for high power WTs 371–4
variable structure control (VSC) 337
vector switching converters (VeSCs) 507, 522
vehicular power systems, multi-port converters for 103–6
vertical axis wind turbines (VAWTs) 352–3
Vienna rectifier 453–4
voltage and/or current back-to-back converters 147
voltage back-to-back converter
matrix converters versus 167
voltage-balancing control 125–9, 132
voltage-driven soft-switched rectifier topologies 591–2
voltage-fed Z-source converter 207
voltage feedforward path 269
voltage levels–based algorithms 61–9
multicarrier PWM 61–2
nearest level control (NLC) 63–9
voltage link systems 473–7
voltage-mode control 246, 268–70
voltage oriented control (VOC) 382–4
voltage overshoot 13, 20, 372
voltage sag correction 518–21

voltage source converter (VSC) 39, 355, 363–4, 371, 489, 503
voltage source inverter (VSI) 44, 151, 170, 205, 474
voltage source rectification (VSR) 151, 153
voltage total harmonic distortion (THD_v) 481
VSC-HVDC transmission 524

Watkins–Johnson converters 77
whole isolated full bridge converter 82
wind-based DG 550
wind energy-based DGs 537
wind energy conversion system (WECS) 137, 351, 363
  current standard designs of 356–61
  emerging configurations of 361
  types of generators used in 354–6
WinDrive concept 361
wind turbines (WTs) 351
  common PEC topologies for 366–9
  conventional power electronic building blocks 363–6
  emerging PEC topologies for 369–70
  generators for 354
  PEC topologies for high power 371–4
  power electronic converters (PECs), control algorithms for 374
    control for stand-alone mode of operation 386–8
    dc/dc boost converters, control for 377–8
    direct power control-space vector modulated (DPC-SVM) 384–5
    direct torque control-space vector modulated (DTC-SVM) 382
    field oriented control (FOC) 380–2
    IS converters, control for 378–80
    maximum power point tracking 375–7
    single-phase grid converter control 385
    voltage oriented control (VOC) 382–4
  power electronic interfaces for variable speed WTs 361
  types of 352–4
  wind energy conversion system (WECS)
    current standard designs of 356–61
    types of generators used in 354–6
  wind energy technology 351
wireless charging 456–64
  advantages of EV wireless charging 464
  control strategies for SS and SP topology 463–4
  necessity of compensation for 458–61
  series–parallel topology, analysis of 461–2
    peak efficiency of 462–3
  series–series topology, analysis of 461
    peak efficiency of 462–3
Wireless Power Consortium (WPC) 578
wireless power transfer (WPT) 577–97
  basic principles 579–80
  different forms of WPT systems 582
    four-coil systems 584–5
    relay and domino resonators, WPT systems with 585
    three-coil systems 583–4
    two-coil systems 582
  maximum energy efficiency principle 581
  maximum power transfer theorem 580–1
  operating principles, choice of 582
  power electronics and control 585
    control 592–4
    practical circuit topologies 588–92
    series and parallel tuned primary and secondary circuits 586–8

safety regulations 594
electromagnetic compatibility 594–5
human exposure 595–6
wound rotor induction generator (WRIG) 355
wound rotor synchronous generator (WRSG) 355, 359, 361

Y-source converter 217
Y-source network 233

Zero Current Switching Quasi-Resonant Converters (ZCS-QRCs) 181–2
full-wave 185–6
half-wave 183–5
zero voltage switching (ZVS) technique 25–33
turn-off characteristics 28–9
12 kV, 2 mm buffer layer N-IGBT 29–33
12 kV, 5 mm buffer layer N-IGBT 33
turn-on characteristics 26–9
Zero Voltage Switching Quasi-Resonant Converters (ZVS-QRCs) 181–2
full-wave 188
half-wave 186–8
Zeta converter 78
Z-source converters 76, 205
basic principles 207–9
basic *Z*-source impedance network 206
categories of impedance source power converters 215
ac–ac converter topologies 226–7
dc–ac topologies 222–6
dc–dc converter topologies 215–21
impact of parameter variations on 213
impedance source network topologies 227
non-transformer based 227–33
with transformer or magnetic coupling 233–6
modeling and control 209–15
shoot-through state 208–9
simplified *Z*-source converter model 211
small-signal models 211–12
voltage-fed 207
Z-source inverter (ZSI) 208, 214
closed-loop voltage control methods of 215–16
ZVS-full-bridge converter 199
ZVT boost converters with an auxiliary switch 199
ZVT Buck–boost converter 196, 198–9